Chronobiology and Chronomedicine
From Molecular and Cellular Mechanisms to Whole Body Interdigitating Networks

Chemical Biology

Editor-in-chief:
Kira J. Weissman, *Lorraine University, France*

Series editors:
Sarah Slavoff, *Yale University, USA*
Eriko Takano, *University of Manchester, UK*

Titles in the series:

1: High Throughput Screening Methods: Evolution and Refinement
2: Chemical Biology of Glycoproteins
3: Computational Tools for Chemical Biology
4: Mass Spectrometry in Chemical Biology: Evolving Applications
5: Mechanisms of Primary Energy Transduction in Biology
6: Cyclic Peptides: From Bioorganic Synthesis to Applications
7: DNA-targeting Molecules as Therapeutic Agents
8: Protein Crystallography: Challenges and Practical Solutions
9: Oxidative Folding of Proteins: Principles, Biological Regulation and Design
10: Chemical and Biological Synthesis: Enabling Approaches for Understanding Biology
11: Synthetic Glycomes
12: The Chemical Biology of Sulfur
13: The Chemical Biology of Phosphorus
14: DNA Damage, DNA Repair and Disease: Volume 1
15: DNA Damage, DNA Repair and Disease: Volume 2
16: The Discovery and Utility of Chemical Probes in Target Discovery
17: Inhibitors of Protein–Protein Interactions
18: The Chemical Biology of Nitrogen
19: RNA Polymerases as Molecular Motors: On the Road
20: Prebiotic Chemistry and Life's Origin
21: The Chemical Biology of Carbon
22: Quorum Quenching: A Chemical Biological Approach for Microbial Biofilm Mitigation and Drug Development
23: Chronobiology and Chronomedicine: From Molecular and Cellular Mechanisms to Whole Body Interdigitating Networks

How to obtain future titles on publication:
A standing order plan is available for this series. A standing order will bring delivery of each new volume immediately on publication.

For further information please contact:
Book Sales Department, Royal Society of Chemistry, Thomas Graham House, Science Park, Milton Road, Cambridge, CB4 0WF, UK
Telephone: +44 (0)1223 420066, Fax: +44 (0)1223 420247,
Email: booksales@rsc.org
Visit our website at books.rsc.org

Chronobiology and Chronomedicine

From Molecular and Cellular Mechanisms to Whole Body Interdigitating Networks

Edited by

Germaine Cornelissen
University of Minnesota, USA
Email: corne001@umn.edu

and

Tsuyoshi Hirota
Nagoya University, Japan
Email: thirota@itbm.nagoya-u.ac.jp

ROYAL SOCIETY OF CHEMISTRY

Chemical Biology No. 23

Print ISBN: 978-1-83916-200-8
PDF ISBN: 978-1-83916-755-3
EPUB ISBN: 978-1-83916-756-0
Print ISSN: 2055-1975
Electronic ISSN: 2055-1983

A catalogue record for this book is available from the British Library

The Royal Society of Chemistry is a charity, registered in England and Wales, Number 207890, and a company incorporated in England by Royal Charter (Registered No. RC000524), registered office: Burlington House, Piccadilly, London W1J 0BA, UK, Telephone: þ44 (0)20 7437 8656.

For further information see our website at www.rsc.org

For general enquiries, please contact books@rsc.org

For EU product safety enquiries, please email books@rsc.org or contact Royal Society of Chemistry Worldwide (Germany) GmbH, Römischer Hof, Unter den Linden 10, 10117 Berlin.

Preface

Just about 50 years ago, researchers were still debating whether circadian (about 24 hour) rhythms had a genetic basis or not. Discoveries of the suprachiasmatic nuclei (SCN) and the presence of clock genes in almost every cell answered this question unambiguously. Meanwhile, technological advances made it possible to investigate the role of circadian rhythms at all levels, from the whole organism to organs and tissues, as well as at the cellular and molecular level. These advances, in turn, resulted in a growing number of investigations that linked clock genes to the overt manifestation of circadian rhythms. More importantly, these studies also led to the realization of the critical role played by the circadian system in health and disease. While this fact was already apparent from early work at the organismic level, which showed how rhythm characteristics changed in the presence of disease, or even in the presence of an elevated disease risk, it now became possible to follow specific molecular pathways involved in the development of a given disease condition. Likewise, chronobiological studies have led and continue to lead to a better understanding of the role played by clock genes in clinical health across different physiological systems. A large body of work in the past two decades or so clearly indicates their ubiquitous involvement in every physiological system.

As the field of chronobiology exploded, the role of the circadian system related to interactions among different physiological systems also came into focus. By bringing experts from each branch of physiology, this book focuses on providing both facts and methods to help the reader grasp the breadth and depth of the fields of chronobiology and chronomedicine. It reviews how molecular and cellular mechanisms account for the observations made at the organismic level, and how interdigitating networks contribute to the physiological functioning of whole organisms. Above all, this book is

Chemical Biology No. 23
Chronobiology and Chronomedicine: From Molecular and Cellular Mechanisms to Whole Body Interdigitating Networks
Edited by Germaine Cornelissen and Tsuyoshi Hirota

Published by the Royal Society of Chemistry, www.rsc.org

about **bridges. Bridges** across species, **bridges** across organizational levels, **bridges** across physiological systems, **bridges** across time scales, **bridges** across the lifespan in health and disease, and **bridges** from basic science to clinical practice, all are important aspects and recurring themes covered in this book.

- Detailed studies of the molecular mechanisms responsible for circadian phenotypes of animals have shown that particular circadian proteins and their interactions are shared across evolutionary distant species. These highly conserved canonical clock genes and their rhythmic expression across diverse life forms represent a **bridge** across evolutionary time that guarantees the continuity of processes that are the essence of life itself.
- Despite inter-species differences, notably related to the rest–activity cycle, the evolutionary conserved mechanism of the circadian clock makes it possible to carry out experiments in the laboratory to gain knowledge which is applicable to humans, a translational **bridge** from bench to bedside.
- From clinical health to disease risk elevation, and eventually to overt disease, the interplay of multiple networks **bridging** several physiological systems illustrates the merit of a broad overview across all physiological systems, serving as an indispensable complement to the increasing number of excellent published reviews focusing on any given special field of interest.
- Another **bridge** discussed in several chapters relates to the intersection between the circadian system, metabolism, and cell cycle checkpoints (regulated by circadian clock genes), which are essential for processes such as growth, repair, and replacement. Similarly there are **bridges** across multiple domains, including the microbiome (which contributes to host homeostasis by regulating circadian rhythmicity, metabolism, and the immune system) and mitochondria (the power house of cells, also involved in calcium homeostasis, programmed cell death, and innate immunity and stem cell regulation).
- Although the concept of circadian medicine has gained momentum, our choice of using chronomedicine in the title of this book conveys the existence of a much broader time structure **bridging** circadian rhythms to both higher- and lower-frequency rhythms, which are also amenable to treatment optimization, as illustrated in some of the chapters.
- Finally, for chronomedicine to be successful and enter mainstream clinical practice, **bridging** results from cohort studies into the provision of optimal healthcare to individual patients is imperative. This book offers ideas, concepts, facts, and methods to enable those working in the field to develop novel therapeutic approaches.

The book is structured into four parts. After a general introduction to the field, the first set of three chapters focuses on basic principles of chronobiology.

They include a review of circadian clocks in mammals, general aspects of oscillating systems, and their role in medicine by Dr Ken-ichi Honma (Hokkaido University, Sapporo, Japan). Dr Yu Tahara and and Dr Shigenobu Shibata (Waseda University, Tokyo, and Hiroshima University, Hiroshima, Japan) have reviewed the entrainment systems in the central and peripheral clocks together with their molecular mechanisms. Dr Kuniaki Otsuka (Women's Medical University, Tokyo, Japan), Larry A Beaty and Dr Germaine Cornelissen (University of Minnesota, Minneapolis, MN, USA) have provided an overview of inter-modulating multi-frequency rhythms.

The second part systematically covers circadian rhythms and the role of clock genes in each of the twelve physiological systems. While an exhaustive coverage of current knowledge in each of these fields is beyond the scope of this book, each chapter introduces major players and important mechanisms, also pointing to challenges and prospects, thereby inviting and motivating the readers to engage in this exciting area of research.

- Dr Sato Honma (Sapporo Hanazono Hospital, Sapporo, Japan) opens this part with a focus on the nervous system, starting with the primary clock in the suprachiasmatic nuclei, which orchestrate clocks in almost every cell throughout the body in a hierarchical multi-oscillator system that enables temporal coordination of all physiological functions.
- In the chapter on the cardiovascular system, Dr Yoshimasa Oyama (Oita University, Oita, Japan), Dr Lori A. Walker (University of Colorado, Aurora, CO, USA) and Dr Tobias Eckle (University of Colorado, Denver, CO, USA) focus on the circadian–hypoxia link, revealing new insights into cardiovascular therapeutics.
- The chapter on the respiratory system, written by Dr Colleen M. Bartman and Dr Y.S. Prakash (Mayo Clinic, Rochester, MN, USA), delves into the importance of circadian clocks in lung function and respiratory diseases, pointing to several avenues of future research aimed at optimizing patient care.
- The renal system reviewed from a practicing physician's perspective by Dr Martin S. Knapp (University of Tasmania, Australia) provides a great illustration of the intersection across multiple physiological systems. Since the kidneys play a substantial role in the elimination of many drugs from the body, and renal impairment may affect drug absorption, this chapter also offers some insights into the field of pharmacology and discusses how the timing of drug administration can be optimized for safety and efficacy.
- Dr Vincent M. Cassone (University of Kentucky, Lexington, KY, USA) examines the role of circadian rhythms related to the gastrointestinal system, providing another example of multiple coordinated physiological systems working in concert to maintain the proper functioning of an organism. This chapter also reviews the role of the gastrointestinal microbiome and its interaction with host signals, and further considers disease conditions associated with altered clock function.

- To illustrate the role of mammalian circadian rhythms related to the endocrine system, Dr Keiko Tominaga (Osaka University, Osaka, Japan) and Dr Hitoshi Okamura (Kyoto University, Kyoto, Japan) have provided a detailed discussion of pineal melatonin and of adrenal glucocorticoids and mineralocorticoids. The authors thereby convey the role played by many endocrine hormones as intermediate players in time coordination within the body, as another great example of communication across multiple physiological systems.
- Dr Annie M. Curtis (RCSI University of Medicine and Health Sciences, Dublin, Ireland) and Dr Martin S. Knapp (University of Tasmania, Australia) review the immune system. This complex network of organs, cells and proteins dispersed throughout the body is involved in a number of physiological systems, as observed in conditions with underlying inflammatory features such as bronchial asthma, rheumatoid arthritis, and cardiovascular disease. The authors also discuss how rhythmicity inherent to the immune system can be harnessed to optimize pharmacological responses.
- In the chapter on the haematopoietic system, Dr Ole D. Laerum (University of Copenhagen, Copenhagen, Denmark, and University of Bergen, Bergen, Norway) describes circadian variations in the different cell lineages and stages of development, from stem and progenitor cells to mature leukocytes and erythrocytes, and discusses how in these cells, clock genes and their proteins interact with other regulatory mechanisms.
- A summary of the current understanding of the circadian timing mechanisms in cartilage and bone provided by Dr Cátia F. Gonçalves and Dr Qing-Jun Meng (University of Manchester, United Kingdom) focuses on the evidence of circadian rhythms in physiology, their molecular underpinnings, tissue-specific targets, and disease links. This chapter also offers a discussion on the clinical relevance of circadian rhythms within the skeleton system.
- The chapter on the muscular system, contributed by Dr Julien Delezie and Dr Katja Lamia (The Scripps Research Institute, La Jolla, CA, USA), provides an overview of the mammalian circadian clock and its influence on the skeletal muscle transcriptome. It discusses the circadian clock's role in muscle growth and regeneration, and outlines ways by which muscle cells regulate glucose, lipid, and mitochondrial metabolism.
- An overview of circadian clock mechanisms within the integumentary system provided by Dr Germaine Cornelissen (University of Minnesota, Minneapolis, MN, USA) and Dr Kuniaki Otsuka (Women's Medical University, Tokyo, Japan) illustrates how multiple independent but coordinated clocks function within distinct anatomic compartments in skin. It also shows how the circadian clock gates the cell cycle of these rapidly differentiating cells, and discusses implications related to various skin diseases.

- Drawing primarily from their own research, Dr Takahiro J. Nakamura (Meiji University, Kawasaki, Kanagawa, Japan), Dr Nana N. Takasu and Dr Wataru Nakamura (Nagasaki University, Nagasaki, Japan) write about the interplay between female reproductive function and the circadian system. They document how circadian rhythms and oestrus cycles affect each other, providing an example of the interrelation between two systems operating at different frequencies. A supplement provided by Dr Germaine Cornelissen (University of Minnesota, Minneapolis, MN, USA) and Dr Kuniaki Otsuka (Women's Medical University, Tokyo, Japan) covers some additional recent work on the role of the circadian system within the broader reproductive system.

The third part includes four chapters related to certain important biological processes. Related to growth, Dr Yoshiki Tsuchiya and Dr Kazuhiro Yagita (Kyoto Prefectural University of Medicine, Kyoto, Japan) review recent progress made in the field of circadian clock ontogeny, documenting the functional coupling between the development of the circadian clock and cellular differentiation during embryogenesis, as well as the dedifferentiation-coupled disruption of the circadian clock during carcinogenesis. The chapter on disease conditions by Dr Germaine Cornelissen (University of Minnesota, Minneapolis, MN, USA), Dr Denis Gubin (Medical University, Tyumen, Russia) and Dr Kuniaki Otsuka (Women's Medical University, Tokyo, Japan) illustrates the involvement of clock genes in pathology, focusing on sleep disorders, mental disorders, neurodegenerative conditions and cancer. It also highlights the interdigitating network of interactions taking place within multiple physiological systems in any given disease condition. Metabolism is the topic of the chapter by Dr Etienne Challet and Dr Patrick Vuillez (Université de Strasbourg, Strasbourg, France), who focus on the many reciprocal interactions occurring between metabolic intracellular pathways and the molecular clockwork. The authors also emphasize the bidirectional relationship between gut microbiota and the host's feeding/fasting cycle, and how nutritional interventions could lend themselves to personalized chronomedicine. Dr Pei Zhu and Dr Clara B. Peek (Northwestern University Feinberg School of Medicine, Chicago, IL, USA) present an overview of the role of circadian clocks in ageing and age-related processes. Their work aimed at understanding the interrelation between circadian decline and ageing and determining consequences of circadian disruption is key to finding potential new treatment modalities targeting age-related pathologies.

The next four chapters constitute the last fourth part, which relates to methodological aspects of chronobiology. Its aim is to provide tools and other resources to the reader interested in conducting research in the field and to those interested in applying chronobiological principles to their respective work. In the chapter on chronotherapeutic approaches, Dr Denis Gubin (Medical University, Tyumen, Russia) reviews current approaches to personalized chronodiagnosis and chronotherapy, discusses how pharmacokinetics and pharmacodynamics vary along the 24 hour scale, and considers several therapeutic modalities, from non-pharmacologic approaches to drugs targeting the circadian clock machinery directly. Dr Dietmar Weinert (Martin Luther University, Halle-Wittenberg, Germany) and Dr Denis Gubin (Medical University, Tyumen, Russia) introduce chronobiological study designs and discuss the relative merits of each approach. They review the benefit of relying on a number of different marker rhythms and describe ways to estimate the endogenous component of overt rhythms. In their easy-to-understand and clearly written chapter, Dr Alexander Yoo and Dr Ron Anafi (University of Pennsylvania, Philadelphia, PA, USA) outline a number of different and complex methods used to detect and characterize rhythms, carefully pointing out the strengths and weaknesses of each approach. As new technologies underlying recent advances in molecular biology are rapidly evolving, this chapter addresses the mathematical and computational challenges associated with accurately identifying and interpreting biological rhythms. In the last chapter, Dr Germaine Cornelissen (University of Minnesota, Minneapolis, MN, USA) and Dr Tsuyoshi Hirota (Nagoya University, Nagoya, Japan) offer some insights into the pitfalls and prospects for the integration of chronomedicine into mainstream clinical practice and discuss future avenues for basic research in chronobiology.

This book project was initiated thanks to an email exchange with Drew Gwilliams, the Series Editor for the Chemical Biology series at the Royal Society for Chemistry. He was interested in an in-depth overview of chronobiology at the cellular and molecular level, which is expected to serve as a professional reference targeting an audience of postgraduates and individuals with more advance knowledge in the field. We are very grateful to him and to all authors who contributed their expertise to bring this endeavor to fruition. We also thank all external reviewers for their positive comments and suggestions.

Germaine Cornelissen
University of Minnesota, Minneapolis, MN, USA

Tsuyoshi Hirota
Nagoya University, Nagoya, Japan

Contents

Chemical Biology No. 23
Chronobiology and Chronomedicine: From Molecular and Cellular Mechanisms to Whole Body Interdigitating Networks
Edited by Germaine Cornelissen and Tsuyoshi Hirota

Published by the Royal Society of Chemistry, www.rsc.org

Part III: Chronopharmaceutics and Chronotheranostics Relevant to Biological Processes

CHAPTER 1

Introduction

GERMAINE CORNELISSEN,*[a] KUNIAKI OTSUKA[a,b] AND TSUYOSHI HIROTA[c]

[a] Halberg Chronobiology Center, University of Minnesota, Minneapolis, Minnesota, USA; [b] Executive Medical Center, Totsuka Royal Clinic, Tokyo Women's Medical University, Tokyo, Japan; [c] Nagoya University, Nagoya, Japan
*Email: corne001@umn.edu

Once dependent on set points, feedback, and regulation, our understanding of the concept of homeostasis is evolving. Currently defined as a self-regulating process through which an organism can maintain internal stability while adjusting to changing external conditions, this cornerstone of physiology is no longer static and unvarying. Rather, it is a dynamic process that can change internal conditions as required to survive external challenges. Homeostatic regulation is no longer merely the product of a single negative feedback loop but instead reflects the complex interaction of multiple feedback systems that can be modified by higher control centers.[1]

Circadian and other rhythms are at the helm of these higher control centers. They play a central role in achieving homeostasis. Even the theory that a biological system is at a steady state or cycles with a single frequency is currently being challenged. It is now appreciated that most biological systems have no homeostatic "set point" but rather fluctuate as composite rhythms consisting of superimposed oscillations.[2] Throughout this book, you will discover how rhythms are the essence of life, and how their status defines health and disease. They also serve as the link to integrate

Chemical Biology No. 23
Chronobiology and Chronomedicine: From Molecular and Cellular Mechanisms to Whole Body Interdigitating Networks
Edited by Germaine Cornelissen and Tsuyoshi Hirota

Published by the Royal Society of Chemistry, www.rsc.org

function from molecules to intact organisms as the chapters in this book illustrate.

What are circadian rhythms? What is their molecular mechanism? How are they orchestrated for the organism to function in harmony? How do they stay synchronized with the external environment? Chapters in this book will provide answers to these questions, given our current state of knowledge, as active research in chronobiology continues to refine the current understanding of the circadian clock architecture, consisting of inter-nested transcriptional loops, and update the fundamental mechanisms of our complex, versatile, and interesting timekeeping system.[3] Deciphering the molecular core clock mechanism in *Drosophila* led to the 2017 Nobel Prize in Physiology or Medicine being awarded to Drs. Jeffrey C. Hall, Michael Rosbash and Michael W. Young "for their discoveries of molecular mechanisms that control circadian rhythms". Fundamentally conserved within the animal kingdom, this mechanism must be important for survival.[4] What used to be a specialist niche within physiology research became a topic influencing every aspect at all levels of the discipline. Genetics and genomics are powerful approaches to enhance biological understanding of circadian and sleep disorders, and related diseases in humans.[5] Understanding biological timing at multiple levels of molecular, cellular, and neural organization will be important for translating chronobiology to humans.[4]

One recurrent theme throughout this book concerns the relationship between circadian disruption and disease risk, affecting a wide range of conditions, from cancer and cardiovascular disease to cognitive impairment, and ways in which the biological clock can serve to alleviate problems arising from our 24-hour society.[6–10] Solutions beyond the use of bright light therapy for social jetlag, the discrepancy in a person's sleep pattern between weekdays and the weekend, will be presented, including meal timing,[11–16] time-restricted feeding,[9,17–20] and exercise,[21–28] while considering the role of the microbiome,[29–35] also from the perspective of "precision medicine".[29,31,32,34,36,37] You will learn about the importance of sleep and sleep quality, and about the role of astrocytes in the brain region that orchestrates circadian rhythms,[38–41] and about the coordination between the biological clock and brain functional networks such as the default mode network.[42] You will further learn about how the timing of medication matters, how drug infusion can be optimized and how new drugs target the circadian clock directly.[43–48]

Rhythms are the essence of life, from the brain waves and the beating of the heart to about-monthly and yearly rhythms. While circadian rhythms have taken central stage now that a molecular mechanism accounts for their manifestation, interest is growing in some other frequency rhythms, as elaborated in some of the following chapters. Among them, interest in about 7-day[49,50] and 12-hour[2,51–56] rhythms has emerged, notably since the latter may play a role in the adaptation to a novel environment, as may be the case in planned missions to Mars.[57–63]

From a practical point of view, recognizing the importance of rhythms leads to a shift in the kind of questions that can be addressed, with important clinical implications not yet fully realized. In addition to asking whether a biomarker is too high or too low, questions like whether its extent of daily change is too large or too small and whether the timing of its recurring high and low values occurs too early or too late bring novel information that is clinically relevant. One example relates to the temperature-based method of natural birth control. Tracking basal body temperature each day to predict the time of ovulation, however, only has limited accuracy. New studies investigate the merit of using wearable devices to measure skin temperature automatically to assess changes in the 24-hour pattern of temperature during the menstrual cycle for a more accurate prediction of the fertility window. A combination of changes in the 24-hour amplitude and phase in addition to the average was shown to be effective for detecting and predicting physiologic events of fertility, including ovulation, pregnancy, parturition, and spontaneous abortions.[64]

Quite generally, rhythm characteristics serve as important biomarkers that can also provide early signs of changes indicative of pre-disease at a stage when prophylactic interventions can be instituted before the disease becomes overt.[65] As documented in relation to each physiological system and some important biological processes in this book, a molecular understanding of circadian (and other) rhythms opens therapeutic insights for treating disease conditions, and most importantly, for helping prevent them. Indeed, circadian rhythm disruptions correlate with, and often precede, many symptoms of various conditions, including neuropsychiatric disorders such as sleep disorders, affective disorders, addiction-related disorders, and neurodegeneration.[66]

Few medical fields have evolved as rapidly as chronobiology and its offshoot, chronomedicine. Initial predictions of an intrinsic timekeeping system were based on the persistence of about 24-hour variations in the absence of external synchronizers. These variations were documented to recur at intervals deviating slightly but statistically significantly from exactly 24 hours and to be synchronized to 24 hours by the external light–dark cycle,[67–70] with their free-running period not being affected by temperature.[71] Most importantly, these periodicities were shown to anticipate, rather than follow entraining environment signals. As such, they were considered to be critical for health and survival as early as the late 1950s and early 1960s, as shown by pioneering work by Aschoff, Halberg, and Pittendrigh.[67–72]

One reason for the rapid changes in the field is that chronobiological applications are far-reaching, touching all physiological systems and all biological processes. Another reason is that breakthroughs in chronobiology benefitted from technical advances in a number of other fields, from genetics, cellular and molecular biology to biomedical engineering, bioinformatics and biostatistics, metabolomics, proteomics and transcriptomics, as well as chemistry, crystallography and drug development.

There has never been a better time to summarize the status of the field and to ask experts in their respective fields to describe past achievements and offer their perspective on future avenues of research. From the original pure description and quantification of rhythmic behavior at the organismic level, new tools have made it possible to better understand the origin of circadian (and other) rhythms and to probe the mechanisms of endogenous biological clocks and their interactions with other important molecular pathways.

Many challenges, however, remain to realize the full potential of chrono-medicine, notably by translating insights regarding the role of circadian clocks in cells and tissues into the practice of clinical medicine.[73] While some general principles highlighted throughout this book point to the bidirectional association between circadian disruption and disease risk, indicating just how essential the circadian clock is for health while its disruption causes pathologies, how to best harness this knowledge to determine the mode of action of chronotherapies has not yet been sufficiently studied and validated in many disease conditions. Such potential exists given that peripheral clocks govern a vast array of molecular and cellular processes at virtually every level of organization and that large fractions of the genome are clock-controlled. More than half of protein-encoding genes show circadian oscillation in distinct patterns across tissues.[74] Detecting defined alterations in physiological rhythms holds promise for identifying individuals at increased disease risk. Many of the best-selling drugs and World Health Organization essential medicines directly target the products of rhythmic genes. Many of these drugs have short half-lives and may benefit from timed dosage.[75] New small molecules are being discovered that target the circadian clock directly.[48]

Perhaps the most important challenge lies in recognizing just how critical a personalized approach is to health maintenance and eventual treatment when needed, as alluded to in several chapters. Circadian patterns can differ greatly from one individual to another, as can their responses to treatment.[76] Large inter-individual variations stem from several factors including genetics, age, sex hormones, and intensity, duration, and timing of light exposure.[77] The idea of adjusting the treatment to individual needs emerged as molecular biology tools became available to provide rapid and accurate diagnostic assays. Not only can the use of biomarkers for personalized medicine help reduce drug risks,[78] the danger of relying on averages that can hide individual differences in clinical trials has also been pointed out.[79] Making such precision medicine possible is also likely to yield the most benefits. It will require an integrative approach to identify and characterize the nature of any circadian disruption, to understand the relative contribution by the different physiological mechanisms that result in the abnormal behavior of specific biomarkers, and, accordingly, to determine the most appropriate intervention to adapt treatment to circadian (or other) rhythms.

In parallel with pursuing a mechanistic understanding of the link between the molecular clock and disease, longitudinal monitoring of physiological variables has a role to play in achieving personalized medicine. As outlined in some of the following chapters, time-specified reference values in clinical health best replace our current fixed limits, allowing for the interpretation of particular elements of a patient's broad time structure in their light. Such an approach shifts part of the responsibility for health maintenance to the patient. It is made possible by the availability of wearable devices and modifications that are currently being implemented in electronic health records to accommodate the handling of time series data.[80] As discussed in this book, it will be critical to incorporate algorithms for the as-one-goes chronobiological analysis of the multitude of data thus collected from personal wearable devices.

New opportunities abound to elevate the exciting fields of chronobiology and chronomedicine to the next level, and we hope that what has fascinated us will stimulate you.

References

1. G. E. Billman, Homeostasis: the underappreciated and far too often ignored central organizing principle of physiology, *Front. Physiol.*, 2020, **11**, 200.
2. B. Zhu, C. C. Dacso and B. W. O'Malley, Unveiling "Musica Universalis" of the cell: a brief history of biological 12 hours rhythms, *J. Endocr. Soc.*, 2018, **2**, 727–752.
3. J. S. Takahashi, Transcriptional architecture of the mammalian circadian clock, *Nat. Rev. Genet.*, 2017, **18**, 164–179.
4. C. R. Cederroth, U. Albrecht, J. Bass, S. A. Brown, J. Dyhrfjeld-Johnsen, F. Gachon, C. B. Green, M. H. Hastings, C. Helfrich-Förster, J. B. Hogenesch, F. Lévi, A. Loudon, G. B. Lundkvist, J. H. Meijer, M. Rosbash, J. S. Takahashi, M. Young and B. Canlon, Medicine in the fourth dimension, *Cell Metab.*, 2019, **30**, 238–250.
5. J. M. Lane, J. Qian, E. Mignot, S. Redline, F. A. J. L. Scheer and R. Saxena, Genetics of circadian rhythms and sleep in human health and disease, *Nat. Rev. Genet.*, 2023, **24**, 4–20.
6. T. Kubo, K. Ozasa, K. Mikami, K. Wakai, Y. Fujino, Y. Watanabe, T. Miki, M. Nakao, K. Hayashi, K. Suzuki, M. Mori, M. Washio, F. Sakauchi, Y. Ito, T. Yoshimura and A. Tamakoshi, Prospective cohort study of the risk of prostate cancer among rotating-shift workers: findings from the Japan collaborative cohort study, *Am. J. Epidemiol.*, 2006, **164**, 549–555.
7. S. M. Abbott, R. G. Malkani and P. C. Zee, Circadian disruption and human health: A bidirectional relationship, *Eur. J. Neurosci.*, 2020, **51**, 67–583.
8. M. Münch, A. Wirz-Justice, S. A. Brown, T. Kantermann, K. Martiny, O. Stefani, C. Vetter, K. P. Wright Jr, K. Wulff and D. J. Skene, The role of daylight for humans: gaps in current knowledge, *Clocks Sleep*, 2020, **2**, 61–85.
9. L. Pickel and H. K. Sung, Feeding rhythms and the circadian regulation of metabolism, *Front. Nutr.*, 2020, **7**, 39.
10. A. I. Socaciu, R. Ionuţ, M. A. Socaciu, A. P. Ungur, M. Bârsan, A. Chiorean, C. Socaciu and A. G. Râjnoveanu, Melatonin, an ubiquitous metabolic regulator: functions, mechanisms and effects on circadian

disruption and degenerative diseases, *Rev. Endocr. Metab. Disord.*, 2020, **1**, 465–478.

11. B. I. Sundararaj, P. Nath and F. Halberg, Circadian meal timing in relation to lighting schedule optimizes catfish body weight gain, *J. Nutr.*, 1982, **12**, 1085–1097.
12. F. Halberg, E. Haus and G. Cornelissen, From biologic rhythms to chronomes relevant for nutrition, in *Not Eating Enough: Overcoming Underconsumption of Military Operational Rations*, ed. B. M. Marriott, National Academy Press, Washington DC, 1995, pp. 361–372.
13. D. Jakubowicz, J. Wainstein, B. Ahrén, Y. Bar-Dayan, Z. Landau, H. R. Rabinovitz and O. Froy, High-energy breakfast with low-energy dinner decreases overall daily hyperglycaemia in type 2 diabetic patients: a randomised clinical trial, *Diabetologia*, 2015, **58**, 912–919.
14. E. Challet, The circadian regulation of food intake, *Nat. Rev. Endocrinol.*, 2019, **15**, 393–405.
15. R. Adafer, W. Messaadi, M. Meddahi, A. Patey, A. Haderbache, S. Bayen and N. Messaadi, Food timing, circadian rhythm and chrononutrition: a systematic review of time-restricted eating's effects on human health, *Nutrients*, 2020, **12**, 3770.
16. S. Aoyama and S. Shibata, Time-of-day-dependent physiological responses to meal and exercise, *Front. Nutr.*, 2020, **7**, 18.
17. K. Gabel, K. K. Hoddy, N. Haggerty, J. Song, C. M. Kroeger, J. F. Trepanowski, S. Panda and K. A. Varady, Effects of 8 hours time restricted feeding on body weight and metabolic disease risk factors in obese adults: A pilot study, *Nutr. Healthy Aging*, 2018, **4**, 345–353.
18. W. Chen, X. Liu, L. Bao, P. Yang and H. Zhou, Health effects of the time-restricted eating in adults with obesity: A systematic review and meta-analysis, *Front. Nutr.*, 2023, **10**, 1079250.
19. X. Wei, B. Lin, Y. Huang, S. Yang, C. Huang, L. Shi, D. Liu, P. Zhang, J. Lin, B. Xu, D. Guo, C. Li, H. He, S. Liu, Y. Xue, Y. Xu and H. Zhang, Effects of time-restricted eating on nonalcoholic fatty liver disease: The TREATY-FLD Randomized Clinical Trial, *JAMA Network Open*, 2023, **6**, e233513.
20. M. Hatori, C. Vollmers, A. Zarrinpar, L. DiTacchio, E. A. Bushong, S. Gill, M. Leblanc, A. Chaix, M. Joens, J. A. Fitzpatrick, M. H. Ellisman and S. Panda, Time-restricted feeding without reducing caloric intake prevents metabolic diseases in mice fed a high-fat diet, *Cell Metab.*, 2012, **15**, 848–860.
21. L. K. Barger, K. P. Wright Jr, R. J. Hughes and C. A. Czeisler, Daily exercise facilitates phase delays of circadian melatonin rhythm in very dim light, *Am. J. Physiol.: Regul., Integr. Comp. Physiol.*, 2004, **286**, R1077–R1084.
22. S. Aoyama and S. Shibata, The role of circadian rhythms in muscular and osseous physiology and their regulation by nutrition and exercise, *Front. Neurosci.*, 2017, **11**, 63.
23. D. Duglan and K. A. Lamia, Clocking in, working out: circadian regulation of exercise physiology, *Trends Endocrinol. Metab.*, 2019, **30**, 347–356.

24. M. Savikj, B. M. Gabriel, P. S. Alm, J. Smith, K. Caidahl, M. Björnholm, T. Fritz, A. Krook, J. R. Zierath and H. Wallberg-Henriksson, Afternoon exercise is more efficacious than morning exercise at improving blood glucose levels in individuals with type 2 diabetes: a randomised crossover trial, *Diabetologia*, 2019, **62**, 233–237.
25. C. A. Wolff and K. A. Esser, Exercise timing and circadian rhythms, *Curr. Opin. Physiol.*, 2019, **10**, 64–69.
26. Y. Choi, J. Cho, M. H. No, J. W. Heo, E. J. Cho, E. Chang, D. H. Park, J. H. Kang and H. B. Kwak, Re-setting the circadian clock using exercise against sarcopenia, *Int. J. Mol. Sci.*, 2020, **21**, 3106.
27. R. Mancilla, B. Brouwers, V. B. Schrauwen-Hinderling, M. K. C. Hesselink, J. Hoeks and P. Schrauwen, Exercise training elicits superior metabolic effects when performed in the afternoon compared to morning in metabolically compromised humans, *Physiol. Rep.*, 2021, **8**, e14669.
28. S. Mendt, H. C. Gunga, D. Felsenberg, D. L. Belavy, M. Steinach and A. C. Stahn, Regular exercise counteracts circadian shifts in core body temperature during long-duration bed rest, *NPJ Microgravity*, 2021, **7**, 1.
29. X. Liang, F. D. Bushman and G. A. FitzGerald, Rhythmicity of the intestinal microbiota is regulated by gender and the host circadian clock, *Proc. Natl. Acad. Sci. U. S. A.*, 2015, **112**, 10479–10484.
30. J. K. Paulose, J. M. Wright, A. G. Patel and V. M. Cassone, Human gut bacteria are sensitive to melatonin and express endogenous circadian rhythmicity, *PLoS One*, 2016, **11**, e0146643.
31. A. E. Rosselot, C. I. Hong and S. R. Moore, Rhythm and bugs: circadian clocks, gut microbiota, and enteric infections, *Curr. Opin. Gastroenterol.*, 2016, **32**, 7–11.
32. R. M. Voigt, C. B. Forsyth, S. J. Green, P. A. Engen and A. Keshavarzian, Circadian rhythm and the gut microbiome, *Int. Rev. Neurobiol.*, 2016, **131**, 193–205.
33. X. Liang and G. A. FitzGerald, Timing the microbes: the circadian rhythm of the gut microbiome, *J. Biol. Rhythms*, 2017, **32**, 505–515.
34. F. Bishehsari, R. M. Voigt and A. Keshavarzian, Circadian rhythms and the gut microbiota: from the metabolic syndrome to cancer, *Nat. Rev. Endocrinol.*, 2020, **16**, 731–739.
35. D. Song, C. S. Yang, X. Zhang and Y. Wang, The relationship between host circadian rhythms and intestinal microbiota: A new cue to improve health by tea polyphenols, *Crit. Rev. Food Sci. Nutr.*, 2021, **61**, 139–148.
36. E. A. Mayer, R. Knight, S. K. Mazmanian, J. F. Cryan and K. Tillisch, Gut microbes and the brain: paradigm shift in neuroscience, *J. Neurosci.*, 2014, **34**, 15490–15496.
37. G. A. Bilkey, B. L. Burns, E. P. Coles, T. Mahede, G. Baynam and K. J. Nowak, Optimizing precision medicine for public health, *Front. Public Health*, 2019, **7**, 42.
38. M. Brancaccio, M. D. Edwards, A. P. Patton, N. J. Smyllie, J. E. Chesham, E. S. Maywood and M. H. Hastings, Cell-autonomous clock

of astrocytes drives circadian behavior in mammals, *Science*, 2019, **363**, 187–192.
39. J. Clasadonte, E. Scemes, Z. Wang, D. Boison and P. G. Haydon, Connexin 43-mediated astroglial metabolic networks contribute to the regulation of the sleep-wake cycle, *Neuron*, 2017, **95**, 1365–1380.
40. S. Mederos, C. Gonzalez-Arias and G. Perea, Astrocyte-neuron networks: A multilane highway of signaling for homeostatic brain function, *Front. Synaptic Neurosci.*, 2018, **10**, 45.
41. M. H. Hastings, M. Brancaccio, M. F. Gonzalez-Aponte and E. D. Herzog, Circadian Rhythms and Astrocytes: The Good, the Bad, and the Ugly, *Annu. Rev. Neurosci.*, 2023, **46**, 123–143.
42. K. Otsuka, G. Cornelissen, Y. Kubo, K. Shibata, M. Hayashi, K. Mizuno, H. Ohshima, S. Furukawa and C. Mukai, Circadian challenge of astronauts' unconscious mind adapting to microgravity in space, estimated by heart rate variability, *Sci. Rep.*, 2018, **8**, 10381.
43. *Chronopharmaceutics: Science and Technology for Biological Rhythm-Guided Therapy and Prevention of Diseases*, ed. B. C. Youan, Wiley, Hoboken, NJ, 2009.
44. F. Levi, A. Okyar, S. Dulong, P. F. Innominato and J. Clairambault, Circadian timing in cancer treatments, *Annu. Rev. Pharmacol. Toxicol.*, 2010, **50**, 377–421.
45. S. Ohdo, Chronotherapeutics: rhythm monitoring, manipulation and disruption, *Adv. Drug Delivery Rev.*, 2010, **62**, 859–875.
46. S. Ohdo, S. Koyanagi and N. Matsunaga, Chronopharmacological strategies focused on chrono-drug discovery, *Pharmacol. Ther.*, 2019, **202**, 72–90.
47. R. J. W. Hill, P. F. Innominato, F. Levi and A. Ballesta, Optimizing circadian drug infusion schedules towards personalized cancer chronotherapy, *PLoS Comput. Biol.*, 2020, **16**, e1007218.
48. E. S. Rasmussen, J. S. Takahashi and C. B. Green, Time to target the circadian clock for drug discovery, *Trends Biochem. Sci.*, 2022, **47**, 745–758.
49. A. E. Reinberg, L. Dejardin, M. H. Smolensky and Y. Touitou, Seven-day human biological rhythms: An expedition in search of their origin, synchronization, functional advantage, adaptive value and clinical relevance, *Chronobiol. Int.*, 2017, **34**, 162–191.
50. T. G. Bromage, Y. Idaghdour, R. S. Lacruz, T. D. Crenshaw, O. Ovsiy, B. Rotter, K. Hoffmeier and F. Schrenk, The swine plasma metabolome chronicles "many days" biological timing and functions linked to growth, *PLoS One*, 2016, **11**, e0145919.
51. B. Zhu, Q. Zhang, Y. Pan, E. M. Mace, B. York, A. C. Antoulas, C. C. Dacso and B. W. O'Malley, A cell-autonomous mammalian 12 hr clock coordinates metabolic and stress rhythms, *Cell Metab.*, 2017, **25**, 1305–1319.e9.
52. Y. Pan, H. Ballance, H. Meng, N. Gonzalez, S. M. Kim, L. Abdurehman, B. York, X. Chen, Y. Schnytzer, O. Levy, C. C. Dacso, C. A. McClung,

B. W. O'Malley, S. Liu and B. Zhu, 12 h clock regulation of genetic information flow by XBP1s, *PLoS Biol.*, 2020, **18**, e3000580.

53. H. Meng, N. M. Gonzales, D. M. Lonard, N. Putluri, B. Zhu, C. C. Dasco, B. York and B. W. O'Malley, XBP1 links the 12 hours clock to NAFLD and regulation of membrane fluidity and lipid homeostasis, *Nat. Commun.*, 2020, **11**, 6215.
54. H. Ballance and B. Zhu, Revealing the hidden reality of the mammalian 12 hours ultradian rhythms, *Cell. Mol. Life Sci.*, 2021, **78**, 3127–3140.
55. B. Zhu, Decoding the function and regulation of the mammalian 12 hours clock, *J. Mol. Cell. Biol.*, 2021, **12**, 752–758.
56. K. Otsuka, S. Murakami, K. Okajima, K. Shibata, Y. Kubo, D. G. Gubin, L. A. Beaty and G. Cornelissen, Appropriate circadian-circasemidian coupling protects blood pressure from morning surge and promotes human resilience and wellbeing, *Clin. Interventions Aging*, 2023, **18**, 755–769.
57. K. Otsuka, G. Cornelissen, Y. Kubo, M. Hayashi, N. Yamamoto, K. Shibata, T. Aiba, S. Furukawa, H. Ohshima and C. Mukai, Intrinsic cardiovascular autonomic regulatory system of astronauts exposed long-term to microgravity in space: observational study, *NPJ Microgravity*, 2015, **1**, 15018.
58. S. Yang, Y. Liu, Y. Yang, Z. Yang, S. Cheng, W. Hou, Y. Wang, Z. Jiang, J. Xiao, H. Guo and Z. Wang, Simulated microgravity influences circadian rhythm of NIH3T3 cells, *Biol. Rhythm Res.*, 2016, **47**, 897–907.
59. F. E. Garrett-Bakelman, M. Darshi, S. J. Green, R. C. Gur, L. Lin, B. R. Macias, M. J. McKenna, C. Meydan, T. Mishra, J. Nasrini, B. D. Piening, L. F. Rizzardi, K. Sharma, J. H. Siamwala, L. Taylor, M. H. Vitaterna, M. Afkarian, E. Afshinnekoo, S. Ahadi, A. Ambati, M. Arya, D. Bezdan, C. M. Callahan, S. Chen, A. M. K. Choi, G. E. Chlipala, K. Contrepois, M. Covington, B. E. Crucian, I. De Vivo, D. F. Dinges, D. J. Ebert, J. I. Feinberg, J. A. Gandara, K. A. George, J. Goutsias, G. S. Grills, A. R. Hargens, M. Heer, R. P. Hillary, A. N. Hoofnagle, V. Y. H. Hook, G. Jenkinson, P. Jiang, A. Keshavarzian, S. S. Laurie, B. Lee-McMullen, S. B. Lumpkins, M. MacKay, M. G. Maienschein-Cline, A. M. Melnick, T. M. Moore, K. Nakahira, H. H. Patel, R. Pietrzyk, V. Rao, R. Saito, D. N. Salins, J. M. Schilling, D. D. Sears, C. K. Sheridan, M. B. Stenger, R. Tryggvadottir, A. E. Urban, T. Vaisar, B. Van Espen, J. Zhang, M. G. Ziegler, S. R. Zwart, J. B. Charles, C. E. Kundrot, G. B. I. Scott, S. M. Bailey, M. Basner, A. P. Feinberg, S. M. C. Lee, C. E. Mason, E. Mignot, B. K. Rana, S. M. Smith, M. P. Snyder and F. W. Turek, The NASA Twins Study: A multidimensional analysis of a year-long human spaceflight, *Science*, 2019, **364**, eaau8650.
60. J. C. Nwanaji-Enwerem, U. Nwanaji-Enwerem, L. Van Der Laan, J. M. Galazka, N. S. Redeker and A. Cardenas, A longitudinal epigenetic aging and leukocyte analysis of simulated space travel: The Mars-500 Mission, *Cell Rep.*, 2020, **33**, 108406.

61. S. I. Fujita, L. Rutter, Q. Ong and M. Muratani, Integrated RNA-seq analysis indicates asynchrony in clock genes between tissues under spaceflight, *Life*, 2020, **10**, 196.
62. B. T. Kassahun, M. Bier and J. Ding, Perturbing circadian oscillations in an *in vitro* suprachiasmatic nucleus with magnetic stimulation, *Bioelectromagnetics*, 2020, **41**, 63–72.
63. K. Otsuka, G. Cornelissen, S. Furukawa, K. Shibata, Y. Kubo, K. Mizuno, T. Aiba, H. Ohshima and C. Mukai, Unconscious mind activates central cardiovascular network and promotes adaptation to microgravity possibly anti-aging during 1 year-long spaceflight, *Sci. Rep.*, 2022, **12**, 11862.
64. W. W. Webster and B. Smarr, Using circadian rhythm patterns of continuous core body temperature to improve fertility and pregnancy planning, *J. Circadian Rhythms*, 2020, **18**, 5.
65. G. Cornelissen and K. Otsuka, Chronobiology of aging: a mini-review, *Gerontology*, 2017, **63**, 118–128.
66. H. K. Cha, S. Chung, H. Y. Lim, J. W. Jung and G. H. Son, Small molecule modulators of the circadian molecular clock with implications for neuropsychiatric diseases, *Front. Mol. Neurosci.*, 2018, **11**, 496.
67. F. Halberg, Physiologic 24 hours periodicity; general and procedural considerations with reference to the adrenal cycle, *Int. Z Vitaminforsch Beih*, 1959, **10**, 225–296.
68. F. Halberg, E. Halberg, C. P. Barnum and J. J. Bittner, Physiologic 24 hours periodicity in human beings and mice, the lighting regimen and daily routine, in *Photoperiodism and Related Phenomena in Plants and Animals*, ed. R. B. Withrow, American Association for the Advancement of Science, Ed. Publ. No. 55, Washington DC, 1959, pp. 803–878.
69. F. Halberg, Temporal coordination of physiologic function, *Cold Spring Harbor Symp. Quant. Biol.*, 1960, **25**, 289–310.
70. F. Halberg, Chronobiology, *Annu. Rev. Physiol.*, 1969, **31**, 675–725.
71. C. S. Pittendrigh, On temperature independence of the clock system controlling emergence time in Drosophila, *Proc. Natl. Acad. Sci. U. S. A.*, 1954, **40**, 1018–1029.
72. J. Aschoff, Exogenous and endogenous components in circadian rhythms, *Cold Spring Harbor Symp. Quant. Biol.*, 1960, **25**, 11–28.
73. A. Kramer, T. Lange, C. Spies, A. M. Finger, D. Berg and H. Oster, Foundations of circadian medicine, *PLoS Biol.*, 2022, **20**, e3001567.
74. R. Allada and J. Bass, Circadian mechanisms in Medicine, *N. Engl. J. Med.*, 2021, **384**, 550–561.
75. R. Zhang, N. F. Lahens, H. I. Ballance, M. E. Hughes and J. B. Hogenesch, A circadian gene expression atlas in mammals: implications for biology and medicine, *Proc. Natl. Acad. Sci. U. S. A.*, 2014, **111**, 16219–16224.
76. A. Ballesta, P. F. Innominato, R. Dallmann, D. A. Rand and F. A. Levi, Systems chronotherapeutics, *Pharmacol. Rev.*, 2017, **69**, 161–199.

77. B. Dose, M. Yalçin, S. P. M. Dries and A. Relógio, TimeTeller for timing health: The potential of circadian medicine to improve performance, prevent disease and optimize treatment, *Front. Digital Health*, 2023, **5**, 1157654.
78. M. Mukhtar, Evolution of biomarkers: drug discovery to personalized medicine, *Drug Discovery Today*, 2005, **10**, 1216–1218.
79. D. Kent and R. Hayward, When averages hide individual differences in clinical trials, *Am. Sci.*, 2007, **95**, 60–68.
80. G. Cornelissen, Applications of cosinor rhythmometry in pharmacology, *J. Pharmacokinet. Pharmacodyn.*, 2021, **48**, 339–359.

Part I: Basic Principles of Chronobiology

CHAPTER 2

Circadian Clocks in Mammals: General Aspects of Oscillating Systems and Roles in Medicine

KEN-ICHI HONMA

Department of Physiology, Hokkaido University Graduate School of Medicine, Sapporo 060-8638, Japan
Email: kenhonma2011@gmail.com

2.1 Endogenous Nature of Circadian Rhythms

2.1.1 Circadian Period

The strongest evidence for the endogenous nature of circadian rhythms is the persistence of rhythmicity under constant environmental conditions with a period slightly deviating from exactly 24 hours. The circadian period varies from species to species and from individual to individual. These variabilities also indicate the endogenous nature of circadian rhythms.

The circadian period is affected by internal as well as external factors such as the intensity of environmental light, physical activity and endocrinological manipulation.[1] In nocturnal animals, the circadian period is positively correlated with light intensity, whereas in diurnal animals, the period is negatively correlated with light intensity under constant light (LL) conditions (Aschoff's rule). In humans, the circadian period under temporal isolation does not significantly correlate with light intensity ranging from total darkness to constant light of 1500 lx[2]. However, the average periods under total darkness (24.48 ± 0.08 hours) or in blind individuals (24.50 ± 0.42 hours) are significantly

Chemical Biology No. 23
Chronobiology and Chronomedicine: From Molecular and Cellular Mechanisms to Whole Body Interdigitating Networks
Edited by Germaine Cornelissen and Tsuyoshi Hirota

Published by the Royal Society of Chemistry, www.rsc.org

shorter than the overall average (25.00 ± 0.50 hours), suggesting that the illuminance response curve is saturated at dim lights of around 100 lx, or alternatively, behavioural restrictions under total darkness shorten the circadian periods.[3] The circadian period shows history dependence. The after-effect of prior entrainment on circadian periods is observed in rodents[4] and humans,[5] and the circadian period in humans under temporal isolation is influenced by the season.[6] This could be an after-effect of the photoperiod to which the volunteers have entrained before isolation.

Endocrinological manipulations such as gonadectomy and sex hormone replacement are known to alter the circadian period and activity time in nocturnal rodents,[7,8] a phenomenon that appears to correlate with the change in the strength of behavioural activities. A sex difference in the circadian period is evident in humans[9] but is not clear in rodents.[10] In humans, the circadian period is significantly longer in males than in females. Specific receptors for testosterone and estradiol are detected in the neurons of the suprachiasmatic nucleus (SCN), the central clock, suggesting that circadian oscillation is affected by these hormones.[11]

The method used to measure behavioural rhythms is known to influence the circadian period.[12] The so-called stress-type actograph such as that obtained from a running-wheel shortens the circadian period in rats when compared with non-stress type actographs such as those stemming from a condenser sensor or passive infrared sensor;[13] the mechanism of which is not elucidated. It could involve a feedback signal from behavioural activity to the circadian pacemaker. Self-selected light–dark (LD) cycles lengthen the circadian period in humans.[2] The mechanism is not known but it could be related to motivation rather than a slightly delayed (by a few minutes) light onset and offset.

2.1.2 Entrainability

The circadian rhythm is entrained by time cues (Zeitgebers) in the environment. Entrainment means the establishment of a stable phase-relationship between the circadian rhythm and the time cue. The strongest and most important time cue is the LD cycle for many creatures. Entrainment takes place not only to the time cue of exactly 24 hours but also to that deviating from 24 hours. However, there is a limit to entrainable periods, which is determined by the sensitivity of the circadian rhythm (oscillator) to light and/or light intensity. Light sensitivity is expressed by the amount of phase-shift of the circadian rhythm induced by light. It depends on the circadian phase of light exposure. The sensitivity is described as a phase response curve (PRC), which is expressed as the amount of phase shift (Y axis) with respect to the circadian phase (X axis). The PRC has two parts: one is when light induces phase delays (D) and the other is when light induces phase-advances (A).[14] When light intensity is not strong, or alternatively the circadian oscillation is strong enough, there is a dead zone between the two parts (D and A), when light does not produce a significant phase shift. When light intensity is very strong, or alternatively the circadian oscillation is relatively weak, light produces phase shifts of up to 12 hours and there is no dead zone.

The PRC to light is a key concept to understand the entrainment of the circadian rhythm to the LD cycle. The PRC is constructed experimentally by giving a single light pulse of 0.5 to 1 hour every 2 to 3 weeks to animals under continuous dark (DD) conditions while recording their free-running behavioural rhythms. The phase shift thus produced is measured either on the next day (immediate PRC) or after a steady state free-run is established (steady-state PRC). The shape of the immediate PRC depends on the phase reference of the circadian behavioural rhythm used (see Figure 2.1). When the onset of activity is used as a reference, the immediate PRC lacks the A portion, whereas when the end of activity (activity offset) is used, the immediate PRC possesses both the D and A portions.[15] However, the D portion could be a result of phase delays in activity onset.

A similar difference in PRC shape is detected for the circadian rhythm in the pineal *N*-actyltransferase activity of hamsters.[16] The PRC shape also depends on the circadian period[4] and the length of activity time.[15]

To maintain a stable phase relationship with the environmental time cue, the circadian rhythm or oscillator phase shifts twice a day, once as a phase

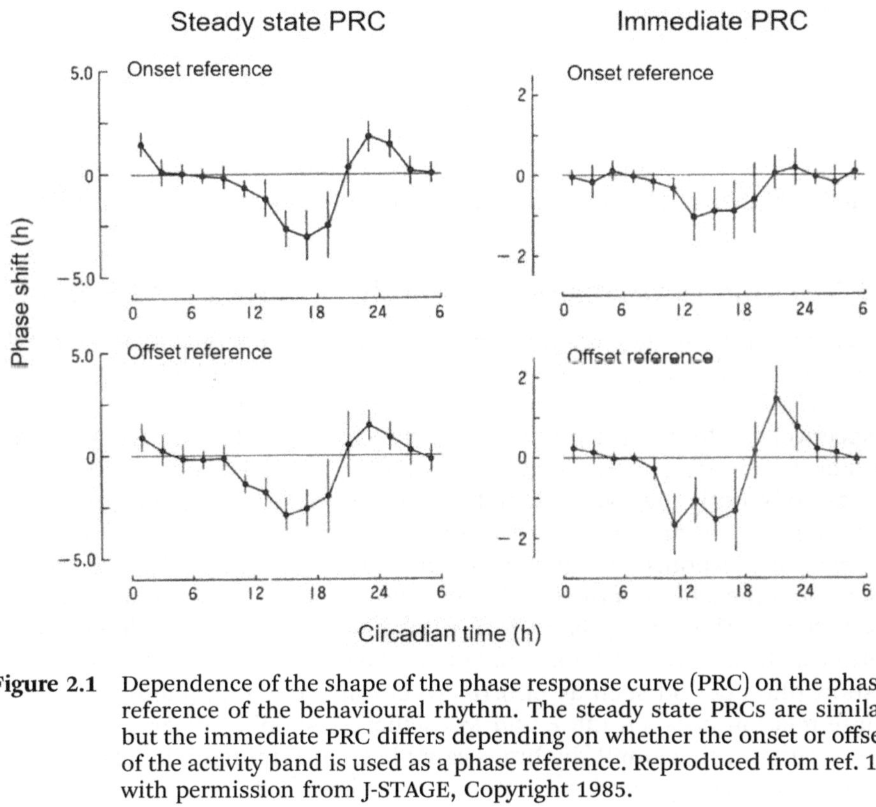

Figure 2.1 Dependence of the shape of the phase response curve (PRC) on the phase reference of the behavioural rhythm. The steady state PRCs are similar but the immediate PRC differs depending on whether the onset or offset of the activity band is used as a phase reference. Reproduced from ref. 15 with permission from J-STAGE, Copyright 1985.

advance (φ_1) due to light hitting the A portion and once as a phase delay (φ_2) due to light hitting the D portion.[14] Entrainment is achieved when the overall phase shift, the sum of φ_1 (positive sign) and φ_2 (negative sign), is equal to the difference between the circadian period (τ) and the period of the time cue (T).

$$\tau - T = \varphi_1 + \varphi_2$$

The circadian rhythm also entrains to a time cue with a period different from 24 hours. According to the above formula, the limit of entrainment is determined by $\tau - \varphi_{1\max} < T < \tau + \varphi_{2\max}$. The phase relation ($\psi$) between the circadian rhythm and the LD cycle systematically changes as T is increased or decreased. The shape of the PRC also explains this phenomenon. Outside the limits of LD entrainment, the circadian rhythm shows the so-called relative coordination, where the oscillation speeds up and down according to the phase relation between the circadian rhythm and the light signal.[1] The entrainment mentioned so far is referred to as non-parametric entrainment. However, non-parametric entrainment does not explain all phenomena associated with light entrainment. There are differences in entrainment to the complete photoperiod and to the skeleton photoperiod in which the LD cycle consists of two short light pulses on a DD background.[17] Continuous light in the light phase may have additional effects on the circadian period. To explain the effect of light, Pittendrigh[17] introduced the velocity response curve (VRC), assuming that light accelerates or decelerates the circadian oscillation; in other words, light changes the angular velocity of the circadian oscillation. The VRC is constructed from the PRC. The VRC could also explain Aschoff's rule under LL. An asymmetric PRC with a larger D portion than A lengthens the circadian period in LL and *vice versa*.

2.1.3 E and M Oscillator Hypothesis — Clock for All Seasons

Based on a unique phenomenon referred to as rhythm splitting, Pittendrigh proposed the evening (E) and morning (M) oscillators, which regulate the onset and offset components of the activity band of behavioural rhythms, respectively.[18] When nocturnal rodents are exposed to LL, the activity onset (or bout) and offset components occasionally split and free-run with different periods until the two activity bouts become antiphasic. An oscillatory coupling is assumed between the two oscillators, the strength of which changes in response to environmental conditions. The photoperiodical changes in the length of the activity band are explained through the E and M oscillators.[19] The activity band is significantly shorter under a long photoperiod such as LD18:6 (18 hours of light and 6 hours of darkness) than under a short photoperiod such as LD6:18. To entrain to a long photoperiod, the phase relation between the E and M oscillators should be changed so that the circadian behavioural rhythm is capable of entraining to the LD cycle. As a result, the activity band is shortened (see Figure 2.2).

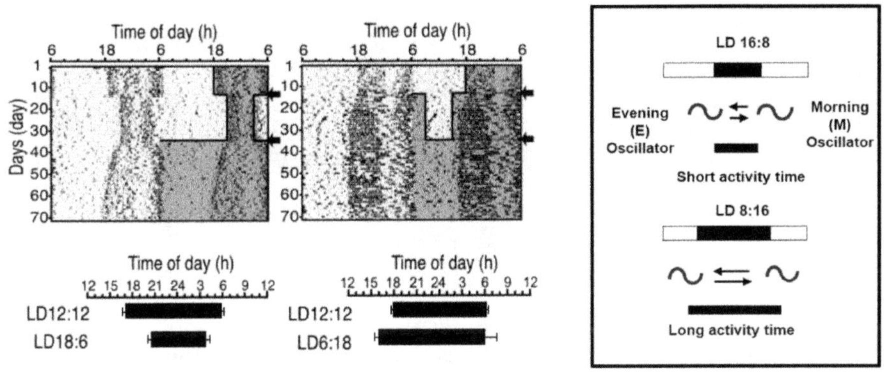

Figure 2.2 Photoperiodic changes in the length of the activity band. The length of the activity band depends on the photoperiod (left), which is explained by the E and M oscillator hypothesis (right). Reproduced from ref. 22 with permission from National Academy of Sciences.

2.1.4 Two-oscillator Hypothesis Explains the Aftereffects of Prior Entrainment and Re-entrainment to Shifted LD Cycles

When the phase of the LD cycle is suddenly shifted by several hours, the circadian behavioural rhythm re-entrains to the new LD cycle with a transient period. The days needed for re-entrainment depend on the direction of the LD shift and on the reference phase of the circadian rhythm. Through double pulse experiments, the circadian oscillator is regarded as immediately shifted after the shift of LD, whereas the slave oscillator regulating the behavioural rhythm takes longer to shift, resulting in a temporal dissociation between the two oscillators and the display of a transient period.[19] The transient period required for re-entrainment is the period needed for the re-coupling of the slave oscillator to the master. An alternative interpretation for the transient period was recently addressed.[20] In this study, the LD cycle is phase-advanced by 8 hours, and 4 days later, the animal is released into DD. As a result of the phase-advance of the LD cycle, the onset of activity moves only slightly, whereas the activity offset phase-advances substantially, resulting in α compression (shortening of activity time). On the following day in DD, the onset of activity starts to phase-advance and the activity offset phase-delays (decompression) until the length of the active span reaches a steady state. During the transient period, the phase relation (ψ) between the two oscillators significantly changes before recovering gradually. Analyses of circadian rhythms in clock gene expression in the SCN support this interpretation.[21]

The after-effects of prior entrainment to a photoperiod or unusual LD on the circadian period are also understood in terms of the E and M oscillators.[19] Establishment of a stable entrainment of the circadian pacemaker requires a change in the coupling state of the two oscillators. Following

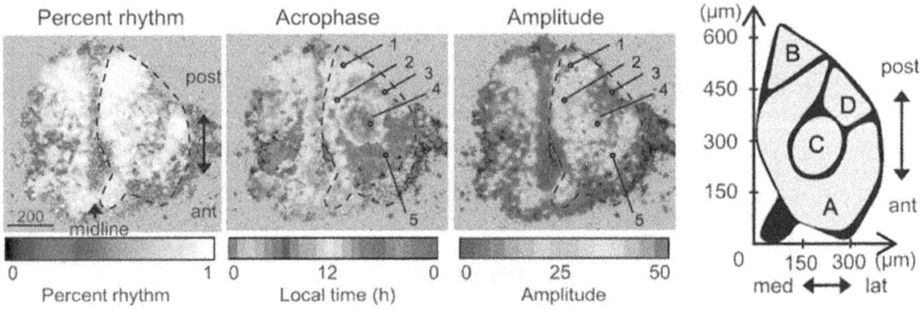

Figure 2.3 Regional pacemakers in the SCN. Circadian oscillators with different characteristics are found in different SCN regions; the E oscillator is located in region A of the horizontal plane of the SCN, the M oscillators are located in region B, the light-induced dumping oscillator is located in region C and the oscillator at the gate of photic input in C. Reproduced from ref. 22 with permission from National Academy of Sciences.

exposure to DD, the altered coupling changes the circadian period until the coupling returns to the previous state.

Plausible sites of the E and M oscillators are identified in the suprachiasmatic nucleus (SCN) in rats and mice.[22,23] The E oscillator is located in the rostral part of the SCN shell and the M oscillator is located in the caudal part of the SCN shell (see Figure 2.3).

The oscillatory structure of the circadian behavioural rhythm seems complex. There is evidence that the circadian behavioural rhythm is composed of multiple activity bouts, which are directly driven by non-SCN circadian oscillators located outside the SCN.[24,25] The E and M oscillators may share these peripheral oscillators and regulate them respectively by oscillatory coupling.

2.2 Molecular Mechanisms of Clock Works

2.2.1 Autoregulatory Feedback Loop

The circadian rhythm in mammals is believed to be generated by the transcription–translation autoregulatory feedback loop (TTFL) involving the clock genes *Periods* (*Per1* and *Per2*), *Cryptochromes* (*Cry1* and *Cry2*), *Bmal1*, and *Clock*, and their protein products.[26] This loop is referred to as the core loop. However, through more sophisticated genetic and biochemical analyses, the molecular mechanism of circadian rhythm generation is found to be more complex.[27] In this loop, the heterodimer of BMAL1/CLOCK, an activator, binds to the E-box enhancer elements on the *Per(s)* and *Cry(s)* promoter zone to stimulate their transcription and increase the respective mRNA levels. The resulting translation enhances the cytoplasmic levels of PER(s) and CRY(s) proteins, which fuse together and with casein kinase 1 ε, δ. The protein complex translocates into the nucleus and suppresses the transactivation effect of BMAL1/CLOCK. The mRNA levels of *Per*(s) and *Cry*(s) decline, resulting in a decrease in PER(s) and CRY(s) levels in the cytosol and subsequently in the nucleus. The suppression of transactivation ceases and a new cycle starts. One turn takes about 24 hours.

The molecular core loop is interlocked with at least two additional molecular feedback loops (see Figure 2.4); one is the *Bmal1* loop and the other is the DEC loop. The *Bmal1* loop regulates BMAL1 production.[28,29] *Bmal1* and *Clock* expressions are activated by RORα,β,γ and repressed by REV-ERBα,β, which compete with each other at the ROR binding elements (RORE) located on the promotor of both clock genes. *Bmal1* expression peaks in the middle of subjective night, 180 degrees out of phase with the *Per1* circadian rhythm, and *Clock* expression peaks in the early subjective day with a shallow rhythmicity.[30] Double knockout of *RevErbα,β* compromises the circadian behavioural rhythm and clock gene expressions in the mouse liver.[31] *RevErbα,β*, together with *E4bp4*, is regarded as a key factor in regulating the circadian output.[32,33]

The DEC loop produces DEC1 and DEC2, basic helix–loop–helix (bHLH) transcription factors, which suppress the transcription of *Per1* by directly interacting with E-box enhancer elements. Their mRNA levels show circadian rhythms peaking in the early and middle subjective day.[34,35] A single knockout of *Dec1* or a double knockout of *Dec1* and *Dec2* slightly but significantly lengthens the free-running period in DD.[36,37] DEC1 is suggested to be involved in resetting the circadian clock in the kidney and adrenal glands.[38]

The dynamics of TTFL could be modified at several points in the feedback loop, resulting in a change in the circadian period. Possible sites are the transcription of *Per(s)/Cry(s)*, the trafficking of their mRNA or clock proteins across the nuclear membrane, and the production (translation) and degradation of clock proteins. One of the excellent examples is the tau mutant hamster, which was discovered from a colony of laboratory animals. Tau mutants have an extremely short tau of the running-wheel activity rhythm in DD.[39] It turned out later that in this animal, a point mutation occurs in the enzyme casein kinase 1 ε (CK1ε), which is involved in the stabilization of PER2 protein.[40] The altered CK1ε enzyme is unable to bind to PER2 and facilitate its degradation.[41] Similarly, casein kinase 1δ (CK1δ) is involved in the degradation of PER2[42] and the inhibitors of CK1δ could prolong the turnover of PER2, resulting in the lengthening of the circadian period.[43] The intracellular concentrations of clock gene products thus seem to be a determinant of the circadian period. Genetic or pharmacological manipulations demonstrate changes in the circadian period and reveal possible sites of action. However, it is not known whether these enzymes are really involved in the physiological regulation of the circadian system.

2.2.2 Input and Output of TTFL

2.2.2.1 TTFL Input

The SCN has three major neural afferents, the signals of which regulate the circadian oscillation. In addition, the SCN receives the humoral information

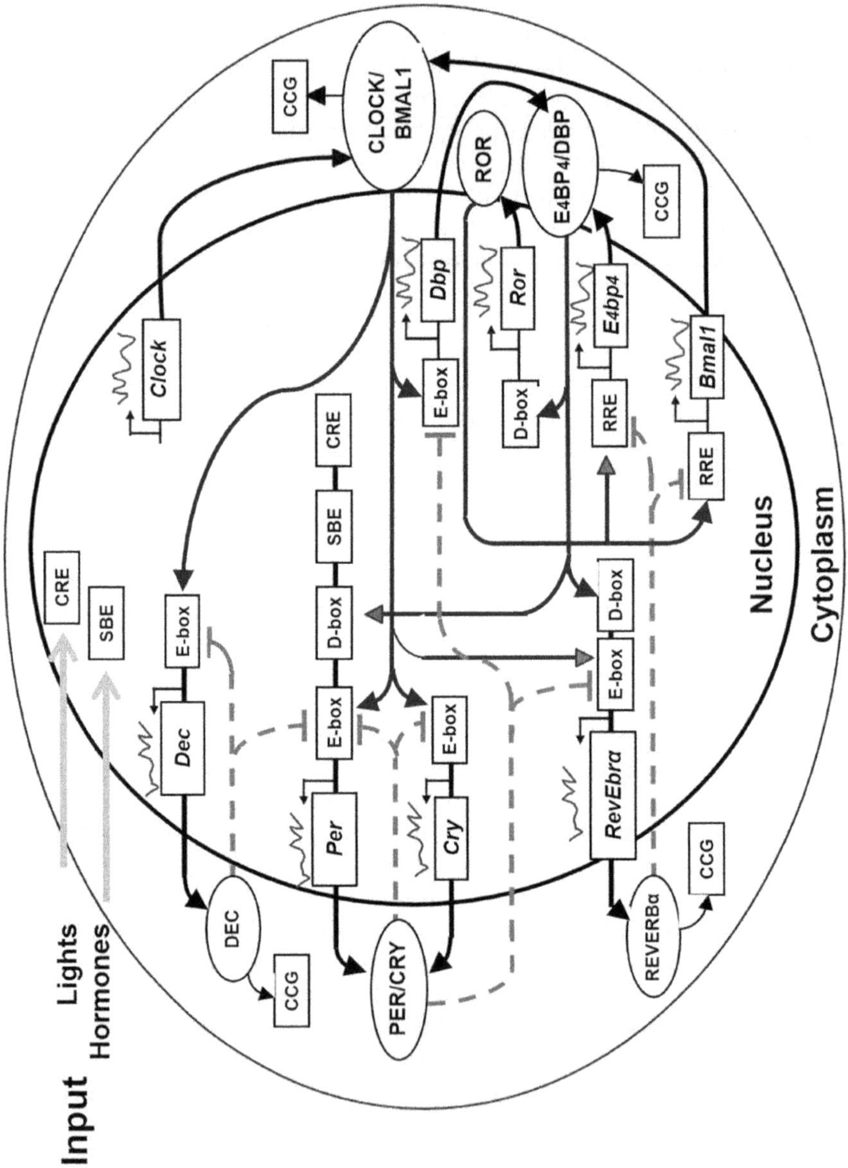
Input
Lights
Hormones
CRE
SBE
Clock
CLOCK/
BMAL1
CCG
Dec
E-box
DEC
Per
Cry
PER/CRY
D-box
Dbp
Ror
ROR
E4BP4/DBP
E4bp4
RRE
Bmal1
RevErbα
REVERBα
Nucleus
Cytoplasm

from peripheral organs such as the gonads. These afferent signals may finally affect the TTFL in the oscillating SCN neurons.

The circadian pacemaker receives photic signals from the retinohypothalamic tract (RHT), the effects of which are phase-dependent as described previously. The light signals are first transmitted to neurons containing the vasoactive intestinal polypeptide (VIP) and gastrin releasing peptide (GRP) in the SCN core, and then to neurons containing arginine vasopressin (AVP) in the SCN shell. Exposure to LL attenuates the expression of *Per2*[44] and prolonged LL exposure constitutively elevates PER2 protein levels in the mouse SCN.[45] The principal neurotransmitter of the RHT is glutamate, which is potentiated by pituitary adenylate cyclase-activating peptide (PACAP). These substances activate the mitogen-activated protein kinase (MAPK) cascade and phosphorylate the cAMP response element-binding (CREB) protein to enhance the transcription of *Per1* and *Per2* through CRE elements in the VIP- and GRP-containing neurons. Subsequently, the secreted VIP and GRP stimulate *Per(s)* expression through vasoactive intestinal peptide receptor 2 (VPAC2) and bombesin 2 (BB2) receptors of the AVP-containing neurons. The MAPK cascade and CREB phosphorylation are involved in these responses.[46]

The circadian pacemaker also entrains to non-photic periodic signals such as vigorous activity through wheel-running[47] or treadmill exercise.[48] Neuropeptide Y (NPY) from the intergeniculate leaflet (IGL) *via* the geniculohypothalamic tract (GHT) and serotonin from the midbrain raphe nuclei are likely involved in the entrainment to timed exercise.[49,50] NPY is known to phase-advance the circadian rhythm *in vivo* and *ex vivo* when applied during the subjective day[51] and it has been observed to reduce *Per1* and *Per2* mRNA levels in the SCN.[52] Electrical stimulation of the midbrain dorsal raphe nucleus increases serotonin release in the SCN and phase-advances the circadian behavioural rhythm through the 5-HT_7 receptor.[53,54]

Gonadectomy lengthens the circadian period in DD and shortens the activity time in hamsters and rats.[55–57] Testosterone replacement restores the period and activity time. Androgen receptors are located in the GRP-containing neurons in the SCN core[58] and phase-dependently affect the light-induced *Per1* and *Per2* expression in the SCN,[59] probably through steroid responsive elements (SRE) in the promotors of *Per1* and *Per2*.[60]

Figure 2.4 TTFL and interlocked loops for circadian rhythm generation. In addition to the core molecular loop of PER, CRY, CLOCK and BMAL1, the interlocked *Bmal1* and DEC loops are illustrated. The right side represents the positive components and the left side the negative components. The inputs to the molecular loop are illustrated in the upper left and the outputs are denoted as Clock-controlled genes (CCG). The *Bmal1* loop consists of the complex feedback loops. BMAL1 production is positively regulated by ROR and negatively by REVERBα, both of which are positively regulated by the E4BP4/DBP complex and negatively by PER/CRY.

On the other hand, the circadian peak of *Per1* expression in the SCN of female rats is detected at night regardless of the stage of the estrous cycle and is abolished by the removal of the ovaries. However, ovarian steroids do not significantly affect the rhythm in *Per1* expression in the SCN, suggesting that brain regions outside the SCN mediate the effects of steroids.[61]

2.2.2.2 TTFL Output

The circadian oscillation generated in an SCN neuron is transmitted to other functions in the same neuron and to peripheral clocks outside the SCN as well as to other SCN neurons. CLOCK and BMAL1 bind to the E box elements in the promotor of other genes such as AVP[62] to transactivate them. CRY is thought to directly interact with the G-protein α-subunit (Gsα) to inhibit cyclic adenosine monophosphate (cAMP) accumulation in the mouse liver.[63] There are various neural projections from the SCN to other sites of the brain.[64] Among them, the hypothalamic projection is the most robust. These efferent neurons contain AVP, VIP or GRP.[65] Evidence for the involvement of humoral factors is obtained through SCN transplantation, where the *de novo* synthesis of synaptic connections is blocked.[66] AVP in the SCN is thought to be released directly into the cerebrospinal fluid.[67]

2.2.3 Clock Gene Mutations

2.2.3.1 Per1 and Per2

The circadian behavioural rhythm in mice is not abolished by a single knockout of *Per1* or *Per2,* but the knockout of both clock genes together seriously compromises the circadian behavioural rhythm.[68,69] The circadian period in DD is shortened by the knockout of the *Per1*[68,69] and *Per2*[69] genes, or it remains unchanged.[70,71] A *Per2* E′-box mutation compromises the circadian behavioural rhythm in DD.[72] Interestingly, the response to LL of the circadian behavioural rhythm differs between *Per1* and *Per2* mutations.[70] *Per1* mutant mice show circadian periods longer than *Per2* mutants, possibly reflecting differences in the PRC shape.[71] These findings suggest that the dynamics differ between the *Per1*- and *Per2*-associated TTFLs. A question arises how the circadian system in the SCN integrates these two different TTFLs. Is it a matter of cellular oscillation or neural network? There are three types of neurons in the SCN in terms of *Per1* and *Per2* expression.[73] Almost half of the SCN neurons express only *Per2*, while a very few neurons express only *Per1*. The other half of the SCN neurons express both *Per1* and *Per2*. About half of dual-expressing neurons show distinct in-phase circadian rhythms, which are similar to either of the neurons that express only *Per1* or *Per2*. These findings suggest an important TTFL characteristic: the functional TTFL in a single cell involves either *Per1* or *Per2*, but not both. Even when *Per1* and *Per2* are simultaneously expressed, the TTFL is regulated by only one of them, either *Per1* or *Per2*.

2.2.3.2 Cry1 and Cry2

A single knockout of *Cry1* shortens the circadian period of the behavioural rhythm, whereas that of *Cry2* lengthens the period.[74] A double knockout of *Cry1* and *Cry2* abolishes the circadian behavioural rhythm.[74] However, the circadian PER2::LUC rhythm is detected in SCN cultures from neonatal *Cry1*/*Cry2*-deficient mice, which is as robust as in wild-type (WT) mice[75,76] (see Figure 2.5, left). The circadian rhythm disappears around weaning, at postnatal day 21. The SCN neurons of *Cry1*/*Cry2*-deficient mice show coherent circadian rhythms in neuronal activity, indicating that the loss of circadian rhythm in adult mice is due to the desynchronization of cellular circadian rhythms.[76] Actually, the adult SCN retains the circadian PER2::LUC rhythm when co-cultured with the SCN of neonatal WT mice. The responsible factor secreted from the donor SCN is AVP.[77] AVP production in *Cry1*/*Cry2*-deficient mice is substantially reduced.[76]

On the other hand, VIP is responsible for the circadian rhythm in the neonatal SCN of *Cry1*/*Cry2*-deficient mice. Knocking out the VIP receptor gene (*Vipr2*) abolishes the circadian PER2::LUC rhythm in *Cry1*/*Cry2*-deficient neonatal mice.[77] VIP is a unique molecule, which is regulated by the circadian pacemaker during the neonatal period but not in adulthood.[78,79] Non-rhythmic release of VIP in the SCN of *Cry1*/*Cry2*-deficient adult mice results in a failure to express coherent circadian rhythmicity. Importantly, the expression of *Per1* and *Per2* in *Cry1*/*Cry2*-deficient mice is substantially enhanced in adults as well as in neonates, indicating that the E-box is kept disinhibited in mutant mice, therefore preventing the TTLF from functioning normally.[80] Despite this, a coherent circadian rhythm is detected in the SCN.

Finally, continuous exposure to LL from birth to several weeks after weaning restores the behavioural rhythm in *Cry1*/*Cry2*-deficient adolescent mice[81] (see Figure 2.5, right). Neonatal LL is known to strengthen the circadian behavioural rhythms in WT rats[82] and to induce circadian rhythms in clock gene mutant arrhythmic mice.[83,84] CRY1 and CRY2 are dispensable in the generation of circadian oscillations.

2.2.3.3 Clock

The clock gene *Clock* is identified by the forward genetics of mutagenized mice.[85] A single mutation in the clock gene, *Clock*$^{-\Delta 19/-\Delta 19}$ exhibits dominant-negative effects and is responsible for the disrupted circadian behavioural rhythm.[86] The *Clock* mutation lengthens the circadian rhythm to up to 28 hours *in vivo* and *ex vivo*.[87,88] The circadian *Per1* expression in the *Clock*-mutant SCN peaks about 4 hours later than the peak in the WT SCN,[89] which is attributed to the lengthening of the circadian period.

Surprisingly, *Clock*-deficient mice continue to express robust circadian rhythms in locomotor activity and show an unaffected TTFL.[90] It was discovered that NPAS2 (MOP4), a paralog of CLOCK, can form a dimer with

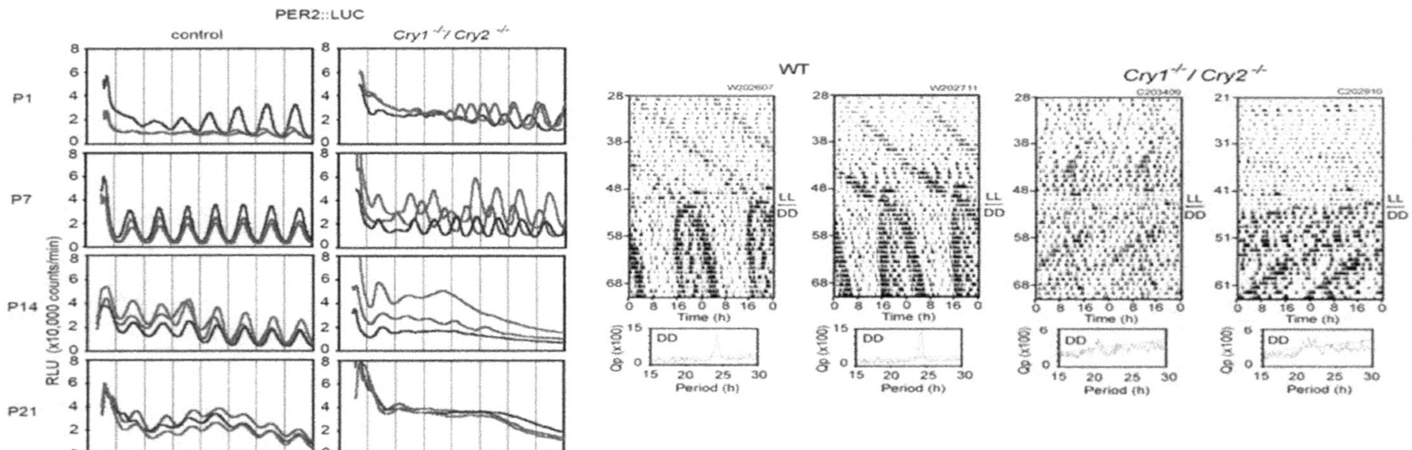

Figure 2.5 CRY1/CRY2 are dispensable for circadian rhythm generation. Circadian rhythms in PER2::LUC expression in cultured SCN slices from WT and *Cry1/Cry2*-deficient neonatal mice (left) and circadian rhythms in locomotor activity (right). Circadian rhythms are detected in PER2::LUC in the early postnatal period but disappear around the day of weaning. Circadian locomotor rhythms appear in adolescent mice released into DD after prolonged exposure to LL from the day of birth. Periodograms are shown for each actograph. Reproduced from ref. 76 and ref. 81 with permission from Springer Nature and PLOS ONE, Copyright 2013 and 2013, respectively.

BMAL1 and compensate for the role of CLOCK.[91] The deficiency of both *Clock* and *Npas2* eliminates the circadian rhythms in *Per1* and *Per2* expression.

2.2.3.4 Bmal1

Knocking out *Bmal1* (*Mop3*) disrupts the circadian behavioural rhythm in mice in DD.[92] However, a circadian rhythm in PER2::LUC is detected in the cultured SCN.[93] The rhythmicity is unstable with a small amplitude: it is therefore called a quasi-circadian rhythm. A mathematical model simulates the quasi-circadian rhythm by assuming intercellular coupling and molecular noise without circadian oscillation. On the other hand, functional replacement of *Bmal1* with its paralog, *Bmal2*, restores the circadian rhythm.[94] More recently, a number of transcripts in the SCN and peripheral organs of *Bmal1*-knockout mice were demonstrated to exhibit a circadian rhythm comparable to that of WT mice.[95]

The circadian *Bmal1* expression rhythm is temporally dissociated from the circadian *Per1* expression rhythm in cultured SCN slices of neonatal mice,[96] suggesting that the core loop and *Bmal1* loop are capable of oscillating independently. In addition, the shape of the PRC for medium exchange differs between the circadian *Bmal1* and *Per1* expression rhythms. A similar dissociation is observed during a phase-shift of the behavioural rhythm induced by a light pulse.[97]

2.3 Biological Clock: A Hierarchical Multi-oscillator System

2.3.1 Clock in the SCN

2.3.1.1 Cellular Circadian Rhythms

The SCN consists of about 10 000 neurons on one side in rats and mice. In dispersed SCN cell culture, 50–90% neurons show circadian rhythms in neuronal activity on the multiple electrode dish (MED), the period of which follows a Gaussian distribution extending from 20 to 28 hours with a mean period of 24.2 hours.[98,99] In a dispersed cell culture, the circadian neuronal rhythms seem to be weakly coupled because the period distribution becomes narrower (from 22 to 26 hours) without changing the mean period in sliced SCN tissue culture.[100,101] More than 80% of solitary mouse SCN cells in a dispersed cell culture show a circadian rhythm in intracellular Ca^{2+} concentration on a physically isolated micro-island without glial cells.[102] The distribution of circadian periods extends from 22.7 to 27.5 hours. These findings indicate that circadian periods in single SCN neurons are diverse and that SCN neural networks integrate cellular circadian rhythms to express a coherent rhythm.

SCN neurons in a dispersed cell culture show circadian rhythms in clock gene expression. A lower cell density is associated with fewer rhythm positive cells.[102,103] The findings suggest that not all SCN neurons possess the circadian rhythm generating machinery and non-oscillating neurons are driven through the neural networks in the SCN.

2.3.1.2 *Core and Shell with Regional Oscillations*

As mentioned already, circadian rhythms in the core (ventral) and shell (dorsal) of the SCN have different functions. The VIP-positive neurons are abundant in the core, whereas the AVP-positive neurons are rich in the shell. They, especially AVP, define the core–shell boundaries.[104] Gamma-aminobutyric acid (GABA) coexists in most SCN neurons.[105] The circadian AVP rhythm is phase-advanced in cultured SCN relative to VIP, probably due to its shorter circadian period.[106] Surgical separation[107] or tetrodotoxin (TTX) treatment[108] induces a dissociation of circadian rhythms in the core and shell. There are gradients of circadian phase in PER2::LUC expression rhythms with the earliest phase at the dorsomedial and the latest at the ventrolateral part, which looks like a wave.[109] However, there is no evidence for signal transduction along with the wave of phases.

2.3.1.3 *Clusters of Cellular Circadian Oscillations*

There are at least three different clusters of cellular circadian oscillations in the SCN (see Figure 2.6). These clusters appear in the adult SCN of *Cry1*/*Cry2*-deficient mice.[77] Co-culture of neonatal WT SCN restores the circadian PER2::LUC rhythm in the adult *Cry1*/*Cry2*-deficient SCN. This restoration is associated with the appearance of a single cluster. These findings indicate that the integration of three clusters into one is important for expressing a coherent circadian rhythm. Actually, there is a single cell cluster in the neonatal SCN of *Cry1*/*Cry2*-deficient mice, which shows a coherent circadian rhythm.[76] Additional knockout of the VIP receptor gene abolishes the circadian PER2::LUC rhythm and results in the expression of three clusters of cellular rhythms. VIP is an integrator of cell clusters in neonatal *Cry1*/*Cry2*-deficient mice.

2.3.2 Clocks in Peripheral Organs

2.3.2.1 *Circadian Body Clock as a Hierarchal Multi-oscillator System*

By means of *in situ* hybridization of clock genes, a circadian rhythm is detected not only in the SCN but also in brain areas outside the SCN.[110,111] Even before the identification of clock genes, several studies documented the presence of circadian rhythms in peripheral tissues isolated from the body.[112,113] The concept of peripheral clocks was established by the discovery of circadian rhythms of clock gene expression in cultured peripheral tissues.[114] The circadian period and the phase differ from organ to organ *ex vivo*.[115]

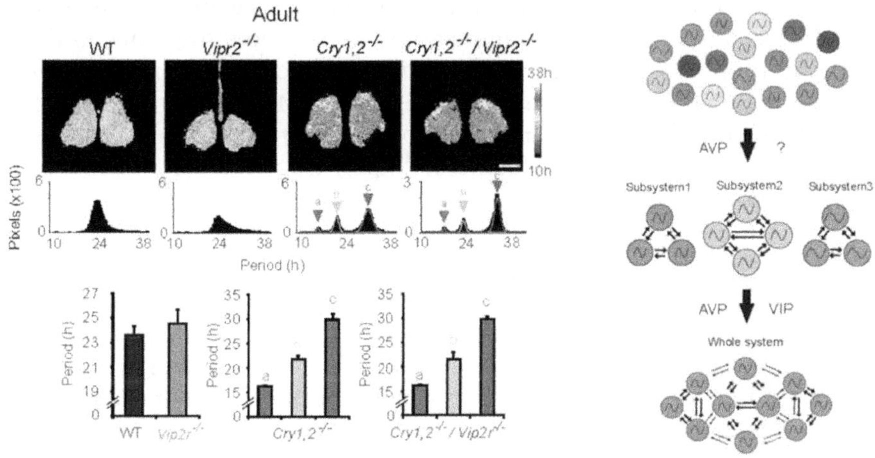

Figure 2.6 Cell clusters with different periodicities. Three cell clusters with mean periodicities of 16, 22 and 28 hours are detected in aperiodic *Cry1/Cry2*- and *Cry1/Cry2/Vipr2*-deficient adult mice (left). A single cluster is observed in periodic WT and *Vipr2*-deficient mice. Illustration of the three-cluster hypothesis for coherent circadian rhythm generation in the mouse SCN (right). Reproduced from ref. 77 with permission from AAAS, Copyright 2016, The Authors.

The circadian body clock consists of the central clock entrainable to the LD cycle and peripheral clocks driving the circadian rhythms specific to peripheral tissues and organs,[116] forming a multi-oscillatory circadian system.

However, the phase relation between the central and peripheral clocks is variable and depends on the tissues and on lifestyle. For example, the ontogeny of peripheral clocks is organ-specific.[116] Most peripheral clocks are influenced by the developmental changes in feeding patterns and physical activity. For neonates, the maternal rhythm is a strong synchronizer for several days after birth, and then the LD cycle becomes the major time cue. Around the time of weaning, the feeding habits of pups change from milk to pellets and their physical activities become more robust and rhythmic.

2.3.2.2 Entrainment to Non-photic Time Cues

Peripheral clocks are entrained by non-photic time cues such as meals, physical exercise and methamphetamine (MAP) administration.[117] These non-photic time cues are known to entrain a fraction of behavioural rhythms in the presence of an intact SCN and all behavioural rhythms in the absence of the SCN. However, it has not yet been determined which factors mediate and reset peripheral clocks. One possible mechanism is body temperature. Most non-photic time cues are associated with changes in physical activity, which alter body temperature. Changes in cellular temperature reset peripheral circadian clocks but not the central clock in the SCN.[118]

Body temperature altered by non-photic time cues is a candidate mediator. Another possibility is the reorganization of the coupling mechanisms among peripheral oscillators to build a *de novo* oscillation such as a food-entrainable (FEO)[119] or MAP-induced oscillation (MAO).[120] Prolonged MAP treatment in rats or mice changes the phase relations among peripheral oscillators in several brain structures,[121] pointing to a MAP-induced re-organization of brain oscillators.

2.3.2.3 Relation Between the Central and Peripheral Clocks

The mechanism of circadian signal transduction from central to peripheral clocks is not well understood. The mediator could be neuronal or humoral, and the transduction may come directly from the SCN clock or indirectly by means of a particular overt rhythm such as the sleep–wake cycle. The transplantation of the neonatal SCN to SCN-lesioned arrhythmic animals is well documented to restore circadian behavioural rhythms[122,123] but it does not restore rhythms in other variables such as adrenocortical hormone levels in blood.[124] In addition, characteristics of the donor SCN are transferred to the recipient's behavioural rhythm.[39] These findings suggest that humoral factor(s) rhythmically released from the SCN send(s) the circadian signal to the structure(s) responsible for behavioural rhythms. However, the site of action of humoral factor(s) is unknown. More recently, peripheral clocks in the liver, kidney and skeletal muscle were demonstrated to entrain to the transplanted SCN circadian rhythm, but those in the heart, spleen and adrenal medulla failed to entrain, suggesting that the responsiveness to the SCN signal differs among peripheral clocks.[125]

Neural outputs from the SCN are well described.[105,126] The major pathway leaves the SCN through the sub-paraventricular zone (SPZ). On the other hand, light signals through the SCN affect autonomic nervous functions such as melatonin synthesis[127] and neuronal activities of the sympathetic and parasympathetic nerves.[128,129] However, roles of the autonomic nervous system relative to central-peripheral coordination are not well understood.

2.3.3 Central and Peripheral Clocks in Humans

2.3.3.1 Circadian Clock in Humans

The circadian clock in humans is essentially the same as that in other mammals but unique in the following aspects, which seem to be exceptional in the animal kingdom. First, behavioural rhythms such as the sleep–wake cycle spontaneously desynchronize from circadian rhythms in deep body temperature or plasma melatonin and show periods deviating markedly from 24 hours. Second, behavioural rhythms occasionally show rhythms with periods of about 48 or 72 hours. Third, behavioural rhythms easily entrain to non-photic time cues. Fourth, sleep and wakefulness are consolidated, except in newborn babies and elderly people.[130]

These characteristics of human behavioural rhythms led us to propose a two-oscillator model, in which the mutually coupled circadian pacemaker and the sleep–wake cycle regulate specific peripheral rhythms.[131]

A two-process model was developed to explain the clock-dependent changes in sleep length and internal desynchronization.[132] However, the important concepts for the model, such as the S-process through which the sleep propensity increases during waking, the thresholds for falling asleep and waking-up, and the interval between falling asleep and awakening have not been materialized yet. The sleep–wake cycle entrains to non-photic time cues independently of the circadian pacemaker.[133] Non-SCN oscillations in the circadian domain have been shown to exist in the brain and are closely related to the behavioural rhythm in rodents.[117] Internal desynchronization is also demonstrated in MAP-treated model animals.[24] The two-process model does not easily explain these findings. Recently, the two-process model was extensively revised by the proposers themselves. The newly advanced model, based on relaxation oscillations, is essentially the same as the two-oscillator model.[134]

2.3.3.2 Internal Desynchronization

Spontaneous internal desynchronization is a unique phenomenon specific to the human circadian clock, which occurs mostly under free-running conditions.[2] During internal desynchronization, the circadian rhythms in core body temperature and plasma melatonin show circadian periods close to 25.0 hours, whereas the sleep–wake cycle exhibits a period up to 34 hours (long internal desynchronization; see Figure 2.7 left) or down to 18 hours (short internal desynchronization). Despite the long waking period in the long internal desynchronization case, the participants view it as a usual day and have three meals during the prolonged waking period. The sense of time is substantially changed and correlates positively with the duration of the waking period.[135] The opposite is observed in the case of short internal desynchronization. The long and short desynchronizations are predicted by the period of the free-running rhythm and the daily mean body temperature before desynchronization occurs.[2,136] During internal desynchronization, the circadian rhythm in core body temperature and the sleep–wake cycle show phase-trapping and relative coordination to each other, a sign of mutual interaction of two oscillators[137] (see Figure 2.7, right)

2.3.3.3 Significance of Human Peripheral Clocks

In humans, circadian rhythms in clock gene expression are detected in many peripheral organs, but the implication of these circadian rhythms is still unclear.[138] Are they synchronized with the sleep–wake cycle or with the circadian pacemaker in the SCN? Rhythms in plasma melatonin and deep body temperature are good markers of the circadian pacemaker. The circadian rhythm of plasma cortisol mostly reflects the circadian pacemaker.

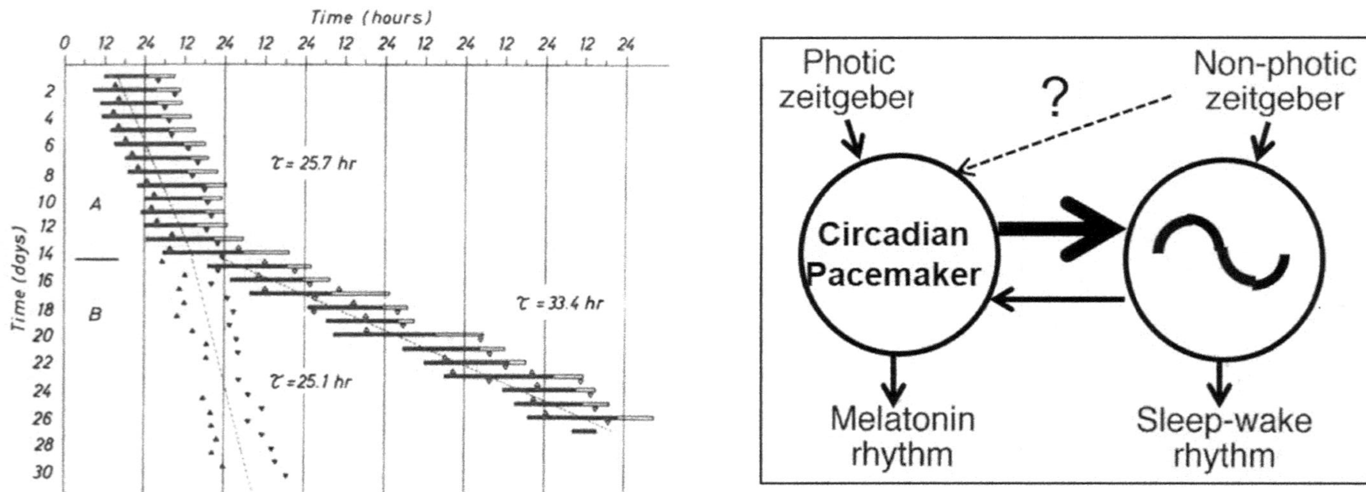

Figure 2.7 Spontaneous internal desynchronization in the human circadian system. Long internal desynchronization between the circadian rhythm in deep body temperature and the sleep/wake cycle is illustrated in a subject free-running under temporal isolation (left). The circadian period of body temperature is 25.1 hours, whereas the period of the sleep/wake cycle is 33.4 hours. Two-oscillator hypothesis for the human circadian system (right). The circadian pacemaker for body temperature and melatonin rhythms and the oscillator for the sleep/wake cycle are mutually coupled. The former oscillator entrains to photic time cues and the latter to non-photic time cue such as forced sleep schedule and exercise.

Other circadian rhythms, such as plasma growth hormone and prolactin, are mainly driven by the sleep–wake cycle or by the oscillator that drives it. Very few studies examined circadian rhythms in peripheral clocks under spontaneous internal desynchronization, which may answer the above question.

Circadian rhythms in clock gene expression could be useful for diagnosing clock-related diseases such as familial advanced sleep-phase syndrome (FASPS).[139] Blood samples are used to screen for mutations in clock genes. Patients suffering from FASPS show altered circadian periods in cultured peripheral tissues.[140]

2.4 Circadian Rhythms in Susceptibility and Responsiveness: Implication for Screening, Diagnosis and Treatment

Circadian rhythms are also known to characterise various responses of body functions to internal as well as external perturbations. The circadian susceptibility to drugs and medical treatments such as radiation therapy has been known for many years and is of special importance for screening beneficial materials, medical diagnosis and treatment. For example, the anesthetic effect of lidocaine is much greater in mid-afternoon than in the early morning or the late evening.[141] Most best-selling drugs in the USA such as Tamiflu for influenza and Ritalin for narcolepsy directly target the product of rhythmic genes, suggesting that these drugs may benefit from timed dosage.[142] The principles and procedures of chronotherapy are reviewed elsewhere in this book. Hrushesky and März[143] systematically surveyed the time dependency of anticancer agents and clinical trials based on timed dosing.

2.4.1 Principle of Timed Chemotherapy for the Treatment of Malignant Tumours

Cell proliferation in malignant tumours has been reported to be a function of time of day, but this is not invariably the case. While the reason is not precisely known, tumour clock disruption is suspected to result in the temporal deregulation of the cell cycle among malignant cells. On the other hand, the immune system of patients, a strong biological defence, and the tolerance of the body to chemical toxicities also show circadian rhythms. Therefore, the best time for chemotherapy or radiation should be selected in order to minimize the toxicity and maximize favorable effects. For example, the toxicity of cisplatin, a widely used anti-cancer drug, is maximal when administered at the transition from the rest to active phase and minimal in the late active phase in nocturnal animals.[144] On the other hand, doxorubicin, another anti-cancer drug, showed maximal toxicity in the middle of the active phase and minimal toxicity in the rest phase.[145] Therefore, the lowest toxicity and the highest therapeutic benefit are, at least theoretically, expected with daytime administration of doxorubicin and nighttime application of cisplatin in experimental animals.[146]

2.4.2 Tumour Response and Patient Survival

Based on animal experiments, a clinical trial was performed. Ovarian cancer patients treated with a combination of cisplatin and doxorubicin were assigned to one of four different schedules: (A) no consistent sequence or intervals between the two drugs; (B) doxorubicin at 0600 followed by cisplatin at 1800; (C) doxorubicin at 18:00 followed by cisplatin at 0600; and (D) alternation between (B) and (C).[144] Results showed that nobody in group A survived more than 5 years. Patients in group B had a 5-year survival rate of 44% and those in group C had a survival rate of 11%. Patients in group D had a survival rate of 75%, indicating that not only the therapeutic benefit but also the toxicity should be considered. If these results are confirmed, timed chemotherapy would be the first choice for ovarian cancer treatment.

2.4.3 New Frontiers for Cancer Treatment

Recently, anticancer approaches proposed by Suili *et al.*[147] use drugs that directly or indirectly target the circadian clock components (clock-drugs). The rationale for this treatment is threefold. First, clock drugs can have pleiotropic effects on multiple cancer hallmarks. Second, clock drugs can sustain rhythms in normal tissue in which circadian function may be disrupted by paracrine factors from tumours. Third, due to the nature of the interlocked feedback loop, a normal circadian clock is resilient to wide fluctuations in the cellular biochemical environment. The molecular resilience of a normal circadian clock may attenuate any adverse effect or toxicity of clock drugs.

The proposal is an attempt to search for new mechanisms to conquer cancers. However, our knowledge is still limited regarding the link between the circadian oscillation in cells or tissues and the proliferation of cancer cells.

References

1. J. Aschoff, *Handbook of behavioral neurobiology: Volume 4: Biological Rhythms*, Plenum Press, New York, 1981, pp. 81–93.
2. R. A. Wever, *The Circadian System of Man: Results of Experiments Under Temporal Isolation*, Springer-Verlag, Berlin, 1979.
3. S. S. Campbell, D. Dawson and J. Zulley, *Sleep*, 1993, **16**, 638.
4. C. S. Pittendrigh and S. Daan, *J. Comp. Physiol.*, 1976, **106**, 223.
5. T. Endo, S. Honma, S. Hashimoto and K. Honma, *Jpn. J. Physiol.*, 1999, **49**, 425.
6. A. Wirz-Justice, R. A. Wever and J. Aschoff, *Naturewissenschaften*, 1984, **71**, 316.

7. S. Daan, D. Damassa, C. S. Pittendrigh and E. R. Smith, *Proc. Natl. Acad. Sci. U. S. A.*, 1975, **72**, 3744.
8. L. P. Morin, K. M. Fitzgerald and I. Zucker, *Science*, 1977, **196**, 305.
9. R. Wever, *Experientia*, 1984, **40**, 1226.
10. F. C. Davis, J. M. Darrow and M. Menaker, *Am. J. Physiol.*, 1983, **244**, R93.
11. Z. Model, M. P. Butler, J. LeSauter and R. Silver, *Horm. Behav.*, 2015, **73**, 1.
12. J. Aschoff, J. Figala and E. Pőppel, *J. Comp. Physiol. Psychol.*, 1973, **85**, 20.
13. N. Yamada, K. Shimoda, K. Takahashi and S. Takahashi, *Physiol. Behav.*, 1986, **36**, 357.
14. C. S. Pittendrigh and S. Daan, *J. Comp. Physiol., A*, 1976, **106**, 253.
15. K. Honma, S. Honma and T. Hiroshige, *Jpn. J. Physiol.*, 1985, **35**, 643.
16. H. Illnerová and J. Vaněček, *J. Comp. Physiol., A*, 1982, **145**, 539.
17. C. S. Pittendrigh and S. Daan, *J. Comp. Physiol., A*, 1976, **106**, 291.
18. C. S. Pittendrigh, *The Neurosciences: Third Study Program*, MIT Press, Cambridge, 1974.
19. C. S. Pittendrigh, *Handbook of behavioral neurobiology: Volume 4: Biological Rhythms*, Plenum Press, 1981, pp. 81–93.
20. Y. Yamanaka, S. Honma and K. Honma, *J. Biol. Rhythms*, 2018, **33**, 614.
21. Y. Yamanaka, S. Honma and K. Honma, *Sci. Rep.*, 2023, under review.
22. N. Inagaki, S. Honma, D. Ono, Y. Tanahashi and K. Honma, *Proc. Natl. Acad. Sci. U. S. A.*, 2007, **104**, 7664.
23. T. Yoshikawa, N. F. Inagaki, S. Takagi, S. Kuroda, M. Yamasaki, M. Watanabe, S. Honma and K. Honma, *Sci. Rep.*, 2017, **7**, 8210.
24. S. Masubuchi, S. Honma, H. Abe, K. Ishizaki, M. Namihiara, M. keda and K. Honma, *Eur. J. Neurosci.*, 2000, **12**, 4206.
25. A. Natsubori, K. Honma and S. Honma, *Eur. J. Neurosci.*, 2013, **37**, 251.
26. S. M. Reppert and D. R. Weaver, *Nature*, 2002, **418**, 935.
27. J. S. Takahashi, *Nat. Rev. Genet.*, 2017, **18**, 164.
28. N. Preitner, F. Damiola, L. Lopez-Molina, J. Zakany, D. Duboule, U. Albrecht and U. Schibler, *Cell*, 2002, **110**, 251.
29. H. R. Ueda, W. Chen, A. Adachi, H. Wakamatsu, S. Hayashi, T. Takasugi, N. Nagano, K. Nakahama, Y. Suzuki, S. Sugano, M. Iino, Y. Shigeyoshi and S. Hashimoto, *Nature*, 2002, **418**, 534.
30. H. Abe, S. Honma, M. Namihira, Y. Tanahashi, M. Ikeda, W. Yu and K. Honma, *Mol. Brain Res.*, 1999, **66**, 104.
31. H. Cho, X. Zhao, M. Hatori, R. T. Yu, G. D. Barish, M. T. Lam, L.-W. Chong, L. DiTacchio, A. R. Atkins, C. K. Glass, C. Liddle, J. Auwerx, M. Downess, S. Panda and R. M. Evans, *Nature*, 2012, **485**, 123.
32. R. Ikeda, Y. Tsuchiya, N. Koike, Y. Umemura, H. Inokawa, R. Ono, M. Inoue, Y. Sasawaki, T. Grieten, N. Okubo, K. Ikoma, H. Fujiwara, T. Kubo and K. Yagita, *Sci. Rep.*, 2019, **9**, 10171.
33. H. Yoshitane, Y. Asano, A. Sagami, S. Sakai, Y. Suzuki, H. Okamura, W. Iwasaki, H. Ozaki and Y. Fukada, *Commun. Biol.*, 2019, **82**, 300.
34. S. Honma, T. Kawamoto, Y. Takagi, K. Fujimoto, M. Noshiro, Y. Kato and K. Honma, *Nature*, 2002, **419**, 841.

35. T. Kawamoto, M. Noshiro, F. Sato, K. Maemura, N. Takeda, R. Nagai, T. Iwata, K. Fujimoto, M. Furukawa, K. Miyazaki, S. Honma, K. Honma and Y. Kato, *Biochem. Biophys. Res. Commun.*, 2004, **313**, 117.
36. M. J. Rossnr, H. Oster, S. P. Wichert, L. Reinecke, M. C. Wehr, J. Reineck, G. Eichele, R. Tanja and K.-A. Nave, *PLoS One*, 2008, **3**, e2762.
37. A. Nakashima, T. Kawamoto, K. K. Honda, T. Ueshima, M. Noshiro, T. Iwata, K. Fujimoto, H. Kubo, S. Honma, N. Yorioka, N. Kohno and Y. Kato, DEC1 modulates the circadian phase of clock gene expression, *Mol. Cell Biol.*, 2008, **28**, 4080.
38. N. Kon, T. Hirota, T. Kawamoto, Y. Kato, T. Tsubota and Y. Fukada, *Nat. Cell Biol.*, 2008, **10**, 1463.
39. M. R. Ralph and M. Menaker, *Science*, 1988, **241**, 1225.
40. P. L. Lowrey, K. Shimomura, M. P. Antoch, S. Yamazaki, P. D. Zemenides, M. R. Ralph, M. Menaker and J. S. Takahashi, *Science*, 2000, **288**, 483.
41. Q. J. Meng, L. Logunova, E. S. Maywood, M. Gallego, J. Lebiechi, T. M. Brown, M. Sladek, A. S. Semikohodskii, N. R. J. Glossop, H. D. Piggins, J. E. Chesham, D. A. Bechtold, S. H. Yoo, J. S. Takahashi, D. M. Virshup, R. P. Boot-Handford, M. H. Hastings and A. S. I. Loudon, *Neuron*, 2008, **58**, 78.
42. Y. Xu, Q. S. Padiath, R. E. Shapiro, C. R. Jones, S. C. Wu, N. Saigoh, K. Saigoh, L. J. Patcek and Y. H. Fu, *Nature*, 2005, **434**, 640.
43. J. P. Etchegaray, K. K. Machida, E. Noton, C. M. Constance, R. Dallmann, M. N. Di Napoli, J. P. DeBruyne, C. M. Lambert, E. A. Yu, S. M. Reppert and D. R. Weaver, *Mol. Cell. Biol.*, 2009, **29**, 3853.
44. M. Sudo, K. Sasahara, T. Moriya, M. Akiyama, T. Hamada and S. Shibata, *Neuroscience*, 2003, **121**, 493.
45. M. Muñoz, S. N. Peirson, M. W. Hankins and R. G. Foster, *J. Biol. Rhythms*, 2005, **20**, 3.
46. R. Hamnett, P. Crosby, J. E. Chesham and M. H. Hastings, *Nat. Commun.*, 2019, **10**, 542.
47. D. M. Edgar and W. C. Dement, *Am. J. Physiol.*, 1991, **261**, R928.
48. E. G. Marchant and R. E. Mistlberger, *Physiol. Behav.*, 1996, **60**, 657.
49. E. G. Marchant, N. V. Watson and R. E. Mistlberger, *J. Neurosci.*, 1997, **17**, 7974.
50. J. D. Glass, G. H. Grossman, L. Farnbauch and L. DiNardo, *J. Neurosci.*, 2003, **23**, 7451.
51. P. C. Yannielli and M. E. Harrington, *Peptides*, 2001, **22**, 547.
52. C. Fukuhara, J. M. Brewer, J. C. Dirden, E. L. Bittman, G. Tosini and M. E. Harrington, *Neurosci. Lett.*, 2001, **314**, 119.
53. T. E. Dudley, L. A. Dinardo and J. D. Glass, *J. Neurophysiol.*, 1999, **81**, 1469.
54. J. D. Glass, L. A. DiNardo and J. C. Ehlen, *Brain Res.*, 2000, **859**, 224.
55. S. Daan, D. Damassa, C. S. Pittendrigh and E. R. Smith, *Proc. Natl. Acad. Sci. U. S. A.*, 1975, **72**, 3744.

56. I. N. Karatsoreos, A. Wang, J. Sasanian and R. Silver, *Endocrinology*, 2007, **148**, 5487.
57. M. P. Butler, I. N. Karatsoreos, J. LeSauter and R. Silver, *Endocrinology*, 2012, **153**, 2344.
58. Z. Model, M. P. Butler, J. LeSauter and R. Silver, *Horm. Behav.*, 2015 **73**, 1.
59. I. N. Karatsoreos, M. P. Butler, J. Lesauter and R. Silver, *Endocrinology*, 2011, **152**, 1970.
60. Z. Travnickova-Bendova, N. Cermakian, S. M. Reppert and P. Sassone-Corsi, *Proc. Natl. Acad. Sci. U. S. A.*, 2002, **99**, 7728.
61. Z. C. Murphy, P. Pezuk, M. Menaker and M. T. Sellix, *Biol. Reprod.*, 2013, **89**, 35.
62. J. M. Coulson, C. E. Fiskerstrand, P. J. Woll and J. P. Quinn, *Biochem. J.*, 1999, **344**, 961.
63. E. E. Zhang, Y. Liu, R. Dentin, P. Y. Pongsawakul, A. C. Liu, T. Hirota, D. A. Nusinow, X. Sun, S. Landais, Y. Kodama, D. A. Brenner, M. Montminy and S. A. Kay, *Nat. Med.*, 2010, **16**, 1152.
64. L. P. Morinl, *Exp. Neurol.*, 2013, 243**4**.
65. A. Kalsbeek and R. M. Buijs, *Cell Tissue Res.*, 2002, **309**, 109.
66. R. Silver, J. LeSauter, P. A. Tresco and M. N. Lehman, *Nature*, 1996, **382**, 810.
67. W. J. Schwartz and S. M. Reppert, *J. Neurosci.*, 1985, **5**, 2771.
68. K. Bae, X. Jin, E. S. Maywood, M. H. Hastings, S. M. Reppert and D. R. Weaver, *Neuron*, 2001, **30**, 525.
69. B. Zheng, U. Albrecht, K. Kaasik, M. Sage, W. Lu, S. Vaishnav, Q. Li, Z. S. Sun, G. Eichele, A. Bradley and C. C. Lee, *Cell*, 2001, **105**, 683.
70. J. S. Pendergast, R. C. Friday and S. Yamazaki, *J. Neurosci.*, 2010, **30**, 12179.
71. S. Steinlechner, B. Jacobmeier, F. Scherbarth, H. Dernbach, F. Kruse and U. Albrecht, *J. Biol. Rhythms*, 2002, **17**, 202.
72. M. Doi, H. Shimatani, Y. Atobe, I. Murai, H. Hayashi, Y. Takahashi, J. M. Fustin, Y. Yamaguchi, H. Kiyonari, N. Koike, K. Yagita, C. Lee, M. Abe, K. Sakimura and H. Okamura, *Nat. Commun.*, 2019, **10**, 2563.
73. T. Yoshikawa, Y. Nakajima, Y. Yamada, R. Enoki, K. Watanabe, M. Yamazaki, K. Sakimura, S. Honma and K. Honma, *Eur. J. Neurosci.*, 2016, **42**, 2678.
74. G. T. van der Horst, M. Muijtjens, K. Kobayashi, R. Takano, S. Kanno, M. Takao, J. de Wit, A. Verkerk, A. P. Eker, D. van Leenen, R. Buijs, D. Bootsma, J. H. Hoeijmakers and A. Yasui, *Nature*, 1999, **398**, 627.
75. E. S. Maywood, E. J. Chesham, J. A. O'Brien and M. H. Hastings, *Proc. Natl. Acad. Sci. U. S. A.*, 2011, **108**, 14306.
76. D. Ono, S. Honma and K. Honma, *Nat. Commun.*, 2013, **4**, 1666.
77. D. Ono, S. Honma and K. Honma, *Sci. Adv.*, 2016, **2**, e1600960.
78. Y. Ban, Y. Shigeyoshi and H. Okamura, *J. Neurosci.*, 1997, **17**, 3920.
79. K. Shinohara, T. Funabashi and F. Kimura, *Mol. Brain Res.*, 1999, **63**, 262.

80. D. Ono, S. Honma, C. Schmal, T. Takumi, T. Kawamoto, K. Fujimoto, Y. Kato and K. Honma, *Sci. Rep.*, 2021, **11**, 19240.
81. D. Ono, S. Honma and K. Honma, *PLoS One*, 2013, **8**, e80615.
82. M. M. Canal-Corretger, T. Cambras, J. Vilaplana and A. Díez-Noguera, *Am. J. Physiol.*, 2000, **278**, R201.
83. K. Spoelstra, M. Oklejewicz and S. Daan, *J. Biol. Rhythms*, 2002, **17**, 520.
84. D. Abraham, R. Dallmann, S. Steinlechner, U. Albrecht, G. Eichele and H. Oster, *J. Biol. Rhythms*, 2006, **21**, 169.
85. D. P. King, Y. Zhao, A. M. Sangoram, L. D. Wilsbacher, M. Tanaka, M. P. Antoch, T. D. Steeves, M. H. Vitaterna, J. M. Kornhauser, P. L. Lowrey, F. W. Turek and J. S. Takahashi, *Cell*, 1997, **89**, 641.
86. N. Gekakis, D. Staknis, H. B. Nguyen, F. C. Davis, L. D. Wilsbacher, D. P. King, J. S. Takahashi and C. J. Weitz, *Science*, 1998, **280**, 1564.
87. M. H. Vitaterna, D. P. King, A. M. Chang, J. M. Kornhauser, P. L. Lowrey, J. D. McDonald, W. F. Dove, L. H. Pinto, F. W. Turek and J. S. Takahashi, *Science*, 1994, **264**, 719.
88. W. Nakamura, S. Honma, T. Shirakawa and K. Honma, K, *Nat. Neurosci.*, 2002, **5**, 399.
89. M. P. Butler, S. Honma, T. Fukumoto, T. Kawamoto, K. Fujimoto, M. Noshiro, Y. Kato and K. Honma, *J. Biol. Rhythms*, 2004, **19**, 126.
90. J. P. Debruyne, E. Noton, C. M. Lambert, E. S. Maywood, D. R. Weaver and S. M. Reppert, *Neuron*, 2006, **50**, 465.
91. J. P. DeBruyne, D. R. Weaver and S. M. Reppert, *Nat. Neurosci.*, 2007, **10**, 543.
92. M. K. Bunger, L. D. Wilsbacher, S. M. Moran, C. Clendenin, L. A. Radcliffe, J. B. Hogenesch, M. C. Simon, J. S. Takahashi and C. A. Bradfield, *Cell*, 2000, **103**, 1009.
93. C. H. Ko, Y. R. Yamada, D. K. Welsh, E. D. Buhr, A. C. Liu, E. F. Zhang, M. R. Ralph, S. A. Kay, D. B. Forger and J. S. Takahashi, *PLoS Biol.*, 2010, **8**, e1000513.
94. S. Shi, A. Hida, O. P. McGuinness, D. H. Wasserman, S. Yamazaki and C. H. Johnson, *Curr. Biol.*, 2010, **20**, 316.
95. S. Ray, U. K. Valekunja, A. Stangherlin, S. A. Howell, A. P. Snijders, G. Damodaran and A. B. Reddy, *Science*, 2020, **367**, 800.
96. S. Nishide, S. Honma and K. Honma, *Sci. Rep.*, 2018, **8**, 14765.
97. D. Ono, S. Honma, Y. Nakajima, S. Kuroda, R. Enoki and K. Honma, *Proc. Natl. Acad. Sci. U. S. A.*, 2017, **114**, E3699.
98. D. K. Welsh, D. E. Logothetis, M. Meister and S. M. Reppert, *Neuron*, 1995, **14**, 697.
99. S. Honma, T. Shirakawa, Y. Katsuno, M. Namihira and K. Honma, *Neurosci. Lett.*, 1998, **250**, 157.
100. S. Honma, W. Nakamura, T. Shirakawa and K. Honma, *Neurosci. Lett.*, 2004, **358**, 173.
101. S. Honma, D. Ono, Y. Suzuki, N. Inagaki, T. Yoshikawa, W. Nakamura and K. Honma, *Progress in Brain Research*, Elsevier, B.V., 2012, vol. 199, pp. 129–141.

102. Y. Hirata, R. Enoki, K. Kuribayashi-Shigetomi, Y. Oda, S. Honma and K. Honma, *Sci. Rep.*, 2019, **9**, 18271.
103. E. D. Herzog, T. Hermanstyne, N. J. Smyllie and M. H. Hastings, *Cold Spring Harbor Perspect. Biol.*, 2017, **9**, a027706.
104. A. N. van den Pol and K. L. Tsujimoto, *J. Neurosci.*, 1985, **15**, 1049.
105. R. Y. Moore and R. K. Leak, *Handbook of Behavioural Neurobiology, volume 12: Circadian Clocks*, Kluwer Academic/Plenum Publishers, New York, 2001, pp. 141–179.
106. W. Nakamura, S. Honma, T. Shirakawa and K. Honma, *Eur. J. Neurosci.*, 2001, **14**, 666.
107. S. Koinuma, T. Asakawa, M. Nagano, K. Furukawa, M. Sujino, K. H. Masumoto, Y. Nakajima, S. Hashimoto, K. Yagita and Y. Shigeyoshi, *Eur. J. Neurosci.*, 2013, **38**, 2832.
108. R. Enoki, S. Kuroda, D. Ono, M. T. Hasan, T. Ueda, S. Honma and K. Honma, *Proc. Natl. Acad. Sci. U. S. A.*, 2012, **109**, 21498.
109. H. Fukuda, I. Tokuda, S. Hashimoto and N. Hayasaka, *PLoS One*, 2011, **6**, e23568.
110. S. Honma, M. Ikeda, H. Abe, Y. Tanahashi, M. Namihira, K. Honma and M. Nomura, *Biochem. Biophys. Res. Commun.*, 1998, **250**, 83.
111. M. Namihira, S. Honma, H. Abe, Y. Tanahashi, M. Ikeda and K. Honma, *Neurosci. Lett.*, 1999, **267**, 69.
112. E. Bunning, Das Weiterlaufen der "physiologischen Uhr" im Saugerdarm ohne zentrale Steuerung, *Naturewissenschaften*, 1958, **45**, 68.
113. R. V. Andrews, *Gegenbaurs Morphol. Jahrb.*, 1971, 117**89**.
114. A. Balsalobre, F. Damiola and U. Schibler, *Cell*, 1998, **93**, 929.
115. S. Y. Nishide, S. Honma, Y. Nakajima, M. Ikeda, K. Baba, Y. Ohmiya and K. Honma, *Genes Cells*, 2006, **11**, 1173.
116. S. Y. Nishide, K. Hashimoto, T. Nishio, K. Honma and S. Honma, *Am. J. Physiol.*, 2014, **306**, R67.
117. K. Honma and S. Honma, *Eur. J. Neurosci.*, 2009, **30**, 1707.
118. E. D. Buhr, S. H. Yoo and J. S. Takahashi, *Science*, 2010, **330**, 379.
119. F. K. Stephan, *Physiol. Bahav.*, 1984, **32**, 663.
120. K. Honma, S. Honma and T. Hiroshige, *Physiol. Behav.*, 1987, **40**, 767.
121. A. Natsubori, K. Honma and S. Honma, *Eur. J. Neurosci.*, 2014, **39**, 229.
122. Y. Sawaki, I. Nihonmatsu and H. Kawamura, *Neurosci. Res.*, 1984, **1**, 67.
123. R. Drucker-Colín, R. Aguilar-Roblero, F. García-Hernández, F. Fernández-Cancino and F. B. Rattoni, *Brain Res.*, 1984, **311**, 353.
124. E. L. Meyer-Bernstein, A. E. Jetton, S. I. Matsumoto, J. F. Markuns, M. N. Lehman and E. L. Bittman, *Endocrinology*, 1999, **140**, 207.
125. H. Guo, J. M. Brewer, M. N. Lehman and E. L. Bittman, *J. Neurosci.*, 2006, **26**, 6406.
126. L. P. Morin, *Exp. Neurol.*, 2013, **243**, 4.
127. R. J. Wurtman, J. Axelrod and J. E. Fischer, *Science*, 1964, **143**, 1328.
128. A. Niijima, K. Nagai, N. Nagai and H. Nakagawa, *Physiol. Behav.*, 1993, **54**, 555.

129. A. Ishida, T. Mutoh, T. Ueyama, H. Bando, S. Masubuchi, D. Nakahara, G. Tsujimoto and H. Okamura, *Cell Metab.*, 2005, **2**, 297.
130. K. Honma, S. Hashimoto, A. Natsubori, S. Masubuchi and S. Honma, *Indian Sleep Med*, 2013, **8**, 51.
131. M. C. Moore-Ede, F. M. Sulzman and C. A. Fuller, *The Clocks That Times Us*, Harvard University Press, Cambridge, 1982.
132. S. Daan, D. Beersma and A. Borbély, *Am. J. Physiol.*, 1984, **246**, R161.
133. S. Hashimoto, K. Nakamura, S. Honma and K. Honma, *Sleep Biol. Rhythms*, 2004, **2**, 29.
134. A. Borbély, S. Daan, A. Wirz-Justice and T. Deboer, *J. Sleep Res.*, 2016, **25**, 131.
135. J. Aschoff, *Human Neurobiology*, 1985, **4**, 41.
136. S. Daan, S. Honma and K. Honma, *J. Biol. Rhythms*, 2013, **28**, 403.
137. R. E. Kronaue, C. A. Czeisler, S. F. Pilato, M. C. Moore-Ede and E. D. Weitzman, *Am. J. Physiol.*, 1982, **242**, R3.
138. G. H. Son, H. K. Cha, S. Chung and K. Kim, *J. Endocr. Soc.*, 2018, **2**, 444.
139. C. R. Jones, S. S. Campbell, S. E. Zone, E. Cooper, A. DeSano, P. J. Murphy, B. Jones, L. Czajkowski and L. Ptácek, *Nat. Med.*, 1999, **5**, 1062.
140. A. Hirano, G. Shi, C. R. Jones, A. Lipzen, L. A. Pennacchio, Y. Xu, W. C. Hallows, T. McMahon, M. Yamazaki, L. J. Ptáček and Y. H. Fu, *eLife*, 2016, **5**, e16695.
141. A. Reinberg and M. Reinberg, *Naunyn Schmiedebergs Arch. Pharmacol.*, 1977, **297**, 149.
142. R. Zhang, N. F. Lahens, H. I. Ballance, M. E. Hughes and J. B. Hogenesch, *Proc. Natl. Acad. Sci. U. S. A.*, 2014, **111**, 16219.
143. W. J. M. Hrushesky and W. J. März, *Biologic Rhythms Clinical and Laboratory Medicine*, Springer-Verlag, New York, 1992, pp. 611–634.
144. W. J. M. Hrushesky, F. Levi, F. Halberg and B. J. Kennedy, *Cancer Res.*, 1982, **42**, 945.
145. W. J. M. Hrushesky, I. Dell, J. Eaton and F. Halberg, *Proc. Am. Assoc. Cancer. Res.*, 1982, **23**, 12.
146. W. M. Hryniuk, M. N. Levine and L. Levin, *NCI monogr.*, 1986, **1**, 87.
147. G. Sulli, M. T. Y. Lam and S. Panda, *Trends Cancer*, 2019, **5**, 475.

CHAPTER 3

Synchronizers of Circadian Rhythms

YU TAHARA[a,b] AND SHIGENOBU SHIBATA*[a,b]

[a] School of Advanced Science and Engineering, Waseda University, Wakamatsu-cho 2-2, Shinjuku-ku, Tokyo, 162-8480, Japan; [b] Laboratory of Public Health, Graduate School of Biomedical and Health Sciences, Hiroshima University, Hiroshima, Japan
*Email: shibatas@waseda.jp

3.1 Introduction

Since the period of our body clock is not exactly 24 hours and may be slightly longer or shorter individually, it has to be adjusted to the actual clock time every day by using environmental factors (*i.e.*, sunlight, food, physical activity, or chemicals).[1] This adjustment is called "entrainment", and the environmental factors are called "synchronizers, Zeitgebers ("time-giver" in German), or entraining factors". Importantly, light is the strongest synchronizer for the central clock, which is the suprachiasmatic nucleus (SCN) located in the hypothalamus. Light information detected by the retina is transmitted to the SCN through the retino-hypothalamic tract.[2] Time information is transferred to the periphery through many physiological pathways, including the neural and endocrine systems, to regulate the peripheral clocks located in all the peripheral tissues. Hence, there are internal synchronizers also for synchronizing the body clocks. Instead of light entrainment in the SCN, peripheral clocks are directly controlled by factors such as food and physical activity, which operate independently of the light–dark cycle. For instance, classic experiments demonstrated that the

Chemical Biology No. 23
Chronobiology and Chronomedicine: From Molecular and Cellular Mechanisms to Whole Body Interdigitating Networks
Edited by Germaine Cornelissen and Tsuyoshi Hirota

Published by the Royal Society of Chemistry, www.rsc.org

locomotor activity or body temperature rhythm can be entrained to the daily feeding time (*i.e.*, mice receive food during a specific 4-hour window every day). In this case, clock gene expression and serum corticosterone rhythms are also entrained to the feeding cycle.[3] However, the SCN clock is not altered in this situation, suggesting that the light–dark signal is dominant in the SCN, but not in the peripheral clocks. Taken together, the internal and external synchronizers are the principal regulators for the central and peripheral clocks to maintain the appropriate time every day. In this chapter, the molecular mechanisms and roles of each synchronizer of the SCN and peripheral tissues are explained, and we will discuss the relationships between each synchronizer and physiological functions, including sleep, mood, memory, and metabolism.

3.2 Light Entrainment in the Central Clock

3.2.1 Photoreception in the Retina

The central clock is entrained not only by photic stimulation, such as environmental light through the retinal–hypothalamic tract (RHT), but also by non-photic stimulation, such as exercise and melatonin administration or other drug treatments to eliminate the difference between the external (environmental and social) clock and our self-oscillating circadian clock. Subsequently, the central clock activates neural signals, hormonal signals, locomotor activity, and other pathways to synchronize the brain and peripheral clocks with the central clock.[4] In photic stimulation-induced entrainment, the following three steps are involved: retinal photoreception, transmission of retinal signals to the SCN, and cellular signal transduction in the SCN neurons.

Rods and cones (classic photoreceptors) were believed to be the only cells responsible for all image and non-image forming processes; however, melanopsin has been discovered as a third new type of photoreceptor in the retina.[5] Melanopsin, belonging to the family of G protein-coupled receptors (GPCRs), is a photoreceptor that is sensitive to 480 nm blue light within intrinsically photosensitive retinal ganglion cells (ipRGCs).[6] Melanopsin plays a significant role in important physiological functions, especially circadian-related events, such as the pupillary light response, circadian entrainment to light stimulation, and the negative "masking" (masking defines a factor that influences the rhythmic output of the circadian clocks, independently of clock processes) of locomotor activity (bright light-induced suppression).[6] If the pupillary light response is lost, bright light causes strong light-induced entrainment. Although the pupillary light response is strong, bright light-induced entrainment exhibits a ceiling effect because of the occurrence of miosis. The negative and positive masking effects of light are important for maintaining a daily activity rhythm under light–dark conditions in mice with clock gene mutations, such as the *Cry1/2* double mutation. These mice acutely lose their activity rhythm under constant light conditions.[6]

There are several innervations from ipRGCs to brain areas, in both nocturnal and diurnal species; common targeted brain areas are involved in the circadian entrainment, pupillary light reflex, and higher visual functions.[6] Bilateral ipRGC innervation was found in the SCN of the hypothalamus, while contralateral ipRGC innervation was found in the intergeniculate leaflet (IGL) in the thalamus; this IGL area is known to be responsible for the dark pulse-induced entrainment (detailed explanation later).[7–11] In addition, the anterior/lateral nucleus and ventrolateral preoptic nucleus of the hypothalamus also receive projections from ipRGCs in mice and rats.[7,9,12] These hypothalamic nuclei may play a role in regulating sleep and instinctive behaviours. The innervation of ipRGCs to the olivary pretectal nucleus plays a critical role in light-induced pupil constriction, while the superior colliculus, which is associated with non-image-forming higher visual function is also innervated by ipRGCs.[7–11] Notably, the brain areas involved in emotional and cognitive regulation such as the amygdala complex, raphe nucleus, and lateral habenula receive innervation from ipRGCs.[13–15]

To determine the role of melanopsin in ipRGCs in light-induced SCN neuronal activity, rodless/coneless mice (rd/rd cl mice) and melanopsin knockout mice (*Opn4*$^{-/-}$ mice) were studied. The early transient phase of activation of SCN neurons is regulated by rods/cones, and the activity of the late sustained phase is regulated by melanopsin.[16–20] Other experiments have confirmed these results through slow and prolonged activation of SCN neurons in rd/rd cl mice.[20] The light-induced phase shift of locomotor activity and light-induced suppression of melatonin release were not impaired in rd/rd cl mice, suggesting the existence of other photoreceptors. Conversely, *Opn4*$^{-/-}$ mice show impairments in both light-induced phase shift (approximately 40% reduction compared to wild type mice) and negative masking effects on the locomotor activity rhythm, while showing a normal circadian rhythm under constant darkness conditions.[21–23] These results indicate that the classical photoreceptors play an important role in transmitting the irradiance information to the mouse SCN;[24] additionally, the results indicate that melanopsin can encode both transient and sustained responses to light. If we simplify the function of rods and cones and melanopsin on photoentrainment, rods may work at a lower light level, cones at an intermediate irradiance level, and melanopsin at a higher irradiance level.

Mice with genetically ablated or chemically damaged ipRGCs exhibit a free-running rhythm under light–dark conditions, suggesting the absence of photic entrainment in these mice. These mice lose light-induced phase-advance and -delay and the light-induced masking effect.[25–27] These studies have strongly demonstrated that the blue light photoreceptor melanopsin in ipRGCs plays an important role in light-induced phase entrainment and the masking of locomotor activity rhythm in animals.

3.2.2 Phase-response Curve (PRC) and Phase Transition Curve (PTC)

We know that circadian clocks do not have a cycle length of precisely 24 hours and exhibit a slightly longer or shorter free-running period. The internal clock in mice usually exhibits a shorter cycle; however, humans show a longer cycle. Therefore, the clock in mice requires a daily phase-delay adjustment, and the clock in humans needs a phase-advance adjustment to ensure that the clock is set to the "real" astronomical day.[28] The daily phase resetting signal is called the Zeitgeber, and the SCN is usually entrained to external light–dark signals. Light signals are not the only factors that provide entrainment. Exercise and hormones such as melatonin can also entrain the SCN clock as non-photic entrainment (see section below).

When mice are maintained under constant darkness, they exhibit free-running locomotor activity. Under such free-running conditions, if mice are exposed to a pulse of light at different times during the subjective daytime or nighttime, the phase shift effects on the endogenous clock vary between advances, delays, or no effect, depending on the timing of light exposure. Light exposure during the subjective daytime induces no or minute effects, while light exposure at the early phase of subjective nighttime (between CT12 and CT18) causes a phase-delay, and light exposure at the later phase of subjective nighttime (CT18-CT24) causes a phase-advance. CT stands for circadian time (where 1 CT hour equals the circadian period/24 hours); CT0 represents the beginning of a subjective day, and CT12 marks the initiation of activity in a nocturnal animal. These differential effects of light are described by the "PRC". Figure 3.1 illustrates how to obtain the PRC in the case of mice. The PRCs of animals, including rodents and humans, exhibit a similar pattern. Light exposure between CT12 and CT18 causes a delay in next day's activity onset, whereas light exposure between CT18 and CT24

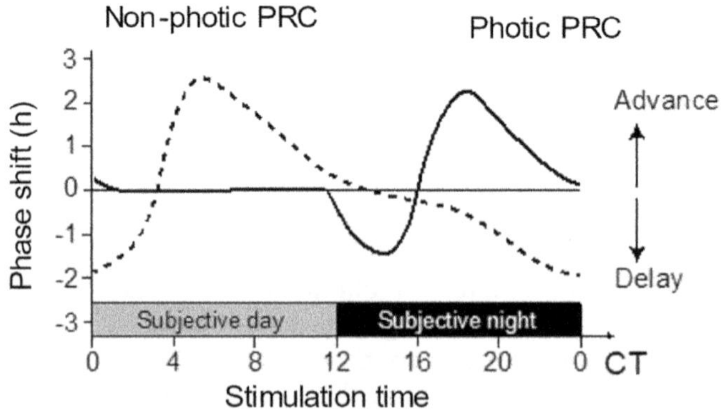

Figure 3.1 Example of the phase-response-curve (PRC) of photic (solid line) and non-photic (dashed line) entrainment in nocturnal animals. The vertical axis indicates phase shift values with positive arms representing phase advances and negative arms representing phase delays.

causes an advance in activity onset on the subsequent day. Notably, the PRCs of diurnal animals, including humans, show small delays with large advances, and those of nocturnal animals show large delays with small advances.[29] In addition, the shape of such a PRC is somehow dependent on the free-running period, whether it is shorter or longer than 24 hours. Undeniably, the size of the phase shift is affected by the duration of exposure and the intensity of light. Additionally, nocturnal animals show a higher sensitivity to light compared to diurnal animals, including humans, suggesting the requirement of stronger light signals for entrainment in humans. Another characteristic is that the speed of re-entrainment to a new light–dark cycle is different for an advance or delay of the environmental light–dark (LD) cycle. Additional days are required to adjust the internal clock in the case of a phase-advance (similar to an eastward trip, which represents a phase advance of the light–dark cycle), compared to that of a phase delay (similar to a westward trip, which represents a phase delay of the light–dark cycle). Considering the PRC and these characteristics, a world trip might be less stressful with less jetlag.[30]

The PTC is also obtained by plotting the new phase, which results from the perturbation of the Zeitgeber stimulus (y-axis) against the internal phase at which the pulse was given (x-axis). A type 0 PRC is explained by a slope of zero of the PTC, and type 1 is explained by a slope near one of the PTC. Whether organisms show a type 1 or type 0 PRC depends on the duration and/or magnitude of Zeitgeber strength and clock sensitivity. Many peripheral clocks and cellular clocks could show type 0 phase resetting, whereas the phase response at the organism level is exhibited as a type 1 PRC.

Abrupt changes from light to dark (dusk) and from dark to light (dawn) serve as entraining time cues in non-parametric entrainment. Continuous cyclic light exposure is considered an important time cue for entrainment in parametric entrainment. Depending on the niche of an organism or depending on the time of year, a combination of parametric and non-parametric entrainment will eventually explain how the circadian clock is entrained. However, only limited information is currently available regarding parametric entrainment, which needs further investigation in the future.

3.2.3 SCN Sub-regions and Retinal Inputs

The SCN has been divided into two sub-regions, the dorsal shell (DM-SCN) and ventral core areas (VL-SCN), based on inputs from the retina and other brain areas, and on neuropeptide distributions (Figure 3.2).[31] The topographical organization of specific neuronal clusters suggests a differential functional division of the VL- and DM-SCN.[32,33] The VL-SCN is located above the optic chiasm and is characterized by neurons that synthesize vasoactive intestinal polypeptides (VIPs) and gastrin-releasing peptides (GRPs). The DM-SCN contains arginine vasopressin (AVP) and calretinin neurons. VIP, GABA, GRP, enkephalin, and calretinin are known to be involved in the

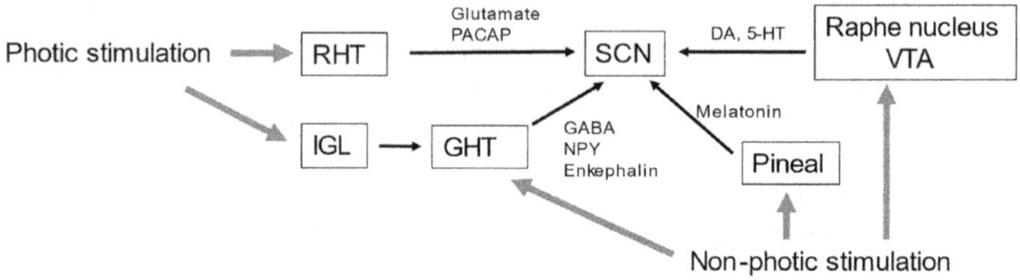

Figure 3.2 Schematic summary of photic and non-photic input pathways of the suprachiasmatic nucleus (SCN).

coupling of the VL- and DM-SCN. The light signal is transduced by photoreceptive cells located in the retina and transmitted to the VL-SCN *via* the RHT. The VL-SCN also receives inputs from the geniculohypothalamic tract (GHT), which originates in the thalamic IGL that receives direct projections from the retina. Transmitters of the RHT are glutamate and pituitary adenylyl cyclase-activating peptide (PACAP), and those of the GHT pathway are neuropeptide Y and GABA.[34,35]

Glutamate activates *N*-methyl-D-aspartate (NMDA) receptors, increases intracellular Ca^{2+} ions, and triggers the activation of diverse signal transduction cascades, including calmodulin kinase II (CaMKII), neuronal nitric oxide synthase activity, cAMP-dependent protein kinase (PKA), and mitogen-activated protein kinase (MAPK). The most common mechanism involved in this pathway is the phosphorylation of cAMP response element-binding protein (CREB). pCREB activates *Per1* and *Per2* transcription by binding to the CRE element in the promoter regions of these genes (Figure 3.3). Glutamate receptors (NMDA, AMPA, and metabotropic types) have been found in the SCN. Functional blockade of these receptors reduces light-induced and glutamate-induced phase shifts of circadian rhythms.

VIP and PACAP are members of a superfamily of structurally related peptide hormones, including glucagon, glucagon-like peptides, secretin, gastric inhibitory peptides, and growth hormone-releasing hormone. PACAP exerts its action through three G-protein-coupled receptors: PAC(1), VPAC(1), and VPAC(2). This family comprises receptors for all peptides structurally related to VIP and PACAP. PAC(1) receptors are selective for PACAP, whereas VPAC(1) and VPAC(2) respond to both VIP and PACAP with high affinity. VIP and PACAP play diverse and important roles in the control of circadian rhythms. Null or overexpression of PACAP or its receptors has demonstrated the fundamental role of PACAP in photic entrainment.[36–39] Both PAC1 and VPAC2 receptors are involved in photic entrainment; the activation of these receptors elicits AC-PKA activation and MAPK activation, causing a phase shift. The optogenetic activation of VIP neurons in the SCN causes phase shifts in the cellular circadian PER2::LUC oscillation *in vitro* and entrains the behavioural rhythm *in vivo*.[40,41] The impairment of VIP neurons attenuated the light-induced phase shifts of behavioural rhythms.[42]

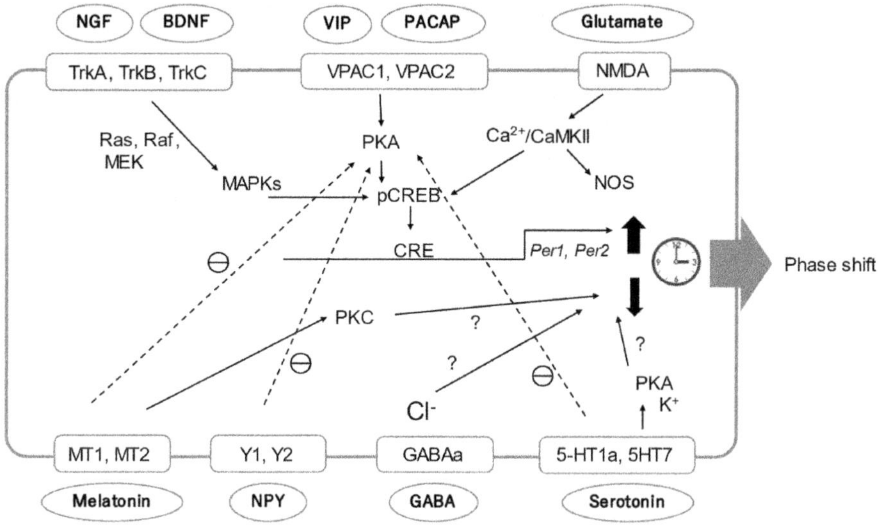

Figure 3.3 Schematic summary of the input signalling pathways in the suprachiasmatic nucleus (SCN), especially for the *Per1/2* induction and inhibition by neuropeptides or hormones.

In mammals, MAPKs include c-jun (JNK), p38, and ERK. ERKs are involved in the photic resetting of the SCN.[43,44] In the SCN, several peptides, such as PACAP, VIP, GRP, nerve growth factor (NGF), and epidermal growth factor (EGF) participate in photic activation through MAPKs, including ERKs.[45–47] However, each MAPK pathway subfamily plays an important role in the input pathway to the clock in different ways. The blocking of PKA activation decreases the phase-delay caused by glutamate administration *in vitro* and glutamate-induced *Per1* gene expression.[48] In addition, the blockade of PKA activation attenuates the light-induced and GRP-induced phase shift in locomotor behaviour, supporting the importance of PKA, not only in the photic signalling pathway, but also in the output signalling pathway from the VL-SCN to the DM-SCN circadian rhythm.[49] The pharmacological blockade of protein kinase C (PKC) attenuates the photic response in the mouse wheel-running rhythm,[50] and circadian variation is observed in the mRNA expression of PKC-α.[50] It has been observed that mice with PKC-α deficit exhibit a 50% decrease in their maximal behavioural response to light.[51] BDNF and NGF activate MAPKs and RAS pathways through tyrosine receptor kinase B (*TrkB*) activation.[52,53] Mice with an enhanced RAS activity revealed the potentiation of both light-induced phase shifts, and pERK1/2 and pCREB pathways.[54]

The phase shifts induced by light exposure are influenced by other factors, such as food, drugs, and non-photic entrainment signals. Such interactions of photic and non-photic entrainment will be mentioned in the following section.

3.2.4 Circadian Light Entrainment in Humans

In humans, similar to rodents, there is an integration of light signals at the level of ipRGCs,[28] and the concept of ipRGCs/rod/cone interaction helps to explain why melanopsin-like responses can be deduced from the action spectra.[55,56] In humans, light exposure early at night produces a phase delay. Additionally, light exposure in the early morning causes a phase advance in the circadian rhythm; these phase delays and phase advances are similar to those observed in rodent experiments, supported by the presence of similar PRC patterns in humans and rodents. However, compared to mice, humans require light stimulation of high irradiance (more than 100 lx) and long duration (>30 minutes) to achieve entrainment.[28] Since the discovery of artificial lighting, exposure to artificial light, especially during the late nighttime, instead of adapting to the natural environmental alternation between light and darkness, has led to specific circadian disruptions. Indeed, chronic circadian disruptions, such as in shift work and social jet lag, lead not only to sleep–wakefulness impairment, but also to impaired performance, as well as to a higher risk of poor health. The use of blue-enriched light-emitting devices immediately before bedtime suppresses melatonin secretion.[57] Melatonin levels were significantly suppressed after 2 hours of Apple iPad tablet usage.[58] Compared to reading a printed book, reading a light-emitting-eBook suppresses evening levels of melatonin and causes a 1.5-hour delay of the dim light melatonin onset (DLMO),[59] suggesting that the use of smartphones and PC tablets during the late night should be restricted. Regarding the use of light-emitting devices late at night, exposure to artificial light at night (LAN) affects various physiological variables, such as metabolic parameters (food intake, lipid/glucose metabolism, body weight, melatonin, and glucocorticoids) in both rodents and humans. Exposure to LAN has been shown to impact glucose homeostasis in humans and has been associated with an increased incidence of overweight, obesity, and atherosclerosis.[60] Individuals who are blind show poor entrainment of circadian rhythms, in addition to suffering from sleep disturbances and a depressed mood.[61] Several ophthalmic factors are associated with an age-related reduction in light transmission, such as senile miosis, reduction of pupil diameter, and reduction of retinal light transmission.[62]

3.3 Non-photic Entrainment in the Central Clock

3.3.1 Exercise and Arousal

The SCN can be entrained not only by light stimulation, but also by other stimuli, such as exercise, chemicals (including melatonin), and the feeding schedule. Neurotransmitters involved in photic entrainment include glutamate and PACAP, while those involved in non-photic entrainment include GABA, neuropeputide Y (NPY), 5-HT, enkephalin, melatonin, and dopamine. The SCN afferents involved in the non-photic modulation of circadian rhythms are divided into two major pathways. One pathway is GHT, including NPY, GABA, and enkephalin, originating from the IGL and terminating in the VL-SCN.[50,63,64] The other pathway is 5-HT, originating from the raphe nucleus and terminating in the VL-SCN.[65]

In an experiment on hamsters, triazolam, a short-acting hypnotic, increased the rate of adjustment of activity rhythms, a feat also achieved by making the animals active on a single occasion in the middle of their normal rest span, immediately after the shift in the LD cycle.[66,67] Entrainment by single or periodic social interactions is also possible, but these effects are often weak and appear to be mediated by social stimulation of arousal.[68,69] Additionally, there is evidence that physical exercise can change the phase of locomotor activity in humans.[70–72] The shape of the PRC under non-photic stimuli is different from that under a photic stimulus. The PRC with phase advances during the subjective day and small delays during the late subjective night is referred to as the non-photic PRC.[53] The PRCs for both triazolam[73] and novel wheel-running locomotor activity pulses,[74] similarly define a typical non-photic profile, with maximal phase advances of 2 to 3 hours when activity is induced during the mid-to-late subjective day.[75] Both novel running and gentle handling have shown an increase in c-Fos expression in IGL neurons; additionally, ablation of the IGL area blocked the phase shifts induced by these arousal treatments.[76,77] Hence, these data suggest the importance of the IGL to SCN pathways for arousal-induced non-photic entrainments.

3.3.2 NPY/GABA

The GHT from the IGL to the SCN contains NPY positive neurons,[78] and the injection of NPY into the SCN mimics a non-photic-induced PRC in hamsters and mice,[79] whereas antiserum against NPY can attenuate a wheel-running-induced phase-advance.[80] *In vitro* treatment of NPY to a SCN slice during the subjective day caused a phase advance of the circadian firing rhythm of SCN neurons.[81] The application of NPY *in vitro* and *in vivo* suppressed *Per1* and *Per2* gene expression in the SCN,[82,83] although induction with light increased *Per1* and *Per2* gene expression in the SCN.[84] Photic and non-photic entrainments interact because NPY injection into the SCN completely blocks the light-induced phase advance.[85] These studies strongly suggest the involvement of NPY activation during non-photic entrainment.

GABA-positive neurons are abundant in both the SCN and IGL.[78,86] Muscimol, a GABAa receptor agonist, injected into the SCN around midday caused a phase advance of the activity rhythm,[87,88] and the *in vitro* administration of muscimol caused a phase advance of electrical activity rhythms in the SCN.[89] The GABAa receptor modifier brotizolam induced a phase advance in the hamster's wheel-running rhythm with a reduction in *Per1* and *Per2* gene expression in the SCN.[90] The GABAergic stimulation by muscimol inhibits light-induced phase shifts,[91] and brotizolam administration attenuates a light-induced behavioural phase shift, and *Per1* and *Per2* gene expression in the SCN.[90] Thus, there are clear interactions between photic and non-photic entrainment influencing both the clock gene expression and behaviour.

3.3.3 Serotonin, Dopamine, and Melatonin

Serotonin (5-HT) reportedly modulates circadian rhythms by means of behavioural arousal phase shifts and photic entrainment.[92,93] Arousal-induced phase shifts involve the release of 5-HT[75] because the 5-HT receptor antagonist blocks phase shifts induced by arousal stimulation.[94] The median raphe innervates the SCN, while the dorsal raphe innervates the IGL.[95] 5-HT neurons from the dorsal raphe thus play an important role in arousal-induced phase shifts.[96] *In vitro* administration of 5-HT1a/7 receptor agonists induced a non-photic type of PRC in the SCN electrical activity rhythm,[81,97] and *in vivo* administration of 5-HT1a/7 receptor agonists resulted in typical non-photic phase shifts.[98–100] Both *in vitro* and *in vivo* injection during the mid-subjective day of 5-HT into the SCN induced phase advances of the SCN clock.[101,102] In nocturnal rodents, the administration of 5-HT1a/7 receptor agonists induced phase advances of the behavioural rhythm with a reduction in *Per1* and *Per2* gene expression in the SCN.[90,103] The activation of 5-HT7 receptors induces an increase in cAMP production,[104] and cAMP activates PKA. A phase shift of the electrical activity rhythm in the SCN induced by a 5-HT1a agonist can be blocked by PKA inhibitors.[105]

Dopamine (DA) neurotransmission controls behaviours important for survival, including voluntary movement and reward processing. Methamphetamine activates DA transmission, changes locomotor activity and motivational behaviour, and induces circadian-like behaviour.[106] Dopaminergic activation thus influences circadian physiology and behaviour. There is evidence identifying a novel, direct connection between DA neurons of the ventral tegmental area (VTA) and the SCN.[107] A recent study demonstrated that dopamine D1 receptor signalling within the SCN is necessary for properly timed resynchronization of activity rhythms to phase-shifted LD cycles. Additionally, the elevation of DA tone through the selective activation of VTA DA neurons accelerates photoentrainment. These findings demonstrate a clear role for the direct DA input to the SCN and the importance of the relationship between the circadian system and the neuromodulatory circuits that govern motivational behaviours.

As nocturnal animals, mice eat 80% of their food during the night when fed *ad libitum* with a healthy diet. In contrast, mice fed a high-calorie diet eat roughly 40% of their food during the day. Rewarding, high-energy foods induce dopamine release, and an increase in SCN dopaminergic tone attenuates the overall SCN excitability *via* dopamine D1 receptor signalling, resulting in the overconsumption of calories, reduced synchrony between the central and peripheral clocks, obesity and metabolic disorders. Remarkably, dopamine D1 receptor signalling in the SCN plays a role in enabling the feeding and metabolic defects associated with chronic exposure to high-calorie food. The dopamine system in the SCN plays a critical role in circadian, feeding and locomotor behaviours.

Melatonin released from the pineal gland conveys dark information to the brain, including the SCN, and to peripheral organs.[108] Melatonin application during the late daytime causes a phase-advance of the SCN clock through the activation of PKC *via* the melatonin type-2 receptor.[109–111] Melatonin induces *Per1* and *Per2* gene expression in the SCN when applied late during the daytime, but not during the middle of daytime.[108] This effect is significantly different from that of the novel wheel;[90] additionally, GABA agonists induce day-active non-photic entrainment, reducing *Per1* and *Per2* gene expression in the SCN.[90]

It is well known that the SCN clock is entrained by light stimulation (photic entrainment), while peripheral clocks are entrained by time-restricted feeding (non-photic entrainment). Food entrainment, notably of the liver clock occurs more rapidly in SCN-lesioned mice, suggesting that the SCN possesses an inhibitory control on non-photic entrainment.[112] Under free-feeding conditions, time-restricted feeding does not affect the phase of the SCN clock. However, under energy restricted conditions, the SCN clock shows a phase advance due to time-restricted feeding.[113] In a human experiment, 5-hour phase-delay of food timing, similar to time-restricted feeding from noon to midnight, revealed a phase delay of the autonomic activity rhythm, and a phase delay of *Per1* and *Per2* gene expression rhythms in subcutaneous adipocytes, but no effect on the rhythm of melatonin release.[114]

3.4 Food Entrainment in the Peripheral Clocks

3.4.1 Signalling Pathway

Peripheral tissues (*i.e.*, liver, fat, and intestine) respond to food intake and change their peripheral clocks. The PRC for entrainment to a single daily meal displays a phase advance caused by food provided in the early part to the end of the light span (at around Zeitgebar Times ZT3–ZT12). ZT0 is defined as the time of light onset in the animal room and ZT12 is the time of lights off. A phase delay occurs when food is given in the middle to the end of the dark span (ZT15–ZT24) in nocturnal mice.[115] Overnight fasting combined with food access limited to the middle of the light span leads to a phase advance of clock gene expression in the liver and kidney.[116] To detect clock entrainment by feeding stimuli, mice need to fast overnight to enhance their eating motivation and then be fed at an odd time, such as during the light span. The overnight fasting itself modulates peripheral clocks (Figure 3.4). The expression of *Per1* and *Per2* genes is activated by fasting-induced glucagon signalling in the liver.[117,118] An elevated level of glucagon activates the transcription factor CREB, and the phosphorylated CREB induces Per expression directly through the CRE site located in the promotor region of the Per genes. Additionally, CREB activation up-regulates other

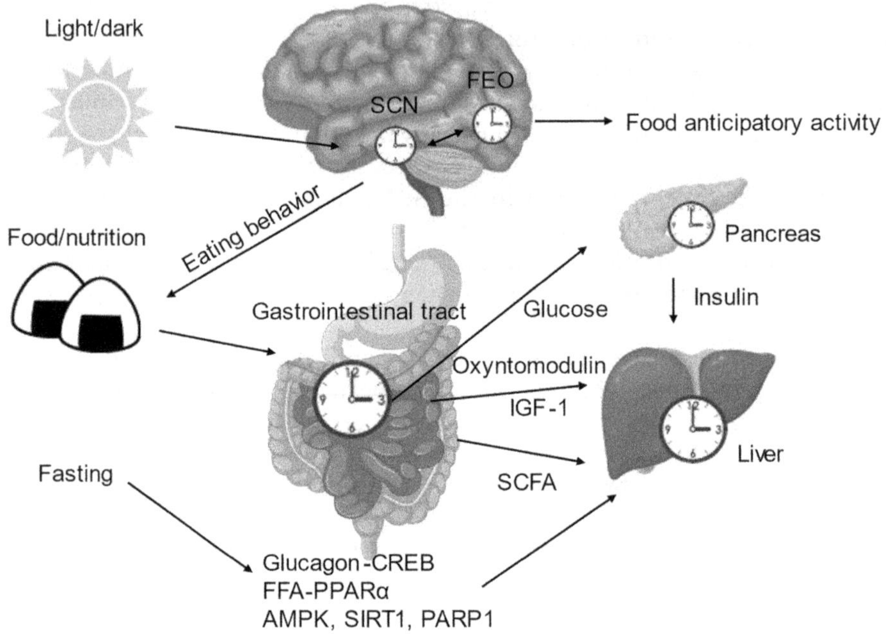

Figure 3.4 Schematic summary of food entrainment in peripheral tissues.

important clock-related genes, including *Pgc-1α*, *Sirt1*, and *Nampt*. This *Per1* elevation is diminished by re-feeding; however, *Per2* is observed to be further activated by re-feeding.[116,117] The fasting-induced glucagon-CREB-*Per* signal can be detected in *Cry1/2* double knockout mice,[118] which means that the molecular clockwork is not necessary in this effect. Moreover, glucagon up-regulates *Bmal1* gene expression in the liver if fasting lasts over 24 hours, which occurs through the CREB/CRTC1 pathway; re-feeding or insulin injection reduces this elevation.[119] Besides, *Rev-erbα* is up-regulated by fasting, which is caused by PPARα activation due to free fatty acid secretion caused by fasting-induced insulin down-regulation.[117] The up-regulated cellular metabolic sensors, *Sirt1* and *Nampt*, are also candidates for the signalling of fasting-re-feeding-induced peripheral clock entrainment.[120] The NAD^+/NADH feedback loop displays day–night variation.[121] Histone deacetylase *Sirt1*, whose activity depends on the NAD^+/NADH ratio, regulates CLOCK: BMAL1 transcriptional activity. PARP1, an ADP-ribosyltransferase, shares NAD^+-related common metabolic signalling pathways and adjusts the DNA binding of CLOCK: BMAL1. Notably, PARP1 KO mice display a delayed entrainment to the daytime-restricted feeding schedule in the liver.[120] As *Sirt1* and *Nampt* gene expression changes after fasting, NAD^+/NADH signalling also plays an important role in feeding entrainment in peripheral clocks. Another candidate is the cellular energy sensor, adenosine monophosphate-activated protein kinase (AMPK), which is phosphorylated and activated by fasting or lower glucose levels. P-AMPK stabilizes the CRY protein and lengthens the period of clock oscillation.[122] Conversely, overnight fasting advances the phase of peripheral clocks,[115,123] and hypocaloric feeding decreases the free-running period of the locomotor activity rhythm in constant darkness.[113] In contrast, mice fed a high-fat diet display a longer free-running period with dampened rhythmicity in their locomotor behaviour in constant darkness.[124] Taken together, the fasting-induced AMPK activation and its effect on the molecular clock might be a compensatory function for fasting-induced molecular clock changes.

Fasting-related hormonal changes partially regulate "food anticipatory activity," which is detected immediately before feeding time (2 to 3 hours before feeding) if the feeding time is restricted to a specific time window in mice.[125,126] Food anticipatory behaviour accounts for an internal clock system; however, the behaviour appears in SCN-lesioned mice, suggesting that there is another novel clock center for food entrainment outside of the SCN. Although research for finding the novel brain center of the food anticipatory activity has been a hot topic in the last two decades, researchers have failed to find it thus far. To date, several brain regions (arcuate nucleus, dorsomedial hypothalamic nucleus (DMH), ventromedial hypothalamic nucleus, and striatum), neural transmitters (dopamine, serotonin, *etc.*), and the core clock genes have been reported to be involved in this system.[127–131] Additionally, fasting-induced ketone (β-hydroxybutyrate) secretion from the liver and ghrelin secretion from the stomach, both up-regulate this anticipatory behaviour.[132,133] In contrast, the light-entrained SCN clock competes with food entrainment and food anticipatory activity through the neural connection between the SCN and the arcuate nucleus.[127] Although the mechanism of food anticipatory activity is still unclear, peripheral clock entrainment to food is completely independent of this behaviour. For example, DMH-lesioned mice or clock gene mutant mice attenuate the anticipatory behaviour; however, their peripheral clock can be entrained normally.[128,134,135] Thus, the peripheral clock directly responds to the feeding pattern.

Re-feeding after overnight fasting produces a powerful synchronizing effect on the molecular clock in peripheral tissues. Insulin, secreted from the β cells of the pancreas in response to a food-induced blood glucose increase, plays the main role in this entrainment.[116,136–138] Insulin directly modifies clock genes *Per2* and *Rev-erbα* in insulin-sensitive tissues, such as the liver and fat, but not in the lungs. Insulin binds to the insulin receptor and activates the PI3K/AKT/mTOR and MAPK cellular signalling pathways, which enhance *Per2* gene expression and promote the rapid translation of PER2.[136–138] Insulin-induced entrainment is detected in fibroblasts, cultured tissues, and peripheral clocks *in vivo* in a time-dependent manner. The enhancement of this insulin signal by re-feeding with incretins, such as exenatide, accelerates a food-induced phase shift in the peripheral clocks.[139] In contrast, inhibition by a competitive peptide inhibitor attenuates the phase shift.[137] The insulin signalling cascade of the insulin receptor depends on the fasting duration before re-feeding; longer fasting results in stronger insulin signalling. Food after longer fasting thus produces a more powerful entrainment in the liver.[123] Additionally, food intake stimulates oxyntomodulin secretion from the gut, and the gut hormone accelerates *Per1* and *Per2* entrainment in the liver after re-feeding. Oxyntomodulin is an anorexigenic incretin hormone produced from preproglucagon that controls energy metabolism in several tissues, including the brain and liver.[140] The inhibition of oxyntomodulin by antibody injection attenuates the food-induced entrainment in the liver. *Angptl8*, a hepatokine has also been identified to have a mechanistic role in food-induced peripheral clock entrainment.[141] *Angptl8* shows a day–night variation in the liver, controlled by glucocorticoids and the LXRα pathway. Adenovirus-mediated inhibition of *Angptl8* in the liver attenuates food-induced *Per* expression induction in the liver. IGF-1, the mediator of the growth hormone produced mainly from the liver, is also an important hormone for the clock entrainment by food.[138,142] The IGF-1 level is up-regulated by protein intake; additionally, IGF-1 synchronizes *Per2* gene expression rhythms *in vitro* and *in vivo*.[142] The inhibition of IGF-1 signalling by the receptor blocker attenuates the protein-induced peripheral clock entrainment. In addition to IGF-1, glucagon is also secreted in response to protein intake and enhances clock entrainment.[142] However, glucagon secretion is up-regulated by fasting and decreased by re-feeding. Therefore, the protein-related entrainment through IGF-1 and glucagon seems to be specific to the protein-enriched diets, an only protein diet, or specific amino acid supplementation.

Taken together, there are several candidates for the regulation of food-induced circadian clock entrainment in peripheral tissues. In addition to the peripheral clocks, food intake modulates the molecular clock in various brain areas, such as the hippocampus and DMH.[128] The fasting re-feeding paradigm immediately enhances *Per* expression in those brain areas.[128] However, the mechanism of these changes in the clock gene in the brain has not been investigated. The abovementioned mediators, such as insulin, oxyntomodulin, and IGF-1 work in the brain with the help of their receptors. Thus, these mediators might regulate clock genes in the brain after re-feeding, which needs to be verified in the future.

3.4.2 Functional Foods

The 2017 Nobel prize in Physiology or Medicine, awarded to J. C. Hall, M. Rosbash and M. W. Young for their discoveries of molecular mechanisms controlling the circadian clock, has accelerated translational studies from the basic understanding to the clinical level.[143] Since the circadian clock can be adjusted by food and nutrition, research focusing on the timing and components of food intake for clock entrainment is now a hot topic. This research is called "chrono-nutrition".[125,144] The detection of functional nutrients for clock entrainment is one of the main ideas for chrono-nutrition. In contrast, as the circadian clock controls metabolism and various physiological functions, the optimal timing of food/nutrient intake needs to be evaluated. Here, recent progress in research seeking functional foods for clock entrainment is reviewed.

From the mechanism of food entrainment in peripheral tissues (previous section), the idea of enhancing hormonal signals through functional nutrients presents an avenue for finding novel candidates. Insulin secretion is dependent on the postprandial glucose level; therefore, the rapidly digestible carbohydrates (*i.e.*, corn and rice starches) show a stronger phase shifting effect on the peripheral clock than the slower digestible carbohydrates (*i.e.*, potato starch) in re-fed mice.[145] Incretin secretion is promoted by omega-3 fatty acids (*i.e.*, docosahexaenoic acid and eicosapentaenoic acid, which are enriched in fish oil) through the orphan receptors GPR41, 43, and 120.[146] Re-feeding these fatty acids accelerates the food-induced phase shift in peripheral tissues in mice.[139] Additionally, insulin secretion is up-regulated by L-ornithine. The injection of L-ornithine itself during the early daytime induces a phase advance of peripheral clocks.[147] Incretin secretion is also enhanced by the short-chain fatty acids secreted from microbiota, and acids are reported to enhance food entrainment. Microbiota ferments soluble indigestible fibers and produces short-chain fatty acids, such as acetic acid, butyric acid, and propionic acids. Since these fatty acids have critical roles in metabolism and immune function, the soluble indigestible fibers consumed are called "prebiotics." Prebiotics might thus have a positive effect on peripheral clocks.[148,149]

The screening of functional nutrients or chemical compounds using an *in vitro* luciferase assay system that monitors clock gene expression has been conducted in several laboratories.[150,151] The automated assay system can screen thousands of chemicals at the same time, and it identified nobiletin as a nutritional synchronizer candidate. Nobiletin, a polymethoxyflavonoid abundant in citrus fruits has been reported to have anti-inflammatory, anti-oxidative stress, and anti-obesity functions. Nobiletin lengthens the circadian period, increases the amplitude and changes the phase of the PER2::LUC rhythm of embryonic fibroblasts *in vitro*, and prevents high-fat or aging-induced attenuation of the locomotor activity rhythm.[152,153] However, the SCN clock is not affected by nobiletin, evaluated through *ex vivo* SCN slice cultures. A similar effect is seen in caffeine; however, caffeine affects the SCN and peripheral clocks. The continuous consumption of coffee induces a longer free-running period in constant darkness in mice. Caffeine lengthens the period, increases the amplitude, and changes the phase of the molecular clock in mouse fibroblasts.[154–156] Caffeine consumption in the early night delays the circadian clock in mice and humans, suggesting that caffeine at night should be prohibited.[156] In addition, triterpenoids (*i.e.*, corosolic acid, cucurbitacin B, and celastrol), cinnamic acid, resveratrol, and harmine (harmala alkaloid) have been reported as synchronizers.[157–159] The modulation of the circadian clock by food or nutrition helps to counteract jet lag or prevent the direct effects of night-shift work. Future studies thus need to investigate further novel candidates and determine the best timing of nutrient intake.

3.5 Physical Activity and Stress Entrainment in Peripheral Clocks

3.5.1 Signalling Pathways

Exercise has the power to adjust the central and peripheral clocks in mice and humans. Several days of time-restricted access to the running wheel or providing treadmill exercise to mice around the middle of the day resulted in a phase advance of peripheral clocks.[160] In contrast, wheel access at the end of the dark span delays the peripheral clock. This evidence suggests that physical activity is an important synchronizer for the circadian system. Treadmill exercise notably produces a larger phase shift compared to running-wheel exercise, suggesting that forced exercise is more powerful than voluntary exercise.[160] This is because the mechanism of this entrainment is mainly through the hypothalamic–pituitary–adrenal (HPA) and sympatho–adrenal–medullary (SAM) axes (Figure 3.5).[161] Physical or psychological stress exposure also activates the HPA and SAM axes, and induces peripheral clock entrainment with circadian stage-dependence. Dexamethasone, a synthesized glucocorticoid receptor agonist, is commonly used for the synchronization of the cellular molecular clock in *in vitro* experiments.[162] Glucocorticoids secreted from adrenal glands induce *Per1/2* and *E4bp4* expression in peripheral tissues through the transcriptional activity of the nuclear glucocorticoid receptor.[163] The injection of dexamethasone in mice early during the daytime causes a phase advance of the peripheral clocks, which is consistent with the effect of exercise and stress exposure.[164] In humans, the administration of hydrocortisone (Cortef, Pfizer, NY, USA) in the late afternoon changes clock gene expression rhythms in blood monocellular cells.[165] However, in this experiment, the rhythm of endogenous melatonin or cortisol was not altered, suggesting that the central clock does not respond to hydrocortisone treatment. The SCN does not express the glucocorticoid receptor in

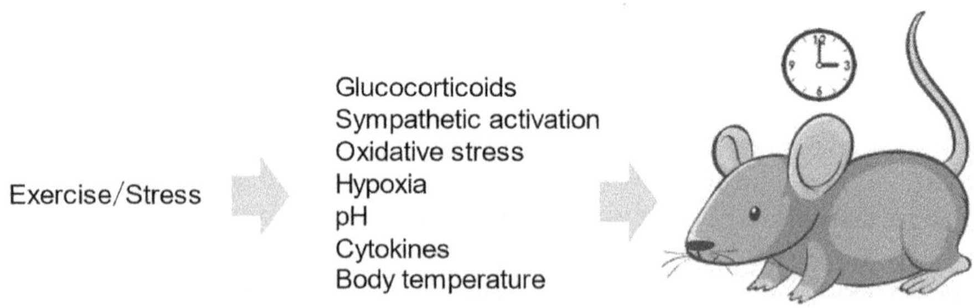

Figure 3.5 Schematic summary of exercise and stress entrainment in peripheral tissues. There are several possible pathways to entrain the circadian clock through exercise and stress. The detailed experimental results are reviewed in the text.

mice.[166] Importantly, the SCN itself regulates the 24-hour rhythm of glucocorticoid production with a peak around the time of awakening by controlling the HPA axis. Additionally, there is an internal clock in the adrenal gland, which coordinates the rate-limiting enzyme StAR (steroidogenic acute regulatory protein) to produce the rhythm of glucocorticoids.[167] A similar phase entrainment in the peripheral clock is seen by the injection of epinephrine, norepinephrine, or adrenergic receptor agonist in mice.[168,169] Sympathetic epinephrine or norepinephrine activation directly modulates clock genes through the transcription factor CREB.[168,170] Phosphorylated CREB binds to the promoter of *Per1* and *Per2*, inducing gene expression.

In addition to the HPA and SAM axes, there are several possible pathways for exercise- and stress-induced entrainment (Figure 3.5). Exercise and stress alter oxidative stress, body temperature, hypoxia, pH, and cytokines; these factors are reported to be synchronizers.[161] Reactive oxygen species (ROS), such as superoxide, hydroxyl radicals, and hydrogen peroxide, are elevated in peripheral tissues as a result of vigorous exercise or physical stress. ROS potently induce the expression of several clock genes, including *Per1*, *Per2*, *Cry1*, *Cry2*, *Clock*, *Bmal1*, *Rev-erbα*, and *Rev-erbβ* in cultured fibroblasts. Additionally, ROS activation through hydrogen peroxide treatment creates a "type 0" phase resetting of clock gene expression rhythms.[171,172] In addition, continuous ROS activation causes a dampened and lengthened oscillation of clock gene expression *in vitro*.[171] *Bmal1* KO mice display higher ROS in tissues and shorter longevity, suggesting a negative feedback between the disrupted circadian clock and ROS accumulation.[173] Mice treated with hydrogen peroxide show a phase shift of peripheral clocks in a time-dependent manner, and the PRC of hydrogen peroxide is similar to the PRC of exercise and stress exposure.[171] The response of ROS is also regulated by the circadian clock through the main transcriptional mediator NRF1, which displays day–night variation controlled by CLOCK:BMAL1.[174,175]

Environmental and internal body temperature changes are also believed to be synchronizers. A 24-hour cycle of temperature, alternating between lower and higher temperatures at 12-hour intervals, entrains the molecular clock gene expression in cultured cells and in *ex vivo* tissue cultures.[176] The SCN clock in hypothalamic slices is not synchronized with the temperature cycle of the incubator observed in brain slices, suggesting that the SCN is resistant to the synchronizing effects of temperature changes.[177] Impaired communication among SCN neurons through tetrodotoxin (a voltage-gated Na^+ channel blocker) or nimodipine (L-type calcium channel blocker) treatment renders the SCN sensitive to temperature resetting. Thus, neural communication in the SCN is implicated in maintaining the SCN's phase in the presence of environmental temperature changes (contrary to light input).[177] Since the inhibition of heat shock factor (*Hsf*) 1 by knockout in fibroblasts attenuates the temperature entrainment of clock genes, it suggests that the heat-HSF1 signalling pathway is involved.[178] Additionally, the inhibition of cold-inducible RNA-binding protein (*Cirp*) also reduces the response to the temperature cycle.[179] CIRP is activated by the temperature cycle and transfers *Clock* mRNA to the cytoplasm to promote translation. *In vivo*, putting mice in a warm water bath for 2 hours early during the light span for several days advances the phase of peripheral clocks, and the effect shows temperature dependence.[180] Taken together, exercise-induced heat production in the body may encourage peripheral clock adjustment.

Hypoxia (decreased oxygen level in tissues) caused by vigorous exercise might be another mediator of peripheral clock entrainment. Hypoxia activates the transcription factors HIF1α/HIF1β heterodimer and, notably, the complex promotes *Per* gene expression through its PER-ARNT-SIM (PAS)-domain binding to the E-box located in the promoter region of *Per* genes.[181,182] Additionally, HIF1α collaborates with BMAL1 protein to initiate *Per2* gene expression. Therefore, hypoxia modulates clock gene expression directly and resets the phase or amplitude. Similar to temperature entrainment, experimentally oscillating the oxygen concentration in the incubator (5% for 12 hours, alternating with 8% for 12 hours) induces cellular clock entrainment.[183] Conversely, chemically-induced continuous hypoxia down-regulates the amplitude and lengthens the period of clock gene expression rhythms in fibroblasts. In mice, a 2-hour exposure to low oxygen levels (14%) at the beginning of the active (dark) span accelerates the recovery of LD shift-induced experimental jetlag. Although it is still unknown whether the SCN itself responds to hypoxia, the data from experimental jetlag suggest that exercise-induced hypoxia may help to reset the circadian clock *in vivo*.[183]

Finally, pH or cytokines are also candidates accounting for the mechanism of exercise-induced entrainment. A lower or higher pH activates TGF-β, which in turn induces one of the clock genes *Dec1* to alter clock oscillation.[184] In addition to the TGF-1 pathway, activated apoptosis signal-regulating kinase (ASK) by cellular stress stimuli, including pH or osmolarity changes, and oxidative stress modify the phosphorylation of CLOCK to reset the clock oscillation.[185] Triple knockout of the ASK family (*Ask1/2/3*) in mice displays an attenuated light-induced phase shift of their locomotor activity rhythms. Additionally, cultured fibroblasts harvested from this triple knockout mouse completely diminished the osmotic stress-induced phase entrainment. Physical and psychological stress up-regulates the pro-inflammatory cytokines IL-6 and TNF-α in peripheral tissues. Inflammation due to these cytokines also modulates clock gene expression. Lipopolysaccharide treatment dampens the amplitude of clock gene expression in peripheral tissues and changes clock gene mRNA levels in the SCN of mice.[186] TNF-α and IL-1β inhibit the transcriptional activity of CLOCK:BMAL1 through E-box, and down-regulate the expression of clock genes, including *Per1/2/3* and *Dbp*.[187,188] Thus, pH changes and inflammation due to physical stress or exercise may promote entrainment.

3.6 Interaction of Synchronizers with Physiological Functions

3.6.1 Impact of Timing of Food Intake on Metabolism

The timing of food intake is important for preventing diet-induced obesity. Obesity resulting from a high-fat diet is not observed in mice on time-restricted feeding (TRF), with food access during a specific time window of 4 to 12 hours, either during the daytime or nighttime.[189,190] The amount of food per body weight was similar between mice fed with *ad libitum* and TRF mice. TRF resets peripheral clocks every day and produces a higher amplitude of clock gene expression in peripheral tissues. These clock modifications may involve the effect of TRF in preventing obesity. However, these changes are not necessary. Mice with whole-body clock gene knockout or liver-specific knockout F show a similar effect of TRF in preventing obesity.[191] TRF (*i.e.*, a 12-hour eating window during the dark span) also includes a 12-hour fasting window every day. Daily fasting results in various benefits not only for metabolism, but also for anti-aging.[192] The main players of fasting signalling are NRF2 (oxidative stress), AMPK and mTOR (autophagy, protein synthesis), PGC-1α (mitochondrial biogenesis), FOXO (cell survival), and SIRT1 (anti-aging, metabolism, inflammation, *etc.*). These pathways are also activated by calorie restriction, which is a 20–30% reduction in food intake. In humans, TRF (eating restricted to 8–12 hours per day) also counteracts obesity or helps pre-diabetic patients in losing body weight and improving blood parameters.[193] TRF is easier to handle for patients, compared with the classic calorie restriction treatment.[192] In addition to the anti-obesity effect, TRF also offers benefits in aging, cancer, and neurodegenerative disorders.[144] The application of TRF in a mouse model of Huntington's disease prevented motor disfunction.[194] Daily food intake with an appropriate fasting window thus helps to maintain health and rhythmicity of circadian clocks and prevents disease progression in humans and rodents.

3.6.2 Impact of Timing of Exercise on Metabolism

From the PRC of exercise-induced peripheral clock entrainment in mice, exercise in the early morning induces a phase advance, and exercise in the evening produces a phase delay.[161] In humans, the PRC of exercise is slightly complicated. Moderate-intensity exercise (60 minutes for 3 days) in the morning or afternoon displays a phase advance of salivary melatonin onset, and exercise at night (7–11 pm) produces a phase delay.[195] The morning–evening preference (chronotype) also affects the PRC.[196] The morning chronotype shows a similar phase response to exercise as described above; however, the evening chronotype shows a phase advance by both morning and late-night exercise. Taken together, the timing of exercise needs to be considered while maintaining the circadian clock; nevertheless, further research is needed to verify the above human evidence.

The timing of exercise also influences the benefits of exercise-related body weight change. "When to exercise" for losing weight is an important issue in the recent worldwide pandemic of obesity, and research on the circadian clock may answer the question. The optimal timing of exercise for losing weight might be in the morning compared to the evening. Two intervention studies (6 weeks or 10 months of intervention) for overweight and obese individuals showed the same outcome, namely, exercise in the morning leads to greater weight reduction than exercise in the evening.[197,198] Another study suggests that exercise before breakfast increases fat oxidation under overnight fasting conditions and may promote weight reduction.[199] In mice, exercise in the early active phase promotes the HIF1α signalling pathway, which induces glycolytic activation and lipid oxidation.[200] To prevent muscle atrophy, exercise in the morning is more powerful than in the evening. A mouse study using hindlimb-unloading-induced muscle atrophy demonstrates that weight bearing in the morning shows a lower reduction of muscle mass.[201] Atrogin1, the main player in muscle atrophy, shows day–night variation under the hindlimb-unloading condition; however, exercise in the morning strongly reduces this activation compared with exercise in the evening. In contrast, better sports performance is achieved in the evening than in the morning, because of the higher body temperature, sympathetic activity, and heart rate.[202] ZMP, as an endogenous AMPK activator, is higher in the body after evening exercise than after morning exrecise.[203] ZMP activates glycolysis and fatty acid oxidation to facilitate exercise performance in the evening. Taken together, the best timing of exercise depends on the purpose. Future studies need to find the optimal timing for each sport and purpose.

3.7 Conclusion

Entrainment is an important function of the circadian clock that is regulated under specific rules (*e.g.*, PRC or cell- or tissue-specific factors). Timing, duration, and size of external and internal Zeitgebers must be considered carefully to keep our body clock in synch. However, detailed mechanisms of entrainment need to be better understood in the future. In addition, the combined effects of each Zeitgeber should be examined to understand their effects in the real world for improving translational research.

Abbreviations

SCN	Suprachiasmatic nucleus
RHT	Retinal–hypothalamic tract
GPCR	G protein-coupled receptors
ipRGCs	Intrinsically photosensitive retinal ganglion cells
IGL	Intergeniculate leaflet

rd/rd cl mice	Rodless/coneless mice
PRC	Phase-response curve
CT	Circadian time
LD	Light–dark
PTC	Phase transition curve
DM-SCN	Dorsal shell
VL-SCN	Ventral core areas
VIP	Vasoactive intestinal polypeptide
GRP	Gastrin-releasing peptide
AVP	Arginine vasopressin
GHT	Geniculohypothalamic tract
PACAP	Pituitary adenylyl cyclase-activating peptide
NMDA	*N*-Methyl-D-aspartate
CaMKII	Calmodulin kinase II
PKA	cAMP-dependent protein kinase
MAPKs	Mitogen-activated protein kinase
CREB	cAMP response element-binding protein
NGF	Nerve growth factor
EGF	Epidermal growth factor
PKC	Protein kinase C
DLMO	Dim light melatonin onset
LAN	Light at night
NPY	Neuropeputide Y
5-HT	Serotonin
DA	Dopamine
VTA	Ventral tegmental area
ZT	Zeitgeber time
AMPK	Adenosine monophosphate-activated protein kinase
HPA	Hypothalamic–pituitary–adrenal
SAM	Sympathoadrenal-medullary
ROS	Reactive oxygen species
Hsf	Heat shock factor
Cirp	Cold-inducible RNA-binding protein
PAS	PER-ARNT-SIM
ASK	Apoptosis signal-regulating kinase
TRF	Time-restricted feeding

Acknowledgements

This work was partially supported by a Grant-in-Aid for Scientific Research (A) (19H01089) and Grant-in-Aid for Exploratory Research (19K21597) from the Japan Society for the Promotion of Science, and by the "Mirai Program" and "FOREST Program" from the Japan Science and Technology Agency (JST). The authors have no conflicts of interest to declare.

References

1. J. Bass and J. S. Takahashi, *Science*, 2010, **330**, 1349–1354.
2. C. S. Colwell, *Nat. Rev. Neurosci.*, 2011, **12**, 553–569.
3. M. Girotti, M. S. Weinberg and R. L. Spencer, *Am. J. Physiol.*, 2009, **296**, E888–E897.
4. P. Lewis, H. W. Korf, L. Kuffer, J. V. Groß and T. C. Erren, *BMJ Open Sport Exerc. Med.*, 2018, **4**, e000443.
5. I. Provencio and R. G. Foster, *Brain Res.*, 1995, **694**, 183–190.
6. M. Duda, A. Domagalik, P. Orlowska-Feuer, O. Krzysztynska-Kuleta, E. Beldzik, M. K. Smyk, A. Stachurska, H. Oginska, J. S. Jeczmien-Lazur, M. Fafrowicz, T. Marek, M. H. Lewandowski and T. Sarna, *Neurosci. Biobehav. Rev.*, 2020, **113**, 190–203.
7. J. J. Gooley, J. Lu, D. Fischer and C. B. Saper, *J. Neurosci.*, 2003, **23**, 7093–7106.
8. J. Hannibal, L. Kankipati, C. E. Strang, B. B. Peterson, D. Dacey and P. D. Gamlin, *J. Comp. Neurol.*, 2014, **522**, 2231–2248.
9. S. Hattar, M. Kumar, A. Park, P. Tong, J. Tung, K. W. Yau and D. M. Berson, *J. Comp. Neurol.*, 2006, **497**, 326–349.
10. J. L. Langel, L. Smale, G. Esquiva and J. Hannibal, *Front. Neuroanat.*, 2015, **9**, 93.
11. L. P. Morin, J. H. Blanchard and I. Provencio, *J. Comp. Neurol.*, 2003, **465**, 401–416.
12. A. Delwig, D. D. Larsen, D. Yasumura, C. F. Yang, N. M. Shah and D. R. Copenhagen, *PLoS One*, 2016, **11**, e0149501.
13. J. L. Ecker, O. N. Dumitrescu, K. Y. Wong, N. M. Alam, S. K. Chen, T. LeGates, J. M. Renna, G. T. Prusky, D. M. Berson and S. Hattar, *Neuron*, 2010, **67**, 49–60.
14. D. C. Fernandez, P. M. Fogerson, L. Lazzerini Ospri, M. B. Thomsen, R. M. Layne, D. Severin, J. Zhan, J. H. Singer, A. Kirkwood, H. Zhao, D. M. Berson and S. Hattar, *Cell*, 2018, **175**, 71–84.e18.
15. X. Li, C. Ren, L. Huang, B. Lin, M. Pu, G. E. Pickard and K. F. So, *Invest. Ophthalmol. Visual Sci.*, 2015, **56**, 8373–8381.
16. P. M. Smallwood, B. P. Olveczky, G. L. Williams, G. H. Jacobs, B. E. Reese, M. Meister and J. Nathans, *Proc. Natl. Acad. Sci. U. S. A.*, 2003, **100**, 11706–11711.
17. G. H. Jacobs and G. A. Williams, *Doc. Ophthalmol.*, 2007, **115**, 137–144.
18. L. S. Mure, C. Rieux, S. Hattar and H. M. Cooper, *J. Biol. Rhythms*, 2007, **22**, 411–424.
19. E. Drouyer, C. Rieux, R. A. Hut and H. M. Cooper, *J. Neurosci.*, 2007, **27**, 9623–9631.
20. T. M. Brown, J. Wynne, H. D. Piggins and R. J. Lucas, *J. Physiol.*, 2011, **589**, 1173–1194.
21. N. Mrosovsky and S. Hattar, *Chronobiol. Int.*, 2003, **20**, 989–999.
22. S. Panda, T. K. Sato, A. M. Castrucci, M. D. Rollag, W. J. DeGrip, J. B. Hogenesch, I. Provencio and S. A. Kay, *Science*, 2002, **298**, 2213–2216.

23. S. Panda, I. Provencio, D. C. Tu, S. S. Pires, M. D. Rollag, A. M. Castrucci, M. T. Pletcher, T. K. Sato, T. Wiltshire, M. Andahazy, S. A. Kay, R. N. Van Gelder and J. B. Hogenesch, *Science*, 2003, **301**, 525–527.
24. H. C. van Diepen, A. Ramkisoensing, S. N. Peirson, R. G. Foster and J. H. Meijer, *FASEB J.*, 2013, **27**, 4204–4212.
25. D. Göz, K. Studholme, D. A. Lappi, M. D. Rollag, I. Provencio and L. P. Morin, *PLoS One*, 2008, **3**, e3153.
26. M. Hatori, H. Le, C. Vollmers, S. R. Keding, N. Tanaka, T. Buch, A. Waisman, C. Schmedt, T. Jegla and S. Panda, *PLoS One*, 2008, **3**, e2451.
27. A. D. Güler, J. L. Ecker, G. S. Lall, S. Haq, C. M. Altimus, H. W. Liao, A. R. Barnard, H. Cahill, T. C. Badea, H. Zhao, M. W. Hankins, D. M. Berson, R. J. Lucas, K. W. Yau and S. Hattar, *Nature*, 2008, **453**, 102–105.
28. R. G. Foster, S. Hughes and S. N. Peirson, *Biology*, 2020, **9**, 180.
29. C. S. Pittendrigh, *Cold Spring Harbor Symp. Quant. Biol.*, 1960, **25**, 159–184.
30. C. I. Eastman and H. J. Burgess, *Sleep Med. Clin.*, 2009, **4**, 241–255.
31. R. Y. Moore, J. C. Speh and R. K. Leak, *Cell Tissue Res.*, 2002, **309**, 89–98.
32. S. Honma, M. Ikeda, H. Abe, Y. Tanahashi, M. Namihira, K. Honma and M. Nomura, *Biochem. Biophys. Res. Commun.*, 1998, **250**, 83–87.
33. L. Yan, I. Karatsoreos, J. Lesauter, D. K. Welsh, S. Kay, D. Foley and R. Silver, *Cold Spring Harbor Symp. Quant. Biol.*, 2007, **72**, 527–541.
34. M. Astiz, I. Heyde and H. Oster, *Int. J. Mol. Sci.*, 2019, **20**, 343.
35. S. Honma, *J. Physiol. Sci.*, 2018, **68**, 207–219.
36. C. S. Colwell, S. Michel, J. Itri, W. Rodriguez, J. Tam, V. Lelièvre, Z. Hu and J. A. Waschek, *Am. J. Physiol.*, 2004, **287**, R1194–R1201.
37. J. Hannibal, P. Brabet and J. Fahrenkrug, *Am. J. Physiol.*, 2008, **295**, R2050–R2058.
38. H. Hashimoto, N. Shintani and A. Baba, *Ann. N. Y. Acad. Sci.*, 2006, **1070**, 75–89.
39. S. Shen, C. Spratt, W. J. Sheward, I. Kallo, K. West, C. F. Morrison, C. W. Coen, H. M. Marston and A. J. Harmar, *Proc. Natl. Acad. Sci. U. S. A.*, 2000, **97**, 11575–11580.
40. J. R. Jones, M. C. Tackenberg and D. G. McMahon, *Nat. Neurosci.*, 2015, **18**, 373–375.
41. C. Mazuski, J. H. Abel, S. P. Chen, T. O. Hermanstyne, J. R. Jones, T. Simon, F. J. Doyle 3rd and E. D. Herzog, *Neuron*, 2018, **99**, 555–563.e555.
42. J. R. Jones, T. Simon, L. Lones and E. D. Herzog, *J. Neurosci.*, 2018, **38**, 7986–7995.
43. G. Q. Butcher, J. Doner, H. Dziema, M. Collamore, P. W. Burgoon and K. Obrietan, *J. Biol. Chem.*, 2002, **277**, 29519–29525.
44. A. N. Coogan and H. D. Piggins, *J. Neurosci.*, 2003, **23**, 3085–3093.
45. G. Q. Butcher, B. Lee, H. Y. Cheng and K. Obrietan, *J. Neurosci.*, 2005, **25**, 5305–5313.
46. H. Hao and J. Schwaber, *Brain Res.*, 2006, **1088**, 45–48.

47. R. Hamnett, P. Crosby, J. E. Chesham and M. H. Hastings, *Nat. Commun.*, 2019, **10**, 542.
48. S. A. Tischkau, E. A. Gallman, G. F. Buchanan and M. U. Gillette, *J. Neurosci.*, 2000, **20**, 7830–7837.
49. R. Sterniczuk, G. R. Yamakawa, T. Pomeroy and M. C. Antle, *Neurosci. Lett.*, 2014, **559**, 24–29.
50. D. R. Bonsall and G. S. Lall, *Chronobiol. Int.*, 2013, **30**, 460–469.
51. V. Jakubcakova, H. Oster, F. Tamanini, C. Cadenas, M. Leitges, G. T. van der Horst and G. Eichele, *Neuron*, 2007, **54**, 831–843.
52. M. S. Alessandro, D. A. Golombek and J. J. Chiesa, *Yale J. Biol. Med.*, 2019, **92**, 241–250.
53. D. A. Golombek and R. E. Rosenstein, *Physiol. Rev.*, 2010, **90**, 1063–1102.
54. T. Serchov, A. Jilg, C. T. Wolf, I. Radtke, J. H. Stehle and R. Heumann, *Mol. Neurobiol.*, 2016, **53**, 1843–1855.
55. G. C. Brainard, J. P. Hanifin, J. M. Greeson, B. Byrne, G. Glickman, E. Gerner and M. D. Rollag, *J. Neurosci.*, 2001, **21**, 6405–6412.
56. K. Thapan, J. Arendt and D. J. Skene, *J. Physiol.*, 2001, **535**, 261–267.
57. M. G. Figueiro, B. Wood, B. Plitnick and M. S. Rea, *Neuroendocrinol. Lett.*, 2011, **32**, 158–163.
58. B. Wood, M. S. Rea, B. Plitnick and M. G. Figueiro, *Appl. Ergon.*, 2013, **44**, 237–240.
59. A. M. Chang, D. Aeschbach, J. F. Duffy and C. A. Czeisler, *Proc. Natl. Acad. Sci. U. S. A.*, 2015, **112**, 1232–1237.
60. G. Fleury, A. Masís-Vargas and A. Kalsbeek, *Obesity*, 2020, **28**(Suppl 1), S18–S28.
61. H. Tabandeh, S. W. Lockley, R. Buttery, D. J. Skene, R. Defrance, J. Arendt and A. C. Bird, *Am. J. Ophthalmol.*, 1998, **126**, 707–712.
62. B. Winn, D. Whitaker, D. B. Elliott and N. J. Phillips, *Invest. Ophthalmol. Visual Sci.*, 1994, **35**, 1132–1137.
63. M. E. Harrington, D. M. Nance and B. Rusak, *Brain Res. Bull.*, 1985, **15**, 465–472.
64. R. Y. Moore and J. P. Card, *J. Comp. Neurol.*, 1994, **344**, 403–430.
65. E. L. Meyer-Bernstein and L. P. Morin, *J. Neurosci.*, 1996, **16**, 2097–2111.
66. N. Mrosovsky and P. A. Salmon, *Nature*, 1987, **330**, 372–373.
67. N. Mrosovsky and P. A. Salmon, *Chronobiol. Int.*, 1990, **7**, 35–41.
68. R. E. Mistlberger and D. J. Skene, *Biol. Rev.*, 2004, **79**, 533–556.
69. N. Mrosovsky, S. G. Reebs, G. I. Honrado and P. A. Salmon, *Experientia*, 1989, **45**, 696–702.
70. O. M. Buxton, M. L'Hermite-Balériaux, U. Hirschfeld and E. Cauter, *J. Biol. Rhythms*, 1997, **12**, 568–574.
71. R. E. Mistlberger and D. J. Skene, *J. Biol. Rhythms*, 2005, **20**, 339–352.
72. O. M. Buxton, S. A. Frank, M. L'Hermite-Balériaux, R. Leproult, F. W. Turek and E. Van Cauter, *Am. J. Physiol.*, 1997, **273**, E536–E542.
73. F. W. Turek and S. Losee-Olson, *Nature*, 1986, **321**, 167–168.
74. S. G. Reebs and N. Mrosovsky, *J. Biol. Rhythms*, 1989, **4**, 39–48.

75. P. K. Jha, H. Bouâouda, A. Kalsbeek and E. Challet, *Neurosci. Biobehav. Rev.*, 2021, **123**, 48–60.
76. R. F. Johnson, R. Y. Moore and L. P. Morin, *Brain Res.*, 1988, **460**, 297–313.
77. M. M. Koletar, H. Y. Cheng, J. M. Penninger and M. R. Ralph, *Chronobiol. Int.*, 2011, **28**, 553–562.
78. L. P. Morin and J. H. Blanchard, *J. Comp. Neurol.*, 2001, **437**, 79–90.
79. S. J. Soscia and M. E. Harrington, *Neurosci. Lett.*, 2005, **373**, 175–178.
80. S. M. Biello, D. Janik and N. Mrosovsky, *Neuroscience*, 1994, **62**, 273–279.
81. S. Shibata and R. Y. Moore, *Brain Res.*, 1993, **615**, 95–100.
82. C. Fukuhara, J. M. Brewer, J. C. Dirden, E. L. Bittman, G. Tosini and M. E. Harrington, *Neurosci. Lett.*, 2001, **314**, 119–122.
83. E. S. Maywood, H. Okamura and M. H. Hastings, *Eur. J. Neurosci.*, 2002, **15**, 216–220.
84. Y. Shigeyoshi, K. Taguchi, S. Yamamoto, S. Takekida, L. Yan, H. Tei, T. Moriya, S. Shibata, J. J. Loros, J. C. Dunlap and H. Okamura, *Cell*, 1997, **91**, 1043–1053.
85. E. T. Weber and M. A. Rea, *Neurosci. Lett.*, 1997, **231**, 159–162.
86. R. Y. Moore and J. C. Speh, *Neurosci. Lett.*, 1993, **150**, 112–116.
87. S. Ebihara, M. Goto and I. Oshima, *J. Biol. Rhythms*, 1988, **3**, 357–364.
88. R. D. Smith, S. Inouye and F. W. Turek, *J. Comp. Physiol*, 1989, **164**, 805–814.
89. K. Tominaga, S. Shibata, T. Hamada and S. Watanabe, *Neurosci. Lett.*, 1994, **166**, 81–84.
90. K. Horikawa, S. Yokota, K. Fuji, M. Akiyama, T. Moriya, H. Okamura and S. Shibata, *J. Neurosci.*, 2000, **20**, 5867–5873.
91. C. F. Gillespie, E. M. Mintz, C. L. Marvel, K. L. Huhman and H. E. Albers, *Brain Res.*, 1997, **759**, 181–189.
92. T. Moriya, S. Yamanouchi, T. Fukushima, T. Shimazoe, S. Shibata and S. Watanabe, *Brain Res.*, 1996, **740**, 261–267.
93. A. Watanabe, T. Moriya, Y. Nisikawa, T. Araki, T. Hamada, S. Shibata and S. Watanabe, *Brain Res.*, 1996, **740**, 329–336.
94. A. Sumova, E. S. Maywood, D. Selvage, F. J. Ebling and M. H. Hastings, *Brain Res.*, 1996, **709**, 88–96.
95. A. T. L. Hughes and H. D. Piggins, *Prog. Brain Res.*, 2012, **199**, 305–336.
96. J. D. Glass, G. H. Grossman, L. Farnbauch and L. DiNardo, *J. Neurosci.*, 2003, **23**, 7451–7460.
97. R. A. Prosser, J. D. Miller and H. C. Heller, *Brain Res.*, 1990, **534**, 336–339.
98. K. Tominaga, S. Shibata, S. Ueki and S. Watanabe, *Eur. J. Pharmacol.*, 1992, **214**, 79–84.
99. K. Horikawa and S. Shibata, *Neurosci. Lett.*, 2004, **368**, 130–134.
100. E. Challet, K. Scarbrough, P. D. Penev and F. W. Turek, *J. Biol. Rhythms*, 1998, **13**, 410–421.
101. S. Shibata, A. Tsuneyoshi, T. Hamada, K. Tominaga and S. Watanabe, *Brain Res.*, 1992, **582**, 353–356.

102. K. J. Bobrzynska, M. H. Godfrey and N. Mrosovsky, *Physiol. Behav.*, 1996, **59**, 221–230.
103. K. Horikawa, K. Fuji, Y. Fukazawa and S. Shibata, *J. Pharmacol. Sci.*, 2013, **123**, 402–406.
104. T. W. Lovenberg, M. G. Erlander, B. M. Baron and J. G. Sutcliffe, *Int. Clin. Psychopharmacol.*, 1993, **8**(Suppl 2), 19–23.
105. R. A. Prosser, H. C. Heller and J. D. Miller, *Brain Res.*, 1994, **644**, 67–73.
106. T. Nikaido, M. Akiyama, T. Moriya and S. Shibata, *Mol. Pharmacol.*, 2001, **59**, 894–900.
107. R. M. Grippo, A. M. Purohit, Q. Zhang, L. S. Zweifel and A. D. Güler, *Curr. Biol.*, 2017, **27**, 2465–2475.e2463.
108. P. C. Kandalepas, J. W. Mitchell and M. U. Gillette, *PLoS One*, 2016, **11**, e0157824.
109. A. J. McArthur, A. E. Hunt and M. U. Gillette, *Endocrinology*, 1997, **138**, 627–634.
110. A. E. Hunt, W. M. Al-Ghoul, M. U. Gillette and M. L. Dubocovich, *Am. J. Physiol.*, 2001, **280**, C110–C118.
111. M. Pfeffer, A. Rauch, H. W. Korf and C. von Gall, *Chronobiol. Int.*, 2012, **29**, 415–429.
112. C. Saini, A. Liani, T. Curie, P. Gos, F. Kreppel, Y. Emmenegger, L. Bonacina, J. P. Wolf, Y. A. Poget, P. Franken and U. Schibler, *Genes Dev.*, 2013, **27**, 1526–1536.
113. J. Mendoza, C. Graff, H. Dardente, P. Pevet and E. Challet, *J. Neurosci.*, 2005, **25**, 1514–1522.
114. S. M. T. Wehrens, S. Christou, C. Isherwood, B. Middleton, M. A. Gibbs, S. N. Archer, D. J. Skene and J. D. Johnston, *Curr. Biol.*, 2017, **27**, 1768–1775.e1763.
115. A. Hirao, H. Nagahama, T. Tsuboi, M. Hirao, Y. Tahara and S. Shibata, *Am. J. Physiol.*, 2010, **299**, G1045–G1053.
116. Y. Tahara, M. Otsuka, Y. Fuse, A. Hirao and S. Shibata, *J. Biol. Rhythms*, 2011, **26**, 230–240.
117. A. Mukherji, A. Kobiita and P. Chambon, *Proc. Natl. Acad. Sci. U. S. A.*, 2015, **112**, E6683–E6690.
118. C. Vollmers, S. Gill, L. DiTacchio, S. R. Pulivarthy, H. D. Le and S. Panda, *Proc. Natl. Acad. Sci. U. S. A.*, 2009, **106**, 21453–21458.
119. X. Sun, F. Dang, D. Zhang, Y. Yuan, C. Zhang, Y. Wu, Y. Wang and Y. Liu, *J. Biol. Chem.*, 2015, **290**, 2189–2197.
120. G. Asher, H. Reinke, M. Altmeyer, M. Gutierrez-Arcelus, M. O. Hottiger and U. Schibler, *Cell*, 2010, **142**, 943–953.
121. Y. Nakahata, M. Kaluzova, B. Grimaldi, S. Sahar, J. Hirayama, D. Chen, L. P. Guarente and P. Sassone-Corsi, *Cell*, 2008, **134**, 329–340.
122. K. A. Lamia, U. M. Sachdeva, L. DiTacchio, E. C. Williams, J. G. Alvarez, D. F. Egan, D. S. Vasquez, H. Juguilon, S. Panda, R. J. Shaw, C. B. Thompson and R. M. Evans, *Science*, 2009, **326**, 437–440.
123. H. Kuroda, Y. Tahara, K. Saito, N. Ohnishi, Y. Kubo, Y. Seo, M. Otsuka, Y. Fuse, Y. Ohura, A. Hirao and S. Shibata, *Sci. Rep.*, 2012, **2**, 711.

124. A. Kohsaka, A. D. Laposky, K. M. Ramsey, C. Estrada, C. Joshu, Y. Kobayashi, F. W. Turek and J. Bass, *Cell Metab.*, 2007, **6**, 414–421.
125. Y. Tahara and S. Shibata, *Neuroscience*, 2013, **253**, 78–88.
126. R. E. Mistlberger, *Eur. J. Neurosci.*, 2009, **30**, 1718–1729.
127. G. Acosta-Galvan, C. X. Yi, J. van der Vliet, J. H. Jhamandas, P. Panula, M. Angeles-Castellanos, M. Del Carmen Basualdo, C. Escobar and R. M. Buijs, *Proc. Natl. Acad. Sci. U. S. A.*, 2011, **108**, 5813–5818.
128. T. Moriya, R. Aida, T. Kudo, M. Akiyama, M. Doi, N. Hayasaka, N. Nakahata, R. Mistlberger, H. Okamura and S. Shibata, *Eur. J. Neurosci.*, 2009, **29**, 1447–1460.
129. C. M. Gallardo, M. Darvas, M. Oviatt, C. H. Chang, M. Michalik, T. F. Huddy, E. E. Meyer, S. A. Shuster, A. Aguayo, E. M. Hill, K. Kiani, J. Ikpeazu, J. S. Martinez, M. Purpura, A. N. Smit, D. F. Patton, R. E. Mistlberger, R. D. Palmiter and A. D. Steele, *eLife*, 2014, **3**, e03781.
130. N. N. Takasu, G. Kurosawa, I. T. Tokuda, A. Mochizuki, T. Todo and W. Nakamura, *PLoS One*, 2012, **7**, e48892.
131. J. Delezie, S. Dumont, C. Sandu, S. Reibel, P. Pevet and E. Challet, *Sci. Rep.*, 2016, **6**, 29386.
132. R. Chavan, C. Feillet, S. S. Costa, J. E. Delorme, T. Okabe, J. A. Ripperger and U. Albrecht, *Nat. Commun.*, 2016, **7**, 10580.
133. J. LeSauter, N. Hoque, M. Weintraub, D. W. Pfaff and R. Silver, *Proc. Natl. Acad. Sci. U. S. A.*, 2009, **106**, 13582–13587.
134. Y. Tahara, A. Hirao, T. Moriya, T. Kudo and S. Shibata, *J. Biol. Rhythms*, 2010, **25**, 9–18.
135. C. A. Feillet, J. A. Ripperger, M. C. Magnone, A. Dulloo, U. Albrecht and E. Challet, *Curr. Biol.*, 2006, **16**, 2016–2022.
136. D. Yamajuku, T. Inagaki, T. Haruma, S. Okubo, Y. Kataoka, S. Kobayashi, K. Ikegami, T. Laurent, T. Kojima, K. Noutomi, S. Hashimoto and H. Oda, *Sci. Rep.*, 2012, **2**, 439.
137. M. Sato, M. Murakami, K. Node, R. Matsumura and M. Akashi, *Cell Rep.*, 2014, **8**, 393–401.
138. P. Crosby, R. Hamnett, M. Putker, N. P. Hoyle, M. Reed, C. J. Karam, E. S. Maywood, A. Stangherlin, J. E. Chesham, E. A. Hayter, L. Rosenbrier-Ribeiro, P. Newham, H. Clevers, D. A. Bechtold and J. S. O'Neill, *Cell*, 2019, **177**, 896–909.e820.
139. A. Furutani, Y. Ikeda, M. Itokawa, H. Nagahama, T. Ohtsu, N. Furutani, M. Kamagata, Z. H. Yang, A. Hirasawa, Y. Tahara and S. Shibata, *PLoS One*, 2015, **10**, e0132472.
140. D. Landgraf, A. H. Tsang, A. Leliavski, C. E. Koch, J. L. Barclay, D. J. Drucker and H. Oster, *eLife*, 2015, **4**, e06253.
141. S. Chen, M. Feng, S. Zhang, Z. Dong, Y. Wang, W. Zhang and C. Liu, *Nat. Commun.*, 2019, **10**, 3518.
142. Y. Ikeda, M. Kamagata, M. Hirao, S. Yasuda, S. Iwami, H. Sasaki, M. Tsubosaka, Y. Hattori, A. Todoh, K. Tamura, K. Shiga, T. Ohtsu and S. Shibata, *EBioMedicine*, 2018, **28**, 210–224.
143. A. Sehgal, *Cell*, 2017, **171**, 1232–1235.

144. Y. Tahara and S. Shibata, *Nat. Rev. Gastroenterol. Hepatol.*, 2016, **13**, 217–226.
145. M. Itokawa, A. Hirao, H. Nagahama, M. Otsuka, T. Ohtsu, N. Furutani, K. Hirao, T. Hatta and S. Shibata, *Nutr. Res.*, 2013, **33**, 109–119.
146. D. Y. Oh, S. Talukdar, E. J. Bae, T. Imamura, H. Morinaga, W. Fan, P. Li, W. J. Lu, S. M. Watkins and J. M. Olefsky, *Cell*, 2010, **142**, 687–698.
147. T. Fukuda, A. Haraguchi, M. Kuwahara, K. Nakamura, Y. Hamaguchi, Y. Ikeda, Y. Ishida, G. Wang, C. Shirakawa, Y. Tanihata, K. Ohara and S. Shibata, *Sci. Rep.*, 2016, **6**, 34665.
148. Y. Tahara, M. Yamazaki, H. Sukigara, H. Motohashi, H. Sasaki, H. Miyakawa, A. Haraguchi, Y. Ikeda, S. Fukuda and S. Shibata, *Sci. Rep.*, 2018, **8**, 1395.
149. V. Leone, S. M. Gibbons, K. Martinez, A. L. Hutchison, E. Y. Huang, C. M. Cham, J. F. Pierre, A. F. Heneghan, A. Nadimpalli, N. Hubert, E. Zale, Y. Wang, Y. Huang, B. Theriault, A. R. Dinner, M. W. Musch, K. A. Kudsk, B. J. Prendergast, J. A. Gilbert and E. B. Chang, *Cell Host Microbe*, 2015, **17**, 681–689.
150. Z. Chen, S. H. Yoo, Y. S. Park, K. H. Kim, S. Wei, E. Buhr, Z. Y. Ye, H. L. Pan and J. S. Takahashi, *Proc. Natl. Acad. Sci. U. S. A.*, 2012, **109**, 101–106.
151. T. Hirota, J. W. Lee, W. G. Lewis, E. E. Zhang, G. Breton, X. Liu, M. Garcia, E. C. Peters, J. P. Etchegaray, D. Traver, P. G. Schultz and S. A. Kay, *PLoS Biol.*, 2010, **8**, e1000559.
152. B. He, K. Nohara, N. Park, Y. S. Park, B. Guillory, Z. Zhao, J. M. Garcia, N. Koike, C. C. Lee, J. S. Takahashi, S. H. Yoo and Z. Chen, *Cell Metab.*, 2016, **23**, 610–621.
153. A. Shinozaki, K. Misawa, Y. Ikeda, A. Haraguchi, M. Kamagata, Y. Tahara and S. Shibata, *PLoS One*, 2017, **12**, e0170904.
154. H. Oike, M. Kobori, T. Suzuki and N. Ishida, *Biochem. Biophys. Res. Commun.*, 2011, **410**, 654–658.
155. S. Narishige, M. Kuwahara, A. Shinozaki, S. Okada, Y. Ikeda, M. Kamagata, Y. Tahara and S. Shibata, *Br. J. Pharmacol.*, 2014, **171**, 5858–5869.
156. T. M. Burke, R. R. Markwald, A. W. McHill, E. D. Chinoy, J. A. Snider, S. C. Bessman, C. M. Jung, J. S. O'Neill and K. P. Wright Jr, *Sci. Transl. Med.*, 2015, **7**, 305ra146.
157. C. Suzuki, S. Fukumitsu and H. Oike, *Phytochemistry*, 2021, **181**, 112539.
158. K. Oishi, S. Yamamoto, H. Oike, N. Ohkura and M. Taniguchi, *Biochem. Biophys. Rep.*, 2017, **9**, 232–237.
159. H. Oike and M. Kobori, *Biosci., Biotechnol., Biochem.*, 2008, **72**, 3038–3040.
160. H. Sasaki, Y. Hattori, Y. Ikeda, M. Kamagata, S. Iwami, S. Yasuda, Y. Tahara and S. Shibata, *Sci. Rep.*, 2016, **6**, 27607.
161. Y. Tahara and S. Shibata, *Free Radical Biol. Med.*, 2018, **119**, 129–138.
162. A. Balsalobre, L. Marcacci and U. Schibler, *Curr. Biol.*, 2000, **10**, 1291–1294.

163. S. Cheon, N. Park, S. Cho and K. Kim, *Nucleic Acids Res.*, 2013, **41**, 6161–6174.
164. Y. Tahara, T. Shiraishi, Y. Kikuchi, A. Haraguchi, D. Kuriki, H. Sasaki, H. Motohashi, T. Sakai and S. Shibata, *Sci. Rep.*, 2015, **5**, 11417.
165. M. Cuesta, N. Cermakian and D. B. Boivin, *FASEB J.*, 2015, **29**, 1360–1370.
166. A. Balsalobre, S. A. Brown, L. Marcacci, F. Tronche, C. Kellendonk, H. M. Reichardt, G. Schutz and U. Schibler, *Science*, 2000, **289**, 2344–2347.
167. A. Y. So, T. U. Bernal, M. L. Pillsbury, K. R. Yamamoto and B. J. Feldman, *Proc. Natl. Acad. Sci. U. S. A.*, 2009, **106**, 17582–17587.
168. H. Terazono, T. Mutoh, S. Yamaguchi, M. Kobayashi, M. Akiyama, R. Udo, S. Ohdo, H. Okamura and S. Shibata, *Proc. Natl. Acad. Sci. U. S. A.*, 2003, **100**, 6795–6800.
169. Y. Tahara, Y. Takatsu, T. Shiraishi, Y. Kikuchi, M. Yamazaki, H. Motohashi, A. Muto, H. Sasaki, A. Haraguchi, D. Kuriki, T. J. Nakamura and S. Shibata, *npj Aging Mech. Dis.*, 2017, **3**, 16030.
170. Z. Travnickova-Bendova, N. Cermakian, S. M. Reppert and P. Sassone-Corsi, *Proc. Natl. Acad. Sci. U. S. A.*, 2002, **99**, 7728–7733.
171. Y. Tahara, A. Yokota, T. Shiraishi, S. Yamada, A. Haraguchi, A. Shinozaki and S. Shibata, *J. Circadian Rhythms*, 2016, **14**, 1–7.
172. T. Tamaru, M. Hattori, Y. Ninomiya, G. Kawamura, G. Vares, K. Honda, D. P. Mishra, B. Wang, I. Benjamin, P. Sassone-Corsi, T. Ozawa and K. Takamatsu, *PLoS One*, 2013, **8**, e82006.
173. R. V. Khapre, A. A. Kondratova, O. Susova and R. V. Kondratov, *Cell Cycle*, 2011, **10**, 4162–4169.
174. V. Pekovic-Vaughan, J. Gibbs, H. Yoshitane, N. Yang, D. Pathiranage, B. Guo, A. Sagami, K. Taguchi, D. Bechtold, A. Loudon, M. Yamamoto, J. Chan, G. T. van der Horst, Y. Fukada and Q. J. Meng, *Genes Dev.*, 2014, **28**, 548–560.
175. J. Lee, M. Moulik, Z. Fang, P. Saha, F. Zou, Y. Xu, D. L. Nelson, K. Ma, D. D. Moore and V. K. Yechoor, *Mol. Cell. Biol.*, 2013, **33**, 2327–2338.
176. S. A. Brown, G. Zumbrunn, F. Fleury-Olela, N. Preitner and U. Schibler, *Curr. Biol.*, 2002, **12**, 1574–1583.
177. E. D. Buhr, S. H. Yoo and J. S. Takahashi, *Science*, 2010, **330**, 379–385.
178. C. Saini, J. Morf, M. Stratmann, P. Gos and U. Schibler, *Genes Dev.*, 2012, **26**, 567–580.
179. J. Morf, G. Rey, K. Schneider, M. Stratmann, J. Fujita, F. Naef and U. Schibler, *Science*, 2012, **338**, 379–383.
180. N. Ohnishi, Y. Tahara, D. Kuriki, A. Haraguchi and S. Shibata, *PLoS One*, 2014, **9**, e100272.
181. C. B. Peek, D. C. Levine, J. Cedernaes, A. Taguchi, Y. Kobayashi, S. J. Tsai, N. A. Bonar, M. R. McNulty, K. M. Ramsey and J. Bass, *Cell Metab.*, 2017, **25**, 86–92.
182. Y. Wu, D. Tang, N. Liu, W. Xiong, H. Huang, Y. Li, Z. Ma, H. Zhao, P. Chen, X. Qi and E. E. Zhang, *Cell Metab.*, 2017, **25**, 73–85.
183. Y. Adamovich, B. Ladeuix, M. Golik, M. P. Koeners and G. Asher, *Cell Metab.*, 2017, **25**, 93–101.

184. N. Kon, T. Hirota, T. Kawamoto, Y. Kato, T. Tsubota and Y. Fukada, *Nat. Cell Biol.*, 2008, **10**, 1463–1469.
185. K. Imamura, H. Yoshitane, K. Hattori, M. Yamaguchi, K. Yoshida, T. Okubo, I. Naguro, H. Ichijo and Y. Fukada, *Proc. Natl. Acad. Sci. U. S. A.*, 2018, **115**, 3646–3651.
186. Y. Yamamura, I. Yano, T. Kudo and S. Shibata, *Chronobiol. Int.*, 2010, **27**, 213–232.
187. G. Cavadini, S. Petrzilka, P. Kohler, C. Jud, I. Tobler, T. Birchler and A. Fontana, *Proc. Natl. Acad. Sci. U. S. A.*, 2007, **104**, 12843–12848.
188. S. Chen, K. K. Fuller, J. C. Dunlap and J. J. Loros, *Front. Immunol.*, 2020, **11**, 867.
189. M. Hatori, C. Vollmers, A. Zarrinpar, L. DiTacchio, E. A. Bushong, S. Gill, M. Leblanc, A. Chaix, M. Joens, J. A. Fitzpatrick, M. H. Ellisman and S. Panda, *Cell Metab.*, 2012, **15**, 848–860.
190. H. Sherman, Y. Genzer, R. Cohen, N. Chapnik, Z. Madar and O. Froy, *FASEB J.*, 2012, **26**, 3493–3502.
191. A. Chaix, T. Lin, H. D. Le, M. W. Chang and S. Panda, *Cell Metab.*, 2019, **29**, 303–319.e304.
192. A. Di Francesco, C. Di Germanio, M. Bernier and R. de Cabo, *Science*, 2018, **362**, 770–775.
193. M. J. Wilkinson, E. N. C. Manoogian, A. Zadourian, H. Lo, S. Fakhouri, A. Shoghi, X. Wang, J. G. Fleischer, S. Navlakha, S. Panda and P. R. Taub, *Cell Metab.*, 2020, **31**, 92–104.e105.
194. D. S. Whittaker, D. H. Loh, H. B. Wang, Y. Tahara, D. Kuljis, T. Cutler, C. A. Ghiani, S. Shibata, G. D. Block and C. S. Colwell, *J. Biol. Rhythms*, 2018, **33**, 535–554.
195. S. D. Youngstedt, J. A. Elliott and D. F. Kripke, *J. Physiol.*, 2019, **597**, 2253–2268.
196. J. M. Thomas, P. A. Kern, H. M. Bush, K. J. McQuerry, W. S. Black, J. L. Clasey and J. S. Pendergast, *JCI Insight*, 2020, **5**, e134270.
197. E. A. Willis, S. A. Creasy, J. J. Honas, E. L. Melanson and J. E. Donnelly, *Int. J. Obes.*, 2020, **44**, 114–124.
198. Z. Alizadeh, S. Younespour, M. Rajabian Tabesh and S. Haghravan, *Clin. Obes.*, 2017, **7**, 157–165.
199. K. Iwayama, R. Kurihara, Y. Nabekura, R. Kawabuchi, I. Park, M. Kobayashi, H. Ogata, M. Kayaba, M. Satoh and K. Tokuyama, *EBioMedicine*, 2015, **2**, 2003–2009.
200. S. Sato, A. L. Basse, M. Schönke, S. Chen, M. Samad, A. Altõntaş, R. C. Laker, E. Dalbram, R. Barrès, P. Baldi, J. T. Treebak, J. R. Zierath and P. Sassone-Corsi, *Cell Metab.*, 2019, **30**, 92–110.e114.
201. S. Aoyama, S. Kojima, K. Sasaki, R. Ishikawa, M. Tanaka, T. Shimoda, Y. Hattori, N. Aoki, K. Takahashi, R. Hirooka, M. Takizawa, A. Haraguchi and S. Shibata, *EBioMedicine*, 2018, **37**, 499–508.
202. E. Facer-Childs and R. Brandstaetter, *Curr. Biol.*, 2015, **25**, 518–522.
203. S. Ezagouri, Z. Zwighaft, J. Sobel, S. Baillieul, S. Doutreleau, B. Ladeuix, M. Golik, S. Verges and G. Asher, *Cell Metab.*, 2019, **30**, 78–91.e74.

CHAPTER 4

Inter-modulating Multi-frequency Rhythms

KUNIAKI OTSUKA,*[a,b] LARRY A. BEATY[b] AND GERMAINE CORNELISSEN[b]

[a] Executive Medical Center, Totsuka Royal Clinic, Tokyo Women's Medical University, Tokyo, Japan; [b] Halberg Chronobiology Center, University of Minnesota, Minneapolis, Minnesota, USA
*Email: frtotk99@ba2.so-net.ne.jp

4.1 Introduction

The study of circadian rhythms in physiology is part of chronobiology, a term coined in 1950 by Halberg (from the Greek "chronos" = time, "bios" = life, and "logos" = science).[1] In 1959, he also introduced the term circadian (from Latin "*circa*" = about, and "dies" = day).[2] It is now known that the circadian system plays a vital role in the coordination of physiological processes, including cell cycle progression, cytokine release, hormonal secretion, sleep and wakefulness, neurobehavioural performance, cardiovascular activity, energy metabolism, and immunity.[3]

Very early on, it was recognized that alongside circadian rhythms, physiological processes are orchestrated by a broad time structure (chronome),[4] consisting of multi-frequency rhythms and trends as a function of development and ageing. The brain's action potentials are faster than those of the heart. The gut's refilling and evacuation are much slower. Reproduction proceeds at an even slower pace. Living matter functions concomitantly at vastly different frequencies, as a system open to its environment. Physiological processes rely on periodicities with different frequencies for a division of labor

Chemical Biology No. 23
Chronobiology and Chronomedicine: From Molecular and Cellular Mechanisms to Whole Body Interdigitating Networks
Edited by Germaine Cornelissen and Tsuyoshi Hirota

Published by the Royal Society of Chemistry, www.rsc.org

over time, all of which are essential to life. Genetic factors known to influence circadian rhythms likely influence the entire chronome, which changes with age and during the course of disease, as illustrated herein. In this chapter, we provide a glimpse of biomedical applications that can benefit from a comprehensive chronobiological approach that considers not only the important and ubiquitous circadian rhythms but also other elements of a broad time structure.

4.2 Characterization of Rhythms

4.2.1 Definitions

Physiological processes fluctuate over time according to partly predictable recurring cycles of different frequencies, which are the combined result of genetics and external influences. Rhythmic variations in a biological system usually only partially reflect the response to a periodic input from the environment. They often relate to programmed and genetically determined periodic processes that anticipate the organism's needs across different time scales, from the electroencephalographic rhythms and the beating of the heart to many circadian rhythms and the menstrual cycle. In this sense, a chronobiological view of physiology replaces the conventional homeostatic view with an approach that accounts for rhythmic variation.

Mean values are thus complemented by amplitudes and acrophases of each periodic component, which are measures of half the predictable change within one cycle and of the timing of overall high values recurring in each cycle, respectively. When non-sinusoidal rhythms are modeled using multiple harmonic terms, the amplitudes and acrophases of all terms in the model describe the rhythm's waveform. The magnitude, the range of predictable change within one cycle, and the orthophase and bathyphase, times of overall maximum and minimum recurring in each cycle, are useful endpoints characteristic of the rhythm's shape (Figure 4.1). They serve as additional endpoints to define health, and to signal a heightened disease risk when they deviate from established norms.

The period of a rhythm is defined as the duration of one cycle. Circadian rhythms were originally defined empirically as rhythms with a period of 24 ± 4 hours to account for the limits of entrainment by the environmental light–dark cycle.[5] Ultradian and infradian rhythms are rhythms with a frequency higher or lower than one cycle in about 24 hours, respectively. Ultradian, circadian, and infradian rhythms continuously undergo modulations, modifications, and synchronization due to periodicities in the external as well as internal environment. Examples include respiratory sinus arrhythmia[6] and the marked circannual modulation of cardiorespiratory functions, partly related to changes in temperature with seasons.[7]

Under usual conditions, most physiological variables undergo spontaneous circadian variations. They also respond in a circadian stage-dependent manner to stimuli ranging from the action of ACTH on corticosterone[8] to the effect

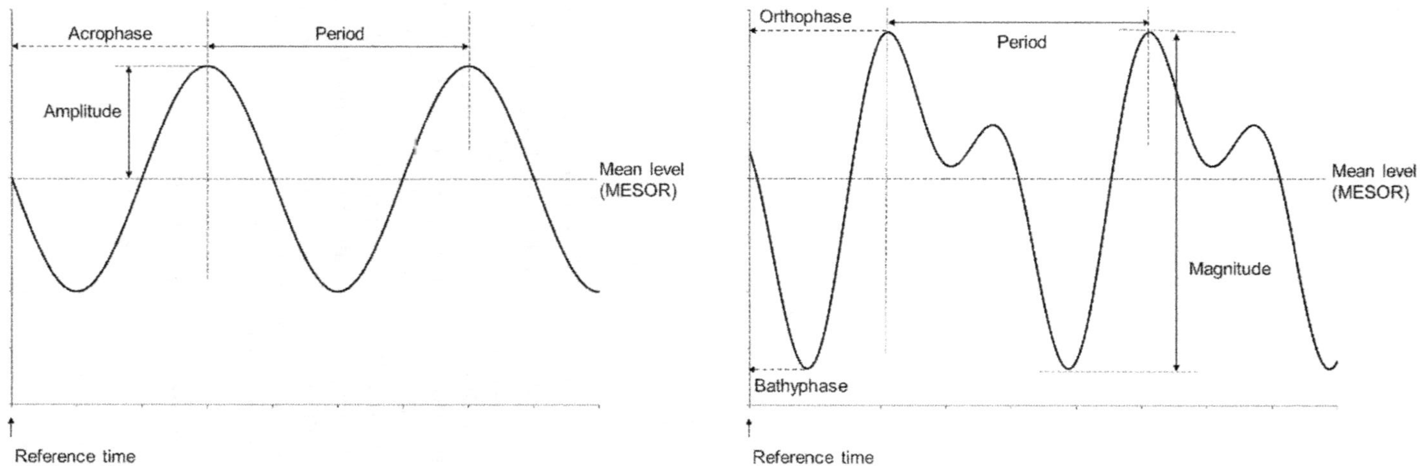

$$100 + 10.\cos(2\pi t/24h + \pi/2)$$

$$100 + 10.\cos(2\pi(t\text{-}10h)/24h + 3*\pi/2) + 7.5.\cos(2\pi(t\text{-}10h)/12h + 3*\pi/1.75)$$

Figure 4.1 Definition of rhythm characteristics. Left: the single-component cosine model with a period of 24 hours, 90° (−270°) phase, and excursion of ±10 units around a mean value (MESOR) of 100. Right: two-component cosine model, including a second harmonic term with a period of 12 hours and an amplitude slightly smaller than the fundamental 24-hour component.

of periodontal surgery on blood pressure.[9] These response rhythms constitute the foundation of chronotherapy. For instance, the incidence of major cardiac events is significantly lower when aortic valve replacement surgery takes place in the afternoon than in the morning. The difference is likely mediated by circadian variations in the expression of *Rev-Erbα*, a unique nuclear receptor family gene regulating the molecular circadian clock.[10] A third entity can further modulate the response of one entity to another in the complex hierarchical system that physiology represents. An example of such "feedsidewards" as these modulations have been named, is the amplification of the corticosterone response to ACTH by the pineal.[11]

4.2.2 Ultradian Rhythms

Ultradian rhythms are broadly defined and cover a wide frequency range. Different frequency domains are related with different physiological processes. Among them are brain waves (0.5–3 Hz, δ; 4–7 Hz, θ; 8–13 Hz, α; 14–25 Hz, β; and 26–70 Hz, γ), respiratory cycles (0.15–0.40 Hz), baroreflex activity (0.04–0.15 Hz), brain thalamic astrocyte activity (infraslow oscillation, ISO) (0.01–0.10 Hz),[12] and the about 90-minute basic rest–activity cycle (BRAC).[13] Ultradian components with periods longer than about 3 to 4 hours are usually difficult to distinguish from harmonics of the circadian system, which describe the rhythm's waveform. However, there is evidence suggesting that about 12-hour components may also have a genetic basis (see below).

Ultradian rhythms are a widely occurring necessity for the coordination and coherence of living processes.[14] Evidence suggests that ultradian rhythms of unknown origin may also regulate clock gene expression in the SCN. Continuous bioluminescence monitoring of the expression of clock gene reporters in the SCN of freely moving mice for up to 3 weeks in constant darkness showed that episodic bursts occurring with a period of approximately 3 hours superimpose circadian rhythms.[15] Synchronous ultradian intracellular calcium rhythms with periods of 0.5 to 4.0 hours may originate from the subparaventricular zone–paraventricular nucleus of the hypothalamus (SPZ–PVN) region, and be transmitted to the SCN, as found in cultured slices from neonatal mice.[16] Suppression of the ultradian rhythm by *N*-methyl-D-aspartate (NMDA)-type glutamate receptor blockers indicates the critical involvement of the glutamatergic mechanism in ultradian rhythm generation, probably mediated by astrocytes in the SPZ–PVN region.[16] Ultradians in different frequency regions are reviewed below.

4.2.2.1 Temporal Co-modulation of Heart–Brain and Brain–Heart Interactions During Sleep

In the high-frequency region of the spectrum, ultradians characterize brain–heart interactions, illustrated in the concomitant EEG and ECG recordings during sleep from an otherwise clinically healthy 14-year boy who complained of palpitations at night, Figure 4.2. Continuous polysomnography

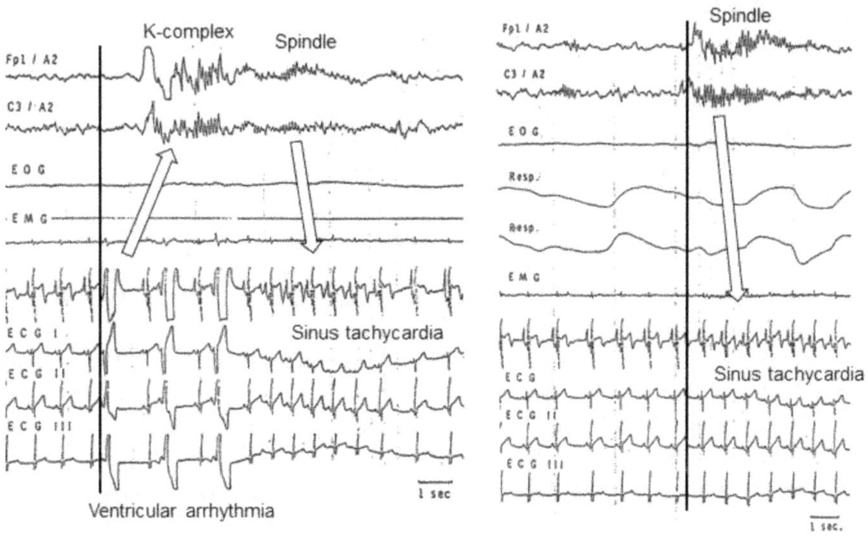

Figure 4.2 Temporal co-modulation of the heart and brain (left) or the brain and heart (right). Left: human EEG during sleep responding to an occurrence of cardiac arrhythmia and showing K-complex with sleep spindle rhythms. Right: Human ECG during sleep responding to an occurrence of sleep spindle in EEG and showing sinus tachycardia.

for four nights showed an intimate correlation in the bidirectional temporal co-modulation between the heart and brain. Every night, the boy's brain responded to each cardiac arrhythmic episode during stage 2 sleep by reproducibly expressing K-complexes in the EEG, while the occurrence of spindles in stage 2 sleep reproducibly induced sinus tachycardia in his ECG.

The brain controls the heart directly through the sympathetic and parasympathetic branches of the autonomic nervous system. Cardiac function can be profoundly altered by the reflex activation of cardiac autonomic nerves in response to inputs from baroreceptors, chemoreceptors, nasopharyngeal and other receptors, and by central autonomic commands, including those associated with stress, physical activity, arousal and sleep.[17] While a deeper understanding of brain–heart interactions is lacking, the question may be raised whether astrocyte–neuron networks are involved.[18–20] Astrocytic *Bmal1* has been reported to play an important role in the coordination of neuronal clocks, indicating that astrocytes could be a new cellular target for neuropharmacology of transient or chronic perturbation of circadian rhythms.[21]

Astrocytes are cells characterized by circadian rhythms, much like neurons.[22,23] Surprisingly, the circadian rhythm of neuronal activity peaks during the daytime, whereas that of astrocytes peaks during the night, in antiphase with SCN neurons. During the night, SCN astrocytes inhibit SCN neurons by releasing gliotransmitters into the extracellular space, which may serve as an inhibitory astrocytic-neuronal coupling signal.[24–26] Multiple experiments indeed suggest that astrocytic networks contribute significantly to the generation (or spreading) of brain oscillations (slow-wave sleep waves in particular) across the SCN, which have broad brain relevance.[17] In addition to glia within the brain, glial cells within multiple autonomic ganglia in the peripheral nervous system, including the heart and the vasculature, also play a role in modulating brain–heart interactions.[27–32]

4.2.2.2 Basic Rest–Activity Cycle (BRAC)

Several studies show that the sleep–dream cycle of human sleep is a general activity pattern of the brain.[33] The BRAC is manifested in several physiological processes. Regular oscillations with periods of 1 to 2 hours in oxygen consumption have been reported,[34] as well as an about 90-minute periodicity in the heart rate (HR), reflecting the performance of a complex vigilance task.[35] Ultradians in the same frequency range reportedly characterize gastric motility,[35] urine flow in awake subjects, which were out of phase with those of electrolyte concentrations and osmolarity,[36] and vigilance,[37] as well as the ability to fall asleep throughout the day.[38] Evidence that the BRAC reflects an intrinsic rhythm in brain activity is provided by the fact that in Sprague–Dawley rats, eating is not only preceded by increases in body and brain temperature but is also integrated into the BRAC.[39] Under conditions of 32 hours of bed rest in a sound–light attenuated room, an ultradian tendency to sleep predominated in narcoleptic patients, with evidence of a strong BRAC.[40] The BRAC in specific endpoints of HR variability (HRV) is amplified in astronauts during long-term missions in space, raising the question whether such amplification may be a sign of early adaptation to microgravity.[41]

The BRAC has been referred to as the basic signature of life,[42] as it is involved in the functioning of the central nervous system, which integrates many somatic, visceral, and neurobehavioural functions.[33–38,43] It manifests itself most strikingly in the alternation of non-REM and REM sleep. In sedentary laboratory rats fed *ad libitum*, rather than being triggered by changes in the amounts of body fuels or other meal-associated factors, eating was integrated into the BRAC in a temporally precise manner, occurring only after an increase in brown adipose tissue temperature.[39] Since the function of brown adipose tissue is to transfer energy from food to heat, this BRAC behaviour points to its survival advantage.

4.2.2.3 About 12-hour Cycles

Although the origin of the 12-hour rhythm has remained elusive, two classes of genes cycling with periods of 12 and 8 hours were identified in the mouse liver, in addition to the 24-hour rhythm.[44] The about 12-hour cycling genes in the mouse liver were enriched in endoplasmic reticulum (ER) and mitochondria-associated metabolism pathways.[45,46] The 12-hour rhythm of gene expression was confirmed to be cell-autonomous, coordinated independently from the circadian clock. It has been proposed that the 12-hour clock is transcriptionally regulated by 12-hour rhythms of XBP1s chromatin recruitment to ER response element.[47,48] This mechanism differs from the circadian clock, which is transcriptionally regulated through a revised TTFL model that includes activation phase-facilitated repression mainly *via* the retinoic acid-related orphan receptor response element.[46] Harmonic components of the circadian clock and the 12-hour clock, synchronizing to different environmental cues, including a response to fasting and feeding, and entrained by metabolic and ER stress cues, confer adaptive advantages by allowing organisms to anticipate daily changes in their environment.

Under usual conditions, circadian rhythms in humans often depart from sinusoidality; the addition of a 12-hour component in the model usually greatly improves the fit. In part, the second harmonic accounts for the fact that the waking span is about twice as long as the sleep span. The 12-hour component contributes significantly to the characterization of the circadian waveform for variables such as core body temperature, blood pressure (BP), cognitive performance, and endocrine functions, among others.[41,49] An example is shown in Figure 4.3 for the case of BP monitored around-the-clock for 7 days by 140 Japanese hypertensive patients (57 ± 10 years of age, 45.4% men).[50] Two peaks are clearly seen, one in the morning occurring 0 to 3 hours after awakening, and another in the early evening, about 9 hours later. It has been suggested that unlike the role played by platelet aggregability, neurohormonal

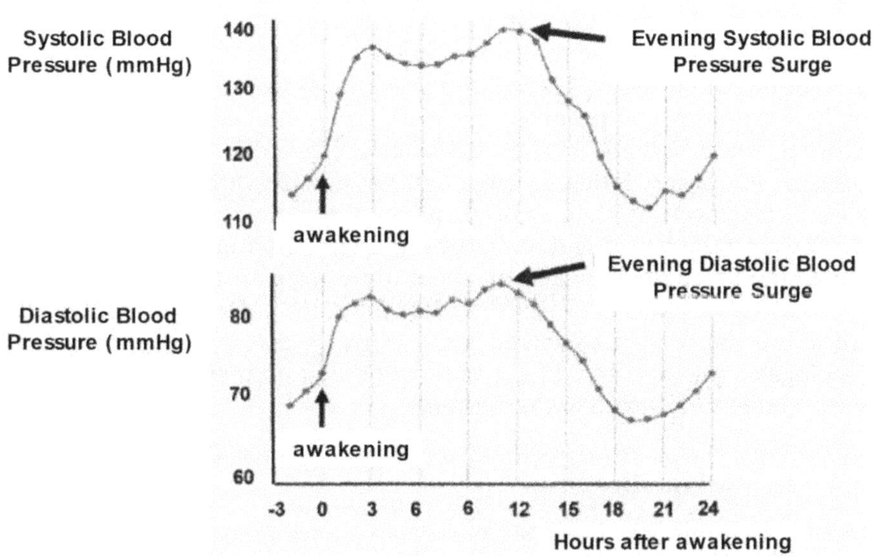

Figure 4.3 Bimodal circadian variation in systolic and diastolic blood pressure, peaking 0 to 3 hours after awakening, and again 9 to 12 hours after awakening. Hourly averages of blood pressure, stacked over a single 24-hour day, computed from 7-day/24-hour ABPM records of 140 study participants. Reproduced from ref. 50 with permission from John Wiley & Sons, Copyright 2019 Wiley Periodicals, Inc.

factors, and fibrinolytic activity in the morning BP peak, the evening rise in BP might reflect the endogenous component of the circadian BP rhythm.[51]

The 12-hour component significantly contributes to describing the circadian rhythm of circulating cortisol, melatonin, basic fibroblast growth factor, and epidermal growth factor in hospitalized patients with metastatic breast cancer. The larger prominence of the 12-hour compared to the 24-hour component in the case of basic fibroblast growth factor raises questions about the role of ultradian variations in the pathophysiology of breast cancer.[52]

4.2.3 Circadian Rhythms

Circadian clocks play a key role in coordinating a vast array of biological processes, which have significant implications for human health. "Circadian" was coined in the early 1950s,[53] and endorsed by international nomenclature committees.[54] The term indicates both partial endogenicity and variability inherent in its characterization. "Circadian" was introduced to replace "diurnal", a term used in biomedicine to distinguish activity or morbidity patterns during the daily photofraction from those during the daily scotofraction, referred to as diurnality *vs.* nocturnality, respectively. While they are obvious signatures of the photic day, circadians are also influenced non-photically by natural and artificial electromagnetic fields,[55,56] and other synchronizers such as the feeding schedule and temperature.

Circadian rhythms are ubiquitous, characterizing almost all physiological variables across most species. They also impact the incidence of morbidity and mortality from different conditions. For instance, sleep is a vulnerable time for respiratory compromise, especially in asthma patients experiencing nocturnal exacerbations.[57] The incidence of potentially life-threatening cardiovascular events follows a circadian periodicity, with a tendency to be higher in the morning than at other times of the day.[58] Fever incidence is much lower in the morning than in the evening. Consequently, fever-causing diseases tend to be overlooked or underappreciated in the morning, especially for sepsis cases and during screenings for infectious disease outbreaks.[59] It is now appreciated that clock genes are involved in all physiological systems, and that a bidirectional relationship exists between disease states and circadian disruption.

4.2.4 Infradian Rhythms

Infradian rhythms are broadly defined and encompass a wide range of frequencies. Because of their lower frequencies, they are harder to study on a longitudinal basis since they need to be observed over long spans. Non-invasive physiological monitoring has made it possible to collect serial data for their investigation, and their assessment has benefitted from the complementary information provided by population studies. Selected infradian components are reviewed below.

4.2.4.1 About 7-day Cycles: Circaseptans

The presence of statistically significant weekly and half-weekly variations in BP and HR has been documented in several populations of different ethnicities, synchronized by the social week.[60,61] Using 7-day/24-hour ABPM (ambulatory BP monitoring), we confirmed a weekly variation in the morning BP surge, which peaked on Mondays.[62] An abrupt and dramatic increase in BP in the morning is a possible trigger for cardiovascular events, which occur more frequently in the morning, between 6 AM and noon.[63,64] Weekly variations in the incidence of cardiovascular events have also been reported, with the highest incidence occurring on Mondays.[62–66]

The weekly distribution of the morning BP surge, which showed a distinct peak on Mondays might play an important causative role in the reported increased incidence of cardiovascular events on Mondays. Indeed, neurohormonal pathophysiologic mechanisms, sympathetic activation, and systemic vasoconstriction associated with the arousal reaction have been postulated to account for the morning BP surge, which is associated with target organ damage and an increased risk of stroke. Performing ABPM for at least 7 days has been recommended accordingly, the current practice of 24-hour ABPM being insufficient for the proper diagnosis of hypertension and for a precise evaluation of treatment efficacy.[62,67–69]

While the social schedule clearly synchronizes about-weekly rhythms, some studies have considered about 7-day rhythms to be independent of the social week, with some possibly related to harmonics of the lunar about-monthly cycle.[70–72] There is evidence, however, that circaseptans may also have an endogenous component. The first observation came from the analysis of a 15-year record of almost daily measurements of urine volume and urinary excretion of 17-ketosterids (17KS) by Christian Hamburger (1904–1992), who was head of the Hormone Department of Statens Serum Institut in Copenhagen, Denmark. His data showed a 7-day synchronized component in both variables for the first 12 years, after which 17-KS started to free-run with a period of about 6.935 days while urine volume remained synchronized to a 7-day cycle.[73] Such internal desynchronization occurred when Christian Hamburger started self-administering testosterone.

Circaseptans are prominently expressed in BP and HR in early extra-uterine life, when they are synchronized by the time of birth rather than by the day of the week, indicating that they are a feature of human development, notably in low-birth-weight infants.[74] About 7-day patterns have long been observed for the span from the onset of illness to the "critical day" when the patient recovers or dies, and are apparent after the application of single stimuli as mild as balneotherapy or as severe as organ transplantation.[75]

4.2.4.2 About-monthly Rhythms

About-monthly cycles are well documented in reproductive hormones, such as estrogen and progesterone, as they play an important role in the menstrual cycle, fertility, and pregnancy. Recent studies have started to examine molecular mechanisms of lunar-controlled rhythms and clocks.[76–78] Reproductive behaviour in many species synchronizes with the lunar cycle, and in humans, the menstrual cycle has a period close to the lunar cycle length. Women's menstrual cycles with a period longer than 27 days were found to be intermittently synchronous with the Moon's luminance and/or gravimetric cycles, with the onset of menses occurring around the time of the new moon, while earlier studies reported that onsets of women's menses occurred around the time of the full moon.[79] Light pollution, which has increased in the past decades, was speculated to account for the change since night light has been shown to shorten the length of the menstrual cycle, which reduces the probability of synchronization with the lunar cycle.[79] Using wrist actimetry, the timing of nocturnal sleep was shown to be synchronized with the lunar cycle in individuals living in a rural area and urban citizens.[80] Sleep was reported to start later and to be of shorter duration on the nights before the full moon when moonlight is available during the hours following dusk.[80]

4.2.4.3 About-yearly Rhythms

Circannual rhythms have evolved as genetically programmed adaptive timing mechanisms that allow organisms to take advantage of favorable seasons for reproduction and growth, and survive through unfavorable seasons. They characterize phenomena as diverse as reproduction, migration patterns, and hibernation. Circannual rhythms are endogenously generated. Evidence suggests that they are independent of the circadian pacemaker system since circannual cycles of body mass and reproductive phenotypes persist after SCN ablation.[81] Several concepts are emerging to account for their basic underlying mechanisms.[81] Specialised thyrotroph cells located in the pars tuberalis of the pituitary gland and adjacent tanycyte cells located in the ependymal wall of the third cerebral ventricle of the brain are thought to act as putative central circannual pacemakers.[82] The pars tuberalis-specific thyrotroph may act as a seasonal calendar cell, driving both hypothalamic and pituitary endocrine circuits.[83]

Many physiological functions follow about-yearly changes, as does the incidence of adverse events from different conditions. While changes in photoperiod serve as a synchronizer of circannual rhythms, ambient temperature is another factor underlying the manifestation of circannual rhythms, notably at higher latitudes. As an example, morbidity and mortality from respiratory and cardiovascular disease peak during the coldest months of the year in both the Northern and Southern Hemispheres. Their large-amplitude circannual rhythm follows the well-recognized and undisputed seasonal pattern affecting human health.[84] The prominence of the circannual rhythm in overall and cardiovascular mortality depends on latitude, as shown in data compiled from 19 countries, amounting to over 54 million deaths.[85] Several risk factors underlying adverse cardiovascular events such as body mass index, BP, low-density lipoprotein (LDL) cholesterol, triglycerides, and glucose concentrations also follow a circannual rhythm.[86] Ambient temperature has been reported to be the dominant factor underlying the higher incidence of mortality from cardiovascular disease and myocardial infarction in the coldest months,[87] while atmospheric pressure is also a contributing factor.[88] Other factors also need to be considered to account for the paradox of excess winter mortality in that the circannual variation in all-cause mortality in Northern countries unexpectedly have a weaker circannual rhythm compared to countries with a more moderate climate.[89]

Other infradian components have been observed to characterize human pathophysiology. Among them are the about half-year[90] and some other long-period components with no known environmental synchronizer, which have periods that are less stable than those of the about-weekly, about-monthly, and about-yearly variations, which may resonate with wobbly non-photic signals present in space weather.[91] For instance, the about 1.3-year oscillation characterizing solar wind speed has been detected in most long-term records of human BP and HR.[92] About 10.5-year cycles similar to the solar activity cycle have been documented in longitudinal records of BP, HR and HRV and transversally in short around-the-clock neonatal records of BP and HR.[92]

4.3 Reference Values in Health

It is not always practical or efficient to uncover the endogenous characteristics of circadian rhythms before applying chronobiologic principles, notably in the clinic. Some investigators have tried to compensate for circadian variation by sampling at a fixed clock hour. As reviewed elsewhere in this book, this practice can be greatly misleading because it does not account for inter-individual variability, and because any difference in the rhythm's amplitude, phase, and/or period when assessing the effect of a given intervention is likely to yield different and even opposite results in studies performed at different clock hours. For these reasons, the advocated approach has been to collect data around the clock from clinically healthy men and women of different ages and to map how circadian rhythm characteristics change throughout the lifespan, while accounting for gender differences. Reference values in terms of time-specified limits (chronodesms) and rhythm parameters can thus be derived, as done for many hormones[93,94] and illustrated herein for endpoints of HRV, and for BP and HR.

4.3.1 Endpoints of Heart Rate Variability

HRV endpoints have become an important tool in cardiology as a non-invasive gauge of autonomic nervous system activity and as fundamental indices reflecting people's health and well-being.[95–99] In humans, rhythmic features of HR and HRV include circadian variations and circaseptan and circannual components, among others. HR and HRV are also characterized by changes, which may be nonlinear in nature, some of them pertaining to what has been called deterministic chaos.[97,100–103]

Using 24-hour ambulatory two-channel Holter monitoring, circadian changes in HRV endpoints were mapped in health as a function of age and development in 285 males (aged from 25 days to 92 years) and 117 females (aged 3 to 91 years). Both time-domain and frequency-domain measures were determined over consecutive 5-min intervals. Fractal scaling was assessed based on the slope β of $1/f$ fluctuations of NN intervals in the frequency range of 0.0001–0.01 Hz.

The 24-hour average normal-to-normal inter-beat interval (NN) is shorter in females than in males, and its standard deviation (SDNN) decreases with age after about 20 years of age. Similar changes are found for indices of parasympathetic tone (r-MSSD and pNN50). Both LF and HF (spectral power of NN in frequency ranges of 0.04–0.15 Hz and 0.15–0.40 Hz, respectively) peak around 15 years of age. Their ratio LF/HF, an index of sympathetic tone, is higher in males than in females and increases with age until about 25–45 years of age, decreasing thereafter.[104] Some HRV endpoints, such as RR50, SDmean (1 min), and 24-hour HF reach a minimum around 66.5, 72.0, and 59.5 years of age, respectively, and increase thereafter, raising the question whether the oldest age group represents a category of long-lived individuals who, by the virtue of their age, may have been healthier than the average person in the younger age groups.[105]

All HRV endpoints undergo marked circadian rhythms, as illustrated in Figure 4.4. HF peaks in the middle of the night and the LF/HF ratio peaks in the afternoon, based on circadian profiles from 218 healthy males aged 3 to 92 years.[105] Time-specified reference values, computed as 90% prediction limits,[106] separately for males and females in different age groups, can be used to screen for abnormal circadian variation, even when it occurs inside the conventional physiological range.

The fractal scaling of HRV, gauged by the correlation dimension of RR intervals, is also characterized by a circadian rhythm, as documented for groups of 11 males in four age groups between 5–10 and 60–65 years of age.[100] The average correlation dimension was found to be largest for men 20 to 25 years of age, and to decrease thereafter, notably during the night. In all age groups, the correlation dimension is higher during the night than during the day. In addition to a decrease in the average value with age, the circadian amplitude is also reduced and the phase is advanced with increasing age.[100]

4.3.2 Blood Pressure Variability

In clinical health, women have a lower BP and a higher HR than men. While diastolic (D) BP reaches a maximum around 50 years of age, systolic (S) BP continues to increase at least until around 80 years of age. As a consequence,

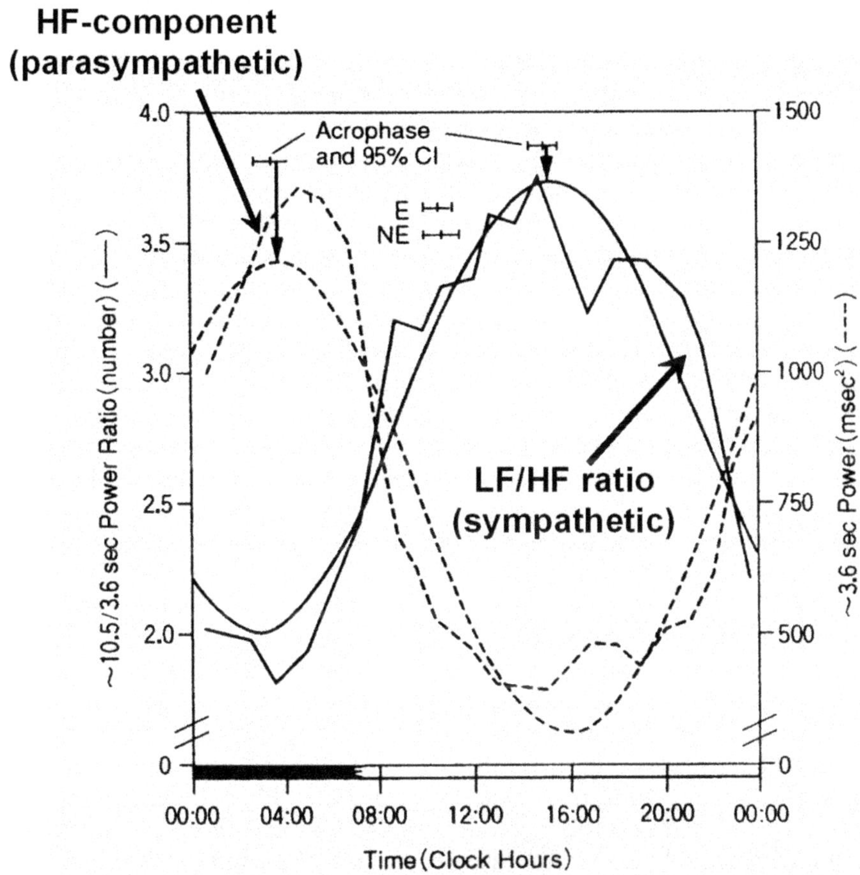

	"Tone"	
	"Sympathetic"	"Parasympathetic"
P:	<0.001	<0.001
MESOR±SE:	2.88±0.04	769±24
Double Amplitude (95% CI):	1.70 (1.46, 1.96)	904 (763, 1045)
Acrophase (95% CI):	14:51 (14:26; 15:24)	03:36 (03:00; 04:12)

Figure 4.4 Circadian variation of HF and the LF/HF ratio. Reproduced from ref. 97 with permission from Springer Nature, Copyright 2016, Springer Japan.

pulse pressure starts increasing after 50 years of age.[107] Both BP and HR are characterized by a large-amplitude circadian rhythm. BP reaches a nadir in the middle of the night, already increasing before awakening, when it increases more rapidly to maintain higher values during the day and starting to slowly decrease in the evening. A circadian rhythm in BP and HR is already discernable during the first week of life, but it has a very small amplitude and its acrophase occurs shortly after midnight, which is opposite to that of the mother.[108] With increasing age, the post-prandial dip in early afternoon becomes more pronounced.[107,109] In view of these results, time-specified reference values, qualified by gender and age were derived from ABPM records from clinically healthy adults 20 years and older.[110] Similar reference values were obtained for pregnant women,[111] and for children and adults practicing manual self-measurements.[112]

In view of the large day-to-day variability in all characteristics of the circadian rhythm in BP, in normotensive as well as hypertensive individuals, a consensus[113] was reached to recommend that ABPM be performed for a minimum of 7 days rather than the current practice of limiting it to 24 hours. Because results from 24-hour ABPM may lead to a diagnosis of normotension on one day and hypertension on another day, the recommendation was extended to everybody rather than being limited to special cases, as currently done in practice. Since circadian rhythm characteristics of BP differ between men and women, and change as a function of age, the consensus advocates that data be analyzed to estimate circadian parameters rather than limiting the computation to mean values over 24 hours, daytime, and nighttime. While it is recognized that current ABPM using an arm cuff may not be readily acceptable to all, thus hampering the implementation of the recommendations made in the consensus report, new technologies are emerging[114–116] that may facilitate the implementation of a chronobiologic approach in the near future.

Utilizing the chronobiological reference values in health for the circadian variation in BP and HR, a computer-generated form, called a sphygmochron, was designed to provide information on a given person's rhythm characteristics and compare them with those of clinically healthy peers matched by gender and age.[117] Results consist of two parts. First, estimates of the MESOR and 24-hour amplitude and acrophase from the least squares fit of a model consisting of cosine curves with periods of 24 and 12 hours to the SBP, DBP, and HR data are listed for comparison with corresponding 90% prediction limits in clinical health. This model approximates the circadian waveform of BP and HR well in most cases. Second, the person's average circadian waveform determined after stacking all data over an idealized 24-hour day is compared to the time-specified 90% prediction limits in health. Using numerical integration, this complementary approach yields estimates of the percentage time elevation above acceptable limits, the extent of excess gauged by the area delineated by the profile when it exceeds the upper limit of acceptability, and the timing of overall excess. It accounts for cases when the circadian waveforms are not approximated well by the two-component model. The sphygmochron readily indicates whether the circadian variation in BP and HR is within acceptable limits or not. If not, it also indicates in which way the circadian variation of BP and/or HR deviates from norms, whether the MESOR is elevated, whether the 24-hour amplitude is excessive, and/or the 24-hour acrophase occurs too early or too late. The combined information from the parametric and non-parametric approaches is useful for screening and diagnosis, and serves as a guide to treatment timing.

4.4 Circadian Disruption

Deriving time-specified reference values in health is particularly important in many instances when the assessment of the endogenous component of the circadian rhythm is not practical. Now that the continuous monitoring of some physiological variables has become feasible, reference values can be personalized. As a minimum, their time structure can serve as markers in preference to reliance on clock hours (or calendar dates).

Chronobiological reference values are indispensable for diagnosing circadian disruption and helping in determining whether it relates to impending disease or healthy ageing. Circadian disruption can be associated with genetic mutations, such as in the case of the Smith–Magenis syndrome, or it can have a behavioural origin, known as social jetlag. Poor circadian hygiene in modern society (suboptimal daylight in office buildings and excessive exposure to artificial light at night) also contributes to circadian disruption. Circadian disruption associated with disease conditions is discussed in Sections 4.5 and 4.6, in relation to chronodiagnosis and chronotherapy.

4.4.1 Social Jetlag

Social jetlag is the name given to the condition in which sleep patterns vary widely between workdays and weekends. It is a consequence of the discrepancy between the individual's circadian rhythm and the social schedule. Social jetlag is associated with chronic sleep loss. A large-scale epidemiological study showed that, beyond sleep duration, social jetlag is also associated with an increased body mass index.[118] Due to differences in the sleep schedule between workdays and free days, the estimated circadian amplitude of physiological variables such as activity, temperature, or HR monitored for a week or longer will be reduced to an extent that is commensurate to the amount of social jetlag present.

Social jetlag is clinically relevant because disrupted phase alignment among clocks in the suprachiasmatic nucleus (SCN) and the periphery is thought to underlie adverse outcomes related to shift work.[119–121] Social jetlag, the most frequent type of circadian misalignment, correlates with a higher risk of developing obesity, diabetes and cardiovascular diseases.[122–125]

Circadian misalignment rather than sleep duration has been associated with fatty liver disease that requires the presence of metabolic dysfunction (MAFLD). In a study using the NHANES 2017–2018 database, circadian misalignment was found to be an independent risk factor for MAFLD, whereas short sleep duration was not an independent risk factor for MAFLD or liver fibrosis after adjusting for circadian misalignment.[126] Regularity of the sleep schedule rather than sleep duration also correlates with academic performance.[127]

4.4.2 Ageing

A biological circadian clock orchestrating most biological variables, is also involved in the circadian coordination of BP.[128] Disruption of circadian rhythms can seriously affect the overall health and increase the risk of adverse cardiovascular outcomes.[129] In mammals, including humans, ageing is associated with alterations in circadian rhythm characteristics, usually involving a reduced amplitude and a greater variability in the phase, which tends to be advanced.[130] As discussed elsewhere in this book, deficient melatonin production and impaired transmission of photic information by retinal ganglion cells are candidate mechanisms underlying age-related changes in circadian function.

Age-related alterations in the circadian rhythm of BP measured automatically around the clock for 7 days were examined as part of a comprehensive community-based medical study.[97] The 283 citizens living in a rural Japanese town (Tosa) (170 women and 113 men) were 39 to 74 years of age. The circadian period of BP, estimated by the maximum entropy method, ranged from 22.5 to 25.5 hours, averaging 24.04 ± 0.46 (SD) hours. Circadian parameters were estimated by fitting a model consisting of cosine curves with periods of 24 and 12 hours.[68] Considering four age groups, the incidence of circadian SBP disruption (defined as an estimated period outside the average period $\pm$ 3 SD) was highest in the oldest age group (>65 years) ($P < 0.05$). SBP was also higher and circadian acrophases of SBP and DBP were statistically significantly advanced in that group as compared to the younger groups.

Although there were no inter-group differences in depressive mood, or subjective quality of life (health and happiness), sleep duration was longer in the elderly. Citizens with a disrupted circadian rhythm (period differing from 24 hours) exhibited an earlier acrophase, a reduced circadian amplitude, and a smaller nightly BP dip as compared to those who were synchronized to a 24-hour rhythm. They were older and more likely to take anti-hypertensive medication. Ageing was thus associated with changes in circadian rhythm characteristics and with an increased incidence of rhythm disruption.[97]

Analysis of 7-day/24-hour BP and HR records from another study of 72 individuals ranging in age from 12 to 106 years showed that the weakening of the circadian rhythm with advancing age was accompanied by an increased prominence of both infradian and ultradian components.[131] While in the elderly (≥75 years), their circadian amplitude of SBP decreased by 33%, their weekly and half-weekly amplitudes increased by 53% and 96%, respectively, and ultradians with periods of 6 and 3 hours were also considerably increased (87% and 76%, respectively).

4.5 Principles of Chronodiagnosis

The derivation of time-specified reference values allows new, more refined questions to be answered. In addition to knowing whether a test value is too low, too high, or acceptable, additional questions are whether the circadian amplitude, phase, and/or period are within acceptable limits, and whether there may have been a transposition of the variance from the circadian to the ultradian and/or infradian domain. By considering the physiological range, even a change in the mean value can be detected earlier, leading the way to primary prevention (pre-habilitation) by the institution of prophylactic, in preference to curative or palliative treatment, including lifestyle modifications.

Rhythm alteration detected before symptoms become overt offers warning signs of pre-disease, which is defined as an increase in disease risk. The goal of chronomedicine aimed at pre-habilitation is to prevent the further development of overt disease.[68,97,132] A better understanding of the mechanisms underlying the chronodiagnosis is likely to lead to novel therapeutic interventions for conditions, including cardiovascular and metabolic disorders and cancer, arising from circadian disruption.[133–136]

4.5.1 Chronodiagnosis Involving Rhythms Other Than Circadian

While chronobiological reference values have not yet been derived for other-than-circadian rhythms, several disease conditions have been linked to ultradian and infradian oscillations, illustrated here for the BRAC and circaseptans. REM sleep is a fundamental state of consciousness of the mammalian brain. REM sleep is marked by wake-like brain activity of low-voltage fast waves occurring in combination with temporary muscle paralysis and vivid dreaming. It is also characterized by muscle twitching and rapid eye movements. Cardiorespiratory control and autonomic tone typically exhibit significant variability in REM sleep.[137] Untimely activation of REM sleep circuits leads to involuntary muscle weakness or paralysis, a major symptom of narcolepsy,[137] a condition where BRAC is prominently expressed.[138] Degeneration of the REM circuit is associated with excessive muscle activation in REM sleep behaviour disorder,[137] which may be associated with other neurological conditions, such as Lewy body dementia, Parkinson's disease, or multiple system atrophy. Inappropriate arousal from sleep during obstructive sleep apnea repeatedly disturbs the activity of sleep circuits, particularly the REM sleep circuit.[137]

The circulating concentrations of major reproductive hormones fluctuate on a short-term basis. It has long been proposed that pituitary–adrenal activation is associated with REM sleep stages.[139] A regularly occurring periodicity of approximately 100–120 min, characterizing the pulsatility in growth hormone, was reported in clinically healthy short boys.[140] In patients with African trypanosomiasis (sleeping sickness), a study of hormonal pulsatility and sleep stages, determined by polygraphic sleep recordings, points to the existence of a robust link between hormonal release and the internal sleep structure.[141] The pulsatility of glucocorticoid hormones is altered in obesity during pregnancy.[142]

From an epidemiologic viewpoint, Reimann[143] points to the dominant incidence of a weekly rhythm in so-called periodic diseases. From a clinical viewpoint, Derer[144] described a 6-day rhythm in human blood leukocytes, reticulocytes, excretion of uric acid, glucocorticoids, and other infradian periodicity more broadly (for review of circaseptans in physiopathology, see ref. 145–147). The about half-weekly variation is enhanced in the HR and temperature of cancer patients,[148] and their larger prominence among treated patients predicts cancer recurrence.

In a study involving 192 citizens of Uraus, Japan, we observed a prominent circaseptan variation in SBP and DBP in mild depression, peaking on Thursdays, which was not found in non-depressed individuals.[149] This result is in agreement with the amplification of the circaseptan component during depressive episodes in a longitudinal ABPM record of a manic-depressive patient,[150] and with a prominent weekly variation in the negative affect, but not in the positive affect among 196 clinically healthy students who repeatedly completed the PANAS questionnaire.[151]

A circaseptan rhythm characterizes SDNN and other time-domain HRV endpoints in middle-aged and elderly individuals.[152] SDNN, which is decreased on Mondays, with a secondary trough on Fridays, is a good predictor of long-term survival, both in patients with myocardial infarction and in a cohort of elderly subjects. It could account in part for the circaseptan pattern of incidence of myocardial infarction peaking on Mondays with a secondary peak on Thursdays, documented in a meta-analysis of 47 studies.[66]

The alterations in ultradian or infradian time structure associated with different disease conditions presented above deserve further investigation to determine whether they may also serve as early warning signs. Unveiling their underlying mechanisms may also lead to a better understanding of the etiologies involved and perhaps also to novel therapies.

4.5.2 Chronodiagnosis Related to Circadian Rhythms

Circadian rhythm alteration is almost invariably seen in the presence of disease, as reviewed for each physiological system in Parts II and III of this book. While emphasis has been placed on cellular and molecular mechanisms, it is worth noting that from an empirical perspective, chronobiological reference values and rhythm parameters used as biomarkers can also play an important role in early and differential diagnosis of disease. For instance, a shortened circadian period of breast skin temperature has been associated with the tumor growth rate and metabolic heat production in mammary carcinomas.[153]

As reviewed above, all HRV endpoints exhibit circadian periodicity, differ between males and females, and change as a function of age. Conventional reference ranges in adulthood, however, rely on the same thresholds irrespective of time, gender, or age. Refined reference standards accounting for these factors may thus lead to a more reliable interpretation of data obtained during screening. For this purpose, short ECG records are best replaced with around-the-clock monitoring for a few days. While this practice has entered clinical care, a chronobiologic interpretation of the data has not yet entered mainstream clinical practice. Certain circadian stages may also be preferable to others for the assessment of disease or disease risk, as alterations from the norm may be greater at one time than at another.[100]

In the case of BP and HR, ABPM records have been obtained from clinically healthy people to derive reference values as 90% prediction limits for circadian rhythm parameters in men and women of different age groups.[106,110] Using these reference values as a basis, several abnormalities in the circadian patterns of BP and HR were identified, called vascular variability disorders (VVDs).[113] A MESOR of SBP and/or DBP above their respective upper 95% prediction limit in health is known as MESOR-hypertension, which indicates that the diagnosis is made based on around-the-clock measurements rather than on casual single values. A pulse pressure exceeding 60 mmHg is considered to be too large, as is a pulse-pressure product above 10 000 mmHg.beats per min. A 24-hour amplitude of SBP and/or DBP that exceeds the upper 95% prediction limit in health has been called CHAT, which stands for circadian hyper-amplitude-tension. A 24-hour acrophase of BP differing by more than 6 hours from the 24-hour acrophase of HR, known as ecphasia, reflects circadian misalignment. This condition is diagnosed in some patients with non-insulin-dependent diabetes mellitus who have impaired autonomic nervous function. A global standard deviation of HR less than 7.5 beats min^{-1} is considered to be too low. A non-24-hour period, defined as a period's 95% confidence interval (estimated by spectral analysis) not covering 24 hours is yet another VVD requiring records spanning at least 7 days. In each case, estimates from a given record are interpreted in the light of reference values from clinically healthy peers matched by gender and age.[117] Illustrative case reports related to the different conditions are presented in detail elsewhere.[97]

Several outcome studies have validated the merit of a chronobiological diagnosis based on VVDs.[68] In Japan,[154,155] a 6-year prospective study involved 297 patients with no initial history of morbid cardiovascular events who underwent 48-hour ABPM. As compared to uncomplicated MESOR-hypertension, the relative risk (RR) and 95% confidence interval (CI) of adverse cardiovascular events were 6.3 [CI: 2.1–18.8] for DBP CHAT, 8.2 [CI: 3.6–18.8] for an excessively large pulse pressure, and 8.2 [CI: 2.8–23.8] for a deficient HRV.[107] VVDs are mostly independent and additive. The incidence of adverse events in uncomplicated MESOR-hypertension was less than 10%. By comparison, the presence of any one or two additional VVDs was associated with a 29% or 53% incidence of adverse events.[68] In Taiwan,[156] the left ventricular mass index (LVMI) served as a surrogate outcome measure in a study involving 1179 untreated patients who underwent a 24-hour ABPM. As compared to the LVMI of $87.5 \pm 0.6\ g\,m^{-2}$ characterizing 1039 patients with an acceptable circadian variation of DBP, the LVMI of 67 patients with DBP CHAT was 11.2% larger and that of the 73 patients with DBP ecphasia was 11.6% larger ($P < 0.001$), a sign of increased cardiovascular disease risk.

A classification in terms of dipping, a method based on the day–night ratio often used to interpret ABPM records, was not discriminatory, pointing to the merit of using chronobiologic reference values qualified for gender and age, and the consideration that some VVDs have a nonlinear relationship with cardiovascular disease risk.

Reliance on chronobiological reference values also helps in surveillance and health maintenance purposes, as illustrated in community-wide studies undertaken in rural areas of Japan.[97] In Uraus (Hokkaido), 207 ordinary citizens, aged 24 to 79 years, who underwent 7-day/24-hour ABPM, were followed-up for 5 years to observe the occurrence of new onset cardiovascular outcomes. Survival analysis showed that the presence of abnormal circadian variability in BP and HR considerably decreased disease-free survival.[97]

4.6 Principles of Chronotherapy

Predictable circadian changes in the physiological state were shown experimentally as early as the 1950s to make the difference between life and death following exposure to noise, bacterial endotoxins, ethanol, or ouabain,[2] as well as librium, acetylcholine, pentobarbital and a host of other agents, including whole-body or partial-body X-ray irradiation.[157] These results led to the concept of "hours of diminished resistance" to the effects of exogenous agents, which occur predictably along the 24-hour scale, and whose timing differs from one agent to another. Since then, numerous studies have shown optimal circadian stages for administering a given treatment to patients with a wide range of conditions, from asthma and endocrine disorders to cancer and cardiovascular disease, thereby increasing treatment efficacy and/or reducing side effects.[158] Specific timing of drug administration can have significant impact on the efficacy and toxicity of drugs. RNA expression profiles in mice have demonstrated that 43% of protein coding genes are rhythmic. Since most commonly prescribed medications target one of these rhythmic genes, chronotherapy is important to optimize the treatment of many health conditions.[159] While chronotherapy is primarily associated with circadian rhythms, ultradian and infradian components are also worthy of consideration.

4.6.1 Chronotherapy Involving Rhythms Other Than Circadian

The pacemaker is perhaps the best example of an application of chronotherapy based on ultradian variation. It restores a regular heartbeat in patients with heart failure and in those with arrhythmia, such as a slow heartbeat after a heart attack. High frequencies are also used for deep brain stimulation, a neurosurgical procedure aimed at changing brain activity in a controlled manner. Through implanted electrodes, electrical impulses are sent to specific targets in the brain for the treatment of movement disorders, including Parkinson's disease, essential tremor, dystonia, obsessive–compulsive disorder, and epilepsy.[160] In clinical trials, deep brain stimulation is studied as a potential treatment for chronic pain and for various affective disorders, including major depression.[160]

In addition to their circadian behaviour, glucocorticoids are secreted in a pulsatile pattern that emerges from delays in the feedforward–feedback interaction between the anterior pituitary and adrenal glands. A combined circadian and ultradian schedule of glucocorticoid delivery should provide safer and more effective outcomes, with major implications for patients with adrenal insufficiency.[161] Dynamic oscillations of glucocorticoids, critical for normal cognitive and metabolic function have been shown in the rat to modulate the pattern of glucocorticoid-sensitive gene expression, modify synaptic activity, and maintain stress responsiveness. Such a pulsatile signal of the glucocorticoid secretion dynamics is of crucial importance also in humans in relation to cognitive and behavioural processes, including optimal responses to emotional stimulations, quality of sleep, and working memory performance.[162] Suppressing the ultradian signal by the administration of long-acting synthetic glucocorticoids impairs glucocorticoid-dependent bodily systems, resulting in the disruption of behaviour, affective state, and cognitive functions. Chronotherapy can now take into consideration the characteristic ultradian behaviour of glucocorticoids, thanks to portable pulsatile continuous subcutaneous delivery systems using a portable infusion pump.[163]

Several studies have also indicated that optimizing the weekly schedule of treatment administration could lead to benefits. In line with the circaseptan periodicity in organ rejection induced or amplified after kidney transplants in Minnesota, Paris, and Milan,[164] segmental pancreatic graft function in rats was prolonged by optimizing both the circadian and circaseptan scheduling of cyclosporine.[165] The longest times to rejection were observed in rats treated around 20 HALO (hours after light onset), with varying daily doses of cyclosporine according to a circaseptan sinusoidal pattern with largest daily doses administered either on the third or fifth day after surgery and at 7-day intervals thereafter. As compared to rats receiving equal daily doses of cyclosporine, those treated according to an optimal circaseptan schedule experienced a 30% prolongation of graft function. A 30% increase in treatment efficacy associated with circaseptan optimization is not negligible, first because it can be achieved in addition to the benefit from circadian optimization, and second because it may be practical to implement.

Circadian and circaseptan stage-dependence was also documented for treatment with Lentinan, a polysaccharide extracted from edible mushrooms, defined as T-oriented adjuvant or specific T-helper factor, administered to LOU rats prior to immunocytoma inoculation.[166] While tumor growth inhibition, survival rate, and survival times were increased in rats treated during their rest span according to a weekly schedule, they were decreased in rats treated near the middle or late activity span with equal daily doses.[166] Tumor cells of four different types of cancer (B14, C3H, glioma, L1210), gauged by the β-ATP peak, were also shown to grow according to both circadian and circaseptan rhythms, the weekly component being much more prominent than the circadian variation.[167] Moreover, when radiation was applied at the time of highest cell division (maximal β-ATP) within these 24-hour and 7-day cycles, tumor response to radiation was statistically significantly more effective than when it was given at the time of lowest cell division (minimal β-ATP).[168]

4.6.2 Chronotherapy Involving Circadian Rhythms

Bright light and melatonin are used to treat conditions such as depression and sleep disorders.[169–175] Many more patients can be helped this way since disorders of sleep are an integral part of neurodegenerative diseases.[170] Many studies have also reported that exogenous melatonin has analgesic and neuroprotective effects in chronic pain,[174] among others.

Circadian chronotherapy also extends to many other medications used to treat a wide variety of different conditions. Treatment has been optimized for inhaled medication, adrenergic and anticholinergic bronchodilators, anti-inflammatory corticoids and cromoglycate derivatives in relation to allergic diseases. The recommendation was made to ingest corticoids at the beginning of activity and long-acting theophylline when diurnal activity ceases.[176] Taking non-sedative anti-histamines (mequitazine and terfenadine) in the evening was reported to be more efficient and better tolerated for treating rhinitis.[175] Since the 1950s, chronotherapy of rheumatoid arthritis in the United States and Europe has involved several non-steroid anti-inflammatory drugs, certain disease modifying anti-rheumatic drugs, and various synthetic corticosteroid medications.[177] Chronotherapy of anticancer drugs considering the patient's circadian changes in xenobiotic metabolism and detoxification, cell cycle events, DNA repair, apoptosis, and angiogenesis can modify the tolerability of anticancer medications 2- to 10-fold in experimental models and in cancer patients.[178] Applications related to cancer and metabolism, which are important in all aspects of pharmacology (absorption, distribution, metabolism, and elimination) are reviewed elsewhere in this book.

The importance of relying on marker rhythmometry to guide chronotherapy is worth noting here. Doubling of 2-year disease-free survival of patients with cancer of the oral cavity was achieved by irradiating the tumor at the time of circadian peak tumor temperature, by comparison to treatment administered either 4 or 8 hours before or after the time of peak tumor temperature, or treatment given as usual, without consideration of timing.[179]

4.6.3 Toward Personalized Chronotherapy

In the case of cardiovascular diseases, BP and HR are marker variables that are relevant not only for guiding the timing of treatment, but also for assessing the response to treatment. In view of the different kinds of abnormal variations in BP and HR known as VVDs, which all increase cardiovascular disease risk, it becomes imperative not to seek an average best time for the population in large clinical trials, but rather to optimize both the timing of treatment and the choice of medication for each individual patient.[180] To this effect, it is worth noting that not all anti-hypertensive drugs have the same effect on the circadian amplitude of BP, that some may increase rather than decrease HRV, and that lowering BP to a larger extent is not necessarily associated with the best outcome, notably in populations where the incidence of CHAT is high.[117,181]

Accurate mathematical models are being developed to optimize drug administration through infusion pumps in order to achieve precision medicine for cancer patients.[182] Modeling the hypothalamic–pituitary–adrenal network similarly leads to refined diagnosis and a personalized approach to treatment in conditions such as depression[183] and congenital adrenal hyperplasia.[184] In patients with type 1 diabetes, use of a sensor-augmented-insulin-pump with a low-glucose suspend feature led to a significant and sustained decrease in glycated hemoglobin, and a significant reduction in severe hypoglycemia and hypoglycemia unawareness.[185]

An understanding of circadian biology is important for personalized and precision medicine. Studies focusing on genetics[186–190] and epigenetics[191–194] improved our understanding of how rhythms sustain health. The discovery that the circadian clock coordinates half the mammalian protein-coding genome indicates the importance of time considerations in the implementation of lifestyle interventions for effectively preventing and/or treating disease.[124,188,192] Histone modifications have been shown to be responsive to various environmental stimuli, including exercise and diet. Collectively, a better understanding of the mechanisms through which the circadian clock function can be compromised will lead to novel therapeutic approaches to address various human diseases, including cardiovascular and metabolic disorders and cancer arising from circadian disruption.[133–136]

4.7 Inter-modulations Among Multi-frequency Rhythms and Chronomics Medicine

Variables related to cellular and humoral immunity share with the nervous system a broad spectrum of rhythms interwoven with age trends. The term "modulation" is often used to refer to the process of fitting or regulating according to a certain measurable proportion; a tempering or toning down". This definition does not do justice to modulators that may lead predictably to both increases and decreases of a given (modulated) response, *i.e.*, to both stimulation and inhibition in physiological terms. Modulation then means no more than (unpredictable) interaction. An alternate definition of "modulation" is the variation of a characteristic (amplitude, phase, or frequency) of a carrier or signal in a periodic or intermittent manner for the transmission of information (as in telegraphy, telephony, radio or television).

The complexity of neuro-immune interactions can perhaps explain why we are far from fully understanding the mechanisms underlying the coordination of biological rhythms.[100,194–198] The molecular clocks present in individual cells involve approximately fifteen clock genes interconnected in regulatory feedback loops.[199] Hence, many mechanisms of diseases and drug effects undergo inter-modulations, not only among circadian variations of the different entities involved, but also involving multi-frequency components. Inter-modulations also arise from effects of competing synchronizers, such as the lighting regimen and the feeding schedule. Time-restricted feeding offered during the early or late part of the active span is associated with large shifts of the circadian acrophase of variables such as insulin, glucagon and growth hormone, but not so much of cortisol.[200–203] The timing of meals can thus alter the internal constellation of phase relations among physiological variables. Time-restricted feeding can be used to decrease body weight and SBP, and improve sleep quality.[204,205]

Biphasic dose responses (hormesis) to environmental conditions are characterized by beneficial stimulation at a low dose and undesired inhibition at a high dose.[206,207] The cellular signaling pathways and molecular mechanisms that mediate hormetic responses have recently been elucidated as typically involving enzymes such as kinases and deacetylases, and transcription factors.[208–211] As the result of multiple interactions among rhythmic variables, chronomodulation is another process accounting for qualitatively as well as quantitatively different effects of the same drug or other stimulus. For example, in children with asthma, the circadian acrophase of the clinically important peak expiratory flow was advanced by a corticosteroid given at one circadian stage, while it was delayed when given at another circadian stage.[212] The multi-frequency components of biologic time structure are coordinated through multi-level interactions, called feedsidewards, in a network of spontaneous, reactive, and modulatory rhythms. An example of feedsidewards relates to serum corticosterone secreted from the bisected adrenal of mice, which undergoes a prominent spontaneous circadian rhythm peaking at the end of the light (rest) span.[11] Its response to a fixed dose of a synthetic ACTH follows circadian periodicity. The largest response is found in the last part of the dark (active) span, indicating that adrenals are most responsive to the pituitary when corticosterone concentrations are minimal. This response of the adrenal to ACTH is amplified by the addition to pineal homogenate, irrespective of the circadian stage at which pineal glands are harvested.[11] In other words, the modulation of the adrenal response to the pituitary by the pineal involves enhancing the response when it is large and reducing it when it is small.

The response of a physiological variable to an external stimulus varies predictably according to the rhythm stage of its exposure or administration. Other factors acting on all rhythm parameters, not just on the mean value, can further modify this response. Much novel information can hence be gained by mapping the biologic time structure and estimating all its characteristics. The information to be derived from this approach is much richer than what can be learned from a two-timepoint design or by fixing the time of day.

Abbreviations

17KS	17-Ketosteroids
ACTH	Adrenocorticotropic hormone
ABPM	Ambulatory blood pressure monitoring
ATP	Adenosine triphosphate
β	Fractal scaling; $1/f$ slope of log(power) *versus* log(frequency) of NN intervals (in 0.0001–0.01 Hz range)
B14	Hamster fibroblast cell line
BP	Blood pressure
BRAC	Basic rest-activity cycle (about 90-minute)
C3H	Mouse embryo cell line
CHAT	Circadian hyper-amplitude-tension
DBP	Diastolic blood pressure
ECG	Electrocardiogram
EEG	Electroencephalogram
ER	Endoplasmic reticulum
HALO	Hours after light onset
HF	HRV power in the high-frequency band (0.15–0.4 Hz)
HR	Heart rate
HRV	Heart rate variability
Hz	Hertz (1/sec)
ISO	Infraslow oscillation (0.01–0.10 Hz)
L1210	Mouse lymphocytic leukemia cell line
LDL	Low-density lipoproteins
LF	HRV power in the low-frequency band (0.04–0.15 Hz)
LOU rats	Inbred albino Louvain rats
LVMI	Left ventricular mass index
MAFLD	Metabolic-associated fatty liver disease
MESOR	Midline estimating statistic of Rhythm
NHANES	National health and nutrition examination survey
NMDA	*N*-Methyl-D-aspartate
NN	Inter-beat interval of normal-to-normal beats (from ECG)
NN50	Number of successive NN intervals that differ by more than 50 ms
PANAS	Positive and negative affect schedule
pNN50	Percentage of successive NN intervals that differ by more than 50 ms
PVN	Paraventricular nucleus
REM	Rapid eye movement
r-MSSD	Root mean square of successive NN interval differences
RNA	Ribonucleic acid

RR	Inter-beat interval (R wave to next R wave in ECG)
SBP	Systolic blood pressure
SCN	Suprachiasmatic nuclei
SDmean (1 min)	Mean over 24 hours of the 1-min SD of NN intervals
SDNN	Standard deviation of NN intervals
SPZ	Subparaventricular zone
TTFL	Transcription-translation feedback loop
VVD	Vascular variability disorder

References

1. F. Halberg and M. B. Visscher, *Proc. Soc. Exp. Biol.*, 1950, **75**, 846.
2. F. Halberg and A. N. Stephens, *Proc. Minn. Acad. Sci.*, 1959, **27**, 139.
3. J. Bass and J. S. Takahashi, *Science*, 2010, **330**, 1349.
4. F. Halberg and G. Cornelissen, *Biochim. Clin.*, 1991, **15**, 159.
5. F. Halberg, F. Carandente, G. Cornelissen and G. S. Katinas, *Chronobiologia*, 1977, **4**(Suppl. 1), 189.
6. P. D. Larsen, Y. C. Tzeng, P. Y. Sin and D. C. Galletly, *Respir. Physiol. Neurobiol.*, 2010, **174**, 111.
7. E. Kristal-Boneh, G. Harari, M. S. Green and J. Ribak, *Blood Press. Monit.*, 1996, **1**, 87.
8. H. Brown, F. Halberg, E. Haus, D. Lakatua, H. Berg, L. Sackett, J. Melby and T. Wilson, *Chronobiologia*, 1980, **7**, 21.
9. F. J. Raab, E. M. Schaffer, G. Guillaume-Cornelissen and F. Halberg, *J. Am. Dent. Assoc.*, 1998, **129**, 461.
10. D. Montaigne, X. Marechal, T. Modine, A. Coisne, S. Mouton, G. Fayad, S. Ninni, C. Klein, S. Ortmans and C. Seunes, *et al.*, *Lancet*, 2018, **391**, 59.
11. F. Halberg, F. Guillaume, S. S. de la Peña, M. Cavallini and G. Cornelissen, *Chronobiologia*, 1986, **13**, 137.
12. B. O. Watson, *Front. Syst. Neurosci.*, 2018, **12**, 44.
13. N. Kleitman, *Physiol. Rev.*, 1949, **29**, 1.
14. D. Lloyd and D. B. Murray, *FEBS Lett.*, 2006, **580**, 2830.
15. D. Ono, K. I. Honma and S. Honma, *Sci. Rep.*, 2015, **5**, 12310.
16. Y. E. Wu, R. Enoki, Y. Oda, Z. L. Huang, K. I. Honma and S. Honma, *Proc. Natl. Acad. Sci. U. S. A.*, 2018, **115**, E9469.
17. A. Silvani, G. Calandra-Buonaura, R. A. Dampney and P. Cortelli, *Philos. Trans. R. Soc., A*, 2016, **374**, 20150181.
18. K. E. Poskanzer and R. Yuste, *Proc. Natl. Acad. Sci. U. S. A.*, 2016, **113**, E2675.
19. S. Mederos, C. González-Arias and G. Perea, *Front. Synaptic Neurosci.*, 2018, **10**, 45.
20. M. G. Frank, *Handb. Exp. Pharmacol.*, 2019, **253**, 83.
21. O. Barca-Mayo, M. Pons-Espinal, P. Follert, A. Armirotti, L. Berdondini and D. De Pietri Tonelli, *Nat. Commun.*, 2017, **8**, 14336.
22. M. D. Ruben and J. B. Hogenesch, *Curr. Biol.*, 2017, **27**, R339.
23. R. Costa and S. Montagnese, *J. Neurochem.*, 2021, **157**, 42.

24. M. A. Banaclocha, *Brain Res. Bull.*, 2007, **73**, 21.
25. M. Brancaccio, A. P. Patton, J. E. Chesham, E. S. Maywood and M. H. Hastings, *Neuron*, 2017, **93**, 1420.
26. M. H. Hastings, E. S. Maywood and M. Brancaccio, *Biology*, 2019, **8**, 13.
27. J. E. Stern, S. Son, V. C. Biancardi, H. Zheng, N. Sharma and K. P. Patel, *Hypertension*, 2016, **68**, 1483.
28. N. Marina, A. G. Teschemacher, S. Kasparov and A. V. Gourine, *Exp. Physiol.*, 2016, **101**, 565.
29. J. Enes, M. Haburčák, S. Sona, N. Gerard, A. C. Mitchell, W. Fu and S. J. Birren, *PLoS One*, 2020, **15**, e0218643.
30. N. Marina, I. N. Christie, A. Korsak, M. Doronin, A. Brazhe, P. S. Hosford, J. A. Wells, S. Sheikhbahaei, I. Humoud, J. F. R. Paton, M. F. Lythgoe, A. Semyanov, S. Kasparov and A. V. Gourine, *Nat. Commun.*, 2020, **11**, 131.
31. N. L. Kikel-Coury, J. P. Brandt, I. A. Correia, M. R. O'Dea, D. F. DeSantis, F. Sterling, K. Vaughan, G. Ozcebe, P. Zorlutuna and C. J. Smith, *PLoS Biol.*, 2021, **19**, e3001444.
32. A. Tedoldi, L. Argent and J. M. Montgomery, *Am. J. Physiol.*, 2021, **320**, C1.
33. E. Othmer, M. P. Hayden and R. Segelbaum, *Science*, 1969, **164**, 447.
34. W. C. Orr, H. J. Hoffman and F. W. Hegge, *Chronobiologia*, 1976, **3**, 293.
35. J. F. Hiatt and D. F. Kripke, *Psychosom. Med.*, 1975, **37**, 320.
36. P. Lavie and D. F. Kripke, *Nature*, 1977, **269**, 142.
37. M. Okawa, M. Matousek, A. L. Nueth and I. Petersen, *Electroencephalogr. Clin. Neurophysiol.*, 1981, **52**, S17.
38. P. Lavie and A. Scherson, *Electroencephalogr. Clin. Neurophysiol.*, 1981, **52**, 163.
39. W. Blessing, M. Mohammed and Y. Ootsuka, *Physiol. Behavior.*, 2012, **105**, 966.
40. L. Nobili, F. Ferrillo, A. Besset, G. Rosadini, G. Schiavi and M. Billiard, *Neurophysiol. Clin.*, 1996, **26**, 51.
41. K. Otsuka, G. Cornelissen, S. Furukawa, Y. Kubo, M. Hayashi, K. Shibata, K. Mizuno, T. Aiba, H. Ohshima and C. Mukai, *Heliyon*, 2016, **2**, e00211.
42. F. E. Yates and L. B. Yates, in *Ultradian Rhythms from Molecules to Mind*, ed. D. Lloyd and E. L. Rossi, Springer, London, 2008, pp. 249–260.
43. D. Bailey, D. Harry, R. E. Johnson and I. Kupprat, *J. Appl. Physiol.*, 1973, **34**, 467.
44. M. E. Hughes, L. DiTacchio, K. R. Hayes, C. Vollmers, S. Pulivarthy, J. E. Baggs, S. Panda and J. B. Hogenesch, *PLoS Genet.*, 2009, **5**, e1000442.
45. M. E. Hughes, H. K. Hong, J. L. Chong, A. A. Indacochea, S. S. Lee, M. Han, J. S. Takahashi and J. B. Hogenesch, *PLoS Genet.*, 2012, **8**, e1002835.
46. B. Zhu, C. C. Dacso and B. W. O'Malley, *J. Endocr. Soc.*, 2018, **2**, 727.
47. B. Zhu, Q. Zhang, Y. Pan, E. M. Mace, B. York, A. C. Antoulas, C. C. Dacso and B. W. O'Malley, *Cell Metab.*, 2017, **25**, 1305.e9.

48. Y. Pan, H. Ballance, H. Meng, N. Gonzalez, S. M. Kim, L. Abdurehman, B. York, X. Chen, Y. Schnytzer and O. Levy, *et al.*, *PLoS Biol.*, 2020, **18**, e3000580.
49. K. Otsuka, S. Murakami, Y. Kubo, T. Yamanaka, G. Mitsutake, S. Ohkawa, K. Matsubayashi, S. Yano, G. Cornelissen and F. Halberg, *Biomed. Pharmacother.*, 2003, **57**(Suppl.1), 1.
50. S. Murakami, K. Otsuka and T. Kono, *J. Clin. Hypertens.*, 2019, **2**, 1675.
51. S. A. Shea, M. F. Hilton, K. Hu and F. A. Scheer, *Circ. Res.*, 2011, **108**, 980.
52. E. Haus, L. Dumitriu, G. Y. Nicolau, S. Bologa and L. Sackett-Lundeen, *Chronobiol. Int.*, 2001, **18**, 709.
53. F. Halberg, G. Cornelissen, G. Katinas, E. V. Syutkina, R. B. Sothern, R. Zaslavskaya, F. Halberg, Y. Watanabe, O. Schwartzkopff and K. Otsuka, *et al.*, *J. Circadian Rhythms*, 2003, **1**, 2.
54. F. Halberg, *Z. Vitamin-, Hormon-u. Fermentforsch.*, 1959, **10**, 225.
55. R. A. Wever, *The Circadian System of Man: Results of Experiments under Temporal Isolation*, Springer-Verlag, New York, 1979.
56. F. Halberg, G. Cornelissen, R. P. Sonkowsky, C. Lanzoni, A. Galvagno, M. Montalbini and O. Schwartzkopff, *New Trends Exp. Clin. Psychiatr.*, 1998, **14**, 15.
57. H. Greenberg and R. I. Cohen, *Curr. Opin. Pulm. Med.*, 2012, **18**, 57.
58. T. D. Giles, *J. Hypertens. Suppl.*, 2006, **24**, S11.
59. C. Harding, F. Pompei, S. F. Bordonaro, D. C. McGillicuddy, D. Burmistrov and L. D. Sanchez, *West. J. Emerg. Med.*, 2020, **21**, 909.
60. M. Shinagawa, K. Otsuka, S. Murakami, Y. Kubo, G. Cornelissen, K. Matsubayashi, S. Yano, G. Mitsutake, K. Yasaka and F. Halberg, *Blood Press. Monit.*, 2002, **7**, 69.
61. K. Otsuka, G. Yamanaka, M. Shinagawa, S. Murakami, T. Yamanaka, K. Shibata, S. Yano, S. Ishizuka, R. B. Singh, G. Cornelissen and F. Halberg, *Biomed. Pharmacother.*, 2004, **58**(Suppl 1), S48.
62. S. Murakami, K. Otsuka, Y. Kubo, M. Shinagawa, T. Yamanaka, S. Ohkawa and Y. Kitaura, *Am. J. Hypertens.*, 2004, **17**, 1179.
63. J. E. Muller, P. H. Stone, Z. G. Turi, J. D. Rutherford, C. A. Czeisler, C. Parker, W. K. Poole, E. Passamani, R. Roberts and T. Robertson, *et al.*, *N. Engl. J. Med.*, 1985, **313**, 1315.
64. H. R. Arntz, S. N. Willich, C. Schreiber, T. Brüggemann, R. Stern and H. P. Schultheiss, *Eur. Heart J.*, 2000, **21**, 315.
65. S. N. Willich, H. Löwel, M. Lewis, A. Hörmann, H. R. Arntz and U. Keil, *Circulation*, 1994, **90**, 87.
66. G. Cornelissen, T. K. Breus, C. Bingham, R. Zaslavskaya, M. Varshitsky, B. Mirsky, M. Teibloom, B. Tarquini, E. Bakken and F. Halberg, *Chronobiologia*, 1993, **20**, 87.
67. S. Murakami, K. Otsuka, T. Kono, A. Soyama, T. Umeda, N. Yamamoto, H. Morita, G. Yamanaka and Y. Kitaura, *Hypertens. Res.*, 2011, **34**, 70.
68. F. Halberg, D. Powell, K. Otsuka, Y. Watanabe, L. A. Beaty, P. Rosch, J. Czaplicki, D. Hillman, O. Schwartzkopff and G. Cornelissen, *Am. J. Physiol. Heart Circ. Physiol.*, 2013, **305**(3), H279.

69. K. Otsuka, K. Okajima, T. Yamanaka, S. Oinuma, J. Sasaki, Y. Taniwaki, K. Otsuka and G. Cornelisen, *J. Am. Geriatr. Soc.*, 2014, **62**, 1602.
70. V. B. Meyer-Rochow and P. J. Brown, *Acta Neurobiol. Exp.*, 1998, **58**, 287.
71. N. Muir and R. Pownall, *J. Pharm. Pharmacol.*, 1983, **36**, 185.
72. L. Pollmann, *Chronobiol. Int.*, 1984, **1**, 151.
73. F. Halberg, M. Engeli, C. Hamburger and D. Hillman, *Acta Endocrinol.*, 1965, **50**(Suppl 103), 5.
74. G. Cornelissen, J. Siegelova and F. Halberg, in *The Importance of Chronobiology in Diagnosing and Therapy of Internal Diseases*, ed. F. Halberg, T. Kenner and B. Fiser, Masaryk University, Brno, 2002, pp. 58–96.
75. G. Cornelissen and F. Halberg, *Percept. Mot. Skills*, 2000, **90**, 579.
76. F. Raible, H. Takekata and K. Tessmar-Raible, *Front. Neurol.*, 2017, **8**, 189.
77. G. Andreatta and K. Tessmar-Raible, *J. Mol. Biol.*, 2020, **432**, 3525.
78. T. S. Kaiser and J. Neumann, *BioEssays*, 2021, **43**, e2100074.
79. C. Helfrich-Förster, S. Monecke, I. Spiousas, T. Hovestadt, O. Mitesser and T. A. Wehr, *Sci. Adv.*, 2021, **7**, eabe1358.
80. L. Casiraghi, I. Spiousas, G. P. Dunster, K. McGlothlen, E. Fernández-Duque, C. Valeggia and H. O. de la Iglesia, *Sci. Adv.*, 2021, **7**, abe0465.
81. V. Kumar and I. Mishra, in *Encyclopedia of Reproduction*, ed. M. K. Skinner, Academic Press, Elsevier, vol. 1, 2018, pp. 442–450.
82. G. Lincoln, *J. Neuroendocrinol.*, 2019, **31**, e12694.
83. S. Wood and A. Loudon, *Gen. Comp. Endocrinol.*, 2018, **258**, 222.
84. M. Smolensky, F. Halberg and F. Sargent II, in *Advances in Climatic Physiology*, ed. S. Itoh, K. Ogata and H. Yoshimura, Igaku Shoin, Tokyo, 1972, pp. 281–318.
85. H. Marti-Soler, S. Gonseth, C. Gubelmann, S. Stringhini, P. Bovet, P. C. Chen, B. Wojtyniak, F. Paccaud, D. H. Tsai, T. Zdrojewski and P. Marques-Vidal, *PLoS One*, 2014, **9**, e113500.
86. H. Marti-Soler, C. Gubelmann, S. Aeschbacher, L. Alves, M. Bobak, V. Bongard, E. Clays, G. de Gaetano, A. Di Castelnuovo and R. Elosua, *et al.*, *Heart*, 2014, **100**, 1517.
87. J. Yang, M. Zhou, C. Q. Ou, P. Yin, M. Li, S. Tong, A. Gasparrini, X. Liu, J. Li, L. Cao, H. Wu and Q. Liu, *Environ. Pollut.*, 2017, **224**, 400.
88. S. Danet, F. Richard, M. Montaye, S. Beauchant, B. Lemaire, C. Graux, D. Cottel, N. Marecaux and P. Amouyel, *Circulation*, 1999, **100**, E1.
89. J. D. Healy, *J. Epidemiol. Community Health*, 2003, **57**, 784.
90. G. Cornelissen, F. Halberg, L. Pöllmann, B. Pöllman, G. S. Katinas, H. Minne, T. Breus, R. B. Sothern, Y. Watanabe and R. Tarquini, *et al.*, *Biomed. Pharmacother.*, 2003, **57**, 45s.
91. F. Halberg, G. Cornelissen, P. Regal, K. Otsuka, Z. R. Wang, G. S. Katinas, J. Siegelova, P. Homolka, P. Prikryl and S. M. Chibisov, *et al.*, *Biomed. Pharmacother.*, 2004, **58**, 50s.
92. G. Cornelissen, D. Gubin, L. A. Beaty and K. Otsuka, *Int. J. Environ. Res. Public Health*, 2020, **17**, 3083.

93. F. Halberg, G. Cornelissen, R. B. Sothern, L. A. Wallach, E. Halberg, A. Ahlgren, M. Kuzel, A. Radke, J. Barbosa, F. Goetz *et al.*, in *Neoplasms-Comparative Pathology of Growth in Animals, Plants and Man*, ed. H. Kaiser, Williams and Wilkins, Baltimore, 1981, pp. 553–596.
94. Y. Touitou, E. Haus, *Biological Rhythms in Clinical and Laboratory Medicine*, Springer-Verlag, Berlin, 1992.
95. J. F. Thayer and R. D. Lane, *Biol. Psychol.*, 2007, **74**, 224.
96. F. Shaffer, R. McCraty and C. L. Zerr, *Front. Psychol.*, 2014, **5**, 1040.
97. K. Otsuka, G. Cornelissen, and F. Halberg, *Chronomics and Continuous Ambulatory Blood Pressure Monitoring – Vascular Chronomics: From 7 Day/24 Hour to Lifelong Monitoring*, Springer Japan, Tokyo, 2016.
98. F. Shaffer and J. P. Ginsberg, *Front. Public Health*, 2017, **5**, 258.
99. K. Otsuka, G. Cornelissen, Y. Kubo, K. Shibata, K. Mizuno, H. Ohshima, S. Furukawa and C. Mukai, *Sci. Rep.*, 2019, **9**, 8995.
100. K. Otsuka, G. Cornelissen and F. Halberg, *Clin. Sci.*, 1997, **93**, 299.
101. K. Otsuka, G. Cornelissen, M. Shinagawa, Y. Kubo, T. Yamanaka, K. Omori, S-i. Ohkawa, Z. Y. Zhao, A. Delyukov and Y. Gorgo, *et al.*, *Ther. Res.*, 1999, **20**, 388.
102. K. Otsuka, G. Cornelissen, M. Shinagawa, Y. Nishimura, Y. Kubo, N. Hotta, C. Fujii, T. Ishii, K. Omori and Y. Watanabe, *et al.*, *Comput. Cardiol.*, 1999, **26**, 587.
103. K. Otsuka, *Chronome & Janus-medicine: Heart Rate Variability (HRV) and BP Variability (BPV) from a viewpoint of chronobiology and ecology*, Medical Review, Tokyo, 1998.
104. K. Otsuka, Y. Kubo, M. Shinagawa, T. Yamanaka, T. Hasebe, K. Omori, Y. Watanabe, N. Kazuma, H. Kawada, M. Kaneko, G. Cornelissen and F. Halberg, *Ther. Res.*, 1997, **18**, 457.
105. K. Otsuka, Y. Nishikawa, Y. Kubo, G. Cornelissen and F. Halberg, *Comput. Cardiol.*, 1997, **24**, 49.
106. W. Nelson, G. Cornelissen, D. Hinkley, C. Bingham and F. Halberg, *Chronobiologia*, 1983, **10**, 179.
107. G. Cornelissen, F. Halberg, K. Otsuka and R. B. Singh, *World Heart J.*, 2008, **1**, 223.
108. J. Schuh, A. Sinaiko, G. Cornelissen, U. Sitka, D. Weinert, B. Tarquini, S. Sensi, G. Mainardi, G. Mello, C. Panero *et al.* in *Proceedings 2nd Anniversary IEEE Symposium on Computer-Based Medical Systems*, Computer Society Press, Washington DC, 1989, pp. 279–287.
109. G. Cornelissen and K. Otsuka, *Gerontology*, 2017, **63**, 118.
110. G. Cornelissen, K. Otsuka, F. Halberg, in *Chronocardiology and Chronomedicine: Humans in Time and Cosmos*, ed. K. Otsuka, G. Cornelissen and F. Halberg, Life Science, Tokyo, 1993, pp. 16–48.
111. G. Cornelissen, F. Halberg, R. Kopher, J. Kato, C. Maggioni, K. Tamura, K. Otsuka, Y. Miyake, M. Ohnishi and K. Satoh, *et al.*, *Chronobiologia*, 1991, **18**, 72.

112. F. Halberg, L. E. Scheving, E. Lucas, G. Cornelissen, R. B. Sothern, E. Halberg, J. Halberg, F. Halberg, J. Carter, K. D. Straub and D. P. Redmond, *Chronobiologia*, 1984, **11**, 217.
113. F. Halberg, G. Cornelissen, K. Otsuka, J. Siegelova, B. Fiser, J. Dusek, P. Homolka, S. Sanchez de la Pena, R. B. Singh and BIOCOS project, *World Heart J.*, 2010, **2**, 279.
114. K. Kario, D. Shimbo, N. Tomitani, H. Kanegae, J. E. Schwartz and B. Williams, *J. Clin. Hypertens.*, 2020, **22**, 135.
115. J. R. Sempionatto, M. Lin, L. Yin, E. De la paz, K. Pei, T. Sonsa-ard, A. N. de Loyola Silva, A. A. Khorshed, F. Zhang, N. Tostado, S. Xu and J. Wang, *Nat. Biomed. Eng.*, 2021, **5**, 737.
116. G. Cornelissen, Y. Watanabe, J. Siegelova, L. A. Beaty, R. K. Singh, R. Singh, R. B. Singh, A. Delcourt, L. Gumarova, D. Gubin, C. H. Chen and K. Otsuka, *Biol. Rhythm Res.*, 2019, **50**, 46.
117. G. Cornelissen, F. Halberg, E. E. Bakken, R. B. Singh, K. Otsuka, B. Tomlinson, A. Delcourt, G. Toussaint, S. Bathina and O. Schwartzkopff, *et al.*, *Biomed. Pharmacother.*, 2004, **58**, 69s.
118. T. Roenneberg, K. V. Allebrandt, M. Merrow and C. Vetter, *Curr. Biol.*, 2012, **22**, 939.
119. F. A. Scheer, M. F. Hilton, C. S. Mantzoros and S. A. Shea, *Proc. Natl. Acad. Sci. U. S. A.*, 2009, **106**, 4453.
120. C. J. Morris, T. E. Purvis, K. Hu and F. A. Scheer, *Proc. Natl. Acad. Sci. U. S. A.*, 2016, **113**, E1402.
121. G. D. Potter, D. J. Skene, J. Arendt, J. E. Cade, P. J. Grant and L. J. Hardie, *Endocr. Rev.*, 2016, **37**, 584.
122. T. Roenneberg, A. Wirz-Justice and M. Merrow, *J. Biol. Rhythms*, 2003, **18**, 80.
123. M. Wittmann, J. Dinich, M. Merrow and T. Roenneberg, *Chronobiol. Int.*, 2006, **23**, 497.
124. K. A. Dyar, D. Lutter, A. Artati, N. J. Ceglia, Y. Liu, D. Armenta, M. Jastroch, S. Schneider, S. de Mateo and M. Cervantes, *et al.*, *Cell*, 2018, **174**, 1571.
125. Á. R. Sűdy, K. Ella, R. Bódizs and K. Káldi, *Front. Neurosci.*, 2019, **13**, 950.
126. Z. Weng, W. Ou, J. Huang, M. Singh, M. Wang, Y. Zhu, R. Kuma and S. Lin, *Nat. Sci. Sleep*, 2021, **13**, 103–111.
127. A. J. K. Phillips, W. M. Clerx, C. S. O'Brien, A. Sano, L. K. Barger, R. W. Picard, S. W. Lockley, E. B. Klerman and C. A. Czeisler, *Sci. Rep.*, 2017, **7**, 3216.
128. A. M. Curtis, Y. Cheng, S. Kapoor, D. Reilly, T. S. Price and G. A. Fitzgerald, *Proc. Natl. Acad. Sci. U. S. A.*, 2007, **104**, 3450.
129. J. Richards and M. L. Gumz, *Am. J. Physiol.*, 2013, **304**, R1053.
130. F. Halberg, in *Hormones in Development and Aging*, ed. A. Vernadakis and P. Timiras, Spectrum Publications, New York, 1982, ch. 13, pp. 451–476.
131. D. Gubin, G. Cornélissen, F. Halberg, G. Gubin, K. Uezono and T. Kawasaki, *In Vivo*, 1997, **11**, 485.

132. K. Otsuka, *Clinical Chronomedicine; Glocalization in Time and Space*, Nakayama Shoten, Tokyo, 2013.
133. S. Panda, *Nat. Rev. Endocrinol.*, 2019, **15**, 67.
134. C. Winter, C. Silvestre-Roig, A. Ortega-Gomez, P. Lemnitzer, H. Poelman, A. Schumski, J. Winter, M. Drechsler, R. de Jong and R. Immler, *et al.*, *Cell Metab.*, 2018, **28**, 175.e5.
135. G. Sulli, A. Rommel, X. Wang, M. J. Kolar, F. Puca, A. Saghatelian, M. V. Plikus, I. M. Verma and S. Panda, *Nature*, 2018, **553**, 351.
136. E. Maury, *Int. J. Mol. Sci.*, 2019, **20**, 1597.
137. J. J. Fraigne, K. P. Grace, R. L. Horner and J. Peever, *Curr. Opin. Pulm. Med.*, 2014, **20**, 527.
138. P. Passouant, F. Halberg, R. Genicot, L. Popoviciu and M. Baldy-Moulinier, *Rev. Neurol.*, 1969, **121**, 155.
139. M. P. Mandell, A. J. Mandell, R. I. Rubin, P. Brill, J. Rodnick, R. Sheff and B. Chaffey, *Life Sci.*, 1966, **5**, 583.
140. K. Goji, *J. Clin. Endocrinol. Metab.*, 1993, **76**, 667.
141. G. Brandenberger, A. Buguet, K. Spiegel, A. Stanghellini, G. Muanga, P. Bogui and M. Dumas, *J. Biol. Rhythms.*, 1996, **11**, 258.
142. L. I. Stirrat, J. J. Walker, K. Stryjakowska, N. Jones, N. Z. M. Homer, R. Andrew, J. E. Norman, S. L. Lightman and R. M. Reynolds, *Clin. Endocrinol.*, 2018, **88**, 592.
143. H. A. Reimann, *Arch. Int. Med.*, 1953, **92**, 494.
144. L. Derer, *Neoplasma*, 1960, **7**, 117.
145. G. Cornelissen, E. Halberg, H. J. Long III, K. Prem, E. Bakken, Y. Touitou, S. Elg, E. Haus and F. Halberg, *Chronobiologia*, 1991, **18**, 153.
146. E. Halberg, H. J. Long III, G. Cornelissen, M. A. Blank, S. Elg, Y. Touitou, E. Bakken, P. Delmore, E. Haus, L. Sackett-Lundeen, K. Prem and F. Halberg, *Chronobiologia*, 1992, **19**, 17.
147. F. Halberg, in *Long- and Short-Term Variability in Sun's History and Global Change*, ed. W. Schröder, Science Edition, Bremen, 2000, pp. 271–301.
148. M. Blank, O. Denisova, G. Cornelissen and F. Halberg, *Anticancer Res.*, 1999, **19**, 853.
149. S. Murakami, K. Otsuka, G. Yamanaka, Y. Kubo, O. Matsuoka, T. Yamanaka, M. Shinagawa, S. Nunoda, S. Ohkawa and T. Wada, *et al.*, *Biomed. Pharmacother.*, 2004, **58**(Suppl 1), S45.
150. K. Otsuka, T. Yamanaka, G. Cornelissen, T. Breus, S. M. Chibisov, R. Baevsky, F. Halberg, J. Siegelova and B. Fiser, *Scr. Med.*, 2000, **73**, 111.
151. G. Cornelissen, D. Watson, G. Mitsutake, B. Fiser, J. Siegelova, J. Dusek, L. Vohlidalova, H. Svacinova and F. Halberg, *Scr. Med.*, 2005, **78**, 89.
152. K. Otsuka, G. Cornelissen and F. Halberg, *Biomed. Pharmacother.*, 2001, **55**, 7.
153. M. Gautherie and C. Gros, *Chronobiologia*, 1977, **4**, 1.
154. K. Otsuka, G. Cornélissen and F. Halberg, *Clin. Drug Invest.*, 1996, **11**, 20.
155. K. Otsuka, G. Cornélissen, F. Halberg and G. Oehlert, *J. Med. Eng. Technol.*, 1997, **21**, 23.

156. G. Cornelissen, F. Halberg, K. Otsuka, R. B. Singh and C. H. Chen, *Hypertension*, 2007, **49**, 237.
157. F. Halberg, E. Haus, S. S. Cardoso, L. E. Scheving, J. F. W. Kühl, R. Shiotsuka, G. Rosene, J. E. Pauly, W. Runge, J. F. Spalding, J. K. Lee and R. A. Good, *Experientia*, 1973, **29**, 909.
158. G. Cornelissen, *J. Pharmacokinet. Pharmacodyn.*, 2021, **48**, 339.
159. R. Zhang, N. F. Lahens, H. I. Ballance, M. E. Hughes and J. B. Hogenesch, *Proc. Natl. Acad. Sci. U. S. A.*, 2014, **111**, 16219.
160. M. L. Kringelbach, N. Jenkinson, S. L. F. Owen and T. Z. Aziz, *Nat. Rev. Neurosci.*, 2007, **8**, 623.
161. S. L. Lightman, M. T. Birnie and B. L. Conway-Campbell, *Endocr. Rev.*, 2020, **41**, 470.
162. K. Kalafatakis, G. M. Russell, C. J. Harmer, M. R. Munafo, N. Marchant, A. Wilson, J. C. Brooks, C. Durant, J. Thakrar, P. Murphy, N. J. Thai and S. L. Lightman, *Proc. Natl. Acad. Sci. U. S. A.*, 2018, **115**, E4091.
163. G. M. Russell, C. Durant, A. Ataya, C. Papastathi, R. Bhake, W. Woltersdorf and S. Lightman, *Clin. Endocrinol.*, 2014, **81**, 289.
164. A. DeVecchi, F. Carandente, D. S. Fryd, F. Halberg, D. E. Sutherland, R. J. Howard, R. L. Simmons, J. S. Najarian, in *Chronopharmacology, Proceedings of the Satellite Symposium of the 7th International Congress on Pharmacology*, ed. A. Reinberg and F. Halberg, Pergamon Press, New York, 1979, pp. 193–202.
165. T. Liu, M. Cavallini, F. Halberg, G. Cornelissen, J. Field and D. E. R. Sutherland, *Experientia*, 1986, **42**, 20.
166. F. Levi, F. Halberg, M. Nesbit, E. Haus and H. Levine, in *Neoplasms—Comparative Pathology of Growth in Animals, Plants and Man*, ed. H. Kaise, Williams and Wilkins, Baltimore, 1981, pp. 267–316.
167. W. Ulmer, G. Cornelissen, M. Revilla, J. Siegelova, J. Dusek and F. Halberg, *Scr. Med.*, 2001, **74**, 87.
168. Halberg F., G. Cornelissen, F. Halberg, W. Ulmer, S. Sanchez de la Peña, J. Siegelova, O. Schwartzkopff and BIOCOS project, *Proceedings, Noninvasive Methods in Cardiology*, Brno, 2008, pp. 63–73.
169. C. Humpston, F. Benedetti, M. Serfaty, S. Markham, J. Hodsoll, A. H. Young and D. Veale, *J. Affect. Disord.*, 2020, **261**, 91.
170. A. Iranzo, *Sleep Med. Clin.*, 2016, **11**, 1.
171. M. Dumont and C. Beaulieu, *Sleep Med.*, 2007, **8**, 557.
172. P. A. Geoffroy and L. Palagini, *Prog. Neuro-Psychopharmacol. Biol. Psychiatry*, 2021, **106**, 110158.
173. S. R. Pandi-Perumal, J. M. Monti, D. Burman, R. Karthikeyan, A. S. BaHammam, D. W. Spence, G. M. Brown and M. Narashimhan, *Psychiatry Res.*, 2020, **291**, 113239.
174. T. Kaur and B. C. Shyu, *Adv. Exp. Med. Biol.*, 2018, **1099**, 229.
175. M. A. Quera Salva, S. Hartley, F. Barbot, J. C. Alvarez, F. Lofaso and C. Guilleminault, *Curr. Pharm. Des.*, 2011, **17**, 1459.
176. P. Gervais and A. Reinberg, *Pathol. Biol.*, 1987, **35**, 969.

177. E. Haus, L. Sackett-Lundeen and M. H. Smolensky, *Bull. NYU Hosp. Jt. Dis.*, 2012, **70**(Suppl 1), 3.
178. F. Lévi, A. Okyar, S. Dulong, P. F. Innominato and J. Clairambault, *Annu. Rev. Pharmacol. Toxicol.*, 2010, **50**, 377.
179. Halberg F., J. Halberg, E. Halberg and F. Halberg, in *Cancer Growth and Progression*, ed. A. L. Goldson, H. Kaiser series ed., Kluwer Academic Publishers, Dordrecht, 1989, ch.19, vol. 9, pp. 227–253.
180. Y. Watanabe, F. Halberg, K. Otsuka and G. Cornelissen, *Clin. Exp. Hypertens.*, 2013, **35**, 257.
181. M. Shinagawa, Y. Kubo, K. Otsuka, S. Ohkawa, G. Cornelissen and F. Halberg, *Biomed. Pharmacother.*, 2001, **55**(Suppl 1), 125s.
182. R. J. W. Hill, P. F. Innominato, F. Lévi and A. Ballesta, *PLoS Comput. Biol.*, 2020, **16**, e1007218.
183. E. O. Bangsgaard and J. T. Ottesen, *Math. Biosci.*, 2017, **287**, 24.
184. M. Al-Kofahi, M. A. Ahmed, M. M. Jaber, T. N. Tran, B. A. Willis, C. L. Zimmerman, M. T. Gonzalez-Bolanos, R. C. Brundage and K. Sarafoglou, *Br. J. Clin. Pharmacol.*, 2020, 1.
185. A. M. Gomez, L. F. Marin Carrillo, O. M. Munoz Velandia, M. A. Rondon Sepulveda, C. M. Arevalo Correa, E. Mora Garzon, M. C. Cuervo Diaz and D. C. Henao Carrillo, *Diabetes Technol. Ther.*, 2017, **19**, 109.
186. A. Charrier, B. Olliac, P. Roubertoux and S. Tordjman, *Int. J. Mol. Sci.*, 2017, **18**, 938.
187. J. Wefers, D. van Moorsel, J. Hansen, N. J. Connell, B. Havekes, J. Hoeks, W. D. van Marken Lichtenbelt, H. Duez, E. Phielix and A. Kalsbeek, *et al.*, *Proc. Natl. Acad. Sci. U. S. A.*, 2018, **115**, 7789.
188. M. D. Ruben, G. Wu, D. F. Smith, R. E. Schmidt, L. J. Francey, Y. Y. Lee, R. C. Anafi and J. B. Hogenesch, *Sci. Transl. Med.*, 2018, **10**, eaat8806.
189. L. S. Mure, H. D. Le, G. Benegiamo, M. W. Chang, L. Rios, N. Jillani, M. Ngotho, T. Kariuki, O. Dkhissi-Benyahya, H. M. Cooper and S. Panda, *Science*, 2018, **359**, eaao0318.
190. C. M. Depner, E. L. Melanson, A. W. McHill and K. P. Wright Jr, *Proc. Natl. Acad. Sci. U. S. A.*, 2018, **115**(23), E5390.
191. S. Masri, K. Kinouchi and P. Sassone-Corsi, *Curr. Opin. Oncol.*, 2015, **27**, 50.
192. D. Molina-Serrano, D. Kyriakou and A. Kirmizis, *Front. Genet.*, 2019, **10**, 192.
193. I. Pacheco-Bernal, F. Becerril-Pérez and L. Aguilar-Arnal, *Clin. Epigenet.*, 2019, **11**, 79.
194. K. Otsuka, T. Yamanaka and Y. Kubo, *J. Ambulatory Monit.*, 1994, **7**, 219.
195. Y. Kubo, S. Murakami, O. Matsuoka, N. Hotta, S. Oinuma, M. Shinagawa, K. Omori, S. Nunoda, K. Otsuka and S. Ohkawa, *et al.*, *Biomed. Pharmacother.*, 2003, **57**(Suppl 1), 110.
196. Y. Kubo, S. Murakami, O. Matsuoka, S. Nunoda, K. Otsuka and S. Ohkawa, *Jpn. J. Clin. Physiol.*, 2003, **33**, 169.
197. N. Hotta, K. Otsuka, S. Murakami, G. Yamanaka, Y. Kubo, O. Matsuoka, T. Yamanaka, M. Shinagawa, S. Nunoda and Y. Nishimura, *et al.*, *Biomed. Pharmacother.*, 2005, **59**(Suppl 1), 45s.

198. N. Yamamoto, Y. Kubo, K. Ishizawa, G. Kim, T. Moriya, T. Yamanouchi and K. Otsuka, *Diabetes Technol. Ther.*, 2010, **12**, 775.
199. A. Ballesta, P. F. Innominato, R. Dallmann, D. A. Rand and F. A. Lévi, *Pharmacol. Rev.*, 2017, **69**, 161.
200. F. Halberg, *Chronobiologia*, 1974, **1**(Suppl.1), 27.
201. B. I. Sundararaj, P. Nath and F. Halberg, *J. Nutr.*, 1982, **12**, 1085.
202. F. Halberg, E. Haus and G. Cornelissen, in *Not Eating Enough: Overcoming Underconsumption of Military Operational Rations*, ed. B. M. Marrioo, National Academy Press, Washinton DC, 1995, pp. 361–372.
203. F. Halberg, G. Cornélissen, Z. Wang, C. Wan, W. Ulmer, G. Katinas, R. Singh, R. K. Singh, R. K. Singh and B. D. Gupta, *et al.*, *J. Exp. Ther. Oncol.*, 2003, **3**, 223.
204. S. Gill and S. Panda, *Cell Metab.*, 2015, **22**, 789.
205. K. Gabel, K. K. Hoddy, N. Haggerty, J. Song, C. M. Kroeger, J. F. Trepanowski, S. Panda and K. A. Varady, *Nutr. Healthy Aging*, 2018, **4**, 345.
206. E. J. Calabrese, K. A. Bachmann, A. J. Bailer, P. M. Bolger, J. Borak, L. Cai, N. Cedergreen, M. G. Cherian, C. C. Chiueh and T. W. Clarkson, *et al.*, *Toxicol. Appl. Pharmacol.*, 2007, **222**, 122.
207. M. P. Mattson, *Ageing Res. Rev.*, 2008, **7**, 1.
208. D. M. Yellon and and J. M. Downey, *Physiol. Rev.*, 2003, **83**, 1113.
209. L. Wu, M. H. Noyan Ashraf, M. Facci, R. Wang, P. G. Paterson, A. Ferrie and B. H. Juurlink, *Proc. Natl. Acad. Sci. U. S. A.*, 2004, 1017**094**.
210. S. A. Kim, Y. M. Lee, J. Y. Choi, D. R. Jacobs Jr. and D. H. Lee, *Environ. Pollut.*, 2018, **233**, 725.
211. E. S. Epel, *Ageing Res. Rev.*, 2020, **63**, 101167.
212. K. Reindl, C. Falliers, F. Halberg, H. Chai, D. Hillman and W. Nelson, *Rass. Neurol. Veg.*, 1969, **23**, 5.

Part II: Systems Approach to Chronopharmaceutics and Chronotheranostics

CHAPTER 5

The Nervous System

SATO HONMA

Center for Sleep and Circadian Rhythm Disorders, Sapporo Hanazono Hospital, 1-30, S-15, W-15, Chuo-ku, Sapporo 064-0915, Japan
Email: sato.honma@gmail.com

5.1 Mammalian Circadian System in the Central Nervous Sytem

5.1.1 The Circadian System Composed of a Master Clock and Peripheral Clocks

The mammalian circadian system comprises a master clock located in the hypothalamic suprachiasmatic nucleus (SCN) and peripheral clocks situated throughout the body, including extra-SCN brain tissues.[1,2] The SCN was long considered to be the exclusive site of the circadian clock in the mammalian body because almost all overt rhythms are abolished after lesioning of the SCN.[3,4] Identification of peripheral clocks in 2000 altered the understanding of the SCN's role in circadian rhythms. Almost all peripheral tissues and organs exhibit circadian rhythms in their clock gene expression *ex vivo*.[5] This discovery indicates that the SCN is not the only clock driving circadian rhythms in physiological functions but rather serves as the master clock, which orchestrates the activities of peripheral clocks throughout the body. Individual cells have their own molecular circadian clock machinery,[6] namely, transcriptional and translational feedback loops that involve several clock genes and their protein products.[7] Cellular clocks must be synchronized at the tissue or organ level by the SCN to function as a peripheral clock. The SCN receives photic signals from the retina to entrain itself to the environmental

Chemical Biology No. 23
Chronobiology and Chronomedicine: From Molecular and Cellular Mechanisms to Whole Body Interdigitating Networks
Edited by Germaine Cornelissen and Tsuyoshi Hirota

Published by the Royal Society of Chemistry, www.rsc.org

light–dark (LD) cycle[4] and sends LD-entrained circadian signals to peripheral clocks so that neurological behaviours and bodily functions are expressed in a temporally coordinated manner throughout the day.

The circadian system, comprising an input system, oscillator, and an output system, is summarized in a so-called Eskinogram[8] (Figure 5.1A). In mammals, the primary input system, the retina, has a circadian clock, while all the output systems, the peripheral organs, contain peripheral clocks. Therefore, the circadian system is a hierarchical multi-oscillator system, with the master clock, the SCN, at the top and several extra-SCN clocks serving as second-level clocks that control peripheral clocks. Also, the SCN itself contains a multi-level oscillator structure (Figure 5.1B).

5.1.2 Multiple Brain Clocks Regulating Behaviour and Physiology

Two SCN-independent rhythms are known to drive functions and behaviours with circadian periodicity even in animals who have lost their circadian rhythms in behaviour and hormone levels after SCN-lesioning. First, food-entrainable rhythms are expressed when food availability is restricted periodically,[9] and, second, methamphetamine-induced rhythms are expressed through the chronic administration of a central stimulant, methamphetamine.[10] Both of these rhythms affect the sleep–wake cycle, body temperature regulation, and several hormones other than melatonin. Extra-SCN brain

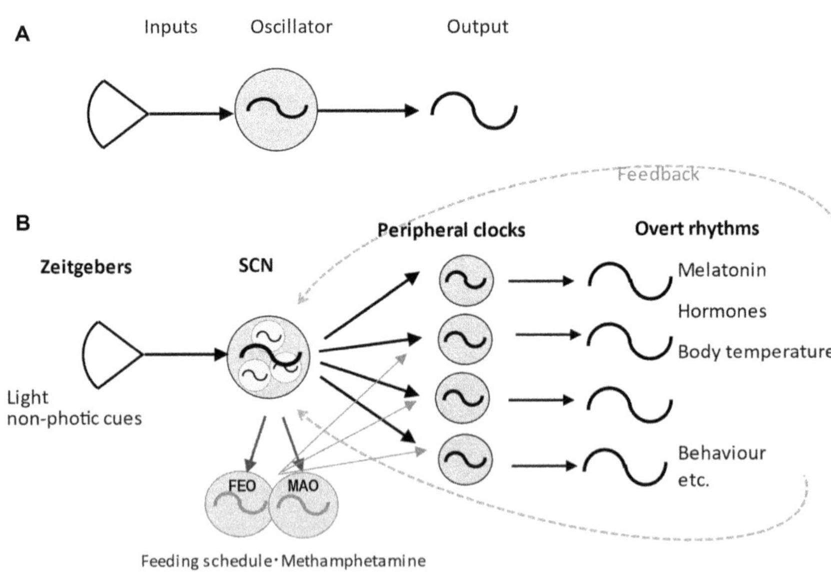

Figure 5.1 Schematic diagram of the circadian system. (A) Eskinogram. (B) Hierarchical multi-oscillator system of mammals with input pathways, output pathways and feedback of output functions to the oscillator (see text in Section 5.3.2).

circadian oscillators, including the food-entrainable oscillator (FEO), and the methamphetamine-induced oscillator (MAO) might underlie the mechanisms responsible for food-entrainable and methamphetamine-induced rhythms. Both of these rhythms exhibit circadian features, including circadian periodicity, entrainment to external (non-photic) time cues, maintaining a free-running pattern under constant conditions following the phase-set by the entraining cue, and exhibiting systematic changes in the phase angle difference between the overt rhythm and Zeitgeber period (T cycle) within the circadian range. These findings suggest that the FEO and MAO are under the control of the SCN, but can dissociate and exhibit food-entrained or methamphetamine-induced rhythms under certain conditions. Although there are several theoretical locations, the exact site of FEO[11] or MAO[12] is not yet identified.

On the other hand, explants of brain regions such as the olfactory bulb (OB)[13] and median eminence[14] exhibit circadian rhythms in the expression of clock genes, which are as robust as those measured in the SCN. While many brain structures other than the SCN have area-specific circadian clocks, the FEO and MAO are capable of driving rhythms for peripheral functions such as behaviour and glucocorticoid production (Figure 5.1B).[2]

5.2 Suprachiasmatic Nucleus: The Master Clock

5.2.1 Identification of the SCN as the Master Clock

In 1972, two studies demonstrating that lesioning of the SCN eliminated circadian rhythms in behaviour[3] and in plasma corticosterone[4] showed that the SCN was the master circadian clock. Also, the afferent pathway from the retinohypothalamic tract was sufficient to entrain the circadian clock in the SCN and the efferent pathway projecting caudally from the SCN was responsible for corticosterone rhythms.[4] Rats with a hypothalamic island, including the SCN, exhibited circadian rhythms only within the island, whereas circadian rhythms were abolished outside the island, including those associated with behaviour.[15] Transplantation of neonatal or fetal SCN tissues to SCN-lesioned animals restored circadian rhythms of behaviour with the donor's genotype.[16,17] These findings indicate that the SCN is a necessary, as well as a sufficient region for controlling rhythmic behaviours.

Apart from these *in vivo* studies, various circadian parameters were examined in cultured SCN tissue. Cultured SCN tissue exhibits circadian rhythms of neuronal activity,[18] neuropeptide release,[19] and clock gene expression.[5] The rhythms continue for days, weeks, and months at the levels of both the single cell[18] and tissues.[5,19]

5.2.2 Functional Anatomy and Histochemistry

The SCN is formed of paired oval nuclei located in the anterior hypothalamus, just above the optic chiasm, in mammals, including humans.

In rats and mice, each nucleus is composed of approximately 10 000 neurons containing various neurotransmitters and neuropeptides.[20,21] The SCN is a highly neuron-rich structure in the central nervous system, and the ratio between neurons and astrocytes is about 3 : 1.[22,23]

The cytoarchitecture of the SCN, including its afferents and efferents, is schematically shown in Figure 5.2. Histochemically and anatomically, the SCN consists of two major areas. The core predominantly comprises neurons containing the vasoactive intestinal polypeptide (VIP) and gastrin-releasing peptide (GRP). The shell predominantly comprises neurons containing arginine vasopressin (AVP). Also, most neurons found in the SCN contain γ-amino butyric acid (GABA).[21,24] Other transmitters and neuropeptides are expressed in specific areas within the SCN, including the regulator of G protein signalling (Rgs16), which is found mainly in the dorsal border of the SCN,[25] prokineticin (PK) 2 and somatostatin (SS), which are found in the shell, and cholecystokinin (CCK), which is located at the edge of the shell.[20,21,23,26] These peptides and transmitters often co-localize with other peptides. Neuromedin S is expressed in about 60% of SCN neurons but not

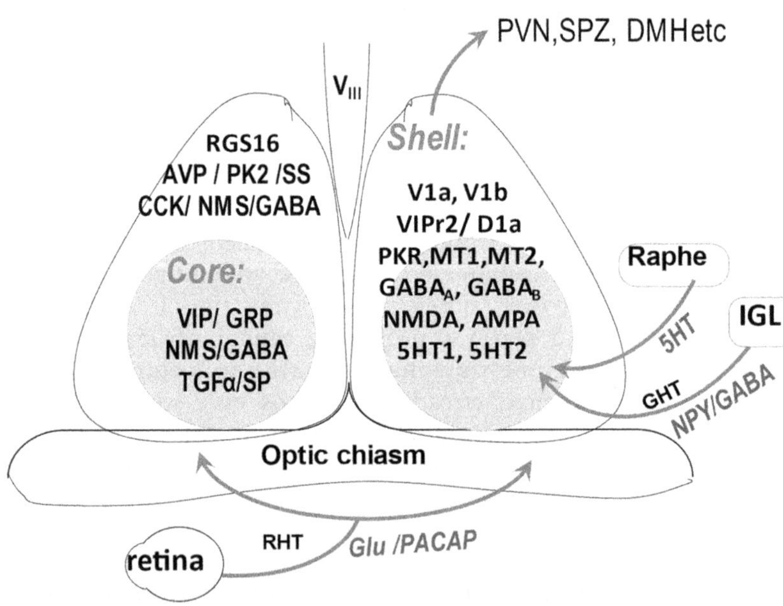

Figure 5.2 Schematic drawing of the SCN in the coronal plane composed of a shell (white area) and a core (gray area). Representative neuropeptides and transmitters are shown on the left side, and receptors for intrinsic and extrinsic transmitters are shown on the right side. Transmitters of the photic (retina) and non-photic (raphe and IGL) input pathways to the SCN are expressed in *italics*. Major output pathways from the SCN are also shown. AMPA: α-amino-3-hydroxy-5-methyl-4-isoxazolepropionic acid receptor, D:dopamine, 5HT: 5 hydroxytryptamine, MT: melatonin, NMDA: *n*-methyl-D-aspartate receptor, and PKR: PK2 receptor. For other abbreviations, see text.

in those expressing GRP.[27] GRP sometimes colocalizes with VIP,[28] and GABA co-localizes with other peptides in almost all SCN neurons.[29] Species differences are also known to exist concerning the neuropeptides found in the SCN. For example, calbindin- and substance P (SP)-containing neurons are found only in the SCN core in hamsters, while in mice, calbindin is found in both the core and shell, whereas SP is rarely found.[21,26] Recently, Wen *et al.*[23] classified mouse SCN neurons into five subtypes according to their circadian and light-induced gene expression patterns, using single-cell RNA-sequencing of the whole SCN.

Some of these neurotransmitters and neuropeptides have physiological roles in circadian timekeeping. Mice lacking VIP[30] or the SCN-specific VIP receptor, VIPR2,[31] often become arrhythmic in their behaviour in constant darkness (DD), but the behavioural changes vary from being almost indistinguishable from the wild-type to complete arrhythmia.[32] Loss of AVP signalling does affect behavioural rhythms. Mice lacking AVP receptors, VP1a and VP1b, specifically in the SCN, show rapid entrainment to an LD shift,[33] while AVP neuron-specific BMAL1 deletion results in lengthening of the circadian period and decoupling the onset and offset activity components.[34]

Almost all SCN neurons are GABAergic,[29] but GABA's role in circadian timekeeping remains poorly understood.[35] Although GABA is a major inhibitory neurotransmitter in the central nervous system (CNS), it acts as an excitatory transmitter during early development[36] and in adults in some CNS regions, including the SCN. However, previous reports have been controversial, indicating that GABA in the SCN is not excitatory,[37] or could be excitatory depending on the circadian phase,[38] developmental stage,[39] photoperiod,[40,41] or SCN sub-region.[42] Recent observations of the GABA knock-out SCN demonstrated that the absence of GABA does not affect PER2 rhythms but induces burst firing and Ca^{2+} increases throughout the 24-hour day. These observations indicate that GABA's action is basically inhibitory and dispensable to circadian timekeeping but essential for refining the circadian output signal from the SCN.[43] Also, GABA seems to be involved in the coupling of regional oscillators within the SCN, which is discussed in Section 5.2.4. Neuromedin S is also expressed in most SCN neurons, whereas a lack of this peptide shows no significant change in the circadian rhythms.[27]

The SCN plays a pivotal role in photic entrainment. Circadian rhythms either phase-advance or phase-delay depending on the timing of light exposure. The relationship between the magnitude of phase shifts and the circadian phase of the light pulse is summarized in a phase-response curve (PRC). The shapes of the PRC to single light pulses are similar in all observed organisms so far, with a phase-delay caused by light at early subjective night, and a phase-advance caused by light at the late subjective night. And the light PRCs of most mammals have a dead zone, the phase when the photic signals do not induce any phase shift in the circadian clock, during the subjective day. The mechanism by which the SCN shuts down light inputs at a certain time of day, namely, the gating mechanism, is still not well understood.

Several reports have predicted mechanisms such as (1) circadian changes in glial functions, including the network formation during daytime enabling the active uptake of extracellular glutamate released from the retinal terminal;[44] (2) circadian increases in the intracellular Ca^{2+} ion level during the subjective day, preventing further increases in Ca^{2+} levels due to photic signals;[45,46] and (3) circadian rises in the expression of clock genes *Period (Per)1* and *Per2* during the subjective day, preventing the photic induction of these genes.[47]

5.2.3 Molecular Mechanisms in the SCN

Interlocked transcriptional and translational feedback loops, composed of several clock genes and their protein products, are considered the molecular machinery for circadian rhythm generation.[7] In a core loop, the transcription of negative elements, *Per1, Per2, Cryptochrome (Cry)1*, and *Cry2,* is activated by heterodimeric transcription factors CLOCK and BMAL1 *via* E-box enhancers to increase PER and CRY levels during the subjective day. PER and CRY molecules, in turn, translocate into the nucleus and bind with the CLOCK/BMAL1 heterodimer, thus inhibiting the transactivation of transcription during the subjective night. Thus, one turn of this loop generates one cycle of about 24 hours. Interlocking with this core loop, *Bmal1* transcription is also regulated by a molecular feedback loop in which transcription factor RORα activates and REV-ERB suppresses *Bmal1* transcription during the subjective night and day, respectively.[1] In addition, the auto-feedback loop of rhythmic *Dec* transcription also interacts with the core loop, whereby the bHLH transcription factors DEC1 and DEC2 suppress the transcription of *Pers* and *Crys*, while PER and CRY molecules suppress the transcription of *Dec1* and *Dec2.* These interlocked multiple molecular feedback loops function to stabilize periodicity in cells and enhance the amplitude when cellular oscillators are synchronized.[48]

At the molecular level, SCN cells are similar to other peripheral cells. The only differences are the high levels of clock gene expression and specific spatiotemporal patterns seen in the phase, amplitude, and photic responses in the SCN. Figure 5.3 shows the circadian rhythms of clock gene expression in the SCN. The phase relationships among expression rhythms of *Per1*, *Per2* and *Bmal1* are similar between *in vivo* recordings using the fiber optic method and *ex vivo* recordings of cultured slices[49] (Figure 5.3A). The circadian phases of different variables in the SCN are the same between nocturnal and diurnal species, including those of clock gene expression, neuronal firing, intracellular calcium ion levels, metabolic rate, and synthesis and release of AVP. Outside the SCN, the rhythms of nocturnal and diurnal species are 180 degrees out of phase (Figure 5.3B). A recent report indicated that nocturnality and diurnality are changeable, depending on seasonal changes in photoperiods and ambient temperature in the field.[50] Hence, the phase-reversal of behavioural rhythms takes place downstream of the SCN.

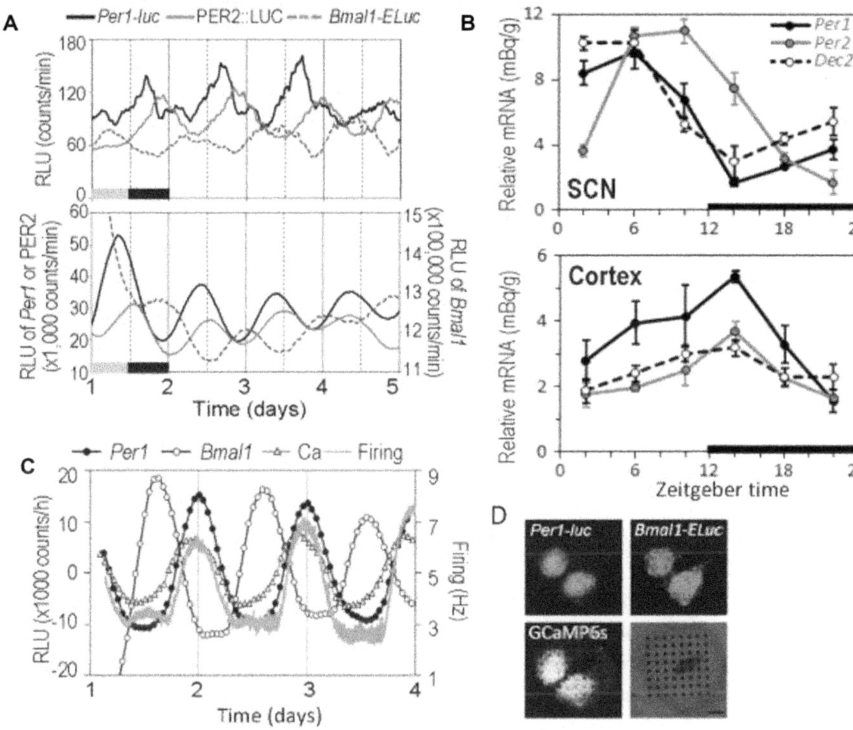

Figure 5.3 Circadian rhythms in the SCN and their phase relations. (A) Clock gene expression rhythms monitored by luciferase reporters. *Per1-luc*, PER2::LUC, and *Bmal1-ELuc* rhythms in the SCN *in vivo* (upper) and *ex vivo* (lower) are shown. *In vivo* bioluminescence rhythms are plotted as a 4 h moving average. *Ex vivo* bioluminescence records are plotted at 10 min intervals as a 50 min moving average. Vertical lines in each panel indicate local times (solid line, 06:00 h; broken line, 18:00 h). Gray and black bars at the bottom of each record indicate the antecedent LD cycle of animals. Reproduced from ref. 49, https://doi.org/10.1038/srep12310, under the terms of the CC BY 4.0 license http://creativecommons.org/licenses/by/4.0/. (B) Phase relation of *Per1*, *Per2* and *Dec2* expression rhythms in the SCN (upper panel) and cortex (lower panel) examined through a quantitative *in situ* hybridization method. Data are expressed as the mean and SE of 5 mice at each time point. Light onset of the LD cycle is designated as Zeitgeber time 0. (C) Phase relation of four rhythms in the SCN simultaneously recorded from a single neonatal SCN slice cultured on a multi-electrode probe. Bioluminescence of *Per1-luc* and *Bmal1-ELuc* was separately measured using a long-pass filter. Intracellular Ca^{2+} was fluorometrically measured through GCaMP6s, a genetically encoded calcium sensor transduction. Spontaneous firing is expressed as the mean firing rate (Hz) from all electrodes covered by bilateral SCN. (D) Four images of the SCN slice shown in (C). Bioluminescence images of *Per1-luc*, *Bmal1-ELuc*, fluorescence image of GCaMP6s, and bright field image (lower right) of cultured SCN slice on a multi-electrode probe. Scale bars are 200μm. (C and D) Reproduced from ref. 130 with permission from National Academy of Sciences).

5.2.4 Multi-oscillator Structure and Networks

The multi-oscillator network structure makes the SCN distinct from the peripheral clocks found throughout the body. SCN neurons in dispersed cell cultures exhibit circadian rhythms in neuronal firing, as well as clock gene expression, with cell-specific circadian periodicity.[51] Therefore, synchronization among these cells is critical for maintaining coherent circadian rhythm expression from the SCN. Whether all SCN cells have cell-autonomous clocks remains a matter of debate. The percentage of cells with stable circadian rhythms reportedly decreases as the *in vitro* cell density decreases,[52] suggesting that the number of SCN cells with cell-autonomous clocks is limited, and mutual coupling in the intact SCN architecture enables synchronization. However, a recent study[53] demonstrated that most solitary SCN neurons, cultured in micro-islands, exhibit stable circadian rhythms in intracellular Ca^{2+} concentrations and clock gene expression when cultured individually, without any contact with other cells, suggesting that most SCN cells exhibit circadian rhythms with cell-specific circadian periods.

On the other hand, genetic and mathematical approaches suggested that clusters of SCN cells with different periodicities form the fundamental components of oscillators.[54,55] SCN neurons from *Cry1*- and *Cry2*-deficient mice form three-period clusters, creating a synchronized rhythm through the action of diffusible factors such as AVP and VIP.[54] Cluster analysis of PER2:LUC bioluminescence records demonstrated the presence of several clusters with different spatiotemporal patterns, which are differentially affected by a lack of clock genes.[55] Several mutations, including those of *tau*,[56] VIPR2 deficiency,[32] and CRY1/CRY2 double deficiency,[54] affect the coupling of cellular rhythms, leading to the loss of overt circadian rhythms.

Multiple regional oscillators in the SCN have been reported by demonstrating dissociations among them. The uncoupling of oscillators in the ventral and dorsal SCN has been observed *in vivo*, following an abrupt phase-shift in the LD cycle. Clock gene expression rhythms in the ventral SCN were observed to change immediately after the LD shift, while it took several days for gene expression to change in the dorsal region.[57] Interestingly, rhythms of clock gene expression and neuronal activity shift differently. In a multi-unit extracellular activity, both ventral and dorsal SCN cells exhibit circadian rhythms with bimodal peaks during the first few days after an LD shift, but the two peaks uncouple to show differentially phased single peaks following either surgical dissection or the administration of a GABA receptor blocker.[58] These data indicate that there are two separate but mutually coupled oscillators in the ventral (core) and dorsal (shell) SCN. In response to light signals, the oscillator in the core, where the light signals are innervated directly and densely, shifts faster than the oscillator in the shell. Neuronal activities are connected *via* GABAergic signalling, contributing to the rhythms of both areas. A cultured slice of the SCN exhibited synchronized circadian rhythms in the release of AVP and VIP, whereas the two rhythms were dissociated by antimitotic treatment.[59] These findings suggest that there are two separate sub-regional oscillators in the shell and core, composed of an AVPergic oscillator cell network and a VIPergic oscillator cell network, respectively. The two sub-regional oscillators are further coupled by an unidentified mechanism. The findings indicate the presence of at least three different synchronizers, one for the coupling among AVP cells, one for VIP cells, and another for the coupling between the two groups.

Three different regional oscillators in the SCN are involved in the adaptation to photoperiods. One in the rostral SCN is phase-set by the light-off signal and regulates the onset of nocturnal behavioural activity; another in the caudal SCN is phase-set by the light-on signal and regulates the end of activity.[60,61] With these oscillators, animals living in middle and high latitudes alter their activity time to optimize the timing of behaviour according to seasonal changes in day length. The third regional oscillator, located in the central areas of the SCN, uncouples and exhibits differently phased rhythms only under long photoperiods. The third oscillator may be involved in the transfer of photic signals to the two oscillators in the rostral and caudal SCN.

When hamsters are under constant light conditions, their behavioural activity sometimes splits into two parts that free-run in parallel but are 180 degrees out of phase with each other. Two coupled oscillators were assumed to explain this splitting phenomenon.[62] During behavioural splitting, clock gene expression rhythms are found to show phase-reversal in the right and left SCN regions, demonstrating the presence of different types of multi-oscillators in the SCN, one in the left region and the other in the right region.[63]

5.2.5 Roles of SCN Glia in the Central Clock

Recent studies have demonstrated that SCN glia, especially astrocytes, are indispensable partners of neurons in circadian pacemaking within the mammalian master clock. Previously, circadian changes in cellular shapes, identified by glial fibrillary acidic protein (GFAP) immunostaining, suggested that SCN glia have a specific role in circadian timekeeping of the hamster.[44] Reduced GFAP signals in rats exposed to DD from birth,[64] and in adult mice following 80 days of exposure to constant light,[65] suggest a physiological role of SCN glia in photic responses. In addition, the significant role of SCN astrocytes in determining circadian periodicity was shown in 2017. Tso *et al.*[66] demonstrated that circadian rhythms of astrocyte-specific *Bmal1* knockout mice had a significantly longer period in behaviour and clock gene expression in the SCN compared to wild type mice. Similarly, astrocyte-specific excision of the *tau*-mutation (short tau CK1ε mutation) in the SCN lengthened the circadian period in mice carrying this mutation. These findings suggest that astrocytes are involved in determining the circadian period in the SCN. Also, Brancaccio *et al.* reported that SCN astrocytes show robust circadian rhythms in intracellular Ca^{2+} levels, which are complementary to neuronal Ca^{2+} rhythms.[67] These authors further demonstrated that glutamate released from astrocytes during the subjective night suppresses glutamate-uptake-related neuronal activity through a presynaptic glutamate transporter. These findings clearly demonstrated another type of circuit, a network associated with gliotransmission in the SCN. A more recent study by this group demonstrated that *Cry1/Cry2* double knockout mice show a phenotype similar to *Cry2* single knockouts, by rescuing *Cry1* specifically in SCN astrocytes.[68] Furthermore, they found that the circadian period was even shorter than that observed following neuron-specific *Cry1* rescue, supporting the observation made by Tso *et al.*[66] that SCN glia have shorter circadian periods than SCN neurons. In contrast to these reports that suggest a supporting role for SCN glia in circadian pacemaking, a suppressing effect has also been observed.[53] Solitary SCN neurons cultured in micro-islands show robust circadian rhythms in Ca^{2+} levels as well as in PER2:LUC, without cohabitation with other cells. However, cohabitation with glia interferes with neuronal Ca^{2+} rhythms. Since the number of glia in a micro-island is always higher than in the SCN *in vivo*, the balance between glia and neurons might be important for maintaining stable circadian rhythm expression from the SCN. Although glia may not affect rhythm generation, they may mask the overt expression of a Ca^{2+} rhythm since solitary SCN neurons exhibit stable clock gene expression rhythms, even when cultured with several glia.

5.3 Input to the SCN

5.3.1 Photic System

5.3.1.1 Functional Structure of the Retinal Circadian Clock

Retinal photoreception is the only pathway through which the mammalian circadian system receives photic inputs to entrain the circadian pacemaker to the environmental LD cycle. In contrast, in animals other than mammals, light is perceived in various tissues such as the pineal gland, deep brain photoreceptors, skin, and cells covering the body.[69] The retina is composed of neurons and glia that are highly organized in stratified layers, which not only sense light and transfer photic signals to the brain but also function as a circadian clock (Figure 5.4). All retinal cells exhibit circadian rhythms in clock gene expression. Circadian rhythms are detected in various retinal functions, such as disc shedding of photoreceptor cells and dopamine and melatonin release from amacrine cells and photoreceptor cells, respectively.[70] The retinal clock exhibits self-sustained rhythms and entrains the rhythms to the LD cycle so that the retina can prepare for LD changes. The retinal clock can oscillate without SCN inputs in mammals. And in vertebrates other than mammals, the retina can act as a master circadian clock, regulating physiological rhythms, including behaviour.[69,70]

5.3.1.2 Different Functions of Photopigments and the Photic Input Pathway to the SCN

The mammalian retina contains three groups of photopigments, rhodopsins for dim-light vision, cone opsins for color vision, and melanopsins for

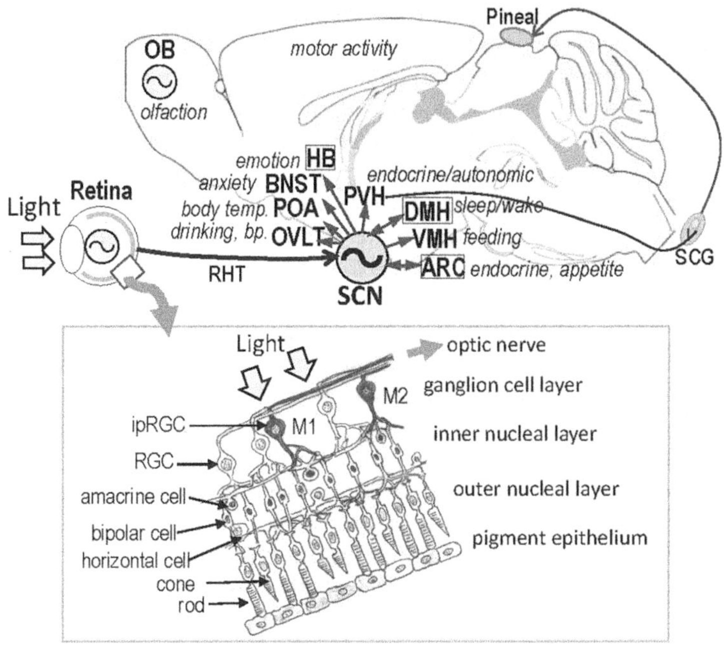

Figure 5.4 Retinal input and major direct projection sites of the SCN. Main functions of the SCN targets are listed in *italics*. bp: blood pressure. Other abbreviations are shown in the text. ⦾: autonomous circadian oscillator. Rectangles: semi-autonomous oscillator that slowly dampens *ex vivo*. Black letters: peripheral oscillator that immediately dampens without a synchronizer. Double arrow indicates bidirectional projections. Schematic diagram in the lower box indicates the cell composition in the retina (a rectangle in the upper diagram). M1 and M2: type 1 and type 2 melanopsin cell, respectively. Other abbreviations are shown in the text.

various physiological functions, including photoentrainment in the SCN. The presence of photoreceptors other than rhodopsin or cone opsin was suggested because mice lacking both rods and cones can entrain to an LD cycle.[71] And a specific photopigment, melanopsin, was found in a small subset (2–5%) of retinal ganglion cells called intrinsically photosensitive retinal ganglion cells (ipRGCs)[72,73] (Figure 5.4). Further studies demonstrated that there are at least six types of ipRGCs (M1–M6 cells), each with different morphological characteristics and retinal distribution.[74,75] Of these, M1 cells are the most comprehensively understood and are involved in non-image-forming photic responses, including photoentrainment. They respond to photic signals with an extremely slow time constant[73] and are involved in various subconscious visual behaviours in addition to photoentrainment such as light adaptation, light-induced pupil constriction, light-induced melatonin suppression, light masking of behaviour, light-induced alertness, and mood.[75–78] ipRGCs have varied physiological functions other than direct light perception, due to the presence of diverse innervations in the CNS such as the anterior hypothalamus (temperature regulation), preoptic area (POA), including the ventrolateral preoptic area (sleep center), bed nucleus of the stria terminalis (BNST) (stress responses), lateral hypothalamic nucleus (feeding, alertness), medial amygdala (mood), periaqueductal gray, and perihabenular zone (anger). M2 cells, which project into the SCN, also innervate the dorsal geniculate body and olivary pretectal nucleus, whereas M3–M6 cells do not innervate the SCN. In addition, M2–M6 cells seem to be involved in different kinds of visual imaging, such as contrast detection,[79] color contrast (M5), and discrimination of spatial details (M6).[74] Since M1 (and M2) cells innervate CNS structures other than the SCN, these pathways may play a role in managing the efferent flow of the circadian signals, in addition to light-based effects such as sleep and temperature regulation.[80]

The retinohypothalamic tract projects mainly into the SCN core. VIP[81] and GRP[82] neurons in the core receive glutamatergic and pituitary adenylate cyclase-activating polypeptide inputs from the retina. These neurons respond to light in a phase-dependent manner to induce *c-fos*[83] and *Per1*[47] expression. However, a recent tracing study clarified that single ipRGCs innervate both right and left SCN regions, and their target cells are not restricted to VIP or GRP cells in the SCN core. In fact, the axon terminal even innervates AVP cells in the SCN shell.[84]

5.3.2 Non-photic Inputs to the SCN

Compared with the vast amount of information on photic inputs, little is known about non-photic inputs. The geniculohypothalamic tract from the intergeniculate leaflets (IGL) and serotonergic inputs from the raphe nucleus both innervate the SCN core.[21] They are the major non-photic projections, and their neurotransmitters (NPY, GABA, and 5HT) induce phase responses in circadian rhythms (PRCs), which are classified as non-photic.[85–87] Non-photic PRCs for dark pulses, exercise (humans) or behavioural arousal (animals), melatonin, NPY, and triazolam are also called dark-pulse-type PRCs, which exhibit a large phase-advance zone in the subjective day. Phase responses following triazolam injection were demonstrated, not directly by a GABAergic stimulus but due to a drug-induced hyper locomotion.[88] Timed exercise has also been reported to phase-set the circadian clock in animals, including hamsters and humans.[86,87] These non-photic inputs may, therefore, be targets for drug discovery.

5.4 Output from the SCN

The neural afferents from the SCN cover most of the hypothalamic areas and several areas in the thalamus and midbrain (Figure 5.4). Tracing studies demonstrated that the projection sites are located dorsal to the anterior of the various hypothalamic areas; dorsally to the subparaventricular zone, paraventricular nucleus (PVH), periventricular nucleus, and then to the anterior hypothalamic area, parvicellular paraventricular nucleus, midline thalamic nuclei, thalamic paraventricular nucleus, dorsomedial nucleus (DMH), ventromedial nucleus (VMH), and posterior hypothalamic area; rostrally to the medial preoptic area, and anteroventral periventricular nucleus; anterodorsally to the medial preoptic nucleus and intermediate lateral septal nucleus; caudally to the POA, BNST, parataenial nucleus and rostral part of the thalamic paraventricular nucleus, and the anterior hypothalamic and retrochiasmatic areas; and laterally to the ventral lateral geniculate nucleus.[24,89] In addition, an immunohistochemical study suggested that the efferents from the SCN also project to the organum vasculosum of the lamina terminalis (OVLT), interpeduncular nucleus, and habenula (HB).[20] The projection sites overlap those of M1 ipRGC, and include brain areas that regulate body temperature, fluid and electrolyte homeostasis, sleep–wakefulness, hunger and satiety, thirst, metabolic state, emotion, motivation, mood, anxiety, and stress responses. By sending circadian signals, the SCN orchestrates these psychophysiological functions throughout the day.

When the SCN was found to be transplantable,[16] many researchers expected that the discrete SCN area might provide the efferent neural pathway from the SCN to particular brain areas that regulate overt rhythms. However, not all of the SCN functions are transplantable. Behavioural rhythms are thought to be driven *via* diffusible factors from the SCN because SCN tissues in a semipermeable capsule have been shown to rescue circadian behavioural rhythms.[90] However, the recovery of circadian hormonal rhythms has not been reported following SCN transplantation,[91] suggesting that different efferent pathways exist for behavioural activity and endocrine rhythms.

The master circadian clock in the SCN orchestrates peripheral clocks throughout the body. Although the precise mechanisms of this orchestration remain poorly understood, there are at least two mechanisms: the autonomic nervous system and the endocrine system. These are not independent downstream of the master clock but interact with one another to integrate information, resulting in coordinated circadian rhythms.

5.4.1 Autonomic Regulation of Peripheral Clocks

5.4.1.1 Output Pathways of the Autonomic Nervous System from the SCN

Autonomic functions in the body, such as the heartbeat, blood pressure, and body temperature, along with digestive functions, including movement, secretion, and muscle tone, all exhibit robust circadian rhythms in a coordinated manner under the control of the circadian pacemaker entrained to the LD cycle.

The paraventricular hypothalamic nucleus (PVH) plays pivotal roles in regulating the autonomic nervous and neuroendocrine systems,[92] and the parvocellular PVH is one of the major projection sites of SCN neurons.[20,24,89] Major descending pathways from the parvocellular PVH run through the medial forebrain bundle and ventral tegmental area. Some of the pathways project to the nucleus ambiguus, the nucleus of the solitary tract, and the dorsal motor nucleus of the vagus, from where the parasympathetic nervous system can be regulated in a circadian manner. Other parts of the descending pathways project to the intermediolateral cell column of the thoracic cord,[93,94] from where sympathetic nervous systems are regulated, according to a circadian rhythm.

5.4.1.2 *Autonomic Inputs Depending on the Peripheral Clocks*

Probably the most well-studied downstream pathway of the SCN is the regulation of the pineal melatonin rhythm; in addition, its immediate suppression by light during the subjective night is a well-studied light-masking pathway of circadian rhythms. The efferent projections reach the pineal gland *via* the PVH, the intermediolateral nucleus of the spinal cord, and the superior cervical ganglion (SCG), from where the post-ganglionic sympathetic nerve returns to the brain and projects into the pineal gland[95] (Figure 5.4). In rodents, the nocturnal release of noradrenaline activates arylalkylamine-*N*-acetyltransferase (AA-ANT), the rate-limiting enzyme for melatonin synthesis *via* the β1 receptor-cAMP pathway, which is also enhanced *via* α1 receptor-protein kinase C activation.[96,97] Post-translational processing of AA-NAT, *i.e.*, cAMP-induced stabilization of AA-NAT through interaction with 14-3-3 proteins to protect proteasomal degradation, is also essential since, in melatonin proficient C3H mice, the absence of CRY1 and CRY2 leads to steady increases in *Aa-nat* mRNA throughout the day, yet the levels of AA-NAT protein remain continually low.[98] However, ungulates and primates show very little change in daily mRNA levels of *Aa-nat,* suggesting that they primarily utilize post-transcriptional mechanisms. Although both nocturnal and diurnal mammals exhibit nocturnal rises in melatonin release, the regulation of the melatonin rhythm by circadian inputs is species-dependent.[99]

Several peripheral clocks, including the one in the liver, are reset by noradrenaline.[100] In contrast, circadian rhythms in clock gene expression and secretory functions in the respiratory systems are mainly regulated by the vagal nerve.[101] These findings suggest that both sympathetic and parasympathetic nervous systems are important pathways through which the central clock regulates peripheral clocks. The predominant system used for each peripheral clock depends on the tissue and function. In the rat submaxillary salivary gland, the expression rhythm of *Per1* exhibits an LD-entrained rhythm even under a restricted daily feeding schedule. Meanwhile, the rhythm entrains to the feeding schedule after surgical sympathectomy.[102] Thus, peripheral clocks may have multiple input pathways, and each clock responds to the secondary input only after the primary input is lost.

5.4.2 Endocrine Regulation of Peripheral Clocks

Various hormone levels exhibit robust and LD-entrained circadian rhythms, reflecting the control of peripheral clocks by the central clock in the SCN.

Each peripheral clock optimizes hormone secretion rhythms by receiving various rhythmic signals and the LD-entrained signal from the SCN is not necessarily the predominant time cue. For some peripheral clocks, feeding or the sleep/wake rhythm is the predominant time cue. Thus, the alignment of time cues, such as environmental light, mealtimes, sleep/wake schedules, and physical exercise, are important in coordinating bodily functions and health. In modern society, these time cues are sometimes dissociated from the LD cycle in humans, due to demanding work schedules, inadequate lifestyles, or poor sleep hygiene. Under such misalignment of various time cues, circadian orchestration is easily lost. Indeed, various hormonal disturbances, such as reduced glucose tolerance and increased glucocorticoid levels, result from sleep deprivation[103] and shiftwork.[104]

The output pathway from the SCN to the PVH is one of the major downstream pathways of hormonal rhythms. Among the three sub-regions in the PVH,[105] the parvocellular PVH contains neuroendocrine cells that produce corticotropin-releasing hormone, thyrotropin-releasing hormone, somatostatin, and AVP. The robust circadian rhythms in the synthesis and/or content of these neuroendocrine hormones strongly suggest that SCN efferents to the PVH are responsible for the circadian rhythms in downstream hormones, including adrenocorticotropic hormone, thyrotropin-releasing hormone, and growth hormone. Among them, circadian rhythms in the hypothalamo–pituitary–adrenocortical (HPA) axis have been well-documented due to the important roles of glucocorticoids in physiological functions, such as stress responses, glucose and lipid metabolism, immune responses, cognition and mood.[106] The circadian peak of glucocorticoids is detected at the wake time in both nocturnal and diurnal animals, which is lost by SCN lesioning. Importantly, glucocorticoids are major synchronizers of peripheral clocks.[107] Thus, special attention should be paid to the timing of glucocorticoid therapy. Also, the hypothalamo–pituitary–thyroid axis is critical for seasonal breeding patterns in animals.[108] Furthermore, long-projecting neurons composed of oxytocin neurons and melanocortin receptor 4-expressing neurons are involved in appetite control and metabolism. The SCN regulates the circadian rhythms of appetite and metabolic state not only *via* the PVH but also by sending direct projections to various hypothalamic areas, including the lateral hypothalamic nucleus, VMH, and arcuate nucleus (ARC).[109] These areas form networks, and the peripheral hormones released under the control of these areas, such as leptin and ghrelin, provide feedback to the CNS.

The efferent pathway of the SCN is linked to another important neuroendocrine rhythm – that of the hypothalamo–gonadal-axis. Reproductive rhythms are lost by lesioning of the SCN, even if the downstream functions are intact,[110] indicating the critical role of the SCN. Reproductive rhythms are regulated by gonadotropin-releasing hormone cells scattered widely in the POA and kisspeptin neurons in the anteroventral periventricular nucleus. AVP cells in the SCN are regarded as responsible output neurons,[111] and direct AVP innervation onto kisspeptin cells has been demonstrated, which is involved in estrogen-positive feedback.[112] VIP might be involved in gonadotropin-releasing hormone cell regulation,[113] and the SCN is the major source of VIP in the hypothalamus. Nevertheless, direct evidence for innervation of SCN VIP neurons has not yet been reported. Interestingly, in hamsters exhibiting behavioural splitting under constant light exposure, daily luteinizing hormone surges in estrogen-primed ovariectomized animals show two peaks per day,[114] each of which is phase-locked to the activation of the ipsilateral SCN,[115] indicating neuronal regulation of reproductive cycles by each SCN region.

5.4.3 Glymphatic System and Cerebrospinal Fluid

The brain's glymphatic system is a lymph-like network of perivascular space involved in replenishing and clearing the brain *via* movements of the cerebrospinal fluid (CSF). The glymphatic system moves metabolic waste such as amyloid-beta and tau-protein out of the brain, especially during deep sleep when the CSF flow rate is increased.[116] The CSF is mainly produced by the choroid plexus (CP) that lines the ventricle and immerses the whole brain. Since the SCN is situated at the bottom of the 3rd ventricle, it has long been considered that the CSF may function as the circadian output pathway of the diffusible signals produced in the SCN. Previously, AVP was shown to be released into the CSF in a circadian fashion, which was disrupted by SCN lesioning.[117] Recent studies revealed the involvement of the circadian clock in the glymphatic system and a novel role for the CSF. The CP exhibits robust circadian rhythms in clock gene expression and has been considered one of the peripheral clocks. On the other hand, the disruption of circadian rhythms in the CP alters the circadian period of the SCN and behaviour, suggesting that circadian rhythms in the CP can provide feedback to the SCN, probably *via* some diffusible factor(s) in the CSF.[118] In addition, both influx and clearance of the glymphatic system show circadian rhythms independent of the LD cycle, and the localization of water channel aquaporin (AQP)-4 in astrocytes plays a critical role in this process.[119] AQP-4 shows increased polarization around vascular structures during the rest phase, thus demonstrating circadian clearance rhythms in the glymphatic system.

5.4.4 Sleep–Wake System

Humans under prolonged temporal isolation often show internal desynchronization between sleep–wake and circadian rhythms in melatonin production and core body temperature. The findings suggest that sleep–wake rhythms are not simply induced by the SCN's circadian clock. Although it is still debatable whether the sleep–wake rhythm is induced by another oscillatory mechanism[120] or non-oscillating homeostatic mechanism,[121] sleep–wake rhythms are phase-set by strict sleep schedules.[122] Furthermore, after the end of the schedule, they start to free-run and eventually synchronize with the melatonin rhythm, which does not entrain to the sleep schedule,[122] indicating the oscillatory nature of sleep–wake rhythms.[123]

Apart from the circadian rhythm, sleep and wakefulness are separately regulated by discrete brain regions. A group of wake-promoting neurons, *i.e.*, monoaminergic neurons in the upper brainstem, such as noradrenergic neurons in the locus coeruleus and serotonergic neurons in the raphe nuclei, as well as histaminergic neurons in the tuberomammillary nucleus and cholinergic neurons in the pedunculopontine and laterodorsal tegmental nuclei, project to the thalamus and the entire cerebral cortex. Orexin neurons in the lateral hypothalamic nucleus exert stabilizing effects on these wake-promoting neurons, while GABAergic sleep-promoting neurons in the ventrolateral preoptic nucleus (VLPO) exert inhibitory influences, inducing sleep. Interestingly, lesioning of any of the above areas affects the amount of sleep or wakefulness.[124] However, animals with SCN-lesions or surgical isolation using a Halasz knife totally lose circadian sleep/wake rhythms and exhibit frequent sleep and wake alterations throughout the day, even under an LD cycle.[3,15,125] While innervation from the SCN to the VLPO or orexin neurons is limited, sleep–wake rhythms are affected by lesioning of the subparaventricular zone[126] and dorsomedial hypothalamic nucleus,[127] where the SCN neurons send a vast number of projections. Thus, Saper *et al.*[124] proposed the neural regulation of circadian sleep–wake rhythms *via* the SCN-subparaventricular zone-dorsomedial hypothalamic nucleus. Although this neural network model has not been fully confirmed, the restoration of behavioural rhythms in SCN-lesioned animals through diffusible factors from transplanted SCN tissue suggests that paracrine signals may be the output pathway through which sleep–wake rhythms are regulated.[90] Indeed, the diffusible factors PK2[128] and transforming growth factor-α (TGF-α)[129] secreted from the SCN are reportedly candidate output signals that regulate sleep–wake rhythms. Since sleep–wake rhythms are so important to animal life, the regulating mechanisms may be redundant, and both humoral and neural factors may be involved.

5.5 Conclusion

Cloning of mammalian clock genes and live monitoring of gene expression or protein levels using luciferase and fluorescent reporters have revealed that circadian clocks are present throughout the body at the cellular, organ, and system levels. These clocks affect behaviour, sleep, and wakefulness. Mammals have developed a circadian clock system with a master clock in the brain, which receives the only source of environmental light, and peripheral clocks throughout the body, thus forming a hierarchical multi-oscillator system that enables physiological functions to be temporally organized by the master clock. The master clock does not drive rhythms in a stimulation and reaction manner but by phase resetting of oscillators. Circadian oscillation is a slow but remarkably stable and precise process. It does not require reactions on the order of milliseconds; instead, a different strategy is necessary for the orchestration of the circadian system, namely, a molecular clock system with transcriptional and translational feedback loops and various post-transcriptional and post-translational modifications, multiple oscillator networks, and interactions with diffusible factors.

Acknowledgements

The author declares no conflicts of interest. All applicable international, national, and institutional guidelines for the care and use of animals were followed. This review article does not contain any studies involving human participants.

References

1. D. R. Weaver and S. M. Reppert, Coordination of circadian timing in mammals, *Nature*, 2002, **418**, 935.
2. S. Honma, The mammalian circadian system: a hierarchical multi-oscillator structure for generating circadian rhythm, *J. Physiol. Sci.*, 2018, **68**, 207.
3. F. K. Stephan and I. Zucker, Circadian rhythms in drinking behavior and locomotor activity of rats are eliminated by hypothalamic lesions, *Proc. Natl. Acad. Sci. U. S. A.*, 1972, **69**, 1583.
4. R. Y. Moore and V. B. Eichler, Loss of a circadian adrenal corticosterone rhythm following suprachiasmatic lesions in the rat., *Brain Res.*, 2973, **42**, 201.
5. S. Yamazaki, R. Numano, M. Abe, A. Hida, R. Takahashi, M. Ueda, G. D. Block, Y. Sakaki, M. Menaker and H. Tei, Resetting central and peripheral circadian oscillators in transgenic rats, *Science*, 2000, **288**, 682.
6. E. Nagoshi, C. Saini, C. Bauer, T. Laroche, F. Naef and U. Schibler, Circadian gene expression in individual fibroblasts: cell-autonomous and self-sustained oscillators pass time to daughter cells, *Cell*, 2004, **119**, 693.
7. N. Gekakis, D. Staknis, H. B. Nguyen, F. C. Davis, L. D. Wilsbacher, D. P. King, J. S. Takahashi and C. J. Weitz, Role of the CLOCK protein in the mammalian circadian mechanism, *Science*, 1998, **280**, 1564.
8. A. Eskin, Identification and physiology of circadian pacemakers, *Fed. Proc.*, 1979, **38**, 2570.
9. F. K. Stephan, The "other" circadian system: food as a Zeitgeber., *J. Biol. Rhythms*, 2003, **17**, 284.
10. K. Honma and S. Honma, The SCN-independent clocks, methamphetamine and food restrition., *Eur. J. Neurosci.*, 2009, **30**, 1707.
11. J. J. Gooley, A. Schomer and C. B. Saper, The dorsomedial hypothalamic nucleus is critical for the expression of food-entrainable circadian rhythms, *Nat. Neurosci.*, 2006, **9**, 398.
12. A. Natsubori, K. Honma and S. Honma, Dual regulation of clock gene Per2 expression in discrete brain areas by the circadian pacemaker and methamphetamine-induced oscillator in rats, *Eur. J. Neurosci.*, 2014, **39**, 229.

13. D. Ono, S. Honma and K. Honma, Circadian PER2::LUC rhythms in the olfactory bulb of freely moving mice depend on the suprachiasmatic nucleus but not on behaviour rhythms, *Eur. J. Neurosci.*, 2015, **42**, 3128.
14. C. Guilding, A. T. L. Hughes, T. M. Brown, S. Namvar and H. D. Piggins, A riot of rhythms: neuronal and glial circadian oscillators in the mediobasal hypothalamus, *Mol. Brain*, 2009, **2**, 28.
15. S. T. Inouye and H. Kawamura, Persistence of circadian rhythmicity in a mammalian hypothalamic "island" containing the suprachiasmatic nucleus, *Proc. Natl. Acad. Sci. U. S. A.*, 1979, **76**, 5962.
16. M. R. Ralph, R. G. Foster, F. C. Davis and M. Menaker, Transplanted suprachiasmatic nucleus determines circadian period, *Science*, 1990, **247**, 975.
17. M. Sujino, K. Masumoto, S. Yamaguchi, G. T. J. van der Horst, H. Okamura and S. T. Inouye, Suprachiasmatic Nucleus Grafts Restore Circadian Behavioral Rhythms of Genetically Arrhythmic Mice, *Curr. Biol.*, 2003, **13**, 664.
18. D. K. Welsh, D. E. Logothetis, M. Meister and S. M. Reppert, Individual neurons dissociated from rat suprachiasmatic nucleus express independently phased circadian firing rhythms, *Neuron*, 1995, **14**, 697.
19. K. Shinohara, S. Honma, Y. Katsuno, H. Abe and K. Honma, Two distinct oscillators in the rat suprachiasmatic nucleus *in vitro*, *Proc. Natl. Acad. Sci. U. S. A.*, 1995, **92**, 7396.
20. A. N. van den Pol and K. L. Tsujimoto, Neurotransmitters of the hypothalamic suprachiasmatic nucleus: immunocytochemical analysis of 25 neuronal antigens, *Neuroscience*, 1985, **15**, 1049.
21. E. E. Abrahamson and R. Y. Moore, Suprachiasmatic nucleus in the mouse: retinal innervation, intrinsic organization and efferent projections, *Brain Res.*, 2001, **916**, 172.
22. F. H. Güldner, Numbers of neurons and astroglial cells in the suprachiasmatic nucleus of male and female rats, *Exp. Brain Res.*, 1983, **50**, 373.
23. S. Wen, D. Ma, M. Zhao, L. Xie, Q. Wu, L. Gou, C. Zhu, Y. Fan, H. Wang and J. Yan, Spatiotemporal single-cell analysis of gene expression in the mouse suprachiasmatic nucleus, *Nat. Neurosci.*, 2020, **23**, 456.
24. R. Y. Moore, The Suprachiasmatic Nucleus and the Circadian Timing System, *Prog. Mol. Biol. Transl. Sci.*, 2013, **119**, 1.
25. M. Doi, I. Murai, S. Kunisue, G. Setsu, N. Uchio, R. Tanaka, S. Kobayashi, H. Shimatani, H. Hayashi, H.-W. Chao, Y. Nakagawa, Y. Takahashi, Y. Hotta, J. Yasunaga, M. Matsuoka, M. H. Hastings, H. Kiyonari and H. Okamura, Gpr176 is a Gz-linked orphan G-protein-coupled receptor that sets the pace of circadian behaviour, *Nat. Commun.*, 2016, **7**, 10583.
26. R. Silver, A. I. Sookhoo, J. LeSauter, P. Stevens, H. T. Jansen and M. N. Lehman, Multiple regulatory elements result in regional specificity in circadian rhythms of neuropeptide expression in mouse SCN., *NeuroReport*, 1999, **10**, 3165.

27. I. T. Lee, A. S. Chang, M. Manandhar, Y. Shan, J. Fan, M. Izumo, Y. Ikeda, T. Motoike, S. Dixon, J. E. Seinfeld, J. S. Takahashi and M. Yanagisawa, Neuromedin s-producing neurons act as essential pacemakers in the suprachiasmatic nucleus to couple clock neurons and dictate circadian rhythms, *Neuron*, 2015, **85**, 1086.
28. W. D. Todd, A. Venner, C. Anaclet, R. Y. Broadhurst, R. De Luca, S. S. Bandaru, L. Issokson, L. M. Hablitz, O. Cravetchi, E. Arrigoni, J. N. Campbell, C. N. Allen, D. P. Olson and P. M. Fuller, Suprachiasmatic VIP neurons are required for normal circadian rhythmicity and comprised of molecularly distinct subpopulations, *Nat. Commun.*, 2020, **11**, 4410.
29. R. Y. Moore and J. C. Speh, GABA is the principal neurotransmitter of the circadian system, *Neurosci. Lett.*, 1993, **150**, 112.
30. C. S. Colwell, S. Michel, J. Itri, W. Rodriguez, J. Tam, V. Lelievre, Z. Hu, X. Liu and J. A. Waschek, Disrupted circadian rhythms in VIP- and PHI-deficient mice, *Am. J. Physiol.: Regul., Integr. Comp. Physiol.*, 2003, **285**, R939.
31. A. J. Harmar, H. M. Marston, S. Shen, C. Spratt, K. M. West, W. J. Sheward, C. F. Morrison, J. R. Dorin, H. D. Piggins, J. C. Reubi, J. S. Kelly, E. S. Maywood and M. H. Hastings, The VPAC2 receptor is essential for circadian function in the mouse suprachiasmatic nuclei, *Cell*, 2002, **109**, 497.
32. S. J. Aton, C. S. Colwell, A. J. Harmar, J. Waschek and E. D. Herzog, Vasoactive intestinal polypeptide mediates circadian rhythmicity and synchrony inmammalian clock neurons, *Nat. Neurosci.*, 2005, **8**, 476.
33. Y. Yamaguchi, T. Suzuki, Y. Mizoro, H. Kori, K. Okada, Y. Chen, J. M. Fustin, F. Yamazaki, N. Mizuguchi, J. Zhang, X. Dong, G. Tsujimoto, Y. Okuno, M. Doi and H. Okamura, Mice genetically deficient in vasopressin V1a and V1b receptors are resistant to jet lag, *Science*, 2013, **42**, 85.
34. M. Mieda, D. Ono, E. Hasegawa, H. Okamoto, K. Honma, S. Honma and T. Sakurai, Cellular clocks in AVP neurons of the SCN are critical for interneuronal coupling regulating circadian behavior rhythm, *Neuron*, 2015, **85**, 1103.
35. D. Ono, K. Honma and S. Honma, GABAergic mechanisms in the suprachiasmatic nucleus that influence circadian rhythm, *J. Neurochem.*, 2021, **157**, 31.
36. Y. Ben-Ari, Excitatory actions of gaba during development: the nature of the nurture, *Nat. Rev. Neurosci.*, 2002, **3**, 728.
37. R. Mason, S. M. Biello and M. E. Harrington, The effects of GABA and benzodiazepines on neurones in the suprachiasmatic nucleus (SCN) of Syrian hamsters, *Brain Res.*, 1991, **552**, 53.
38. S. Wagner, M. Castel, H. Gainer and Y. Yarom, GABA in the mammalian suprachiasmatic nucleus and its role in diurnal rhythmicity, *Nature*, 1997, **387**, 598.

39. M. Ikeda, T. Yoshioka and C. N. Allen, Developmental and circadian changes in Ca^{2+} mobilization mediated by GABAA and NMDA receptors in the suprachiasmatic nucleus, *Eur. J. Neurosci.*, 2003, **17**, 58.
40. S. Farajnia, T. L. van Westering, J. H. Meijer and S. Michel, Seasonal induction of GABAergic excitation in the central mammalian clock, *Proc. Natl. Acad. Sci. U. S. A.*, 2014, **111**, 9627.
41. K. E. Rohr, H. Pancholi, S. Haider, C. Karow, D. Modert, N. J. Raddatz and J. Evans, Seasonal plasticity in GABAA signaling is necessary for restoring phase synchrony in the master circadian clock network, *eLife*, 2019, **8**, e49578.
42. J. Schaap, H. Albus, H. T. van der Leest, P. H. C. Eilers, L. Détári and J. H. Meijer, Heterogeneity of rhythmic suprachiasmatic nucleus neurons: Implications for circadian waveform and photoperiodic encoding, *Proc. Natl. Acad. Sci. U. S. A.*, 2003, **100**, 15994.
43. D. Ono, K. Honma, Y. Yanagawa, A. Yamanaka and S. Honma, GABA in the suprachiasmatic nucleus refines circadian output rhythms in mice, *Commun. Biol.*, 2019, **2**, 232.
44. M. Lavialle and J. Servière, Circadian fluctuations in GFAP distribution in the Syrian hamster suprachiasmatic nucleus, *NeuroReport*, 1993, **4**, 1243.
45. C. S. Colwell, Linking neural activity and molecular oscillations in the SCN, *Nat. Rev. Neurosci.*, 2011, **12**, 553.
46. R. Enoki, S. Kuroda, D. Ono, M. T. Hasan, T. Ueda, S. Honma and K. Honma, Topological specificity and hierarchical network of the circadian calcium rhythm in the suprachiasmatic nucleus, *Proc. Natl. Acad. Sci. U. S. A.*, 2012, **109**, 21498.
47. Y. Shigeyoshi, K. Taguchi, S. Yamamoto, S. Takekida, L. Yan, H. Tei, T. Moriya, S. Shibata, J. J. Loros, J. C. Dunlap and H. Okamura, Light-induced resetting of a mammalian circadian clock is associated with rapid induction of the mPer1 transcript, *Cell*, 1997, **91**, 1043.
48. S. Honma, T. Kawamoto, Y. Takagi, K. Fujimoto, F. Sato, N. Noshiro, Y. Kato and K. Honma, *Dec1* and *Dec2* are regulators of mammalian molecular clock, *Nature*, 2002, **419**, 841.
49. D. Ono, K. Honma and S. Honma, Circadian and ultradian rhythms of clock gene expression in the suprachiasmatic nucleus of freely moving mice, *Sci. Rep.*, 2015, **5**, 12310.
50. S. Daan, K. Spoelstra, U. Albrecht, I. Schmutz, M. Daan, B. Daan, F. Rienks, I. Poletaeva, G. Dell'Omo, A. Vyssotski and H.-P. Lipp, Lab mice in the field: unorthodox daily activity and effects of a dysfunctional circadian clock allele, *J. Biol. Rhythms*, 2011, **26**, 118.
51. S. Honma, D. Ono, Y. Suzuki, N. Inagaki, T. Yoshikawa, W. Nakamura and K. Honma, Suprachiasmatic nucleus: cellular clocks and networks, in *Progress in Brain Research*, ed. A. Kalsbeek, M. Merrow, T. Roenneberg and R. G. Foster, The Neurobiology of Circadian Timing, Elsevier, Amsterdam, 2012, vol. 199, p. 129.

52. A. B. Webb, N. Angelo, J. E. Huettner and E. D. Herzog, Intrinsic, nondeterministic circadian rhythm generation in identified mammalian neurons, *Proc. Natl. Acad. Sci. U. S. A.*, 2009, **106**, 16493.
53. Y. Hirata, R. Enok, K. Kuribayashi-Shigetomi, Y. Oda, S. Honma and K. Honma, Circadian rhythms in Per1, PER2 and Ca^{2+} of a solitary SCN neuron cultured on a microisland, *Sci. Rep.*, 2019, **9**, 18271.
54. D. Ono, S. Honma and K. Honma, Differential roles of AVP and VIP signaling in the postnatal changes of neural networks for coherent circadian rhythms in the SCN, *Sci. Adv.*, 2016, **2**, e1600960.
55. N. C. Foley, T. Y. Tong, D. Foley, J. Lesauter, D. K. Welsh and R. Silver, Characterization of orderly spatiotemporal patterns of clock gene activation in mammalian suprachiasmatic nucleus, *Eur. J. Neurosci.*, 2011, **33**, 1851.
56. K. Shimomura and M. Menaker, Light-induced phase shifts in tau mutant hamsters, *J. Biol. Rhythms*, 1994, **9**, 97.
57. M. Nagano, A. Adachi, K. Nakahama, T. Nakamura, M. Tamada, E. Meyer-Bernstei, A. Sehgal and Y. Shigeyoshi, An abrupt shift in the day/night cycle causes desynchrony in the mammalian circadian center, *J. Neurosci.*, 2003, **23**, 6141.
58. H. Albus, M. J. Vansteensel, S. Michel, G. D. Block and J. H. Meijer, A GABAergic mechanism is necessary for coupling dissociable ventral and dorsal regional oscillators within the circadian clock, *Curr. Biol.*, 2005, **15**, 886.
59. K. Shinohara, S. Honma, Y. Katsuno, H. Abe and K. Honma, Two distinct oscillators in the rat suprachiasmatic nucleus *in vitro*, *Proc. Natl. Acad. Sci. U. S. A.*, 1995, **92**, 7396.
60. N. Inagaki, S. Honma, D. Ono, Y. Tanahashi and K. Honma, Separate oscillating cell groups in mouse suprachiasmatic nucleus couple photoperiodically to the onset and end of daily activity, *Proc. Natl. Acad. Sci. U. S. A.*, 2007, **104**, 7664.
61. T. Yoshikawa, N. Inagaki, S. Takagi, S. Kuroda, M. Yamasaki, M. Watanabe, S. Honma and K. Honma, Localization of photoperiod responsive circadian oscillators in the mouse suprachiasmatic nucleus, *Sci. Rep.*, 2017, **7**, 820.
62. C. S. Pittendrigh and S. Daan, A functional analysis of circadian pacemakers in nocturnal rodents. V. Pacemaker structure: A clock for all seasons, *J. Comp. Physiol.*, 1976, **106**, 333.
63. H. O. de la Iglesia, J. Meyer, A. Carpino Jr and W. J. Schwartz, Antiphase oscillation of the left and right suprachiasmatic nuclei, *Science*, 2000, **290**, 799.
64. T. Ikeda, N. Iijima, K. Munekawa, A. Ishihara, Y. Ibata and M. Tanaka, Functional retinal input stimulates expression of astroglial elements in the suprachiasmatic nucleus of postnatal developing rat, *Neurosci. Res.*, 2003, **47**, 39.
65. T. Moriya, Y. Yoshinobu, Y. Kouzu, A. Katoh, H. Gomi, M. Ikeda, T. Yoshioka, S. Itohara and S. Shibata, Involvement of glial fibrillary

acidic protein (GFAP) expressed in astroglial cells in circadian rhythm under constant ighting conditions in mice, *J. Neurosci. Res.*, 2000, **60**, 212.
66. C. F. Tso, T. Simon, A. C. Greenlaw, T. Puri, M. Mieda and E. D. Herzog, Astrocytes regulate daily rhythms in the suprachiasmatic nucleus and behavior, *Curr. Biol.*, 2017, **27**, 1055.
67. M. Brancaccio, A. P. Patton, J. E. Chesham, E. S. Maywood and M. H. Hastings, Astrocytes control circadian timekeeping in the suprachiasmatic nucleus *via* glutamatergic signaling, *Neuron*, 2017, **93**, 1420.
68. M. Brancaccio, M. D. Edwards, A. P. Patton, N. J. Smyllie, J. E. Chesham, E. E. S. Maywood and M. H. Hastings, Cell-autonomous clock of astrocytes drives circadian behavior in mammals, *Science*, 2019, **363**, 187.
69. H. Underwood, C. T. Steele and B. Zivovic, Circadian organization and the roles of the pineal in the birds, *Microsc. Res. Tech.*, 2001, **53**, 48.
70. G. Y. P. Ko, Circadian regulation in the retina: From molecules to network, *Eur. J. Neurosci.*, 2020, **51**, 194.
71. S. Hattar, R. J. Lucas, N. Mrosovsky, S. Thompson, R. H. Douglas, M. W. Hankins, J. Lem, M. Biel, F. Hofmann, R. G. Foster and K. W. Yau, Melanopsin and rod-cone photoreceptive systems account for all major accessory visual functions in mice, *Nature*, 2003, **424**, 76.
72. S. Hattar, H. W. Liao, M. Takao, D. M. Berson and K. W. Yau, Melanopsin-containing retinal ganglion cells: architecture, projections, and intrinsic photosensitivity, *Science*, 2002, **295**, 1065.
73. D. M. Berson, F. A. Dunn and M. Takao, Phototransduction by retinal ganglion cells that set the circadian clock, *Science*, 2002, **295**, 1070.
74. M. T. H. Do, Melanopsin and the intrinsically photosensitive retinal ganglion cells: Biophysics to behavior, *Neuron*, 2019, **104**, 205.
75. T. M. Schmidt, M. T. H. Do, D. Dacey, R. Lucas, S. Hattar and A. Matynia, Melanopsin-positive intrinsically photosensitive retinal ganglion cells: from form to function, *J. Neurosci.*, 2011, **31**, 16094.
76. R. J. Lucas, A. E. Allen, N. Milosavljevic, R. Storchi and T. Woelders, Can we see with melanopsin?, *Ann. Rev. Vis. Sci.*, 2020, **6**, 453.
77. R. J. Lucas, S. Hattar, M. Takao, D. M. Berson, R. G. Foster and K. W. Yau, Diminished pupillary light reflex at high irradiances in melanopsin-knockout mice, *Science*, 2003, **299**, 245.
78. D. C. Fernandez, P. M. Fogerson, L. L. Ospri, M. B. Thomsen, R. M. Layne, D. Severin, J. Zhan, J. H. Singer, A. Kirkwood, H. Zhao, D. M. Berson and S. Hattar, Light affects mood and learning through distinct retina-br andin pathways, *Cell*, 2018, **175**, 71.
79. T. Sonoda, S. K. Lee, L. Birnbaumer and T. M. Schmid, Melanopsin phototransduction is repurposed by ipRGC subtypes to shape the function of distinct visual circuits, *Neuron*, 2018, **99**, 754.
80. A. C. Rupp, M. Ren, C. M. Altimus, D. C. Fernandez, M. Richardson, F. Turek, S. Hattar and T. M. Schmidt, Distinct ipRGC subpopulations mediate light's acute and circadian effects on body temperature and sleep, *eLife*, 2019, **8**, e44358.

81. Y. Ibata, Y. Takahashi, H. Okamura, F. Kawakami, H. Terubayashi, T. Kubo and N. Yanaihara, Vasoactive intestinal peptide (VIP)-like immunoreactive neurons located in the rat suprachiasmatic nucleus receive a direct retinal projection, *Neurosci. Lett.*, 1989, **97**, 1.
82. M. Tanaka, S. Hayashi, Y. Tamada, T. Ikeda, Y. Hisa, T. Takamatsu and Y. Ibata, Direct retinal projections to GRP neurons in the suprachiasmatic nucleus of the rat, *NeuroReport*, 1997, **8**, 2187.
83. J. M. Kornhauser, D. E. Nelson, K. E. Mayo and J. S. Takahashi, Photic and circadian regulation of c-fos gene expression in the hamster suprachiasmatic nucleus, *Neuron*, 1990, **5**, 127.
84. D. C. Fernandeza, Y.-T Chang, S. Hattara and S.-K. Chen, Architecture of retinal projections to the central circadian pacemaker, *Proc. Natl. Acad. Sci. U. S. A.*, 2016, **113**, 6047.
85. I. C. Webb, M. C. Antle and R. E. Mistlberger, Regulation of circadian rhythms in mammals by behavioral arousal, *Behav. Neurosci.*, 2014, **128**, 304.
86. N. Mrosovsky, Locomotor activity and non-photic influences on circadian clocks, *Biol. Rev. Biol. Proc. Cambridge Philos. Soc.*, 1996, **71**, 343.
87. R. E. Mistlberger and D. J. Skene, Social influences on mammalian circadian rhythms: animal and human studies, *Biol. Rev.*, 2004, **79**, 533.
88. C. Wickland and F. W. Turek, Phase-shifting effect of triazolam on the hamster's circadian rhythm of activity is not mediated by a change in body temperature, *Brain Res.*, 1991, **560**, 12.
89. A. G. Watts, L. W. Swanson and G. Sanchez-Watts, Efferent projections of the suprachiasmatic nucleus: I. Studies using anterograde transport of Phaseolus vulgaris leucoagglutinin in the rat, *J. Comp. Neurol.*, 1987, **258**, 204.
90. R. Silver, J. LeSauter, P. A. Tresco and M. N. Lehman, A diffusible coupling signal from the transplanted suprachiasmatic nucleus controlling circadian locomotor rhythms, *Nature*, 1996, **382**, 810.
91. E. L. Meyer-Bernstein, A. E. Jetton, S. I. Matsumoto, J. F. Markuns, M. L. Lehman and E. L. Bittman, Effects of suprachiasmatic transplants on circadian rhythms of neuroendocrine function in golden hamsters, *Endocrinology*, 1999, **140**, 207.
92. L. W. Swanson and P. E. Sawchenko, Paraventricular nucleus: a site for the integration of neuroendocrine and autonomic mechanisms, *Neuroendocrinology*, 1980, **31**, 410.
93. L. W. Swanson and H. G. J. M. Kuypers, The paraventricular nucleus of the hypothalamus: Cytoarchitectonic subdivisions and organization of projections to the pituitary, dorsal vagal complex, and spinal cord as demonstrated by retrograde fluorescence double-labeling methods, *J. Comp. Neurol.*, 1980, **194**, 555.
94. J. C. Geerling, J.-W Shin, P. C. Chimenti and A. D. Loew, Paraventricular hypothalamic nucleus: axonal projections to the brainstem, *J. Comp. Neurol.*, 2010, **518**, 1460.

95. R. Teclemariam-Mesbah, G. J. Terhorst, F. Postema, J. Wortel and R. M. Bujis, Anatomical demonstration of the suprachiasmatic nucleus–pineal pathway, *J. Comp. Neurol.*, 1999, **406**, 171.
96. D. C. Klein, D. Sugden and J. L. Weller, Postsynaptic alpha-adrenergic receptors potentiate the beta-adrenergic stimulation of pineal serotonin N-acetyltransferase, *Proc. Natl. Acad. Sci. U. S. A.*, 1983, **80**, 599–603.
97. A. K. Ho and D. C. Klein, Activation of alpha 1-adrenoceptors, protein kinase C, or treatment with intracellular free Ca^{2+} elevating agents increases pineal phospholipase A2 activity. Evidence that protein kinase C may participate in Ca^{2+}-dependent alpha 1-adrenergic stimulation of pineal phospholipase A2 activity, *J. Biol. Chem.*, 1987, **262**, 11764.
98. Y. Yamanaka, Y. Yamada, K. Honma and S. Honma, Cryptochrome deficiency enhances transcription but reduces protein levels of pineal Aanat, *J. Mol. Endocrinol.*, 2018, **61**, 219.
99. J. D. Johnston and D. J. Skene, Regulation of mammalian neuroendocrine physiology and rhythms by melatonin, *J. Endocrinol.*, 2015, **226**, T187.
100. H. Terazono, T. Mutoh, S. Yamaguchi, M. Kobayashi, M. Akiyama, R. Udo, S. Ohdo, H. Okamura and S. Shibata, Adrenergic regulation of clock gene expression in mouse liver, *Proc. Natl. Acad. Sci. U. S. A.*, 2003, **100**, 6795.
101. H. Bando, T. Nishio, G. T. J. van der Horst, S. Masubuchi, Y. Hisa and H. Okamura, Vagal Regulation of Respiratory Clocks in Mice, *J. Neurosci.*, 2007, **27**, 4359.
102. N. Vujović, A. J. Davidson and M. Menaker, Sympathetic input modulates, but does not determine, phase of peripheral circadian oscillators, *Am. J. Physiol.*, 2008, **295**, R355.
103. K. Spiegel, R. Leproult and E. Van Cauter, Impact of sleep debt on metabolic and endocrine function, *Lancet*, 1999, **354**, 1435.
104. F. A. J. L. Scheer, K. Hu, H. Evoniuk, E. E. Kelly, A. Malhotra, M. F. Hilton and S. A. Shea, Impact of the human circadian system, exercise, and their interaction on cardiovascular function, *Proc. Natl. Acad. Sci. U. S. A.*, 2010, **107**, 20541.
105. L. W. Swanson and P. E. Sawchenko, Paraventricular nucleus: a site for the integration of neuroendocrine and autonomic mechanisms, *Neuroendocrinology*, 1980, **31**, 410.
106. S. Chung, G. H. Son and K. Kim, Circadian rhythm of adrenal glucocorticoid: Its regulation and clinical implications, *Biochim. Biophys. Acta*, 2011, **1812**, 581–591.
107. K. Oishi, N. Amagai, H. Shirai, K. Kadota, N. Ohkura and N. Ishida, Genome-wide expression analysis reveals 100 adrenal gland-dependent circadian genes in the mouse liver, *DNA Res.*, 2005, **12**, 191–202.
108. K. Ikegami, S. Refetoff, E. Van Cauter and T. Yoshimura, Interconnection between circadian clocks and thyroid function, *Nat. Rev. Endocrinol.*, 2019, **15**, 590.

109. R. Méndez-Hernández, C. Escobar and R. M. Buijs, Suprachiasmatic nucleus-arcuate nucleus axis: Interaction between time and metabolism essential for health, *Obesity*, 2020, **28**, S10–S17.
110. K. Brown-Grant and G. Raisman, Abnormalities in reproductive function associated with the destruction of the suprachiasmatic nuclei in female rats, *Proc. R. Soc. London*, 1977, **198**, 279.
111. E. L. Bittman, Circadian function in multiple cell types is necessary for proper timing of the preovulatory LH surge, *J. Biol. Rhythms*, 2019, **34**, 622.
112. B. Vida, L. Deli, E. Hrabovszky, T. Kalamatianos, A. Caraty, C. W. Coen, Z. Liposits and I. Kalló, Evidence for suprachiasmatic vasopressin neurones innervating kisspeptin neurones in the rostral periventricular area of the mouse brain: regulation by oestrogen, *J. Neuroendocrinol.*, 2010, **22**, 1032.
113. E. Vijayan, W. K. Samson, S. I. Said and S. M. McCann, Vasoactive Intestinal Peptide: Evidence for a Hypothalamic Site of Action to Release Growth Hormone, Luteinizing Hormone, and Prolactin in Conscious Ovariectomized Rats, *Endocrinology*, 1979, **104**, 53.
114. J. M. Swann and F. W. Turek, Multiple circadian oscillators regulate the timing of behavioral and endocrine rhythms in female golden hamsters, *Science*, 1985, **228**, 898.
115. H. O. de la Iglesia, J. Meyer and W. J. Schwartz, Lateralization of circadian pacemaker output: activation of left- and right-sided luteinizing releasing hormone neurons involves a neural rather than a humoral pathway, *J. Neurosci.*, 2003, **23**, 7412.
116. L. Xie, H. Kang, Q. Xu, M. J. Chen, Y. Liao, M. Thiyagarajan, J. O'Donnell, D. J. Christensen, C. Nicholson, J. J. Iliff, T. Takano, R. Deane and M. Nedergaard, Sleep drives metabolite clearance from the adult brain, *Science*, 2013, **342**, 373–377.
117. W. J. Schwartz and S. M. Reppert, Neural regulation of the circadian vasopressin rhythm in cerebrospinal fluid: a pre-eminent role for the suprachiasmatic nuclei, *J. Neurosci.*, 1985, **5**, 2771–2887.
118. J. Myung, C. Schmal, S. Hong, Y. Tsukizawa, P. Rose, Y. Zhang, M. J. Holtzman, E. D. Schutter, H. Herzel, G. Bordyugov and T. Takumi, *Nat. Commun.*, 2018, **9**, 1062.
119. L. M. Hablitz, V. Plá, M. Giannetto, H. S. Vinitsky, F. F. Stæger, T. Metcalfe, R. Ngyuyen, A. Benrais and M. Nedergaard, Circadian control of brain glymphatic and lymphatic fluid flow, *Nat. Commn.*, 2020, **11**, 4411.
120. R. Wever, *The circadian system of Man: Results of Experiments under Temporal Isolation*, Springer-Verlag, New York, 1979.
121. S. Daan, D. G. Beersma and A. A. Borbély, Timing of human sleep: recovery process gated by a circadian pacemaker, *Am. J. Physiol.*, 1984, **246**, R161.
122. S. Hashimoto, K. Nakamura, S. Honma and K. Honma, Non-photic entrainment of human rest-activity cycle independent of circadian pacemaker, *Sleep Biol. Rhythms*, 2004, **2**, 29.

123. A. A. Borbély, S. Daan, A. Wirz-Justice and T. Deboer, The two-process model of sleep regulation: a reappraisal, *J Sleep Res.*, 2016, **25**, 31.
124. C. B. Saper, T. E. Scammell and J. Lu, Hypothalamic regulation of sleep and circadian rhythms, *Nature*, 2005, **437**, 1257.
125. N. Ibuka and H. Kawamura, Loss of circadian rhythm in sleep-wakefulness cycle in the rat by suprachiasmatic nucleus lesions, *Brain Res.*, 1975, **96**, 76.
126. J. Lu, Y. H. Zhang, T. C. Chou, S. E. Gaus, J. K. Elmquist, P. Shiromani and C. B. Saper, Contrasting effects of ibotenate lesions of the paraventricular nucleus and subparaventricular zone on sleep-wake cycle and temperature regulation, *J. Neurosci.*, 2001, **21**, 4864.
127. T. C. Chou, T. E. Scammel, J. J. Gooley, S. E. Gaus, C. B. Saper and J. Lu, Critical role of dorsomedial hypothalamic nucleus in a wide range of behavioral circadian rhythms, *J. Neurosci.*, 1069, 2003**23**.
128. M. Y. Cheng, C. M. Bullock, C. Li, A. G. Lee, J. C. Bermak, J. Belluzzi, D. R. Weaver, F. M. Leslie and Q. Y. Zhou, Prokineticin 2 transmits the behavioural circadian rhythm of the suprachiasmatic nucleus, *Nature*, 2002, **417**, 405.
129. A. Kramer, F. C. Yang, P. Snodgrass, X. Li, T. E. Scammell, F. C. Davis and C. J. Weitz, Regulation of Daily Locomotor Activity and Sleep by Hypothalamic EGF Receptor Signaling, *Science*, 2001, **294**, 2511.
130. D. Ono, S. Honma, Y. Nakajima, S. Kuroda, R. Enoki and K. Honma, Dissociation of Per1 and Bmal1 circadian rhythms in the suprachiasmatic nucleus in parallel with behavioral outputs, *Proc. Natl. Acad. Sci. U. S. A.*, 2017, **114**, E3699.

CHAPTER 6

The Circadian–Hypoxia Link

YOSHIMASA OYAMA,[a] LORI A. WALKER[b] AND TOBIAS ECKLE*[c]

[a] Department of Anesthesiology and Intensive Care Medicine, Oita University Faculty of Medicine, Oita, Japan; [b] Division of Cardiology, Department of Medicine, University of Colorado Anschutz Medical Campus, Aurora, CO, USA; [c] Department of Anesthesiology, University of Colorado Denver School of Medicine, Aurora, CO 80045, USA
*Email: tobias.eckle@cuanschutz.edu

6.1 The Circadian–Hypoxia Link

Sunlight and oxygen caused dramatic changes in our environment during evolution. Crucial elements in the evolution of adaptation to sunlight and oxygen were the so-called cyanobacteria. These prokaryotes were some of the first organisms to appear on Earth.[1] When cyanobacteria appeared on Earth, these photosynthetic organisms had to adapt to the consistent oscillations of sunlight and darkness. In doing so, cyanobacteria that developed an internal clock were preferentially selected throughout evolutionary time. Cyanobacteria would therefore express photosynthetic genes in anticipation of a light phase to optimally target metabolic processes during a time when solar resources would be available, compared to an inopportune time when it would be energetically wasteful to express genes involved in photosynthesis.[2,3]

The photosynthetic property of cyanobacteria led to the accumulation of oxygen as a by-product in the atmosphere, which contained minimal to no oxygen. Oceans and rock absorbed much of the initial free oxygen released by cyanobacteria until oxygen gas reached its maximum levels, and it was then absorbed by the land until about 850 million years ago when oxygen gas accumulated in the atmosphere to unprecedented levels, triggering

Chemical Biology No. 23
Chronobiology and Chronomedicine: From Molecular and Cellular Mechanisms to Whole Body Interdigitating Networks
Edited by Germaine Cornelissen and Tsuyoshi Hirota

Published by the Royal Society of Chemistry, www.rsc.org

'The Great Oxygenation Event'.[1] Due to these environmental changes, almost all living organisms have developed molecular mechanisms allowing them to sense light and oxygen.[1–5]

6.1.1 Oxygen-sensing Pathways: Hypoxia Inducible Factor

Throughout the early stages of life on Earth, when 'The Great Oxygenation Event' killed off anaerobic bacteria from this planet, subsequent organisms that appeared and evolved throughout evolutionary time carried with them the adaptive ability to sense environmental oxygen levels in the form of hypoxia-inducible factors (HIFs).[6–9] The ability of cells to adapt to hypoxia is an evolutionarily conserved, innate, and adaptive response that is advantageous for almost all organisms to rely on when environmental conditions demand the cells to do so for survival.

HIFs consist of two subunits, an oxygen-regulated subunit hypoxia-inducible factor 1 α (HIF1A) and a constitutively active subunit hypoxia-inducible factor 1 β (HIF1B).[10] In typical oxygen environments, HIF1A is regulated by oxygen-sensing prolyl hydroxylase domain proteins (PHDs). These PHDs sense the oxygen status of the cell by using iron as a co-factor to reduce oxygen to carbon dioxide in the prolyl hydroxylation of HIF1A. This post-translational modification allows for HIF1A ubiquitination by the von Hippel–Lindau tumor suppressor (pVHL) and subsequent HIF1A degradation. Due to the iron-dependent nature of PHDs, the lack of oxygen thereby inhibits PHD functionality, leading to HIF1A stabilization. In hypoxia, HIF1A enters the nucleus, where it binds to hypoxia response elements (HREs, consensus sequence 5′-RCGTG-3′) in the promoter of hypoxia-dependent genes that are crucial for adaptation to low oxygen availability (Figure 6.1).[11–13] HIF1A and hypoxia-sensing pathways are evolutionarily essential for environmental adaptability to changes in oxygen availability and redox status of the cell.[6,7,9,14–20]

6.1.2 PAS Domains: An Evolutionary Link between Circadian and Hypoxia Pathways

In mammals, light- and oxygen-sensing are interconnected on a molecular level.[7,13,17,21,22] HIF1A is part of the same protein family as specific core circadian proteins such as period (PER, Figure 6.2). It is the superfamily of PAS domain-containing proteins that are sensors for oxygen, light, or metabolism.[7,13,17,22,23] PAS domain-containing proteins are present in all kingdoms of life, *Bacteria*, *Archaea*, and *Eucarya*, and are present in a vast array of proteins and enzymes, including kinases, chemo- and photoreceptors, circadian clock proteins, ion channels, cyclic nucleotide phosphodiesterases, and hypoxia response proteins.[22] The 250–300 amino acid PAS domains are named after the homologous region between the three *Drosophila* genes: period (PER), aryl hydrocarbon receptor nuclear translocator (ARNT), and the single-minded gene homolog (SIM). PER was initially

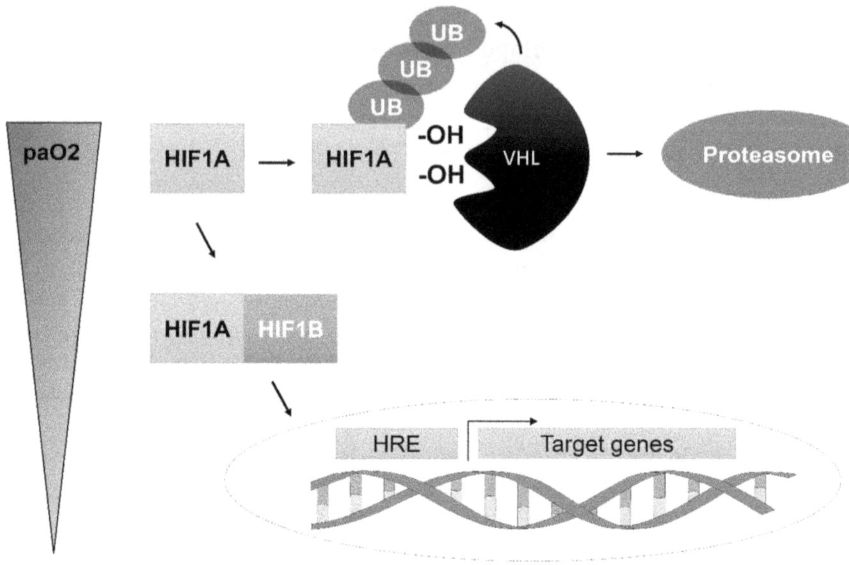

Figure 6.1 HIF1A degradation during normoxia and HIF1A stabilization and transcription during hypoxia. Hypoxia inducible factor 1 α (HIF1A) is ubiquitinated (UB) *via* the Von-Hippel–Lindau protein (VHL) and thereby undergoes proteasomal degradation under normoxia. In hypoxia, HIF1A binds to the constitutively expressed hypoxia inducible factor 1 β (HIF1B), which translocates to the nucleus to bind to hypoxia response elements (HRE) in the promoter regions of its target genes.

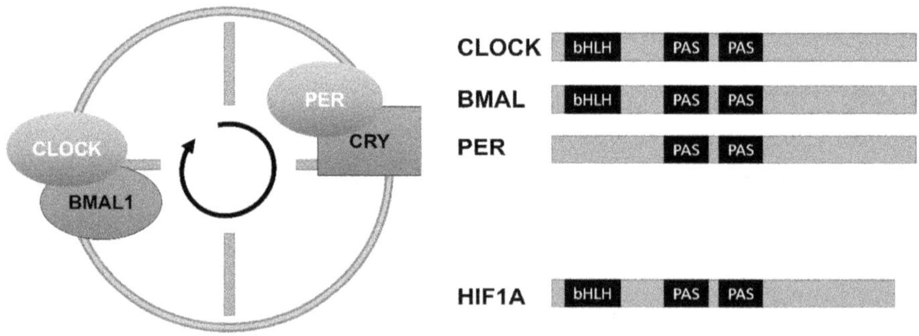

Figure 6.2 The PAS domain connects the circadian system with HIF1A. Core circadian rhythm proteins share the PAS domain with HIF1A.

described in *Drosophila and* identified as a vital regulator of the circadian clock in *Drosophila*. ARNT is an alias for HIF1B, and as mentioned before, it interacts with HIF1A to initiate transcriptional adaptation to hypoxia. While PAS domain proteins have been identified for their function in signal transduction pathways[22] and their ability to bind to basic helix–loop–helix (bHLH) DNA binding domains as transcription factors,[13] some PAS domains can be used for protein–protein interactions. In particular, the PAS domains in PER proteins are thought to be involved in protein–protein interactions, forming heterodimers and homodimers.[13,17,23] Indeed, PER-PAS domains can act as a dimerization surface so that PER proteins can interact at a protein–protein interaction level with other PAS domain-containing proteins.[13,20] As PER2 and HIF1A have both PAS domains, one might assume that they have similar functions in adaptation to environmental changes under oxygen or light conditions.[5]

6.1.3 PER2-HIF1A Interconnection in Cardioprotection from Myocardial Ischaemia

Based on the evolutionarily conserved relationship between HIF1A and PER2, investigations in the field of adenosine-dependent adaptation to low oxygen availability identified a reciprocal relationship between PER2 and HIF1A in cardioprotection. Extracellular adenosine is a signalling molecule critical for cells to adapt to hypoxia.[24–27] In the extracellular compartment, ecto-5′-nucleotidase (CD73) converts 5′-adenosine monophosphate (AMP) to adenosine, which then signals through four distinct adenosine receptors: ADORA1, ADORA2A, ADORA2B or ADORA3.[24,28,29]

The crucial role of extracellular adenosine signalling during low oxygen events has been well established through various studies.[30] For instance, hypoxia or ischaemia conditions significantly enhance extracellular adenosine production and signalling,[31–33] providing protection against hypoxia *via* inhibition of vascular leakage,[34] or hypoxia-driven inflammation.[35,36] Furthermore, adenosine signalling has been well established in providing organ protection from ischaemia.[37–41] As such, studies using pharmacological or genetic approaches to prevent adenosine production or signalling revealed that the cardioprotective responses elicited by cardiac ischaemic preconditioning were entirely abolished.[30,42,43] Cardiac ischaemic preconditioning was discovered in the early 1980s when researchers began to explore nucleotide metabolism during myocardial ischaemia. Here, Murry *et al.* demonstrated in dogs that short, non-lethal ischaemic periods before a longer myocardial ischaemia time resulted in a 70% infarct size reduction when compared to non-preconditioned hearts.[44] After this initial observation, ischaemic precondition has been repeatedly confirmed in various species and organs. As such, ischaemic preconditioning has been recognised as "the strongest form of *in vivo* protection against myocardial ischaemic injury other than early reperfusion".[45]

While all four adenosine receptors have been implicated in mediating the effects of ischaemic preconditioning, the ADORA2B receptor was found to be the most critical for the effects of cardiac ischaemic preconditioning, as evidenced by studies comparing all four adenosine receptors using gene targeted mice.[31] To gain better insight into the mechanisms of ADORA2B-dependent cardio-adaptation to ischaemia, a microarray study from ischaemic-preconditioned murine hearts with genetic ablation of the

ADORA2B receptor was performed.[46] These studies suggested a previously unrecognised role for the circadian rhythm protein PER2 in mediating adenosine-elicited cardio-adaptive responses during ischaemia.[46] Consistent with previous studies implicating the molecular network of circadian rhythm proteins in the regulation of cellular metabolism,[47,48] PER2 was found to initiate metabolic reprogramming of the ischaemic heart *via* an enhanced use of oxygen efficient glycolysis[46] (Figure 6.3).

Further mechanistic studies revealed that ischaemic preconditioning upregulated cardiac PER2 in wildtype mice, which was abolished in *Adora2b*-deficient mice.[46] Furthermore, *Per2* mRNA was found to be upregulated in heart tissue from patients with myocardial ischaemia-induced heart failure.[46] *In vitro* studies using human endothelial cells demonstrated that ADORA2B signalling increased *Per2* mRNA *via* phospho-CREB (cyclic AMP response element binding protein) binding to the PER2 E-box promoter region. Furthermore, it was found that ADORA2B signalling prevented PER2 protein degradation through cullin-1 deneddylation.[46] Considering the known role of HIF1A in cardiac ischaemic preconditioning and adenosine signalling as previously described,[42] continued studies in human endothelial cells overexpressing oxygen-stable HIF1A, with or without PER2 or an ADORA2B agonist, provided insights into metabolic adaptation. Human endothelial cells exposed to an ADORA2B agonist and overexpression of oxygen-stable HIF1A increased the gene expression of glycolytic enzymes, a result that was abolished in *Per2*-deficient cells,[46] suggesting that PER2 is the master regulator of hypoxic adaptation.

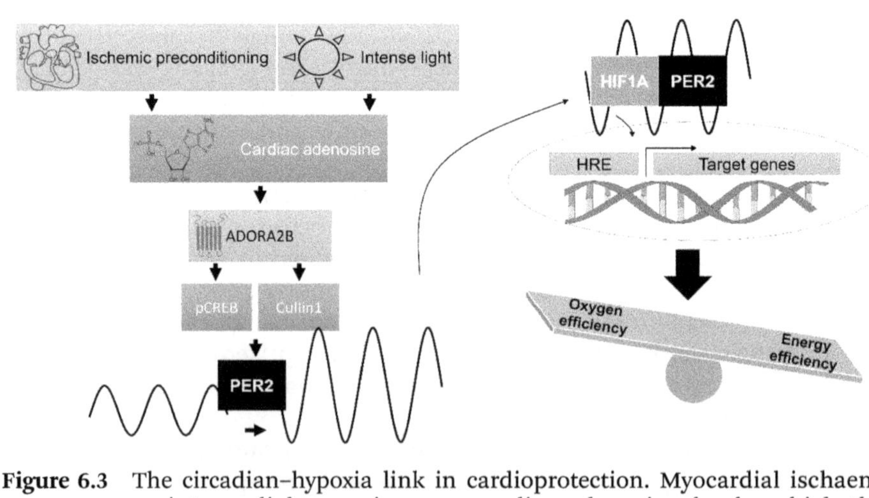

Figure 6.3 The circadian–hypoxia link in cardioprotection. Myocardial ischaemia or intense light can increase cardiac adenosine levels, which then signals *via* the adenosine A2B receptor (ADORA2B). Cardiac ADORA2B signalling then increases PER2 protein and transcript levels *via* phospho CREB or CULLIN1. Increased cardiac PER2 signalling also increases HIF1A signalling, resulting in an oxygen efficient metabolism, which is cardioprotective.

Studies deciphering the connection between PER2 and HIF1A confirmed the hypothesis that PER2 is an upstream regulator of HIF1A. *Per2*$^{-/-}$ mice expressed less *Hif1α* mRNA, a result also found in *Adora2b*$^{-/-}$ mice. Furthermore, HIF1A was not stabilised in *Adora2b*$^{-/-}$ mice upon oxygen depletion, and *Adora2b*$^{-/-}$ mice were also unable to induce the gene expression of glycolytic enzymes during ischaemic preconditioning, an effect similarly seen in *Per2*$^{-/-}$ mice.[46] Furthermore, it was found that cardiac HIF1A protein levels oscillate in a circadian pattern along with protein levels of the HIF1A downstream targets, pyruvate dehydrogenase kinase and lactate dehydrogenase, both key enzymes of glycolysis. Notably, PER2 was necessary for this HIF1A oscillation, and without PER2, HIF1A was not stabilised during the ischaemic preconditioning of the heart.[46] While HIF1A stabilization required PER2 during hypoxia, PER2 circadian oscillation did not need HIF1A, together suggesting that (1) PER2 is dependent on ADORA2B signalling, (2) HIF1A is dependent on PER2 for glycolytic gene expression, and (3) PER2 may be a master regulator of metabolic adaptation to low oxygen conditions.[46]

In support of a circadian oscillation pattern in cardiac HIF1A protein levels dependent on PER2,[46] another research group reported the necessity of the circadian clock for HIF1A's ability to sense the cellular oxygen status.[49] In fact, the circadian clockwork has been found to regulate HIF1A nuclear protein level localization and oscillation.[23] In a direct link between circadian and hypoxia pathways, HIF1A was found at the HRE in the PER2 promoter and its binding enhanced the amplitude of PER2 oscillation.[49] Additionally, ChIP-seq studies revealed that BMAL1-CLOCK binding to an E-box in the HIF1A promoter regulated HIF1A transcription.[50] Indeed, there is a growing body of evidence supporting a bi-directional link between circadian and hypoxia pathways, including PER2-dependent HIF1A binding to HREs in promoter regions of hypoxic target genes from *in vitro* experiments in HeLa cells[51] to zebrafish experiments discovering hypoxia-mediated expression of the PER2 gene through regulating HIF1A binding at E-box regions in promoters.[52] Investigations into this circadian–hypoxia link revealed daily rhythmic oxygen levels in the blood and tissue of rodents within a physiological range that peaked during the dark phase compared to the light phase. Mimicking rhythmic oxygenation fluctuations was sufficient to reset the circadian clock through synchronization in a HIF1A-dependent fashion.[22,53] These studies support the hypothesis that PER2 is an effector protein for HIF1A and suggest a role for HIF1A in the core circadian loop and a bi-directional relationship between light- and oxygen-sensing pathways.

6.2 Circadian Entrainment and Amplitude Enhancement

Maintaining circadian entrainment is critical for overall health and therefore targeting circadian biology to prevent or treat disease is a promising strategy.

A key feature of the circadian system is its synchronization by light.[5,54] In fact, sunlight is the dominant 'Zeitgeber' (time giver) for circadian rhythms and the primary mechanism of synchronization or entrainment. Without light, biological systems revert to their internal free-running clock, which is generally slightly shorter or slightly longer than the 24-hour day but will most likely fade over time, as indicated by isolated human cell lines.[55] In fact, research using non-intrusive techniques to monitor mouse activity up to 470 days has shown that mice exhibit circannual oscillations with an overall decline in activity over time.[56]

Zeitgebers like sunlight synchronize the internal circadian clock to the environment and maintain a 24-hour oscillation. Sunlight activates melanopsin receptors on the retina and signals to the suprachiasmatic nuclei (SCN) *via* the retinohypothalamic tract. In the SCN, the signal is transmitted to the molecular clockwork.[57,58] Melanopsin receptors in retinal ganglion cells are in particular sensitive to blue light. Thus, blue light at 470 nm shifts the circadian phase in a melanopsin-dependent manner.[59] Interestingly, only light with intensities higher than 180 LUX can entrain the human circadian system,[60] whereas intense light (>10 000 LUX) appears to be most effective.

PER2 and other circadian protein levels oscillate in peripheral tissues like the heart, similar to the brain,[46,61] and it is believed that secreted neurohormonal signalling molecules are responsible for this phenomenon.[57,62,63] Interestingly, adenosine and 5′-AMP plasma levels exhibited circadian patterns in studies investigating the effects of light on liver metabolism.[62] Previous investigations into ischaemic preconditioning-mediated cardioprotective mechanisms revealed that adenosine signalling through the ADORA2B receptor was cardioprotective in a PER2-dependent manner.[46] In these studies, intense light was used to stabilize cardiac PER2 and to mimic ischaemic preconditioning of the heart. In follow-up studies, intense light exposure increased cardiac adenosine levels.[64] Furthermore, intense light-mediated cardioprotection was abolished in *Adora2b*$^{-/-}$ mice, and intense light was found to increase the circadian amplitude of PER2 in mice and humans, supporting the notion that increasing the robustness of the circadian system can be organ protective[64] (Figure 6.4).

A form of circadian disruption is the dampening of the circadian amplitude, which is observed as part of the aging process.[65] Therefore, researchers believe that a robust circadian timekeeping system and amplitude enhancement are essential for human health and well-being.[4,5,66–69] Indeed, approaches to increase entrainment and thereby the amplitude of the clock have been found to be beneficial in certain disease states. One study found that restricted feeding increased the circadian amplitude and could prevent mice from becoming obese when exposed to a high-fat diet.[70] Another study found short-term caloric restricted feeding in mice to be cardioprotective. Also, cardiac microRNA profiling of short-term caloric restricted feeding was associated with the circadian clock.[71] Similarly, an independent study tested whether physical activity would restore a disrupted circadian rhythm.

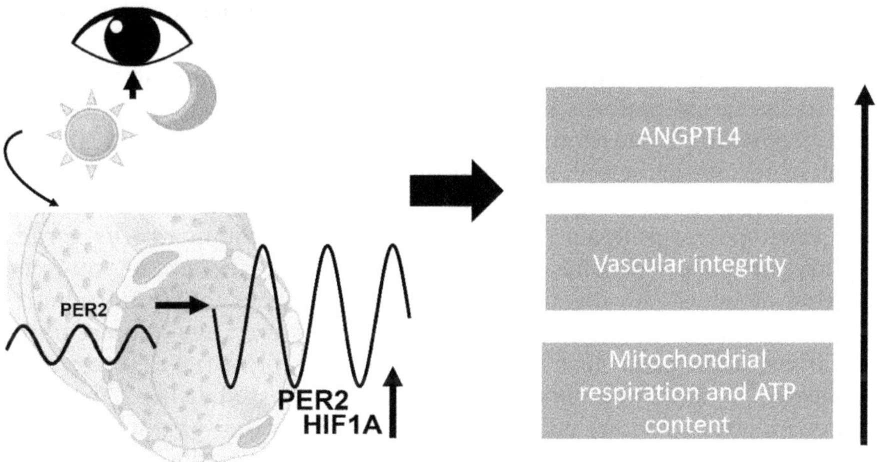

Figure 6.4 Intense light-elicited endothelial PER2 boosts vascular integrity. Intense light increases the amplitude of PER2 and protects the heart from ischaemia *via* endothelial expressed PER2. Intense light-elicited PER2 amplitude enhancement then boosts the vasculature *via* upregulation of the endothelial barrier protective protein ANGPTL4. PER2 further protects the cardiac endothelium *via* improved mitochondrial respiration and increased ATP content under low oxygen conditions.

Here, vasointestinal polypeptide (VIP)-deficient mice exhibiting a disrupted circadian rhythm were studied using a wheel running system.[72] Indeed, voluntary scheduled exercise increased the amplitude of PER2 expression and rescued the disrupted circadian system in VIP-deficient mice. Studies on blue-enriched light exposure in rats found similar effects on circadian rhythmicity and amplitude. Besides, animal health was found to be significantly improved, which was shown through analyses of metabolic and physiological parameters.[73] Similarly, one study demonstrated that physiological cycling of light and darkness is crucial for cardiac remodeling, and disruption of day/night rhythms exacerbated disease pathophysiology.[74] Here, they used a murine model of pressure overload, which caused cardiac hypertrophy, and demonstrated that a 20-hour light cycle caused circadian rhythm disruption associated with disease progression. Using echocardiography, they could prove that circadian rhythm disruption reduced cardiac contractility. Thus, it is compelling that strategies directed towards increasing the robustness of circadian rhythms, such as light exposure, restricted feeding, or scheduled exercise, would be cardioprotective.

Interestingly, entrainment strategies are the same between diurnal and nocturnal animals, like humans and mice. It is a common misperception that a diurnal animal would have circadian clockwork opposite to that of a nocturnal animal. In fact, wheel running, which significantly changes mouse behaviour is still being used as the gold standard to determine day/night activity in rodents. However, modern video camera technology has revealed that mice are not only active at night but also show significant activity during the light phase,[75] suggesting that activity levels are not an ideal output of the circadian rhythm. In addition, certain rodent species will switch from diurnal to nocturnal patterns of activity upon the presentation of a running wheel.[76] Nocturnal and diurnal creatures maintain nearly identical oscillations of the fundamental circadian biology machinery through the same mechanisms to achieve light-elicited entrainment. The difference between these two types of animals lies in how the SCN is coupled to its various outputs, particularly activity levels, behaviour, and eating, among other behavioural activities. We use the behavioural differences to identify whether an animal is diurnal or nocturnal, which is, however, not representative of their circadian phase and therefore does not imply a 180-degree change in phase.[77–80] Metabolic rhythms in the SCN assessed using ^{14}C-labeled 2-deoxyglucose in various animals with either nocturnal or diurnal nature revealed consistent peaks during the day even though the animals have varying levels of activity at different times of the day.[80] Moreover, studies investigating the phase response curve in response to light in both diurnal and nocturnal animals found essentially the same phase relative to the daylight cycle in both kinds of mammals.[80,81]

6.2.1 Intense Light in Circadian Amplitude Enhancement and Cardioprotection

The onset of myocardial infarction is notoriously circadian in nature, with most of the cases occurring in the early morning hours after a long period without daylight,[82] and the severity of infarction is much higher in the morning compared to the evening.[83] During the darker winter with its shorter light periods, it has been found that myocardial infarction cases significantly increase in all US states.[84] While the increase in the winter had initially been attributed to low temperatures, the fact that this phenomenon is observed in all states instead suggests that a lack of efficient circadian entrainment *via* light may be the underlying mechanism.

The full release of epidemiological studies in circadian biology emphasised deciphering underlying mechanisms of this link to ultimately gain understanding and identify targets for the prevention or treatment of cardiovascular diseases.[5,63] In a recent study, intense light increased the circadian amplitude of PER2, which reduced infarct sizes following myocardial ischaemia.[64] Indeed, the mentioned epidemiological studies reinforce the widely accepted association that light is the most dominant Zeitgeber for circadian entrainment in mammals[85] and that this effect on the suprachiasmatic nucleus is synchronised with peripheral organs.[86] It is believed that the light signal targeting the SCN is transmitted to the peripheral organs *via* secreted signalling molecules, or possibly *via* the sympathetic nervous system.[87,88] Early studies investigating the effects of light on liver metabolism discovered that plasma adenosine levels were circadian,[62] suggesting that adenosine could be the signalling molecule released from the SCN to synchronize the peripheral organs.

If mice were housed under intense light conditions instead of ambient light conditions for up to one week (14 h light: 10 h darkness, 10 000 LUX, full-spectrum with UV filter), infarct sizes following myocardial ischaemia decreased, which was, however, abolished in *Per2*$^{-/-}$ mice. Investigation of the most cardioprotective light exposure strategy revealed enhancement of the PER2 amplitude.[89,90] Furthermore, it was shown that PER2 amplitude enhancement was abolished in blind mice and that intense light increased HIF1A transcription in the uninjured heart before any ischaemic event.[64,91] Circadian amplitude enhancement has been found to offer protection in different settings[70,90] and is, thus, an intense research area.[89,92,93] As circadian amplitude enhancement remains a black box, a wide genome array profiling of intense light-dependent cardiac gene expressions was performed. In this array, angiopoietin-like 4 (ANGPTL4) was identified as the gene that is most robustly upregulated by intense light-elicited PER2. Interestingly, ANGPTL4 is known to be upregulated in cardiac endothelial cells *via* HIF1A[94] and to protect endothelial barrier function during myocardial ischaemia (MI) and reperfusion injury.[95] Although endothelial disruption is a recognised critical event in MI[96] leading to functional abnormality, cellular edema, and apoptosis,[97] clinically applicable strategies to protect the endothelial barrier have not been established yet.

Since ANGPTL4 is an endothelial-secreted protein that has been found to maintain vascular integrity during acute myocardial ischaemia and reperfusion (IR) injury,[95,98] mice with an endothelial-specific *Per2* deletion were exposed to myocardial ischaemic injury. Here, endothelial-specific *Per2*$^{-/-}$ mice had significantly increased infarct sizes and abolished light-elicited cardioprotection.[64] Subsequent studies using Evans blue dye to determine the extent of vascular leakage confirmed that intense light considerably improved endothelial barrier function during myocardial ischaemia and reperfusion injury, which, however, was abolished in mice with an endothelial-specific deletion of *Per2*. Thus, these data suggest that intense light can boost vascular function and protect from a subsequent myocardial ischaemic event.[64] Further studies revealed that intense light induced the endothelial barrier protective protein ANGPTL4 in a PER2-dependent manner. Most interestingly, these studies were further able to demonstrate that intense light increased HIF1A binding to the promoter region of *Angptl4*. These data illustrate that intense light-elicited amplitude enhancement of endothelial PER2 boosts vascular integrity *via* induction of HIF1A-ANGPTL4, which is cardioprotective.

Studies on tight junction expression and cell permeability *in vitro* using a stable lentiviral-mediated PER2 knockdown human microvascular endothelial cell line[99] revealed reduced HIF1A-transcription of the tight junction protein claudin-1 and increased cell permeability during hypoxia.[64]

Affinity purification–mass spectrometry-based proteomic analysis of endothelial PER2 protein interactions under hypoxic conditions further revealed a prominent metabolic role for PER2 and suggested that PER2 controls mitochondrial function *via* mitochondrial complex 4 under hypoxia.[64]

As ATP levels play a critical role in maintaining a functional endothelial barrier,[100–102] it was hypothesised that endothelial PER2 controlled mitochondrial oxygen consumption rates and, consequently, ATP production under hypoxia. Indeed, oxygen consumption rates and thereby ATP production were significantly reduced in human endothelial cells with a PER2 knockdown. Similar to studies showing that HIF1A increases mitochondrial complex 4 subunit COX4.2 over COX4.1 in hypoxia to enhance oxygen efficiency and provide adaptation to hypoxia by means of maintaining ATP levels,[103] hypoxia-mediated increases of *COX4.2* mRNA were abolished in PER2 knockdown endothelial cells. Furthermore, myocardial ischaemia-induced increases in complex 4 activity were dampened in $Per2^{-/-}$ mice.

Previous studies identified that PER2 overexpression increased the hypoxia-mediated transactivation and binding of HIF1A to the promoter region of its target genes.[51] Published data also demonstrated abolished HIF1A binding to the promoter region of its target gene LDHA (lactate dehydrogenase A) in hypoxic PER2 knockdown endothelial cells.[64] Interestingly, it could be shown that global HIF1A-HRE (hypoxia response element) binding following intense light-mediated amplitude enhancement of PER2 was significantly increased in murine cardiac tissue. Intense light also significantly increased HIF1A binding to the *Angptl4* promoter region and significantly increased the circadian amplitude of cardiac *Claudin-1* or *Cox4.2* mRNA levels.[64] These studies illustrate that PER2 amplitude enhancement *via* intense light is a critical control point of HIF1A-ANGPTL4-regulated pathways to maintain vascular integrity in myocardial IR-injury.

6.3 Light as Circadian Rhythm Targetting Therapy in Humans

Intense or bright light therapy [10 000 LUX] has been used to treat seasonal mood disorders in humans for a long time.[104] Recent studies further investigated the impact of intense light on healthy humans.[64] Healthy human volunteers were exposed to 30 min of intense light in the morning from 08:00 until 08:30 for five consecutive days. Each day, blood was drawn after light therapy at 08:30. As control, plasma samples were collected in the morning from human volunteers without any light treatment. These studies revealed that intense light was able to increase PER2 protein levels in collected buccal or plasma samples at 09:00 or 21:00. The increase in the evening was higher than in the morning. It showed (a) that PER2 induction was possible after a more extended period following light exposure and (b) that light probably increased the circadian amplitude of PER2. Interestingly this phenomenon was found in two different tissues simultaneously (plasma and buccal swaps).[64]

As the efficacy of intense light on the circadian system is measured through melatonin levels,[60] melatonin plasma levels were also analyzed. These analyses revealed that intense light exposure in the morning significantly suppressed plasma melatonin levels immediately following light therapy at 08:30, which was not seen in the control group exposed to room light at the same time.[64]

Plasma samples from healthy human volunteers were further analyzed, and it was found that intense light significantly decreased plasma triglycerides. Plasma triglycerides are a clinical marker for insulin sensitivity and functional carbohydrate metabolism.[105] Since light reduced plasma triglyceride levels, these studies suggested that intense light therapy could increase insulin sensitivity and improve glucose metabolism. A targeted metabolomics screen was performed on these human plasma samples to gain a better insight into the effects of intense light on human metabolism. These studies found that intense light regulated metabolic pathways. Critical intermediates of glycolysis or the TCA cycle were significantly decreased, indicating that light eventually increased metabolism.[64]

It has been well described that there is a strong correlation between sleep deprivation and decreased insulin sensitivity, resulting in a diabetic phenotype in young, healthy human volunteers.[69] Thus, if intense light therapy could also improve sleep, this would further support the notion that intense light could ameliorate a diabetic phenotype in humans. In the aforementioned studies, the effects of light therapy on sleep were studied using actigraphy.[106] Here, the studies demonstrated that intense light was able to reduce wake-up episodes after the sleep onset, which resulted in overall improved sleep efficiency.[64] Further actigraphy analyses revealed that intense light increased healthy human volunteers' activity during the day, suggesting that light indeed increased the circadian amplitude. Although these studies indicate that intense light may present a promising strategy for the treatment of metabolic disorders, it remains unclear if intense light would also be effective in treating myocardial ischaemia.

6.3.1 Light in the Regulation of Circadian MicroRNAs in Humans

Studies have indicated a role for circadian microRNAs in cardioprotection. However, in general, studies on cardiac circadian microRNAs are rare.[71] One study investigated the effect of cardiac ischaemic preconditioning on PER2-dependent microRNAs and found miR-21 to be one of the top PER2-regulated microRNAs.[91] Subsequent *in vitro* studies demonstrated that miR-21 was essential for regulating PER2- and HIF1A-dependent anaerobic glycolysis.[46] In fact, myocardial ischaemia heavily relies on anaerobic glycolysis, which remains the only energy source during low oxygen conditions. It has been shown that anaerobic glycolysis allows the heart to function during ischaemia, and if anaerobic glycolysis is inhibited, immediate myocardial cell death will occur.[107]

Further analysis of miR-21 in these processes revealed that *miR-21*$^{-/-}$ mice had larger infarct sizes following myocardial ischaemia and reperfusion injury.[46] Furthermore, intense light elicited cardioprotection, and PER2-dependent cardioprotection was abolished in *miR-21*$^{-/-}$ mice. As suggested by PER2-dependent regulation, miR-21 was also found to be upregulated upon intense light exposure in wild type mice.

Interestingly, the studies mentioned above on miR-21 also used intense light in healthy humans. Here, intense light exposure for 30 minutes over five days in the morning between 08:00 and 08:30 increased miR-21 or PER2-dependent phosphofructokinase activity in plasma samples.[91] These findings illustrate that light could become a promising strategy to activate PER2-elicited cardioprotection[46] *via* circadian amplitude enhancement and activation of cardioprotective pathways such as miR-21.[108,109] More detailed studies on intense light therapy in humans will hopefully enable the dissection of these mechanisms further.

6.3.2 Intense Light as Anticoagulant Therapy in Humans

Acute coronary thrombosis is a common cause of myocardial infarction.[110] It is well described in patients with heart failure or patients with coronary artery disease. The involvement of circadian rhythms in thrombosis generation has been reported in mice and men. Still, many studies in mice are not well defined due to the lack of tissue-specific studies.[111,112]

Indeed, a study using a mouse model with tissue-specific deletion of *Per2* in the megakaryocyte lineage demonstrated that *Per2*-deficient platelets aggregate faster than platelets from wild type animals.[113] These studies also found increased infarct sizes following myocardial ischaemia in mice with a *Per2* deletion in the megakaryocyte lineage. Furthermore, protein array analyses using plasma from healthy human volunteers, who were exposed to intense light for five days for 30 minutes in the morning (08:00–08:30), revealed that light inhibited procoagulant pathways. Also, using an aggregometer, it was found that intense light exposure reduced the clotting rate in humans.[113]

In contrast, most other studies on the role of circadian rhythm proteins in platelet function or thrombus formation have used whole body knockout mice. For instance, it has been reported that *Per2*$^{-/-}$ mice had reduced platelet counts, and platelets were compromised in their ability to aggregate.[114] Another study demonstrated that the expression of thrombopoietin followed a circadian pattern,[111] and *Clock* mutant mice were found to have increased thrombopoietin, a significant increase in megakaryocyte numbers, and significantly higher platelet counts than wild type controls. Unfortunately, *Clock*-deficient platelets were not analyzed in these studies. In line with the study on *Clock* mutant mice, whole-body *Bmal1*$^{-/-}$ showed enhanced platelet aggregation upon ADP stimulation.[115] Regardless, these data emphasize that results from whole-body null mice affecting circadian core clock proteins are challenging to interpret and that future effort should be directed toward analyzing the tissue-specific role of circadian proteins in disease development.

The importance of light as a regulator of the circadian system has been well described.[4,5,63,64,91] Interestingly, studies on platelet turnover found that megakaryopoiesis is regulated by light *via* the circadian machinery in the suprachiasmatic nuclei.[116] As such, it is not fully surprising that intense light creates an anti-thrombotic signature in plasma samples and that the clotting rate, which is platelet-dependent, is significantly reduced after intense light therapy in healthy humans. Despite these promising initial results from healthy humans, further research will be necessary to fully understand the intense light mechanisms in inhibiting coagulation.

6.4 Intense Light Mimicking Drugs

Adenosine-3′, 5′- cyclic monophosphate (cAMP) signalling is an essential part of the oscillatory network.[117] cAMP signalling is rhythmic and strengthens the transcriptional loop of the suprachiasmatic nuclei. This characteristic ultimately regulates the amplitude, phase, and period of circadian rhythms. It became evident in studies where cAMP was a vital regulator of the *Per2* transcript and PER2 protein.

ADORA2B adenosine receptors, like other Gs-protein-coupled receptors, signal mainly *via* cAMP-dependent protein kinase A (PKA). Signalling downstream from PKA proceeds through phosphorylation of the transcription factor CREB, resulting in its activation. Activated CREB can mediate gene expression directly by interacting with gene promoters. Indeed, it has been found that (1) a specific ADORA2B agonist mimics cardiac ischaemic preconditioning,[31] (2) PER2 is central to ADORA2B-mediated cardiac ischaemic preconditioning[46] and (3) light elicited cardioprotection and PER2-dependent cardioprotection require the ADORA2B receptor.[64] Interestingly, forskolin, a cAMP-increasing drug, has long been used as a light mimicking drug *in vitro* to synchronize and enhance circadian rhythms. As such, drugs capable of increasing cAMP signalling, such as ADORA2B agonists, might be exciting candidates to mimic the beneficial effects of light in humans.

Based on a comprehensive search for drugs that could further manipulate the circadian rhythm, several circadian amplitude enhancers have been identified.[90] One promising drug is flavonoid nobiletin, which has been found to enhance the circadian amplitude and restore a metabolic syndrome in obese mice. Nobiletin was further found to be cardioprotective in a PER2-dependent manner.[118] Future research will have to evaluate the impact of such drugs that directly affect the circadian machinery and assess positive and negative effects.

6.5 Circadian Rhythms in the Cardiovascular System

While the hypoxia–circadian link reveals a deeper understanding of evolutionary developed circadian mechanisms in the heart, circadian rhythms play an important role in the control of many other physiological and pathophysiological functions of the heart. In fact, almost all cardiovascular functions have a day/night pattern, including blood pressure,[119] heart rate,[120] circulating catecholamines,[121] and blood coagulation markers,[122] and vascular endothelial function.[123] Moreover, epidemiological data revealed a robust morning increase in adverse cardiovascular events, including stroke,[124] myocardial infarction,[82] serious ventricular arrhythmias,[125] and sudden cardiac death.[126] Furthermore, approximately 13% of genes and approximately 8% of proteins are rhythmic in the heart.[127–129]

However, while many cardiac functions such as blood pressure or the heart rate have a 24-hour pattern, there are no clean genetic mouse models revealing that these mechanisms are under the direct control of the circadian machinery. Just because the mammalian system is circadian, it is difficult to know if a cardiac function is indeed circadian-controlled or simply the output of a circadian system. For instance, is the heart rate slower at night because certain circadian proteins regulate the heart rate or is the heart rate slow as the metabolism is under the direct control of the circadian system,[130] which ultimately slows down the heart rate? In a study using *Bmal1*$^{-/-}$ mice, it was found that the 24-hour variation in the heart rate and blood pressure was lost in those animals. Remarkably, these genetic models revealed slower heart rates and lower blood pressures than wildtype controls.[131] Considering circadian disruption and circadian deficiency being involved in the development of cardiovascular disease, the observed phenotype would rather support protection of the cardiovascular system than being harmful. Regardless, additional mechanistic studies identified that the clock regulates the heart rate and blood pressure by modulating the synthesis and disposition of catecholamines, suggesting that the heart rate and blood pressure are more indirectly controlled *via* neurohormonal signalling pathways.

In general, mice lacking the core circadian clock gene *Bmal1* are behaviourally arrhythmic and die prematurely. In fact, *Bmal1*$^{-/-}$ mice developed age-associated dilated cardiomyopathy, which was associated with myofilament disorganization.[132] These studies illustrated the crucial role of BMAL1 in the heart. In fact, studies have shown that BMAL1 is critical for normal mitochondrial activity and cardiac function using human embryonic stem cells.[133] Studies using heart specific *Bmal1*$^{-/-}$ mice also found significant cardiac mitochondrial defects and further confirmed severe progressive heart failure with age.[134]

As indicated by the *Bmal1* deficient mouse model, the circadian clock not only regulates the blood pressure or heart rate but also plays a central role in cardiac metabolism. This in fact has been highlighted over the last two decades using clock mutant mice but also has been repeatedly confirmed using other genetic circadian mouse models.[135] As such, mice that overexpressed a dominant negative *Clock* mutant exhibited a reduced fatty acid response of the heart during fasting.[136] Cardiomyocyte-specific circadian *Clock* mutant mice had increased myocardial oxygen consumption and fatty acid oxidation rates but decreased cardiac efficiency.[137] Furthermore, cardiomyocyte *Clock* mutant mice had a changed myocardial response to a chronic high-fat diet. In these studies, it was found that the cardiomyocyte clock inactivates a hormone-sensitive lipase during the active/awake phase both at the transcriptional and post-translational level, ultimately leading to marked cardiac steatosis.[138] Like *Bmal1*$^{-/-}$ mice, *Clock* mutant mice also developed cardiovascular disease with age resulting in cardiac hypertrophy and fibrosis.[139] Also, heart-specific *Bmal1*-deficient mice were not only found to have a decreased cardiac function but to exhibit systemic insulin resistance.[140] Collectively, studies have demonstrated that BMAL1 influences metabolism signalling, and contractile function of the heart.[134] While CLOCK and BMAL1 are clearly central to cardiac metabolism, PER2 has also repeatedly been found to regulate cardiac glucose and lipid metabolism.[46,64,141]

In addition to these physiological functions, several genetic models point towards a strong circadian involvement under pathophysiological conditions. As such, it was found that mice that are deficient in circadian proteins cryptochrome 1 and cryptochrome 2 exhibit abnormal production of the mineralocorticoid aldosterone. While these mice did not show any alterations in blood pressures at the baseline, a high-salt diet consisting of 3.15% sodium resulted in a significant blood pressure increase.[142]

Krüppel-like factor 15 (Klf15), a transcriptional regulator, was identified to be under circadian control and to target KChIP2, the regulatory β-subunit for the repolarizing transient outward potassium current in the heart. While Klf15-null mice did not exhibit spontaneous arrhythmias on ECG telemetry, intra-cardiac programmed electrical stimulation revealed a marked increase in ventricular arrhythmias when compared to wildtype mice.[143] These studies were the first to suggest KLF15 as a molecular mechanism for sudden cardiac death, which exhibits a strong 24-hour occurrence pattern.

Per2-deficient mice suffer from multiple abnormalities but do not exhibit a significant cardiovascular phenotype at the baseline. However, *Per2*-deficient mice exposed to hind limb ischaemia had impaired blood flow recovery and developed auto-amputation of the distal limb. These findings were attributed to an impaired endothelial progenitor cell function in *Per2*$^{-/-}$ mice.[144] Endothelial-specific Per2 knockout animals also showed increased vascular permeability and larger infarct sizes following myocardial ischaemia.[64] Furthermore, conditional deletion of *Bmal1* in the endothelium resulted in increased microvascular and macrovascular injury upon an ischaemic insult.[145] In fact, endothelial cells of *Bmal1*$^{-/-}$ or *Per2*$^{-/-}$ mice have significantly decreased endothelial nitric oxide synthase activation and, consequently, reduced nitric oxide production and increased superoxide levels.[146,147]

While there are many epidemiological studies that align with the above findings in mice, a genetic equivalent in men has yet to be identified. In fact, while putative deficiency of circadian genes has been labeled as 'circadian disruption' in humans, no technology exists yet that is able to determine such a state with certainty in a clinical setting.

Regardless, the evidence that circadian rhythms are crucial for heart health is striking. Furthermore, the evidence regarding circadian amplitude enhancement and its beneficial effects on human health cannot be ignored. However, future studies using intense light or circadian amplitude enhancing drugs will be necessary to further decipher these mechanisms in humans. Incorporating novel medical technologies such as mRNA therapies, CRIPSR, RNA-Seq, precision medicine, or even health wearables will be necessary in these endeavors to finally provide insights into novel cardioprotective therapies through targeting circadian rhythms.

References

1. A. L. Zerkle, S. W. Poulton, R. J. Newton, C. Mettam, M. W. Claire, A. Bekker and C. K. Junium, *Nature*, 2017, **542**, 465–467.
2. S. B. Gaudana, S. Krishnakumar, S. Alagesan, M. G. Digmurti, G. A. Viswanathan, M. Chetty and P. P. Wangikar, *Front. Microbiol.*, 2013 **4**, 374.
3. R. Saha, D. Liu, A. Hoynes-O'Connor, M. Liberton, J. Yu, M. Bhattacharyya-Pakrasi, A. Balassy, F. Zhang, T. S. Moon, C. D. Maranas and H. B. Pakrasi, *mBio*, 2016, **7**, 1–14.
4. J. Brainard, M. Gobel, K. Bartels, B. Scott, M. Koeppen and T. Eckle, *Semin. Cardiothorac. Vasc. Anesth.*, 2015, **19**, 49–60.
5. J. Brainard, M. Gobel, B. Scott, M. Koeppen and T. Eckle, *Anesthesiology*, 2015, **122**, 1170–1175.
6. G. L. Semenza, *Science*, 2007, **318**, 62–64.
7. B. E. McIntosh, J. B. Hogenesch and C. A. Bradfield, *Annu. Rev. Physiol.*, 2010, **72**, 625–645.
8. G. L. Semenza, *J. Mol. Med.*, 2007, **85**, 1293–1294.
9. G. L. Semenza, *J. Appl. Physiol.*, 2000, **88**, 1474–1480.
10. H. K. Eltzschig and P. Carmeliet, *N. Engl. J. Med.*, 2011, **364**, 656–665.
11. G. L. Semenza, B. H. Jiang, S. W. Leung, R. Passantino, J. P. Concordet, P. Maire and A. Giallongo, *J. Biol. Chem.*, 1996, **271**, 32529–32537.
12. R. H. Wenger and M. Gassmann, *Biol. Chem.*, 1997, **378**, 609–616.
13. Y. Z. Gu, J. B. Hogenesch and C. A. Bradfield, *Annu. Rev. Pharmacol. Toxicol.*, 2000, **40**, 519–561.
14. G. L. Semenza, *Exp. Physiol.*, 2007, **92**, 988–991.
15. G. L. Semenza, *Biochem. J.*, 2007, **405**, 1–9.
16. G. L. Semenza, *Sci. STKE*, 2007, **2007**, cm8.
17. J. B. Hogenesch, Y. Z. Gu, S. Jain and C. A. Bradfield, *Proc. Natl. Acad. Sci. U. S. A.*, 1998, **95**, 5474–5479.
18. L. E. Huang, Z. Arany, D. M. Livingston and H. F. Bunn, *J. Biol. Chem.*, 1996, **271**, 32253–32259.
19. S. Salceda and J. Caro, *J. Biol. Chem.*, 1997, **272**, 22642–22647.
20. N. S. Chandel, E. Maltepe, E. Goldwasser, C. E. Mathieu, M. C. Simon and P. T. Schumacker, *Proc. Natl. Acad. Sci. U. S. A.*, 1998, **95**, 11715–11720.
21. R. S. Edgar, E. W. Green, Y. Zhao, G. van Ooijen, M. Olmedo, X. Qin, Y. Xu, M. Pan, U. K. Valekunja, K. A. Feeney, E. S. Maywood, M. H. Hastings, N. S. Baliga, M. Merrow, A. J. Millar, C. H. Johnson, C. P. Kyriacou, J. S. O'Neill and A. B. Reddy, *Nature*, 2012, **485**, 459–464.

22. B. L. Taylor and I. B. Zhulin, *Microbiol. Mol. Biol. Rev.*, 1999, **63**, 479–506.
23. E. L. Button, D. C. Bersten and M. L. Whitelaw, *Exp. Cell Res.*, 2017, **356**, 141–145.
24. B. B. Fredholm, *Cell Death Differ.*, 2007, **14**, 1315–1323.
25. A. Ohta and M. Sitkovsky, *Nature*, 2001, **414**, 916–920.
26. M. V. Sitkovsky, D. Lukashev, S. Apasov, H. Kojima, M. Koshiba, C. Caldwell, A. Ohta and M. Thiel, *Annu. Rev. Immunol.*, 2004, **22**, 657–682.
27. M. Thiel, A. Chouker, A. Ohta, E. Jackson, C. Caldwell, P. Smith, D. Lukashev, I. Bittmann and M. V. Sitkovsky, *PLoS Biol.*, 2005, **3**, e174.
28. H. K. Eltzschig, *Anesthesiology*, 2009, **111**, 904–915.
29. G. Hasko, J. Linden, B. Cronstein and P. Pacher, *Nat. Rev. Drug Discovery*, 2008, **7**, 759–770.
30. H. K. Eltzschig, *Anesthesiology*, 2009, **111**, 904–915.
31. T. Eckle, T. Krahn, A. Grenz, D. Kohler, M. Mittelbronn, C. Ledent, M. A. Jacobson, H. Osswald, L. F. Thompson, K. Unertl and H. K. Eltzschig, *Circulation*, 2007, **115**, 1581–1590.
32. T. Eckle, A. Grenz, S. Laucher and H. K. Eltzschig, *J. Clin. Invest.*, 2008, **118**, 3301–3315.
33. H. K. Eltzschig, J. C. Ibla, G. T. Furuta, M. O. Leonard, K. A. Jacobson, K. Enjyoji, S. C. Robson and S. P. Colgan, *J. Exp. Med.*, 2003, **198**, 783–796.
34. T. Eckle, M. Faigle, A. Grenz, S. Laucher, L. F. Thompson and H. K. Eltzschig, *Blood*, 2008, **111**, 2024–2035.
35. H. K. Eltzschig, L. F. Thompson, J. Karhausen, R. J. Cotta, J. C. Ibla, S. C. Robson and S. P. Colgan, *Blood*, 2004, **104**, 3986–3992.
36. P. Rosenberger, J. M. Schwab, V. Mirakaj, E. Masekowsky, A. Mager, J. C. Morote-Garcia, K. Unertl and H. K. Eltzschig, *Nat. Immunol.*, 2009, **10**, 195–202.
37. A. Grenz, H. Osswald, T. Eckle, D. Yang, H. Zhang, Z. V. Tran, K. Klingel, K. Ravid and H. K. Eltzschig, *PLoS Med.*, 2008, **5**, e137.
38. M. L. Hart, C. Much, I. C. Gorzolla, J. Schittenhelm, D. Kloor, G. L. Stahl and H. K. Eltzschig, *Gastroenterology*, 2008, **135**, 1739–1750 e1733.
39. C. M. Lappas, Y. J. Day, M. A. Marshall, V. H. Engelhard and J. Linden, *J. Exp. Med.*, 2006, **203**, 2639–2648.
40. J. Linden, *J. Clin. Invest.*, 2006, **116**, 1835–1837.
41. Z. Yang, Y. J. Day, M. C. Toufektsian, S. I. Ramos, M. Marshall, X. Q. Wang, B. A. French and J. Linden, *Circulation*, 2005, **111**, 2190–2197.
42. T. Eckle, D. Kohler, R. Lehmann, K. El Kasmi and H. K. Eltzschig, *Circulation*, 2008, **118**, 166–175.
43. D. Kohler, T. Eckle, M. Faigle, A. Grenz, M. Mittelbronn, S. Laucher, M. L. Hart, S. C. Robson, C. E. Muller and H. K. Eltzschig, *Circulation*, 2007, **116**, 1784–1794.
44. C. E. Murry, R. B. Jennings and K. A. Reimer, *Circulation*, 1986, **74**, 1124–1136.
45. R. A. Kloner, R. Bolli, E. Marban, L. Reinlib and E. Braunwald, *Circulation*, 1998, **97**, 1848–1867.

46. T. Eckle, K. Hartmann, S. Bonney, S. Reithel, M. Mittelbronn, L. A. Walker, B. D. Lowes, J. Han, C. H. Borchers, P. M. Buttrick, D. J. Kominsky, S. P. Colgan and H. K. Eltzschig, *Nat. Med.*, 2012, **18**, 774–782.
47. R. D. Rudic, P. McNamara, A. M. Curtis, R. C. Boston, S. Panda, J. B. Hogenesch and G. A. Fitzgerald, *PLoS Biol.*, 2004, **2**, e377.
48. F. W. Turek, C. Joshu, A. Kohsaka, E. Lin, G. Ivanova, E. McDearmon, A. Laposky, S. Losee-Olson, A. Easton, D. R. Jensen, R. H. Eckel, J. S. Takahashi and J. Bass, *Science*, 2005, **308**, 1043–1045.
49. C. B. Peek, D. C. Levine, J. Cedernaes, A. Taguchi, Y. Kobayashi, S. J. Tsai, N. A. Bonar, M. R. McNulty, K. M. Ramsey and J. Bass, *Cell Metab.*, 2017, **25**, 86–92.
50. Y. Wu, D. Tang, N. Liu, W. Xiong, H. Huang, Y. Li, Z. Ma, H. Zhao, P. Chen, X. Qi and E. E. Zhang, *Cell Metab.*, 2017, **25**, 73–85.
51. M. Kobayashi, A. Morinibu, S. Koyasu, Y. Goto, M. Hiraoka and H. Harada, *FEBS J.*, 2017, **284**, 3804–3816.
52. M. Egg, L. Koblitz, J. Hirayama, T. Schwerte, C. Folterbauer, A. Kurz, B. Fiechtner, M. Most, W. Salvenmoser, P. Sassone-Corsi and B. Pelster, *Chronobiol. Int.*, 2013, **30**, 510–529.
53. Y. Adamovich, B. Ladeuix, M. Golik, M. P. Koeners and G. Asher, *Cell Metab.*, 2017, **25**, 93–101.
54. V. Jakubcakova, H. Oster, F. Tamanini, C. Cadenas, M. Leitges, G. T. van der Horst and G. Eichele, *Neuron*, 2007, **54**, 831–843.
55. S. M. Reppert and D. R. Weaver, *Nature*, 2002, **418**, 935–941.
56. K. Pernold, E. Rullman and B. Ulfhake, *Sci. Rep.*, 2021, **11**, 4961.
57. J. S. Takahashi, H. K. Hong, C. H. Ko and E. L. McDearmon, *Nat. Rev. Genet.*, 2008, **9**, 764–775.
58. A. D. Guler, J. L. Ecker, G. S. Lall, S. Haq, C. M. Altimus, H. W. Liao, A. R. Barnard, H. Cahill, T. C. Badea, H. Zhao, M. W. Hankins, D. M. Berson, R. J. Lucas, K. W. Yau and S. Hattar, *Nature*, 2008, **453**, 102–105.
59. V. Pilorz, S. K. Tam, S. Hughes, C. A. Pothecary, A. Jagannath, M. W. Hankins, D. M. Bannerman, S. L. Lightman, V. V. Vyazovskiy, P. M. Nolan, R. G. Foster and S. N. Peirson, *PLoS Biol.*, 2016, **14**, e1002482.
60. A. J. Lewy, T. A. Wehr, F. K. Goodwin, D. A. Newsome and S. P. Markey, *Science*, 1980, **210**, 1267–1269.
61. Z. Bendova and A. Sumova, *Physiol. Res.*, 2006, **55**, 623–632.
62. J. Zhang, K. Kaasik, M. R. Blackburn and C. C. Lee, *Nature*, 2006, **439**, 340–343.
63. S. Bonney, K. Hughes, P. N. Harter, M. Mittelbronn, L. Walker and T. Eckle, *Int. J. Biochem. Cell Biol.*, 2013, **45**, 667–671.
64. Y. Oyama, C. M. Bartman, S. Bonney, J. S. Lee, L. A. Walker, J. Han, C. H. Borchers, P. M. Buttrick, C. M. Aherne, N. Clendenen, S. P. Colgan and T. Eckle, *Cell Rep.*, 2019, **28**, 1471–1484 e1411.

65. A. M. Schroeder and C. S. Colwell, *Trends Pharmacol. Sci.*, 2013, **34**, 605–619.
66. H. K. Ritchie, E. R. Stothard and K. P. Wright, *Curr. Pharm. Des.*, 2015, **21**, 3438–3442.
67. D. M. Arble, K. M. Ramsey, J. Bass and F. W. Turek, *Best Pract. Res., Clin. Endocrinol. Metab.*, 2010, **24**, 785–800.
68. R. G. Foster, S. N. Peirson, K. Wulff, E. Winnebeck, C. Vetter and T. Roenneberg, *Prog. Mol. Biol. Transl. Sci.*, 2013, **119**, 325–346.
69. C. M. Depner, E. R. Stothard and K. P. Wright Jr, *Curr. Diabetes Rep.*, 2014, **14**, 507.
70. M. Hatori, C. Vollmers, A. Zarrinpar, L. DiTacchio, E. A. Bushong, S. Gill, M. Leblanc, A. Chaix, M. Joens, J. A. Fitzpatrick, M. H. Ellisman and S. Panda, *Cell Metab.*, 2012, **15**, 848–860.
71. H. Noyan, O. El-Mounayri, R. Isserlin, S. Arab, A. Momen, H. S. Cheng, J. Wu, T. Afroze, R. K. Li, J. E. Fish, G. D. Bader and M. Husain, *PLoS One*, 2015, **10**, e0130658.
72. A. M. Schroeder, D. Truong, D. H. Loh, M. C. Jordan, K. P. Roos and C. S. Colwell, *J. Physiol.*, 2012, **590**, 6213–6226.
73. R. T. Dauchy, M. A. Wren-Dail, A. E. Hoffman, J. P. Hanifin, B. Warfield, G. C. Brainard, S. M. Hill, V. P. Belancio, E. M. Dauchy and D. E. Blask, *Comp. Med.*, 2016, **66**, 373–383.
74. T. A. Martino, N. Tata, D. D. Belsham, J. Chalmers, M. Straume, P. Lee, H. Pribiag, N. Khaper, P. P. Liu, F. Dawood, P. H. Backx, M. R. Ralph and M. J. Sole, *Hypertension*, 2007, **49**, 1104–1113.
75. R. S. Bains, S. Wells, R. R. Sillito, J. D. Armstrong, H. L. Cater, G. Banks and P. M. Nolan, *J. Neurosci. Methods*, 2018, **300**, 37–47.
76. M. J. Kas and D. M. Edgar, *J. Neurosci.*, 1999, **19**, 328–333.
77. A. M. Vosko, M. H. Hagenauer, D. L. Hummer and T. M. Lee, *Am. J. Physiol.*, 2009, **296**, R353–R361.
78. L. Smale, T. Lee and A. A. Nunez, *J. Biol. Rhythms*, 2003, **18**, 356–366.
79. N. Kronfeld-Schor, G. Bloch and W. J. Schwartz, *Proc. Biol. Sci.*, 2013, **280**, 20131354.
80. L. Smale, A. A. Nunez and M. D. Schwartz, *Biol. Rhythm Res.*, 2008, **39**, 305–318.
81. M. Mahoney, A. Bult and L. Smale, *J. Biol. Rhythms*, 2001, **16**, 149–162.
82. J. E. Muller, P. H. Stone, Z. G. Turi, J. D. Rutherford, C. A. Czeisler, C. Parker, W. K. Poole, E. Passamani, R. Roberts and T. Robertson, *et al.*, *N. Engl. J. Med.*, 1985, **313**, 1315–1322.
83. A. Suarez-Barrientos, P. Lopez-Romero, D. Vivas, F. Castro-Ferreira, I. Nunez-Gil, E. Franco, B. Ruiz-Mateos, J. C. Garcia-Rubira, A. Fernandez-Ortiz, C. Macaya and B. Ibanez, *Heart*, 2011, **97**, 970–976.
84. F. A. Spencer, R. J. Goldberg, R. C. Becker and J. M. Gore, *J. Am. Coll. Cardiol.*, 1998, **31**, 1226–1233.
85. T. Roenneberg, C. J. Kumar and M. Merrow, *Curr. Biol.*, 2007, **17**, R44–45.

86. R. Zhang, N. F. Lahens, H. I. Ballance, M. E. Hughes and J. B. Hogenesch, *Proc. Natl. Acad. Sci. U. S. A.*, 2014, **111**, 16219–16224.
87. J. S. Takahashi, *Nat. Rev. Genet.*, 2017, **18**, 164–179.
88. D. Yuan, R. D. Collage, H. Huang, X. Zhang, B. C. Kautza, A. J. Lewis, B. S. Zuckerbraun, A. Tsung, D. C. Angus and M. R. Rosengart, *Proc. Natl. Acad. Sci. U. S. A.*, 2016, **113**, 5239–5244.
89. H. Wang, E. van Spyk, Q. Liu, M. Geyfman, M. L. Salmans, V. Kumar, A. Ihler, N. Li, J. S. Takahashi and B. Andersen, *Cell Rep.*, 2017, **20**, 1061–1072.
90. B. He, K. Nohara, N. Park, Y. S. Park, B. Guillory, Z. Zhao, J. M. Garcia, N. Koike, C. C. Lee, J. S. Takahashi, S. H. Yoo and Z. Chen, *Cell Metab.*, 2016, **23**, 610–621.
91. C. M. Bartman, Y. Oyama, K. Brodsky, L. Khailova, L. Walker, M. Koeppen and T. Eckle, *PLoS One*, 2017, **12**, e0176243.
92. G. F. Gloston, S. H. Yoo and Z. J. Chen, *Front Neurol.*, 2017, **8**, 100.
93. J. Gile, B. Scott and T. Eckle, *Crit. Care Med.*, 2018, **46**, e600–e608.
94. T. Inoue, T. Kohro, T. Tanaka, Y. Kanki, G. Li, H. M. Poh, I. Mimura, M. Kobayashi, A. Taguchi, T. Maejima, J. Suehiro, A. Sugiyama, K. Kaneki, H. Aruga, S. Dong, J. F. Stevens, S. Yamamoto, S. Tsutsumi, T. Fujita, X. Ruan, H. Aburatani, M. Nangaku, Y. Ruan, T. Kodama and Y. Wada, *Genome Biol.*, 2014, **15**, R63.
95. A. Galaup, E. Gomez, R. Souktani, M. Durand, A. Cazes, C. Monnot, J. Teillon, S. Le Jan, C. Bouleti, G. Briois, J. Philippe, S. Pons, V. Martin, R. Assaly, P. Bonnin, P. Ratajczak, A. Janin, G. Thurston, D. M. Valenzuela, A. J. Murphy, G. D. Yancopoulos, R. Tissier, A. Berdeaux, B. Ghaleh and S. Germain, *Circulation*, 2012, **125**, 140–149.
96. A. K. Singhal, J. D. Symons, S. Boudina, B. Jaishy and Y. T. Shiu, *Vasc. Dis. Prev.*, 2010, **7**, 1–14.
97. M. Sezer, N. van Royen, B. Umman, Z. Bugra, H. Bulluck, D. J. Hausenloy and S. Umman, *J. Am. Heart Assoc.*, 2018, **7**, e009949.
98. A. Xu, M. C. Lam, K. W. Chan, Y. Wang, J. Zhang, R. L. Hoo, J. Y. Xu, B. Chen, W. S. Chow, A. W. Tso and K. S. Lam, *Proc. Natl. Acad. Sci. U. S. A.*, 2005, **102**, 6086–6091.
99. Y. Oyama, C. M. Bartman, S. Bonnie, J. S. Lee, L. A. Walker, J. Han, C. H. Borchers, P. M. Buttrick, N. Clendenen, S. P. Colgan and T. Eckle, *bioRxiv*, 2019, 561340, DOI: 10.1101/561340.
100. I. A. Kolosova, T. Mirzapoiazova, D. Adyshev, P. Usatyuk, L. H. Romer, J. R. Jacobson, V. Natarajan, D. B. Pearse, J. G. Garcia and A. D. Verin, *Circ. Res.*, 2005, **97**, 115–124.
101. P. A. Gajwani, L. Wang, A. M. A. Ray, S. A. Krantz, A. A. Karginov and J. Rehman, *FASEB J.*, 2019, **33**, 788.782.
102. L. M. JanssenDuijghuijsen, S. Grefte, V. C. J. de Boer, L. Zeper, D. A. M. van Dartel, I. van der Stelt, M. Bekkenkamp-Grovenstein, K. van Norren, H. J. Wichers and J. Keijer, *Front. Physiol.*, 2017, **8**, 794.
103. R. Fukuda, H. Zhang, J. W. Kim, L. Shimoda, C. V. Dang and G. L. Semenza, *Cell*, 2007, **129**, 111–122.

104. N. Yorguner Kupeli, N. S. Bulut, G. Carkaxhiu Bulut, E. Kurt and K. Kora, *Psychiatry Res.*, 2017, **260**, 432–438.
105. H. N. Ginsberg, Y. L. Zhang and A. Hernandez-Ono, *Arch. Med. Res.*, 2005, **36**, 232–240.
106. P. H. Lee and L. K. Suen, *Sleep Breath.*, 2017, **21**, 209–215.
107. J. Aragones, P. Fraisl, M. Baes and P. Carmeliet, *Cell Metab.*, 2009, **9**, 11–22.
108. H. Ritchie, E. R. Stothard and K. P. Wright, *Curr. Pharm. Des.*, 2015, 3438–3442.
109. K. P. Wright Jr, A. W. McHill, B. R. Birks, B. R. Griffin, T. Rusterholz and E. D. Chinoy, *Curr. Biol.*, 2013, **23**, 1554–1558.
110. A. Takada, K. Saito, A. Ro, M. Kobayashi, A. Hamamatsu, T. Murai and N. Kuroda, *Leg. Med.*, 2003, **5**(Suppl 1), S292–S294.
111. C. J. Tracey, X. Pan, J. H. Catterson, A. J. Harmar, M. M. Hussain and P. S. Hartley, *J. Thromb. Haemostasis*, 2012, **10**, 662–669.
112. B. Zheng, D. W. Larkin, U. Albrecht, Z. S. Sun, M. Sage, G. Eichele, C. C. Lee and A. Bradley, *Nature*, 1999, **400**, 169–173.
113. Y. Oyama, S. Shuff, P. Davizon-Castillo, N. Clendenen and T. Eckle, *PLoS One*, 2020, **15**, e0244792.
114. Y. Zhao, Y. Zhang, S. Wang, Z. Hua and J. Zhang, *Thromb. Res.*, 2011, **127**, 122–130.
115. P. R. Somanath, E. A. Podrez, J. Chen, Y. Ma, K. Marchant, M. Antoch and T. V. Byzova, *J. Cell. Physiol.*, 2011, **226**, 132–140.
116. P. S. Hartley, J. Sheward, E. Scholefield, K. French, J. M. Horn, M. C. Holmes and A. J. Harmar, *Br. J. Haematol.*, 2009, **146**, 185–192.
117. J. S. O'Neill, E. S. Maywood, J. E. Chesham, J. S. Takahashi and M. H. Hastings, *Science*, 2008, **320**, 949–953.
118. Y. Oyama, C. M. Bartman, J. Gile, D. Sehrt and T. Eckle, *Curr. Pharm. Des.*, 2018, **24**, 3376–3383.
119. M. A. Weber, J. I. Drayer, D. K. Nakamura and F. A. Wyle, *Am. J. Cardiol.*, 1984, **54**, 115–119.
120. J. P. Degaute, P. van de Borne, P. Linkowski and E. Van Cauter, *Hypertension*, 1991, **18**, 199–210.
121. M. B. Turton and T. Deegan, *Clin. Chim. Acta*, 1974, **55**, 389–397.
122. P. Angleton, W. L. Chandler and G. Schmer, *Circulation*, 1989, **79**, 101–106.
123. M. E. Otto, A. Svatikova, R. B. Barretto, S. Santos, M. Hoffmann, B. Khandheria and V. Somers, *Circulation*, 2004, **109**, 2507–2510.
124. W. J. Elliott, *Stroke*, 1998, **29**, 992–996.
125. N. Twidale, S. Taylor, W. F. Heddle, B. F. Ayres and A. M. Tonkin, *Am. J. Cardiol.*, 1989, **64**, 1204–1206.
126. S. N. Willich, R. J. Goldberg, M. Maclure, L. Perriello and J. E. Muller, *Am. J. Cardiol.*, 1992, **70**, 65–68.
127. T. Martino, S. Arab, M. Straume, D. D. Belsham, N. Tata, F. Cai, P. Liu, M. Trivieri, M. Ralph and M. J. Sole, *J. Mol. Med.*, 2004, **82**, 256–264.
128. K. F. Storch, O. Lipan, I. Leykin, N. Viswanathan, F. C. Davis, W. H. Wong and C. J. Weitz, *Nature*, 2002, **417**, 78–83.

129. P. Podobed, W. G. Pyle, S. Ackloo, F. J. Alibhai, E. V. Tsimakouridze, W. F. Ratcliffe, A. Mackay, J. Simpson, D. C. Wright, G. M. Kirby, M. E. Young and T. A. Martino, *Am. J. Physiol.*, 2014, **307**, R121–R137.
130. J. Bass, *Nature*, 2012, **491**, 348–356.
131. A. M. Curtis, Y. Cheng, S. Kapoor, D. Reilly, T. S. Price and G. A. Fitzgerald, *Proc. Natl. Acad. Sci. U. S. A.*, 2007, **104**, 3450–3455.
132. M. Lefta, K. S. Campbell, H. Z. Feng, J. P. Jin and K. A. Esser, *Am. J. Physiol. Heart Circ. Physiol.*, 2012, **303**, H475–H485.
133. E. Li, X. Li, J. Huang, C. Xu, Q. Liang, K. Ren, A. Bai, C. Lu, R. Qian and N. Sun, *Protein Cell*, 2020, **11**, 661–679.
134. M. E. Young, R. A. Brewer, R. A. Peliciari-Garcia, H. E. Collins, L. He, T. L. Birky, B. W. Peden, E. G. Thompson, B. J. Ammons, M. S. Bray, J. C. Chatham, A. R. Wende, Q. Yang, C. W. Chow, T. A. Martino and K. L. Gamble, *J. Biol. Rhythms*, 2014, **29**, 257–276.
135. L. Zhang and M. K. Jain, *J. Clin. Invest.*, 2021, **131**, e148276.
136. D. J. Durgan, N. A. Trexler, O. Egbejimi, T. A. McElfresh, H. Y. Suk, L. E. Petterson, C. A. Shaw, P. E. Hardin, M. S. Bray, M. P. Chandler, C. W. Chow and M. E. Young, *J. Biol. Chem.*, 2006, **281**, 24254–24269.
137. M. S. Bray, C. A. Shaw, M. W. Moore, R. A. Garcia, M. M. Zanquetta, D. J. Durgan, W. J. Jeong, J. Y. Tsai, H. Bugger, D. Zhang, A. Rohrwasser, J. H. Rennison, J. R. Dyck, S. E. Litwin, P. E. Hardin, C. W. Chow, M. P. Chandler, E. D. Abel and M. E. Young, *Am. J. Physiol.*, 2008, **294**, H1036–1047.
138. J. Y. Tsai, P. C. Kienesberger, T. Pulinilkunnil, M. H. Sailors, D. J. Durgan, C. Villegas-Montoya, A. Jahoor, R. Gonzalez, M. E. Garvey, B. Boland, Z. Blasier, T. A. McElfresh, V. Nannegari, C. W. Chow, W. C. Heird, M. P. Chandler, J. R. Dyck, M. S. Bray and M. E. Young, *J. Biol. Chem.*, 2010, **285**, 2918–2929.
139. F. J. Alibhai, J. LaMarre, C. J. Reitz, E. V. Tsimakouridze, J. T. Kroetsch, S. S. Bolz, A. Shulman, S. Steinberg, T. P. Burris, G. Y. Oudit and T. A. Martino, *J. Mol. Cell. Cardiol.*, 2017, **105**, 24–37.
140. T. Nakao, A. Kohsaka, T. Otsuka, Z. L. Thein, H. T. Le, H. Waki, S. S. Gouraud, H. Ihara, M. Nakanishi, F. Sato, Y. Muragaki and M. Maeda, *Chronobiol. Int.*, 2018, **35**, 499–510.
141. S. Bonney, D. Kominsky, K. Brodsky, H. Eltzschig, L. Walker and T. Eckle, *PLoS One*, 2013, **8**, e71493.
142. M. Doi, Y. Takahashi, R. Komatsu, F. Yamazaki, H. Yamada, S. Haraguchi, N. Emoto, Y. Okuno, G. Tsujimoto, A. Kanematsu, O. Ogawa, T. Todo, K. Tsutsui, G. T. van der Horst and H. Okamura, *Nat. Med.*, 2010, **16**, 67–74.
143. D. Jeyaraj, S. M. Haldar, X. Wan, M. D. McCauley, J. A. Ripperger, K. Hu, Y. Lu, B. L. Eapen, N. Sharma, E. Ficker, M. J. Cutler, J. Gulick, A. Sanbe, J. Robbins, S. Demolombe, R. V. Kondratov, S. A. Shea, U. Albrecht, X. H. Wehrens, D. S. Rosenbaum and M. K. Jain, *Nature*, 2012, **483**, 96–99.
144. C. Y. Wang, M. S. Wen, H. W. Wang, I. C. Hsieh, Y. Li, P. Y. Liu, F. C. Lin and J. K. Liao, *Circulation*, 2008, **118**, 2166–2173.

145. A. D. Bhatwadekar, E. Beli, Y. Diao, J. Chen, Q. Luo, A. Alex, S. Caballero, J. M. Dominguez 2nd, T. E. Salazar, J. V. Busik, M. S. Segal and M. B. Grant, *Am. J. Pathol.*, 2017, **187**, 1426–1435.
146. C. B. Anea, M. Zhang, D. W. Stepp, G. B. Simkins, G. Reed, D. J. Fulton and R. D. Rudic, *Circulation*, 2009, **119**, 1510–1517.
147. H. Viswambharan, J. M. Carvas, V. Antic, A. Marecic, C. Jud, C. E. Zaugg, X. F. Ming, J. P. Montani, U. Albrecht and Z. Yang, *Circulation*, 2007, **115**, 2188–2195.

CHAPTER 7

The Respiratory System

COLLEEN M. BARTMAN*[a] AND Y. S. PRAKASH[a,b]

[a] Departments of Anesthesiology and Perioperative Medicine, Mayo Clinic, Rochester, Minnesota, USA; [b] Departments of Physiology & Biomedical Engineering, Mayo Clinic, Rochester, Minnesota, USA
*Email: bartman.colleen@mayo.edu

7.1 Circadian Rhythm in Respiratory Physiology

7.1.1 History of Rhythms in the Lung

Changes in breathing (*e.g.*, lung function, breathing rate, tidal volume) during the day *versus* the night is not a new phenomenon. Greek–Roman physicians were some of the first to note fluctuations in breathing based on the time of day. In the 2nd century, Soranus of Ephesus documented coughing and breathing difficulties, particularly "at the beginning and the end of the night", in people infected with what is now recognized as pulmonary *Myobacterium tuberculosis*.[1] In the 5th century, Caelius Aurelianus described rhythmic changes in breathing and worsened wheezing or asthma attacks at night (translated as "*On the heavy breath and wheeze which is called asthma by the Greeks, this disease is a burden, and men suffer more than women, and the elderly more than the young... during winter and at night more than during the day or in the spring,*" *Caelius Aurelianus reprinted 1722, as cited in* ref. 2 and 3). These observations were documented again in the 15th and 16th centuries by Wirsung and Castello Petro Vasco, respectively, who described what we now know as 'nocturnal asthma' – severe worsening of asthmatic symptoms in the middle of the night and during the dark winter months compared to the daytime and sunny summer months.[3]

Chemical Biology No. 23
Chronobiology and Chronomedicine: From Molecular and Cellular Mechanisms to Whole Body Interdigitating Networks
Edited by Germaine Cornelissen and Tsuyoshi Hirota

Published by the Royal Society of Chemistry, www.rsc.org

Castello Petro Vasco may have been the first to term 'nocturnal asthma' as he used the phrase '*asthma noctu*' in his textbook from 1616. Documentations such as these have continued to be noted since these initial observations. 18th century physician Sir John Floyer documented his personal struggle with asthma that worsened during sleep and French physician Murat classified asthma as a nighttime disorder.[3,4] These documented observations made it clear that there was a link between breathing, asthma, and daytime *versus* nighttime, but the extent of circadian rhythmicity in lung physiology and pathophysiology and underlying mechanisms were centuries away from being uncovered.

7.1.2 Functional Rhythms

7.1.2.1 Airway Caliber

The normal healthy lung exhibits diurnal variation in airway caliber, which involves the distensibility and dimensions of both bronchial and parenchymal elements. Airway caliber can be assessed through multiple variables, each measured through functional tests such as spirometry, plethysmography, or a peak flow meter. Pulmonary function tests help determine whether changes in airway/lung function are obstructive (air flow is decreased due to airway resistance) or restrictive (air flow is limited due to the inability of the lower airways to sufficiently expand). Several important airway caliber measurements are defined below:

Tidal volume: amount of air inhaled and exhaled during normal breathing
Vital capacity: total volume exhaled after maximum inhale
Total lung capacity: total maximum volume in the lungs after inhaling
Functional residual capacity (FRC): amount of air remaining in lungs after exhaling normally
Forced vital capacity (FVC): maximum amount of air exhaled forcefully
Forced expiratory volume in one second (FEV1): amount of air that can be forced from the lungs in one second
Peak expiratory flow (PEF): maximum flow rate of expiration from forced exhalation
Specific airway conductance (sG_{aw}): intrinsic airway resistance measured during tidal breathing.

Measurements of airway caliber have been documented to fluctuate over 24 hours in healthy individuals. For example, there is a dip in FEV1 and PEF around approximately 4:00 AM. Studies using environmentally controlled chambers with normal healthy volunteers revealed diurnal variations in sG_{aw} (lowest at 4:00 AM and highest at 12:00 PM) and diurnal patterns in FRC (lowest at 12:00 AM and highest at 8:00 AM).[5] A study examining the FEV1 of normal healthy men on a shift work schedule found diurnal variation in FEV1 with an increase in FEV1 for shift workers who both started and ended their shift in the morning, and decreased FEV1 for shift workers who both started and ended their shifts in the afternoon.[6] These findings indicate that environmental exposures may have a more detrimental effect on ventilatory capacity depending on shift-work schedules. Measurements of PEF in normal individuals showed a diurnal rhythm with a similar time of peak and dip in function, although the same PEF measurements in patients with asthma showed significantly greater amplitude in their rhythms.[7] An exaggerated amplitude in lung function may be an explanation for more severe symptoms occurring during the night in individuals with asthma (Figure 7.1).

7.1.2.2 Bronchial Responsiveness and Allergen Challenges

Bronchial responsiveness to allergen challenges shows patterns over a 24-hour period. Furthermore, 24-hour variation in PEF has been associated with altered bronchial responsiveness to bronchoconstrictor agonists such as histamine inhalation. A lower PEF indicates narrowing of the airways and a slower speed of air throughout the lungs. When PEF is at its lowest point in a 24-hour day (typically around 4:00 AM), bronchial hyper-responsiveness is at its highest. This indicates that the greater amplitude of the 24-hour variation results in a greater dip in PEF, which may be a significant contributor to bronchial hyper-responsiveness.[8] Bronchial challenge tests, in which nebulized methacholine or histamine is inhaled at incremental doses, can be helpful in diagnosing airway diseases through the measurement of the degree by which there is narrowing or constricting of airways. Studies have investigated methacholine challenge at twelve-hour intervals (4:00 AM and 4:00 PM) in healthy individuals compared to those with nocturnal asthma. Although both groups had increased responses to methacholine at

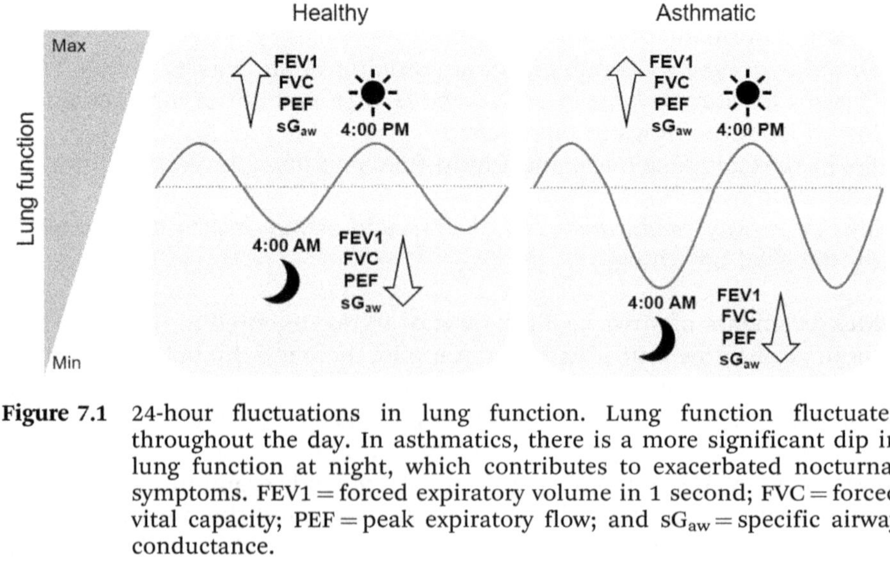

Figure 7.1 24-hour fluctuations in lung function. Lung function fluctuates throughout the day. In asthmatics, there is a more significant dip in lung function at night, which contributes to exacerbated nocturnal symptoms. FEV1 = forced expiratory volume in 1 second; FVC = forced vital capacity; PEF = peak expiratory flow; and sG_{aw} = specific airway conductance.

4:00 AM compared to 4:00 PM, those with asthma had a larger increase in hyperreactivity during the night compared to during the day.[8,9] Additionally, individuals with greater dips in PEF during the night (for both asthmatic and normal individuals) had heightened bronchial responsiveness at that time.[8] Finally, measurements of bronchial responsiveness through bronchial challenge with house dust exposure showed hyper-responsiveness at night in asthma patients.[10] In combination with a dip in lung function in the middle of the night, allergen exposure elicits a more robust response during that time, which, together, may help explain nocturnal symptoms of airway diseases.

7.1.2.3 Ventilatory Response and Sleep Apnea

Pulmonary ventilation (V_E), tidal volume, and breathing rate have shown to follow circadian patterns. These changes are not necessarily linked to the normal variation in physical activity. The V_E response to hypercapnia or hypoxia has a 24-hour pattern, as well. In humans, it is known that the metabolic rate (O_2 consumption [VO_2] and CO_2 production [VCO_2]) and V_E decrease during sleep, accompanied by increased alveolar (P_ACO_2) and arterial pressures of carbon dioxide (P_aCO_2).[11,12] Experiments with humans under constant light exposure remove the 24-hour metabolic fluctuations in terms of VO_2 and blunt daily variation and amplitude of V_E.[13,14] Chemoreceptors in the lung, which aid in maintaining tidal volume and normal V_E, do so by sensing changes in O_2 and CO_2. Hypoxia or hypercapnia therefore increase chemoreceptor activity and alter breathing patterns. 24-hour fluctuations in V_E chemosensitivity thus may be affected by the hypoxic ventilatory response. Studies involving humans or rats exposed to hypobaric hypoxia show that respiratory gases have less of an effect on the timing of circadian patterns of lung function (*i.e.*, V_E) and more of an effect on the amplitude of the rhythmicity. Failure to coordinate fluctuations in VO_2 and hypoxic ventilatory response would lead to disruption of blood gas homeostasis.[15] Additionally, there is a V_E response to hypercapnia in a time-of-day-dependent manner.[15] With the circadian pattern of V_E and response to hypoxia or hypercapnia, studies suggest humans are more adaptive to changes in pulmonary gases when blood gas homeostasis would be least disrupted (*i.e.*, it is advantageous to coordinate with VO_2).[15] Furthermore, failure to take advantage of endogenous circadian patterns in V_E can lead to heightened disease symptoms at certain times of the day.[15]

7.1.3 Clinical Presentations of Circadian Component of Respiratory Disease

7.1.3.1 Asthma in the Clinic

Asthma is a chronic lung disease involving obstruction of airflow, underlying inflammation, and bronchial hyper-responsiveness. Understanding the natural rhythms of lung function enables researchers and physicians to decipher the pathophysiology of nocturnal asthma. Rhythmic patterns in asthma symptoms are nearly universal experiences with symptoms worsening during the night. This coincides with the deepest dip in lung function in patients with airway diseases, therefore exacerbating symptom severity. In healthy humans, circadian rhythmicity in FEV1 and PEF has been established with their lowest point around 3:00 AM to 4:00 AM (and their peak about twelve hours later, 3:00 PM to 4:00 PM).[9,16,17] Interestingly, studies have shown that although the times of peak and trough of PEF remain unchanged, the amplitude of such rhythms are exaggerated in asthma patients[7] (Figure 7.1). Importantly, this appears to be only a one-sided exacerbation, as worsened lung function at night does not necessarily correlate with increased (higher than normal) lung function during the day. Rather, it seems that the exaggerated dip in lung function at night hinders improved lung function during the day. A potential hypothesis for this phenomenon could be that failure to reach optimal lung function during the day triggers a larger dip during the night as a compensatory response. In support of this, studies on asthmatic patients found that the greater dip in PEF during the nighttime (*i.e.*, a deeper dip in lung function) is associated with increased limits in airflow during the day.[9] Furthermore, worsened symptoms at night lead to higher rates of emergency room visits and hospitalization during this time compared to the early morning hours of the day. Asthma-related hospitalization and deaths from respiratory arrest occur more frequently during the night and are associated with a wider variability in daytime PEF.[18–20] Variations in lung function and the efficacy of treatments in a time-of-day-dependent manner (with consideration of individual chronotypes, as well) have been proposed as important factors to consider to prevent increased risk of death from nocturnal asthma.[21] The high amplitude dips in lung function such as PEF seen in the middle of the night have been attributed to asthma mortality.[22]

7.1.3.2 COPD in the Clinic

Chronic obstructive pulmonary disease (COPD) is characterized by irreversible obstruction of airflow and abnormal inflammatory response. Circadian variation in COPD has also been noted.[23] Patients with chronic bronchitis and emphysema who were hospitalized with acute exacerbations of symptoms are advised to be monitored at night due to the higher risk of overnight death from associated hypoxaemia and hypercapnia.[24] Deaths from chronic bronchitis and emphysema in patients were also reported to be highest in the middle of the night (between 1 and 5 AM).[25] Studies have shown that patients arriving at the emergency department in the middle of the night have more severe symptoms than those arriving during the day, and these patients subsequently require intensive treatment and experience poorer outcomes.[26] For example, patients who presented with COPD symptoms during the night had a higher chance of being intubated upon admission.[26]

7.1.3.3 Lung Cancer in the Clinic

The most common type of lung cancer is non-small cell lung cancer in which there are adenocarcinomas, squamous cell cancers, and large cell carcinomas. Patients with non-small-cell lung cancer have disrupted circadian rhythms, so targeting rhythmicity and optimizing the timing of therapeutic treatments may play an important role in improving the overall quality of life in this population. For example, melatonin supplementation for patients with advanced lung cancer has shown to be a potential consideration for improving the overall quality of life.[27,28] Researchers and physicians have proposed that considering and integrating circadian rhythmicity as a component of patient care can provide a multitude of improvements and potentially extend survival of patients with advanced lung cancer.[29] Mechanistically, studies have found core clock proteins such as PER1 and PER2 to act as tumour suppressors and deletion of these genes promotes tumour growth and progression, which has been investigated both *in vivo* and *in vitro*.[29–32] These studies suggest that the circadian clock, at either the molecular or physiological level, could indeed be an advantageous target for controlling cancer.

7.1.3.4 Shift Work and Lung Disease

The effect of shift work on overall health has long been a focus for many circadian rhythm researchers because it has been associated with a wide variety of diseases. Multiple studies have investigated the effect of circadian disruption arising from chronic shift work on lung physiology and pathophysiology (Table 7.1). Studies examining lung function in underground coal workers documented disrupted rhythmicity in lung function, which was dependent on shift schedules.[33] Cohort studies involving humans with a history of shift work and subsequent altered sleep patterns found that this environmental circadian disruption elevated the risk of developing pulmonary fibrosis later in life.[34] Additionally, cohort studies involving nurses who consistently worked rotating night shifts found an increased risk of developing lung cancer, but only in those who also smoked.[35] These studies suggest that circadian disruption may increase susceptibility to additional insults that promote tumorigenesis. In another study, people who are permanent shift workers are at an elevated risk of developing moderate to severe asthma.[36] Without a doubt, environmental circadian disruption has detrimental effects on the lung and therefore presents a potential strategy for optimizing therapeutic approaches through targeting the circadian clock.

Table 7.1 Environmental circadian disruption and lung pathology.

Circadian disruption model	Clock and physiological effects on the lung	Ref.
Chronic jet lag paradigm in Fischer rats using injected NK-sensitive adenocardinoma cells	Altered expression of *Per2* and *Bmal1* genes Promoted lung tumour development	67
Chronic jet lag paradigm in mice using a mouse model of lung adenocarcinoma	Promoted lung tumour growth and progression Decreased survival	69
Chronic jet lag paradigm in male and female mice	Airway resistance at PEEP lower in males in the chronic jet lag group Clock gene expression decreased in males in the chronic jet lag group *Per2* and *Reverba* gene expression increased in males in chronic jet lag group *Bmal1* and *Reverba* gene expression decreased in females in chronic jet lag group *Per2* and *Cry2* gene expression increased in females in the chronic jet lag group	41
Chronic jet lag paradigm or free-running conditions (constant darkness) in male and female PER2::luc mice	Lung rhythmicity was phase-advanced under both chronic jet lag and free-running conditions	42
Phase advance protocol in PER2::luc mice	Lung rhythmicity and kinetics shifted	40
Chronic jet lag paradigm in mice inoculated with Lewis lung carcinoma cells	Increased tumour proliferation rate and metastases in the chronic jet lag group Altered expression of genes involved in cell cycle, apoptosis, immunity, and tumour suppression in the chronic jet lag group	114
Human cohorts from UK Biobank	Markers of circadian disruption (sleep length, chronotype, shift work) were associated with pulmonary fibrosis Altered clock gene expression in patients with idiopathic pulmonary fibrosis Increased REV-ERBα protein expression in lung tissue from humans with idiopathic pulmonary fibrosis	34
Cohorts of nurses working rotating night shifts for more than 15 years, smokers *vs.* non-smokers	Increased lung cancer risk in night-shift group who also smoked	35

Table 7.1 (*Continued*)

Circadian disruption model	Clock and physiological effects on the lung	Ref.
Human cohorts from UK Biobank	Consistent night shift workers had increased risk of all severities of asthma Morning chronotypes on irregular shift patterns showed increased risk of moderate-severe asthma	36
Chronic jet lag protocol in mice alongside Sendai virus and influenza A	Increased weight loss, airway resistance, methacholine sensitivity in circadian disrupted mice following SeV infection	57

7.2 Peripheral Molecular Clocks in the Lung

7.2.1 Environmental or Genetic Circadian Disruption in Allergic and Immune Response

The term 'nocturnal asthma' has long been in the literature, and modern technologies have allowed researchers to understand the underlying mechanisms behind why asthmatics experience worsened symptoms at night. Additionally, asthmatics are particularly sensitive to allergic triggers that, upon exposure, promote asthma attacks. Researchers have investigated the role of circadian networks in the allergic response to such triggers. Two common models for investigating the involvement of circadian rhythm and circadian disruption are environmental and genetics: chronic jet lag exposure protocols (altered light-dark cycles; Table 7.1) and genetic deletion of core clock genes in mice (Table 7.2).

The first observation of rhythmicity in immune response was shown in mice exposed to endotoxin and it was found that inflammatory responses varied depending on when the endotoxin was administered in a 24-hour period.[37] Chronic jet lag-mediated circadian disruption or free-running conditions of constant darkness revealed significant shifts in circadian rhythms of core clock genes in the mouse lung.[38] Circadian disruption through both chronic jet lag and genetic deletion of *Bmal1* (*Arntl*) in mice combined with exposure to Sendai virus and influenza A resulted in an exacerbated inflammatory response, increased airway resistance, and decreased survival – factors relevant to asthma pathophysiology.[39] Influenza A inhalation was shown to alter the phase and amplitude of core clock genes in the lung.[40] The conditional deletion of *Bmal1* in pulmonary airway epithelial cells not only disrupted circadian rhythmicity of transcriptional targets in these cells, but also resulted in worsened lung function and recovery following influenza A infection with increased neutrophil infiltration and inflammatory response in the lung.[41] Notably, the timing of influenza A infection administered to mice influences the inflammatory response in the lung, injury, viral clearance, and survival.[42] Influenza A administration

Table 7.2 Molecular circadian clock and lung pathophysiology.

Mode of clock disruption and/or insult/exposure	Effect on lung clock	Physiological effects in the lung	Ref.
Cigarette smoke			
C57BL6/J adult mice Lung epithelial cell-specific *Bmal1* KO mice Lung tissue from patients with COPD	BMAL1 acetylated and degraded due to reduced levels of SIRT1, an NAD^+-dependent deacetylase, upon cigarette smoke exposure	Abnormal clock protein expression and oscillation (due to altered posttranslational modifications) promotes proinflammatory response in the lung, emphysema, and worse lung function	51
Bronchial epithelia *Reverba* or *Reverbb* KO mice	Increased chemokines CXCL1, CXCL2, G-CSF in BAL	Exaggerated response to cigarette smoke exposure, earlier mortality in *Reverba* KO mice	39
Sprague-Dawley rats	Shift in expression of clock genes and phase shift after cigarette smoke exposure	Increased inflammatory response	53
Tumour-resistant C57BL6/J adult mice Tumour-susceptible AJ adult mice	Decreased *Reverba* and *Bmal1* gene expression in the lungs after cigarette smoke exposure Increased *Per2* and *Cry1* gene expression in the lungs after cigarette smoke exposure		54
Influenza A			
Bmal1 KO mice infected with Sendai or influenza A virus		Exaggerated asthma-like changes in the airway following infection in *Bmal1* KO mice (increased mucus production, increased airway resistance to methacholine challenge)	57
Bmal1 KO mice infected with influenza A virus at different times of day		Time of day mice infected with influenza determines inflammation, viral clearance, and survival, which is dependent on BMAL1	58

Bmal1 KO in club cell secretory protein expressing pulmonary airway epithelial cells and influenza viral infection	*Bmal1* KO in club cell secretory protein expressing pulmonary airway epithelial cells impacted metabolic and inflammatory gene expression patterns and revealed new transcriptomic rhythms	Increased bronchoalveolar neutrophils, reduced lung function, and poor response to influenza infection	59
Cigarette smoke and influenza A infection			
C57BL/6J adult mice Whole-body *Bmal1* KO mice	Altered phase and amplitude Lung rhythm of BMAL1, PER1, PER2, CRY1, and REV-ERBβ Lack of rhythm of CLOCK, CRY2, REV-ERBα, RORα	Increased inflammatory response in the lung Pro-fibrotic phenotype Reduced survival Changes in normal fluctuations in lung function Development of emphysema	52
LPS			
Bronchial epithelial cell specific *Bmal1* KO mice *Bmal1* KO in myeloid cells (macrophage and monocyte) and neutrophils	CXCL5 under circadian control in bronchiolar epithelial cells, which determines circadian variation in response to endotoxin and bacterial infection	Deletion of *Bmal1* from bronchial epithelial cells disrupts glucocorticoid signalling and subsequently altered CXCL5 and neutrophil infiltration in the lung and ultimately decreased efficacy in bacterial clearance from the lung	47
C57BL/6J mice	Endotoxin exposure alters circadian gene periodicity and differentially altered core clock gene expression in that some oscillations were blunted while others became more exaggerated	Altered rhythmic leukocyte and granulocytes in the lung in a BMAL1-dependent manner	48
Per2 KO mice administered LPS		*Per2* KO mice more resistant to LPS-induced endotoxic shock than controls *Per2* KO mice have decreased serum levels of proinflammatory cytokines (gamma interferon, interleukin-1b) following LPS challenge	60

Table 7.2 (*Continued*)

Mode of clock disruption and/or insult/exposure	Effect on lung clock	Physiological effects in the lung	Ref.
Bleomycin-induced fibrosis			
C57BL/6J adult mice *Nrf2* KO mice PER2::Luc mice *Clock*$^{\Delta 19}$ mice	Rhythmic expression of antioxidant genes in mouse lung, which are dependent on NRF2	When NRF2 is at its lowest point in the circadian cycle, mice have more severe oxidative injury and fibrotic phenotype	55
TGF-β1-adenovirus-infected mice Human lung epithelial cells and fibroblasts with *BMAL1* KD	TGF-β1 increases BMAL1 expression *in vitro* and *in vivo* BMAL1 deacetylated with TGF-β1 treatment *BMAL1* KD in culture blunts TGF-β1 signalling pathways and disrupts epithelial-to-mesenchymal transition (epithelial cells) and differentiation (fibroblasts)		56
Bleomycin-induced fibrosis in *Reverba* mutated mesenchymal, club, or myeloid cells from mouse lungs	Myofibroblasts have circadian oscillations in fibrotic lung that are asynchronous	Mutated REV-ERBα in fibroblastic cells: increased fibrotic response in the lung *Reverba* knockdown in culture: increased myofibroblast differentiation	34
Hypoxia or hyperoxia			
Dose response hyperoxic acute lung injury in mice	Dose-response increase in *Clock* gene expression with increasing O_2	Increases in *Clock* gene expression associated with increases in inflammatory markers and alveolar congestion in hyperoxia-induced lung injury	43
Neonatal hyperoxia exposure or neonatal mouse lung fibroblasts	Increased *Reverba* gene expression *in vivo* and altered rhythmicity of REV-ERBα *in vitro*	Neonatal oxidative stress (hyperoxia) increases *Reverba* gene expression *via* NRF2 Inflammation (NFκB pathway) inhibits *Reverba* gene expression *via* NFκB	73

Adult mice exposed to influenza A virus after hyperoxia exposure as neonates Mice with *Bmal1* KO in alveolar type 2 cells with lung function measured as adults		Mice exposed to hyperoxia as neonates did not have time-of-day protection from influenza A infection as adults *Bmal1* KO in alveolar type 2 cells similar effect on lung function and response to infection in adulthood compared to adult mice exposed to hyperoxia as neonates	71
Ventilator-induced lung injury			
Sprague-Dawley rats with endotracheal intubation and mechanical ventilation	Decreased *Reverba* gene and REV-ERBα protein expression	Agonism of REV-ERBα activity attenuated injury in terms of lung oedema, inflammatory cell infiltration, and pro-inflammatory cytokines	44
Lung cancer			
Human lung adenocarcinomas	*PER1, PER2, PER3, CRY1, CRY2, and BMAL1* gene expression decreased in adenocarcinomas *vs.* normal lung Clock gene expression increased in adenocarcinomas *vs.* normal lung		69
Per2 or *Bmal1* KO mouse model with lung adenocarcinoma		Promoted lung tumour growth progression Increased c-MYC expression, increased proliferation, dysregulated metabolic activity Decreased survival	69
Mice injected with tumoural cells		Lung exhibited diurnal patterns of neutrophil infiltration Lung metastatic tumour development and neutrophil infiltration were dependent on time of day mice were injected with tumoural cells	68
Hospitalized patients receiving chemotherapy for advanced non-small-cell lung cancer	Significantly abnormal circadian function compared to ambulatory patients (actigraphy data)	Poor sleep quality, high fatigue	27

immediately before the animal's active phase (lights off) resulted in worse outcomes compared to administration immediately before the animal's rest phase (lights on).[43] Furthermore, the variable effect of influenza A infection throughout a 24-hour day was dependent on BMAL1.[44]

In an ovalbumin mouse model, myeloid specific *Bmal1* knockout (KO) mice showed increased eosinophil infiltration in the lung following exposure to ovalbumin, providing a potential explanation for asthma symptoms at night at a time when BMAL1 expression is at its lowest.[38] However, exposure to ovalbumin in *Reverba (Nr1d1)* KO mice did not exacerbate allergic inflammatory response, suggesting a distinct and specific role of core clock components in innate and adaptive immune responses.[39]

The magnitude of 24-hour fluctuations in immune cell functions and inflammatory responses has been shown in macrophages.[45,46] More recently, macrophage cytokine secretion in response to endotoxin was measured in mice throughout a 24-hour period. The results showed that *Bmal1* or *Reverba* KO blunts the rhythmicity of the endotoxin response and increases macrophage cytokine secretion.[46] Rhythmicity in endotoxin response in conditional *Bmal1* KO models was only evident in airway epithelial club cells, but not in myeloid cells.[46,47] Furthermore, in the airway epithelial cell specific *Bmal1* KO mouse model, increased endotoxin response was due to disrupted recruitment of chemokine/neutrophil CXCL5 to the lung. Under normal conditions, BMAL1 would inhibit gene expression of *Cxcl5* at night, but deficiency in BMAL1 results in increased response to endotoxin.[47] Additionally, endotoxaemic mice elicit a rhythmic gene signature in the whole mouse lung, a transcriptome of which is involved in immune response, and these mice had disrupted core clock gene expression or rhythmicity. Additionally, endotoxaemic mice had altered rhythms of lung leukocyte counts. Since endotoxin both elicited unique gene expression rhythms and disrupted core clock components, these researchers investigated whether *Bmal1* deletion would be impactful. Importantly, *Bmal1* deletion was shown to have a selective effect on the endotoxaemic mouse lung: BMAL1 was important for rhythmic granulocyte recruitment and gene expression, but it was not as vital for rhythms in gene expression in pneumocytes and endothelial cells. This differential role of BMAL1 in immune response highlights the importance of cell-type specific roles for the clock in respective cellular functions.[48] Furthermore, REV-ERBα protein stability may serve as a key link to immunity in the lung since the proinflammatory cytokine response to endotoxin degrades REV-ERBα, leading to increased endotoxin response in bronchial epithelial cells.[39] These studies identify REV-ERBα as a signalling molecule between the core clock and degree of inflammatory response in bronchial epithelial cells through which cytokine signalling leads to phosphorylation, SUMOylation, ubiquitination, and degradation of REV-ERBα, which stops the inhibition of the inflammatory response. Since BMAL1 is involved in circadian regulation of downstream immune functions through positive transcriptional regulation of *Reverba*, targeting REV-ERBα/β stability is a viable option for modulating immune responses.[39]

7.2.2 Cigarette Smoke and COPD

Similar to asthma, COPD symptoms have been noted to be exacerbated at night.[26] While smokers and patients with COPD exhibit circadian variation in PEF with peaks occurring at a similar time of day as controls and those receiving second-hand smoke, the average PEF variability is lower while the amplitude of PEF is exaggerated.[49] This has also been shown in children exposed to second-hand cigarette smoke where there is a decrease in average PEF and an increase in the amplitude of PEF compared to children who were not exposed.[49,50]

Experiments *in vivo* have shown decreased amplitude and altered clock gene expression in mice exposed to cigarette smoke; BMAL1 gets degraded in mice exposed to cigarette smoke with epithelial-specific *Bmal1* KO showing an increased inflammatory response to cigarette smoke.[51] In lung tissue from COPD patients, BMAL1 expression was also dampened in the smoking group compared to those who did not smoke.[51] Additionally, mice exposed to cigarette smoke and influenza A infection showed altered clock gene expression and an increased inflammatory response in the lung, changes in normal fluctuations in lung function, and development of emphysema.[52] *Bmal1* KO mice that only received influenza A infection (not cigarette smoke) developed similar phenotypes with increased inflammatory response in the lung and remodeling.[52] An exaggerated response to cigarette smoke exposure in circadian disrupted mouse models revealed that in *Reverba* KO mice, exposure to cigarette smoke resulted in earlier mortality and increased CXCL1 in bronchoalveolar lavage (BAL) from the lung.[39] In rats, cigarette smoke exposure altered the clock gene expression and led to a shift in the rhythm in the lungs.[53] In the lungs of mice, both whole smoke and filtered smoke decreased *Reverba* expression.[54]

7.2.3 Bleomycin Model of Fibrotic Lung Disease and the Clock

Little is known about the circadian clock in the fibrotic lung, but the bleomycin mouse model of lung fibrosis has begun to uncover captivating mechanisms with respect to inflammatory response in this model system. In particular, researchers have investigated how the circadian system and NRF2/glutathione-mediated antioxidant defence pathway are involved in the development of the fibrotic lung. Mice administered bleomycin when NRF2 levels are lowest experience a more severe fibrotic outcome than mice administered bleomycin when NRF2 levels are highest.[55] More importantly, *NRF2* is under circadian control, with CLOCK and BMAL1 positively regulating its gene expression *via* an E-box element in its promoter.[55] Therefore, not only does NRF2 accumulate in a circadian fashion in the lung, but so do other antioxidant genes involved in the glutathione antioxidant defence pathway.[55] This effect is consistent with the finding that NRF2 rhythmicity is altered in *Clock*$^{\triangle 19}$ mutant mice, leading to oxidative damage and a pro-fibrotic phenotype, even without bleomycin challenge.[55] Additionally, BMAL1 appears to play a role in development of the fibrotic lung. While TGFβ1 can upregulate BMAL1 expression *in vitro* and *in vivo*, *ARNTL* knockdown in lung epithelial cells and lung fibroblasts in culture revealed attenuation of TGFβ1 signalling as well as TGFβ1-induced epithelial-to-mesenchymal transition.[56]

Additional findings have shown REV-ERBα to be part of a pro-fibrotic phenotype and a potential target for treatment. Researchers have found that pulmonary fibrotic tissue exhibits increased amplitude and desynchrony resulting from mesenchymal/fibroblast cell infiltration in the lung.[34] Additionally, mutant REV-ERBα in mesenchymal cells exacerbated the fibrotic phenotype in the bleomycin-induced fibrosis mouse model.[34] Human studies found that circadian disruption through altered environmental factors (shift work, sleep length, and chronotype), elevates the risk of developing pulmonary fibrosis[34] (Table 7.1).

7.2.4 Pneumonia and Sepsis

Mouse models of pneumonia induced by *Streptococcus pneumoniae* infections revealed that the time of day when the infection was administered affected the severity of pneumococcal pneumonia, which was reflected by variation in bacterial load in the lung and neutrophilia infiltration in the lung.[47] Targeted *Bmal1* KO in airway epithelial cells of mice resulted in increased lung neutrophil infiltration in response to bacterial infection but it did not result in an improved response or clearance of bacteria from the lung.[47] In wild-type (WT) mice infected with Sendai virus and influenza A, the inflammatory response fluctuates over a 24-hour period and depends on the time of inoculation,[57] which determines survival, viral clearance, and inflammatory response to influenza A infection. Studies in *Bmal1* KO mice revealed heightened acute lung injury and inflammation in response to parainfluenza or influenza A.[52,57,58] Further mechanistic studies have pinpointed airway epithelial cells as a potential major factor in this response to infection.[59] Finally, mice lacking BMAL1 exposed to viral infections exhibit a prolonged response to and recovery from pneumonia.[52,57–59] When mice housed on a 12 : 12 L : D cycle were inoculated in the middle of the dark period (6 hours following lights-off), almost all of the mice survived (>90%).[37] However, the majority of mice inoculated with endotoxin around 06:00 did not survive following septic shock (<20%).[37] This effect potentially involves PER2 since *Per2* KO mice, housed under 12 : 12 L : D conditions, lack diurnal fluctuations in survival following endotoxin infection and sepsis.[60]

Studies measuring melatonin variation in ICU patients with and without sepsis found that the endogenous rhythm of melatonin was abolished in patients with sepsis compared to those without sepsis.[61] Experimental studies in humans showed endotoxin-driven dampening of clock gene expression in human peripheral blood leukocytes, neutrophils, and monocytes.[62] One mechanism that may help our understanding of the relationship between circadian rhythms and sepsis is through studies involving melatonin. Genetic deletion of melatonin receptors results in increased inflammatory response in the mouse lung following infection and sepsis.[63] In support of this observation, administration of exogenous melatonin attenuated effects of sepsis and increased survival time, an effect not seen in animals deficient in melatonin receptors.[64]

7.2.5 Lung Cancer

Circadian disruption *via* shift work has been statistically associated with a variety of cancers, including lung adenocarcinoma.[65] Additionally, small-molecule agonists of REV-ERBα have exhibited antitumor activity.[66] Circadian disruption in rats using jet-lag protocols of light exposure promoted lung tumour development when rats were induced with natural killer (NK)-sensitive adenocarcinoma cells.[67] Furthermore, studies have shown diurnal accumulation of neutrophils in the mouse lung, which has downstream effects on the expression of pulmonary genes involved in carcinogenesis and migration.[68] Diurnal neutrophil infiltration in the mouse lung also led to invasion of melanoma cells. This group found that deletion of *Bmal1* in neutrophils not only ablated the diurnal infiltration of neutrophils in the lungs, but neutrophil-specific *Bmal1* deletion also reduced rhythmicity of melanoma cell invasion in the mouse lung.[68] Mouse models of lung adenocarcinoma, featuring genetic mutations in *p53* and *Kras*, have frequently been used since this model mimics human non-small cell lung cancer. A group of researchers used this lung adenocarcinoma mouse model to determine the effects of the circadian clock and circadian disruption on lung tumorigenesis. They found that environmental circadian disruption (jet-lag model) and the presence of *Per2* or *Bmal1* whole body KO mice in the context of *p53* and *Kras* mutations promoted lung tumour growth and progression and death[69] (Table 7.1). Genetic circadian disruption increased c-MYC expression, proliferation, and metabolic dysfunction in the lung[69] (Table 7.2). Additionally, another group used the lung adenocarcinoma mouse model to show that the tumor itself could function as an endogenous Zeitgeber, since this model showed tumor-driven changes in hepatic clock gene expression and metabolic function.[70]

7.2.6 Early Exposures, Aging, and Airway Disease Later in Life

Maintaining circadian rhythmicity has been shown to elicit protection from influenza infection in adults. However, this can be influenced by early life exposures. Neonatal hyperoxia exposure has been recently demonstrated to alter the circadian system and inflammatory response in adulthood; neonatal mice exposed to hyperoxia demonstrate a lack of protection from influenza A infection in adulthood, which is usually circadian-gated.[71] This effect was also shown in adult mice lacking BMAL1 in alveolar type 2 cells.[71] A model of hyperoxic acute lung injury in mice revealed that oxygen exposure increased *Clock* gene expression, which correlated with the exposure time and dose of oxygen along with the level of inflammatory response in the lung.[43,46] In modelling obstructive sleep apnea, mice given brief exposure to intermittent hypoxia have been found to show altered circadian and immune pathways in the lung.[72] Additionally, neonatal mice exposed to hyperoxia showed an increase in the expression of *Reverba* and altered REV-ERBα rhythmicity *in vitro* through oxidative and inflammatory stress pathways, such as NFκB.[73] Recent *in vitro* work conducted with fetal airway smooth muscle cells has shown the existence, responsiveness, and functionality of the clock and clock networks in the developing airway with a unique role in hyperoxia.[74] Many pathways involved in airway disease pathophysiology are also targets of clock networks.[48] Indeed, response to various oxygen exposures such as hypoxia has been shown to be highly tissue-specific and time-of-day-dependent, causing phase shifts in peripheral clocks. However, hypoxia did not retain synchronicity among organ systems but rather each organ had its own unique response to low O_2, leading to inter-tissue misalignment (*i.e.*, the extent of phase-shift in response to hypoxia was not universal among all tissues). For example, while the lung was phase-advanced in response to hypoxia, the liver was phase-delayed.[75] The relationship between circadian networks and oxygen-sensing pathways is most likely through HIF1α and is bidirectional in nature. Oxygen and carbon dioxide levels in blood and tissue have been shown to exhibit circadian rhythms and pulsing oxygen is sufficient to promote clock gene expression and synchrony *in vitro* in a HIF1α-dependent fashion.[76–78]

A disrupted circadian clock has been implicated in aging and senescence. BMAL1 appears to have a distinct importance relative to the timing of its expression since postnatal deletion of murine *Bmal1* does not result in reduced life span, infertility, altered body weight or blood glucose levels, or age-dependent arthropathy, which are classic phenotypes of whole-body *Bmal1* KO mice in which BMAL1 is not present during fetal development.[79,80] Mechanistic studies found that the premature aging phenotype in *Bmal1* KO mice involves increased senescent cells in multiple tissues, including those of the lung *in vivo*.[81] Furthermore, inhibition of mTORC1 with rapamycin increased the lifespan of *Bmal1* KO mice, and *in vitro* data using lung fibroblasts supported that BMAL1 is a negative regulator of TORC1 activity.[82] *In vitro* studies conducted with human lung fibroblasts also showed that senolytics such as quercetin, caffeic acid, and resveratrol alter clock gene and protein expression in culture.[83] Finally, the transcriptomic response to cigarette smoke exposure has shown significant similarities to the transcriptomic changes that occur in the mouse lung from normal age-related processes.[84]

7.2.7 Sexual Dimorphism

Since there are notable sex differences in the immune system, it is plausible that sex differences also exist in the circadian system and response to shift-work-mediated circadian disruption. For example, mathematical modelling predicts that there are sex differences in pro-inflammatory cytokine levels in the lungs of rats exposed to LPS and a shift work regime, with females having less cytokines than males. Additionally, these studies suggest that there is variation in the circadian nature of the immune response depending on the sex.[85] Finally, studies conducted in male and female mice exposed to a chronic jet lag paradigm displayed disrupted lung function and clock gene expression, indicating evidence of sexual dimorphism.[41]

7.3 Chronotherapy and Chronotherapeutics in Lung Disease

7.3.1 Targeting the Clock and Implications for the Clinic

Large dataset investigations of gene expression in the lung across a series of time points over 24 hours have begun elucidating the oscillatory patterns of common drug targets for the treatment of airway diseases.[86–90] An in-depth lung transcriptome analysis in rat lungs has shed light onto pathways involved in lung physiology and pathophysiology, such as during the circadian day when drug targets are expressed.[88] Investigation of the mouse transcriptome in various organs over time identified patterns of oscillatory genes in the lung; many of these genes are directly targeted by medications listed in the World Health Organization's essential medicines list.[87] Additional studies conducted in both mouse and human lungs have identified multiple drug targets of genes associated with airway diseases.[86] Nuclear receptors REV-ERBα/β and retinoic acid receptor-related orphan receptors (RORs) have also been considered for chronotherapy.[91] Additionally, 24-hour studies in an asthmatic mouse model show appropriate timing of inhaled dexamethasone can prevent significant phase-shifting of the lung clock in asthma treatment.[92] Many products of genes associated with airway diseases have oscillatory patterns in rodent lungs, which has enabled researchers to categorize pathways involved based on their involvement as growth factors, in the extracellular matrix, in signalling pathways, drug metabolism, bronchodilation, and involvement with nuclear receptors (Table 7.3).[86–88] Furthermore, this has helped in identification of potential chronotherapeutic targets (Table 7.3).

7.3.2 Chronotherapy in Asthma

Current treatments for mild-moderate asthma involve inhaled corticosteroids ± long-acting beta agonists, with severe asthma requiring even higher doses of inhaled corticosteroids alongside additional measures

Table 7.3 Genes with oscillating expression patterns that may be targetable for lung disease.[a]

Function	Oscillating target	Drug target, if known	Ref.
Bronchodilation	Adrenergic-β2 receptor, *Adrb2*	Disulfiram; Advair Diskus; Combivent; ProAir HFA; terbutaline; arformoterol; salmeterol	86–88
Chemokines	C-X-C motif chemokine ligand 5, *Cxcl5*		47, 87 and 115
Glucocorticoid receptors	*Nr3c1*	Prednisone; triamcinolone; triamcinolone acetonide; beclomethasone; beclomethasone dipropionate; mometasone furoate; ciclesonide	86 and 115
Growth factors	Hepatocyte growth factor, *Hgf*		88
	Platelet-derived growth factor-α, *Pdgfa*		88
	Platelet-derived growth factor receptor-α, *Pdgfra*		
	Platelet-derived growth factor receptor-β, *Pdgfrb*		
	Connective tissue growth factor, *Ctgf*		88
	Fibroblast growth factor 1, *Fgf1*		88
	Fibroblast growth factor receptor 2, *Fgfr2*		
	c-Fos-induced growth factor, *Figf*		88
	Fms-related tyrosine kinase 1, *Flt1*		88
	Transforming growth factor-β receptor 3, *Tgfbr3*		88
Extracellular matrix	Elastin, *Eln*		88
	Fibulin 5, *Fbln5*		88
	Fibrillin 1, *Fbn1*		86 and 88
	Thrombospondin 1, *Thbs1*		88
Metallopeptidases	Metallopeptidase 14, *Mmp14*		88
	A disintegrin and metalloproteinase metallopeptidase 33, *Adam33*		88
	Adam9		86
	Tissue inhibitor of metalloproteinase, *Timp3*		88

Cytoskeleton proteins	Vasodilator-stimulated phosphoprotein, *Vasp*		88
Signalling ligands	Adrenomedullin, *Adm*		88
	Angiopoietin 1, *Angpt1*		88
	Cardiotrophin 1, *Ctf1*		88
cAMP/cGMP signalling	Phosphodiesterase 5A, cGMP-specific, *Pde5a*		88
Receptor tyrosine kinases	*Efnb2*		86
	Map4k1	Sunitnib; Bosutinib; Crizotinib; Dasatinib; Axitinib; Vanetanib; Sorafenib; Nilotinib;	86 and 87
	Map4k3	Sunitnib; Crizotinib; Dasatinib; Axitinib; Vanetanib;	86 and 87
	Slk		86
	Fyn		86
	Kdr		86
	Pkn2		86
	Taok		86
	Taok2		86
Drug metabolism	Carboxylesterase 3, *Ces3*		88
	Cytochrome P-450 4V3, *Cyp4v3*		88
	P-450 oxidoreductase, *Por*		88
	Flavin-containing monooxygenase 3, *Fmo3*		88
	Cytochrome P-45 2A3, *Cyp2a3*		88
	Carboxylesterase 1, *Ces1*		88
	Flavin-containing monooxygenase 2, *Fmo2*		86 and 88
	Sulfotransferase family, cytosolic, 1A1, *Sult1a1*		88
	Glutathione S-transferase-α3, *Gsta3*		88

[a] Subset of oscillating genes relevant to lung function collected from publications as referenced.[86–90]

to reduce inflammation (*e.g.*, systemic corticosteroids).[93] Considering that endogenous cortisol levels are well known to have circadian rhythmicity within the body, it is not surprising that systemic corticosteroids are more effective when administered at certain times of the day (Table 7.4).

Studies have shown that administering corticosteroids at 3 PM is more efficacious for treating nocturnal asthma and appears to work with endogenous circadian cortisol fluctuations, rather than disrupting its natural oscillation.[94] In support of this, administration of prednisolone was only effective at 3:00 PM (compared to at 8 AM or 8 PM, at which times administration was ineffective) in reducing the nocturnal dip in FEV1 and decreasing neutrophils, eosinophils, lymphocytes, and macrophages in BAL.[95] Infusion of methylprednisolone is more efficacious at certain times of the day compared to others; between 8 AM and 4 PM, methylprednisolone did not suppress adrenals, while infusion between 12 and 4 AM elicited severe adrenocortical suppression (with moderate suppression between 4 and 8 PM and 4 and 8 AM).[96]

Similar time-of-day efficacy has been found for inhaled corticosteroids (Table 7.4). One dose of triamcinolone acetate at 3:00 PM had similar efficacy as the same dose spread out across four times in the day,[97] and both of these schedules improved PEF in the morning and evening, an effect not seen when triamcinolone acetate was given in one dose at 8:00 AM.[98] Ciclesonide given either in the morning or evening did not show differences in alleviating symptoms, but only ciclesonide given in the evening improved morning PEF, better than when given in the morning.[99] Inhaled beclomethasone was found to be similarly efficacious when given once in the afternoon compared to when given twice during the day.[100] When mometasone is given once in the evening, it has been shown to be more effective in preserving lung function in the morning and reducing asthma symptoms throughout the day than when mometasone is given once in the morning or twice during the day.[101,102]

β2-Agonists are frequently used for targeting airway smooth muscle relaxation and promoting bronchodilation and anti-inflammatory effects. Long-acting β2-agonist salmeterol has been shown to be equally efficacious when given in one dose at night compared to being given two times during the day.[103] Clinically, administration of long-acting β2-agonists has been considered. Asthmatics receiving terbutaline at 8:00 AM and 8:00 PM, based on known fluctuations in lung function, experienced increased PEF and FEV1 throughout a 24-hour day and a reduced nocturnal dip that typically occurs in lung function around 4:00 AM.[104–106] In terms of short-acting β2-agonists, procaterol and fenoterol have been shown to induce clock gene *PER1* in human bronchial epithelial cells *in vitro*.[107]

Other treatments for asthma have time-of-day-dependent efficacy as well (Table 7.4). Leukotriene receptor antagonists, such as montelukast, have been shown to be more efficacious when given in the evening compared to the morning, as evidenced by increased FEV1.[108] Anti-inflammatory agents

Table 7.4 Time-of-day dependency in drug efficacy.[a]

Drug	Most efficacious time	Effect	Ref.
Oral corticosteroids			
Prednisone	3:00 PM	Decreased inflammatory response and increased nocturnal FEV1 *vs.* 8:00 AM and 8:00 PM administration	95
Inhaled corticosteroids			
Triamcinolone	3:00 PM	Single dose at 3:00 PM is equally efficacious as the same dose spread out over 4× per day	97 and 98
Beclomethasone	5:00 PM or 10:00 PM	Single dose at 5:00 PM or at 10:00 PM is equally efficacious as $\frac{1}{2}$ the dose given 2× per day (8:00 AM and 10:00 PM)	100
Mometasone	8:00–11:00 PM	Single dose around 10:00 PM preserved FEV1, PEF, $FEF_{25-75\%}$, in morning and improved asthma symptoms	101 and 102
		1× per day in the AM as efficacious as $\frac{1}{2}$ the dose given 2× per day	
Ciclesonide	—	Morning and evening dosing found equally efficacious with respect to spirometry, PEF, and symptoms	99
Intravenous corticosteroids			
Methylprednisolone	3:00 PM	Increased nocturnal PEF (*vs.* 3:00 AM, 7:00 AM, 7:00 PM administration)	94
β2-Adrenergic agonists			
Terbutaline	AM/PM	2× per day dosing when change in lung function expected to occur in circadian day results in increased PEF and FEV1 and lessened nocturnal dip in lung function	104–106
Bambuterol	PM	Single dose in evening provides 24-h effects equivalent to 3× per day terbutaline	116
Fluticasone furoate/Vilanterol	—	Morning and evening dosing found equally efficacious with respect to lung function in asthmatics	117
Combination formoterol + tiotropium	—	Single dose tiotropium more efficacious than 2× per day formoterol in COPD for daytime symptoms (no difference in nighttime symptoms)	112
		Addition of Formoterol in morning to tiotropium improved FEV1, FVC, and IC in COPD	

Table 7.4 (*Continued*)

Drug	Most efficacious time	Effect	Ref.
Combination formoterol + tiotropium and chronotype	Varies	Tiotropium in AM combined with formoterol 2× per day most efficacious for moderate-severe COPD Tiotropium in PM combined with formoterol 2× per day most efficacious for patients with exaggerated nighttime symptoms	118
Salmeterol	Evening	Single dose in the evening is equally efficacious as $\frac{1}{2}$ dose given 2× per day	103
Anticholinergic agents			
Oxitropium bromide	Varies	Some, but not all, patients with nocturnal asthma saw improved symptoms (improved morning dip in PEF) when single dose administered at night	110
Atropine	AM	Increased PEF in patients with asthma when administered at 4 AM and 4 PM, but larger increase in PEF when given in AM	111
Tiotropium alone	N/A	No significant differences noted in morning *versus* evening administration with respect to FEV1 in COPD	119
Leukotriene receptor antagonists			
Montelukast	Evening	Single dose in the evening more efficacious than single dose in the morning by increasing morning FEV1 and PEF	108
Theophyllines			
Theophylline "formulation"	8:00 PM	Single dose in the evening more efficacious than at 8:00 AM in terms of increased FEV1 and PEF, as well as improved symptoms	109
Theostat	Low AM/ high PM	2× per day low dose in AM and high dose in PM improved FEV1 in patients with COPD	113

[a] Time-of-day efficacies collected from publications as referenced.[86–90]

such as theophyllines are more effective when given at night than when given during the morning, and they alleviate symptoms, as shown by improved PEF at night and increased theophylline in the serum during the night when lung function dips.[109] Finally, anticholinergic agents given in high enough doses and given either late at night or very early in the morning can increase nocturnal dips in PEF in asthmatics.[110,111]

7.3.3 Chronotherapy in COPD

While studies have shown circadian fluctuations in lung function such as FEV1 in patients with COPD, and rhythmicity dips around 4:00 AM similar to asthmatics, patients with COPD do not exhibit the same exaggerated amplitude of rhythms that is associated with asthma. Current treatments for patients with COPD include long-acting muscarinic antagonists and long-acting β2-agonists. The addition of long-acting β2-agonist formoterol to tiotropium administered once a day improved FEV1, FVC, and inspiratory capacity in patients with COPD over a 24-hour period[112] with a second dose of formoterol in the evening providing additive beneficial effects in lung function during the night.[112] Theophyllines have also been used to treat patients with COPD with studies showing long-acting theostat providing the most benefit when given at night[113] (Table 7.4).

7.3.4 Goals for Chronotherapy in Airway Disease

There are robust data on variability in efficacy of therapeutics for multiple lung diseases, but there are some pitfalls that create opportunities for improvement in research studies and implementation of chronotherapeutics in the clinic. Although time-of-day-dependent efficacy is evident, it has not necessarily been put into practice. This may be because more precise studies are needed to understand mechanisms and cause-and-effect relationships. Major considerations need to be taken into account in studies investigating efficacy of therapeutics such as individual patient chronotype, the current treatment regime, adherence to the prescribed regime, the duration of the study, and numerous time points to measure lung function and symptom severity/variability. Additionally, mechanistic studies are needed to gain a better understanding of how chronotherapeutics leverage endogenous clock systems to attenuate symptom severity and variability. This must start with understanding the basic mechanisms of the circadian component of lung physiology and immune response. Furthermore, understanding the connection between lung function, immunity, and circadian networks may provide a more personalized approach to treatments by using biomarkers of the clock and clock-controlled pathways to determine when to administer certain treatments for individual patients (Figure 7.2).

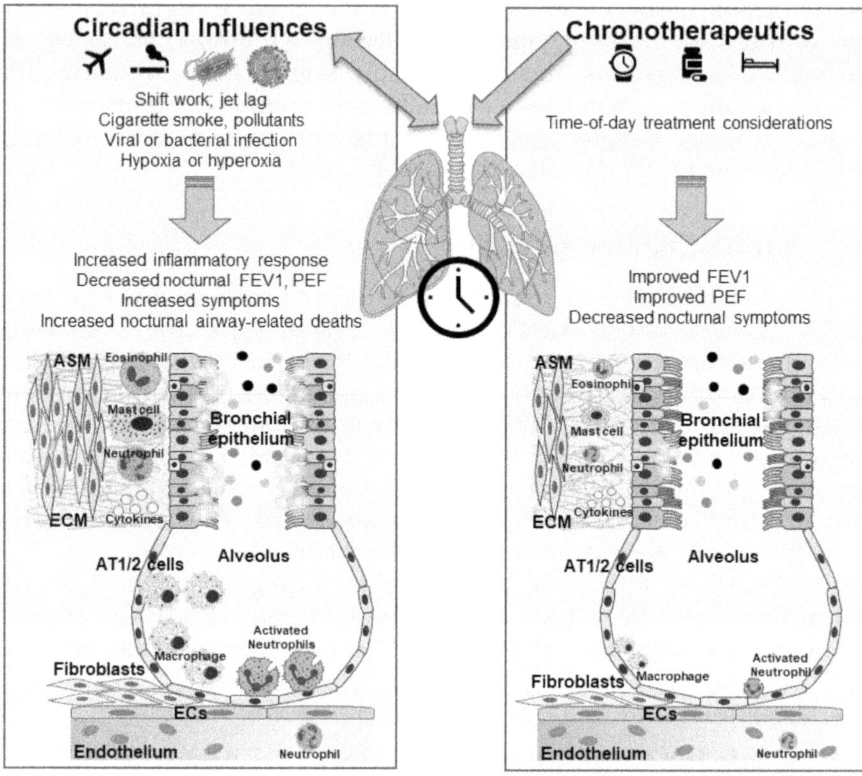

Figure 7.2 The lung clock. Factors that influence the circadian clock in the lung are shift work/jet lag, cigarette smoke/pollutants, viral or bacterial infection, or shifts in oxygen exposure. Reciprocally, disrupted circadian rhythms in the lung can increase susceptibility to infection with poorer recovery outcomes. Chronotherapeutics may maintain lung rhythmicity to provide protection from lung disease. The figure made in part using a Motifolio Scientific Illustration Toolkit.

7.4 Conclusion

From the Greek–Roman physicians in the 2nd century to the robust and flourishing field of circadian rhythms in the lung in the present day, the importance of clocks in lung function and disease is undeniable. Many advances in science and medicine have enabled researchers to investigate underlying mechanisms involved in lung clocks, which of course unveils additional questions:

1. There are over 40 types of cells involved in the pulmonary system. Do molecular clocks have unique functions in each cell type of the lung? How do the various cell types communicate with each other in terms of circadian synchronization?

2. What are the differences between species (and in particular our nocturnal rodent model systems as compared to humans) in lung clocks and how do they impact our understanding of scientific findings?
3. What are the roles of lung clocks in the developing lung and adaptation to postnatal life where oxygen exposure is relatively hyperoxic?
4. Is there sexual dimorphism in lung clocks leading to sex-based differences in airway disease exacerbations and time-of-day responsiveness to treatment?
5. How can our understanding of the circadian component of lung function and disease help make meaningful improvements in chronotherapeutic strategies for patients?

The future of the field of chronobiology with respect to the lung is both vast and intellectually exhilarating. Having readily available tools to investigate circadian rhythm, the lung, and its cellular clocks (*e.g.*, luciferase animals, light exposure regimens, human sample biobanks, large-scale sequencing and bioinformatics technologies, *etc.*) are undoubtedly vital resources that the field can and should leverage. Taking advantage of these opportunities will enable researchers to understand endogenous circadian components of airway physiology that can ultimately shed light on chronotherapeutics that can be used to optimize therapeutic approaches in the clinic.

Abbreviations

FEV1	Forced expiratory volume in 1 second
PEF	Peak expiratory flow
FRC	Functional residual capacity
IC	Inspiratory capacity
COPD	Chronic obstructive pulmonary disease
FVC	Forced vital capacity
V_E	Pulmonary ventilation
P_ACO_2	Alveolar CO_2 pressure
P_aCO_2	Arterial CO_2 pressure
CXCL5	C-X-C motif chemokine ligand 5
NR1D1	Nuclear receptor subfamily 1 group D member 1
sG_{aw}	Specific airway conductance
NK	Natural killer (cells)
RORs	Retinoic acid receptor-related orphan receptors

Acknowledgements

The work was supported by an American Heart Association Postdoctoral Fellowship 20POST35210002 (Bartman), the Mayo Clinic Specialized Center of Research Excellence and Women's Health Research Center Career

Enhancement Core U54 AG044170 (Bartman), and grants from the US National Institutes of Health R01 HL088029 and R01 HL056470 (Prakash). Motifolio Scientific Illustration Toolkit was used for generating the figure.

References

1. M. Karamanou and G. Androutsos, *Am. J. Respir. Crit. Care Med.*, 2012, **186**, 571.
2. B. Lemmer, *Chronobiol. Int.*, 2019, **36**, 883–885.
3. B. Lemmer, *Chronobiol. Int.*, 2009, **26**, 1019–1068.
4. A. Sakula, *Thorax*, 1984, **39**, 248–254.
5. H. D. Kerr, *Arch. Environ. Health*, 1973, **26**, 144–152.
6. E. Guberan, M. K. Williams, J. Walford and M. M. Smith, *Br. J. Ind. Med.*, 1969, **26**, 121–125.
7. M. R. Hetzel and T. J. Clark, *Thorax*, 1980, **35**, 732–738.
8. G. Ryan, K. M. Latimer, J. Dolovich and F. E. Hargreave, *Thorax*, 1982, **37**, 423–429.
9. R. J. Martin, L. C. Cicutto and R. D. Ballard, *Am. Rev. Respir. Dis.*, 1990, **141**, 33–38.
10. P. Gervais, A. Reinberg, C. Gervais, M. Smolensky and O. DeFrance, *J. Allergy Clin. Immunol.*, 1977, **59**, 207–213.
11. E. A. Phillipson, *Am. Rev. Respir. Dis.*, 1978, **118**, 909–939.
12. D. P. White, J. V. Weil and C. W. Zwillich, *J. Appl. Physiol.*, 1985, **59**, 384–391.
13. R. Stephenson, R. M. Mohan, J. Duffin and T. M. Jarsky, *Am. J. Physiol.*, 2000, **278**, R282–286.
14. C. M. Spengler, C. A. Czeisler and S. A. Shea, *J. Physiol.*, 2000, **526**(Pt 3), 683–694.
15. J. P. Mortola, *Eur. J. Appl. Physiol.*, 2004, **91**, 119–129.
16. T. J. Clark, *Chest*, 1987, **91**, 137S–141S.
17. M. Kraft, R. J. Martin, S. Wilson, R. Djukanovic and S. T. Holgate, *Am. J. Respir. Crit. Care Med.*, 1999, **159**, 228–234.
18. M. R. Hetzel, T. J. Clark and M. A. Branthwaite, *Br. Med. J.*, 1977, **1**, 808–811.
19. G. M. Cochrane and J. H. Clark, *Thorax*, 1975, **30**, 300–305.
20. J. R. Bateman and S. W. Clarke, *Thorax*, 1979, **34**, 40–44.
21. J. T. Li and C. E. Reed, *Am. J. Med.*, 1985, **79**, 10–15.
22. L. R. Bagg and D. T. Hughes, *Eur. J. Respir. Dis.*, 1980, **61**, 298–302.
23. T. L. Petty, *Am. J. Med.*, 1988, **85**, 21–23.
24. W. T. McNicholas and M. X. Fitzgerald, *Br. Med. J.*, 1984, **289**, 878.
25. V. G. Tirlapur, *Br. Med. J.*, 1984, **289**, 1540.
26. C. L. Tsai, B. E. Brenner and C. A. Camargo Jr., *Chronobiol. Int.*, 2007, **24**, 699–713.
27. R. D. Levin, M. A. Daehler, J. F. Grutsch, J. Quiton, C. G. Lis, C. Peterson, D. Gupta, K. Watson, D. Layer, S. Huff-Adams, B. Desai, P. Sharma, M. Wallam, M. Delioukina, P. Ball, M. Bryant, M. Ashford,

D. Copeland, M. Ohmori, P. A. Wood and W. J. Hrushesky, *Br. J. Cancer*, 2005, **93**, 1202–1208.
28. S. Carnio, R. F. Di Stefano and S. Novello, *Lung Cancer: Targets Ther.*, 2016, **7**, 73–82.
29. W. J. Hrushesky, J. Grutsch, P. Wood, X. Yang, E. Y. Oh, C. Ansell, S. Kidder, C. Ferrans, D. F. Quiton, J. Reynolds, J. Du-Quiton, R. Levin, C. Lis and D. Braun, *Integr. Cancer Ther.*, 2009, **8**, 387–397.
30. C. Lee, D. R. Weaver and S. M. Reppert, *Mol. Cell. Biol.*, 2004, **24**, 584–594.
31. X. Yang, P. A. Wood, E. Y. Oh, J. Du-Quiton, C. M. Ansell and W. J. Hrushesky, *Breast Cancer Res. Treat.*, 2009, **117**, 423–431.
32. X. Yang, P. A. Wood, C. M. Ansell, D. F. Quiton, E. Y. Oh, J. Du-Quiton and W. J. Hrushesky, *Chronobiol. Int.*, 2009, **26**, 1323–1339.
33. R. G. Love, *Br. J. Ind. Med.*, 1983, **40**, 153–159.
34. P. S. Cunningham, P. Meijer, A. Nazgiewicz, S. G. Anderson, L. A. Borthwick, J. Bagnall, G. B. Kitchen, M. Lodyga, N. Begley, R. V. Venkateswaran, R. Shah, P. F. Mercer, H. J. Durrington, N. C. Henderson, K. Piper-Hanley, A. J. Fisher, R. C. Chambers, D. A. Bechtold, J. E. Gibbs, A. S. Loudon, M. K. Rutter, B. Hinz, D. W. Ray and J. F. Blaikley, *Proc. Natl. Acad. Sci. U. S. A.*, 2020, **117**, 1139–1147.
35. E. S. Schernhammer, D. Feskanich, G. Liang and J. Han, *Am. J. Epidemiol.*, 2013, **178**, 1434–1441.
36. R. J. Maidstone, J. Turner, C. Vetter, H. S. Dashti, R. Saxena, F. Scheer, S. A. Shea, S. D. Kyle, D. A. Lawlor, A. S. I. Loudon, J. F. Blaikley, M. K. Rutter, D. W. Ray and H. J. Durrington, *Thorax*, 2021, **76**, 53–60.
37. F. Halberg, E. A. Johnson, B. W. Brown and J. J. Bittner, *Proc. Soc. Exp. Biol. Med.,*, 1960, **103**, 142–144.
38. Z. Zaslona, S. Case, J. O. Early, S. J. Lalor, R. M. McLoughlin, A. M. Curtis and L. A. J. O'Neill, *Am. J. Physiol.*, 2017, **312**, L855–L860.
39. M. Pariollaud, J. E. Gibbs, T. W. Hopwood, S. Brown, N. Begley, R. Vonslow, T. Poolman, B. Guo, B. Saer, D. H. Jones, J. P. Tellam, S. Bresciani, N. C. Tomkinson, J. Wojno-Picon, A. W. Cooper, D. A. Daniels, R. P. Trump, D. Grant, W. Zuercher, T. M. Willson, A. S. MacDonald, B. Bolognese, P. L. Podolin, Y. Sanchez, A. S. Loudon and D. W. Ray, *J. Clin. Invest.*, 2018, **128**, 2281–2296.
40. A. J. Davidson, O. Castanon-Cervantes, T. L. Leise, P. C. Molyneux and M. E. Harrington, *Eur. J. Neurosci.*, 2009, **29**, 171–180.
41. H. Hadden, S. J. Soldin and D. Massaro, *J. Appl. Physiol.*, 2012, **113**, 385–392.
42. G. Wolff, M. J. Duncan and K. A. Esser, *J. Appl. Physiol.*, 2013, **115**, 373–382.
43. V. Lagishetty, P. T. Parthasarathy, O. Phillips, J. Fukumoto, Y. Cho, I. Fukumoto, H. Bao, R. Cox Jr., L. Galam, R. F. Lockey and N. Kolliputi, *Am. J. Physiol.*, 2014, **306**, C999–C1007.
44. H. Li, C. Wang, J. Hu and J. Tan, *Int. Immunopharmacol.*, 2014, **18**, 249–254.

45. M. Keller, J. Mazuch, U. Abraham, G. D. Eom, E. D. Herzog, H. D. Volk, A. Kramer and B. Maier, *Proc. Natl. Acad. Sci. U. S. A.*, 2009, **106**, 21407–21412.
46. J. E. Gibbs, J. Blaikley, S. Beesley, L. Matthews, K. D. Simpson, S. H. Boyce, S. N. Farrow, K. J. Else, D. Singh, D. W. Ray and A. S. Loudon, *Proc. Natl. Acad. Sci. U. S. A.*, 2012, **109**, 582–587.
47. J. Gibbs, L. Ince, L. Matthews, J. Mei, T. Bell, N. Yang, B. Saer, N. Begley, T. Poolman, M. Pariollaud, S. Farrow, F. DeMayo, T. Hussell, G. S. Worthen, D. Ray and A. Loudon, *Nat. Med.*, 2014, **20**, 919–926.
48. J. A. Haspel, S. Chettimada, R. S. Shaik, J. H. Chu, B. A. Raby, M. Cernadas, V. Carey, V. Process, G. M. Hunninghake, E. Ifedigbo, J. A. Lederer, J. Englert, A. Pelton, A. Coronata, L. E. Fredenburgh and A. M. Choi, *Nat. Commun.*, 2014, **5**, 4753.
49. R. Casale and P. Pasqualetti, *Respiration*, 1997, **64**, 251–256.
50. R. Casale, G. Natali, D. Colantonio and P. Pasqualetti, *Thorax*, 1992, **47**, 801–803.
51. J. W. Hwang, I. K. Sundar, H. Yao, M. T. Sellix and I. Rahman, *FASEB J.*, 2014, **28**, 176–194.
52. I. K. Sundar, T. Ahmad, H. Yao, J. W. Hwang, J. Gerloff, B. P. Lawrence, M. T. Sellix and I. Rahman, *Sci. Rep.*, 2015, **4**, 9927.
53. S. Gebel, B. Gerstmayer, P. Kuhl, J. Borlak, K. Meurrens and T. Muller, *Toxicol. Sci*, 2006, **93**, 422–431.
54. V. T. Vasu, C. E. Cross and K. Gohil, *Integr. Cancer Ther.*, 2009, **8**, 321–328.
55. V. Pekovic-Vaughan, J. Gibbs, H. Yoshitane, N. Yang, D. Pathiranage, B. Guo, A. Sagami, K. Taguchi, D. Bechtold, A. Loudon, M. Yamamoto, J. Chan, G. T. van der Horst, Y. Fukada and Q. J. Meng, *Genes Dev.*, 2014, **28**, 548–560.
56. C. Dong, R. Gongora, M. L. Sosulski, F. Luo and C. G. Sanchez, *Respir. Res.*, 2016, **17**, 4.
57. A. Ehlers, W. Xie, E. Agapov, S. Brown, D. Steinberg, R. Tidwell, G. Sajol, R. Schutz, R. Weaver, H. Yu, M. Castro, L. B. Bacharier, X. Wang, M. J. Holtzman and J. A. Haspel, *Mucosal Immunol.*, 2018, **11**, 97–111.
58. S. Sengupta, S. Y. Tang, J. C. Devine, S. T. Anderson, S. Nayak, S. L. Zhang, A. Valenzuela, D. G. Fisher, G. R. Grant, C. B. Lopez and G. A. FitzGerald, *Nat. Commun.*, 2019, **10**, 4107.
59. Z. Zhang, L. Hunter, G. Wu, R. Maidstone, Y. Mizoro, R. Vonslow, M. Fife, T. Hopwood, N. Begley, B. Saer, P. Wang, P. Cunningham, M. Baxter, H. Durrington, J. F. Blaikley, T. Hussell, M. Rattray, J. B. Hogenesch, J. Gibbs, D. W. Ray and A. S. I. Loudon, *FASEB J.*, 2019, **33**, 6226–6238.
60. J. Liu, G. Malkani, X. Shi, M. Meyer, S. Cunningham-Runddles, X. Ma and Z. S. Sun, *Infect. Immun.*, 2006, **74**, 4750–4756.
61. G. Mundigler, G. Delle-Karth, M. Koreny, M. Zehetgruber, P. Steindl-Munda, W. Marktl, L. Ferti and P. Siostrzonek, *Crit. Care Med.*, 2002, **30**, 536–540.

62. B. Haimovich, J. Calvano, A. D. Haimovich, S. E. Calvano, S. M. Coyle and S. F. Lowry, *Crit. Care Med.*, 2010, **38**, 751–758.
63. A. Kleber, S. Altmeyer, B. Wolf, A. Wolf, T. Volk, T. Fink and D. Kubulus, *Inflammation Res.*, 2014, **63**, 1023–1033.
64. T. Fink, M. Glas, A. Wolf, A. Kleber, E. Reus, M. Wolff, D. Kiefer, B. Wolf, H. Rensing, T. Volk and A. M. Mathes, *Crit. Care Med.*, 2014, **42**, e22–31.
65. S. Masri and P. Sassone-Corsi, *Nat. Med.*, 2018, **24**, 1795–1803.
66. G. Sulli, A. Rommel, X. Wang, M. J. Kolar, F. Puca, A. Saghatelian, M. V. Plikus, I. M. Verma and S. Panda, *Nature*, 2018, **553**, 351–355.
67. R. W. Logan, C. Zhang, S. Murugan, S. O'Connell, D. Levitt, A. M. Rosenwasser and D. K. Sarkar, *J. Immunol.*, 2012, **188**, 2583–2591.
68. M. Casanova-Acebes, J. A. Nicolas-Avila, J. L. Li, S. Garcia-Silva, A. Balachander, A. Rubio-Ponce, L. A. Weiss, J. M. Adrover, K. Burrows, N. A-González, I. Ballesteros, S. Devi, J. A. Quintana, G. Crainiciuc, M. Leiva, M. Gunzer, C. Weber, T. Nagasawa, O. Soehnlein, M. Merad, A. Mortha, L. G. Ng, H. Peinado and A. Hidalgo, *J. Exp. Med.*, 2018, **215**, 2778–2795.
69. T. Papagiannakopoulos, M. R. Bauer, S. M. Davidson, M. Heimann, L. Subbaraj, A. Bhutkar, J. Bartlebaugh, M. G. Vander Heiden and T. Jacks, *Cell Metab.*, 2016, **24**, 324–331.
70. S. Masri, T. Papagiannakopoulos, K. Kinouchi, Y. Liu, M. Cervantes, P. Baldi, T. Jacks and P. Sassone-Corsi, *Cell*, 2016, **165**, 896–909.
71. Y. Issah, A. Naik, S. Y. Tang, K. Forrest, T. G. Brooks, N. Lahens, K. N. Theken, M. Mermigos, A. Sehgal, G. S. Worthen, G. A. FitzGerald and S. Sengupta, *eLife*, 2021, **10**, e61241.
72. G. Wu, Y. Y. Lee, E. M. Gulla, A. Potter, J. Kitzmiller, M. D. Ruben, N. Salomonis, J. A. Whitsett, L. J. Francey, J. B. Hogenesch and D. F. Smith, *eLife*, 2021, **10**, e63003.
73. G. Yang, C. J. Wright, M. D. Hinson, A. P. Fernando, S. Sengupta, C. Biswas, P. La and P. A. Dennery, *Antioxid. Redox Signaling*, 2014, **21**, 17–32.
74. C. M. Bartman, A. Matveyenko, C. Pabelick and Y. S. Prakash, *Am. J. Physiol.*, 2021, **320**, L451–L466.
75. G. Manella, R. Aviram, N. Bolshette, S. Muvkadi, M. Golik, D. F. Smith and G. Asher, *Proc. Natl. Acad. Sci. U. S. A.*, 2020, **117**, 779–786.
76. Y. Adamovich, B. Ladeuix, M. Golik, M. P. Koeners and G. Asher, *Cell Metab.*, 2017, **25**, 93–101.
77. Y. Adamovich, B. Ladeuix, J. Sobel, G. Manella, A. Neufeld-Cohen, M. H. Assadi, M. Golik, Y. Kuperman, A. Tarasiuk, M. P. Koeners and G. Asher, *Cell Metab.*, 2019, **29**, 1092–1103 e1093.
78. E. J. O'Connell, C. A. Martinez, Y. G. Liang, P. A. Cistulli and K. M. Cook, *Am. J. Physiol.*, 2020, **319**, C533–C540.
79. G. Yang, L. Chen, G. R. Grant, G. Paschos, W. L. Song, E. S. Musiek, V. Lee, S. C. McLoughlin, T. Grosser, G. Cotsarelis and G. A. FitzGerald, *Sci. Transl. Med.*, 2016, **8**, 324ra316.

80. R. V. Kondratov, A. A. Kondratova, V. Y. Gorbacheva, O. V. Vykhovanets and M. P. Antoch, *Genes Dev.*, 2006, **20**, 1868–1873.
81. R. V. Khapre, A. A. Kondratova, O. Susova and R. V. Kondratov, *Cell Cycle*, 2011, **10**, 4162–4169.
82. R. V. Khapre, A. A. Kondratova, S. Patel, Y. Dubrovsky, M. Wrobel, M. P. Antoch and R. V. Kondratov, *Aging*, 2014, **6**, 48–57.
83. Y. Okada and M. Okada, *Mol. Biol. Rep.*, 2020, **47**, 1021–1032.
84. M. A. Choukrallah, J. Hoeng, M. C. Peitsch and F. Martin, *BMC Genomics*, 2020, **21**, 291.
85. S. M. C. Abo and A. T. Layton, *PLoS Comput. Biol.*, 2021, **17**, e1008514.
86. R. C. Anafi, L. J. Francey, J. B. Hogenesch and J. Kim, *Proc. Natl. Acad. Sci. U. S. A.*, 2017, **114**, 5312–5317.
87. R. Zhang, N. F. Lahens, H. I. Ballance, M. E. Hughes and J. B. Hogenesch, *Proc. Natl. Acad. Sci. U. S. A.*, 2014, **111**, 16219–16224.
88. S. Sukumaran, W. J. Jusko, D. C. Dubois and R. R. Almon, *J. Appl. Physiol.*, 2011, **110**, 1732–1747.
89. K. Krakowiak and H. J. Durrington, *Pulm. Pharmacol.*, 2018, **4**, 29–43.
90. K. K. Truong, M. T. Lam, M. A. Grandner, C. S. Sassoon and A. Malhotra, *Ann. Am. Thorac. Soc.*, 2016, **13**, 1144–1154.
91. I. K. Sundar, H. Yao, M. T. Sellix and I. Rahman, *Am. J. Physiol.*, 2015, **309**, L1056–1075.
92. N. Hayasaka, T. Yaita, T. Kuwaki, S. Honma, K. Honma, T. Kudo and S. Shibata, *Endocrinology*, 2007, **148**, 3316–3326.
93. K. F. Chung, S. E. Wenzel, J. L. Brozek, A. Bush, M. Castro, P. J. Sterk, I. M. Adcock, E. D. Bateman, E. H. Bel, E. R. Bleecker, L. P. Boulet, C. Brightling, P. Chanez, S. E. Dahlen, R. Djukanovic, U. Frey, M. Gaga, P. Gibson, Q. Hamid, N. N. Jajour, T. Mauad, R. L. Sorkness and W. G. Teague, *Eur. Respir. J.*, 2014, **43**, 343–373.
94. A. Reinberg, F. Halberg and C. J. Falliers, *Chronobiologia*, 1974, **1**, 333–347.
95. W. R. Beam, D. E. Weiner and R. J. Martin, *Am. Rev. Respir. Dis.*, 1992, **146**, 1524–1530.
96. F. Ceresa, A. Angeli, G. Boccuzzi and G. Molino, *J. Clin. Endocrinol. Metab.*, 1969, **29**, 1074–1082.
97. D. J. Pincus, S. J. Szefler, L. M. Ackerson and R. J. Martin, *J. Allergy Clin. Immunol.*, 1995, **95**, 1172–1178.
98. D. J. Pincus, T. R. Humeston and R. J. Martin, *J. Allergy Clin. Immunol.*, 1997, **100**, 771–774.
99. D. S. Postma, C. Sevette, Y. Martinat, N. Schlosser, J. Aumann and H. Kafe, *Eur. Respir. J.*, 2001, **17**, 1083–1088.
100. M. Gagnon, J. Cote, J. Milot, H. Turcotte and L. P. Boulet, *Chest*, 1994, **105**, 1732–1737.
101. M. Noonan, J. P. Karpel, G. W. Bensch, J. W. Ramsdell, D. R. Webb, K. B. Nolop and B. N. Lutsky, *Ann. Allergy, Asthma, Immunol.*, 2001, **86**, 36–43.

102. A. D'Urzo, J. P. Karpel, W. W. Busse, L. P. Boulet, M. E. Monahan, B. Lutsky and H. Staudinger, *Curr. Med. Res. Opin.*, 2005, **21**, 1281–1289.
103. P. Faurschou, A. M. Engel and O. C. Haanaes, *Allergy*, 1994, **49**, 827–832.
104. D. S. Postma, G. H. Koeter, J. J. Keyzer and H. Meurs, *J. Allergy Clin. Immunol.*, 1986, **77**, 471–477.
105. G. H. Koeter, D. S. Postma, J. J. Keyzer and H. Meurs, *Eur. J. Clin. Pharmacol.*, 1985, **28**, 159–162.
106. R. Dahl, H. Harving, L. Sawedal and S. Anehus, *Br. J. Dis. Chest*, 1988, **82**, 237–241.
107. M. Takata, N. Burioka, S. Ohdo, Y. Fukuoka, M. Miyata, M. Endo, H. Suyama and E. Shimizu, *Chronobiol. Int.*, 2005, **22**, 777–783.
108. M. J. Noonan, P. Chervinsky, M. Brandon, J. Zhang, S. Kundu, J. McBurney and T. F. Reiss, *Eur. Respir. J.*, 1998, **11**, 1232–1239.
109. G. E. D'Alonzo, M. H. Smolensky, S. Feldman, L. A. Gianotti, M. B. Emerson, H. Staudinger and V. W. Steinijans, *Am. Rev. Respir. Dis.*, 1990, **142**, 84–90.
110. C. I. Coe and P. J. Barnes, *Chest*, 1986, **90**, 485–488.
111. J. F. Morrison, S. B. Pearson and H. G. Dean, *Br. Med. J.*, 1988, **296**, 1427–1429.
112. J. A. van Noord, J. L. Aumann, E. Janssens, J. J. Smeets, J. Verhaert, B. Disse, A. Mueller and P. J. Cornelissen, *Eur. Respir. J.*, 2005, **26**, 214–222.
113. B. Bruguerolle, F. Philip-Joet, M. Parrel and A. Arnaud, *Chronobiol. Int.*, 1987, **4**, 381–385.
114. M. Wu, J. Zeng, Y. Chen, Z. Zeng, J. Zhang, Y. Cai, Y. Ye, L. Fu, L. Xian and Z. Chen, *Oncol. Rep.*, 2012, **27**, 1417–1428.
115. T. Okabe, R. Chavan, S. S. Fonseca Costa, A. Brenna, J. A. Ripperger and U. Albrecht, *J. Cell Sci.*, 2016, **129**, 4143–4154.
116. A. L. Kuusela, M. Marenk, G. Sandahl, J. Sanderud, K. Nikolajev and B. Persson, *Pediatr. Pulmonol.*, 2000, **29**, 194–201.
117. R. D. Kempsford, A. Oliver, J. Bal, L. Tombs and D. Quinn, *Respir. Med.*, 2013, **107**, 1873–1880.
118. C. Terzano, A. Petroianni, V. Conti, D. Ceccarelli, E. Graziani, A. Sanduzzi and S. D'Avelli, *Respir. Med.*, 2008, **102**, 1701–1707.
119. P. M. Calverley, A. Lee, L. Towse, J. van Noord, T. J. Witek and S. Kelsen, *Thorax*, 2003, **58**, 855–860.

CHAPTER 8

Chronobiology and the Kidney

MARTIN S. KNAPP†

12 Townsend St., Ivanhoe, Melbourne, 3079, Australia
Email: martin@knapp.id.au

8.1 Nephron Function in Health and Chronobiology

In healthy adults, in a stable state of hydration, the concentration of urine at night was observed to increase in early 19th-century studies. In classic studies, Verney in 1947[1] recorded changes in plasma and urine osmolality after altering the osmolality of plasma reaching the cranial contents. He concluded that osmoreceptors existed and communicated with the posterior pituitary to modify the release of a humoral anti-diuretic substance, which is now known to be arginine vasopressin (AVP). Later studies demonstrated a renal response subsequent to a change in AVP on a time scale that indicates the involvement of supplementary mechanisms to that demonstrated by Verney. Measurements of circulating AVP in health demonstrated a circadian rhythm in AVP levels and water excretion, with a peak occurring at the onset of the active phase.[2–5] This would not be expected if only AVP was responsible for the higher concentration and lower volume of urine excreted at night. A circadian rhythm of solute excretion, similar to that observed in healthy individuals, is observed when AVP is absent, as seen in cranial diabetes insipidus (CDI).[5] The collecting tubules of the nephron can increase the concentration of tubular filtrate at night as water reabsorption increases

†Senior Lecturer, University of Tasmania, Australia (retired). Previously: Specialist Physician and Nephrologist, Bristol and Nottingham, UK, Washington University, USA, Jeddah, Saudi Arabia and Universities of Melbourne and Tasmania, Australia.

Chemical Biology No. 23
Chronobiology and Chronomedicine: From Molecular and Cellular Mechanisms to Whole Body Interdigitating Networks
Edited by Germaine Cornelissen and Tsuyoshi Hirota

Published by the Royal Society of Chemistry, www.rsc.org

due to a decreased quantity of solute being excreted by each nephron. An increased volume of urine is excreted at night if AVP is absent. In CDI, the decreased solute levels in nephrons, caused by circadian rhythmicity in solute excretion, increase the ability to generate a certain concentration of the collecting tubular fluid and consequently lead to a decrease in urine output. Additional studies indicate that circadian molecular clocks also modify renal collecting tubular function and contribute to circadian rhythmicity[6–11] (see Figure 8.1).

The kidneys, in addition to the essential contributions by the nephrons, described above, also perform endocrine functions. These include secretion and regulation of erythropoietin (EPO), critical for red blood cell production. Anaemia, requiring administration of EPO, develops in most patients with late-stage kidney failure. It has only recently been elucidated that circadian EPO expression is regulated by master clock proteins (CRY1, CRY2, CLOCK, and BMAL1) with EPO acting distinctively within the complex regulatory network of erythropoiesis. Sciesielski *et al.*[6] recommended administration of

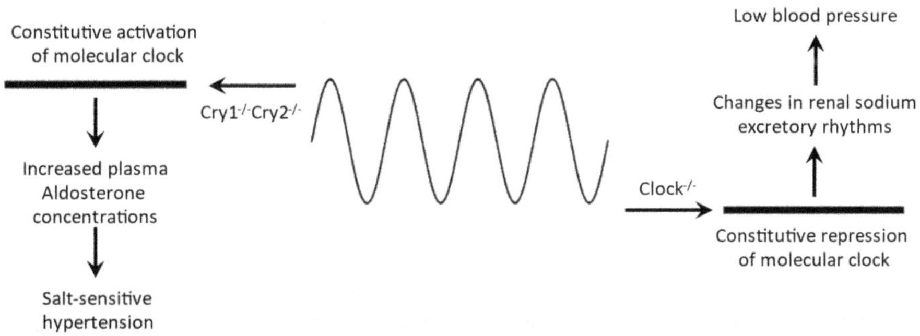

Figure 8.1 Molecular pathways and mechanisms directly controlled in the nephron by the tubular circadian clock. The proximal tubules of the nephron contain sodium-glucose linked transporter isoform1 (SGLT1) and the sodium-hydrogen exchanger isoform3 (NHE3), both located on the apical membrane and regulated by the circadian clock. Sodium is reabsorbed from the filtrate in a quantity appropriate to maintain sodium balance and pumped back into the blood by the basolateral Na, K-ATPase. In the loop of Henle, cells in the thick ascending loop sodium–potassium–chloride cotransporter isoform 2 (NKCC2) are regulated by the nuclear receptor ERRβ. These cells also contain renal outer medullary potassium channel (ROMK) responsible for pumping the appropriate quantity of potassium back into the filtrate, and also have a basolateral Na, K-ATPase. The distal convoluted tubule cells contain apically located sodium chloride cotransporters (NCCs). These cells reabsorb more Na^+ and Cl^- from the filtrate and are also regulated by the circadian clock. When nephron filtrate enters the collecting duct from the distal tubule, the epithelial sodium channel (ENaC) reabsorbs more Na^+ from the filtrate and is also regulated by the clock, as are FXYD5, CAV-1, and ET-1 with FXYD5 positively regulating the Na, K-ATPase, while CAV-1 negatively regulates ENaC. Reproduced from ref. 157 with permission from Elsevier, Copyright 2010.

recombinant human EPO late in the evening to patients with kidney insufficiency, just preceding its physiological circadian maximum around midnight. EPO is only available as an injection, often self-administered by patients, and their suggestion merits consideration as the supply of EPO is a significant expense to those funding the management of chronic renal failure.

Renin output from the kidney, which influences angiotensin levels and aldosterone production, plays an important role in the maintenance of blood pressure (BP) in health and relevant to the development and progression of hypertension (HTN), to the progression of chronic kidney failure (CKF), and potentially to end-stage renal disease (ESRD). Renin, angiotensin and aldosterone (RAA) all exhibit circadian periodicity, and have been studied by several groups considering the role of intracellular clock genes in periodicity. All of them are targets for pharmaceuticals developed for the management of cardiac and renal conditions and HTN. Studies on laboratory rats and mice provided valuable information on the mechanisms and control of the observed circadian variation and its disruption in genetically modified strains after ablation of the suprachiasmatic nuclei (SCN) and through medications, including those influencing the RAA and melatonin (MLT).[7,12–14]

8.2 Renal Function in Clinical Disorders and Chronobiology

8.2.1 Enuresis

In monosymptomatic nocturnal enuresis (MNE) (bedwetting), the central nervous system does not respond by waking when the bladder reaches its emptying volume (BEV). This situation is normal early in life but classified as enuresis at an empirically allocated age of five. Polysymptomatic nocturnal enuresis (PNE), in contrast, occurs when urinary tract symptoms also occur during daytime, with a greater range of possible diagnoses. These, especially a reduced BEV, may not be noticed in the daytime but are relevant during nocturnal sleep as bladder emptying will be earlier.

A loss of circadian rhythmicity in AVP was documented in some individuals with MNE. Nørgaard *et al.*[15] recorded reduced or absent circulating nocturnal anti-diuretic hormone (ADH), a situation that could be considered to be “nocturnal cranial diabetes insipidus”, with nocturnal levels of ADH lower than those in control groups without enuresis. This cohort was responsive to synthetic ADH, desmopressin acetate (DDAVP) (Minirin®), with lower nocturnal urine volumes and, due to slower filling of the bladder during sleep, they demonstrated an improved ability to detect BEV and respond to it, or sleep until morning without micturition. A subgroup of children with enuresis exhibited a reduced renal response to AVP, indicating nocturnal nephrogenic diabetes insipidus; some of them responded positively to evening administration of DDAVP. Others with MNE had hypercalciuria associated with increased sodium retention during daytime.[16]

Management strategies were introduced to establish a central nervous system that responds during sleep, either by waking to a full bladder or by delaying micturition. Experimental studies[17,18] resulted in the development of management strategies based on interventions with wetness-activated alarms for those not responding to education aimed at increasing awareness of when BEV has been reached. Wetness alarms were effective in most cases but not all.[19–21] Many of those not responding had a reduced nocturnal output of AVP. Adults with persisting MNE had a marked reduction in nocturnal renal sensitivity to the antidiuretic effect of AVP.[2] Bedtime administration of DDAVP reduced nocturnal urine flow and enuresis stopped or decreased. A combination of DDAVP and alarm use was found to be effective when success had not been achieved with either alone. Results were attributed to DDAVP delaying the first night-time bedwetting, with the alarm more likely to develop "conditioning" when micturition was delayed.[21]

8.2.2 Nocturia

Nocturia resulting from increased nocturnal urine production is more troublesome when associated with pathology or disturbed function in the lower urinary tract with a reduction in BEV. This is variable among individuals. When the bladder cannot accommodate the urine passed from the kidney during sleep, it results in nocturia or enuresis and, often, a reduction in the time of restorative slow wave sleep. A review of 52 relevant publications[22] concluded that 60% experienced nocturia without enuresis, defined as micturition twice or more on most nights, which is a threshold for nocturia to be "troublesome" and to influence the quality of life. This study established that cardiovascular dysfunction, polyuria, sleep disturbance, mental health, metabolic and inflammatory changes, health status and lower urinary tract symptoms could not be considered independently to be a single cause of nocturia. Polypharmacy, increasing age and body mass index (BMI) all had confounding effects. There were significant interactions between voiding at night and metabolic, cardiovascular, hormonal status, mental health, sleep and inflammatory changes indicating that nocturia is a likely marker of co-morbid poor health. Patients with lower urinary tract symptoms responding to treatment had generally reported only a modest reduction in nocturia episodes, not appreciably different from placebo. Nocturia was associated with compromised glucose homeostasis, reduced nocturnal plasma melatonin levels, increased catecholamine levels and a high likelihood of using sedatives.[22] Sleep disturbance should hence be considered when choosing a therapy for nocturia as well as lower urinary tract function and systemic disorders[22] (see Table 8.1). The possibility that the response of nocturia to medication (for example, with melatonin or with hypnotics) follows a circadian rhythm has been studied both in the laboratory and in randomised clinical trials (RCTs).[14,23–25] Potential toxicity and side effects of pharmaceuticals make it important to initially consider non-pharmaceutical interventions if there is no cause identified with specific

Table 8.1 Conditions associated with reversal of the normal circadian rhythm of urine flow and nocturia.[27]

(1) Associated with salt and water retention	
Congestive heart failure	Brod and Fejfar, 1950[128]
Cirrhosis of the liver	Borst and de Vries, 1950[127]
	Jones *et al.*, 1952[129]
Nephrotic syndrome	Koopman *et al.*, 1994[130]
Under-nutrition	Hutchinson, 1951[131]
(2) Adrenal disorders	
Cushing's syndrome	Doe *et al.*, 1960[132]
	Knapp *et al.*, 1967[5] Knapp, 1969[27]
Hyperaldosteronism	Lennon *et al.*, 1961[133]
	Bartter *et al.*, 1962;[134] Rossier, 2010[136]
	Doi *et al.*, 2010[137]
Steroid therapy	Rosenbaum *et al.*, 1952[135]
	Kowanko *et al.*, 1978[29]
Addison's disease	Doe *et al.*, 1960[132]
(3) Reno-vascular disorders	
Renal artery stenosis	
Anatomical	Rosenheim *et al.*, 1963[46]
Functional	Donaldson *et al.*, 1967[47]
	Ginn and Parry, 1964[138]
Hypertension	Berlyne *et al.*, 1965[139]
	Lemmer, 1994[140]
(4) Associated with abnormal wake/sleep rhythm	
Depressive illness	Elithorn *et al.*, 1966[141]
Trans-world travel	Flink and Doe, 1959[142]
(5) Others	
Chronic Renal Failure	Hillier, Knapp, Cove-Smith, 1980;[26] Hillier and Knapp, 1974[143]
Steatorrhoea	Flear *et al.*, 1959[144]
After head injury	de Wardener, 1958[145]
Parkinson's disease	Chen *et al.*, 2020,[155] Batla *et al.*, 2021[156]
Transplantation kidney	Berlyne *et al.*, 1968[146]
	Albertsen and Petersen, 1968[147]
Anorexia nervosa	Russell and Bruce, 1966[148]
Enuresis	Nørgaard, 1985[15]
Ideopathic Oedema	Kaye M 2008;[34] Dunnigan *et al.*, 2004[33]
Sleep disorders	Papworth *et al.*, 2022[152]

management indicated. Nocturia is intrusive, disrupting sleep and potentially causing falls. Decreasing or redistributing the daily fluid intake may reduce nocturnal urine volume. Daytime rest supine, with legs elevated, can increase daytime output, most likely when conditions are present that have a posture-related reduced daytime urine flow (see Table 8.1).[26,27] Pharmaceuticals with anti-cholinergic effects have been found to be useful for some individuals, as confirmed in RCTs, but they come with side effects, especially mouth dryness, a reason why some do not continue with this therapy. DDAVP is now established as a medication to consider for nocturia when systemic disease is not present, lower urinary tract dysfunction has been excluded, and when records of bladder emptying establish adequate bladder capacity with a normal or excess of urine during the sleep period. Some consider a daily total output exceeding 33% passed overnight as an indication for a trial of bedtime DDAVP. Serum sodium levels or serum osmolality need to be monitored to detect hypo-osmolality and adjust or stop DDAVP as hypo-osmolality and its treatment are potentially dangerous. Limiting the DDAVP effect to the sleep period is facilitated by formulations for mucosal absorption from the mouth. Nocdurna® is a wafer providing an appropriate duration of action of DDVP that restores, in most cases, a more normal circadian pattern of anti-diuretic action output without water retention during the day.[28] Careful monitoring, with dose reductions, especially in females, is recommended as inappropriate retention of water can lead to serious outcomes (the Food and Drug Administration in the USA issued an alert specific to this potential complication).

8.2.3 Nocturia and Glucocorticoids

Nocturia is one of several side effects experienced by patients receiving corticosteroids, notably when ingested as an evening medication. Urine flow was studied both in rats and in patients recording measurements at home, contrasting changes in the timing of once-daily corticosteroid administration.[29] In patients, a change in the circadian rhythm of urine flow was documented when contrasting morning with evening administration in all patients taking doses higher than 10 mg per day of prednisolone. In rat studies, the peak excretion of urine and potassium moved when dexamethasone was administered at the end of activity (lights off) but rhythms remained similar to those prior to injections when dexamethasone was given after light onset (see Figure 8.2).

Animals demonstrating this phase shift in excretion were documented to have some suppression of the adreno–pituitary axis, a possible explanation for the disruption in rhythmicity of urinary excretion. In patients studied to assess adreno–pituitary function by measuring the 24-hour excretion of 11-hydroxycorticosteroid (11-OHCS), those with low 11-OHCS excretion had disturbed urinary excretion patterns, suggesting that this may have been the mechanism for the abnormalities of urinary excretion. Other clinical and laboratory observations, reviewed later, indicate that an excess of aldosterone and other mineralocorticoids, whether administered or associated with pathological conditions, cause clinical syndromes with disturbed circadian rhythmicity, including HTN and disturbances of electrolytes and acid/base balance.

The possibility that nocturia might be a disturbance in the chronobiology of the bladder of some patients has not been considered by most investigators. Chihara *et al.*[30] documented that glucocorticoids are a key cue for a bladder clock *in vitro* and *in vivo*. Analysis of data from a *Bmal1*-dLuc human urothelial cell-line showed significant shifts in gene expression after cortisol

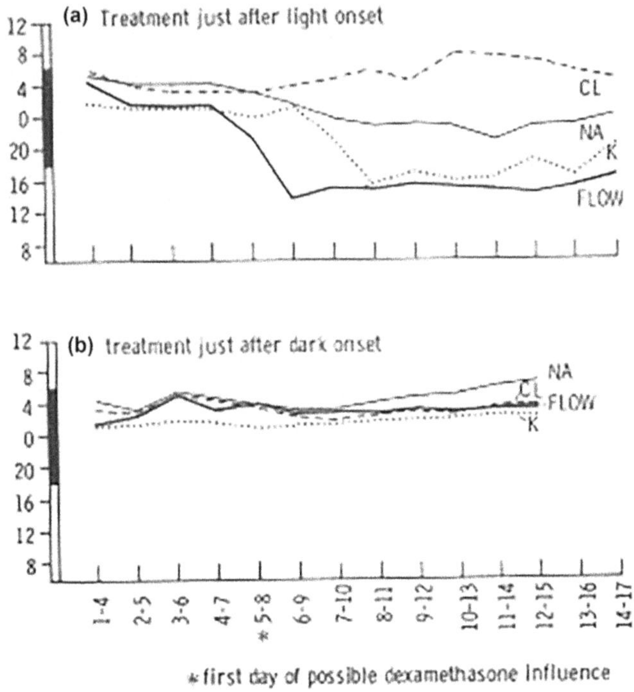

Figure 8.2 Corticosteroid administration and circadian rhythm of urine flow in rats. Once-daily glucocorticoid was administered to a group of rats kept in LD12:12 either just after light onset (a) or just after dark onset (b). Before dexamethasone treatment (on days 1–7 when saline is given), 24-hour acrophases (time of estimated maximum) of urinary flow, Na, K, and Cl were stable, peaking toward the end of the dark phase. Acrophases remained stable when dexamethasone was administered just after dark onset (b) but underwent large changes when it was administered just after light (inactivity) onset (a). Data analysed sequentially by fitting a 24-hour cosine curve over 4 days of data, moved in 1-day increments.[29]

treatment. Rhythmic bladder-clock gene expression *in vivo* remained unchanged by bilateral adrenalectomy, but it shifted 4 hours forward by corticosterone administration during the light (rest) phase. In mice that underwent both bilateral adrenalectomy and corticosterone administration during the inactive phase, the bladder clock shifted 8 to 12 hours. The 24-hour rhythm of volume voided per micturition decreased in these mice, while they maintained 24-hour activity rhythms. The authors concluded that the 24-hour rhythm of glucocorticoid signalling may be a synchronizer that overcomes other bladder-clock entrainment factors and coordinates the 24-hour rhythm of volume voided per micturition.[30]

Clinicians and their patients need to be aware that the time of administration of corticosteroid medication may be responsible for troublesome nocturia when some or all is taken in the evening, and that shifting the time of administration may resolve this issue. Potentially dangerous suppression of the adreno–pituitary axis may also occur with prolonged evening administration of corticosteroids. This should be avoided unless established as being essential to control a disease process or to reduce the risk of rejection in a transplanted organ (see Section 8.5.1). A schedule of administration that nearly simulates the endogenous rhythmicity, while slowly decreasing the prescribed daily dose,[31] reduces side-effects.

8.2.4 Nocturia, Congestive Cardiac Failure (CCF) and Other Systemic Conditions

Nocturia can be the result of disease and can be reduced with management of that disease (see Table 8.1). Nocturia becomes common in older age groups, even when healthy.[32] It also occurs in a range of conditions observed to be associated with polyuria, an increased total 24-hour output of urine, and in those with an increased nocturnal output of urine due to disruption and reduction in the circadian rhythm of urine output or reversal in the phase of that rhythm (see Table 8.1).

Individuals exhibiting inappropriate fluid retention when no pathological process can be identified are classified as having "idiopathic" oedema. This syndrome, found predominantly in females, is characterized by unexplained daytime fluid retention with a nocturnal diuresis. In a study involving 196 patients with symptoms of troublesome fluid retention, but lacking a diagnosis to explain their symptoms of subcutaneous swelling of face, hands, breasts, abdomen and feet and without pitting oedema on examination, there was significantly more evidence of affective disturbance than in controls. Dunnigan *et al.*[33] failed to detect symptoms, signs or abnormal observations to explain the symptoms and the reversal of circadian rhythmicity in urine excretion. Despite a possible genetic component, the pathophysiology remains unexplained. The authors suggested that the swelling resulted from increased capillary permeability with secondary hypovolaemia compensated by oliguria, sodium and water retention and weight gain, aggravated by walking or standing. In older men, Kaye[34,35] observed that individuals with nocturia experienced sodium and water retention during the day and natriuresis and diuresis overnight; nocturia occurred when weight gain was greatest, and was reduced by an afternoon diuretic or daytime recumbency. He concluded that idiopathic nocturia was due to daytime volume expansion associated with the upright position. The literature on this topic, including studies by Birder *et al.*, identifies alternative pathophysiology, including the modulation of bladder smooth muscle by AVP[36,37] (see Section 8.2.5).

8.2.5 Nocturia and Lower Urinary Tract Syndromes (LUTSs)

Reaching the maximal capacity of the bladder (BEV) triggers a central need to empty, often associated with a change from sleep towards alertness.

In some individuals, nocturia is explained only by a reduction in the BEV, with no increase in the volume of nocturnal urine[37] coming from the kidneys. A reduced BEV may be associated with changes in bladder epithelium, including interstitial cystitis, and, in women, it may be associated with pelvic floor prolapse and the laxity of the uterosacral ligament (USL). Conditions that are not explained by these diagnoses are termed "overactive bladder (OAB)". Bladder emptying has received less attention in research than urine production, contributing to the paucity of effective management strategies for OAB. A classification was proposed to include "idiopathic nocturnal detrusor overactivity"[37,38] as a separate entity to nocturnal polyuria syndromes, to bring a focus on the importance of bladder pathophysiology and on the contribution of cellular function, the urinary bladder urothelium and circadian rhythmicity. The bladder is capable of sensory and transducer function, mediated through a network of functional receptors and ion channels.[36,37] Intravenous or *in situ* AVP evokes a huge increase in intravesical pressure, suggesting that AVP may be involved in modulating bladder smooth muscle tone through the regulation of intravesical pressure. Treatment with peripheral nerve blocks along with the addition of local corticosteroids led to a highly significant improvement in patients who did not respond to other treatments for OAB/BPS with nocturia,[39] supporting the hypothesis that neurogenic inflammation was causing the symptoms. Others have proposed that in some cases, the inflammation is a response to urinary pathogens not detected by routine methods of detection[40,41] but responding to long courses of antibiotics.[40,41] Others have experienced improvements on a gluten-free diet, suggesting that non-coeliac gluten sensitivity may account for some individuals with LUTS who do not respond to other treatment strategies.[42] Nocturia should thus be regarded as a type of circadian dysfunction, requiring further study. There is an urgent need for new treatment modalities as there are many older persons with urgency of nocturnal micturition and unexplained low bladder emptying volumes, many with an inadequate response to available therapies and with substantial comorbidity from nocturnal falls associated with nocturia.

8.3 Hypertension and Chronobiology

8.3.1 Blood Pressure and Chronobiology

The association between HTN and kidney disease was identified by Richard Bright in 1827.[43] The importance of HTN in relation to human health has been increasingly recognised to shorten the lifespan and increase comorbidity. In addition to "essential" and "malignant" HTN, some cases can be considered "secondary" HTN, as a reason for HTN is established (such as increased aldosterone), and specific treatment may be available. Patients with HTN and an adrenal adenoma secreting an excess amount of aldosterone than that appropriate for homeostasis were first described by Conn and Louis in 1955.[44] BP often returns to within reference ranges with removal of the adenoma. Other similar patients had an increased secretion of aldosterone, but without neoplastic changes in the adrenal, and this condition was termed idiopathic hyperaldosteronism (IHA). The pharmaceutical spironolactone was developed and found to be effective in assisting diagnosis and in the non-surgical management of patients. Improvements in diagnostic methodology, including the laboratory assays of renin, angiotensin and aldosterone, have identified a variable percentage of those with HTN being associated with hyperaldosteronism[45] and others identified with an alternative hormonal aetiology for their HTN, including some with excess renin production, secondary to stenosis or compression of a renal artery.[46,47]

Some laboratory studies aimed at improving the understanding of the pathophysiology of HTN and at developing and utilising pharmaceuticals to modify HTN consider chronobiological aspects. Lemmer *et al.*, in the first of an elegant series of experiments in transgenic TGR(mREN-2)27 rats, characterized by fulminant HTN with a disturbance in circadian BP regulation and inversed circadian BP profiles, investigated circadian rhythms in several hormones participating in BP regulation.[12,13,48] They observed that peak concentrations of corticosterone at around 1500 h were more than two times higher in TGR rats than in control Sprague-Dawley rats, whereas no differences were observed during this active "lights off" phase. They concluded that in TGR rats, the overall increase in plasma renin activity and aldosterone may contribute to the elevated BP and the high levels of corticosterone and plasma renin activity during daytime may be involved in the inverse circadian BP profiles recorded in transgenic animals.[12,13] Okamura *et al.*[7,49] and Logan[50] reviewed studies conducted in another experimental model, the arrhythmic *Cry1/Cry2* double-knockout (*Cry*-null) mouse, with salt-sensitive HTN and hyperaldosteronism. Earlier, also in *Cry*-null mice, Nugrahaningsih *et al.*[51] observed that decreasing the BP without blocking aldosterone action did not provide beneficial effects to the kidney. They documented the involvement of a new type of aldosterone-synthesising enzyme controlled by the circadian clock and angiotensin II, concluding that *Cry*-null mice can be considered as a model for idiopathic hyperadrenalism (IHA) and its complications, and that the temporal components of an aldosterone-regulating system in the pathophysiology of HTN are promising targets for future research. Bothou *et al.* supported this finding and provided a comprehensive review documenting links between aldosterone excess and metabolic complications.[44,52]

The nocturnal fall in BP was noted in healthy humans soon after the availability of sphygmomanometers to measure BP and was also noted in some patients with HTN. The development of portable devices allowed studies for 24 hours and longer, both during activity and sleep and outside of hospitals and laboratories.[53,54] Debate continues on the role of ambulatory BP monitoring (ABPM) and devices for self-measurements in clinical decision-making, including discussions on the merits of considering nocturnal values when advising on medications and on the selection, frequency and circadian timing of that medication.[50,55]

Table 8.2 Non-dipping nocturnal blood pressure in "malignant" hypertension.[56]

Blood pressure during sleep in malignant and essential hypertension					
		Essential hypertension		Malignant hypertension	
Mean	Blood pressure	Day–sleep difference, Δ	P-value (H_0: $\Delta = 0$)	Day–sleep difference, Δ	P-value (H_0: $\Delta = 0$)
Day	Systolic	15 mmHg	<0.01	0.1 mmHg	>0.90
	Diastolic	9.1 mmHg	<0.01	−2.6 mmHg	0.2 to 0.1
Sleep	Systolic	15 mmHg	<0.01	−1.6 mmHg	0.7 to 0.06
	Diastolic	8.3 mmHg	<0.01	−4.9 mmHg	0.2 to 0.3

In 1963, we reported on our study involving automatic recording of BP over 24 hours[56] in a subgroup of patients with severe ("accelerated" or "malignant") HTN. Patients with malignant HTN (MHTN), with grade 3 or 4 optic fundal changes, treated with the then-available pharmaceuticals, were expected to progress to end-stage kidney failure within two years. We were the first to report on an absent or reversed nocturnal fall in BP associated with sleep in patients with MHTN. This non-dipping or reversed nocturnal BP contrasted with that of patients studied with "essential" HTN, who were known to have a much better renal prognosis and exhibited a nocturnal fall when BP was recorded during sleep (see Table 8.2). It was proposed that medication that provided improved control of HTN throughout the 24-hour period should be an objective of treatment to reverse or prevent MHTN and improve prognosis. We observed that bringing the BP nearer to "normal" with medication for only some part of the 24 hours could reverse the "malignant phase" in some patients (with change in the optic fundal changes to less than grade 3 and better renal outcomes).

8.3.2 Blood Pressure and Evening *vs.* Morning Dosing

Studies published during the subsequent decades were reviewed in 2015 by Stranges *et al.*[57] They concluded that despite evidence and potential to reduce cardiovascular risk, chronotherapy is not uniformly recommended to treat HTN, with further work needed regarding patient selection and use of anti-hypertensive drugs to evaluate the definitive impact of chronotherapy on cardiovascular outcomes. Additional reviews[58,59] concluded that there was substantial evidence that evening administration could reverse non-dipper patterns in hypertensive patients with chronic kidney disease (CKD), but they did not find a significant difference in all-cause or cardiovascular mortality. As there was no documented increased risk associated with night-time administration of anti-hypertensive drugs, this regimen was suggested to be seriously considered for management of HTN,[59] but other reviews supported the need for further RCTs and for more "personalised" prescription for individuals.[60,61]

The Hygia study, an RCT of individuals with HTN in a multi-centre prospective chronotherapy trial in primary care settings in a single Spanish community, provided regular ABPM data with follow-up for a median duration of 6.3 years.[62] This study provided outcome data from 10 084 patients, allocated to either waking or bedtime medication, favouring bedtime over awakening administration time for control of BP. More individuals achieved nocturnal dipping ($p < 0.001$), with long-term follow-up demonstrating reductions in cardiovascular events. Later reviews of published treatment-time trials by Hermida *et al.*[63,64] confirmed earlier conclusions that favoured bedtime administration, finding no RCT supporting morning administration, except when considering compliance. In the systematic review of 155 trials conducted by the Hygia investigators, 23 972 individuals were evaluated using 37 different single and 14 dual-combination therapies.[64] There was no increase in frequency of side effects, including nocturnal neurological symptoms, associated with evening/bedtime dosing. Most (83.9%) trials reported clinically and statistically significant benefits, including decreased incidence of adverse effects, improved kidney function, and reduced cardiac pathology when treated at bedtime/evening rather than upon waking/morning.

Kreutz *et al.*[65] challenged conclusions from the Hygia study,[66] as did Turgeon *et al.*[67] who provided an extensive analysis. These publications documented concerns, among others, with deviations from Cochrane guidelines for RCTs in the Hygia publications, which they considered did not document sufficient auditing of randomisation or of reported side effects. Other concerns[68] included a high risk of bias and questions regarding the extent of decline in cardiovascular events relative to the difference in achieved BP reduction. This later position paper from the International Society of Hypertension[68] did not consider the available evidence adequate to support a recommendation for bedtime rather than morning administration. Their consensus statement is that "complete 24-hour control of BP should be targeted using readily available, long-acting antihypertensive medications as monotherapy or combinations administered in a single morning dose". In a 2022 editorial in The Lancet,[69] commenting on the results from the large HELP study, it was noted that there was no advantage in evening dosing compared to morning dosing, proposing a personalised approach when prescribing while awaiting publication of other trials in progress. One non-randomised prospective study from Japan using MRI imaging[70] did consider potential neurological effects, demonstrating increased abnormalities in elderly Japanese with marked nocturnal dipping. Strokes with MRI changes were noted to be more frequent in those having a morning surge of BP, suggesting that when treating HTN, reduction of the morning surge might be a therapeutic target. Further work is needed to assess the significance of these findings for other populations known to have different patterns of cardiovascular disease.[71] The meta-analysis conducted by Zhao *et al.*[72] reviewed 36 RCTs up to August 2021, including those from the Hygia group with analysis favouring evening administration but emphasized, like other reviews, that further trial results are needed, with several trials in progress that may assist in reaching conclusions.[67–69,72,73]

This lack of consensus has created a problem for those providing guidelines for management and for clinicians advising patients. NICE in the UK did not, for example, in their 2019 updated guidelines, include the prescription of bedtime dosing as part of standard management.[74,75] The Hygia study protocol did include 48-hour ABPM for patients in both morning and bedtime administration groups, and these data were available when adjusting medication for individuals participating in both groups. Clinicians were better able to modify non-dipping by increasing medication at bedtime in that group. This option was not available as a response in the group randomised to morning administration, even when non-dipping persisted. This provides support for bedtime administration when there is simultaneous monitoring *via* ABPM to check whether dipping is restored or steep morning surges reduced.

A day–night BP ratio (DNR) computed from 24-hour ABPM records was initially used to identify dippers and non-dippers, depending on whether the DNR was above or below 10%, respectively. Classification has since been extended to include four classes: reverse dippers when DNR < 0; non-dippers when 0 < DNR < 10%, dippers when 10% < DNR < 20%, and extreme dippers when DNR > 20%.[75–77] In the experience of some individuals using this classification, it is primarily a DNR < 0 (reverse dipping) that is associated with adverse outcomes and non-dipping is much less predictive of an adverse event. Reverse dipping occurs when the phase of the circadian rhythm is abnormal (computed to be outside the reference range identifying normal phase and amplitude of the circadian rhythm in BP).[76] Reverse dipping is often seen in patients with CKD and in patients with diabetes with autonomic nervous dysfunction and also in patients over 70 years with an associated cardiovascular risk.[77] Abnormal timing of the circadian BP rhythm is considered to be better in predicting adverse outcomes than a classification in terms of dipping based on the DNR as documented in diabetic patients in a series of publications from Nakano *et al.*[78] Recent data from experimental studies conducted by this group demonstrate that a non-dipper pattern of BP is associated with the progression of renal injury during the development of salt-dependent hypertension in rats.[79]

8.4 Chronic Kidney Disease, Chronobiology and Chronodisruption

8.4.1 Chronic Renal Failure and Chronobiology

Adjustment to decreasing nephron loss enables an individual with CKD experiencing non-acute progression in the number of functioning nephrons or chronic progressive kidney failure (CPKF) to be unaware of loss of 80–90% of kidney function with little or no impact on daily functioning. As the number of nephrons decreases, there are, however, substantial changes in many organs and tissues. The mechanisms involved in changing function in the residual "intact" nephrons are explained by the "intact nephron hypothesis".[80] Negative "trade off" results from the mechanisms needed to modify transport in residual "intact" nephrons.[81,82] Nocturnal urine, for example, is less concentrated than in a healthy state, explained by greater solute excretion by each remaining nephron, with a reduced ability of each nephron to limit reabsorption of free water in the collecting tubules despite release of ADH.[26]

8.4.2 Chronodisruption in Chronic Renal Failure

Substantial understanding in recent years of the molecular aspects of chronobiology in health and, to a limited extent, in disease states include effects of temporary disturbances of circadian rhythms and of other repetitive patterns of functioning in the renal system. Disturbances that are temporary, such as those associated with jetlag, have been termed chronodisturbances, contrasting with chronodisruptions, which persist and become chronic, possibly leading to a disease or as a feature of a disease state. This progression was documented by those investigating side effects from glucocorticosteroids, documenting the disturbance to adreno–pituitary function when this included evening administration, similar to that noted in patients following removal of an adrenal adenoma, which had been secreting cortisol without rhythmicity in its release and independent of adrenocorticotrophic hormone (ACTH). A minority of patients studied with these temporary disturbances[31,83] experienced prolonged suppression of the adreno–pituitary axis and a risk of a crisis with corticosteroid withdrawal. Erren *et al.*[84,85] proposed the term "chronodisruption", in contrast to chronodisturbances. These authors reviewed the use of their concept over the previous two decades, commenting on interventions preventing chronodisturbances becoming chronodisruptions, such as time-restricted feeding in mice to prevent disruption of the peripheral circadian clocks and their metabolic impact from imposed chronic jetlag.

Lemmer's group studied the molecular biology relating to renal function and its relationship with BP, cardiovascular functions and associated hormonal rhythms in their elegant sequence of investigations in rats. They established that the introduction of a single transgene can result in a great variety of disturbances in body functions and their rhythmic organisation, including HTN. They observed an inverse BP profile with higher BP values in the resting phase than in the activity phase, with profiles of the heart rate and motor activity remaining unchanged. Detailed studies in transgenic TGR(mREN2) rats, a rat model of human secondary HTN, demonstrated disturbed circadian organisation at the level of the brain, heart and kidneys.[12] In these non-transgenic rats, they observed a loss of rhythmicity in cardiovascular functions, including BP after lesions in the SCN, a further indication of a key central role of the SCN in cardiovascular rhythmicity. Rhythmicity in renal function parameters measured was similar in TGR rats and controls, despite the inverted BP rhythms in TGR rats, with only minor differences in relative amplitude and acrophase. The authors concluded that time-dependant changes in systemic blood flow may be of greater importance for circadian regulation of kidney function than systemic BP. They proposed further studies on variables controlling intravascular volume, such as renin, angiotensin II, vasopressin and anti-natriuretic peptides. Studies are needed in both healthy humans and in CKD and HTN patients to validate that rat data can be extrapolated to humans and, especially, to consider modifications to the circadian rhythmic structure due to postural differences in humans. These have been documented to modify some circadian variations in renal and cardiovascular functions.

An extensive literature by Carriazo *et al.*[86] provides evidence that chronodisruption may lead to kidney disease and its progression (CPKD). The authors, however, failed to find publications identifying chronodisrupters, the events that might modify chronodisturbances to become chronodisruptions. If chronodisrupters can be identified, they could be therapeutic targets to prevent or slow down the progression of CKD. Some of them may be present even before there is a decrease in GFR. The most likely scenario is that an accumulation of uraemic toxins might be expected to be active only after a significant loss of renal mass with a decrease in GFR and nephron numbers, resulting in "trade off",[81] with an accumulation of toxins, producing a tissue response before activating alternative renal or other excretory routes. Uric acid is a known example of a circulating "toxin", with relative insolubility under some conditions of pH and hydration. Precipitates of urate in urine can result in stones in the urine collecting system or cause inflammatory changes in renal tissue, interstitial nephritis, when in excess from too much endogenous production or restricted renal excretion. A medication, allopurinol, was developed to arrest the progression from the urate precursor (xanthine) to urate that enables products of cell metabolism to be excreted by both healthy and damaged kidneys. Allopurinol, used to reduce or prevent gout arthritis, might also influence chronic progressive renal failure (CPRF). An RCT that aimed to evaluate this potential[87] enrolled 369 patients with stage 3 or 4 CKD, with 185 assigned to receive allopurinol as a daily titrated dose to reach serum urate targets. Urate-lowering treatment with allopurinol did not slow the decline in eGFR as compared to placebo.[87] Treatment timing was not considered in this RCT, leaving the chronopharmacological question "would the medication have been effective with a different time of administration?". The morbidity, mortality and expense associated with end stage chronic renal failure (ESCRF) remain major problems in all countries. A search for strategies to influence the rate of progression in CPRF merits a more detailed consideration of the role of chronodisrupters and the strategies that may ameliorate their negative impact on rates of progression.

In addition to excretory functions, the kidneys produce hormones with systemic effects and specific functions important for health. These endocrine functions can be disturbed in some renal disorders, including renal failure. Reduction in renal mass results in reduced endocrine output and regulation by clock genes may be disturbed. In studies on mice without renal impairment, Zuber *et al.*[9] found marked circadian rhythms in the expression of a large number of genes involved in homeostatic functions of the kidney. These authors further note that the distal convoluted tubule (DCT), connecting tubule (CNT) and cortical collecting duct (CCD) possess an intrinsic circadian timing system characterized by robust oscillations in the expression of circadian core clock genes (*Bmal1*, *Clock*, *Npas2*, *Per*, *Cry*, and *Nr1d1*) and clock-controlled PAR bZip transcriptional factors (*Dbp*, *Hlf*, and *Tef*). The *Clock* knockout mice or mice devoid of *Dbp/Hlf/Tef* (triple knockout) exhibited significant changes in renal expression of several key regulators of water or sodium balance (vasopressin V2 receptor, aquaporin-2, aquaporin-4, and ENaC). They also noted that the loss of *Clock* led to a complex phenotype characterized by partial diabetes insipidus, dysregulation of sodium excretion rhythms, and a significant decrease in BP.[9] Stow and Gumz[10] also highlight the development of a model for clock-controlled sodium and water transport in renal epithelial cells. Johnston and Pollock[88] provided a further review of the circadian regulation of renal function, noting that kidney disease is associated with a loss of circadian function that may increase the risk of cardiovascular complications.

8.4.3 End-stage Chronic Renal Failure (ESCRF)

In many patients with chronic progressive renal failure, the deterioration is linear when the eGFR, or the reciprocal of plasma creatinine is plotted to monitor renal function[89,90] (see Figure 8.3). When a plot demonstrates deterioration, an estimate of the date when dialysis may be needed (EDD) is

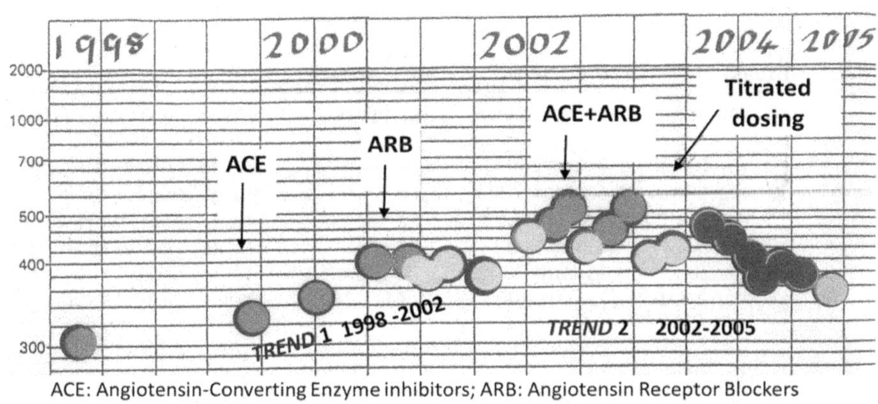

Figure 8.3 This graph of patient 2 (Table 8.3) was updated during clinic visits to motivate and inform the patient. Using this customised template,[90] the reciprocal of plasma creatinine was plotted. The *y* axis is inverted to show deterioration upward (trend 1). The date of probably needing dialysis (EDD)[56] is predicted and confirmed statistically[149] to be in 2006. After the introduction of self-monitoring and titrated evening dosing (Trend 2), deterioration was slowed with a return of renal function close to values in 1998.

provided by extrapolating to the point where plasma creatinine on the most recent trend will be around 1000 mmol l^{-1} (see Table 8.3). In the patients listed in Table 8.3, this was recalculated after intervention, including evening medication. Dialysis was postponed in 7 patients of this cohort of these 13 patients presenting to the author in 2001–2002 with chronic progressive renal failure.

Only four RCTs in the reviews referenced (see Section 8.3.2) provided data on rates of progression in CPKF when contrasting morning with evening/bedtime administration of anti-hypertensive medications, all supporting bedtime as the preferred administration time. Detailed data and analysis needed to associate the timing of medication with arrest or slowing of progressive deterioration in those individuals are lacking in these trials, as in other RCTs.[91,92] Lewis *et al.* conducted studies[92,93] that included data from almost 2000 patients with diabetic nephropathy, with a minimum follow-up period of 2 years, and concluded that many patients achieve regression or apparent remission of clinical evidence of diabetic nephropathy, but without analysis of the influence of treatment timing. Results are becoming available from trials of new agents (for example, SGLT2 inhibitors) that influence outcomes in diabetic nephropathy, with demonstrated changes in many segments of the nephron. RCTs have documented that renal function progression is modified in some patients with once-daily administration of an SGLT2 inhibitor, contrasting with the progression

Table 8.3 Estimated date of dialysis (EDD) estimated from reciprocal of plasma creatinine as a function of time in patients with chronic progressive renal failure.[89,90,124,149,150]

Patient	Breakpoint	EDD (line 1) pre-breakpoint	EDD (line 2) post-breakpoint	Outcome
1	23 Apr 2001	11 Aug 2001	>10 years	Slowed until non-renal death in 2006[a]
2	19 Nov 2002	9 Apr 2006	>10 years	Stable, not on dialysis[b]
3	14 Aug 2000	13 Mar 2001	>10 years	Stable, not on dialysis[b]
4	13 Mar 2001	13 Oct 2002	>10 years	Stable until dialysis[c]
5	12 Sep 2002	6 Sep 2004	>10 years	Stable, not on dialysis[b]
6	23 Apr 2001	11 Aug 2001	>10 years	Stable, not on dialysis[b]
7	16 Mar 2001	18 Jan 2002	28 Mar 2010	Progression slowed with dialysis postponed until 2006
8	12 Dec 2000	23 Mar 2003	18 Sep 2005	Non-renal death in 2004[a]
9	Not detected	21 Nov 2003		Dialysis, 2003[c]
10	Not detected	19 Sep 2004		Dialysis, 2004[c]
11	Not detected	$P>0.05$		Dialysis, 2002[d]
12	Not detected	$P>0.05$		Dialysis, 2003[d]
13	Not detected	$P>0.05$		Dialysis, 2002[d]

[a] Predicted EDD (line 2) not required as non-renal death.
[b] Stable after interventions and breakpoint with dialysis postponed >10 years.
[c] No slowing or arrest of progression and dialysis required as EDD on line 1.
[d] Progression without statistical linearity to enable calculation of EDD ($p>0.05$).

observed in the placebo group based on the analysis of rates of progression to renal failure.[94,95]

As RCTs have not yet provided a consensus, it is important that observational studies are also considered when making treatment decisions. Timio *et al.*,[96] in a prospective non-randomised study, observed a cohort of 48 patients with HTN and decreased renal function, studied with ABPM at the onset of an observation period of 3 years. Patients who did not demonstrate nocturnal dipping at the onset documented more rapid progression of renal dysfunction and of proteinuria (an indicator of a poor renal prognosis), compared to those who did. They concluded that nocturnal BP control should be one objective of treatment in patients with impaired renal function. Hjortknjae *et al.*[97] evaluated nocturnal anti-hypertensive treatment in 24 patients with type 1 diabetes with autonomic neuropathy and non-dipping. In a randomised, placebo-controlled, double-blind cross-over trial, over 12 weeks, they observed that non-dipping treated with an ACE inhibitor at bedtime, as opposed to morning, increased BP dipping and improved abnormal BP profiles, with a potentially beneficial effect on long-term outcomes.

8.5 Renal Transplantation and Chronobiology

8.5.1 Renal Transplantation and Immuno-suppression

The timing and frequency of medication, surveyed in UK transplant units in 1980 and adjacent years, varied between both units and between patients.[98,99] Studies by Halberg *et al.* demonstrated that cyclosporine's immunosuppressive actions could be enhanced by timing its administration, with prolongation of cardiac graft function in rats and of renal transplants in dogs.[100,101] In a study involving 19 pairs of dogs, renal allografting outcomes were contrasted when cylosporine was infused not at a single clock hour but continuously with a programmable Medtronic pump. The delivery rate was programmed to infuse sinusoidally-varying doses, which peaked at 08:30 or at 20:30, with improved outcomes when infusion peaked at 20:30, including prolonged graft survival ($P<0.01$).[101] Circadian rhythms in cell-mediated immunity had been demonstrated to be present in both rats and humans by the author of this chapter and the Nottingham group, with the magnitude of immune responses dependent on the time of exposure to antigens and with documented circadian rhythmicity in the suppression of immune responses with medication. In pilot studies, by this group, prolonged graft survival depended on timing of cyclosporine administration in dogs receiving renal allografts.[102,103] A circaseptan (about 7-day) rhythm was also reported in the rejection of transplanted kidneys in humans.[104] Since the 7-day component had a phase dependent on the time of transplantation, it was not synchronized by environmental cues such as lunar gravity. The Nottingham group later confirmed a circaseptan pattern in the onset of rejection episodes. These observations and others, documented in a review

"Lymphocytes are rhythmic, is this important" in 1984,[105] supported the need to more carefully consider the timing of pharmaceutical administration to suppress rejection.

Initially, many transplant units prescribed morning and evening doses of prednisone, often at higher doses than in common use in other clinical specialties. One UK transplant unit, in Belfast, had a policy of morning administration only. Audits in 1979 and 1982 had shown a wide range of outcomes in different units in the UK.[103,106] The Belfast unit, with 82% one-year graft survival, achieved the best results among units providing transplantation for patients from a defined catchment (Northern Ireland), perhaps related to their use of morning corticosteroid and azathioprine. A retrospective analysis in the Nottingham unit estimated the circadian timing of rejection onset as the time when improving or stable function changed to deterioration, which occurred during the night,[89,99,107–109] supporting the hypothesis that timing of medication influences outcomes.[99]

Support for this result was not, however, provided by two small prospective trials, contrasting morning-only dosing of low-dose prednisolone with twice-daily high-dose glucocorticoids, or with the same total dose divided into morning and evening. While graft survival did not differ, side effects were markedly reduced from excess glucocorticoid and obesity in the lower-dose morning cohort.[106] These results prompted many units to reduce the total dose of corticosteroid and, in some cases, to use a single daily dose taken in the morning. Other immune-suppressive medications have been introduced since then, but with limited consideration given to treatment timing. More emphasis has been given to modifying the structure of established pharmaceuticals or developing new pharmaceuticals to modify the immune response. Tacrolimus has improved results after solid-organ transplantation and is currently in use by most units. A modified formulation, tacrolimusQD (TQD), with modified pharmacokinetics was introduced as a once-daily regimen.[110] Fontova *et al.*[110] found lower whole-blood and intracellular tacrolimus concentrations after nightly doses, presumably influenced by a distinct circadian clock. The significantly lower tacrolimus exposure after the nightly dose was not translated into a significant reduction of the pharmacodynamics effect, providing a conceptual basis for a better understanding of the results from therapeutic monitoring of twice-daily tacrolimus formulations.[110] Reviews of tacrolimus administration reported on observed circadian variations in pharmacodynamics that should be considered when planning AUC-based monitoring.[111,112] On the basis of pharmacokinetic studies showing reduced AUC (area under the curve) and Cmax (peak serum concentration) in the evening, morning administration was selected for TQD. If future reports establish that morning is the preferred time to administer TQD after transplantation for this single dose formulation, it may come into general use after development to utilise circadian pharmacodynamics. To date, guidelines for management after solid-organ transplantation of medications other than corticosteroids have not been influenced by chronobiology to a significant extent.

8.5.2 Renal Transplantation and Disturbed Circadian Rhythms of Urinary Excretion

A disturbance in the circadian rhythm of urine flow after renal transplantation was described in the early literature. In some patients, reversed circadian rhythms were observed to be transient, with rhythms becoming less abnormal as time after the procedure elapsed. Several mechanisms have been suggested, including the role of medication, especially of glucocorticoids administered in the evening. In a small study, Buijsen *et al.*[113] found no difference in renal glomerular clearance in patients 4 to 7 months after renal transplantation compared to healthy controls, except in one patient with stage 3 renal failure (GFR of 39 ml min^{-1}) who had no nocturnal dipping of GFR. All patients were in the prone position for 24 hours. Abnormal post-transplant rhythms may reflect responses to daytime postural change. The similarity between healthy controls and transplant recipients who have absent renal innervation suggests that changes in autonomic function are not the mechanism of the renal circadian rhythm of urine flow in health. A recent report from Iran on 114 patients post-transplantation noted nocturnal polyuria in 68% of them,[114] who were more likely to have non-dipping BP, usually associated with reduced renal function. Restoring nocturnal BP dipping might result in improved renal outcomes after transplantation. Evening medication for HTN significantly restored a nocturnal BP dip in paediatric patients after renal transplantation at 6 months.[115] The impact on the incidence of post-transplant renal function (rate of progression, long-term complications, and graft or patient survival) has not yet been reported. The results suggest, however, that trends of deterioration may be modified, which should encourage clinical teams to give more attention to nocturia and ABPM[116] in post-transplantation management.[117] Reviews in 2018[118] and 2021[119] of post-transplantation care commented on remaining uncertainty about several aspects of management, and expressed concerns about the design, conduct and conclusions of some ongoing trials. They did not comment, however, on the lack of information on chronobiological aspects of medications in common use or on the possible relevance of other aspects of chronobiology, such as the timing of transplant procedures. There is a need for trials with larger numbers and longer follow-up of old and new formulations of pharmaceuticals to utilise the rhythmic variations in transplant recipients to reduce post-transplant rejection and other post-transplant complications.

8.6 Transfer of Renal Chronobiology Into Clinical Practice

The first international symposium focusing on chronopharmacology and chronotherapeutics was held in Tallahassee, Florida, in 1979, with Proceedings published later in 1981.[120] The editors of the Proceedings of that conference indicated in their preface that "the foundation for a new science, chronotherapeutics, had been laid, with great potential for development". Of the 39 presentations, two had a focus on renal aspects, including transplantation, with others on immunology, adrenocortical function and other topics with potential relevance to the management of renal patients. The author of this chapter contributed papers on "Circadian variations in cell-mediated immunity and in the timing of human allograft rejection"[116,121] and on "Transfer of chronobiology from the laboratory to outpatients and community", also providing an invited editorial for the British Medical Journal entitled "Chronotherapeutics a new clinical science"[122] and later a review in 1984.[105] Despite the significant body of chronobiological knowledge accumulated in the subsequent four decades, its influence on clinical practice has been limited, leaving potential for further improvement. Chronic and progressive graft failure in some recipients remains an unsolved problem. To further improve clinical outcomes, it will be critically important for investigators to clearly communicate their chronobiologic findings at the cellular and molecular as well as at the whole-body level to chronopharmacologists and those in different clinical specialties. This should facilitate the incorporation of established and new information into clinical research and, when established, into clinical practice.

There is also a need to evaluate the role of more detailed data analysis in trials and clinics to facilitate the introduction and evaluation of innovative strategies. Traditional and virtual meetings should facilitate communications between the basic and applied scientific communities and between those providing and those using graphical and statistical methods for analysis.[89,109,123–126,150,151] The methods selected should take into account circadian rhythms and other periodic variations, including sleep and sleep disorders. A 2020 consensus provided an assessment of obstructive sleep apnoea with other key sleep disorders, notably insomnia, restless legs syndrome/periodic limb movements of sleep, and parasomnia, including non-REM parasomnia and REM sleep behaviour disorder.[152–154] In the clinics that collect real-time data, rather few provide real-time analysis of time series with only some incorporating the reality of chronobiological variations and sleep-linked changes. Challenges to utilising selected methods of analysis of time series include situations where there are irregular sequences of data, missed data points and episodes, including acute illness that disrupt data collection, all common occurrences when monitoring a chronic condition. Better agreement is needed for the statistical management of laboratory and clinical data, including data from monitoring devices. Standardization of methods of analysis should facilitate the preparation of guidelines and their acceptance, help with the design of future research protocols, and streamline the reporting of results and the compilation of reviews.

Abbreviations

ABPM	Ambulatory blood pressure monitoring
ADH	Antidiuretic hormone

AVP	Arginine vasopressin
BEV	Bladder emptying volume
BP	Blood pressure
CKD	Chronic kidney disease
CKF	Chronic kidney failure
CPKF	Chronic progressive kidney failure
CCF	Congestive cardiac failure
CDI	Cranial diabetes insipidus
DNR	Day–night ratio (from ABPM)
DDAVP	Minirin® Nocturna® desmopressin
GFR	Glomerular filtration rate
eGFR	Estimated glomerular filtration rate
EDD	Expected date of dialysis
ESRF	End stage renal failure
EPO	Erythropoietin
HTN	Hypertension
MHTN	Malignant hypertension
MLT	Melatonin
MNE	Monosymptomatic nocturnal enuresis
LUTS	Lower urinary tract syndrome
OSA	Obstructive sleep apnoea
OAB	Overactive bladder
PCr	Plasma creatinine
PNE	Polysymptomatic nocturnal enuresis
RCT	Randomised controlled trial
REM	Rapid eye movement
RAA	Renin, angiotensin and aldosterone
SCN	Suprachiasmatic nucleus
ACR	Urinary albumin/creatinine ratio

Acknowledgements

I am grateful for the support of many individuals when exploring the potential importance of chronobiology in clinical nephrology and the role of clinical information systems, interactive computer graphics, and Bayesian statistical methods in patient care and research, especially colleagues in Nottingham Hospitals (Professor Roger Blamey, Dr Rodney Cove-Smith, Dr Roy Pownall, Dr Inge Kowanko and Dr Ian Trimble) and in the University of Nottingham Department of Mathematics (Professor AFM Smith, Kerry Gordon and Michael West) and also the foundation staff of Clinical Computing Ltd of London. Our contributions evolved from sharing experiences and technology with others in the British Renal Computing Group, especially Dr EJ Will (Chair, 1982–1988), in the International Society of Chronobiology, and with others who shared similar interests in many countries, with important input from students and post-graduates who made major contributions to many projects.

References

1. E. B. Verney, *Proc. R. Soc. London*, 1947, **135**(878), 25–106.
2. G. Robertson, S. Rittig, L. Kovacs, M. B. Gaskill, P. Zee and J. Nanninga, *Scand. J. Urol.*, 1999, **202**, 36–39.
3. C. P. George, F. H. Messerli, J. Genest, W. Nowaczynski, R. Boucher and M. Kuchel Orofo-Oftega, *J. Clin. Endocrinol. Metab.*, 1975, **41**(2), 332–338.
4. T. Zsoter and S. Sebok, *Acta Med. Scand.*, 1955, **152**(1), 47–52.
5. M. S. Knapp, *The Circadian Rhythm of Urine Excretion*, Bristol University, UK, 1967.
6. L. K. Sciesielski, M. Felten, L. Michalick, K. M. Kirschner, G. Lattanzi, C. L. J. Jacobi, T. Wallach, V. Lang, D. Landgraf, A. Kramer and C. Dame, *Kidney Int.*, 2021, **100**, 1071–1080.
7. H. Okamura, M. Doi, Y. Yamaguchi and J-M. Fustin, *Curr. Hypertens. Rep.*, 2011, **13**, 103–108.
8. D. Firsov and O. Bonny, *Nat. Rev. Nephrol.*, 2018, **14**(10), 626–635.
9. A. Zuber, G. Centeno, S. Pradervand, S. Nikolaeva, L. Maquelin, L. Cardinaux, O. Bonny and D. Firsov, *Proc. Natl. Acad. Sci. U. S. A.*, 2009, **106**(38), 16523–16528.
10. L. Stow and M. L. Gumz, *J. Am. Soc. Nephrol.*, 2011, **22**(4), 598–604.
11. O. Bonny, M. Vinciguerra, M. L. Gumz and G. Mazzoccoli, *Nephrol., Dial., Transplant.*, 2013, **28**(1), 2421–2431.
12. B. Lemmer, K. Witte, H. Enzminger, S. Schiffer and S. Gorbey, *Chronobiol. Int.*, 2003, **20**(4), 711–738.
13. B. Lemmer, K. Witte, A. Schänzer and A. Finedeisen, *Chronobiol. Int.*, 2000, **17**(5), 645–658.
14. A. Fathollahi, F. Daneshgari and A. T. Hanna-Mitchell, *Curr. Urol.*, 2014, **8**, 113–118.
15. J. P. Nørgaard, E. B. Pedersen and J. C. Djurhuus, *J. Urol.*, 1985, **134**(5), 1029–1031.
16. A. Raes, J. Dehoorne, P. Hoebeke, E. Van Laecke, R. Donckerwolcke and J.V. Walle, *J. Urol.*, 2006, **176**(3), 1147–1151.
17. N. D. Crosby, *Med. J. Aust.*, 1950, **2**(15), 533–543.
18. H. Malem, M. S. Knapp and E. J. Hillier, *Br. Med. J.*, 1982, **285**(6334), 22.
19. M. Knapp, *Med. J. Aust.*, 1992, **156**(9), 672.
20. K. E. Fordham and S. R. Meadow, *Arch. Dis. Child.*, 1989, **64**(5), 651–656.
21. M. S. Knapp, *Aust. J. Prim. Health*, 1995, **1**(1), 74–78.
22. W. F. Bower, D. M. Whishaw and F. Khan, *Neurourol. Urodyn.*, 2017, **36**(3), 697–705.
23. M. J. Drake, I. W. Mills and J. G. Noble, *J. Urol.*, 2004, **171**(3), 1199–1202.
24. M. J. Drake, L. Canham, N. Cotterill, D. Delgado, J. Homewood, K. Inglis, L. Johnson, M. C. Kisanga, D. Owen, P. White and D. Cottrell, *BMC Neurol.*, 2018, **18**, 107.
25. K. Sugaya, S. Nishijima, M. Miyazato, K. Kadekawa and Y. Ogawa, *J. Int. Med. Res.*, 2007, **35**(5), 685–691.

26. P. Hillier, M. S. Knapp and R. Cove-Smith, *Q. J. Med.*, 1980, **49**(196), 461–478.
27. M. S. Knapp, *J. R. Coll. Physicians London*, 1969, **3**(4), 398–405.
28. J. P. Weiss, E. A. van der Meulen and K. V. Juul, *Eur. Urol. Focus*, 2020, **6**(5), 1006–1012.
29. I. C. Kowanko, R. Pownall, M. S. Knapp, A. J. Swannell and P. G. C. Mahoney, *Chronobiologia*, 1978, **5**(2), 181–182.
30. I. Chihara, H. Negoro, J. Kono, Y. Nagumo, H. Tsuchiya, K. Kojo, M. Shiga, K. Tanaka, S. Kandori, B. J. Mathis and H. Nishiyama, *Commun. Biol.*, 2023, **6**(1), 81.
31. B. H. B. Robinson, D. Mattingly and C. L. Cope, *Br. Med. J.*, 1962, **1**(5292), 1579–1584.
32. J. F. Duffy, K. Scheuermaier and K. R. Loughlin, *Curr. Aging Sci.*, 2016, **9**(1), 34–43.
33. M. G. Dunnigan, J. B. Henderson, D. Hole and A. J Pelosi, *Q. J. Med.*, 2004, **97**(11), 755–764.
34. M. Kaye, *Nephron Physiol.*, 2008, **109**(1), 8.
35. M. Kaye, *J. Can Urol. Assoc.*, 2008, **2**(6), 604–608.
36. L. A. Birder, A. S. Wolf-Johnston, E. K. Jackson, A. J. Wein and R. Dmochowski, *Neurourol. Urodyn.*, 2019, **38**(1), 393–397.
37. L. A. Birder and P. E. V. Van Kerrebroeck, *Urology*, 2019, **133S**, 14–23.
38. J. P. Weiss, *Rev. Urol.*, 2012, **14**(3–4), 48–55.
39. S. Patil, G. Daniel, R. Vyas, Y. Tailor, M. Howell, T. Ahmed, C. Reutter and A. Shrikhande, *Neurourol. Urodyn.*, 2022, **41**(1), 220–228.
40. K. Gill, H. Horsley, S. Swamy, R. Khasriya and J. Malone-Lee, *BMC Urol.*, 2021, **21**, 39.
41. S. Swamy, A. S. Kupelian, R. Khasriya, D. Dharmasena, H. Toteva, T. Dehpour, L. Collins, L. Rohn and J. Malone-Lee, *Int. Urogynecol. J.*, 2019, **30**, 409–414.
42. K. Rostami and S. Hogg-Kollars, *BMJ*, 2012, **345**, e7982.
43. R. Bright, *Reports of Medical Cases. Illustrating diseases by reference to morbid anatomy*, Longman, London, 1827.
44. J. W. Conn and L. H Louis, *Trans. Assoc. Am. Physicians*, 1955, **68**, 215–231.
45. K. W. Choy, P. J. Fuller, G. Russell, Q. Li, M. Leenaerts and J. Yang, *Br. Med. J.*, 2022, **377**, e065250.
46. M. I. Rosenheim, E. J. Ross, O. M. Wrong, C. J. Hodson, D. R. Davies and J. F. Smith, *Am. J. Med.*, 1963, **34**(5), 735–740.
47. I. M. Donaldson, A. Doig and I. C. Knight, *Am. J. Med.*, 1967, **43**(2), 289–293.
48. M. Pons, A. Schnecko, K. Witte, B. Lemmer, J. M. Waterhouse and J. Cambar, *Am. J. Physiol.: Regul., Integr. Comp. Physiol.*, 1996, **271**(4Pt2), R1002–R1008.
49. H. Okamura, M. Doi, K. Goto and R. Kojima, *Hypertens. Res.*, 2016, **39**(10), 681–687.
50. A. G. Logan, *J. Am. Coll. Cardiol.*, 2022, **79**(17), 1687–1689.

51. D. A. A. Nugrahaningsih, N. Emoto, N. Vignon-Zellweger, E. Purnomo, K. Yagi, K. Nakayama, M. Doi, H. Okamura and K-i. Hirata, *Hypertens. Res.*, 2014, **37**, 202–209.
52. C. Bothou, F. Beuschlein and A. Spyroglou, *Diabetes Metab.*, 2020, **46**(1), 1–7.
53. J. S. Floras, J. V. Jones, M. O. Hassan and P. Sleight, *Br. Med. J.*, 1982, **285**, 1387–1392.
54. G. W. Liddle, *Arch. Intern. Med.*, 1966, **117**(6), 739–743.
55. P. K. Whelton, R. M. Carey, W. S. Aronow, D. E. Casey Jr, K. J. Collins, C. Dennison Himmelfarb, S. M. DePalma, S. Gidding, K. A. Jamerson, D. W. Jones, E. J. MacLaughlin, P. Muntner, B. Ovbiagele, S. C. Smith Jr, C. C. Spencer, R. S. Stafford, S. J. Taler, R. J. Thomas, K. A. Williams Sr, J. D. Williamson and J. T. Wright Jr, *Hypertension*, 2018, **71**(6), e13–e115.
56. D. B. Shaw, M. S. Knapp and D. H. Davies, *Lancet*, 1963, **1**(7285), 797–799.
57. P. M. Stranges, A. M. Drew, P. Rafferty, J. E. Shuster and A. D. Brooks, *Ann. Pharmacother.*, 2015, **49**(3), 323–334.
58. C. Wang, Y. Ye, C. Liu, Y. Zhou, L. Lv, C. Cheng, S. Li, T. Lou and X. Liu, *Intern. Med. J.*, 2017, **47**(8), 900–906.
59. N. P. Bowles, S. S. Thosar, M. X. Herzig and S. A. Shea, *Curr. Hypertens. Rep.*, 2018, **20**(11), 97.
60. M. S. Knapp, *Intern. Med. J.*, 2018, **48**(2), 226–227.
61. R. Refinetti, G. Cornelissen and F. Halberg, *Biol. Rhythm Res.*, 2007, **38**(4), 275–325.
62. R. C. Hermida, J. J. Crespo, M. Domínguez-Sardiña, A. Otero, A. Moyá, M. T. Ríos, E. Sineiro, M. C. Castineira, P. A. Callejas, L. Pousa, J. L. Salgado, C. Duran, J. J. Sanchez, J. R. Fernandez, A. Mojon, D. E. Ayala and for the Hygia Project Investigators, *Eur. Heart J.*, 2020, **41**(48), 4565–4576.
63. R. C. Hermida, J. J. Crespo and M. Domínguez-Sardiña, *Eur. Heart J.*, 2020, **41**(16), 1602–1603.
64. R. C. Hermida, R. G. Hermida-Ayala, M. H. Smolensky, A. Mojón and J. R. Fernández, *Expert Opin. Drug Metab. Toxicol.*, 2020, **16**(12), 1159–1173.
65. R. Kreutz, S. E. Kjeldsen, M. Burnier, K. Narkiewicz, S. Oparil and G. Mancia, *Blood Pressure*, 2020, **29**(3), 135–136.
66. R. C. Hermida, J. J. Crespo, M. Dominguez-Sardina, A. Otero, A. Moya, M. T. Rios, E. Sineiro, M. C. Castineira, P. A. Callejas, L. Pousa, J. Salgado, C. Duran, J. J. Sanchez, J. R. Fernandez, A. Mojon and D. E. Ayala, *Eur. Heart J.*, 2020, **41**(16), 1600.
67. R. D. Turgeon, A. D. Althouse, J. B. Cohen, B. Enache, J. B. Hogenesch, M. E. Johansen, R. Mehta, G. Meyerowitz-Katz, B. Ziaeian and S. Hiremath, *Hypertension*, 2021, **78**(3), 871–878.
68. G. Stergiou, M. Brunstrom, T. MacDonald, K. G. Kyriakoulis, M. Bursztyn, N. Khan, G. Bakris, A. Kollias, A. Menti, P. Muntner, M. Orias, N. Poulter, D. Shimbo, B. Williams, A. M. Adeoye,

A. Damasceno, L. Korostovtseva, Y. Li, E. Muxfeldt, Y. Zhang, G. Mancia, R. Kreutz and M. Tomaszewski, *J. Hypertens.*, 2022, **40**(10), 1847–1858.
69. H. M. Salah and J. L. Mehta, *Lancet*, 2022, **400**(10361), 1383–1385.
70. K. Kario, T. G. Pickering, Y. Umeda, S. Hoshide, Y. Hoshide, M. Morinari, M. Murata, T. Kuroda, J. E. Schwartz and K. Shimada, *Circulation*, 2003, **107**(10), 1401–1406.
71. H. Jacob, A. M. Curtis and C. J. Kearney, *Biochem. Pharmacol.*, 2020, **182**, 114254.
72. J. Zhao, Y. Zeng, J. Weng, J. Zhang, T. Yao, M. Yuan and X. Shen, *Eur. J. Integr. Med.*, 2022, **50**, 102111.
73. R. G. Hermida-Ayala, A. Mojon, J. R. Fernández, M. H. Smolensky and R. C. Hermida, *Chronobiol. Int.*, 2022, **39**(4), 493–512.
74. R. J. Boffa, M. Constanti, C. N. Floyd, A. S. Wierzbicki and on behalf of the Guideline Committee, *BMJ*, 2019, **367**, l5310.
75. R. H. Fagard, *Expert Rev. Cardiovasc. Ther.*, 2009, **7**(6), 599–605.
76. G. Cornelissen, F. Halberg, K. Otsuka, R. B. Singh and C. H. Chen, *Hypertension*, 2007, **49**, 237–239.
77. P. Palatini, P. Verdecchia, L. J. Beilin, K. Eguchi, Y. Imai, K. Kario, T. Ohkubo, S. D. Pierdomenico, F. Saladini, J. E. Schwartz, L. Wing, S. Signorotti and G. Reboldi, *Hypertension*, 2020, **75**(2), 324–330.
78. S. Nakano, M. Ogihara, C. Tamura, M. Kitazawa, M. Nishizawa, T. Kigoshi and K. Uchida, *J. Diabetes Its Complications*, 1999, **13**(4), 224–231.
79. A. Sufiun, A. Rahman, K. Rafiq, Y. Fujisawa, D. Nakano, H. Kobara, T. Masaki and A. Nishiyama, *Int. J. Mol. Sci.*, 2020, **21**(6), 2248.
80. N. S. Bricker, P. A. Morrin and S. W. Kime Jr, *Am. J. Med.*, 1960, **28**, 77–98.
81. N. S. Bricker, *N. Engl. J. Med.*, 1972, **286**(20), 1093–1099.
82. N. S. Bricker, L. G. Fine, M. Kaplan, M. Epstein, J. J. Bourgoignie and A. Light, *N. Engl. J. Med.*, 1978, **299**, 1287–1293.
83. C. L. Cope, *Adrenal Steroids and Disease*, Pitman, London, 1965.
84. T. C. Erren, C. Piekarski and R. J. Reiter, *Nutrients*, 2022, **14**(2), 315.
85. T. C. Erren, R. J. Reiter and C. Piekarski, *Naturwissenschaften*, 2003, **90**(11), 485–494.
86. S. Carriazo, A. M. Ramos, A. B. Sanz, M. D. Sanchez-Niño, M. Kanbay and A. Ortiz, *Toxins*, 2020, **12**(3), 151.
87. S. V. Badve, E. M. Pascoe, A. Tiku, N. Boudville, F. G. Brown, A. Cass, P. Clarke, N. Dalbeth, R. O. Day, J. R. de Zoysa, B. Douglas, R. Faull, D. C. Harris, C. M. Hawley, G. R. D. Jones, J. Kanellis, S. C. Palmer, V. Perkovic, G. K. Rangan, D. Reidlinger, L. Robison, R. J. Walker, G. Walters, D. W. Johnson and for the CKD-FIX Study Investigators, *N. Engl. J. Med.*, 2020, **382**, 2504–2513.
88. J. G. Johnston and D. M. Pollock, *Free Radicals Biol. Med.*, 2018, **119**, 93–107.
89. M. S. Knapp, A. F. Smith, I. M. Trimble, R. Pownall and K. Gordon, *Kidney Int.*, 1983, **24**(4), 474–486.

90. M. S. Knapp, R. Blamey, R. Cove-Smith and M. Heath, *Lancet*, 1977, **2**(8049), 1183.
91. T. G. Feest, E. J. Dunn and C. J. Burton, *Q. J. Med.*, 1999, **92**(5), 275–282.
92. J. B. Lewis, T. Berl, R. P. Bain, R. D. Rohde, E. J. Lewis and the Collaborative Study Group, *Am. J. Kidney Dis.*, 1999, **34**(5), 809–817.
93. E. J. Lewis and J. B. Lewis, *Clin. Exp. Nephrol.*, 2003, **7**, 1–8.
94. C. P. Cannon, R. Pratley, S. Dagogo-Jack, J. Mancuso, S. Huyck, U. Masiukiewicz, B. Charbonnel, R. Frederich, S. Gallo, F. Cosentino, W. J. Shih, I. Gantz, S. G. Terra, D. Z. I. Cherney, D. K. McGuire and for the VERTIS CV Investigators, *N. Engl. J. Med.*, 2020, **383**, 1425–1435.
95. V. Perkovic, M. J. Jardine, B. Neal, S. Bompoint, H. J. L. Heerspink, D. M. Charytan, R. Edwards, R. Agarwal, G. Bakris, S. Bull, C. P. Cannon, G. Capuano, P. L. Chu, D. de Zeeuw, T. Greene, A. Levin, C. Pollock, D. C. Wheeler, Y. Yavin, H. Zhang, B. Zinman, G. Meininger, B. M. Brenner, K. W. Mahaffey and CREDENCE Trial Investigators, *N. Engl. J. Med.*, 2019, **380**(24), 2295–2306.
96. M. Timio, S. Venanzi, S. Lolli, G. Lippi, C. Verdura, C. Monarca and E. Guerrini, *Clin. Nephrol.*, 1995, **43**(6), 382–387.
97. H. O. Hjortkjaer, T. Jensen, K. F. Kofoed, U. M. Mogensen, P. E. Sigvardsen, L. Kober, K. L. Hilsted, H. Corinth, S. Theilade and J. Hilsted, *BMJ Open*, 2016, **6**(12), e012307.
98. *Kidney Transplantation, Principles and Practice*, ed. P. J. Morris, Saunders, Philadelphia, 4th edn, 1994, ch. 37, pp. 516–517.
99. M. S. Knapp, N. P. Byrom, R. Pownall and P. Mayor, *Br. Med. J.*, 1980, **281**(6252), 1382–1385.
100. M. Cavallini, F. Halberg, Sutherland DER, G. Cornelissen, J. Heil and J. S. Najarian, *Transplantation*, 1986, **41**, 654–657.
101. M. Cavallini, G. Magnus, F. Halberg, T. Liu, M. Y. Field, R. Sibley, J. S. Najarian and D. E. R. Sutherland, *Transplant. Proc.*, 1983, **1**(4, Suppl 1), 2960–2966.
102. R. Pownall and M. S. Knapp, *Clin. Sci. Med.*, 1978, **54**, 447–449.
103. J. R. Cove-Smith, P. Kabler, R. Pownall and M. S. Knapp, *Br. Med. J.*, 1978, **2**(6132), 253–254.
104. A. de Vecchi, F. Halberg, R. B. Sothern, A. Cantaluppi and C. Ponticelli, in *Chronopharmacology and Chronotherapeutics*, ed. C. A. Walker, C. M. Winget and K. F. A. Soliman, Florida A & M University Foundation, Tallahassee, Florida, 1981, pp. 339–353.
105. M. S. Knapp and R. Pownall, *Br. Med. J.*, 1984, **289**(6455), 1328–1330.
106. P. J. Griffin and J. R. Salaman, *Br. J. Surg.*, 1985, **72**(1), 10–12.
107. A. F. Smith and M. West, *Biometrics*, 1983, **39**(4), 867–878.
108. I. Trimble, *A quantative approach to monitoring renal allograft function: Mathematical analysis of data. M.Phil*, University of Nottingham, 1980.
109. A. F. M. Smith, M. West, K. Gordon, M. S. Knapp and I. M. G. Trimble, *J. R. Stat. Soc. Series. D Stat.*, 1983, **32**, 46–54.

110. P. Fontova, H. Colom, R. Rigo-Bonnin, L. N. van Merendonk, A. Vidal-Alabro, N. Montero, E. Melilli, M. Meneghini, A. Manonelles, J. M. Cruzado, J. Torras, J. M. Grinyo, O. Bestard and N. Lloberas, *Front. Pharmacol.*, 2021, **12**, 636048.
111. M. T. Gustavsen, K. Midtvedt, I. Robertsen, J. B. Woillard, J. Debord, R. A. Klaasen, N. T. Vethe, S. Bergan and A. Åsberg, *Clin. Transl. Sci.*, 2020, **13**(6), 1327–1335.
112. S. I. Park, C. R. Felipe, P. G. Pinheiro-Machado, R. Garcia, H. Tedesco-Silva Jr. and J. O. Medina-Pestana, *Fundam. Clin. Pharmacol.*, 2007, **21**(2), 191–197.
113. J. G. Buijsen, B. A. van Acker, G. C. Koomen, M. G. Koopman and L. Arisz, *Nephrol., Dial., Transplant.*, 1994, **9**(9), 1330–1333.
114. M. Lebadi, A. Monfared, Z. Atrkar Roshan, A. Babaei Jandghi, S. N. Mozafari Chenijani and S. A. Fakhr Mousavi, *Nephrourol Mon.*, 2018, **10**(1), e60128.
115. C. B. Sethna, L. G. Grossman, P. Dhanantwari, S. Gurusinghe, N. Laney, R. Frank and K. E. Meyers, *Pediatr. Transplant.*, 2020, **24**(8), e13854.
116. M. S. Knapp, in *Chronopharmacology and Chronotherapeutics*, ed. C. A. Walker, C. M. Winget and K. F. A. Soliman, Florida A & M University Foundation, Tallahassee, Florida, 1981, pp. 37–44.
117. R. Pownall and M. S. Knapp, *Int. J. Immunopharmacol.*, 1980, **1**, 293–298.
118. K. J. M. O'Donoghue, R. D. Reed, S. R. Knight, J. M. O'Callaghan, A. A. Ayaz-Shah, S. Hassan, A. Weissenbacher, P. J. Morris and L. H. M. Pengel, *Transplantation*, 2018, **102**(9), 1419–1439.
119. S. Hariharan, A. K. Israni and G. Danovitch, *N. Engl. J. Med.*, 2021, **385**(8), 729–743.
120. *Chronopharmacology and Chronotherapeutics*, ed. C. A. Walker, C. M. Winget and K. F. A. Soliman, Florida A & M University Foundation, Tallahassee, Florida, 1981.
121. M. S. Knapp, R. Pownall and J. R. Cove-Smith, in *Chronopharmacology and Chronotherapeutics*, ed. C. A. Walker, C. M. Winget and K. F. A. Soliman, Florida A & M University Foundation, Tallahassee, Florida, 1981, pp. 329–338.
122. Editorial, *Br. Med. J.*, 1978, 1376.
123. D. B. Morgan and E. J. Will, *Kidney Int.*, 1983, **24**(4), 438–445.
124. Gordon K., *Modelling and monitoring of medical time series*, thesis, University of Nottingham, 1986.
125. M. S. Knapp and K. Gordon, in *Chronobiotechnology and Chronobiological Engineering*, ed. L. E. Scheving, F. Halberg and C. F. Ehret, Martinus Nijhoff, Dordrecht, The Netherlands, 1987, pp. 131–140.
126. *Perspectives on UK Clinical Renal Computing 1979–1994*, ed. E. J. Will, Symposium, Roy Coll Phys, London, 2017.
127. J. G. G. Borst and L. A. De Vries, *Lancet*, 1950, **2**(6618), 1–6.
128. J. Brod and Z. Fejfar, *Q. J. Med.*, 1950, **19**, 75187–200.
129. R. A. Jones, G. O. McDonald and J. H. Last, *J. Clin. Invest.*, 1952, **31**(3), 326–334.

130. M. G. Koopman, G. C. Koomen, B. A. van Acker and L. Arisz, *Q. J. Med.*, 1994, **87**(2), 109–117.
131. A. O. Hutchinson, R. A. McCance and E. M. Widdowson, *Spec. Rep. Ser. – Med. Res. Counc. (G. B.)*, 1951, **275**, 216–225.
132. R. P. Doe, J. A. Vennes and E. B. Flink, *J. Clin. Endocrinol. Metab.*, 1960, **20**, 253–265.
133. E. J. Lennon, P. P. Ruetz and W. W. Engstrom, *Am. J. Med.*, 1961, **30**, 475–485.
134. F. C. Bartter and C. S. Delea, *Ann. N. Y. Acad. Sci.*, 1962, **98**, 969–983.
135. J. D F. B. Rosenbaum, R. K. Davis and E. C. Rossmeisl, *J. Clin. Invest.*, 1952, **5**, 507–520.
136. B. C. Rossier, *Nat. Med.*, 2010, **16**(1), 27–28.
137. M. Doi, Y. Takahashi, R. Komatsu, F. Yamazaki, H. Yamada, S. Haraguchi, N. Emoto, Y. Okuno, G. Tsujimoto, A. Kanematsu, O. Ogawa, T. Todo, K. Tsutsui, G. T. J. van der Horst and H. Okamura, *Nat. Med.*, 2010, **16**(1), 67–74.
138. H. E. Ginn Jr. and W. L. Parry, *South. Med. J.*, 1964, **57**, 739–745.
139. G. M. Berlyne, S. Nilwarangkur, K. Janabi and M. Cooper, *Q. J. Med.*, 1965, **34**(136), 463–479.
140. B. Lemmer, in *Biological Rhythms in Clinical and Laboratory Medicine*, ed. Y. Touitou and E. Haus, Springer-Verlag, Berlin, 1992, pp. 418–427.
141. A. Elithorn, P. K. Bridges, M. C. Lobban and B. E. Tredre, *Br. Med. J.*, 1966, **2**(5530), 1620–1623.
142. E. B. Flink and R. P. Doe, *Proc. Soc. Exp. Biol. Med.*, 1959, **100**(3), 498–501.
143. P. Hillier and M. S. Knapp, *J. Physiol.*, 1974, **242**(2), 111P–112P.
144. C. T. Flear, W. T. Cooke and A. Quinton, *Clin. Sci.*, 1959, **18**, 137–146.
145. H. E. de Wardener, *The kidney: An outline of normal and abnormal function*, Churchill Livingstone, London, 1958.
146. G. M. Berlyne, N. P. Mallick, Y. K. Seedat, E. C. Edwards, R. Harris and W. M. Orr, *Lancet*, 1968, **2**(7565), 435–436.
147. K. Albertsen and V. P. Petersen, *Lancet*, 1968, **2**(7568), 636–637.
148. G. F. Russell and J. T. Bruce, *Am. J. Med.*, 1966, **40**(1), 38–48.
149. P. A. Rowe, R. E. Richardson, P. R. Burton, A. G. Morgan and R. P. Burden, *Am. J. Nephrol.*, 1989, **9**, 38–43.
150. H. C. Rayner, M. E. Thomas and D. V. Milford, *Plot all the dots: graphs reveal the progression of kidney disease: understanding kidney disease*, Springer, Swiitzerland, 2020, p. 387.
151. M. S. Knapp, K. Gordon, A. F. M. Smith and R. Pownall, in *Medical Informatics Europe 85*, ed. F. H. Roger, P. Grönroos, R. Tervo-Pellikka and R. O'Moore, Springer, Berlin, 1985, p. 813.
152. E. Papworth, S. Dawson, E. J. Henderson, S. H. Eriksson, H. Selsick, J. Rees, A. Glimson, E. Strong, N. Cotterill, A. Huntley and M. J. Drake, *Eur. Urol. Focus*, 2022, **8**(1), 42–51.
153. T. J. Ge, J. Vetter and H. H. Lai, Sleep Disturbance and Fatigue Are Associated With More Severe Urinary Incontinence and Overactive Bladder Symptoms, *Urology*, 2017, **109**, 67–73.

154. Y. H. Chung, J. R. Kim, S. J. Choi and E. Y. Joo, *PLoS One*, 2022, **17**(4), e0267441.
155. Z. Chen, G. Li and J. Liu, *Neurobiol. Dis.*, 2020, **134**, 104700.
156. A. Batla, S. Simeoni, T. Uchiyama, L. deMin, J. Baldwin, C. Melbourne, S. Islam, K. P. Bhatia, M. Pakzad, S. Eriksson and J. N. Panicker, *Eur. J. Neurol.*, 2021, **28**(6), 1884–1892.
157. D. Firsov and O. Bonny, Circadian regulation of renal function, *Kidney Int.*, 2010, **78**(7), 640–645.

CHAPTER 9

Circadian Rhythms and the Gastrointestinal System

VINCENT M. CASSONE

Department of Biology, University of Kentucky, Lexington, KY, 40506, USA
Email: Vincent.Cassone@uky.edu

9.1 Biological Clocks and the Gastrointestinal System

Daily and circadian rhythms influence many aspects of physiology and behaviour in humans, non-human animals and nearly all free-living organisms. While daily rhythms, which are the cyclical expression of processes over a 24-hour period, may reflect underlying circadian rhythms, one must experimentally demonstrate the persistence of the cyclical expression of processes with periods of approximately 24 hours (*circa* meaning "about", and *dian* meaning a "day") in organisms placed in continuous environmental conditions such as constant darkness (DD) for it to be considered *circadian.* Otherwise, daily rhythms may merely reflect direct responses to environmental signals, a process called "masking". In contrast, circadian rhythms are the expression of an endogenous oscillator or oscillators, known as the circadian clock, that must synchronize, or entrain, to environmental time cues (Zeitgebers) such as the light:dark cycle (LD), temperature, or food. The ascription of time of day to a particular clock-time can be confusing. Therefore, chronobiologists have, by convention, assigned the 24 hours in a single day "Zeitgeber Times" (ZT) to standardize phase, such that ZT0 represents the beginning of the day, and ZT12 represents the beginning of the night. In organisms expressing endogenous rhythmicity in DD, called

Chemical Biology No. 23
Chronobiology and Chronomedicine: From Molecular and Cellular Mechanisms to Whole Body Interdigitating Networks
Edited by Germaine Cornelissen and Tsuyoshi Hirota

Published by the Royal Society of Chemistry, www.rsc.org

"free-running", circadian time (CT) is designated relative to the internal subjective phase of the organism, such that CT0 represents the beginning of the "subjective day" and CT12 corresponds to the beginning of the "subjective night". These and other formal properties of circadian timing are described elsewhere in this volume.

Among the functions controlled by circadian clocks are digestive processes within the gastrointestinal (GI) system, or alimentary canal, and the accessory digestive organs, such as the teeth, tongue, gallbladder, and several digestive glands, such as the salivary glands, liver, and the exocrine pancreas. As we will see below, these digestive processes are regulated by circadian oscillators within the gastrointestinal system itself and modulated by circadian pacemakers residing in the hypothalamic suprachiasmatic nuclei (SCN), which are in turn entrained to local time *via* Zeitgebers such as timed feeding and light, respectively (Figure 9.1).

Nearly every behaviour exhibits two stages: appetitive *versus* consummatory.[1,2] Much has been written recently about the appetitive aspects of food consumption[3–5] in which anticipation of a meal is regulated by a

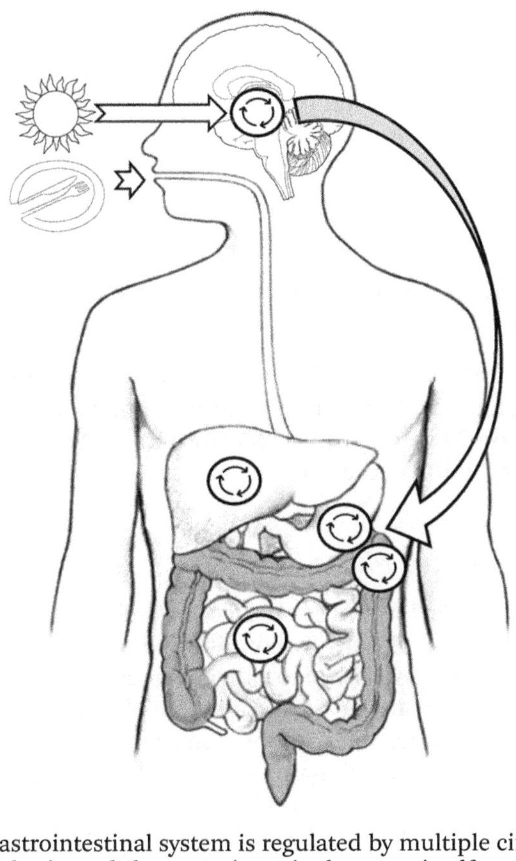

Figure 9.1 The gastrointestinal system is regulated by multiple circadian oscillators in the brain and the gastrointestinal system itself.

"food-entrainable oscillator" (FEO). However, relatively little attention has been paid to the "other" FEO regulating the consumption of food, absorption of nutrients and the elimination of wastes. Of course, digestive processes are instigated by the presence of food entering the gastrointestinal system. However, the gastrointestinal system, like the appetitive FEO, anticipates the presentation of food *via* a "gustatory FEO". Although we will touch on the appetitive FEO, the gustatory aspect of food consumption will be the focus here.

There are several phases of digestion, and each of these is regulated on a basis of daily and circadian timing. The *oral/ingestive phase* of digestion involves breaking down food into smaller pieces to increase the surface area and mixing foodstuffs with saliva, where salivary amylase begins the process of breaking down carbohydrates. Food then progresses into the *gastric phase via* propulsion of food from the oral pharynx into the oesophagus, where food progresses to the stomach *via* peristaltic activity. In the stomach, gastric pepsin breaks proteins into smaller fragments, and gastric acid inactivates pathogenic bacteria and viruses. The *intestinal phase* is a multi-functional digestive process: (1) chemical digestion further breaks down carbohydrates, proteins and lipids *via* enzymes secreted by the intestinal mucosa, exocrine pancreas, and liver, (2) absorption of nutrients by the intestinal epithelia, especially in the small intestine, and (3) continued absorption of nutrients and water by the large intestine. Finally, the *defecation phase* controls the elimination of faeces *via* the anus. Each of these processes is coordinated by circadian clock function as well as by the more proximal mechanisms mentioned above.

9.1.1 Expression of Daily Gastrointestinal Rhythms in Humans

In humans, it is difficult to study physiological mechanisms of digestion under controlled, temporally isolated conditions. However, rhythms of many aspects of digestion have been observed in clinical situations during each phase of digestion delineated above.

9.1.1.1 Oral/Ingestive Phase

It is well-known from Pavlov's early studies that salivation can be induced by the presence of food and classically conditioned to anticipate the prospects of a meal.[6] However, the *propensity* to salivate is regulated by the time of day, presumably under the influence of circadian clocks. Salivary flow rates peak in the early morning as does the composition of salivary sodium and calcium salts.[7,8] Furthermore, salivary α-amylase levels are highest in the early morning[9,10] in anticipation of breakfast. Using long-term manometric measurements, efficacy of oesophageal peristalsis is found to be greater during the day than during the night in children[11] and in healthy adults.[12]

9.1.1.2 *Gastric Phase*

In the human stomach, several aspects of gastric physiology change throughout the day.[13] Gastric acid secretion and volume are lowest during the morning and highest around dusk,[14,15] while pH is lowest during the night, rising from pH 1.5 to 3.0 as the morning progresses.[14,15] Commensurate with the gastric volume data, the rate of gastric emptying is significantly higher in the morning than in the evening,[16,17] thus reducing the level of acid in the stomach.

9.1.1.3 *Intestinal Phase*

During periods of fasting, the gastrointestinal system exhibits a "migrating motor complex" (MMC) in which electromechanical activity in the smooth muscle exhibits coordinated rhythmic contractions from the stomach through the small intestines. The MMC is myogenically generated but is modulated by enteric neurons, parasympathetic afferents *via* the vagus nerve, and sympathetic afferents from the thoracic and lumbar sympathetic chain.

This process is thought to serve a "house-keeping" function by moving undigested food into the next phase of digestion. Typically, the MMC cycles from a period of quiescence of 45–60 minutes in which relatively little electrical or muscular activity can be measured, followed by 30 minutes of progressively increasing frequencies of peristaltic contractions, culminating in a phase of frequent contractile activity before it cycles back to quiescence. The frequency and velocity of the MMC varies depending upon the time of day, with higher propagation velocities observed during the day than during the night.[18]

Similarly, colonic peristaltic activity is expressed rhythmically such that the activity is lowest just before meals and minimal during the night or when individuals are asleep.[19] Contractions are highest in the morning following awakening, and urges for defecation are maximal.[19,20] The force of colonic propulsion is 3 times greater upon waking than during sleep, when it is minimal, while the presentation of a meal induces a smaller increase.[20,21]

Increases in intestinal, biliary, and pancreatic secretions typically accompany the MMC and colonic contractions, and these have been shown to be expressed rhythmically under clinical conditions as well. While intestinal secretions of amylase, chymotrypsin and trypsin are highly correlated with the frequency of MMC, being higher during the day than during the night, the amplitude of those secretions is larger during the night than during the day.[22,23] Similarly, biosynthesis of bile salts and their release into the human intestinal lumen are higher during the night than during the day.[24]

9.1.2 Expression of Gastrointestinal Rhythms in Animal Models

To understand the underlying mechanisms regulating human gastrointestinal rhythmicity, it is critical to develop animal models that can accurately reflect the human condition. There are, of course, limitations to these animal models in this regard. First, most biomedical studies are conducted in nocturnal rodents, which, by definition, are active and feeding during the night. This may seem inconsequential, but the activity of the circadian pacemaker in the SCN is highest during the day in both diurnal and nocturnal animals, and the production of the pineal hormone melatonin is highest during the night in both diurnal and nocturnal animals. Thus, the phase relationship of ingestion to the sleep–wake cycle is approximately 180° out of phase in humans compared to rats and mice. Adding to this, many laboratory strains of mice, such as the commonly employed C57 strains, are incapable of producing melatonin at all due to spontaneous mutations in the biosynthetic pathways producing this hormone.[25,26] Third, dietary differences between humans and animal models can prove to be problematic for generalization. Despite these caveats, animal studies are critical for our understanding of these or any other processes.

9.1.2.1 Endogenous Gastrointestinal Rhythmicity

The GI system, from the pharyngeal mucosa to the rectum, and many other associated GI tissues, such as the liver, the exocrine pancreas and salivary glands exhibit daily and circadian rhythms.[27–32]

9.1.2.2 Oral/Ingestive Phase

As with sleep regulation,[33,34] food consumption is regulated homeostatically as well as by non-homeostatic mechanisms,[35] including a circadian clock.[36] While the homeostatic control of feeding through energy balance at lower levels of nutrition is well regulated, the upper levels of energy balance do not always follow this rule, and as a result, non-homeostatic feeding such as hedonic eating can result in obesity.[36] Circadian control of feeding is evident from the fact that in rodent models maintained in constant darkness (DD), feeding rhythms are expressed with a circadian period of close to, but not exactly, 24 hours in parallel with locomotor activity rhythms,[3] which are entrainable by environmental light–dark cycles (LD) congruently with locomotor behaviour.[3,37] However, the circadian clock underlying the food-anticipatory behaviour can be experimentally separated from the light-entrainable circadian system, since presentation of a timed meal in the light phase of the LD cycle, when nocturnal rodents are not usually active, results in food-anticipatory and feeding behaviour during the presentation of the timed meal (Figure 9.2).[3,4,38] This will be discussed further below.

9.1.2.3 Intestinal Phase

There is strong evidence that the rodent gastrointestinal system itself contains a circadian clock capable of generating circadian patterns of motility, absorption and gene expression independently from central clock structures

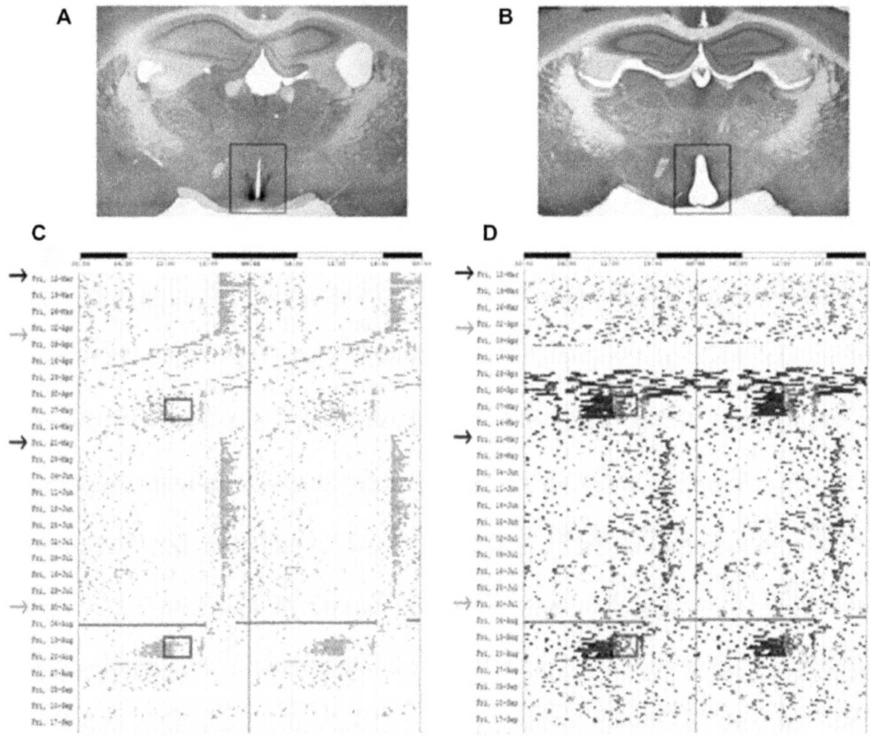

Figure 9.2 Effects of timed feeding on circadian locomotor rhythms in intact and SCN-lesioned mice. (A) Histological section through the SCN in a representative sham-operated mouse. (B) Histological section through the SCN region of a representative SCN-lesioned mouse. (C) Double-plotted actogram of rhythmic circadian locomotor behaviour in a sham-operated mouse. (D) Double-plotted actogram of arrhythmic locomotor behaviour of an SCN-lesioned mouse. First arrow indicates the beginning of the LD cycle. Second arrow indicates placement of mouse into constant darkness (DD). Third arrow represents a timed feeding regimen in DD. Fourth arrow represents placement of mice into *ad lib* feeding in DD. Boxes represent times in which faeces of mice were collected. Reproduced from ref. 38 with permission from American Physiological Society, Copyright 2012.

in the brain.[29,39] First, measures of colonic motility *in vivo* using an ambulatory telemetry system exhibit slow wave rhythms of colonic pressure during the day in LD and subjective day in DD that transitions into higher frequency contractions during the night and subjective night,[29,30] which is inverse to the situation in humans.[20,21]

Second, circadian rhythms in motility are expressed by excised gastrointestinal tissues in culture. Colonic tissue maintained in tissue culture exhibits increased contractility in response to increasing concentrations of the neurotransmitter acetylcholine (Ach). This response to Ach is rhythmic, such that colonic circular muscle contractility in response to Ach is greater during the night in LD and subjective night in DD than during the day and subjective day.[29] In addition, circadian rhythms of gene expression of the *Period* gene, using a bioluminescent luciferase reporter, persist in colonic tissue *in vitro* as well.[38]

Transcriptomic analyses of mouse distal colon tissues revealed a wide array of gene products that are expressed rhythmically in an LD cycle.[30] Genes associated with cell signalling (19%), gene expression (16%), cell death (13%) and cell cycle (12%) are the predominant gene ontology categories that are expressed rhythmically on a daily basis. The times at which gene expression levels peak (acrophase) may vary, even within gene ontology categories. For example, B cell leukemia/lymphoma 2 (*Bcl2*) and beclyn (*Becn*) both regulate autophagy and apoptosis and they are also associated with the diagnosis of colon cancer. Yet, the acrophase for *Bcl2* is midnight (Zeitgeber time ZT17), while the *Becn* acrophase is midday (ZT5).[30] Gene products associated with electrolyte absorption by the rat colon are regulated by the circadian clock *in vivo*.[40,41] The sodium pump (subunits Atp1a1 and Atp1b1), channels (α-, β-, and γ-subunits of Enac and Cftr), transporters (Dra, Ae1, Nkcc1, Kcc1, and Nhe3), and the Na^+/H^+ exchanger (NHE) regulatory factor (Nherf1) are expressed rhythmically in the rat colonic mucosa, peaking at ZT12, or at the beginning of the subjective night.

As a measure of both colonic motility and the defecation phase of digestion, stool weight and number reliably reflect the timing of intestinal motility in mice, since they defecate throughout the circadian day, albeit at variable rates.[29,38] In mice, the stool number and weight are rhythmically produced *in vivo* such that faeces are produced predominantly during the night when the mice are exposed to LD cycles and during the subjective night when they are held in DD.[38]

9.1.3 Mechanisms of Circadian Clock Function

As is described elsewhere in this volume in more detail, circadian rhythms in multicellular animals are generated by a transcription/translation feedback loop.[42,43] Briefly, *Clock* and *Bmal1* are transcribed and then translated in the cytoplasm, where the CLOCK and BMAL1 proteins dimerize and re-enter the nucleus; these dimers then bind to E-box sequences on the promoter regions of so-called "clock-controlled genes", which are activated. Among the clock-controlled genes are output pathways that subserve rhythmic functions within the cell, depending on the tissue in which they reside. In addition, *Period (Per1, Per2,* and *Per3)* and *Cryptochrome (Cry1* and *Cry2)*, which have promoters containing E-box sequences, are transcribed. The mRNA of these genes travel to the cytoplasm, where they are translated into PER1, PER2, PER3, CRY2 and CRY2 proteins. These form oligomers, which re-enter the nucleus, where they interrupt the BMAL1/CLOCK activation of transcription. Interference of the workings of these molecular clockworks by mutation and/or knockout results in disruption and/or abolition of behavioural and physiological circadian rhythms. There are other components of this gene network, which are discussed in detail elsewhere in the literature[42,43] and in this volume.

The mouse gastrointestinal system expresses clock genes in a rhythmic fashion both *in vivo* and *in vitro*.[29,31] Gastrointestinal tissues such as the stomach, proximal, middle and distal colonic tissues express *Clock, Bmal1, Per1, Per2, Per3, Cry1* and *Cry2* mRNA (Figure 9.3), as well as BMAL1, PER1 and PER2 protein. Staining is most intense in epithelial cells along colonic crypts, especially in the bases of the crypts, and is also present in the myenteric plexus of the stomach and throughout the colon.[34] Similarly, Sladek *et al.*[40] have shown that rat colonic epithelial cells rhythmically express clock genes as well. Clock gene rhythms in the rat gastrointestinal epithelium exhibit a "temporal gradient" such that the upper GI tract is phase advanced relative to the lower GI cells.[40–43] In addition, the circadian pattern of clock gene expression persists even outside the animal, since rhythms of *Per2::luciferase* (PER2::LUC) reporter gene expression persist in cultured mouse intestines.[38] Gastrointestinal rhythms are regulated by the clock gene network since a double knockout of *Per1* and *Per2* abolishes circadian rhythms of stool production as well as measures of intestinal motility *in vivo* and *in vitro*[31] (Figure 9.4).

In addition, system-level circadian organization coordinates circadian rhythms in mammals such that the "master pacemaker" residing in the SCN generates a high-amplitude circadian rhythm in gene expression, including the clock genes and several clock-controlled genes, electrical activity and secretion.[43–45] The circadian rhythm in the SCN is synchronized to the LD cycle *via* a specialized retinohypothalamic tract (RHT), and the SCN influences circadian output through multi-synaptic pathways within the brain. Among the outputs are sympathetic efferents that control the pineal gland's secretion of melatonin and other sympathetically-regulated processes. Surgical destruction of the SCN abolishes overt circadian rhythms in the animal, and transplantation of foetal SCN tissue into the lesion site of an SCN-lesioned rodent can restore circadian rhythms.

The SCN is important for gastrointestinal rhythms, since SCN lesions in mice fed *ad libitum* abolish circadian rhythms of stool output.[38] The SCN appears to synchronize the GI rhythms through circadian control of sympathetic tone, since chemical sympathectomy with the neurotoxin 6-hydroxydopamine (6OHDA) also abolishes rhythms in stool output of mice in which food is constantly available, and pulsed administration of the β-adrenergic agonist isoproterenol to cultured colons carrying the PER2::LUC knock-in phase shifts rhythms of bioluminescence.[38] Importantly, timed feeding restores daily patterns of defecation in SCN-lesioned and 6OHDA-treated mice, suggesting that even in the absence of rhythmic input from the SCN *via* the sympathetic nervous activity, the circadian clock in the gastrointestinal tract can entrain to timed meals.

The pineal hormone melatonin has also attracted interest in the context of gastrointestinal health and rhythmicity.[46] The mammalian pineal gland synthesizes and secretes melatonin on a daily and circadian basis such that

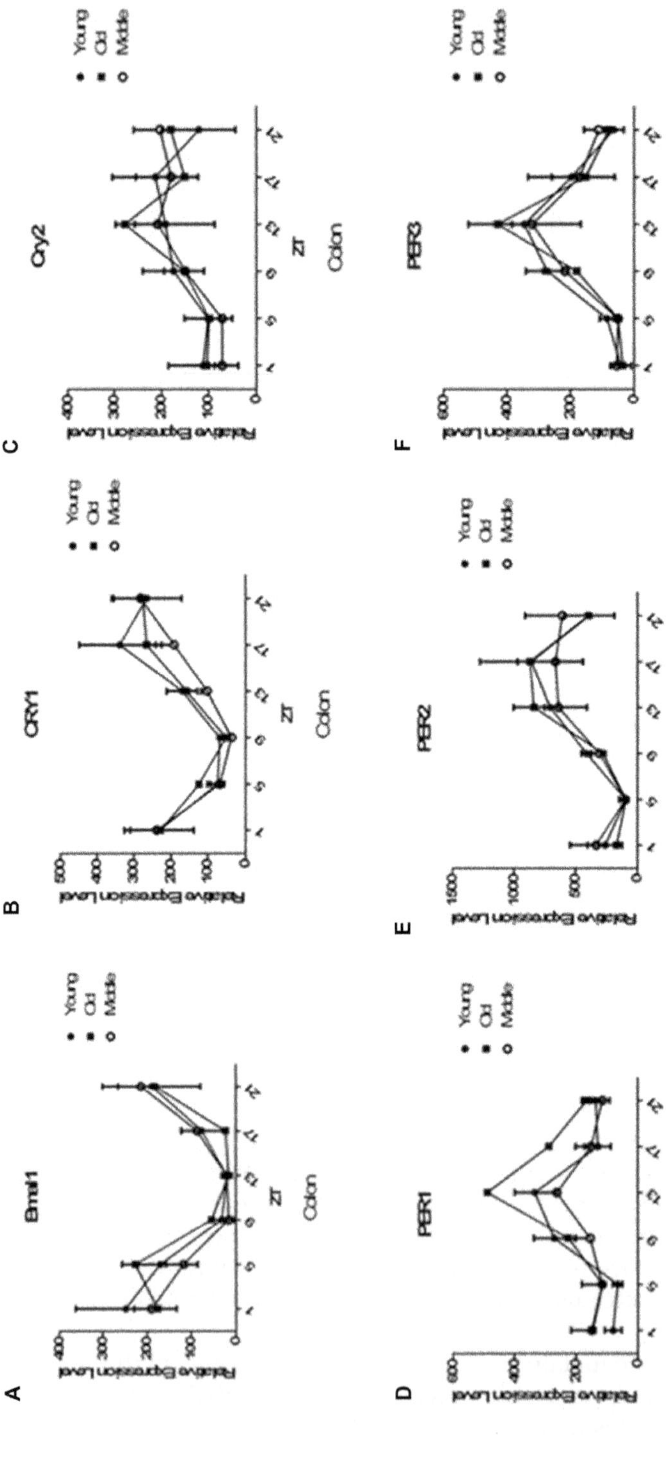

Figure 9.3 Daily patterns in the expression of clock genes in the mouse intestine. Age does not affect daily changes in colonic clock gene content. Rhythms of *Bmal1* (A) *Cry1* (B), *Cry2* (C), *Per1* (D), *Per2* (E) and *Per3* (F) are similar in young (3-month-old), middle-aged (12-month-old) and aged (24-month old) CBA mice. Reproduced from ref. 32 with permission from American Physiological Society, Copyright 2018.

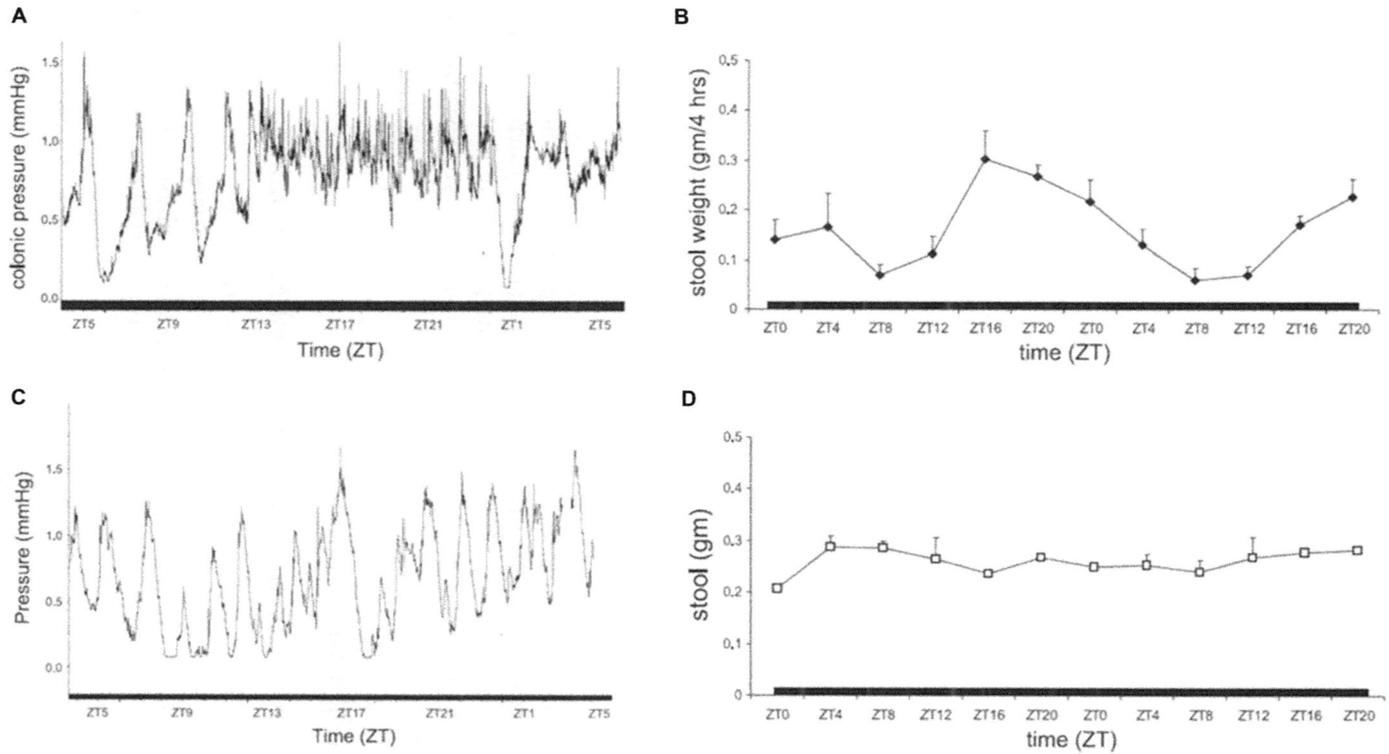

Figure 9.4 Rhythmicity in measures of colonic motility in WT and *Per1*/*Per2* double-knockout mice in constant darkness (DD). (A) Intracolonic pressure changes in a single WT mouse in DD. (B) Stool output in WT mice. (C) Intracolonic pressure changes in *Per1*/*Per2* double-knockout mice in DD. (D) Stool output in *Per1*/*Per2* double-knockout mice ($n=5$). Rhythmic changes in colonic motility are attenuated in *Per1*/*Per2* knockout-mice. Reproduced from ref. 31 with permission from American Physiological Society, Copyright 2010.

blood, brain and tissue levels of the hormone are elevated during the night and subjective night.[46] Pinealocytes and retinal photoreceptors take up the amino acid tryptophan (Trp) and convert Trp to 5-hydroxytryptophan (5-HTrp) *via* tryptophan hydroxylase (TPH). Then, aromatic amino acid decarboxylase (AAADC) converts 5-HTrp to serotonin (5HT). During the night, arylalkylamine-*N*-acetyltransferase (AANAT) converts 5HT to *N*-acetylserotonin (NAS), which, in turn, is converted to melatonin by hydroxyindole-*O*-methyl transferase (HIOMT). In mammals, this rhythm is under the direct control of pacemakers within the SCN *via* a multi-synaptic pathway that includes the sympathetic nervous system.[46]

Interestingly, melatonin is also synthesized within enterochromaffin and/or enterochromaffin-like cells in the gastrointestinal system.[47–49] The evidence for this is primarily derived from immunocytochemical detection of melatonin itself in these cells; estimation of gastrointestinal melatonin levels range from 10–400X serum levels of the hormone.[50,51] However, there are also gene expression data that show the expression of TPH, DDC, AANAT and HIOMT (ASMT) in the small intestine, colon, and pancreas, as well as both melatonin receptors (MT1 and MT2),[52] so it is likely that the biosynthetic pathway for gastrointestinal melatonin is similar, if not identical, to pineal and retinal melatonin biosynthesis. Remarkably, although the gastrointestinal system synthesizes melatonin,[53] gastrointestinal melatonin does not pass into the systemic circulation.[54] It is therefore an open question as to what exactly melatonin does in the gastrointestinal tract.

Several authors have suggested that gastrointestinal melatonin affects gastrointestinal function itself, preventing mucosal ulcerations, reduction of gastric release of hydrochloric acid, and stimulating the gastrointestinal immune system.[47,51] Furthermore, Bubenik[51] reported that melatonin is also secreted into the lumen of the gastrointestinal system, either *via* the hepatic duct or from the enterochromaffin cells themselves. Corroborating this view, Paulose *et al.*[32,55] have detected circadian patterns of melatonin in the faeces of CBA mice, which are genetically capable of synthesizing the hormone.

9.2 Microbiome

9.2.1 Description of the Gastrointestinal Microbiome

It is now well known that within the human body, a large ecosystem of microorganisms resides and interacts with the host tissues and among themselves.[56] Over a period of 10 years and across two phases of research, the National Institutes of Health's Human Microbiome Project[56] established tissue-specific indices of microbial communities using a wide array of approaches, including phenotypic and genotypic analyses of eukaryotic, eubacterial, archaeal, and viral organisms in healthy and some specific disease states. In the first phase, they established community composition in nasal, oral, skin, vaginal and gastrointestinal communities, while in the second phase, longitudinal studies were conducted in preterm birth, inflammatory bowel disease and pre-diabetic patients.

While studies of the gut microbiome in gastrointestinal disease have a long history,[57] recent research has shown that the gut microbiome is a dynamic ecosystem that can have profound effects on human and animal health.[58] The human intestine is home to approximately 10^{14} bacteria comprising approximately 1000 species.[59] Bacteria comprise about 98% of the microbiome under healthy conditions.

Although there are more than 1000 bacterial species in the gastrointestinal system, the human gut microbiome is dominated by two phyla: Bacteroidetes and Firmicutes.[60,61] Although species within the Bacteroidetes phylum usually maintain a complex and beneficial association with the host, they can be significant clinical pathogens and are among the most antibiotic-resistant species in the gut. Similarly, members of the Firmicutes phylum comprise both beneficial and pathogenic species, including the genuses *Bacillus, Lactobacillus, Enterococcus* and *Clostridium.*[60] Among the species of lower density in the gut, the Proteobacteria *Escherichia coli (E.coli)* and *Klebsiella (*formerly named *Enterobacter) aerogenes (K. aerogenes)* are usually beneficial commensals in the gut, but they can be pathogenic under certain circumstances, especially in hospital settings.[59]

The bacteria within the gastrointestinal microbiome interact with the host through a wide array of mechanisms in which they respond to hormonal or cell-surface signals as well as nutrition from the host. The presence of alcohol and many drugs can significantly alter the composition and responses of the gut microbiota as well.[62] Conversely, bacterial components within the microbiome are known to signal many aspects of their physiology to the host, which can significantly affect the health of the host. These signals include short-chain fatty acids and vitamins produced by bacterial metabolism.[62,63]

9.2.2 Circadian Changes in the Microbiome

It is unlikely that the organisms within the gastrointestinal microbiome are exposed to environmental light cycles, at least in larger animals. However, the microbial flora is exposed to a plethora of environmental changes. These include nutrient availability, host-derived hormones, anti-microbial peptides, antibodies, and small changes in body temperature (T_B). As a result, the composition of the microbiome changes depending on the time of day.[64,65]

Using laboratory mice as a model system, several groups[65–67] reported that the relative abundance of several operational taxonomic units (OTUs), a measure of bacterial diversity, in murine faeces changed depending on the time of day. They showed that, while most of the microbiome did not vary, 15%–20% of the detected OTUs exhibited daily oscillations. The relative abundance of some OTUs, notably of Streptococcacea and Oscillospiracea, was higher during the day than during the night, while the relative abundance of others, notably of Clostridiacae and Lactobacillacea, was higher during the night than during the day. Unsurprisingly, the total biomass of foecal bacteria, as was faeces itself, was higher during the night in these nocturnal mice.[65]

These diurnal oscillations in taxonomic abundance represent parallel changes in functional processes.[68] Using the Kyoto Encyclopedia of Genes and Genomes (KEGG) pathway analyses, Thaiss *et al.*[66,69] identified KEGG pathways that predominate at certain times of day. Gene products associated with environmental sensing, detoxification and bacterial chemotaxis are more abundant during the inactive phase (day for nocturnal mice) than during the active phase (night). Gene products associated with energy metabolism, cell growth, DNA repair and protein production are elevated during the night. These rhythmic pathways result in daily oscillations in bacterial metabolites such as short-chain fatty acids, polyamines, and amino acids, which in turn can affect the circadian organization of the host.[63]

Conversely, the host significantly affects the rhythmic patterns in the gastrointestinal microbiome.[70] Double knockout of negative elements *Period 1* and *Period 2* (*Per1/2*$^{-/-}$) abolishes daily oscillations in both the bacterial relative abundance and KEGG pathways.[66] Knockout of the positive element *Bmal1* also abolishes circadian rhythms in gastrointestinal microbiota.[65] Furthermore, subjecting mice to an artificial jet-lag regime shifts the rhythm of relative abundance among bacterial components of the gut and induces dysbiosis among the mice,[66] suggesting that circadian host signals are critical for the maintenance of microbiotic rhythmicity, and that this coordination is important for the host's health. The mechanism(s) through which the host synchronizes microbiotic rhythmicity is(are) yet to be determined.

9.2.3 *Klebsiella* (Formerly Named *Enterobacter*) *Aerogenes*

One hypothesis is that at least one mechanism through which the host influences the gastrointestinal microbiome involves the circadian secretion of melatonin into the lumen of the gut. Paulose *et al.*[71] identified, from metagenomics data in GenBank, several enteric bacteria that expressed sequences with 24–42% identities to known melatonin binding sites in the human genome. These included sequences in *Klebsiella*, *Clostridiales*, and *Lactobacillus* species, but not in *Escherichia coli*. Colonies formed by clinical isolates of *Klebsiella aerogenes*, formerly named *Enterobacter aerogenes,* a Gram negative, indole-negative motile bacterium, proliferated on semi-solid agar significantly more rapidly in the presence of melatonin in a specific, dose-dependent fashion. This effect was specific to melatonin, as *K. aerogenes* spreads further in the presence of melatonin than in the presence of equimolar concentrations of tryptophan, serotonin or *N*-acetylserotonin. RNASeq analysis of planktonic cultures of *K. aerogenes* revealed that 1 nM melatonin affected specific processes in a growth-stage specific fashion.[72] Sequences associated with fimbrial assembly, biofilm formation, metabolism and divalent ion transport were increased primarily during exponential growth, suggesting that the hormone may promote colony formation.

K. aerogenes also expressed a daily pattern of motility. To test the hypothesis that these bacteria possessed a circadian clock, cultures of *K. aerogenes* were transformed to express a reporter construct derived from *Photorhabdus* luciferase (lux) with the promoter region of *MotA*, a part of the flagellar motor of *K. aerogenes.* Using a luminometer, circadian rhythms in bioluminescence were detected that were temperature compensated for the circadian period with a Q_{10} of 0.96, meaning that the period does not change over a span of constant temperature (T_A) from 27 to 40 °C. However, if cultures of *K. aerogenes* are maintained in cycles of T_A simulating natural circadian changes in body temperature (T_B; 1–3 °C in amplitude), the circadian patterns of MotA expression entrain to the temperature cycles,[73] suggesting that this enteric commensal bacterium expresses a temperature-compensated circadian rhythm in motility and gene expression that responds to the melatonin hormone and that can entrain to physiologically relevant cycles of T_A.[74]

9.2.4 *Bacillus Subtilis*

Although it is usually thought of as a soil microorganism, the Proteobacterium *Bacillus subtilis* has also been demonstrated to be present in the human gastrointestinal microbiome.[57,75] *B. subtilis* strains have been obtained from ileum biopsies and human foecal samples, which, when isolated, could form biofilms, secrete antimicrobial proteins and sporulate anaerobically.

Recently, Eelderink-Chen *et al.*[76] have found that *B. subtilis* also exhibits a temperature-compensated circadian clock, which appears to be contingent on the expression of a biofilm. This organism can entrain to both LD cycles and to a 3 °C temperature cycle of 25.5 °C and 28.5 °C. These temperatures do not resemble those found within the gastrointestinal system, but it is possible that this bacterium may be an additional enteric microorganism with a clock. These data, together with the *K. aerogenes* work, raise the exciting possibility that at least some members of the microbiome may synchronize to the circadian rhythms of their host *via* the entrainment of their own circadian clock.

9.3 Effects of Gastrointestinal Rhythmicity on Overt Circadian Organization

As stated above, restricted feeding schedules in which food is available for only an hour or so every 24 hours entrain the wheel-running rhythms of rats, causing the animals to run vigorously for 1 to 3 hours before the timed feeding interval.[3,27,28,37,77] This feeding schedule entrains the circadian system because the phase of the anticipatory activity onset synchronizes with the time of food presentation, even when food is present *ad libitum.* In this situation, the animal free-runs with a circadian period with a phase that is synchronized by the predictable presence of a meal. Yet, although light is a quantifiable Zeitgeber, the presence or absence of a meal is complex, both in the nature of the food substance itself and in the behaviour it elicits. In other words, is it the olfactory or gustatory nature of the meal, the nutritional value, the abdominal distention, and/or the behavioural act of eating that entrains circadian clocks?[37]

Olfaction appears to have little or no role in food anticipatory behaviour because intranasal zinc sulfate administration, which renders rats "behaviourally anosmic," has no effect on the behaviour, nor, for that matter, does complete ablation of the olfactory bulbs affect the behaviour. However, gustation appears to play some role because rats and hamsters will entrain to the presence of "palatable" foods and express anticipatory behaviour even if regular rodent chow is available *ad libitum.* This effect appears to be independent of meal size because rats fed large volumes of non-nutritive mash do not entrain to the regime. In contrast, the nutritional value of the meal does affect anticipation. If rats are restricted to meals containing just two of three major nutrient groups—protein, carbohydrate, or fat—they will anticipate the timed presentation of these macronutrients. Moreover, rats will phase shift to a timed presentation of glucose but not of saccharin, vegetable, or mineral oil, suggesting that glucose holds a special Zeitgeber property for the circadian clock. Ablation of sensory inputs carrying gustatory and/or satiety sensation appears to influence circadian anticipation of food and entrainment to the restricted food schedule. Destruction of the hindbrain parabrachial region, which processes visceral and gustatory information, attenuates the anticipatory behaviour. These data suggest that the biological clock within the SCN, which regulates circadian patterns of appetite and feeding, is itself sensitive to the presence or absence of food.[3,4,27,28]

Surprisingly, although SCN lesions abolish all *spontaneous* circadian rhythmicity, entrainment to restricted food regimens and anticipation to restricted food schedules persist in SCN-lesioned rats (Figure 9.2). Food anticipatory locomotor behaviour persists even in rats[27,28] and mice[38] rendered arrhythmic by destruction of the SCN (Figure 9.2), strongly suggesting an SCN-independent food entrainable oscillator.

The rhythmicity that is entrained by feeding is not a self-sustained rhythm; rats and mice become arrhythmic as soon as the regime is completed. However, food entrains an SCN-independent oscillator because SCN-lesioned rats will entrain to time cycles of less than or greater than 24 hours with a discrete limit of entrainment. However, if there is an SCN-independent oscillator within the brain, where is it?

Lesion studies to identify the SCN-independent oscillator have been largely unsuccessful. Surgical destruction of the olfactory bulbs, area postrema, ventromedial nucleus, hippocampus, arcuate nucleus, and neocortex has little effect on the rhythmic anticipation of food and food entrainment. Only lesions to the afferent parabrachial region reduce the amplitude of the behavioural manifestation of rhythmic appetite, but they does not abolish it altogether. This suggests that the food-associated pacemaker is at least partially peripheral, perhaps in the gastrointestinal system itself.[27,77]

It is not clear to what extent rhythmicity in the gastrointestinal microbiome influences overt circadian behaviour. Nonetheless, the absence of the gut microbiome in germ-free mice alters the expression of clock genes in the intestine and liver.[63,69] This effect in turn alters hepatic levels of cholesterol, free fatty acids, bilirubin, and lactate. Furthermore, dysbiosis may deleteriously affect the microbiotic balance among beneficial and pathogenic bacteria, which may indirectly affect the circadian clock or at least its outputs.[69]

9.4 Disease Conditions Associated with Altered Clock Function

Many disease states have been associated with disorders or disruptions of the circadian clock. These disruptions include mutations or knockouts of core clock genes, surgical manipulation of circadian system structures such as the SCN or pineal gland, or abnormal lighting conditions, such as constant light (LL) or shifting LD cycles as is experienced during transmeridian flight (jet-lag) or shift-work scheduling.

9.4.1 Ageing

It is not clear that it is appropriate to consider ageing as a disease state. However, the process of normal ageing can have moderate to severe effects on gastrointestinal health.[78] Oesophageal contractions and tension decreases,[79] the stomach loses much of its elasticity and is less resistant to damage,[80] and the intestinal lumen can show marked changes in the microflora.[78] In addition, pancreatic and hepatic fibrosis increases as we age.[80] Furthermore, other indirect or secondary factors associated with age can also profoundly upset gastrointestinal homeostasis, including a decrease in physical activity, an increase in the use of several drugs, especially non-steroidal anti-inflammatory drugs (NSAIDS), which can cause GI bleeding, and opiates, which result in constipation.[8]

However, there is little or no evidence that the process of normal ageing has deleterious effects on the circadian clockworks themselves. Although gastrointestinal tissues were not used, Yamazaki *et al.*[81] showed that circadian rhythms of *Per1* bioluminescence in several brain and peripheral tissues of rats, including the retrochiasmatic area (RCA), paraventricular nucleus of the hypothalamus (PVN), pineal gland, pituitary, and arcuate nucleus in the brain, as well as the cornea, kidney, liver, and lung, were largely unaffected by age, although RCA and PVN were phase-advanced in older rats.

Transcriptomics analyses of young (3-month old), middle-aged (12-month old), and old-aged (24-month old) melatonin-competent CBA mice revealed little or no change in the amplitude or phase of rhythms in clock gene expression in the small intestine and colon (Figure 9.3).[32,82] In contrast, diurnal patterns in the expression of nitric oxide synthase 1 (*nos1*) and of immune factors *irak4* and *interleukin*-8 were observed in the colon of young mice, but they were lost in middle-aged and aged animals. Furthermore, the diurnal pattern of melatonin synthesis genes was altered by age, and stool melatonin levels showed a significant decline between young mice and aged cohorts. These data provide additional evidence for the persistence of the circadian clock throughout the ageing process and highlight its importance to health.

While not specific to gastrointestinal function, knockout of the positive element *Bmal1* in laboratory mice results in premature ageing and a decreased lifespan.[83] These animals exhibit many symptoms of premature ageing, including loss of skeletal muscle mass, cataracts, and organ shrinkage. However, one caveat to the idea that loss of circadian rhythm generation in the *Bmal1*$^{-/-}$ mouse is that these mice develop in the absence of this important transcription factor and may be susceptible to pleiotropic effects of the knockout. When conditional *Bmal1*$^{-/-}$ mice are compared to standard *Bmal1*$^{-/-}$ mice, the effects of reduced lifespan, fertility, body weight and age-dependent arthropathy largely go away.[84]

9.4.2 Colorectal Cancer

Colorectal cancer is the most prevalent cancer of the gastrointestinal system and also one of the most commonly diagnosed cancers.[85,86] Most colorectal cancers are adenocarcinomas, which start in cells that synthesize and secrete mucus to lubricate the epithelia of the colon and rectum. There are other, less common types of colorectal tumours. These include carcinoids, gastrointestinal stromal tumours, lymphomas, and sarcomas. By and large, all cancers co-opt the cellular control of cell proliferation, differentiation, DNA damage repair and apoptosis, which are all processes under circadian regulation.[85]

Clock gene expression rhythms have been shown to be altered in human colorectal tumour tissues when compared to matched healthy mucosa taken at single timepoints.[54] Expression of *ARNTL (BMAL1), PER1, PER2, PER3* and *CRY2* was lower in tumour samples than in matched healthy controls. No significant effect of the cancer was found in *CLOCK* and *CRY1* expression. In another clinical study, expression of *CRY1* and *CRY2* in human colorectal cancer tissues was lower than their expression in adjacent healthy tissues.[87] Similarly, reduced expression of *PER3* was observed in colon and rectal cancerous tissue compared to healthy adjacent tissues.[87] In this study, the authors indicated that *PER3* expression was indicative of multiple pathological outcomes, including tumour location, differentiation, and cancer stage, and that the incidence of death was higher in patients with the lowest *PER3* expression. Since these were single time-point datasets, it is not clear if the decrease in expression represents an actual decrease or rather a change in the circadian phase of the circadian clock gene cycle. Thus, from these studies alone, it is not clear whether the *circadian* function of clock genes plays a primary role in carcinogenesis or if these transcription factors have a pleiotropic role.

Even so, animal studies support the notion that circadian clock disruption contributes to gastrointestinal cancers. For example, mice with *Per2* knockout exhibit an increase in neoplasia and mice that are genetically transformed to produce more polyps exhibit increased polyp formation.[88–90] *Bmal1* knockdown with RNAi accelerated the progression of tumour growth in mice *in vivo,* and, in C26 colon cancer cells, *Bmal1* knockdown decreased the efficacy of chemotherapeutics in reducing apoptosis and DNA damage.[91] Environmental disruption of circadian clock function also increases the likelihood of gastrointestinal neoplasia. Experimentally induced jet-lag in mice increases intestinal inflammation and the development of colonic polyps in mice that are genetically predisposed to polyp formation.[88,89]

In contrast, evidence for a direct role of circadian function in human colorectal cancer is controversial. Comparison of shift-workers (350 graveyard shift-workers *versus* 410 early morning controls) revealed no association of graveyard shift with colorectal cancer in female workers who had been in their respective work cycles for over 7.5 years.[85] Furthermore, these authors could find no association of colorectal cancer in 7 different shift-work variables, including light exposure at night or phase-shift.

9.4.3 Inflammatory Bowel Disease

Inflammatory bowel disease (IBD) is an ongoing inflammation of all or part of the gastrointestinal system, causing abdominal pain, cramping, diarrhea, nausea, vomiting and fatigue. Crohn's disease and ulcerative colitis are the two most common forms of IBD; both involve chronic inflammation of the digestive tract. Approximately 10 million adults have IBD worldwide, with 3 million of these residing in the United States. They are both treatable but there is no cure. Crohn's disease can attack any part of the gastrointestinal system from the mouth to the anus, although it is most often associated with the small intestine and colon. It can also manifest in all layers of the bowel wall. In contrast, ulcerative colitis only involves the innermost lining of the colon, forming ulcers in the mucosal membranes in the intestines. The primary causes of IBD are not fully understood, but they likely entail a combination of inherited and environmental factors that increase immune responses to the intestinal lining, which in turn result in localized inflammation.

Patients with IBD frequently experience problems with sleep quantity and quality, and since sleep is regulated by the circadian clock, it is likely that disruptions in sleep may indicate underlying circadian rhythm disruption. Alternatively, the discomfort associated with IBD may mitigate sleep quality.[62] Animal studies support the notion that circadian rhythm disruption worsens IBD, rather than the other way round. Exposure of mice to 12-hour phase shifts every 5 days for 3 months had no effect on body weight in mice compared to non-shifted mice. However, when mice were exposed to dextran sodium sulfate (DSS) in their drinking water and then allowed to drink water without DSS, an IBD-like intestinal syndrome was induced.[70] Non-shifted mice lost 10–15% of their body weight and exhibited significant histopathological damage in their colons. Mice who underwent both the DSS treatment and the phase shift lost 20–25% body weight and exhibited 40% more histopathological damage. These data clearly point to an interaction between circadian organization and the worsening disease state, even if the primary cause of the disease may not be circadian in nature. Even so, the disruption of the circadian clock associated with modern lifestyles may be a contributing environmental factor that increases the pathology.

9.5 Conclusion

Circadian organization comprises both hierarchical and distributed pacemakers and peripheral oscillators that are coordinated by the hypothalamic SCN (Figure 9.1). Oscillators within the SCN are entrained to environmental light cycles *via* the retinohypothalamic tract and coordinate downstream peripheral oscillators *via* neural, paracrine, and endocrine processes. Among the peripheral tissues that are coordinated by the SCN is the gastrointestinal system *via* the sympathetic nervous system.

The gastrointestinal system is a complex system comprising the alimentary tract and several major accessory organs, including the salivary glands, the exocrine pancreas, gall bladder and the liver. At each level of gastrointestinal function, circadian clock function regulates the homeostatic functions of consuming, breaking down, digesting and absorbing nutrients.[92] This system contains a semi-independent circadian clock capable of producing endogenously generated rhythms of gene expression, motility, and absorption *in vivo* and *in vitro*, capable of entraining directly to timed meals.

Within the gastrointestinal lumen, a vibrant ecosystem of microorganisms, the gastrointestinal microbiome, is regulated by the gastrointestinal clock through the presentation of nutrients, hormones, and changes in T_B, among other signals. Some of the bacteria likely directly respond to the host signals, while others, possessing endogenous circadian clocks themselves, may entrain to the host's Zeitgebers, and still others may respond to signals generated by members of the gut microbiome itself.

Intriguingly, the circadian signals derived from the gut microbiome may influence gastrointestinal rhythmicity. Furthermore, circadian signals originating from the gastrointestinal clock may in turn provide feedback to alter overt circadian behaviour as is evidenced by anticipatory behaviour in response to timed feeding. Thus, the circadian system is really an orchestra of multiple oscillators (*e.g.*, the SCN) within multiple oscillators (*e.g.*, the gastrointestinal clock) within multiple oscillators (*e.g.*, the gut microbiome), which in turn feedback at every level.

Disruption of the finely tuned system may be critical in exacerbating or even causing pathological process and disease.[93,94] The presence of artificial lighting systems, ever-present food in modern refrigerators and society's pressures likely expose us all to significant biomedical danger. Recognition of these dangers will be critical for human health in the future.

Acknowledgements

Gastrointestinal research in Cassone's lab has been funded by NIH grants R21 DK 074477 to W Hoogerwerf, and to NIH R01 AG045833 and R01 GM118541 to VM Cassone. The author thanks Ms. Melissa Whitfield-Rucker for assistance with this manuscript.

References

1. W. Craig, *Proc. Natl. Acad. Sci. U. S. A.*, 1917, **3**, 685–688.
2. G. F. Ball and J. Balthazart, *J. Horm. Behav.*, 2008, **53**(2), 307–318.
3. F. K. Stephan, *Physiol. Behav.*, 1986, **38**(4), 537–544.
4. V. M. Cassone and F. K. Stephan, *Nutrition*, 2002, **18**(10), 814–819.
5. J. S. Pendergast and S. Yamazaki, *J. Biol. Rhythms*, 2018, **33**(5), 458–474.
6. R. E. Clark, *Integr. Physiol. Behav. Sci.*, 2004, **39**(4), 279–294.
7. C. Dawes, *J. Physiol.*, 1972, **220**, 529–545.
8. K. Bitar, B. Greenwood-Van Meerveld, R. Saad and J. W. Wiley, *Neurogastroenterol. Motil.*, 2011, **23**(6), 490–501.
9. C. M. Ghiciuc, C. L. Cozma-Dima, V. Pasquali, P. Renzi, S. Simeoni, C. E. Lupusoru and F. R. Patacchioli, *Neuroendocrinol. Lett.*, 2011, **32**(4), 475–480.
10. R. R. Wilcox, D. A. Granger, S. Szanton and F. Clark, *Physiol. Behav.*, 2014, **129**, 11–16.
11. D. K. Chitkara, C. Fortunato and S. Nurko, *J. Pediatr. Gastroenterol. Nutr.*, 2004, **38**, 192–197.
12. H. J. Stein, S. Singh and T. R. Demeester, *Dis. Esophagus*, 2004, **17**(4), 297–303.
13. S. Lanzon-Miller, R. E. Pounder, M. R. Hamilton, N. A. Chronos, S. Ball, J. E. Mercieca, M. Olausson and C. Cederberg, *Aliment. Pharmacol. Ther.*, 1987, **1**(3), 225–237.
14. J. G. Moore and F. Halberg, *Dig. Dis. Sci.*, 1986, **31**, 1185–1191.
15. J. G. Moore and F. Halberg, *Chronobiol. Int.*, 1987, **4**(1), 101–110.
16. R. H. Goo, J. G. Moore, E. Greenberg and N. P. Alazraki, *Gastroenterology*, 1987, **93**(3), 515–518.
17. P. C. Grammaticos, A. Doumas and G. Koliakos, *Hell. J. Nucl. Med.*, 2015, **18**(1), 60–62.

18. D. Kumar, D. Wingate and Y. Ruckebusch, *Gastroenterology*, 1986, **91**, 926–930.
19. F. Narducci, G. Bassotti, M. Gaburri and A. Morelli, *Gut*, 1987, **28**(1), 17–25.
20. S. S. Rao, P. Sadeghi, J. Beaty, R. Kavlock and K. Ackerson, *Am. J. Physiol.: Gastrointest. Liver Physiol.*, 2001, **280**(4), G629–G639.
21. J. J. Auwerda, D. J. Bac and W. R. Schouten, *Dis Colon Rectum*, 2001, **44**(9), 1328–1332.
22. J. Keller and P. Layer, *J. Appl. Physiol.*, 2002, **93**(2), 592–600.
23. J. Keller, G. Groger, L. Cherian, B. Günther and P. Layer, *Am. J. Physiol.: Gastrointest. Liver Physiol.*, 2001, **280**, G273–G278.
24. M. M. Schroor, H. P. Sennels, J. Fahrenkrug, H. L. Jørgensen, J. Plat and R. P. Mensink, *Nutrients*, 2019, **11**(7), 1439.
25. S. Ebihara, T. Marks, D. J. Hudson and M. Menaker, *Science*, 1986, **231**(4737), 491–493.
26. T. Kasahara, K. Abe, K. Mekada, A. Yoshiki and T. Kato, *Proc. Natl. Acad. Sci. U. S. A.*, 2010, **107**(14), 6412–6417.
27. A. J. Davidson, A. S. Poole, S. Yamazaki and M. Menaker, *Genes, Brain Behav.*, 2003, **2**(1), 32–39.
28. A. J. Davidson, O. Castañón-Cervantes and F. K. Stephan, *Liver Int.*, 2004, **24**(3), 179–186.
29. W. A. Hoogerwerf, H. L. Hellmich, G. Cornélissen, F. Halberg, V. B. Shahinian, J. Bostwick, T. C. Savidge and V. M. Cassone, *Gastroenterology*, 2007, **133**(4), 1250–1260.
30. W. A. Hoogerwerf, M. Sinha, A. Conesa, B. A. Luxon, V. B. Shahinian, G. Cornélissen, F. Halberg, J. Bostwick, J. Timm and V. M. Cassone, *Gastroenterology*, 2008, **135**(6), 2019–2029.
31. W. A. Hoogerwerf, V. B. Shahinian, G. Cornélissen, F. Halberg, J. Bostwick, J. Timm, P. A. Bartell and V. M. Cassone, *Am. J. Physiol.: Gastrointest. Liver Physiol.*, 2010, **298**(2), G143–G150.
32. J. K. Paulose, C. V. Cassone and V. M. Cassone, *Physiol. Genomics*, 2019, **51**(1), 1–9.
33. S. Daan, D. G. Beersma and A. A. Borbély, *Am. J. Physiol.*, 1984, **246**(2 Pt 2), R161–R183.
34. A. A. Borbély, S. Daan, A. Wirz-Justice and T. Deboer, *J. Sleep Res.*, 2016, **25**(2), 131–143.
35. C. M. Liu and S. E. Kanoski, *Physiol. Behav.*, 2018, **193**(Pt B), 223–231.
36. E. Challet, *Nat. Rev. Endocrinol.*, 2019, **15**(7), 393–405.
37. F. K. Stephan, *J. Biol. Rhythms*, 2002, **17**(4), 284–292.
38. J. N. Malloy, J. K. Paulose, Y. Li and V. M. Cassone, *Am. J. Physiol.: Gastrointest. Liver Physiol.*, 2012, **303**(4), G461–G473.
39. L. Polidarová, M. Sládek, M. Soták, J. Pácha and A. Sumová, *Chronobiol Int.*, 2011, **28**(3), 204–215.
40. M. Sládek, M. Rybová, Z. Jindráková, Z. Zemanová, L. Polidarová, L. Mrnka, J. O'Neill, J. Pácha and A. Sumová, *Gastroenterology*, 2007, **133**(4), 1240–1249.

41. M. Soták, L. Polidarová, J. Musílková, M. Hock, A. Sumová and J. Pácha, *Am. J. Physiol.: Gastrointest. Liver Physiol.*, 2011, **301**(6), G1066–G1074.
42. J. Pácha and A. Sumová, *Acta Physiol.*, 2013, **208**(1), 11–24.
43. D. Bell-Pedersen, V. M. Cassone, D. J. Earnest, S. S. Golden, P. E. Hardin, T. L. Thomas and M. J. Zoran, *Nat. Rev. Genet.*, 2005, **6**(7), 544–556.
44. A. Patke, M. W. Young and S. Axelrod, *Nat. Rev. Mol. Cell Biol.*, 2020, **21**(2), 67–84.
45. A. M. Rosenwasser and F. W. Turek, *Sleep Med. Clin.*, 2015, 403–412.
46. D. C. Klein, S. L. Coon, P. H. Roseboom, J. L. Weller, M. Bernard, J. A. Gastel, M. Zatz, P. M. Iuvone, I. R. Rodriguez, V. Bégay, J. Falcón, G. M. Cahill, V. M. Cassone and R. Baler, *Recent Prog. Horm. Res.*, 1997, **52**, 307–357.
47. I. M. Kvetnoy, I. E. Ingel, T. V. Kvetnaia, N. K. Malinovskaya, S. I. Rapoport, N. T. Raikhlin, A. V. Trofimov and V. V. Yuzhakov, *Neuroendocrinol. Lett.*, 2002, 121–132.
48. N. T. Raikhlin and I. M. Kvetnoy, *Acta Histochem.*, 1976, 19–24.
49. N. T. Raikhlin, I. M. Kvetnoy and V. N. Tolkachev, *Nature*, 1975, 344–345.
50. G. A. Bubenik, *Horm. Res.*, 1980, 313–323.
51. G. A. Bubenik, *Dig. Dis. Sci.*, 2002, 2336–2348.
52. F. Soderquist, P. M. Hellstrom and J. L. Cunningham, *PLoS One*, 2015, e0120195.
53. G. Huether, *Ann. N. Y. Acad. Sci.*, 1994, 146–158.
54. P. Konturek, T. Brzozowski and S. Konturek, *J. Phys. Pharm.*, 2011, 139–150.
55. J. K. Paulose, C. Wang, B. F. O'Hara and V. M. Cassone, *Nat. Sci. Sleep*, 2019, 113–121.
56. L. M. Proctor, H. H. Creasy, J. M. Fettweis, J. L. Lloyd-Price, A. Mahurhar, W. Zhou, G. A. Buck, M. P. Snyder, J. F. Strauss, G. M. Weinstock, O. White and C. Huttehower, *Nature*, 2019, 641–648.
57. S. Bengmark, *Gut*, 1998, 2–7.
58. R. E. Ley, D. A. Peterson and J. I. Gordon, *Cell*, 2006, 837–848.
59. C. Manichanh, N. Borruel, F. Casellas and F. Guarnaer, *Nat. Rev. Gastroenterol.*, 2012, 599–608.
60. P. B. Eckburg, E. M. Bik, C. N. Bernstein, E. Purdom, L. Dethlefson, M. Sargent, S. R. Gill, K. E. Nelson and D. A. Relman, *Science*, 2005, 1635–1638.
61. H. M. Wexler, *Clin. Microbiol. Rev.*, 2007, 593–621.
62. R. M. Voigt, C. B. Forsyth, S. J. Green, P. A. Engen and A. Keshavarzian, *Int. Rev. Neurobiol.*, 2016, 193–205.
63. V. Leone, S. M. Gibbons, K. Martinez, A. L. Hutchison, E. Y. Huang, C. M. Cham, J. F. Pierre, A. F. Heneghan, A. Nadimpalli, N. Hubert, E. Zale, Y. Wang, Y. Huang, B. Theriault, A. R. Dinner, M. W. Musch, K. A. Kudsk, B. J. Prendergast, J. A. Gilbert and E. B. Chang, *Cell Host Microbe*, 2015, 681–689.
64. X. Liang and G. A. FitzGerald, *J. Biol. Rhythms*, 2017, 505–515.
65. X. Liang, F. D. Bushman and G. A. FitzGerald, *Proc. Natl. Acad. Sci. U. S. A.*, 2015, 10479–10484.

66. C. A. Thaiss, D. Zeevi, M. Levy, G. Zilberman-Schapira, J. Suez, A. C. Tengeler, L. Abramson, M. N. Katz, T. Korem, N. Zmora, Y. Kuperman, I. Biton, S. Gilad, A. Harmelin, H. Shapiro, Z. Halpern, E. Segal and E. Elinav, *Cell*, 2014, **159**(3), 514–529.
67. A. Zarrinpar, A. Chaix, S. Yooseph and S. Panda, *Cell Metab.*, 2014, **20**(6), 1006–1017.
68. A. Montagner, A. Korecka, A. Polizzi, Y. Lippi, Y. Blum, C. Canlet, M. Tremblay-Franco, A. Gautier-Stein, R. Burcelin and Y. C. Yen, *Sci. Rep.*, 2016, 20127.
69. C. A. Thaiss, M. Levy, T. Korem, L. Dohnalová, H. Shapiro, D. A. Jaitin, E. David, D. R. Winter, M. Gury-BenAri, E. Tatirovsky, T. Tuganbaev, S. Federici, N. Zmora, D. Zeevi, M. Dori-Bachash, M. Pevsner-Fischer, E. Kartvelishvily, A. Brandis, A. Harmelin, O. Shibolet, Z. Halpern, K. Honda, I. Amit, E. Segal and E. Elinav, *Cell*, 2016, 1495–1510.
70. F. Preuss, Y. Tang, A. D. Laposky, D. Arble, A. Keshavarzian and F. W. Turek, *Am. J. Physiol.: Regul., Integr. Comp. Physiol.*, 2008, R2034–R2040.
71. J. K. Paulose, J. M. Wright, A. G. Patel and V. M. Cassone, *PLoS One*, 2016, **11**(1), e0146643.
72. K. Graniczkowska, C. L. Shaffer and V. M. Cassone, *Genomics*, 2022, 110321, DOI: 10.1016/j.ygeno.2022.110321.
73. J. K. Paulose, C. V. Cassone, K. B. Graniczkowska and V. M. Cassone, *iScience*, 2019, **19**, 1202–1213.
74. K. B. Graniczkowka and V. M. Cassone, in *Circadian Rhythms in Bacteria and Microbiomes*, ed. C. H. Johnson and M. Rust, Springer Nature, 2021, pp. 351–364.
75. H. A. Hong, R. Khaneja, N. M. K. Tam, A. Cazzato, S. Tan, M. Urdaci, A. Brisson, A. Gasbarrini, I. Barnes and S. M. Cutting, *Res. Microbiol.*, 2009, 134–143.
76. Z. Eelderink-Chen, J. Bosman, F. Sartor, A. N. Dodd, ÁT Kovács and M. Merrow, *Sci. Adv.*, 2021, **7**, eabe2086.
77. C. A. Comperatore and F. K. Stephan, *J. Biol. Rhythms*, 1987, 227–242.
78. A. B. Thomson and M. Keelan, *Can. J. Physiol. Pharmacol.*, 1986, 30–38.
79. M. C. Geokas and B. J. Haverback, *Am. J. Surg.*, 1969, 881–892.
80. M. C. Geokas, C. N. Conteas and A. P. Majumbar, *Clin. Geriatr Med*, 1985, 177–205.
81. S. Yamazaki, M. Straume, H. Tei, Y. Sakaki, M. Menaker and G. D. Block, *Proc. Natl. Acad. Sci. U. S. A.*, 2002, 10801–10806.
82. V. M. Cassone, J. K. Paulose and C. E. Harpole, in *Circadian Rhythms and their Impact on Aging*, ed. S. M. Jazwinski, V. Belancio and S. Hill, Springer, 2017, pp. 83–102.
83. R. V. Kondratov, A. A. Kondratova, V. Y. Gorbacheva, O. V. Vykhovanets and M. P. Antoch, *Genes Dev.*, 2006, 1868–1873.
84. G. Yang, L. Chen, G. R. Grant, G. Paschos, W. L. Song, E. S. Musiek, V. Lee, S. C. McLoughlin, T. Grosser, G. Cotsarelis and G. A. FitzGerald, *Sci. Transl. Med.*, 2016, 324ra16.

85. W. M. Walasa, R. N. Carey, S. Si, L. Fritschi, J. S. Heyworth, R. C. Fernandez and T. Boyle, *Occup. Environ. Med.*, 2018, 344–350.
86. X. Wang, D. Yan, M. Teng, J. Fan, C. Zhou, D. Li, G. Qiu, X. Sun, T. Li, T. Xing, H. Tang, X. Peng and Z. Peng, *Ann. Surg. Oncol.*, 2012, 3081–3088.
87. G. Mazzoccoli, A. Panza, M. R. Valvano, O. Palumbo, M. Carella, V. Pazienza, G. Biscaglia, F. Tavano, P. Di Sebastiano, A. Andriulli and A. Piepoli, *Chronobiol. Int.*, 2011, 841–851.
88. F. Bishehsari, R. M. Voigt and A. Keshavarzian, *Nat. Rev. Endocrinol.*, 2020, 731–739.
89. F. Bishehsari, A. Saadalla, K. Khazaie, P. A. Engen, R. M. Voigt, B. B. Shetuni, C. Forsyth and A. Keshavarian, *Int. J. Mol. Sci.*, 2016, **17**, 2017.
90. X. Yang, P. A. Wood, C. Ansell and W. J. Hrushesky, *Integr. Cancer Ther.*, 2009, 309–316.
91. Z. L. Zeng, M. W. Wu, J. Sun, Y. L. Sun, Y. C. Cai, Y. J. Huang and L. J. Xian, *J. Biochem.*, 2010, 319–326.
92. L. Polidarová, M. Soták, M. Sládek, J. Pacha and A. Sumová, *Chronobiol. Int.*, 2009, **26**(4), 607–620.
93. F. Deng, K. Yang and G. Zheng, *Dis. Markers*, 2020, **2020**, 6486238.
94. A. Mukherji, A. Kobiita, T. Ye and P. Chambon, *Cell*, 2013, 812–827.

CHAPTER 10

Circadian Clock Signals in Endocrine Glands

KEIKO TOMINAGA[a] AND HITOSHI OKAMURA*[b]

[a] Osaka University Graduate School of Frontier Biosciences, Osaka 565-0871, Japan; [b] Kyoto University Graduate School of Medicine, Sakyo-ku, Kyoto 606-8501, Japan
*Email: okamura.hitoshi.4u@kyoto-u.ac.jp

10.1 Mammalian Circadian System and the Endocrine Regulation

Many endocrine functions exhibit a circadian rhythm of approximately 24 hours, similar to behavioural rhythms.[1–4] The discovery of mammalian clock genes and the core clock machinery of transcription–translation of the feedback loop (TTFL) in most organs has raised the possibility that virtually 60 trillion cells of the human body have the oscillating ability using the same transcription-based core loop of clock genes.[1,5–7] The mammalian circadian system is strictly hierarchical, with the SCN as its top. The clock oscillation in the body is abolished when the SCN is lesioned, although the ablation of any other region cannot abolish the rhythm,[8,9] and the rhythm can be restored through SCN transplantation[10,11] or with gene rescue.[12] Thus, unlike birds and lower vertebrates, no endocrine glands can play the role of effective oscillators at system levels in mammals.[13] Moreover, they relay SCN time signals to peripheral organs.

Endocrine signals mediate SCN circadian signals in the entire body. Does this transcriptional rhythm occur in the SCN in living, free-moving animals? Optical fiber insertion into the SCN bearing *Per1*-luciferase or *Per2*-luciferase

Chemical Biology No. 23
Chronobiology and Chronomedicine: From Molecular and Cellular Mechanisms to Whole Body Interdigitating Networks
Edited by Germaine Cornelissen and Tsuyoshi Hirota

Published by the Royal Society of Chemistry, www.rsc.org

genes enables us to monitor *Per1* or *Per2* transcription activity in living animals.[14,15] Using this luminescence monitoring system, luminescence showed a clear circadian fluctuation (see Figure 10.1A) with peaks during the daytime and troughs at night in both mice[14] and rats.[15] Moreover, luciferase peaks in the subjective morning and troughs during the subjective night were very similar to the electrical activity rhythm.[16] Thus, net clock gene expression reflects net electrical activity, which can be widely applied to other brain areas in mice and rats.

SCN neurons are GABAergic.[17,18] These GABAergic neurons also co-store a variety of peptides.[19] Arginine vasopressin (AVP) cells located in the dorsomedial part and vasoactive intestinal peptide (VIP) cells located in the ventral SCN are the two main components. AVP and VIP cells form local circuits inside the SCN,[20,21] and these local circuits generate robust circadian oscillations.

In addition to forming local circuits inside the SCN, both AVP/GABA cells and VIP/GABA cells in SCN neurons extend their neuronal processes outside the SCN, namely, toward the lateral septum, many hypothalamic nuclei, and the paraventricular thalamic nucleus. Among them, the strongest output extends to the broad neighbouring subparaventricular zone (SPVZ) in the anterior hypothalamus,[22,23] through which the SCN and various hypothalamic neuroendocrine nuclei and brainstem autonomic nuclei are connected.

Clear 24-hour day rhythms of serum concentrations of anterior pituitary hormones, including prolactin, thyroid stimulating hormone (TSH), and adrenocorticotropic hormone (ACTH), are observed in humans[24] (see Figure 10.1B). SCN circadian signals, directly or indirectly from the SPVZ, regulate three major neuroendocrine functions of the hypothalamus: the hypothalamus–pituitary–thyroid (HPT) axis important for energy homeostasis, the hypothalamus–pituitary–gonadal (HPG) axis important for reproduction, and the hypothalamus–pituitary–adrenal (HPA) axis for adaptation to stress from the environment.[25,26] The thyrotropin-releasing hormone (TRH) in the periventricular region of the paraventricular nucleus (PVN),[27–29] corticotropin-releasing hormone (CRH) in the parvocellular part of the PVN, and kisspeptin in the arcuate nucleus are involved in these regulations.[25,26,30] The TRH signals enter the pituitary portal circulation through the median eminence and stimulate the release of TSH from the anterior pituitary into the general circulation. For the regulation of the gonadotropin-releasing hormone (GnRH) in the preoptic area, SCN-VIP directly innervates the GnRH neurons,[31] whereas SCN-vasopressin makes contact with kisspeptin neurons and indirectly stimulates GnRH neurons,[31–33] controlling the secretion of luteinizing hormone (LH) and follicle-stimulating hormone (FSH) from the anterior pituitary, and finally controlling ovarian and testicular functions. CRH neurons in the PVN receiving circadian signals control ACTH release from the pituitary, and finally regulate glucocorticoids from the adrenal cortex (see Figure 10.1B).

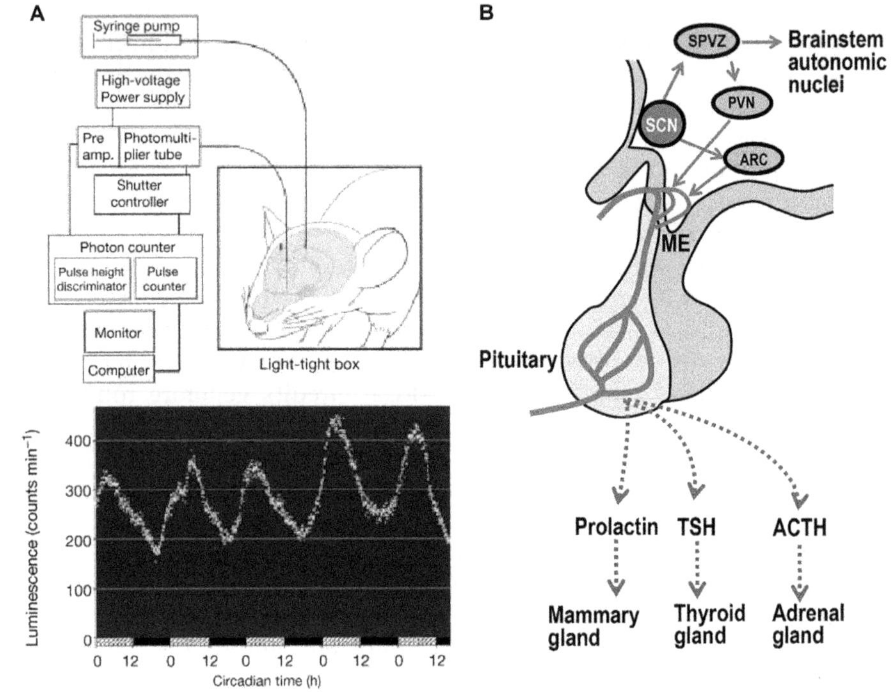

Figure 10.1 Suprachiasmatic nucleus and neuroendocrine hypothalamus. (A) Monitoring gene expression of the suprachiasmatic nucleus (SCN) in freely living animals. Upper fig.: An illustrated schema showing the experiment, monitoring gene expression of the SCN in the living mouse brain. A polymer optical fiber (500 μm in diameter and a numerical aperture of 0.5) was inserted just above the SCN of a Per1-luc transgenic mouse. Luciferin was continuously infused through the cerebral ventricle. The other end of the optical fiber was connected to a photomultiplier as part of a photon-counting apparatus measuring luminescence under constant dark conditions. Lower fig.: Circadian fluctuation of luminescence in the SCN. One dot represents an average of 5 minutes. Hatched and closed bars at the bottom of the figure represent the subjective day and subjective night, respectively. This model system enables real-time gene expression to be monitored in the intact brain under physiological conditions. Lower Figure 10.1A reproduced from ref. 14 with permission from Springer Nature, Copyright 2001. (B) Illustration of how SCN circadian signals regulate hormones in the pituitary anterior lobe. The possible outputs from the SCN to the neuroendocrine hypothalamic nuclei are illustrated. Neurosecretory signals of hypothalamic nuclei enter the portal circulation at median eminence (ME) and regulate the release of hormones of the anterior pituitary, including thyrotropin-stimulating hormone (TSH), prolactin, and adrenocorticotropic hormone (ACTH), to the systemic circulation. ARC, arcuate nucleus; ME, median eminence; PVN, paraventricular nucleus; and SPVZ, subparaventricular zone.

In addition to the hypothalamic nuclei, circadian output signals extend more caudally to the brainstem autonomic nuclei. Circadian signals originating from the SPVZ and PVN polysynaptically pass through the medial forebrain bundle and reticular formation, and project to the central sympathetic and parasympathetic nuclei of the brainstem and spinal cord.[22,23,25] Finally, cholinergic signals from the intermediolateral column of the cervical spinal cord (IML) relay to postganglionic noradrenergic neurons in the sympathetic trunks that innervate visceral organs. The output to parasympathetic nuclei, finally converging at the dorsal motor nucleus of the vagus, is also important, particularly in the gastrointestinal and respiratory organs. Bando *et al.*[34] reported the existence of a clock-gene-mediated oscillating system in the respiratory system, whose rhythm was abolished after ablation of the SCN.

The above mentioned mechanism is the basic mechanism through which endocrine functions exhibit rhythmic patterns (see ref. 25, 26, 30 and 32). Here, we describe three representative hormones with unique circadian manifestations: pineal melatonin as a humoral factor intimately regulated by SCN neuronal signalling, adrenal glucocorticoids serving as integrated circadian mediators in whole body cells, and adrenal mineralocorticoids as circadian regulators of blood circulation. Since this review aims to understand how circadian signals are expressed as endocrine signals, the specific detailed molecular clock mechanisms of each endocrine organ are referenced in excellent published reviews (ovary,[35,36] thyroid gland,[37] adrenal cortex,[38] pituitary gland,[39] pancreas,[40] and all organs[3]).

10.2 Pineal Melatonin as an Endocrine Signal Regulated by Sympathetic Nerves Carrying SCN Signals

The pineal gland is an endocrine gland located in the epithalamus in the uppermost part of the third ventricle, and is classified as a circumventricular organ secreting melatonin. Melatonin is synthesized by the pineal gland during the night,[41] and it is released into both the general circulation and the cerebrospinal fluid of the third ventricle[42] immediately after its synthesis in the pineal gland. Melatonin plays a variety of roles in circadian and seasonal physiology (rest-activity/sleep-wake cycle, metabolism, cardiovascular, immune and reproductive states, among others), through its receptors (MT1 and MT2)[43,44] and non-receptor-mediated actions such as through its antioxidant actions.[45]

Comparative anatomical and endocrinological studies have clarified that the pineal gland originally expresses photoreceptors in lower vertebrates, and thus light signals directly regulate pineal production of melatonin. However, mammals receive sympathetic nerve regulation from the superior cervical ganglion (SCG), resulting in a loss of photosensitivity.[46] In mammals, the circadian production of melatonin is strictly regulated by SCN neural signals.[47] SCN signals are multi-synaptically relayed to preganglionic neurons in the IML of the spinal cord,[25,30] and postganglionic cells in the SCG enter the pineal gland (see Figure 10.2A). TTFL-elicited E-box-driven signals activate SCN neuronal signals in the daytime, and these signals are transmitted to the rest of the brain. The silencing of SCN circadian signals at night finally activates postganglionic noradrenergic neurons, leading to the release of noradrenaline in the pineal gland. Noradrenaline activates both α- and β-adrenergic receptors, leading to an increase in cyclic AMP (cAMP) and intracellular calcium in pinealocytes. cAMP activates the PKA-dependent phosphorylation of cyclic AMP regulatory element-binding protein (CREB). Phosphorylated CREB binds to the CRE sites of the arylalkylamine-*N*-acetyltransferase (AANAT) gene[48] and increases *AANAT* mRNA by 100-fold. The amplitude of this increase is modulated by the inducible cAMP early repressor (ICER), which competes with CREB for CRE sites.[3]

Exposure to light through the eyes is known to decrease the serum level of melatonin, and it is now known that this effect arises from the regulation of

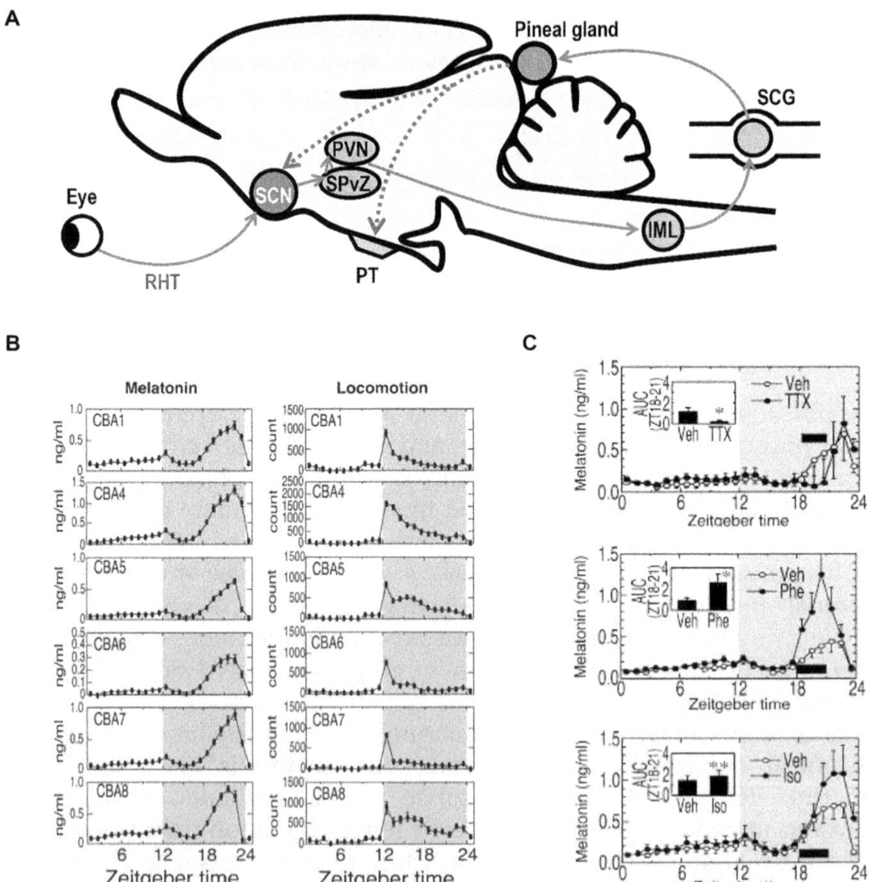

AANAT at the protein level. At night, AANAT is phosphorylated and bound to the chaperone protein 14-3-3,[49] and this complex is not degraded by the proteasome and remains active. Light exposure at night rapidly decreases cAMP levels in the pineal gland, and dephosphorylates AANAT, leading to dissociation of the AANAT/14-3-3 complex, followed by inactivation and proteosomal proteolytic destruction,[48,49] which finally induces rapid reduction of melatonin.

The SCN is necessary for circadian regulation and light inhibition of melatonin synthesis. By applying a 22-hour light–dark (LD) cycle forced desynchrony protocol to monitor melatonin secretion, Schwartz *et al.*[50] proposed that circadian melatonin secretion is controlled by an endogenous light-independent oscillator within the dorsomedial subdivision of the SCN, and the light inhibition of melatonin secretion is controlled by light information *via* the ventrolateral subdivision of the SCN.

How does secreted melatonin act on the SCN? Melatonin secreted into the third ventricle can be used as a sympathetic nerve-conveyed feedback signal to the master clock cells in the SCN, which have high levels of MT1 receptors.[43,44] Since MT1 receptors are also strongly expressed in the pars tuberalis, the upper ridge of the anterior pituitary attached to the median eminence, another circumventricular organ, plays an important role in local TSH secretion.[37] Since MT2 receptors are also expressed in various

Figure 10.2 Sympathetic regulation of pineal melatonin secretion. (A) Schematic representation of the SCN–pineal network. SCN-time signals out to SPVZ/PVN are relayed to the sympathetic intermediolateral cell column of the spinal cord (IML). Cholinergic signals from the IML are relayed to the noradrenergic neurons of the superior cervical ganglion (SCG), which innervate the pineal gland. Produced pineal melatonin enters into general circulation and into the cerebrospinal fluid (CSF) in the third ventricle, through which it activates MT1 receptors in the SCN and TuA (pars tuberalis of the anterior pituitary (PT)). Light information from the eye enters into the brain through the retinohypothalamic tract (RHT)–SCN route. PVN, hypothalamic paraventricular nucleus and SPVZ, subparaventricular zone. (B) 24-hour day profiles of melatonin levels and locomotor activity for six CBA mice maintained in 12L/12D cycles for 10 consecutive days (shown as means $\pm$ SEM). The dominant circadian period of melatonin and locomotor rhythms of each mouse was about 24 hours ($P < 0.05$) according to χ^2-periodgram analysis. A 12-hour rhythmic component was also statistically validated ($P < 0.05$). The dark periods are indicated in grey. (C) Effect of tetrodotoxin (TTX) ($n = 5$), phenylephrine (Phe) ($n = 6$) or isoproterenol (Iso) ($n = 5$) infusion at night (ZT18-21) on dialysate melatonin concentrations sampled using transverse pineal probes. Insets in each graph show the area under the curve (AUC) levels for nighttime melatonin at ZT18–21. Experiments are performed in LD condition, and the dark periods are indicated in grey. $^*P < 0.05$; $^{**}P < 0.01$; significantly different from vehicle (Veh) controls. Figure 10.2B and Figure 10.2C are reproduced from ref. 51 with permission from National Academy of Sciences.

regions, including the amygdala, hypothalamic arcuate nucleus, and thalamic nuclei, a direct effect on these brain regions might occur in relation to sleep function. Melatonin released into the general circulation would also carry an endocrine signal of the central clock in the SCN, providing a nocturnal/circadian message to the whole body through MT2 receptors.[52,53]

By directly measuring the spontaneous neural activity by inserting electrodes into the brain of nocturnal rats after isolating the hypothalamus using a Halasz knife, Inouye and Kawamura[16] reported that the neuronal activity of the hypothalamic SCN was higher at daytime, while extra-SCN brain areas showed higher activity at night. It was demonstrated that this inverted rhythm form was also reflected in the *Per* gene expression levels: SCN *Per1*, *Per2*, and *Per3* expression in mice peaked in the subjective morning and reached their trough during the subjective early night,[54] in contrast to their expression in the cortex, which peaked at night.[55] In conjunction with the finding that deoxyglucose uptake in the SCN is also high during the daytime,[56] TTFL-evoked metabolic and neuronal activity shown as *Per*-gene oscillations show a peak during the daytime with inverted neuronal activity observed in non-SCN brain areas in nocturnal animals. In diurnally active species, the daytime high/nighttime trough pattern of SCN neuronal profile is maintained and in regions outside the SCN, neural activity and expression of *Per* genes also have the same phase as in the SCN.[57–61] This finding demonstrates that the SCN-timekeeping profile is fixed to the daytime-peak pattern, irrespective of diurnality and nocturnality, although cellular clocks in non-SCN brain areas reflect their cell activities.

Melatonin secretion from the pineal gland is high at night in diurnally-active humans and nocturnally-active rodents (rats/mice), which is different from other non-SCN brain sites, such as the cerebral cortex, showing differences according to diurnality/nocturnality. Similar to humans, melatonin secretion in diurnal species (sheep and *degus*) has a night-high pattern.[62] This finding suggests that pineal melatonin reflects SCN function in both diurnal and nocturnal species. However, large phase differences between SCN gene expression and pineal gene expression should be noted in rats;[63] there is an about 10-hour difference in the case of *Per1* gene expression profiles between the SCN and the pineal gland.

Although pineal melatonin production is regulated by the circadian clock in the SCN in mammals, it is difficult to study this process in mice because most strains of inbred mice have lost their ability or are very poor at synthesizing melatonin because of mutations in AANAT and hydroxyindole-*O*-methyltransferase (HIOMT).[64] However, a few strains, CBA, MSM and C3H, which possess the enzymes for melatonin synthesis, display a circadian rhythm of melatonin secretion, although continuous recording has not been reported.[65–68]

To assess a 24-hour pattern of endogenous melatonin secretion in mice, Nakahara *et al.*[51] established a transverse brain microdialysis technique. Interestingly, because melatonin has been shown to be secreted into the ventricular system directly from the pineal gland, melatonin concentrations in the ventricular cerebrospinal fluid (CSF) are several times higher than blood melatonin levels.[42] By applying the transverse microdialysis probe adjacent to the pineal body, a clear and stable 24-hour melatonin rhythm was observed over 100 days in CBA mice. Melatonin secretion showed a bimodal pattern, with the first small peak at the start of darkness and the second most prominent peak occurring near the end of the night (see Figure 10.2B). The free-running rhythms for melatonin concentrations and locomotor activity under completely dark conditions also had nearly identical periods (23.7 hours). The phase relationship between the two peaks, one appearing early for locomotor activity and the other late at night for melatonin secretion under normal LD conditions (see Figure 10.2B), persisted even in constant darkness. Furthermore, the activity at night is decreased when melatonin is high, suggesting that high melatonin levels decrease the activity levels, although not sleep.

Nakahara *et al.*[51] investigated how a brief light pulse affected melatonin levels in dialysates sampled from transverse probes in CBA mice. When the animals were exposed to a 15-minute light pulse at circadian time 20 (CT20), dialysate melatonin levels significantly decreased to basal levels within 2 hours after the start of light exposure. Application of the brief light pulse at CT20 induced a phase advance of locomotor activity after 2 days. The phase advance of the melatonin rhythm was more immediate, and a phase advance of the large nightly peak was observed even on the day after light exposure. These data are consistent with the previously reported pineal and serum melatonin concentrations in rats[69] and mice.[70]

Innervation from the superior cervical ganglion to the pineal gland[47,71–74] differently regulates melatonin release through different receptors. A direct search for α-receptors and β-receptor agonists on the pineal gland revealed that both receptors enhanced nocturnal melatonin secretion, but each receptor acts differently (see Figure 10.2C). In other words, pineal stimulation with the α-adrenergic agonist phenylephrine results in an immediate and transient increase in melatonin levels, suggesting post-translational regulation of protein levels by the AANAT/14-3-3 complex.[48] Furthermore, the injection of the adrenergic β-agonist isoproterenol showed a delayed increase in melatonin, but its elevation lasted over a long period, which suggests its regulation at the transcriptional level. In addition, nocturnal increases in melatonin levels were blocked by the α-adrenergic antagonist prazosin and the β-receptor antagonist propranolol.[51]

10.3 Adrenal Glucocorticoids as Endocrine Signals to the Whole Body Integrating Sympathetic and Neuroendocrine Signals

The robust circadian secretion of glucocorticoids (cortisol in humans[75] and corticosterone in rodents[8]) from the zona fasciculata (ZF) of the adrenal gland represents an early major discovery in circadian biology. The stress-related regulation of glucocorticoid secretion by the hypothalamo–pituitary–adrenal (HPA) axis has been thoroughly investigated; however, the circadian control of its secretion has long remained obscure. The glucocorticoid rhythm is abolished after lesions of the SCN; therefore, it is clear that the central clock is indispensable for driving this secretion rhythm.[8] In contrast to humans, which are diurnally active, serum glucocorticoid levels in nocturnal mice showed a peak at early night and a trough at early daytime. The increase in glucocorticoids occurs before the onset of the active phase,[75] indicating that its rise does not reflect an acute response to arousal; however, it is rather controlled by the SCN in anticipation of the onset of activity. Compared to melatonin, it should be noted that the cortisol phase is approximately 4–6 hours behind melatonin; therefore, the cortisol peak occurs just around the time of awakening in humans.

How does the circadian secretion of glucocorticoids occur? According to the HPA axis, circadian rhythms of corticosterone release are induced by the circadian secretion of ACTH from the pituitary. Supporting this, in human sera sampled continuously,[75] ACTH secretion shows robust 24-hour rhythms, always occurring 30–60 minutes ahead of, and homothetic to glucocorticoid rhythms. Although these data suggest that circadian ACTH changes drive the

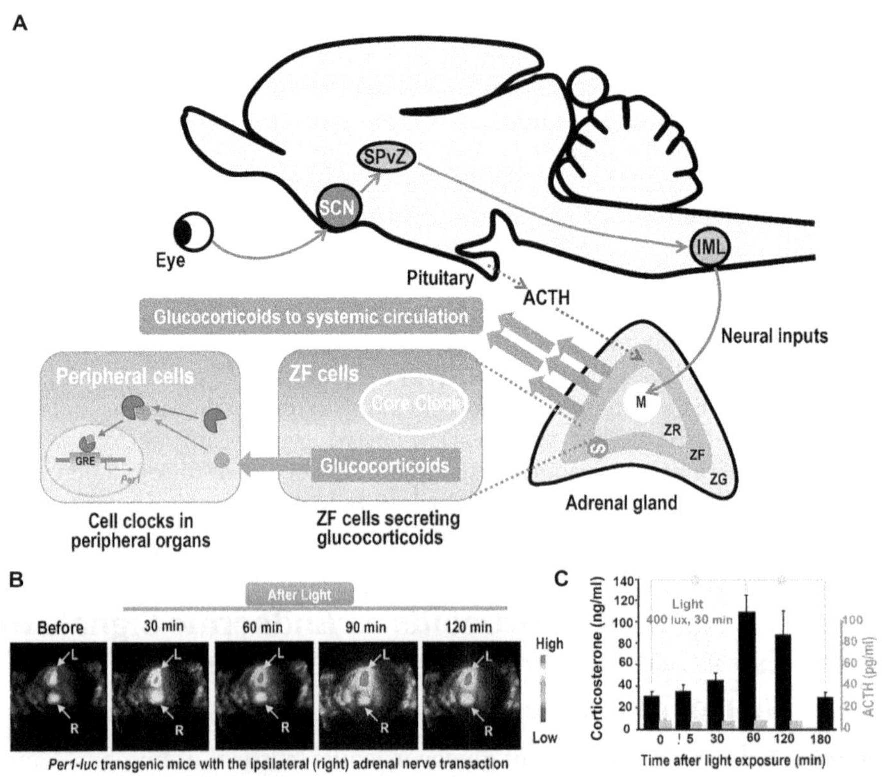

circadian release of glucocorticoids from the adrenal glands, a series of animal experiments in rats and calves suggest that while ACTH seems to be a necessary factor for the circadian secretion of glucocorticoids, the circadian generator *per se* is extra-pituitary.[76–79] In rats, Ottenweller *et al.*[76] demonstrated that hypophysectomy completely abolished the 24-hour rhythmic secretion of corticosterone, and implantation of an ACTH pellet into hypophysectomized animals restored corticosterone rhythms,[77] which were completely abolished upon surgical spinal transection at T7 (7th thoracic vertebra). Since L1 (1st lumbar vertebra) level transection was ineffective in abolishing rhythmic corticosterone, Ottenweller and Meier[77] concluded that the sympathetic preganglionic neurons in spinal T7-L1 are crucial for inducing 24-hour rhythms of glucocorticoids in rats. Later, by performing bilateral transection of the thoracic splanchnic (sympathetic) nerve, Ultrich-Lai *et al.*[80] reported ACTH-independent 24-hour changes in glucocorticoids in rats.

The importance of extra-pituitary regulation of the circadian secretion of glucocorticoids was also speculated from a series of experiments in cows directly measuring steroids in the adrenal veins.[78,79] Edwards and Jones[79] demonstrated that high-frequency stimulation of adrenal sympathetic nerves and/or intra-aortic infusion of acetylcholine strongly potentiated the glucocorticoid response (see Figure 10.3A). This sympathetic nerve

Figure 10.3 Circadian regulation of glucocorticoid secretion. (A) Schematic representation of the SCN–adrenal network. SCN-time signals are transmitted to the adrenal gland *via* the central sympathetic intermediolateral cell column of the spinal cord (IML). In the adrenal gland, ACTH and sympathetic signals are integrated, converted into glucocorticoid signals, and released into the general circulation. In the peripheral tissues, clocks receive glucocorticoid signals that activate the binding of glucocorticoid receptors (GR) to the glucocorticoid responsive elements (GRE) of the *Per1* gene, leading to the resetting of these clocks. M, medulla of the adrenal gland; SPvZ, subparaventricular zone, ZG, zona glomerulosa of the adrenal cortex; ZF, zona fasciculata of the adrenal cortex; and ZR, zona reticularis of the adrenal cortex. (B) Light-induced *Per1-luc* luminescence in visceral organs after hemidenervation of the sympathetic nerve. Although light exposure (2000 lux, 30 minutes) acutely induces luciferase bioluminescence in the abdomen of *Per1* promoter-luciferase (*Per1-luc*) transgenic mice, denervation of the right adrenal nerve silences the ipsilateral *Per1-luc* induction in the right adrenal gland (R), without affecting the induction of *Per1-luc* in the contralateral left adrenal gland (L). (C) Effect of light on plasma levels of corticosterone and ACTH. Mice were exposed to light (400 lux; 30 minutes) at the subjective night (CT16), and plasma levels of corticosterone and ACTH were examined after 5, 30, 60, 120, and 180 minutes (means $\pm$ SEM, $n = 8$ at each time point). Note that corticosterone in plasma strongly increased at 60 and 120 minutes. However, at 180 minutes, it returned to the level before light exposure. Despite of this corticosterone change, ACTH remained unchanged, even after light exposure. The comparison of light-exposed and light-unexposed control was analyzed by Student t test (grey asterisks; $P < 0.05$). (B and C) were reproduced from ref. 92 with permission from Elsevier, Copyright 2005.

stimulation-mediated corticosterone secretion was abolished by hypophysectomy, although it was restored by exogenous and constant administration of low amounts of ACTH. Thus, ACTH is necessary for the secretion of glucocorticoids by stimulating acetylcholine treatment or activation of sympathetic nerves.[78] Although non-ACTH steroidogenic mediators were still unrevealed, there are a couple of data sets indicating that catecholamines in the adrenal medulla increase cortical steroidogenic activity.[81,82] These data strongly suggest that SCN-evoked time signals, transmitted to sympathetic nerves, regulate the circadian secretion of corticosterone.

Oster *et al.*[83] demonstrated that the circadian clock located in the adrenal gland plays a role in the secretion of corticosterone from isolated adrenals *in vitro* and that corticosterone secretion in response to ACTH is circadian time-dependent. This suggests that the adrenal local clock modifies the sensitivity to ACTH. Later, they demonstrated that the knockout of the core clock gene *Bmal1* leads to a lack of responsiveness of the mouse adrenal gland to ACTH, impaired corticosterone production, and inadequate responses to stress.[84] However, since ACTH appears to act only as a prerequisite for the central-preganglionic innervation-induced circadian change in corticosterone release *in vivo*,[77] the role of the local adrenal clock in the integration of these afferent signals for corticosterone synthesis may be limited in living animals.

Light is a powerful synchronizer of circadian rhythms, and 1 hour of exposure to 8000 lux bright light improves the metabolic and hormonal status of circadian rhythm-related sleep–wake disorders, including seasonal affective disorders,[85] although its mechanism is poorly understood. Bright light therapy has now been applied to many circadian sleep disorders, including jet lag syndrome and shift work sleep disorder, which are known to accompany various metabolic issues, including hypertension, cancer, and diabetes.[86–88] In these patients, light therapy improves the psychiatric status, disordered hormones, and metabolism. However, the effects of light have only been established on melatonin,[89] and the remaining powerful and broad effects of light on body metabolism and hormones remain to be clarified.

Although adrenal sympathetic nerves play an important role in the generation of 24-hour glucocorticoid rhythms, their relationship to rhythm generation in the SCN and its entrainment to the light–dark cycle have not been addressed. To reveal the functional output of the SCN, we focused on the perception of light signals. In mammals, light perception is restricted to the eye, and the retinohypothalamic tract conveys retinal light information to the SCN. It is known that the SCN is the only recipient of light signals used for entraining circadian behaviour,[90] and light exposure induces immediate early gene (IEG) expression in retinofugal areas of the SCN; therefore, the light-inducibility of IEG in "blind" visceral organs would be a marker delineating the output pathways from the SCN. In this screening, *Per1* promoter-luciferase (*Per1-luc*) transgenic animals were used because *Per1* also behaves as an IEG.[91] In these experiments, Ishida *et al.*[92] observed reporter gene expression in the visceral organs of *Per1-luc* mice using a two-dimensional photon-counting camera, in which the whole animal could be observed at once (see Figure 10.3B). Brief exposure to light caused a dramatic increase in luminescence from the adrenal gland, whereas other organs such as the liver or kidneys maintained background levels of luminescence, demonstrating that *Per1* expression in the adrenal gland was directly under the control of the SCN.

Moreover, further experiments using microarrays revealed that 156 genes were upregulated and 39 genes were downregulated in the adrenal gland in response to light. This adrenal gene expression response was abolished after SCN-lesioning or by transection of the adrenal sympathetic nerve; therefore, it is clear that light information is conveyed to the adrenal gland *via* the SCN through a sympathetic route. Furthermore, this gene expression response was accompanied by a surge in plasma and brain corticosterone levels after 60–90 minutes, although without concomitant activation of the hypothalamo–hypophyseal–adrenal (HPA) axis (see Figure 10.3C). This activation is circadian time-dependent since the steroid response is higher during the subjective night compared to the subjective day. Therefore, this activation and its mechanisms are completely different from the classical stress reaction, in which glucocorticoids and ACTH levels increase immediately (*i.e.*, 2–5 minutes) to the maximal level.[92]

Ipsilateral adrenal nerve transection abolished the light-induced *Per1-luc* response from the denervated adrenal gland without affecting the response in the contralateral gland (see Figure 10.3B), whereas bilateral transection of adrenal sympathetic nerves completely abolished the light-induced secretion of corticosterone.[92] These experiments demonstrated that corticosterone secretion is directly regulated by the SCN *via* the intermediolateral nucleus (IML) sympathetic nerve route[93] (see Figure 10.3A). Since plasma levels of adrenaline increase within a few minutes after the initiation of light exposure, and since adrenaline treatment (2 mg kg^{-1}, s.c.) significantly increased the adrenal *Per1-luc* luminescence and plasma corticosterone,[92] the intra-adrenal/peri-adrenal ganglion neurons projecting to the adrenal cortex[94,95] might be involved in the process, subsequently activating steroidogenic cells.[96]

An important remaining question is the physiological meaning of SCN-controlled and light-induced glucocorticoid secretion. The released corticosterone may bind to glucocorticoid receptors[97] and activate the *Per1* gene in systemic cells,[98] leading to the resetting of peripheral clocks (see Figure 10.3A). Thus, light-evoked corticosterone secretion may have a role in transmitting environmental light signals to "blind" internal organs. However, even under constant conditions, such as in continuous darkness, the circadian activation of the SCN–sympathetico–adrenal axis is also required to generate circadian plasma glucocorticoid levels peaking at dawn. Thus, the adrenal gland acts as the transducer of SCN-clock time, from nerve impulses to hormonal surges in the circadian timing system. The SCN, by directly controlling the secretion of corticosterone from the adrenal gland, synchronizes metabolism in the whole body *via* ubiquitous glucocorticoid receptors.

Neurons that emerge from the IML of the spinal cord directly innervate the adrenal medulla, without changing synapses at the sympathetic ganglia, releasing acetylcholine from the presynaptic terminals. Thus, anatomically, it can be assumed that the adrenal medulla is equivalent to the post-ganglionic cells, and the sympathetic nerve innervating the adrenal medulla is equivalent to the preganglionic fiber. In fact, the adrenal medulla receives acetylcholine at nicotinic receptors and releases adrenaline/nordadrenaline, just as postganglionic neurons do. However, the effects of light on these hormonal organs, pineal and adrenal cortex, are different: light suppresses melatonin secretion, but oppositely activates adrenal corticosterone secretion. Therefore, the function of sympathetic afferents appears to be regionally different.[99]

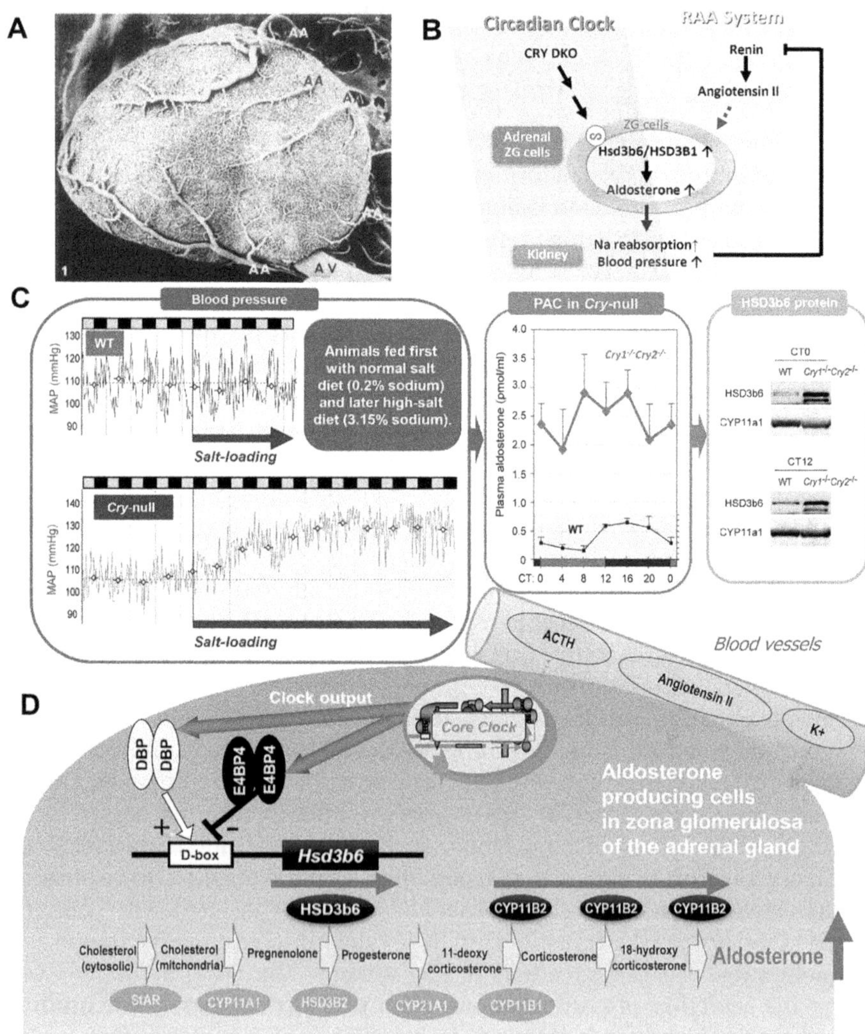

10.4 Adrenal Mineralocorticoids: Renin–Angiotensin–Aldosterone System Meets the Local Clock in the Zona Glomerulosa and Regulates Blood Pressure

Although pineal melatonin and adrenal glucocorticoids are strongly regulated by the central clock in the SCN, circadian regulation of the adrenal mineralocorticoids is not intimately regulated by the SCN; however, it is regulated by the local clock of the adrenal zona glomerulosa (ZG). The outer surface of the adrenal gland is covered with a rich supply of arteries,[100] which run through the capsule and branch repeatedly to form a subcapsular capillary plexus, where the ZG is located (see Figure 10.4A). Thus, the ZG, the

Figure 10.4 Zona glomerulosa of the adrenal cortex and the salt-sensitive hypertension in *Cry1*/*Cry2*-knockout (Cry-null) mice. (A) Scanning electron micrograph of an isolated vascular cast of the right adrenal gland of the rat. The adrenal arteries (AAs) approach the gland and run through the capsule. These arteries branch repeatedly and form a subcapsular capillary plexus. The adrenal vein (AV) emerges from the gland at its medio-inferior surface. Reproduced from ref. 100 with permission from John Wiley & Sons, Copyright © 1982 Wiley-Liss, Inc. (B) Schematic presentation of the encounter of the renin–angiotensin–aldosterone (RAA) system and the circadian clock system in the zona glomerulosa (ZG) of the adrenal cortex. Both local clocks in ZG and angiotensin II regulate the *Hsd3b6*/*HSD3B1* gene, and control the secretion of aldosterone, which promotes Na^+ and water reabsorption in urinary tubules in the kidney. (C) *Cry*-null mice exhibit salt-sensitive hypertension. Left to right, BP, plasma aldosterone concentration (PAC), and protein and enzyme activity of 3β-HSD are shown. (Left) Representative recordings of temporal profiles of mean arterial BP of wild type (WT) (8-day recording) and *Cry*-null mice (15-day recording). In both groups, animals were kept in constant darkness (DD) and fed either normal salt (0.2% sodium) for first 4 days, high salt (3.15% sodium) after 5th day. BP data were collected in 30-s bouts at 5-minute intervals and plotted using a 1-hour moving average. The 24-hour means of each day are shown by open squares. Horizontal dotted lines indicate the mean value of BP on days 1–4 on a normal salt diet. (Middle) 24-hour profiles of plasma aldosterone concentration. (Right) Western analyses of HSD3b6 protein, and enzymatic activity converting pregnenolone to progesterone in the adrenal gland. Note strong expression of HSD3b6 proteins and enzymatic activities in ZG cells. Figures in C were reproduced from ref. 104 with permission from Springer Nature, Copyright 2009. (D) Schematic illustration demonstrating clock interaction with *Hsd3b6*, and aldosterone synthetic pathway. The *Hsd3b6* promoter bears two functional D-box *cis*-elements, through which the transcription factors DBP and E4BP4 activate and repress Hsd3b6 transcription, respectively. In the *Cry*-null adrenal gland, the expression of *Dbp* is constitutively elevated, while *E4bp4* remains suppressed. Together, these data favour a scenario in which a constitutive high expression of DBP leads to the abnormally high transcriptional activation of *Hsd3b6*.

outermost layer of the adrenal gland, is one of the most richly blood-supplied areas of the body.[100,101] Utilizing this anatomical location, ZG cells receive circulatory information related to angiotensin and K^+ levels to maintain blood pressure (BP)[102] by secreting mineralocorticoids[103] (see Figure 10.4B).

Circadian variations in BP are an important physiological feature relevant to human health. BP increases when we get up, peaks between noon and evening, and then decreases during the night. Under a shift-work schedule, BP has been shown to increase.[105,106] Perturbations in the 24-hour rhythms of BP also affect the prognosis of hypertension; non-dippers (whose BP at night does not decrease) tend to show cardiovascular complications later in life.[107] Therefore, BP is a representative output of circadian rhythms; however, its molecular mechanisms are unknown. Hypertension is a long-term medical condition in which arterial BP is persistently elevated. Hypertension is the most important risk factor for stroke, heart disease, and kidney disease, with severe implications for life expectancy and disability.

In humans, epidemiological studies have reported that high salt intake increases the risk of hypertension,[108] and in chimpanzees, Denton *et al.*[109] clearly demonstrated that salt intake dose-dependently increases BP 6–67 weeks after salt loading. This salt-sensitivity is species-specific. Rodents, particularly mice, are reported to have difficulty in developing a hypertensive state.

In studies using *Per1-luc* transgenic mice,[92] we noticed that *Per1* genes were more strongly and rhythmically expressed in the ZG than ZF or ZR. If this strong and rhythmic expression of clock genes in the ZG means an important regulation of aldosterone production/secretion rhythms by the peripheral clock, perturbation of the rhythms could raise abnormalities in BP, because they act on the kidneys to promote sodium and water reabsorption, resulting in an increase in circulating plasma volume, and raise BP. Therefore, we measured the BP of completely arrhythmic *Cry*-null mice[6,110,111] in which central clocks in the SCN and the cellular clocks in whole-peripheral tissues cease to function.

Arterial BP was monitored using telemetry in freely moving animals in constant darkness, with a PA-C10 pressure transmitter (DSI, Boston), in which the catheter was inserted into the aortic arch through the left common carotid artery (see Figure 10.4C). In wild-type (WT) mice, a normal salt diet containing 0.2% Na^+ induced marked circadian fluctuations in mean arterial pressure with a trough during the daytime, the sleeping phase, and a peak during the night, the active phase. The day–night rhythm of arterial pressure did not change with the high-salt diet containing 3.5% Na^+. The 24-hour maxima and minima of BP did not change after the high-salt diet. In *Cry*-null mice fed a normal salt diet, the arterial pressure was similar to that in WT mice; however, their circadian pattern was lost. When a high-salt diet was given, arterial BP in *Cry*-null mice increased day by day, and after one week, it increased by approximately 30 mmHg (see Figure 10.4C, left).

What molecular mechanisms are involved in the hypertension of salt-loaded *Cry*-null mice? BP is determined by a combination of cardiac output and peripheral resistance. Two possible systems are involved: the renin–angiotensin (RA) system with vasoconstriction and sodium and water retention, and the sympathetic nervous system with abnormal baroreceptor reflexes. When we investigated the status of humoral factors in *Cry*-null mice, we found that plasma aldosterone levels were significantly increased in *Cry*-null mice, being 5–10 times higher than in WT mice.[104] Since plasma renin levels were reduced in *Cry*-null mice, it was considered that the increase in plasma aldosterone was not caused by activation of the RA system. Isolated adrenal glands from *Cry*-null mice demonstrated a marked increase in aldosterone release;[104] therefore, it is highly speculated that hyperaldosteronism in the plasma of *Cry*-null mice was caused by an increase in aldosterone synthesis in the adrenal glands (see Figure 10.4D). Therefore, we applied DNA microarrays to the adrenal glands and found that only one steroid synthetic enzyme, Hsd3b6, was strongly expressed in *Cry*-null mice. Hsd3b6 is a 3β-HSD protein that converts pregnenolone to progesterone in the steroid synthetic pathway. Although 3β-HSD is present in all steroid-producing cells, we found that Hsd3b6 was localized in aldosterone-producing cells in the ZG by immunohistochemistry and *in situ* hybridization; however, glucocorticoid or sex-steroid-producing cells did not express this subtype in the adrenal gland.

Next, we asked how is Hsd3b6 regulated by clock genes. Aldosterone-producing cells have a clock that governs thousands of clock-controlled genes (ccg), and Hsb3b6 is one such ccg, regulated by the D-box *cis* element on the promoter (see Figure 10.4D). The *Hsd3b6* promoter bears two functional D-boxes, through which the transcription factor DBP activates *Hsd3b6* transcription and E4BP4 represses it.[112] In the *Cry*-null adrenal gland, the expression of *Dbp* was elevated; however, *E4bp4* remained moderately expressed. The high expression of *Dbp* in *Cry*-null mice represents a feature commonly observed in E-box-controlled genes. These data support a scenario in which a high expression of DBP leads to the abnormally high transcriptional activation of Hsd3b6.

To understand whether these findings are relevant for human hypertensive disease, we first identified the ZG-specific subtype of 3β-HSD. Human 3β-HSD has two subtypes, HSD3B1 and HSD3B2. Previously, it was reported that only the HSD3B2 subtype is present in the adrenal glands.[113] By applying the laser microdissection method combined with the TaqMan MGB probe method to recognize the difference between single nucleotide sequences in the human adrenal gland, we found that HSD3B1 is expressed in aldosterone-producing cells expressing CYP11B2 in the ZG.

The *Hsd3b* genes in human and mouse form a syntenic cluster at chromosomal positions 1p12 and 3F2.2, respectively. The human HSD3B region holds two expressed genes, *HSD3B1* and *HSD3B2*, and five pseudogenes, Ψ1–5. On the other hand, the mouse Hsd3b region contains six expressed genes (types I to VI). The mouse *Hsd3b6* gene is thought to encode a functional homolog of human HSD3B1 since mice Hsd3b6 and human HSD3B1 both show similar regional expression profiles in the placenta, skin, and adrenal ZG.[104,114]

Our *Cry*-null analysis strongly suggests that Hsd3b6 in the ZG regulates aldosterone synthesis and contributes to salt-sensitive hypertension in mice. Hsd3b6/HSD3B1 gene expression in the ZG is regulated by low salt or angiotensin II *via* the AT1 receptor.[115] Thus, HSD3B1 is an enzyme that is controlled by the clock and angiotensin II. However, so far, the role of HSD3B1 in human hypertensive disease remains elusive.

Clinically, adrenal gland-originated hypertension, called primary aldosteronism (PA) or Conn's syndrome, is characterized by overproduction of the aldosterone hormone by the adrenal glands due to hyperplasia or tumours. Excess aldosterone increases the absorption of sodium and water in renal tubules, increasing the blood volume and BP and causing hypertension. PA is classified into two subgroups: aldosterone-producing adenoma (APA), associated with tumours, and idiopathic hyperaldosteronism (IHA), which is characterized by adrenal hyperplasia. Through immunohistochemistry using subtype-specific human 3β-HSD, we found that the hyperproliferating ZG cells in IHA expressed high amounts of HSD3B1, but tumour cells in APA strongly expressed HSD3B2.[114,116] Based on these findings, we speculate that the cause of excess plasma aldosterone levels might be different between IHA and APA. In IHA, excess aldosterone is produced by excess HSD3B1 on both sides of the adrenal gland with hyperplasia, whereas in APA, excess aldosterone is produced by high HSD3B2 production in tumour cells. Abnormal aldosterone synthesis in PA has long been thought to involve the aldosterone synthase CYP11B2, which is the final enzyme in aldosterone synthesis, although CYP11B2 is not always high in adrenals of primary aldosteronism.[117–119] Hyperaldosteronism with bilateral hyperplasia of the adrenal zona glomerulosa in *Cry*-null mice[104,120] suggests that *Cry*-null mice can be used as a new animal model for IHA.

Primary aldosteronism, which accounts for 10% of hypertension cases,[121] has recently been highlighted as a treatable form of hypertension. The incidence of cardiovascular complications, such as atrial fibrillation, is higher than in essential hypertension.[122] In recent years, somatic mutations in potassium channels KCNJ5, ATP1A1, and CACNA1D have been reported as causes of APA.[123–125] The increased cellular activity caused by this mutation is thought to be associated with increased aldosterone production. Multiple enzymes are involved in aldosterone biosynthesis. Among them, StAR is an aldosterone synthase, and CYP11B2/aldosterone synthase mediates the last step. Based on the analysis of clock-disturbed *Cry*-null mice showing hyperaldosteronism, we propose that a new subtype of 3β-HSD, HSD3B1/Hsd3b6, is a new candidate regulator of aldosterone synthesis and can be a cause of IHA. The temporal components of an aldosterone-regulating system in the pathophysiology of hypertension are promising targets for future studies.

10.5 Conclusion

Clock genes are expressed in most cells throughout the body, and time information is generated by all cells controlling cellular metabolism. To survive on Earth, effective timing information at the system level is necessary for efficient physiology and behaviour. In mammals, environmental light-dark information is restricted to the eyes, and a hierarchical circadian system is organized with the SCN in the hypothalamus at its top. Many endocrine hormones play a role as intermediate players in the time signal. The SCN directly controls the hypothalamus–pituitary system and indirectly controls endocrine organs *via* the autonomic nervous system. Pineal melatonin is regulated solely by the sympathetic nervous system, which provides circadian signals back to the SCN and regulates sleep, arousal and seasonal hormones. Adrenal glucocorticoids are regulated by pituitary hormones and the sympathetic nervous system, and play a major role in transmitting central SCN information to peripheral visceral cells. In contrast to the above two systems, adrenal mineralocorticoids are regulated by the circadian clock in the zona glomerulosa. The renin–angiotensin system and potassium levels in the circulation are sensed by ZG cells in the adrenal glands, which secrete aldosterone and control BP. Therefore, various hormones are involved in the generation of 24-hour rhythms using a unique control system that corresponds to their function. Since rotating shift-work and long working hours are common in the 21st century, reflecting the globalization of economic activity, understanding hormonal signals may offer a novel perspective for reducing the risk of life-threatening lifestyle-related diseases.

References

1. J. S. Takahashi, H. K. Hong, C. H. Ko and E. L. McDearmon, *Nat. Rev. Genet.*, 2008, **9**, 764–775.
2. R. Silver and L. J. Kriegsfeld, *Eur. J. Neurosci.*, 2014, **39**, 1866–1880.
3. K. A. Dyar, D. Lutter, A. Artati, N. J. Ceglia, Y. Liu, D. Armenta, M. Jastroch, S. Schneider, S. de Mateo, M. Cervantes, S. Abbondante, P. Tognini, R. Orozco-Solis, K. Kinouchi, C. Wang, R. Swerdloff, S. Nadeef, S. Masri, P. Magistretti, V. Orlando, E. Borrelli, N. H. Uhlenhaut, P. Baldi, J. Adamski, M. H. Tschöp, K. Eckel-Mahan and P. Sassone-Corsi, *Cell*, 2018, **174**, 1571–1585.e1511.
4. M. H. Hastings, E. S. Maywood and J. S. O'Neill, *Curr. Biol.*, 2008, **18**, R805–R815.
5. A. Balsalobre, F. Damiola and U. Schibler, *Cell*, 1998, **93**, 929–937.
6. K. Yagita, F. Tamanini, G. T. van Der Horst and H. Okamura, *Science*, 2001, **292**, 278–281.
7. U. Schibler and P. Sassone-Corsi, *Cell*, 2002, **111**, 919–922.
8. R. Y. Moore and V. B. Eichler, *Brain Res.*, 1972, **42**, 201–206.
9. F. K. Stephan and I. Zucker, *Proc. Natl. Acad. Sci. U. S. A.*, 1972, **69**, 1583–1586.

10. M. R. Ralph, R. G. Foster, F. C. Davis and M. Menaker, *Science*, 1990, **247**, 975–978.
11. M. Sujino, K. H. Masumoto, S. Yamaguchi, G. T. van der Horst, H. Okamura and S. T. Inouye, *Curr. Biol.*, 2003, **13**, 664–668.
12. E. L. McDearmon, K. N. Patel, C. H. Ko, J. A. Walisser, A. C. Schook, J. L. Chong, L. D. Wilsbacher, E. J. Song, H. K. Hong, C. A. Bradfield and J. S. Takahashi, *Science*, 2006, **314**, 1304–1308.
13. C. P. Richter, *Proc. Natl. Acad. Sci. U. S. A.*, 1960, **46**, 1506–1530.
14. S. Yamaguchi, M. Kobayashi, S. Mitsui, Y. Ishida, G. T. van der Horst, M. Suzuki, S. Shibata and H. Okamura, *Nature*, 2001, **409**, 684.
15. Y. Yamaguchi, K. Okada, T. Mizuno, T. Ota, H. Yamada, M. Doi, M. Kobayashi, H. Tei, Y. Shigeyoshi and H. Okamura, *J. Biol. Rhythms*, 2016, **31**, 108–111.
16. S. T. Inouye and H. Kawamura, *Proc. Natl. Acad. Sci. U. S. A.*, 1979, **76**, 5962–5966.
17. H. Okamura, A. Bérod, J. F. Julien, M. Geffard, K. Kitahama, J. Mallet and P. Bobillier, *Neurosci. Lett.*, 1989, **102**, 131–136.
18. B. Gao and R. Y. Moore, *J. Biol. Rhythms*, 1996, **11**, 172–179.
19. E. E. Abrahamson and R. Y. Moore, *Brain Res.*, 2001, **916**, 172–191.
20. K. Tominaga, S. I. Inouye and H. Okamura, *Neuroscience*, 1994, **59**, 1025–1042.
21. Y. Ibata, H. Okamura, M. Tanaka, Y. Tamada, S. Hayashi, N. Iijima, T. Matsuda, K. Munekawa, T. Takamatsu, Y. Hisa, Y. Shigeyoshi and F. Amaya, *Front. Neuroendocrinol.*, 1999, **20**, 241–268.
22. A. G. Watts, L. W. Swanson and G. Sanchez-Watts, *J. Comp. Neurol.*, 1987, **258**, 204–229.
23. A. G. Watts and L. W. Swanson, *J. Comp. Neurol.*, 1987, **258**, 230–252.
24. A. van Coevorden, J. Mockel, E. Laurent, M. Kerkhofs, M. L'Hermite-Balériaux, C. Decoster, P. Nève and E. Van Cauter, *Am. J. Physiol.*, 1991, **260**, E651–E661.
25. R. M. Buijs and A. Kalsbeek, *Nat. Rev. Neurosci.*, 2001, **2**, 521–526.
26. A. M. Neumann, C. X. Schmidt, R. M. Brockmann and H. Oster, *Auton. Neurosci.*, 2019, **216**, 1–8.
27. D. Jordan, M. Veisseire, F. Borson-Chazot and R. Mornex, *Horm. Res.*, 1987, **27**, 216–224.
28. W. Russell, R. F. Harrison, N. Smith, K. Darzy, S. Shalet, A. P. Weetman and R. J. Ross, *J. Clin. Endocrinol. Metab.*, 2008, **93**, 2300–2306.
29. J. Philippe and C. Dibner, *J. Biol. Rhythms*, 2015, **30**, 76–83.
30. A. Kalsbeek, R. van der Spek, J. Lei, E. Endert, R. M. Buijs and E. Fliers, *Mol. Cell. Endocrinol.*, 2012, **349**, 20–29.
31. E. M. van der Beek, V. M. Wiegant, H. A. van der Donk, R. van den Hurk and R. M. Buijs, *J. Neuroendocrinol.*, 1993, **5**, 137–144.
32. R. M. Buijs, M. A. Guzmán Ruiz, R. Méndez Hernández and B. Rodríguez Cortés, *Auton. Neurosci.*, 2019, **218**, 43–50.
33. W. P. Williams III, S. G. Jarjisian, J. D. Mikkelsen and L. J. Kriegsfeld, *Endocrinology*, 2011, **152**, 595–606.

34. H. Bando, T. Nishio, G. T. van der Horst, S. Masubuchi, Y. Hisa and H. Okamura, *J. Neurosci.*, 2007, **27**, 4359–4365.
35. A. Sen and H. M. Hoffmann, *Mol. Cell. Endocrinol.*, 2020, **501**, 110655.
36. M. T. Sellix, *J. Biol. Rhythms*, 2015, **30**, 7–19.
37. K. Ikegami, S. Refetoff, E. Van Cauter and T. Yoshimura, *Nat. Rev. Endocrinol.*, 2019, **15**, 590–600.
38. H. Oster, E. Challet, V. Ott, E. Arvat, E. R. de Kloet, D. J. Dijk, S. Lightman, A. Vgontzas and E. Van Cauter, *Endocr. Rev.*, 2017, **38**, 3–45.
39. H. W. Korf, *Gen. Comp. Endocrinol.*, 2018, **258**, 236–243.
40. M. Perelis, K. M. Ramsey, B. Marcheva and J. Bass, *J. Biol. Rhythms*, 2016, **31**, 323–336.
41. J. Arendt, *J. Biol. Rhythms*, 2005, **20**, 291–303.
42. D. X. Tan, L. C. Manchester and R. J. Reiter, *Med. Hypotheses*, 2016, **86**, 3–9.
43. P. Klosen, S. Lapmanee, C. Schuster, B. Guardiola, D. Hicks, P. Pevet and M. P. Felder-Schmittbuhl, *J. Pineal Res.*, 2019, **67**, e12575.
44. B. Lacoste, D. Angeloni, S. Dominguez-Lopez, S. Calderoni, A. Mauro, F. Fraschini, L. Descarries and G. Gobbi, *J. Pineal Res.*, 2015, **58**, 397–417.
45. J. Cipolla-Neto and F. G. D. Amaral, *Endocr. Rev.*, 2018, **39**, 990–1028.
46. L. Vollrath, *Prog. Brain Res.*, 1979, **52**, 25–38.
47. R. J. Reiter, *Experientia*, 1993, **49**, 654–664.
48. D. C. Klein, *J. Biol. Chem.*, 2007, **282**, 4233–4237.
49. J. A. Gastel, P. H. Roseboom, P. A. Rinaldi, J. L. Weller and D. C. Klein, *Science*, 1998, **279**, 1358–1360.
50. M. D. Schwartz, C. Wotus, T. Liu, W. O. Friesen, J. Borjigin, G. A. Oda and H. O. de la Iglesia, *Proc. Natl. Acad. Sci. U. S. A.*, 2009, **106**, 17540–17545.
51. D. Nakahara, M. Nakamura, M. Iigo and H. Okamura, *Proc. Natl. Acad. Sci. U. S. A.*, 2003, **100**, 9584–9589.
52. V. Lyssenko, C. L. Nagorny, M. R. Erdos, N. Wierup, A. Jonsson, P. Spégel, M. Bugliani, R. Saxena, M. Fex, N. Pulizzi, B. Isomaa, T. Tuomi, P. Nilsson, J. Kuusisto, J. Tuomilehto, M. Boehnke, D. Altshuler, F. Sundler, J. G. Eriksson, A. U. Jackson, M. Laakso, P. Marchetti, R. M. Watanabe, H. Mulder and L. Groop, *Nat. Genet.*, 2009, **41**, 82–88.
53. T. Tuomi, C. L. F. Nagorny, P. Singh, H. Bennet, Q. Yu, I. Alenkvist, B. Isomaa, B. Östman, J. Söderström, A. K. Pesonen, S. Martikainen, K. Räikkönen, T. Forsén, L. Hakaste, P. Almgren, P. Storm, O. Asplund, L. Shcherbina, M. Fex, J. Fadista, A. Tengholm, N. Wierup, L. Groop and H. Mulder, *Cell Metab.*, 2016, **23**, 1067–1077.
54. T. Takumi, K. Taguchi, S. Miyake, Y. Sakakida, N. Takashima, C. Matsubara, Y. Maebayashi, K. Okumura, S. Takekida, S. Yamamoto, K. Yagita, L. Yan, M. W. Young and H. Okamura, *EMBO J.*, 1998, **17**, 4753–4759.

55. S. Masubuchi, S. Honma, H. Abe, K. Ishizaki, M. Namihira, M. Ikeda and K. Honma, *Eur. J. Neurosci.*, 2000, **12**, 4206–4214.
56. W. J. Schwartz and H. Gainer, *Science*, 1977, **197**, 1089–1091.
57. N. Mrosovsky, K. Edelstein, M. H. Hastings and E. S. Maywood, *J. Biol. Rhythms*, 2001, **16**, 471–478.
58. G. Lincoln, S. Messager, H. Andersson and D. Hazlerigg, *Proc. Natl. Acad. Sci. U. S. A.*, 2002, **99**, 13890–13895.
59. A. M. Vosko, M. H. Hagenauer, D. L. Hummer and T. M. Lee, *Am. J. Physiol.: Regul., Integr. Comp. Physiol.*, 2009, **296**, R353–R361.
60. B. B. Otalora, M. H. Hagenauer, M. A. Rol, J. A. Madrid and T. M. Lee, *J. Biol. Rhythms*, 2013, **28**, 249–261.
61. C. Ramanathan, A. Stowie, L. Smale and A. A. Nunez, *Neuroscience*, 2010, **170**, 758–772.
62. S. J. Lee, T. Liu, A. Chattoraj, S. L. Zhang, L. Wang, T. M. Lee, M. M. Wang and J. Borjigin, *J. Pineal Res.*, 2009, **47**, 75–81.
63. S. Takekida, L. Yan, E. S. Maywood, M. H. Hastings and H. Okamura, *Eur. J. Neurosci.*, 2000, **12**, 4557–4561.
64. S. Ebihara, T. Marks, D. J. Hudson and M. Menaker, *Science*, 1986, **231**, 491–493.
65. A. Conti and G. J. Maestroni, *J. Pineal Res.*, 1996, **20**, 138–144.
66. D. J. Kennaway, A. Voultsios, T. J. Varcoe and R. W. Moyer, *Am. J. Physiol.: Regul., Integr. Comp. Physiol.*, 2002, **282**, R358–R365.
67. B. Vivien-Roels, A. Malan, M. C. Rettori, P. Delagrange, J. P. Jeanniot and P. Pévet, *J. Biol. Rhythms*, 1998, **13**, 403–409.
68. C. Zhang, S. J. Clough, E. B. Adamah-Biassi, M. H. Sveinsson, A. J. Hutchinson, I. Miura, T. Furuse, S. Wakana, Y. K. Matsumoto, K. Okanoya, R. L. Hudson, T. Kato, M. L. Dubocovich and T. Kasahara, *J. Pineal Res.*, 2021, **71**, e12748.
69. H. Illnerová, J. Vaněcek, J. Krecek, L. Wetterberg and J. Sääf, *J. Neurochem.*, 1979, **32**, 673–675.
70. D. J. Kennaway, A. Voultsios, T. J. Varcoe and R. W. Moyer, *Am. J. Physiol.: Regul., Integr. Comp. Physiol.*, 2003, **284**, R1231–R1240.
71. D. C. Klein, *Ciba Found. Symp.*, 1985, **117**, 38–56.
72. W. J. Drijfhout, C. J. Grol and B. H. Westerink, *J. Neurochem.*, 1993, **61**, 936–942.
73. D. Sugden, M. A. Namboodiri, D. C. Klein, J. E. Pierce, R. Grady Jr and I. N. Mefford, *Endocrinology*, 1985, **116**, 1960–1967.
74. P. Voisin, G. Van Camp and J. P. Collin, *J. Neurochem.*, 1990, **54**, 1953–1960.
75. D. E. Henley, J. A. Leendertz, G. M. Russell, S. A. Wood, S. Taheri, W. W. Woltersdorf and S. L. Lightman, *J. Med. Eng. Technol.*, 2009, **33**, 199–208.
76. J. E. Ottenweller, A. H. Meier, B. R. Ferrell, N. D. Horseman and A. Proctor, *Endocrinology*, 1978, **103**, 1875–1879.
77. J. E. Ottenweller and A. H. Meier, *Endocrinology*, 1982, **111**, 1334–1338.
78. A. V. Edwards and C. T. Jones, *J. Physiol.*, 1987, **382**, 385–396.
79. A. V. Edwards and C. T. Jones, *J. Anat.*, 1993, **183**(Pt 2), 291–307.

80. Y. M. Ulrich-Lai, M. M. Arnhold and W. C. Engeland, *Am. J. Physiol.: Regul., Integr. Comp. Physiol.*, 2006, **290**, R1128–R1135.
81. S. R. Bornstein, M. Ehrhart-Bornstein, W. A. Scherbaum, E. F. Pfeiffer and J. J. Holst, *Endocrinology*, 1990, **127**, 900–906.
82. S. R. Bornstein, W. C. Engeland, M. Ehrhart-Bornstein and J. P. Herman, *Trends Endocrinol. Metab.*, 2008, **19**, 175–180.
83. H. Oster, S. Damerow, S. Kiessling, V. Jakubcakova, D. Abraham, J. Tian, M. W. Hoffmann and G. Eichele, *Cell Metab.*, 2006, **4**, 163–173.
84. A. Leliavski, A. Shostak, J. Husse and H. Oster, *Endocrinology*, 2014, **155**, 133–142.
85. N. E. Rosenthal, D. A. Sack, J. C. Gillin, A. J. Lewy, F. K. Goodwin, Y. Davenport, P. S. Mueller, D. A. Newsome and T. A. Wehr, *Arch. Gen. Psychiatry*, 1984, **41**, 72–80.
86. A. L. Chesson Jr, M. Littner, D. Davila, W. M. Anderson, M. Grigg-Damberger, K. Hartse, S. Johnson and M. Wise, *Sleep*, 1999, **22**, 641–660.
87. A. Knutsson, *Occup. Med.*, 2003, **53**, 103–108.
88. E. Filipski, F. Delaunay, V. M. King, M. W. Wu, B. Claustrat, A. Gréchez-Cassiau, C. Guettier, M. H. Hastings and L. Francis, *Cancer Res.*, 2004, **64**, 7879–7885.
89. A. J. Lewy, T. A. Wehr, F. K. Goodwin, D. A. Newsome and S. P. Markey, *Science*, 1980, **210**, 1267–1269.
90. K. P. Wright Jr and C. A. Czeisler, *Science*, 2002, **297**, 571.
91. Y. Shigeyoshi, K. Taguchi, S. Yamamoto, S. Takekida, L. Yan, H. Tei, T. Moriya, S. Shibata, J. J. Loros, J. C. Dunlap and H. Okamura, *Cell*, 1997, **91**, 1043–1053.
92. A. Ishida, T. Mutoh, T. Ueyama, H. Bando, S. Masubuchi, D. Nakahara, G. Tsujimoto and H. Okamura, *Cell Metab.*, 2005, **2**, 297–307.
93. R. M. Buijs, J. Wortel, J. J. Van Heerikhuize, M. G. Feenstra, G. J. Ter Horst, H. J. Romijn and A. Kalsbeek, *Eur. J. Neurosci.*, 1999, **11**, 1535–1544.
94. H. Holgert, A. Dagerlind and T. Hökfelt, *J. Comp. Neurol.*, 1996, **371**, 603–620.
95. H. Holgert, A. Dagerlind and T. Hökfelt, *Horm. Metab. Res.*, 1998, **30**, 315–322.
96. S. R. Bornstein, M. Ehrhart-Bornstein, W. A. Scherbaum, E. F. Pfeiffer and J. J. Holst, *Endocrinology*, 1990, **127**, 900–906.
97. A. Balsalobre, S. A. Brown, L. Marcacci, F. Tronche, C. Kellendonk, H. M. Reichardt, G. Schütz and U. Schibler, *Science*, 2000, **289**, 2344–2347.
98. S. Koyanagi, S. Okazawa, Y. Kuramoto, K. Ushijima, H. Shimeno, S. Soeda, H. Okamura and S. Ohdo, *Mol. Endocrinol.*, 2006, **20**, 573–583.
99. S. F. Morrison, *Am. J. Physiol.: Regul., Integr. Comp. Physiol.*, 2001, **281**, R683–R698.
100. A. Kikuta and T. Murakami, *Am. J. Anat.*, 1982, **164**, 19–28.
101. T. Murakami, H. Oukouchi, Y. Uno, A. Ohtsuka and T. Taguchi, *Arch. Histol. Cytol.*, 1989, **52**, 461–476.

102. R. C. Hermida, D. E. Ayala, C. Calvo, F. Portaluppi and M. H. Smolensky, *Adv. Drug Delivery Rev.*, 2007, **59**, 923–939.
103. K. Kamperis, S. Hagstroem, E. Radvanska, S. Rittig and J. C. Djurhuus, *Am. J. Physiol. Renal. Physiol.*, 2010, **299**, F404–F411.
104. M. Doi, Y. Takahashi, R. Komatsu, F. Yamazaki, H. Yamada, S. Haraguchi, N. Emoto, Y. Okuno, G. Tsujimoto, A. Kanematsu, O. Ogawa, T. Todo, K. Tsutsui, G. T. van der Horst and H. Okamura, *Nat. Med.*, 2010, **16**, 67–74.
105. Y. Suwazono, M. Dochi, K. Sakata, Y. Okubo, M. Oishi, K. Tanaka, E. Kobayashi and K. Nogawa, *Hypertension*, 2008, **52**, 581–586.
106. F. A. Scheer, M. F. Hilton, C. S. Mantzoros and S. A. Shea, *Proc. Natl. Acad. Sci. U. S. A.*, 2009, **106**, 4453–4458.
107. K. Kario and K. Shimada, *Clin. Exp. Hypertens.*, 2004, **26**, 177–189.
108. I. C. R. Group, *BMJ*, 1988, **297**, 319–328.
109. D. Denton, R. Weisinger, N. I. Mundy, E. J. Wickings, A. Dixson, P. Moisson, A. M. Pingard, R. Shade, D. Carey and R. Ardaillou, *et al.*, *Nat. Med.*, 1995, **1**, 1009–1016.
110. G. T. van der Horst, M. Muijtjens, K. Kobayashi, R. Takano, S. Kanno, M. Takao, J. de Wit, A. Verkerk, A. P. Eker, D. van Leenen, R. Buijs, D. Bootsma, J. H. Hoeijmakers and A. Yasui, *Nature*, 1999, **398**, 627–630.
111. H. Okamura, S. Miyake, Y. Sumi, S. Yamaguchi, A. Yasui, M. Muijtjens, J. H. Hoeijmakers and G. T. van der Horst, *Science*, 1999, **286**, 2531–2534.
112. S. Mitsui, S. Yamaguchi, T. Matsuo, Y. Ishida and H. Okamura, *Genes Dev.*, 2001, **15**, 995–1006.
113. A. M. Moisan, M. L. Ricketts, V. Tardy, M. Desrochers, F. Mébarki, J. L. Chaussain, S. Cabrol, M. C. Raux-Demay, M. G. Forest, W. G. Sippell, M. Peter, Y. Morel and J. Simard, *J. Clin. Endocrinol. Metab.*, 1999, **84**, 4410–4425.
114. K. Yamamura, M. Doi, H. Hayashi, T. Ota, I. Murai, Y. Hotta, R. Komatsu and H. Okamura, *Mol. Cell. Endocrinol.*, 2014, **382**, 131–138.
115. T. Ota, M. Doi, F. Yamazaki, D. Yarimizu, K. Okada, I. Murai, H. Hayashi, S. Kunisue, Y. Nakagawa and H. Okamura, *Mol. Cell. Biol.*, 2014, **34**, 3880–3894.
116. M. Doi, F. Satoh, T. Maekawa, Y. Nakamura, J. M. Fustin, M. Tainaka, Y. Hotta, Y. Takahashi, R. Morimoto, K. Takase, S. Ito, H. Sasano and H. Okamura, *J. Clin. Endocrinol. Metab.*, 2014, **99**, E257–E262.
117. H. Sasano, *Endocr. J.*, 1994, **41**, 471–482.
118. I. Sakuma, S. Suematsu, Y. Matsuzawa, J. Saito, M. Omura, T. Maekawa, Y. Nakamura, H. Sasano and T. Nishikawa, *Endocr. J.*, 2013, **60**, 329–336.
119. Y. Nakamura, Y. Yamazaki, S. Konosu-Fukaya, K. Ise, F. Satoh and H. Sasano, *J. Steroid Biochem. Mol. Biol.*, 2015, **153**, 57–62.
120. H. Okamura, M. Doi, K. Goto and R. Kojima, *Hypertens. Res.*, 2016, **39**, 681–687.
121. C. Schirpenbach and M. Reincke, *Nat. Clin. Pract. Endocrinol. Metab.*, 2007, **3**, 220–227.

122. J. W. Funder, *Science*, 2011, **331**, 685–686.
123. M. Choi, U. I. Scholl, P. Yue, P. Björklund, B. Zhao, C. Nelson-Williams, W. Ji, Y. Cho, A. Patel, C. J. Men, E. Lolis, M. V. Wisgerhof, D. S. Geller, S. Mane, P. Hellman, G. Westin, G. Åkerström, W. Wang, T. Carling and R. P. Lifton, *Science*, 2011, **331**, 768–772.
124. E. A. Azizan, H. Poulsen, P. Tuluc, J. Zhou, M. V. Clausen, A. Lieb, C. Maniero, S. Garg, E. G. Bochukova, W. Zhao, L. H. Shaikh, C. A. Brighton, A. E. Teo, A. P. Davenport, T. Dekkers, B. Tops, B. Küsters, J. Ceral, G. S. Yeo, S. G. Neogi, I. McFarlane, N. Rosenfeld, F. Marass, J. Hadfield, W. Margas, K. Chaggar, M. Solar, J. Deinum, A. C. Dolphin, I. S. Farooqi, J. Striessnig, P. Nissen and M. J. Brown, *Nat. Genet.*, 2013, **45**, 1055–1060.
125. A. E. Teo, S. Garg, L. H. Shaikh, J. Zhou, F. E. Karet Frankl, M. Gurnell, L. Happerfield, A. Marker, M. Bienz, E. A. Azizan and M. J. Brown, *N. Engl. J. Med.*, 2015, **373**, 1429–1436.

CHAPTER 11

Clock Proteins and Circadian Rhythms in Immunity

ANNIE M. CURTIS*[a] AND MARTIN S. KNAPP[b]

[a] School of Pharmacy and Biomolecular Sciences (PBS) and Tissue Engineering Research Group (TERG), RCSI University of Medicine and Health Sciences, 123 St. Stephens Green, Dublin 2, Ireland; [b] Consultant Physician (retired) Melbourne, Australia, Previously: Universities of Tasmania, Melbourne, Nottingham, Bristol, Director, Chronotherapeutics Research Group, University of Nottingham, UK
*Email: anniecurtis@rcsi.com

11.1 Overview of the Immune System and its Cellular Components

The immune system is a complex network of organs, tissues, cells and molecules, which collectively protect the body against foreign invaders such as microorganisms and repair damage to the body as a result of injury or disease (Figure 11.1). In order to be effective, the immune system must have the following four properties. First, it must swiftly *recognise* infection/damage. Next, it must be able to *contain*, *constrain* and *eliminate* the infection, predominantly through the specialised effector functions of the different leukocytes (also known as white blood cells). All leukocytes arise from stem cells in the bone marrow. Specifically, how the effector function of leukocytes is modulated by the molecular clock will be discussed at length in this chapter. Frequently, infections drive *immunological memory*, so that if individuals are re-exposed to the same infectious agent, they can mount a faster, more tailored response against it. This is termed protective immunity,

Chemical Biology No. 23
Chronobiology and Chronomedicine: From Molecular and Cellular Mechanisms to Whole Body Interdigitating Networks
Edited by Germaine Cornelissen and Tsuyoshi Hirota

Published by the Royal Society of Chemistry, www.rsc.org

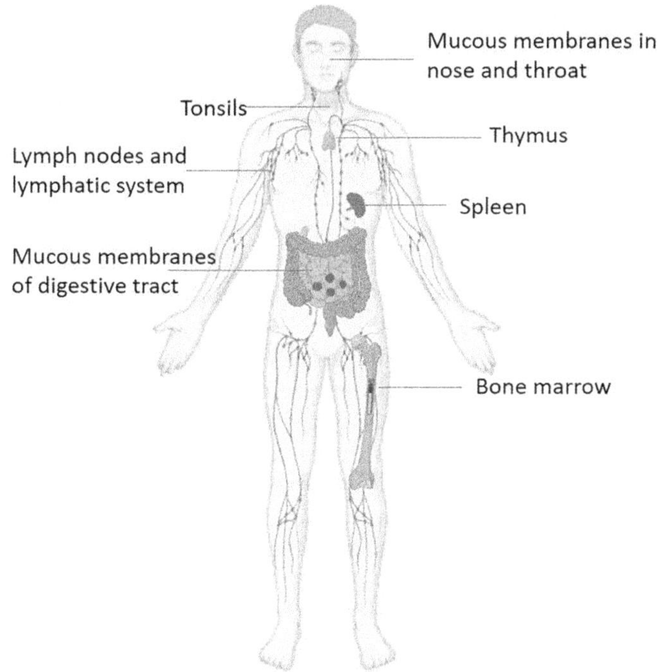

Figure 11.1 The immune system is comprised of a number of organs and their cellular constituents. The bone marrow and thymus are the primary lymphoid organs. The lymph nodes, spleen and mucous membranes are secondary lymphoid tissues. Mucous membranes are membranes that line body cavities, chiefly the respiratory, digestive and urogenital tracts. Created with Biorender.com.

and is the feature of immunity exploited through vaccination. Finally, the immune system must be able to *self-regulate* itself and not *over* or *under* respond to pathogens or endogenous factors. Failure to self-regulate and maintain homeostasis is what contributes to conditions such as allergy, lupus and multiple sclerosis (in which the immune system over responds) or HIV and cancer (which are examples of the immune system under-responding).

11.2 Innate *Versus* Adaptive Immunity and its Cellular Components

The immune system is organised into two principal components, the innate and the adaptive arm. When a pathogen succeeds in breaching one of the host's anatomic barriers such as the skin and mucous membranes, the innate immune system, which is *the first line of defence,* is called into action.[1] The immediate innate response is performed *via* antimicrobial soluble molecules present in blood, fluid or epithelial secretions such as lysozymes, defensins and a system of plasma proteins known as the complement system to target the pathogen. In the second phase, the innate immune cells, particularly cells of the myeloid lineage (neutrophils, basophils, eosinophils and mast cells along with monocytes, which differentiate into macrophages and dendritic cells (DCs)) are called into action. These cells are derived from the common myeloid progenitor cell of the bone marrow (Figure 11.2). These cells use invariant germline encoded receptors called pattern recognition receptors to sense pathogen-associated molecular patterns (PAMPs) on microbes or danger-associated molecular patterns (DAMPs) released from dead or injured cells. Pattern recognition receptors (PRRs) include the family of toll-like receptors (TLRs).[2] Humans express 10 different TLRs whereas mice express 12. In both species, TLR4 is included, which recognises lipopolysaccharide (LPS) found on Gram-negative bacteria. DAMPs, such as members of the S100 family of proteins, are released from cells after trauma and are recognised by TLR2 and TLR4. Thus, these PRRs allow the body to recognise both stranger (foreign pathogens) and danger (self-molecules released upon damage) signals. Ligation of pattern recognition receptors triggers macrophages and neutrophils to engulf foreign pathogens. Upon activation, macrophages and neutrophils secrete proteins such as cytokines and chemokines. Cytokines, like IL-1β and TNF, activate nearby cells bearing appropriate receptors, whereas chemokines, such as CCL2, attract cells

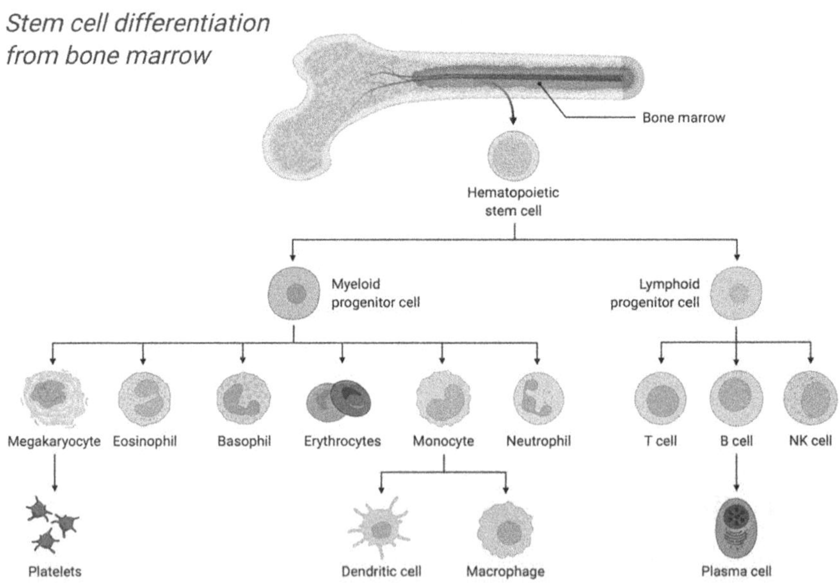

Figure 11.2 Lineage tree representation of cells of the immune system. Immune cells are derived from hematopoietic stem cells. These HSCs can differentiate into either myeloid or lymphoid progenitor cells. These cells can then branch out into specific cell types associated with innate (myeloid) or adaptive (lymphoid) immunity. Created with Biorender.com.

bearing the cognate receptor out of the blood stream and into the infected/damaged tissue. Collectively, this initiates the inflammatory cascade or response, which is characterized by heat, pain, redness, swelling and loss of function.[3]

If the components of the innate immune response are unable to clear and resolve the infection/damage in the first 72 hours post insult, the adaptive immune response is called into play.[4] The adaptive immune response is slower than the innate immune response but is much more specific. While innate immune cells can be activated through their pattern recognition receptors by a broad spectrum of molecular patterns common to pathogens, adaptive immune cells are highly specific for antigens (Ag). An Ag is defined as a molecule or molecular structure, which when present in the body triggers an immune response. Ag may originate within the body or from the external environment. Ag presenting cells (APCs) such as DCs can present Ag to T cells in the form of peptides on major histocompatibility molecules (MHCs). The common lymphoid progenitor in the bone marrow is the precursor for the T and B lymphocytes of the adaptive immune system (Figure 11.2). B lymphocytes mature in the bone marrow, whereas precursor T lymphocytes migrate from the bone marrow to mature in the thymus. The bone marrow and thymus are considered primary lymphoid organs (Figure 11.1). Once matured, T and B lymphocytes enter circulation and travel through the secondary lymphoid organs, the lymph nodes, spleen and mucosal lymphoid tissues. Natural killer (NK) cells are also derived from the common lymphoid progenitor, but NK cells are not antigen specific and are considered part of the innate immune system. Each mature lymphocyte bears a unique variant of a prototype Ag receptor. For T lymphocytes (T cells), this is the T cell receptor (TCR) and for B lymphocytes (B cells), it is the B cell receptor (BCR). This results in a population of lymphocytes expressing a huge repertoire of receptors that are highly diverse in their Ag binding sites. Two billion lymphocytes circulate in the human body. This ensures capacity to recognise a given Ag from a wide variety of pathogens that individuals may encounter throughout their lifetime. Lymphocytes not yet activated by Ag are described as naive, while those, which have met their corresponding Ag are known as effector lymphocytes.

Lymphocytes are presented with Ag in the secondary lymphoid organs. DCs at the site of infection have the capacity to engulf the pathogen and degrade it intracellularly to produce Ag. Ag is then presented on the DC surface *via* MHC proteins. DCs are highly specialised APCs, although macrophages can also perform this function.[5] Therefore, APCs are the bridge between the innate and adaptive immune response. APCs migrate *via* the lymphatic vessels to lymph nodes and the Ag-loaded MHC of the APC finds and binds to a specific TCR. Expression of co-stimulatory molecules on the cell surface of the APC and T cell induces T-cell clonal expansion. Engagement of co-stimulatory molecules on T cells post TCR engagement is required for a robust response. During clonal expansion, lymphocytes are able to duplicate themselves 2–4 times every day for 3–5 days, producing a clone of around 1000 identical fully functional effector daughter cells. B cells have Ag-specific receptors, which enable them to internalize, process and present Ag. Effector T cells are required to activate Ag-specific B cells, which causes differentiation of B cells into plasma cells, which secrete antibodies. For B cells, antibodies are the secreted form of the B cell's Ag receptor. Most plasma cells generated in the lymph node or spleen take up residence in the bone marrow and continue to generate large amounts of antibodies into the blood circulation. Finally, many of these effector T and B cells will die after antigen is eliminated. However, some will persist as memory T and B cells. These are the cells, which are swiftly activated upon secondary challenge and form the basis of long-lasting protective immunity or immunological memory.[6]

11.3 Historical Perspective on Circadian Immunology

The previous section has summarised how the innate and adaptive immune system collectively operate to provide host protection. We will next explore the intersections between circadian rhythms and immunity, and how this information first came about.

Studies in the 1940s, 1950s and 1960s documented 24-hour or daily patterns in the numbers of cell types relevant to immunity. For example, eosinophils in healthy individuals show a circadian pattern, contrasting with disturbance of circadian patterns from individuals with "allergic" disorders such as asthma.[7–9] Studies reported in this era often did not discuss the possible relevance of circadian variations in mammals to human activities and/or their potential application in clinical medicine. In the middle decades of the 20th century, pioneers of circadian research, Pittendrigh and Aschoff, explored the components and organisation of circadian rhythms and participated in focussed discussion on their observations and hypotheses.[10,11] In the subsequent decades, studies continued to report circadian changes observed in healthy mammals with some publications from the 20th century extending their interest to human patients. Halberg, Scheving, Haus, Reinberg, Knapp, with co-workers, demonstrated the temporal influences on several aspects of mammalian immunity as expressed in the intact organism (Figure 11.3). They reported observations identifying variations in immune cell counts,[12] immune cell responses,[13] disease,[14] response to LPS,[15] rejection of skin grafts from a donor, kidney transplant rejection, delayed hypersensitivity reactions,[16] lymphocyte proliferation,[17] serum protein and antibody concentrations,[18] effects of drugs on immune response and more. For the first time, these studies integrated chronobiological knowledge with knowledge of the immune system, and set up the foundations for the field of chronoimmunology. Halberg defined this field as "*describing the results of a chronobiological approach to immunologic phenomena; chronoimmunologic methodology involves less risk of error and/or false*

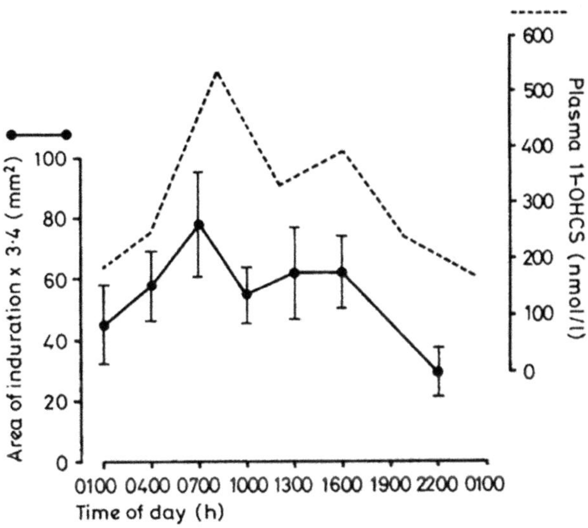

Figure 11.3 Circadian variation in an immune response in humans to intradermal tuberculin.[16] Tuberculin are protein extracts from Mycobacterium tuberculosis, the bacteria which cause tuberculosis or TB. Reproduced from ref. 16 with permission from BMJ, Copyright 1978.

information than the conventional homeostatic approach of sampling at some convenient yet not necessarily pertinent time…".

Franz Halberg, who coined the term circadian,[19] was an early proponent of timing therapy to maximise benefits and minimise toxicity and he suggested "chronopharmacology" and "chronotherapeutics" as collective nouns to identify this medical application of chronobiology. The main findings of this era were that some rhythms within the immune system are of sufficient amplitude to be relevant to the healthy function of the host or can signal disease when they are absent or disturbed.

In the later years of the 20th century, various symposia and textbooks did not reference or document the established rhythmicity in immune function and its potential to modify the management of clinical specialties. The 1990s marked the cloning of the mammalian clock genes,[20] and this heralded a resurgence in interest in circadian immunity, particularly evident from around 2010 onwards. The main findings of this period from *circa* 2010 to present day will be discussed in more detail in the subsequent sections.

11.4 Overview of the Circuitry Imparting Rhythmic Control on Immunity

A key aspect of the immune system is the continuous migration of the cells from the primary to secondary lymphoid organs to the tissues and back, in order to sense invading pathogens and protect the host. Therefore, sophisticated higher-level signals are necessary to ensure correct regulation across the whole body and mount a coordinated whole-body response when required. At present, the main mechanisms for circadian synchronization of the immune system are as follows.

First, akin to other peripheral clocks, the immune clock receives synchronizing signals from the suprachiasmatic nuclei (SCN), and a predominant pathway of immune clock synchronisation is SCN control of endocrine function *via* the adrenal gland. The SCN signals to the adrenal gland *via* two pathways, first *via* direct innervation of the adrenal gland, and second through the hypothalamic–pituitary–adrenal (HPA) axis. The SCN influences the paraventricular nucleus of the hypothalamus to secrete corticotrophin-releasing hormone (CRH) and arginine vasopressin (AVP), stimulating the release of adrenocorticotropic hormone (ACTH) from the pituitary gland in a circadian fashion. ACTH controls the circadian production and secretion of glucocorticoids (cortisol in humans and corticosterone in mice) from the adrenal gland. Glucocorticoids exhibit maximal secretion from the adrenal glands during the early hours of activity. Measurements of cortisol correlate with levels of circulating eosinophils, one of the earliest human circadian immune rhythms to be documented.[8] In addition, local adrenal clocks are thought to mediate the responsiveness to ACTH in a circadian fashion, as circadian glucocorticoid rhythms persist independently of the SCN.[21] This is likely to maintain persistence of circadian glucocorticoids even when the HPA axis is activated under stressful conditions. This results in a rhythmic release of glucocorticoids with levels peaking just prior to the active phase (day in humans and night in rodents). Glucocorticoids exert profound immunosuppressive effects on the immune system. Glucocorticoids, bound to the glucocorticoid receptor (GR), regulate the transcription of target genes to activate a number of anti-inflammatory mechanisms.[22] The SCN also uses neuronal cues of the autonomic nervous system (ANS) in the form of adrenaline and noradrenaline to signal to immune cells.[23]

Metabolic cues resulting from rhythmic feeding/fasting behaviours are also likely to act as immune clock synchronisers.[24] It is also likely that exercise, through mechanical, biochemical and metabolic stimuli can also impart rhythmic immune control.[25] Therefore, multiple interconnections exist through which the environmental factors ("Zeitgebers") such as light, food and activity can signal rhythmicity to the immune system, allowing for temporal coherence with the outside environment. Zeitgebers are defined as external environmental cues that synchronise the endogenous clock to the Earth's diurnal and seasonal cycles. ZT0 is the time of onset of a signal. ZT0–ZT12 represents the subjective day when the organism is exposed to the light during entrainment, whereas ZT12–ZT24 represents the subjective night.

The second mechanism of the immune clock involves the expression of the key circadian timing genes, such as *Bmal1*, *Clock*, *Period*, *Cryptochrome*, *Rev-erbs* and *Ror* genes, which together make up the endogenous circadian oscillator. It appears that most immune cells express the main circadian proteins, but levels and phase differ among the different immune cells.[26]

Collectively, the endogenous oscillator can either directly or indirectly control a range of key immune genes involved in host immunity.[26] Circadian genes operate to regulate the type and magnitude of immune responses in a cell type-specific manner, resulting in very striking 24-hour phenotypes/biology. A classic example of this was demonstrated by Halberg, who first showed a 24-hour variation in survival when mice were treated with LPS.[15] Mice were much more likely to succumb to the LPS challenge when injected prior to the onset of activity *versus* prior to the rest phase. This observation was followed by studies showing daily rhythmicity in response to *Streptococcus pneumoniae* and this periodicity in the response was altered, but not abolished, in blind and adrenalectomized mice.[27,28] The daily variation in response to LPS is dependent on the endogenous oscillator in immune cells, as mice lacking *Bmal1* in the myeloid lineage lose the time-of-day effect and are more susceptible to lethality.[29] However, another study did not observe any impact with myeloid *Bmal1* and *Clock* deletion in response to circadian LPS challenge.[30] Nonetheless, daily rhythmicity in the immune response to infection with dependence on certain clock genes has now been demonstrated in mice challenged with bacteria,[31,32] viruses[33,34] and parasites[35,36] (Table 11.1). However, the phase of the response and impact of clock gene deletion differ depending on the pathogen in question.

Finally, the immune system is sometimes referred to as a liquid organ due to the ability of immune cells to migrate between bone marrow, blood and organs under steady state and activated conditions. Every day, there is a highly regulated distribution of leukocytes around the body, which has been demonstrated in humans,[37] mice[23] and hamsters.[38] When interpreting mouse *versus* human circadian studies, it is important to consider that mice are 'nocturnal', and humans are 'diurnal'. In terms of their peripheral clocks, including the immune clock, these broadly match with their activity and rest phases. For example, the main clock gene *Bmal1* shows peak mRNA expression in mouse peripheral tissues at the onset of the rest phase, the same being true for diurnal species.[39] Therefore, to minimise confusion, throughout this chapter we will refer to clock time in relation to the rest and activity phase of the species in question, rather than referring to day or night. Broadly speaking, it is observed that leukocytes are predominantly in circulation during the rest phase and extravasate into tissues during the active phase. This is achieved *via* rhythmic release of adrenaline by adrenergic nerves, which acts on β-adrenoreceptors of endothelial cells to increase the expression of adhesion molecules on their surface.[23] This rhythmic expression of adhesion molecules such as ICAM1 in muscle and P-selectin, E-selectin and VCAM on bone marrow endothelial cells leads to leukocyte adhesion and migration into these tissues in a circadian manner. These daily fluctuations in leukocytes also underpin the outcome of an inflammatory challenge. The increase in susceptibility of mice to LPS challenge at the onset of their activity is also due to this rhythmic cell migration. Neutrophil migration into tissues is higher when LPS is given prior to the onset of activity. Mice lacking the adhesion molecule *Icam1*, which

Table 11.1 A selection of studies showing the impact of administration time and clock gene deletion on the immunological response to infection[a].

Type of infection and delivery method	Timing of infection	Outcomes in WT	Genetic model	Outcome in genetic model	Ref.
Bacterial infection Salmonella (*S. Typhimurium*)	ZT4 (early rest phase) or ZT10 (early active phase).	Higher pro-inflammatory gene response and higher colonisation of Salmonella at ZT4.	*Clock* mutant and *Clock* deficient mice.	Altered timing of reaction following infection and reduction in pro-inflammatory gene expression.	31
Bacterial infection *Streptococcus pneumoniae*	ZT0 (beginning of rest phase) or ZT12 (beginning of activity phase).	Higher bacterial burden in lungs after ZT0 infection *versus* ZT12.	*Bmal1* deletion in pulmonary epithelial cells.	Increased neutrophil recruitment over WT but no difference in bacterial burden.	32
Viral infection Herpesvirus	ZT0 or ZT10.	Higher viral replication after ZT0 infection *versus* ZT10.	*Bmal1* deficient mice.	Time-of-day difference abolished with increased viral replication with ZT12 infection.	33
Viral infection Influenza A	ZT11 and ZT23.	Decreased survival after ZT11 infection *versus* ZT23.	Deletion of *Bmal1 via* tamoxifen induced CRE recombinase deletion in mice aged 6–8 weeks.	Time-of-day difference abolished. Survival at both time points (ZT11/23) mirrored WT survival at ZT11.	34
Helminth infection *Trichuris muris*	ZT0 and ZT12.	ZT12 infection led to significantly higher worm burden than ZT0.	*Bmal1* deletion in dendritic cells.	Time-of-day difference in worm burden absent.	35
Protozoan infection *Leishmania*	CT3, CT9, CT15 and CT21	CT9 and CT15 infection led to the greatest parasitic load in comparison to CT3.	*Bmal1* deletion in bone marrow.	Time-of-day difference in parasite load was absent.	36

[a] ZT: Zeitgeber Time (ZT0 is time of light onset; ZT4 indicates 4 hours after light onset); CT: Circadian Time (standard marker of time similar to ZT, based on the free-running period).

prevents neutrophil migration into tissues, lose this 24-hour variability in response to LPS and do not display an increase in mortality when challenged at activity onset.[23] These elegant studies demonstrate that the circadian immune response is dependent on the molecular oscillator of the immune cells, the molecular oscillators within cells of the microenvironment and systemic factors.[40] Finally, a key aspect of adaptive immunity involves the migration of dendritic cells along with T and B cells in and out of lymph nodes. There is enhanced guidance of dendritic cells into lymphatic vessels at ZT7, and this is driven by circadian control of the chemokine receptor CCR7 on dendritic cells along with rhythmic gradients of its ligand CCL21 on lymphatic endothelial cells.[41] The highest number of T and B cells in lymph nodes is observed just after they enter their activity phase (ZT13). T and B cells display peak expression of the chemokine receptor 7 (CCR7) in time with the highest expression of CCL21 in lymph nodes, which leads to enhanced recruitment of these cells to the lymph node at this time. Conversely, rhythmic expression of the sphingosine-1-phosphate receptors on T and B lymphocytes, with increased expression around ZT5 drives egress of lymphocytes out of the lymph nodes and into efferent lymph.[42]

Conceptionally, it is useful to think about these systems as circuits in which circadian information flows through different compartments of the circuit, the SCN, the autonomic nervous system, the endocrine system, immune cells and their microenvironment. In many instances, there is reciprocal communication between the different compartments, creating a feedback loop to determine the overall response. The ultimate goal of this system is to orchestrate an appropriate immune response ordered correctly in space and time.

11.5 Clocks within Myeloid Cells

The innate immune system is of ancient origin and is found in vertebrates, invertebrates and also in plants. Granulocytes and monocytes are collectively called myeloid cells, which are the main cells of the innate immune system (Figure 11.2). Granulocytes (cells, which contain cytoplasmic granules) encompass neutrophils, basophils, eosinophils and mast cells. Neutrophils have a characteristic multi-lobed nucleus, whereas basophils and eosinophils have a bi-lobed nucleus. Monocytes are also short-lived, motile cells found in the blood, bone marrow and spleen. They do not contain granules and have a large characteristic kidney shaped nucleus.

11.5.1 Monocytes and Macrophages

Monocytes quickly respond and migrate to sites of infection. They are often considered a systemic reservoir of myeloid precursors, important in the renewal of the long lived tissue macrophages and DCs[43] (Figure 11.3).

Monocyte numbers peak in the blood during the rest phase of mice[40] and humans,[44] and they emigrate into tissues during the active phase. Blood monocytes exhibit a 24-hour variation in clock genes, with their phase broadly aligned with that observed in other peripheral cells. This rhythmicity in migration is due to rhythms in Ly6C^{hi} (inflammatory monocytes referred to as 'classical'), and not Ly6C^{low} (patrolling monocytes referred to as 'non classical'). The number of these Ly6C^{hi} inflammatory monocytes peaks in blood at ZT8. Subsequently, when mice are infected with a low dose of *L. monocytogenes* at ZT8, they clear the infection better than those infected at ZT0. However, when mice are dosed with a higher inoculum of *L. monocytogenes* at ZT8, they more readily succumb than mice infected at ZT0.[45] Importantly, this increased lethality at ZT8 was associated with an enhanced inflammatory response and not higher bacterial burden. The BMAL1/CLOCK complex played a central role in these effects due to the rhythmic recruitment of the histone methylation enzyme and transcriptional repressor EZH2, leading to rhythmic repression in gene expression of the chemokines *Ccl2*, *Ccl8* and *S100a8*.[45] Mice lacking myeloid *Bmal1* were more susceptible to *L. monocytogenes*-induced lethality as well as diet-induced obesity and insulin resistance.[45]

Macrophages either originate as blood monocytes that leave the bloodstream to differentiate into tissues or are seeded into tissues during embryonic development and give rise to long-lasting self-renewing macrophage populations.[46] The term macrophages is aptly derived from two Greek words "makro" and "phagein", which together mean "big eater". Thus, a crucial function of macrophages is the ingestion of pathogens or damaged cells *via* phagocytosis. To date, the circadian control of macrophages has been intensely studied, most likely due to the ease of generating large numbers of bone marrow-derived macrophages *in vitro*. However, bone marrow-derived macrophages should only ever be considered a "model system" and any results in these cells require additional confirmation from macrophages isolated *in vivo*.

A key feature of both monocytes and macrophages is their use of membrane-bound pattern recognition receptors to sense the external environment for infection and damage.

TLR4, one of the most studied toll-like receptors, senses the endotoxin LPS found on Gram-negative bacteria such as *E. coli* and *Salmonella*.[47] TLR4 ligation induces a series of signaling cascades that convert monocytes and macrophages from a quiescent to an immunologically active state.[48] Halberg's historic observation that *E. coli* endotoxin-induced death varied depending on the time of injection suggested that TLR4 receptor signaling was under circadian control.[49] However, circadian oscillations in *Tlr4* expression have not been observed in macrophages,[50] but many genes downstream of TLR4 were found to cycle[51] such as *IkBα*, which negatively regulates NF-κB, and *Adam17*, a metalloproteinase involved in TNFα release.[52] Direct circadian impact on other TLRs, such as TLR9, an intracellular receptor that senses bacterial and viral DNA, has been observed.[50] In a TLR9-dependent model of sepsis, greater lethality was observed at the time-of-day coinciding with the highest TLR9 expression in splenic macrophages and B cells. Therefore, circadian control in some TLRs and downstream signaling pathways appears as an important mechanism directing circadian inflammation in cells of the myeloid lineage. Loss of myeloid *Bmal1* often leads to an enhanced and often detrimental response to the host in models of sepsis such as cecal ligation puncture.[53]

A central role of monocytes and macrophages is the production of inflammatory mediators such as cytokines and chemokines to coordinate the inflammatory response (Figure 11.4). Many studies have investigated the impact of myeloid clock components on induction of these pathways. Myeloid BMAL1 has been shown to suppress the production of the cytokines IL-1β, IL6 and IFN-γ in response to *L.monocytogenes*.[45] BMAL1 can directly regulate LPS-mediated induction of the potent pro-inflammatory cytokine IL-1β through control of NRF2 and ROS[54] or enhancer RNA (eRNA) transcription.[55] TLR4 activation results in increased levels of miR-155, which targets *Bmal1* for degradation, thus potentiating inflammation.[29] BMAL1 has also been shown to sense nutrients such as glucose, acting as a metabolic checkpoint. Macrophages lacking *Bmal1* show alterations in both glycolysis, Krebb cycle intermediates, mitochondrial respiration and amino acid metabolism. Some of these effects are due to aberrant hypoxia-inducible factor 1α (HIF-1α) activation, which results in higher IL-1β and also an immunosuppressive tumour microenvironment.[56,57] Macrophages derived from mice with a *Clock* mutation display reduced expression of pro-inflammatory cytokines and chemokines in response to LPS and Salmonella.[58] REV-ERBα also regulates a host of cytokines and chemokines from macrophages such as *Il6*,[59] *Ccl2*,[60] and inhibits eRNA transcription.[61] Inflammasomes are innate immune system sensors made up of cytosolic multi-protein oligomers. Their role is to

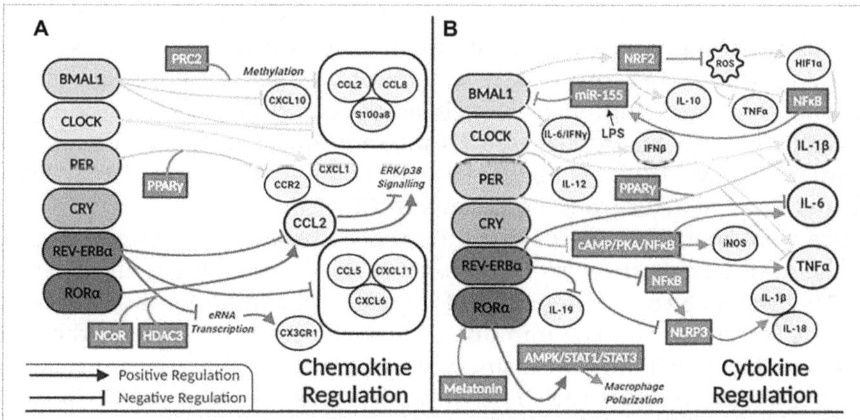

Figure 11.4 Multiple pathways of clock control of cytokine and chemokine expression in monocytes and macrophages. Details circadian clock proteins mediated molecular regulation of (A) chemokines and (B) cytokines. Reproduced from ref. 68, https://doi.org/10.3389/fimmu.2021.700431, under the terms of the CC BY 4.0 license https://creativecommons.org/licenses/by/4.0/.

sense PAMPs or DAMPs and activate the release of pro-inflammatory cytokines such as IL-1β and IL-18. One of the most characterized inflammasomes is NLRP3, and it is regulated by REV-ERBα in mouse and human macrophages.[62] Likewise, the PER and CRY complex suppresses inflammatory components in macrophages.[63,64] Rhythms in macrophage phagocytosis have also been observed and are dependent on BMAL1.[65–67]

Consequently, components of the endogenous molecular clock along with their temporal variation significantly impact many aspects of monocyte and macrophage function. Studies in these innate immune cells have highlighted the broad spectrum of ways by which the circadian system can impact monocyte- and macrophage-dependent immunity. In addition to the pathways discussed above, other pathways linking clock proteins to chemokine and cytokine production from monocytes and macrophages have been uncovered, as illustrated in Figure 11.4, and are detailed more fully in this review.[68]

11.5.2 Dendritic Cells

Dendritic cells (DCs) sit at the interface between innate and adaptive immunity. The main function of DCs is to process antigen intracellularly and present that antigen on its surface to T cells. The term dendritic cells was first coined in 1973 by Steinman and Cohn; Steinman[69] later went on to receive a Nobel prize for this research in 2011. DC subsets are complex. Conventional DCs (cDCs), which originate from a common DC progenitor (CDP) of myeloid origin are subdivided into two populations, cDC1s and cDC2s. Both cDC1s and cDC2s can present antigens on MHCII to CD4+ T cells, whereas only cDC1s can cross-present antigens on MHCI to CD8+ T cells. On the other hand, plasmacytoids (pDCs), which can derive from either myeloid or lymphoid progenitors are highly effective at sensing viruses and rapidly produce large amounts of interferons (IFNs). DCs display robust circadian rhythms,[70] although their circadian features have been studied far less than those of macrophages. DCs migrate to the skin and tissues in a circadian fashion; this was abolished in a model of SCN circadian arrhythmia and is regulated through gradients of CCR7 and CCL21.[38,40,45]

Mice were more capable of expelling the gastrointestinal dwelling parasitic helminth *Trichuris muris* when infected in the morning than the evening. This was due to the induction of a greater T_H2 immune response with morning infection. The T_H2 immune response is partly characterized by the release of IL5, which induces eosinophils, which are highly adept at clearing parasites. The circadian variability in parasite expulsion was lost in mice lacking *Bmal1* in the DC lineage.[35]

11.5.3 Neutrophils

Neutrophils are the most abundant type of leukocytes in the blood and they are the first immune cells recruited to a site of acute inflammation.

Neutrophils quickly respond to trap and kill invading pathogens. They are the predominant white blood cells in humans (50–70%), but comprise a much lower percentage in mice (10–25%).[71] Each day, there is a massive production of neutrophils within the bone marrow and timed release of these cells into blood circulation. This is accomplished by the oscillatory expression of CXCL12 and its cognate receptor CXCR4, which provides a key retention signal for neutrophils and other cells within the bone marrow (BM). During the mouse-resting phase, downregulation of CXCL12 drives the release of neutrophils. Sympathetic nerves innervate the BM to deliver 24-hour adrenergic signals through B3 adrenergic receptors to stromal cells, which inhibit CXCL12. During the active phase, cholinergic signals inhibit this adrenergic activity, and increase CXCL12.[72] While CXCL12 may serve as a BM retention signal, there is evidence that CXCL1/CXCL2, the ligands for CXCR2, act as a release signal for neutrophils. Thus, there is tight regulation of the cues driving circadian neutrophil egress from bone marrow and circadian oscillations in the blood. Once in circulation, neutrophils are short-lived cells (6–10 hours in mice). The neutrophil phenotype changes over this short time window, a process known as neutrophil aging.[73] As neutrophils age, they exhibit increased surface expression of CXCR2 and decreased surface expression of CXCR4. Neutrophil aging is a circadian process controlled at the cell intrinsic level by *Bmal1*. In this instance, it appears that once these cells leave the bone marrow, the molecular clock acts like a timer controlling their aging and clearance in a tightly controlled temporal manner.[74]

Neutrophils infiltrate healthy tissues in a circadian manner, instructing circadian programs in those tissues. Circadian migration of neutrophils into the bone marrow controls hematopoietic niches[75] while their migration into the lung controls the organ's transcriptional activity and function.[76] During bacterial infection in the lung, epithelial cells recruit neutrophils in a circadian manner through the expression of *Cxcl5* and glucocorticoids. Genetic ablation of *Bmal1* in epithelial cells resulted in an exaggerated response.[32] Further work determined that rhythmic pulmonary neutrophilia persisted despite non-rhythmic CXCL5 responses.[77] Thus, neutrophil function and migration are profoundly oscillatory, and further research is required to understand the underlying mechanisms and their consequences for neutrophil-related diseases.

11.5.4 Eosinophils

Eosinophils are largely known for their pathogenic role in allergic asthma, but they also play a role in controlling helminth or worm infections. Eosinophils are recruited to sites of infection and can release cytotoxic proteins such as major basic proteins (MBP) and eosinophil cationic proteins (ECPs). Eosinophil counts display a 24-hour variation in humans[78] and mouse circulation and the rhythm is dependent on adrenal hormones.[12] Rhythmic recruitment of eosinophils to the intestine is mediated by the cytokine IL-5. Feeding induces the hormone vasoactive intestinal peptide (VIP), which innervates vasoactive intestinal polypeptide receptors on type 2 innate lymphoid cells (ILCs) to secrete IL5, which drives the rhythmic accumulation of eosinophils within tissues. This provides a circadian–immune metabolic axis, which may be perturbed in conditions like asthma and allergies.[79] The molecular clock in macrophages and neutrophils dampens IL5 expression, as loss of *Bmal1* in these cells potentiates eosinophil accumulation in allergic lung inflammation.[80]

11.5.5 Mast Cells and Basophils

In addition to eosinophils, mast cells and basophils are the other main effector cells of allergic responses and host defence against parasitic infection. Mast cells are long-lived tissue-resident cells that can release histamine, along with cytokines such as IL-4 and IL-13, which promote T_H2 differentiation, IgE production and lipid mediators such as leukotrienes and prostaglandins. Basophils share many similarities with mast cells. In allergic diseases such as allergic rhinitis, asthma and food allergies, mast cells and basophils are activated by an IgE-dependent mechanism. Therefore, it is interesting that the CLOCK protein modulates the transcription of the high-affinity IgE receptor (FcεRI) in mast cells. Pre-stored histamine and cysteinyl leukotrienes are released in a circadian manner by mast cells following IgE-mediated activation.[81] Pharmacological resetting of the molecular clock in mast cells reduced FcεRI signaling and ameliorated allergic symptoms in a model of allergic rhinitis.[82] Furthermore, factors derived from adrenal glands synchronise the mast cell-intrinsic clockwork, as adrenalectomy abolished mast cell rhythms.[83] Therefore, accruing evidence demonstrates circadian control relating to eosinophils, basophils and mast cell function. Collectively, this provides mechanistic insights into how circadian disruption can promote the development of allergies.[84]

11.5.6 Natural Killer Cells

Natural killer (NK) cells are derived from a common lymphoid progenitor, which gives rise to T and B cells; however, NK cells are classified as part of the innate immune system. They are classed as group 1 innate lymphocytes (ILCs), which will be discussed further below. NK cells, as their name suggests, are best known for killing virally infected cells or cancerous cells. When NK cells make the decision to kill, they do so by releasing cytotoxic granules containing perforin and granzyme, which leads to lysis of the target cell. NK cell numbers display strong 24-hour variations in the circulation, peaking in humans around midday.[85] In rat NK cells, the expression of the key cytokines, IFNγ and TNF, along with the cytolytic factors and activity, also peaks during the middle of the active period.[86–88] Rhythmicity of splenic NK cells is achieved in part through rhythmic neural noradrenergic stimulation.[89] NK cytotoxicity peaks during periods of wakefulness in humans.[90]

Finally, induction of chronic jet lag in rats led to enhanced tumour growth in the lungs due to the dysregulation of the rhythmic pattern of inflammatory mediators and cytotoxic activity of NK cells.[91]

There is now evidence to support significant circadian control of all cells of the innate immune system and across most, if not all, functional aspects of innate immunity. A full account of all the mechanisms underlying these effects still requires significant investigation. However, the mechanisms and their consequences in terms of innate immune function is likely to be unique to each innate immune cell and the microenvironment upon which that cell resides.

11.6 Clock Function within Cells of the Adaptive Immune System

The adaptive immune system is *'younger'* than the innate immune system, as it arose in evolution less than 500 million years ago and is confined to vertebrates. The term lymphoid is typically associated with the adaptive immune system. The main adaptive immune cells or lymphocytes, which we will discuss here, are CD4 and CD8 T cells and B cells (Figure 11.2). The area of circadian adaptive immunity has been reviewed more thoroughly by Downton and colleagues.[92] The adaptive immune system works through clonal selection, wherein both T and B cells express specific receptors to a distinct antigen on MHC and then reproduce to generate a clonal cell population, which eliminates that antigen. Somatic recombination of gene segments, which is essentially a type of stochastic gene rearrangement, ensures broad diversity of these T and B receptors to a range of antigens. Earlier studies clearly showed that the level of T and B cell proliferation in response to concanavalin (a plant lectin, which binds to mannose residues of various glycoproteins and activates lymphocytes) or LPS varied with the timing of stimulation,[93] paving the way for more in-depth studies on circadian function in lymphocytes and adaptive immunity. As previously mentioned, strong circadian oscillations of T and B cell numbers in circulation have been identified. The numbers of both B and T cells peak in the blood in the latter half of the species' respective rest phase.[23] Mouse and human circulating leukocytes exhibit these opposite circadian oscillations in humanized mice, which are mice grafted with the human bone marrow. These rhythms are driven by levels of the chemokine *Cxcr4*, which is controlled by a heterodimer of BMAL1 with HIF-1α[94].

11.6.1 T Cells

T cells are derived from hematopoetic stem cells, which migrate from the bone marrow to the thymus gland (hence their name) to mature. From the thymus, they mature into several distinct subtypes, of which CD8+ "killer" and CD4+ "helper" are the two main subtypes, named due to the presence of CD8 or CD4 on their surface. A key feature of T cells is the presence of the

T cell receptor (TCR). CD8 T cells are often referred to as cytotoxic T lymphocytes (CTLs) and can directly kill virus-infected and cancer cells. Through their TCR, these cells recognise antigen-loaded MHC I molecules on other cells, including DCs. Upon engagement of this receptor, they can kill by releasing cytokines such as TNF and IFNγ, the cytotoxic granules perforin and granzymes and destruction of cells using Fas/FasL interactions.

The CD4+ helper T cells function by indirectly killing cells identified as foreign. They *assist or help* other immune cells, including the activation of CTLs and macrophages along with maturation of B cells into antibody-producing plasma cells. Helper T cells (or T_H cells) become activated when they interact with antigen-presenting cells (APCs) such as dendritic cells expressing cognate antigen on MHC class II receptors. Once activated, they rapidly divide and can differentiate into several subtypes depending on the activating cytokines present. The CD4+ T cell subtypes include T_H1, T_H2, T_H17, T_H22, T_H9, T follicular helper (Tfh) and natural regulatory T cells (nTregs). Each of these subtypes performs specialised functions. For example, T_H1 produces pro-inflammatory cytokines such as IFNγ and TNF, whereas T_H2 produces anti-inflammatory cytokines such as IL10 and promotes the allergic response through secretion of IL4 and IL5. The next two sections will address our current understanding of circadian impacts on CD4+ and CD8+ T cells.

11.6.2 CD4+ T Cells and Subtypes

Both human and mouse CD4+ T cells express clock genes with 24-hour rhythms.[95] However, in terms of T cell development, loss of *Bmal1* does not impact the overall T cell number.[96] A circadian rhythm exists in the response of T cells to antigen. Furthermore, stimulation of human T cells with PMA/ionomycin, which bypasses the T cell membrane receptor complex led to the circadian activation of intracellular signaling pathways and cytokine production (IL-2 and IFNγ).[95,97]

In mice, CD4+ T cells display rhythmic expression of the two transcription factors, NFIL3 and RORγt.[98] Importantly, RORγt, which is encoded by the gene *Rorc*, is the key transcription factor, which drives T_H17 differentiation, whereas NFIL3 is the circadian-regulated basic leucine zipper transcription factor also known as E4BP4. Intriguingly, NFIL3/E4BP4 has been shown to regulate cytokine production in T_H2 and to a lesser extent in T_H1[99,100] and it also plays a critical role in the development of all innate lymphoid cells, including NK cells.[101] Therefore, this axis may be important in terms of circadian control across a number of immune cells and subtypes. In T cells, NFIL3 rhythmically suppresses *Rorc* during the active phase. However, T cells isolated during the rest phase have higher levels of *Rorc* and thus are more prone to differentiate into T_H17 cells. Interestingly, the core circadian clock protein REV-ERBα directly suppresses the transcription of *Nfil3*, thus further connecting the clock circuitry to T_H17 cell differentiation. A chronic jet lag model in mice disturbed this pathway, leading to enhanced proportions of T_H17 cells and rendering the mice more susceptible to experimental colitis. *Clock*$^{(\Delta)19/19}$ mice also displayed reduced frequencies of T_H1 cells.[98] Therefore, it is interesting to note that both CLOCK and REV-ERBα drive the development of T_H1 and T_H17, which are two highly pro-inflammatory T cell subsets. Further work is needed to fully elucidate the role of the endogenous clockwork machinery in terms of CD4 T cell subset development as the outcome will likely depend on the T cell subset in question and the microenvironment in which the cells reside. Both CD4+ and CD8+ T cells enter and exit the lymph nodes in a highly circadian manner that is dependent on endogenous circadian control of the chemokine receptors *Ccr7* (which controls lymphocyte homing to lymph nodes) and *S1pr1* (which controls lymphocyte egress from the lymph nodes). Furthermore, the rise in glucocorticoids ahead of the activity phase is also involved in T cell redistribution into lymph nodes and spleen.[102]

11.6.3 CD8+ T Cells

The CD8+ or cytotoxic T cell response is strongly influenced by the time-of-day and shows 24-hour rhythms in PER2 upon synchronisation.[103] Mice were challenged or immunised with DCs pre-loaded with antigen. A higher response in terms of CD8+ T cell proliferation and IFNγ production was observed when animals were vaccinated in the middle of the rest phase *versus* activity phase.[103] This higher T cell activity translated to higher protection against bacterial challenge to *Listeria monocytogenes*. Transcriptomics of CD8+ T cells harvested from the lymph nodes over two days identified that many rhythmic transcripts were downstream of the TCR, including regulators of the AKT and mTOR pathways. mTOR activation was clearly rhythmic in CD8+ T cells, both basally and in response to vaccination. It is particularly noteworthy that mTOR signaling represents a significant node of circadian T cell control, given that T cells depend on mTOR signaling to sense and integrate immune signals from both cellular and environmental sources.[104]

11.6.4 B Cells

The circadian variation in B cell number in human blood is less pronounced than the other lymphocytes.[105] Nonetheless, there is mounting evidence highlighting the importance of the circadian clock in regulating B cell function. Mice lacking *Bmal1* throughout the body displayed significantly lower numbers of mature B cells in blood, spleen and bone marrow.[106] Adoptive transfer experiments demonstrated that this dysfunction in B cell maturation was not due to the lack of B cell *Bmal1* but actually due to the loss of this circadian gene in the bone marrow microenvironment. There is still a significant gap in our understanding as to how circadian rhythms in the bone marrow niche impacts hematopoeisis. In a separate study, mice deficient in both *Cry1* and *Cry2* (*Cry* double knockout (DKO)) developed an autoimmune phenotype with high levels of serum IgG antibodies and antinuclear antibodies.[107] This was characterized by higher numbers of mature B cells and over-activation of the B cell receptor (BCR) signaling pathway. Further work both in mice and humans is required to understand the circadian control of B cell fate and function and how this might impact disease progression.

11.6.5 Other Immune Cells and Immune Modulation

Over the last decade, the field of cellular immunology has expanded to now include the broad spectrum of innate lymphocyte cells (ILCs). In its simplest definition, ILCs are cells, which belong to the lymphoid lineage, but unlike T and B cells, they do not express antigen-specific receptors. Therefore, ILCs are the innate counterparts of adaptive immune responses, but unlike the adaptive immune system, which takes time to mount a response, ILCs are activated quickly to help fight off invading pathogens. ILCs can now be tentatively classified into five groups. Group 1 are the NK cells discussed above, but the other four groups are ILC1-3 and lymphoid tissue inducer cells.[108] All ILCs are derived from the common lymphoid progenitor (CLP) cell. In simple terms, NK cells are a type of cytotoxic innate lymphocytes, and akin to the CD8+ T cells, and both NK cells and ILC1s guard against intracellular threats and cancer. ILC2s share features with the CD4+ T_H2 cells and protect against extracellular microbes, but can also have pathogenic roles in conditions such as asthma. ILC3s have been shown to express circadian oscillations in clock and cytokine genes and the deletion of *Bmal1* in ILC3s resulted in markedly reduced numbers of ILC3s in the intestine.[109,110] These ILC groups have just recently been discovered; therefore their intersections and connections with the circadian clockwork machinery are relatively unexplored.

Nonetheless, other cells, besides leukocytes such as fibroblasts, epithelial and endothelial cells, perform important immune functions to orchestrate the immune response. For example, fibroblast-like synoviocytes (FLSs) line the inner layer of the synovial membrane within joints. FLSs play a critical role in the pathogenesis of rheumatoid arthritis as they contribute to joint destruction through the production of inflammatory mediators and matrix-degrading molecules. FLSs lacking *Bmal1* produce far greater levels of pro-inflammatory cytokines and chemokines in response to TNF, and lack of *Bmal1* in these cells increases severity of disease in a mouse model of inflammatory arthritis.[111] The molecular clock in epithelial cells of the bronchioles (also known as club cells) modulates bacterial responses by driving rhythmic neutrophil recruitment to the lung.[32] Similarly, the lack of *Bmal1* in these club cells results in an exaggerated inflammatory response.

Endothelial cells also display significant rhythms in adhesion molecules and chemokines, offering a mechanism in which leukocytes are temporally recruited into peripheral tissues.[23] Rhythmic recruitment of leukocytes into tissues was ablated with endothelial-specific deletion of *Bmal1*.[40]

This highlights the vast spectrum of control through which the clockwork mechanism in cells outside of the classical leukocyte lineage works, to assist in the coordination of immune responses.

11.7 Importance of Circadian Immunology in Health and Disease and Potential for Therapeutic Approaches

It is now becoming increasingly clear that clock disruption in immune cells is an aggravating factor in a range of chronic inflammatory conditions such as arthritis, cardiovascular disease and diabetes. This is supported by evidence from animal models of genetic or environmental clock disruption as well as through epidemiological investigations in real-world scenarios. Circadian misalignment is the term used to describe a number of circumstances in which there is misalignment of Zeitgebers, such as inappropriate sleep and wake times, or inappropriately timed feeding with those sleep/wake cycles, or it can occur when there is misalignment of central to peripheral rhythms.[112] Epidemiological studies of shift workers, especially in rotating and permanent night-time shift workers reveal an association of circadian misalignment or disruption with increased susceptibility to chronic inflammatory disease.[113] Long-term night-shift workers had greater numbers of neutrophils, monocytes and lymphocytes in comparison to day-shift workers, strongly suggesting immune dysfunction in this cohort.[114] Transcriptomic analysis of peripheral blood mononuclear cells (PBMCs) from individuals subjected to a 4-day simulated night-shift work protocol showed a significant dampening of rhythmic transcripts along with the dampening of the natural killer cell-mediated response.[115] These studies involving shift workers provide strong evidence of the detrimental effects of circadian misalignment or circadian disruption on our immune system.

11.7.1 Musculoskeletal Disease

Musculoskeletal symptoms such as arthritis are among the most commonly seen in general practice and are well documented to vary with "worst" times[116] and "worst" seasons.[117] In rheumatoid arthritis (RA), the symptoms of early morning stiffness and pain in joints are common (Figure 11.5). This is due to a nightly circadian release of pro-inflammatory cytokines, namely, TNF and IL-6 from synovial macrophages. The collagen-induced arthritis (CIA) murine model of inflammatory arthritis, faithfully demonstrates 24-hour rhythms in disease severity and joint inflammation. Mice show increased symptoms and inflammation at ZT6 ahead of their activity phase, reflecting the human experience. This rhythmic inflammation is driven by a combination of fibroblast-like synoviocytes (FLSs) and infiltrating immune cells. Interestingly, CRY1 and CRY2 repressed inflammation in the FLS, and pharmacological activation of CRY proteins limited the inflammation.[118]

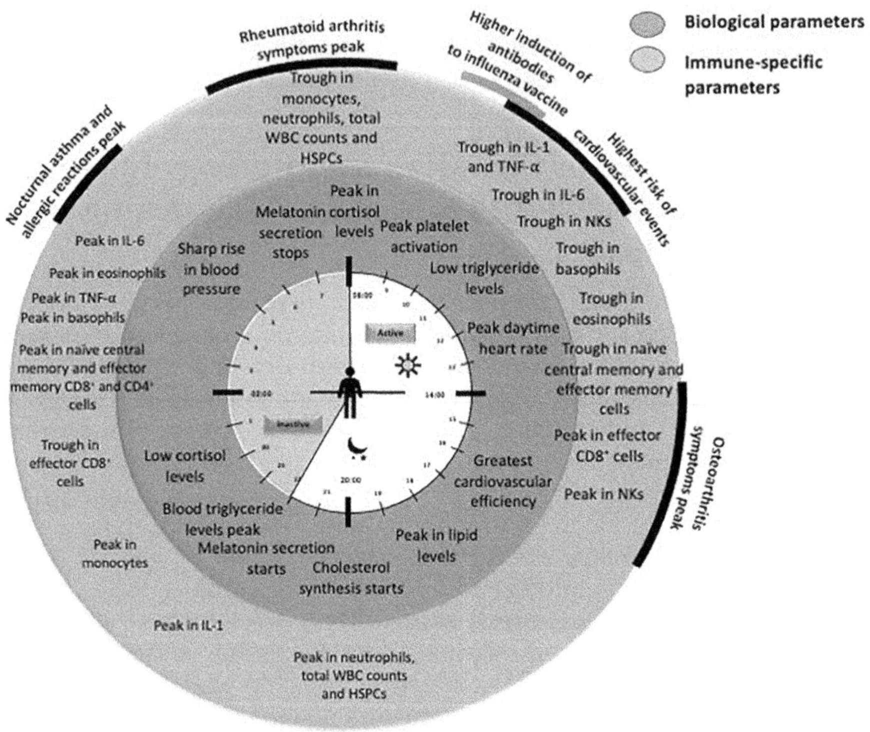

Figure 11.5 Biological parameters, immune parameters and consequent immune-related diseases, which vary with time of day in humans. Reproduced from ref. 122 with permission from Elsevier, Copyright 2020.

As such, direct clock protein modulation in the joint may represent a new therapeutic avenue for RA. Conversely, arthritic joint inflammation is dampened during the active phase, due to an increase in recruitment of anti-inflammatory nTregs into the joint.[119] As mentioned above, nTregs are a subpopulation of T cells that suppress the immune response, and are critical in maintaining homeostasis and self-tolerance. They do this by inhibiting T cell proliferation and cytokine production. The enhanced number of nTregs within the arthritic joint during the active phase suppresses the production of inflammatory mediators from Ly6Chi monocytes. nTregs play a critical role in many autoimmune diseases. Therefore, the identification that these cells are cycling into the joint is particularly intriguing.

The temporal variation in rheumatoid arthritis symptoms prompted early studies to determine if an optimal time for administration of medications existed. Early trials were conducted to explore the chronopharmacology of anti-inflammatory medications such as flurbiprofen and prednisolone in patients with RA.[120,121] Both studies proved the benefit of appropriately timed medication for this disease. Since then, a number of studies have shown the benefit of a chronotherapeutic approach of glucocorticoids, non-steroidal anti-inflammatory drugs (NSAIDs) and methotrexate for RA treatment.[122] A comprehensive review outlining the impact of circadian timing on the treatment of a number of chronic inflammatory diseases, including RA, can be found in ref. 122.

11.7.2 Autoimmune Disease

Multiple Sclerosis (MS) is considered an autoimmune disease in which the immune system destroys myelin. Myelin is the protective fatty substance that coats and wraps around nerve fibres in the brain and spinal cord. Destruction of myelin by the immune system causes nerve damage and this is what leads to issues with mobility. A large population-based study identified that teenagers conducting shift-work have an increased risk to develop MS in later life.[123] Furthermore, incidences of relapse in MS occur most frequently in spring and summer, which has been attributed to lower levels of the clock hormone melatonin.[124] Ligation of melatonin to its receptor (MT1) has two discrete effects. It suppresses the differentiation of T_H17 cells *via* the *Rev-erbα/Nfil3/Rorc* axis, but at the same time promotes the differentiation of IL-10-producing type 1 regulatory Tr1 cells. IL10 is a potent anti-inflammatory cytokine, as it can inhibit the activity of T_H1 cells, NK cells and macrophages. Collectively, levels of melatonin negatively correlate with multiple sclerosis activity in humans.[124] Mouse studies also infer the importance of the circadian immune system in this disease. Experimental autoimmune encephalomyelitis (EAE) is a mouse model of MS in which mice are immunised with myelin/oligodendrocyte glycoprotein (MOG), a minor component of myelin along with complete Freunds adjuvant and pertussis toxin. Immunisation of these components at ZT6 leads to significantly higher levels of disease *versus* ZT18, which corresponds to the time when high numbers of innate and adaptive immune cells are residing in the lymph nodes. The myeloid and T cell clocks were implicated in this response.[42,125]

11.7.3 Infectious Disease

Microorganisms ranging from bacteria, viruses, fungi or parasites can cause disease. We are just beginning to unravel how circadian control of the immune response might impact the host response to infection and consequent disease progression. A range of studies using bacteria, viruses and parasites have shown how the circadian clocks in immune cells influence the host response to infection (Table 11.1).[31,32,35,36,126,127] However, in the case of malaria parasites, they synchronise their replication with the host feeding rhythms, irrespective of the host immune circadian clocks.[128] The importance of our immune circadian clocks and their interactions with pathogens in terms of SARS-CoV-2 infection and COVID-19 disease progression has also been discussed.[129–131] Circadian regulation of ACE2, the receptor, which controls cell entry of SARS-CoV-2, has been observed in epithelial cells. Silencing of *BMAL1* or treatment with the synthetic REV-ERB agonist reduces

ACE2 expression and inhibits viral entry, replication and secretion of infectious particles. Thus, understanding the circadian control of the host response will offer new insights into improving therapeutic targeting strategies against a range of infectious pathogens.

11.7.4 Vaccination

Vaccinations essentially use both the innate and adaptive immune response to generate immunological memory against a specific target pathogen. This provides protection to an individual, as the immune system has been trained to respond to the pathogen, and thus it responds more appropriately and rapidly if the individual is infected with that pathogen. Vaccines have been the single most successful innovation of modern medicine, saving millions of lives against debilitating infectious disease such as small pox, polio, meningitis, and more recently against SARS-CoV-2. There is increasing evidence that the timing of vaccine administration influences the host response. A randomized trial of 276 adults aged above 65 years demonstrated that morning vaccination (9–11 am) for influenza resulted in a stronger antibody response compared to those vaccinated in the afternoon (3–5 pm) (Figure 11.5).[132] In recent years, it has become clear that the innate immune system can create a memory of a pathogen through epigenetic changes.[129] This allows innate immune cells to have an enhanced response to subsequent triggers. An independent study design showed that the time of immunisation influenced the induction of trained immunity. In a cohort of 302 participants, those inoculated with the Bacillus Calmette–Guerin (BCG) vaccine between 8 and 9 am showed stronger trained immunity than those vaccinated between noon and 1 pm.[133] It is also interesting to note that regular sleep after hepatitis A vaccination enhanced the anti-body response.[134,135] Although there are numerous differences between these studies, it is intriguing that both BCG, a live attenuated vaccine, and influenza, a subunit vaccine, showed enhanced efficacy in the morning in humans. More mechanistic work has been performed in mice.[103,136] There is a circadian variation in the T and B cell counts in secondary lymphoid organs, which correlates with the efficacy in response to certain antigens.[42,137] Collectively, these findings provide a biological basis for the circadian influence on vaccine efficacy. It is unlikely that there is one optimal daily administration time for all vaccines. It is more likely that the optimal time of vaccination will depend of the mechanism of action of the vaccine, the route of administration and the patient population. Further investigation is required to fully understand how best we can harness this for vaccine therapeutic gain and this may be relevant to ongoing vaccination programs, including those against COVID-19.[138]

11.8 Final Discussion and Future Directions

Decades of research have now started to uncover the significant contribution of circadian rhythms in immunity, and how this relates to health and disease. The level of coordination controlling this rhythmicity is vast, incorporating the SCN, neuronal/hormonal systems, the innate and adaptive arms of immunity and the surrounding microenvironment. Immune cell molecular clock function and rhythmicity impact every type of immune response and appear important in many, if not all, immune-related conditions and responses (Figure. 11.5). It is now clear that robust synchronised rhythms in our immune system are required for maintaining optimal health and reducing our susceptibility to a number of inflammatory diseases. There is currently intense interest in either using timing as a variable in drug therapy or directly targeting the biological clock and its downstream targets.[122,139] There is a need for more focus on the selection of methodologies to detect rhythmicity and to apply chronobiological knowledge into clinical and other "real-life" situations. To date, the majority of circadian immunology studies have been conducted in mice. Going forward, a more acute focus on translating this knowledge into clinical medicine and practise is needed. This can only be achieved through significant collaboration across a range of disciplines, including circadian biology, immunology, biochemistry, systems biology, and computational biology, along with our clinical colleagues. Promising recent advances such as the ability to track an individual's internal time using gene expression data will aid in translating circadian immunology and may indeed become an integral component of "personalised medicine" in the future.[140,141]

Acknowledgements

This work was funded by a Science Foundation Ireland Career Development Award (17/CDA/4688) and Frontiers for the Future Project Award (20/FFP-P/8636) along with an Irish Research Council Laureate Award (IRCLA/2017/110), all provided to A. M. C.

References

1. J. S. Marshall, R. Warrington, W. Watson and H. L. Kim, An introduction to immunology and immunopathology, *Allergy, Asthma, Clin. Immunol.*, 2018, **14**(Suppl 2), 49.
2. T. Kawasaki and T. Kawai, Toll-like receptor signaling pathways, *Front. Immunol.*, 2014, **5**, 461.
3. A current view on inflammation, *Nat. Immunol.*, 2017, **18**(8), 825.
4. F. A. Bonilla and H. C. Oettgen, Adaptive immunity, *J. Allergy Clin. Immunol.*, 2010, **125**(2 Suppl 2), S33–S40.
5. P. A. Roche and P. Cresswell, Antigen Processing and Presentation Mechanisms in Myeloid Cells, *Microbiol. Spectrum*, 2016, **4**(3), DOI: 10.1128/microbiolspec.MCHD-0008-2015.
6. S. A. Slamanig and M. A. Nolte, The Bone Marrow as Sanctuary for Plasma Cells and Memory T-Cells: Implications for Adaptive Immunity and Vaccinology, *Cells*, 2021, **10**(6), 1508.

7. R. Dahl, Diurnal variation in the number of circulating eosinophil leucocytes in normal controls and asthmatics, *Acta Allergol.*, 1977, **32**(5), 301–303.
8. F. Halberg, M. B. Visscher, E. B. Flink, K. Berge and F. Bock, Diurnal rhythmic changes in blood eosinophil levels in health and in certain diseases, *J. Lancet.*, 1951, **71**(8), 312–319.
9. J. Y. Lee, M. S. Lee, J. S. Lee, F. E. Halberg, J. Halberg and O. Schwartzkopff, *et al.*, A Tribute to Franz Halberg, MD, *Hypertension*, 2015, **66**(6), 1090–1092.
10. J. Aschoff, Exogenous and endogenous components in circadian rhythms, *Cold Spring Harbor Symp. Quant. Biol.*, 1960, **25**, 11–28.
11. C. S. Pittendrigh, Circadian rhythms and the circadian organization of living systems, *Cold Spring Harbor Symp. Quant. Biol.*, 1960, **25**, 159–184.
12. H. E. Brown and T. F. Dougherty, The diurnal variation of blood leucocytes in normal and adrenalectomized mice, *Endocrinology*, 1956, **58**(3), 365–375.
13. G. Fernandes, F. Carandente, E. Halberg, F. Halberg and R. A. Good, Circadian rhythm in activity of lympholytic natural killer cells from spleens of Fischer rats, *J. Immunol.*, 1979, **123**(2), 622–625.
14. R. Pownall and M. S. Knapp, Circadian rhythmicity of delayed hypersensitivity to oxazolone in the rat, *Clin. Sci. Mol. Med.*, 1978, **54**(4), 447–449.
15. F. Halberg, E. A. Johnson, B. W. Brown and J. J. Bittner, Susceptibility rhythm to E. coli endotoxin and bioassay, *Proc. Soc. Exp. Biol. Med.*, 1960, **103**, 142–144.
16. J. R. Cove-Smith, P. Kabler, R. Pownall and M. S. Knapp, Circadian variation in an immune response in man, *Br. Med. J.*, 1978, **2**(6132), 253–254.
17. G. Fernandes, F. Halberg, E. J. Yunis and R. A. Good, Circadian rhythmic plaque-forming cell response of spleens from mice immunized with SRBC, *J. Immunol.*, 1976, **117**(3), 962–966.
18. L. E. Scheving, J. E. Pauly and T. H. Tsai, Circadian fluctuation in plasma proteins of the rat, *Am. J. Physiol.*, 1968, **215**(5), 1096–1101.
19. F. Halberg, G. Cornelissen, G. Katinas, E. V. Syutkina, R. B. Sothern and R. Zaslavskaya, *et al.*, Transdisciplinary unifying implications of circadian findings in the 1950s, *J. Circadian Rhythms*, 2003, **1**(1), 2.
20. D. P. King, Y. Zhao, A. M. Sangoram, L. D. Wilsbacher, M. Tanaka and M. P. Antoch, *et al.*, Positional cloning of the mouse circadian clock gene, *Cell*, 1997, **89**(4), 641–653.
21. J. Husse, A. Leliavski, A. H. Tsang, H. Oster and G. Eichele, The light-dark cycle controls peripheral rhythmicity in mice with a genetically ablated suprachiasmatic nucleus clock, *FASEB J.*, 2014, **28**(11), 4950–4960.
22. S. Vandevyver, L. Dejager, J. Tuckermann and C. Libert, New insights into the anti-inflammatory mechanisms of glucocorticoids: an emerging role for glucocorticoid-receptor-mediated transactivation, *Endocrinology*, 2013, **154**(3), 993–1007.

23. C. Scheiermann, Y. Kunisaki, D. Lucas, A. Chow, J. E. Jang and D. Zhang, *et al.*, Adrenergic nerves govern circadian leukocyte recruitment to tissues, *Immunity*, 2012, **37**(2), 290–301.
24. R. G. Carroll, G. A. Timmons, M. P. Cervantes-Silva, O. D. Kennedy and A. M. Curtis, Immunometabolism around the Clock, *Trends Mol. Med.*, 2019, **25**(7), 612–625.
25. B. Boukelia, E. C. Gomes and G. D. Florida-James, Diurnal Variation in Physiological and Immune Responses to Endurance Sport in Highly Trained Runners in a Hot and Humid Environment, *Oxid. Med. Cell. Longevity*, 2018, **2018**, 3402143.
26. C. Scheiermann, J. Gibbs, L. Ince and A. Loudon, Clocking in to immunity, *Nat. Rev. Immunol.*, 2018, **18**(7), 423–437.
27. R. D. Feigin, V. H. San Joaquin, M. W. Haymond and R. G. Wyatt, Daily periodicity of susceptibility of mice to pneumococcal infection, *Nature*, 1969, **224**(5217), 379–380.
28. P. G. Shackelford and R. D. Feigin, Periodicity of susceptibility to pneumococcal infection: influence of light and adrenocortical secretions, *Science*, 1973, **182**(4109), 285–287.
29. A. M. Curtis, C. T. Fagundes, G. Yang, E. M. Palsson-McDermott, P. Wochal and A. F. McGettrick, *et al.*, Circadian control of innate immunity in macrophages by miR-155 targeting Bmal1, *Proc. Natl. Acad. Sci. U. S. A.*, 2015, **112**(23), 7231–7236.
30. V. Lang, S. Ferencik, B. Ananthasubramaniam, A. Kramer and B. Maier, Susceptibility rhythm to bacterial endotoxin in myeloid clock-knockout mice, *eLife*, 2021, **10**, e62469.
31. M. M. Bellet, E. Deriu, J. Z. Liu, B. Grimaldi, C. Blaschitz and M. Zeller, *et al.*, Circadian clock regulates the host response to Salmonella, *Proc. Natl. Acad. Sci. U. S. A.*, 2013, **110**(24), 9897–9902.
32. J. Gibbs, L. Ince, L. Matthews, J. Mei, T. Bell and N. Yang, *et al.*, An epithelial circadian clock controls pulmonary inflammation and glucocorticoid action, *Nat. Med.*, 2014, **20**(8), 919–926.
33. R. S. Edgar, A. Stangherlin, A. D. Nagy, M. P. Nicoll, S. Efstathiou and J. S. O'Neill, *et al.*, Cell autonomous regulation of herpes and influenza virus infection by the circadian clock, *Proc. Natl. Acad. Sci. U. S. A.*, 2016, **113**(36), 10085–10090.
34. S. Sengupta, S. Y. Tang, J. C. Devine, S. T. Anderson, S. Nayak and S. L. Zhang, *et al.*, Circadian control of lung inflammation in influenza infection, *Nat. Commun.*, 2019, **10**(1), 4107.
35. T. W. Hopwood, S. Hall, N. Begley, R. Forman, S. Brown and R. Vonslow, *et al.*, The circadian regulator BMAL1 programmes responses to parasitic worm infection via a dendritic cell clock, *Sci. Rep.*, 2018, **8**(1), 3782.
36. S. Kiessling, G. Dubeau-Laramee, H. Ohm, N. Labrecque, M. Olivier and N. Cermakian, The circadian clock in immune cells controls the magnitude of Leishmania parasite infection, *Sci. Rep.*, 2017, **7**(1), 10892.
37. E. Haus and M. H. Smolensky, Biologic rhythms in the immune system, *Chronobiol. Int.*, 1999, **16**(5), 581–622.

38. B. J. Prendergast, E. J. Cable, P. N. Patel, L. M. Pyter, K. G. Onishi and T. J. Stevenson, *et al.*, Impaired leukocyte trafficking and skin inflammatory responses in hamsters lacking a functional circadian system, *Brain, Behav., Immun.*, 2013, **32**, 94–104.
39. L. S. Mure, H. D. Le, G. Benegiamo, M. W. Chang, L. Rios and N. Jillani, *et al.*, Diurnal transcriptome atlas of a primate across major neural and peripheral tissues, *Science*, 2018, **359**(6381), eaao0318.
40. W. He, S. Holtkamp, S. M. Hergenhan, K. Kraus, A. de Juan and J. Weber, *et al.*, Circadian Expression of Migratory Factors Establishes Lineage-Specific Signatures that Guide the Homing of Leukocyte Subsets to Tissues, *Immunity*, 2018, **49**(6), 1175–90.e7.
41. S. J. Holtkamp, L. M. Ince, C. Barnoud, M. T. Schmitt, F. Sinturel and V. Pilorz, *et al.*, Circadian clocks guide dendritic cells into skin lymphatics, *Nat. Immunol.*, 2021, **22**(11), 1375–1381.
42. D. Druzd, O. Matveeva, L. Ince, U. Harrison, W. He and C. Schmal, *et al.*, Lymphocyte Circadian Clocks Control Lymph Node Trafficking and Adaptive Immune Responses, *Immunity*, 2017, **46**(1), 120–132.
43. P. Italiani and D. Boraschi, From Monocytes to M1/M2 Macrophages: Phenotypical vs. Functional Differentiation, *Front. Immunol.*, 2014, **5**, 514.
44. J. Born, T. Lange, K. Hansen, M. Molle and H. L. Fehm, Effects of sleep and circadian rhythm on human circulating immune cells, *J. Immunol.*, 1997, **158**(9), 4454–4464.
45. K. D. Nguyen, S. J. Fentress, Y. Qiu, K. Yun, J. S. Cox and A. Chawla, Circadian gene Bmal1 regulates diurnal oscillations of Ly6C(hi) inflammatory monocytes, *Science*, 2013, **341**(6153), 1483–1488.
46. G. Hoeffel and F. Ginhoux, Ontogeny of Tissue-Resident Macrophages, *Front. Immunol.*, 2015, **6**, 486.
47. T. Kawai and S. Akira, The role of pattern-recognition receptors in innate immunity: update on Toll-like receptors, *Nat. Immunol.*, 2010, **11**(5), 373–384.
48. M. K. Vidya, V. G. Kumar, V. Sejian, M. Bagath, G. Krishnan and R. Bhatta, Toll-like receptors: Significance, ligands, signaling pathways, and functions in mammals, *Int. Rev. Immunol.*, 2018, **37**(1), 20–36.
49. F. Halberg, E. A. Johnson, B. W. Brown and J. J. Bittner, Susceptibility Rhythm to E. coli Endotoxin and Bioassay, *Exp. Biol. Med.*, 1960, **103**(1), 142–144.
50. C. Silver Adam, A. Arjona, E. Walker Wendy and E. Fikrig, The Circadian Clock Controls Toll-like Receptor 9-Mediated Innate and Adaptive Immunity, *Immunity*, 2012, **36**(2), 251–261.
51. M. Keller, J. Mazuch, U. Abraham, G. D. Eom, E. D. Herzog and H. D. Volk, *et al.*, A circadian clock in macrophages controls inflammatory immune responses, *Proc. Natl. Acad. Sci. U. S. A.*, 2009, **106**(50), 21407–21412.
52. J. H. Bell, A. H. Herrera, Y. Li and B. Walcheck, Role of ADAM17 in the ectodomain shedding of TNF-α and its receptors by neutrophils and macrophages, *J. Leukocyte Biol.*, 2007, **82**(1), 173–176.

53. W. Deng, S. Zhu, L. Zeng, J. Liu, R. Kang and M. Yang, *et al.*, The Circadian Clock Controls Immune Checkpoint Pathway in Sepsis, *Cell Rep.*, 2018, **24**(2), 366–378.
54. J. O. Early, D. Menon, C. A. Wyse, M. P. Cervantes-Silva, Z. Zaslona and R. G. Carroll, *et al.*, Circadian clock protein BMAL1 regulates IL-1beta in macrophages via NRF2., *Proc. Natl. Acad. Sci. U. S. A.*, 2018, **115**(36), E8460–E8468.
55. Y. Oishi, S. Hayashi, T. Isagawa, M. Oshima, A. Iwama and S. Shimba, *et al.*, Bmal1 regulates inflammatory responses in macrophages by modulating enhancer RNA transcription, *Sci. Rep.*, 2017, **7**(1), 7086.
56. R. K. Alexander, Y. H. Liou, N. H. Knudsen, K. A. Starost, C. Xu and A. L. Hyde, *et al.*, Bmal1 integrates mitochondrial metabolism and macrophage activation, *eLife*, 2020, **9**, e54090.
57. G. A. Timmons, R. G. Carroll, J. R. O'Siorain, M. P. Cervantes-Silva, L. E. Fagan and S. L. Cox, *et al.*, The Circadian Clock Protein BMAL1 Acts as a Metabolic Sensor In Macrophages to Control the Production of Pro IL-1beta, *Front. Immunol.*, 2021, **12**, 700431.
58. M. M. Bellet, E. Deriu, J. Z. Liu, B. Grimaldi, C. Blaschitz and M. Zeller, *et al.*, Circadian clock regulates the host response to Salmonella, *Proc. Natl. Acad. Sci. U. S. A.*, 2013, **110**(24), 9897–9902.
59. J. E. Gibbs, J. Blaikley, S. Beesley, L. Matthews, K. D. Simpson and S. H. Boyce, *et al.*, The nuclear receptor REV-ERBalpha mediates circadian regulation of innate immunity through selective regulation of inflammatory cytokines, *Proc. Natl. Acad. Sci. U. S. A.*, 2012, **109**(2), 582–587.
60. S. S. T. Sato, J. Ogasawara, M. Takahashi, T. Izawa, K. Imaizumi, N. Taniguchi, H. Ohno and T. Kizaki, A Circadian Clock Gene, Reverbα, Modulates the Inflammatory Function of Macrophages through the Negative Regulation of Ccl2 Expression, *J. Immunol.*, 2014, **192**(1), 407–417.
61. M. T. Lam, H. Cho, H. P. Lesch, D. Gosselin, S. Heinz and Y. Tanaka-Oishi, *et al.*, Rev-Erbs repress macrophage gene expression by inhibiting enhancer-directed transcription, *Nature*, 2013, **498**(7455), 511–515.
62. B. Pourcet, M. Zecchin, L. Ferri, J. Beauchamp, S. Sitaula and C. Billon, *et al.*, Nuclear Receptor Subfamily 1 Group D Member 1 Regulates Circadian Activity of NLRP3 Inflammasome to Reduce the Severity of Fulminant Hepatitis in Mice, *Gastroenterology*, 2018, **154**(5), 1449–64.e20.
63. R. Narasimamurthy, M. Hatori, S. K. Nayak, F. Liu, S. Panda and I. M. Verma, Circadian clock protein cryptochrome regulates the expression of proinflammatory cytokines, *Proc. Natl. Acad. Sci. U. S. A.*, 2012, **109**(31), 12662–12667.
64. H. Xu, H. Li, S. L. Woo, S. M. Kim, V. R. Shende and N. Neuendorff, *et al.*, Myeloid cell-specific disruption of Period1 and Period2 exacerbates diet-induced inflammation and insulin resistance, *J. Biol. Chem.*, 2014, **289**(23), 16374–16388.

65. J. Oliva-Ramirez, M. M. Moreno-Altamirano, B. Pineda-Olvera, P. Cauich-Sanchez and F. J. Sanchez-Garcia, Crosstalk between circadian rhythmicity, mitochondrial dynamics and macrophage bactericidal activity, *Immunology*, 2014, **143**(3), 490–497.
66. S. S. Geiger, A. M. Curtis, L. A. J. O'Neill and R. M. Siegel, Daily variation in macrophage phagocytosis is clock-independent and dispensable for cytokine production, *Immunology*, 2019, **157**(2), 122–136.
67. G. B. Kitchen, P. S. Cunningham, T. M. Poolman, M. Iqbal, R. Maidstone and M. Baxter, *et al.*, The clock gene Bmal1 inhibits macrophage motility, phagocytosis, and impairs defense against pneumonia, *Proc. Natl. Acad. Sci. U. S. A.*, 2020, **117**(3), 1543–1551.
68. G. A. Timmons, J. R. O'Siorain, O. D. Kennedy, A. M. Curtis and J. O. Early, Innate Rhythms: Clocks at the Center of Monocyte and Macrophage Function, *Front. Immunol.*, 2020, **11**, 1743.
69. R. M. Steinman and Z. A. Cohn, Identification of a novel cell type in peripheral lymphoid organs of mice. I. Morphology, quantitation, tissue distribution, *J. Exp. Med.*, 1973, **137**(5), 1142–1162.
70. A. C. Silver, A. Arjona, M. E. Hughes, M. N. Nitabach and E. Fikrig, Circadian expression of clock genes in mouse macrophages, dendritic cells, and B cells, *Brain, Behav., Immun.*, 2012, **26**(3), 407–413.
71. A. Hidalgo, E. R. Chilvers, C. Summers and L. Koenderman, The Neutrophil Life Cycle, *Trends Immunol.*, 2019, **40**(7), 584–597.
72. A. Aroca-Crevillen, J. M. Adrover and A. Hidalgo, Circadian Features of Neutrophil Biology, *Front. Immunol.*, 2020, **11**, 576.
73. J. M. Adrover, J. A. Nicolas-Avila and A. Hidalgo, Aging: A Temporal Dimension for Neutrophils, *Trends Immunol.*, 2016, **37**(5), 334–345.
74. J. M. Adrover, C. Del Fresno, G. Crainiciuc, M. I. Cuartero, M. Casanova-Acebes and L. A. Weiss, *et al.*, A Neutrophil Timer Coordinates Immune Defense and Vascular Protection, *Immunity*, 2019, **50**(2), 390–402.e10.
75. M. Casanova-Acebes, C. Pitaval, L. A. Weiss, C. Nombela-Arrieta, R. Chevre and N. A. G., *et al.*, Rhythmic modulation of the hematopoietic niche through neutrophil clearance, *Cell*, 2013, **153**(5), 1025–1035.
76. M. Casanova-Acebes, J. A. Nicolas-Avila, J. L. Li, S. Garcia-Silva, A. Balachander and A. Rubio-Ponce, *et al.*, Neutrophils instruct homeostatic and pathological states in naive tissues, *J. Exp. Med.*, 2018, **215**(11), 2778–2795.
77. L. M. Ince, Z. Zhang, S. Beesley, R. M. Vonslow, B. R. Saer and L. C. Matthews, *et al.*, Circadian variation in pulmonary inflammatory responses is independent of rhythmic glucocorticoid signaling in airway epithelial cells, *FASEB J.*, 2019, **33**(1), 126–139.
78. J. D. Acland and A. H. Gould, Normal variation in the count of circulating eosinophils in man, *J. Physiol.*, 1956, **133**(2), 456–466.
79. J. C. Nussbaum, S. J. Van Dyken, J. von Moltke, L. E. Cheng, A. Mohapatra and A. B. Molofsky, *et al.*, Type 2 innate lymphoid cells control eosinophil homeostasis, *Nature*, 2013, **502**(7470), 245–248.

80. Z. Zaslona, S. Case, J. O. Early, S. J. Lalor, R. M. McLoughlin and A. M. Curtis, *et al.*, The circadian protein BMAL1 in myeloid cells is a negative regulator of allergic asthma, *Am. J. Physiol.: Lung Cell. Mol. Physiol.*, 2017, **312**(6), L855–L860.
81. A. Baumann, S. Gonnenwein, S. C. Bischoff, H. Sherman, N. Chapnik and O. Froy, *et al.*, The circadian clock is functional in eosinophils and mast cells, *Immunology*, 2013, **140**(4), 465–474.
82. Y. Nakamura, N. Nakano, K. Ishimaru, N. Ando, R. Katoh and K. Suzuki-Inoue, *et al.*, Inhibition of IgE-mediated allergic reactions by pharmacologically targeting the circadian clock, *J. Allergy Clin. Immunol.*, 2016, **137**(4), 1226–1235.
83. Y. Nakamura, N. Nakano, K. Ishimaru, M. Hara, T. Ikegami and Y. Tahara, *et al.*, Circadian regulation of allergic reactions by the mast cell clock in mice, *J. Allergy Clin. Immunol.*, 2014, **133**(2), 568–575.
84. A. Nakao, Circadian Regulation of the Biology of Allergic Disease: Clock Disruption Can Promote Allergy, *Front. Immunol.*, 2020, **11**, 1237.
85. C. A. Beam, C. Wasserfall, A. Woodwyk, M. Akers, H. Rauch and T. Blok, *et al.*, Synchronization of the Normal Human Peripheral Immune System: A Comprehensive Circadian Systems Immunology Analysis, *Sci. Rep.*, 2020, **10**(1), 672.
86. A. Arjona and D. K. Sarkar, Circadian oscillations of clock genes, cytolytic factors, and cytokines in rat NK cells, *J. Immunol.*, 2005, **174**(12), 7618–7624.
87. A. Arjona and D. K. Sarkar, The circadian gene mPer2 regulates the daily rhythm of IFN-gamma, *J. Interferon Cytokine Res.*, 2006, **26**(9), 645–649.
88. A. Arjona and D. K. Sarkar, Evidence supporting a circadian control of natural killer cell function, *Brain, Behav., Immun.*, 2006, **20**(5), 469–476.
89. R. W. Logan, A. Arjona and D. K. Sarkar, Role of sympathetic nervous system in the entrainment of circadian natural-killer cell function., *Brain, Behav., Immun.*, 2011, **25**(1), 101–109.
90. G. Gatti, D. Del Ponte, R. Cavallo, M. L. Sartori, A. Salvadori and R. Carignola, *et al.*, Circadian changes in human natural killer-cell activity, *Prog. Clin. Biol. Res.*, 1987, **227A**, 399–409.
91. R. W. Logan, C. Zhang, S. Murugan, S. O'Connell, D. Levitt and A. M. Rosenwasser, *et al.*, Chronic shift-lag alters the circadian clock of NK cells and promotes lung cancer growth in rats, *J. Immunol.*, 2012, **188**(6), 2583–2591.
92. P. Downton, J. O. Early and J. E. Gibbs, Circadian rhythms in adaptive immunity, *Immunology*, 2020, **161**(4), 268–277.
93. A. I. Esquifino, L. Selgas, A. Arce, V. D. Maggiore and D. P. Cardinali, Twenty-four-hour rhythms in immune responses in rat submaxillary lymph nodes and spleen: effect of cyclosporine, *Brain, Behav., Immun.*, 1996, **10**(2), 92–102.
94. Y. Zhao, M. Liu, X. Y. Chan, S. Y. Tan, S. Subramaniam and Y. Fan, *et al.*, Uncovering the mystery of opposite circadian rhythms between

mouse and human leukocytes in humanized mice, *Blood*, 2017, **130**(18), 1995–2005.

95. T. Bollinger, A. Leutz, A. Leliavski, L. Skrum, J. Kovac and L. Bonacina, *et al.*, Circadian clocks in mouse and human CD4+ T cells, *PLoS One*, 2011, **6**(12), e29801.
96. S. Hemmers and A. Y. Rudensky, The Cell-Intrinsic Circadian Clock Is Dispensable for Lymphocyte Differentiation and Function, *Cell Rep.*, 2015, **11**(9), 1339–1349.
97. M. Cuesta, P. Boudreau, G. Dubeau-Laramee, N. Cermakian and D. B. Boivin, Simulated Night Shift Disrupts Circadian Rhythms of Immune Functions in Humans, *J. Immunol.*, 2016, **196**(6), 2466–2475.
98. X. Yu, D. Rollins, K. A. Ruhn, J. J. Stubblefield, C. B. Green and M. Kashiwada, *et al.*, TH17 cell differentiation is regulated by the circadian clock, *Science*, 2013, **342**(6159), 727–730.
99. M. Kashiwada, S. L. Cassel, J. D. Colgan and P. B. Rothman, NFIL3/E4BP4 controls type 2 T helper cell cytokine expression, *EMBO J.*, 2011, **30**(10), 2071–2082.
100. Y. Motomura, H. Kitamura, A. Hijikata, Y. Matsunaga, K. Matsumoto and H. Inoue, *et al.*, The transcription factor E4BP4 regulates the production of IL-10 and IL-13 in CD4+ T cells, *Nat. Immunol.*, 2011, **12**(5), 450–459.
101. T. Kostrzewski, A. J. Borg, Y. Meng, I. Filipovic, V. Male and A. Wack, *et al.*, Multiple Levels of Control Determine How E4bp4/Nfil3 Regulates NK Cell Development, *J. Immunol.*, 2018, **200**(4), 1370–1381.
102. A. Shimba, G. Cui, S. Tani-Ichi, M. Ogawa, S. Abe and F. Okazaki, *et al.*, Glucocorticoids Drive Diurnal Oscillations in T Cell Distribution and Responses by Inducing Interleukin-7 Receptor and CXCR4, *Immunity*, 2018, **48**(2), 286–98.e6.
103. C. C. Nobis, G. Dubeau Laramee, L. Kervezee, D. Maurice De Sousa, N. Labrecque and N. Cermakian, The circadian clock of CD8 T cells modulates their early response to vaccination and the rhythmicity of related signaling pathways, *Proc. Natl. Acad. Sci. U. S. A.*, 2019, **116**(40), 20077–20086.
104. H. Chi, Regulation and function of mTOR signalling in T cell fate decisions, *Nat. Rev. Immunol.*, 2012, **12**(5), 325–338.
105. G. Mazzoccoli, R. B. Sothern, A. De Cata, F. Giuliani, A. Fontana and M. Copetti, *et al.*, A timetable of 24 hours patterns for human lymphocyte subpopulations, *J. Biol. Regul. Homeostatic Agents*, 2011, **25**(3), 387–395.
106. Y. Sun, Z. Yang, Z. Niu, J. Peng, Q. Li and W. Xiong, *et al.*, MOP3, a component of the molecular clock, regulates the development of B cells, *Immunology*, 2006, **119**(4), 451–460.
107. Q. Cao, X. Zhao, J. Bai, S. Gery, H. Sun and D. C. Lin, *et al.*, Circadian clock cryptochrome proteins regulate autoimmunity, *Proc. Natl. Acad. Sci. U. S. A.*, 2017, **114**(47), 12548–12553.
108. E. R. Kansler and M. O. Li, Innate lymphocytes-lineage, localization and timing of differentiation, *Cell. Mol. Immunol.*, 2019, **16**(7), 627–633.

109. F. Teng, J. Goc, L. Zhou, C. Chu, M. A. Shah and G. Eberl, *et al.*, A circadian clock is essential for homeostasis of group 3 innate lymphoid cells in the gut, *Sci. Immunol.*, 2019, **4**(40), eaax1215.
110. Q. Wang, M. L. Robinette, C. Billon, P. L. Collins, J. K. Bando and J. L. Fachi, *et al.*, Circadian rhythm-dependent and circadian rhythm-independent impacts of the molecular clock on type 3 innate lymphoid cells, *Sci. Immunol.*, 2019, **4**(40), eaay7501.
111. L. E. Hand, S. H. Dickson, A. J. Freemont, D. W. Ray and J. E. Gibbs, The circadian regulator Bmal1 in joint mesenchymal cells regulates both joint development and inflammatory arthritis, *Arthritis Res. Ther.*, 2019, **21**(1), 5.
112. K. G. Baron and K. J. Reid, Circadian misalignment and health, *Int. Rev. Psychiatry*, 2014, **26**(2), 139–154.
113. G. Kecklund and J. Axelsson, Health consequences of shift work and insufficient sleep, *BMJ*, 2016, **355**, i5210.
114. M. D. Wirth, M. E. Andrew, C. M. Burchfiel, J. B. Burch, D. Fekedulegn and T. A. Hartley, *et al.*, Association of shiftwork and immune cells among police officers from the Buffalo Cardio-Metabolic Occupational Police Stress study, *Chronobiol. Int.*, 2017, **34**(6), 721–731.
115. L. Kervezee, M. Cuesta, N. Cermakian and D. B. Boivin, Simulated night shift work induces circadian misalignment of the human peripheral blood mononuclear cell transcriptome, *Proc. Natl. Acad. Sci. U. S. A.*, 2018, **115**(21), 5540–5545.
116. M. Cutolo, Glucocorticoids and chronotherapy in rheumatoid arthritis, *RMD Open*, 2016, **2**(1), e000203.
117. H. Mori, T. Sawada, S. Nishiyama, K. Shimada, K. Tahara and H. Hayashi, *et al.*, Influence of seasonal changes on disease activity and distribution of affected joints in rheumatoid arthritis, *BMC Musculoskeletal Disord.*, 2019, **20**(1), 30.
118. L. E. Hand, T. W. Hopwood, S. H. Dickson, A. L. Walker, A. S. Loudon and D. W. Ray, *et al.*, The circadian clock regulates inflammatory arthritis, *FASEB J.*, 2016, **30**(11), 3759–3770.
119. L. E. Hand, K. J. Gray, S. H. Dickson, D. A. Simpkins, D. W. Ray and J. E. Konkel, *et al.*, Regulatory T cells confer a circadian signature on inflammatory arthritis, *Nat. Commun.*, 2020, **11**(1), 1658.
120. I. C. Kowanko, R. Pownall, M. S. Knapp, A. J. Swannell and P. G. Mahoney, Circadian variations in the signs and symptoms of rheumatoid arthritis and in the therapeutic effectiveness of flurbiprofen at different times of day, *Br. J. Clin. Pharmacol.*, 1981, **11**(5), 477–484.
121. I. C. Kowanko, R. Pownall, M. S. Knapp, A. J. Swannell and P. G. Mahoney, Time of day of prednisolone administration in rheumatoid arthritis, *Ann. Rheum. Dis.*, 1982, **41**(5), 447–452.
122. H. Jacob, A. M. Curtis and C. J. Kearney, Therapeutics on the clock: Circadian medicine in the treatment of chronic inflammatory diseases, *Biochem. Pharmacol.*, 2020, **182**, 114254.

123. A. K. Hedstrom, T. Akerstedt, J. Hillert, T. Olsson and L. Alfredsson, Shift work at young age is associated with increased risk for multiple sclerosis, *Ann. Neurol.*, 2011, **70**(5), 733–741.
124. M. F. Farez, I. D. Mascanfroni, S. P. Mendez-Huergo, A. Yeste, G. Murugaiyan and L. P. Garo, *et al.*, Melatonin Contributes to the Seasonality of Multiple Sclerosis Relapses, *Cell*, 2015, **162**(6), 1338–1352.
125. C. E. Sutton, C. M. Finlay, M. Raverdeau, J. O. Early, J. DeCourcey and Z. Zaslona, *et al.*, Loss of the molecular clock in myeloid cells exacerbates T cell-mediated CNS autoimmune disease, *Nat. Commun.*, 2017, **8**(1), 1923.
126. F. Rijo-Ferreira, T. Carvalho, C. Afonso, M. Sanches-Vaz, R. M. Costa and L. M. Figueiredo, *et al.*, Sleeping sickness is a circadian disorder, *Nat. Commun.*, 2018, **9**(1), 62.
127. X. Zhuang, D. Forde, S. Tsukuda, V. D'Arienzo, L. Mailly and J. M. Harris, *et al.*, Circadian control of hepatitis B virus replication, *Nat. Commun.*, 2021, **12**(1), 1658.
128. K. F. Prior, D. R. van der Veen, A. J. O'Donnell, K. Cumnock, D. Schneider and A. Pain, *et al.*, Timing of host feeding drives rhythms in parasite replication, *PLoS Pathog.*, 2018, **14**(2), e1006900.
129. M. G. Netea, J. Dominguez-Andres, L. B. Barreiro, T. Chavakis, M. Divangahi and E. Fuchs, *et al.*, Defining trained immunity and its role in health and disease, *Nat. Rev. Immunol.*, 2020, **20**(6), 375–388.
130. S. Sengupta, T. G. Brooks, G. R. Grant and G. A. FitzGerald, Accounting for Time: Circadian Rhythms in the Time of COVID-19, *J. Biol. Rhythms*, 2021, **36**(1), 4–8.
131. S. Sengupta, L. Ince, F. Sartor, H. Borrmann, X. Zhuang and A. Naik, *et al.*, Clocks, Viruses, and Immunity: Lessons for the COVID-19 Pandemic, *J. Biol. Rhythms*, 2021, **36**(1), 23–34.
132. J. E. Long, M. T. Drayson, A. E. Taylor, K. M. Toellner, J. M. Lord and A. C. Phillips, Morning vaccination enhances antibody response over afternoon vaccination: A cluster-randomised trial, *Vaccine*, 2016, **34**(24), 2679–2685.
133. L. C. J. de Bree, V. P. Mourits, V. A. Koeken, S. J. Moorlag, R. Janssen and L. Folkman, *et al.*, Circadian rhythm influences induction of trained immunity by BCG vaccination, *J. Clin. Invest.*, 2020, **130**(10), 5603–5617.
134. T. Lange, S. Dimitrov, T. Bollinger, S. Diekelmann and J. Born, Sleep after vaccination boosts immunological memory, *J. Immunol.*, 2011, **187**(1), 283–290.
135. T. Lange, B. Perras, H. L. Fehm and J. Born, Sleep enhances the human antibody response to hepatitis A vaccination, *Psychosom. Med.*, 2003, **65**(5), 831–835.
136. A. C. Silver, A. Arjona, W. E. Walker and E. Fikrig, The circadian clock controls toll-like receptor 9-mediated innate and adaptive immunity, *Immunity*, 2012, **36**(2), 251–261.
137. K. Suzuki, Y. Hayano, A. Nakai, F. Furuta and M. Noda, Adrenergic control of the adaptive immune response by diurnal lymphocyte

recirculation through lymph nodes, *J. Exp. Med.*, 2016, **213**(12), 2567–2574.

138. W. Wang, P. Balfe, D. W. Eyre, S. F. Lumley, D. O'Donnell and F. Warren, *et al.*, Time of Day of Vaccination Affects SARS-CoV-2 Antibody Responses in an Observational Study of Health Care Workers, *J. Biol. Rhythms*, 2022, **37**(1), 124–129.
139. R. F. N. Ribeiro, C. Cavadas and M. M. C. Silva, Small-molecule modulators of the circadian clock: Pharmacological potentials in circadian-related diseases, *Drug Discovery Today*, 2021, **26**(7), 1620–1641.
140. N. Wittenbrink, B. Ananthasubramaniam, M. Munch, B. Koller, B. Maier and C. Weschke, *et al.*, High-accuracy determination of internal circadian time from a single blood sample, *J. Clin. Invest.*, 2018, **128**(9), 3826–3839.
141. G. Wu, M. D. Ruben, L. J. Francey, D. F. Smith, J. D. Sherrill and J. E. Oblong, *et al.*, A population-based gene expression signature of molecular clock phase from a single epidermal sample, *Genome Med.*, 2020, **12**(1), 73.

CHAPTER 12

The Haematopoietic System

OLE D. LAERUM[a,b]

[a] The Finsen Laboratory and Department of Radiation Biology, Rigshospitalet/University of Copenhagen, Copenhagen, Denmark;
[b] Institute of Clinical Research, The Gade Laboratory of Pathology, Haukeland University Hospital/University of Bergen, Bergen, Norway
Email: ole.laerum@uib.no

12.1 Introduction

The production of blood cells involves two parallel processes: few cells give rise to many, and few immature cells develop into mature cells of several types. It has been known for many decades that a stem cell can give rise to a progeny of a million cells or more within a week or two, depending on the species. In addition, the pool of stem cells has a high capacity to renew its own population, whereby a steady state can be maintained, even under conditions of high demand for mature blood cells.

12.2 Haematopoiesis: A Large Organ and a Generalized System All Over the Body

Using simple dissection techniques on human cadavers, the magnitude of this "organ" within the red bone marrow in the skeleton was reported in the 1920s to be between 1.6 and 3.7 kg.[1] As an organ, which comprises roughly 4 percent of the body weight, the haematopoietic tissue as a whole seems to be larger than both the liver and the brain.

Once the mature blood cells are produced, they spend a period in the bone marrow before being released into the peripheral blood, from which the

Chemical Biology No. 23
Chronobiology and Chronomedicine: From Molecular and Cellular Mechanisms to Whole Body Interdigitating Networks
Edited by Germaine Cornelissen and Tsuyoshi Hirota

Published by the Royal Society of Chemistry, www.rsc.org

leukocytes later migrate to the tissues. The rate of production of blood cells is so high that the whole bone marrow is replaced by new cells about two times every month (Figure 12.1).

It has long been known that the rate of production of blood cells is not constant, but occurs in rhythmic patterns, related to day and night, month and season and with further alterations according to the age of the individual.[2] Such biological rhythms have even been observed during embryonic life.[3]

12.3 Some Historical Remarks

When conducting longitudinal studies on the cellular composition of peripheral blood in patients with mental diseases, Rud[4] reported in 1947 that eosinophil granulocytes exhibited variations over 24 hours with the maximum in the early morning and the minimum in the afternoon and early evening. A few years later, the occurrence of such patterns was verified in

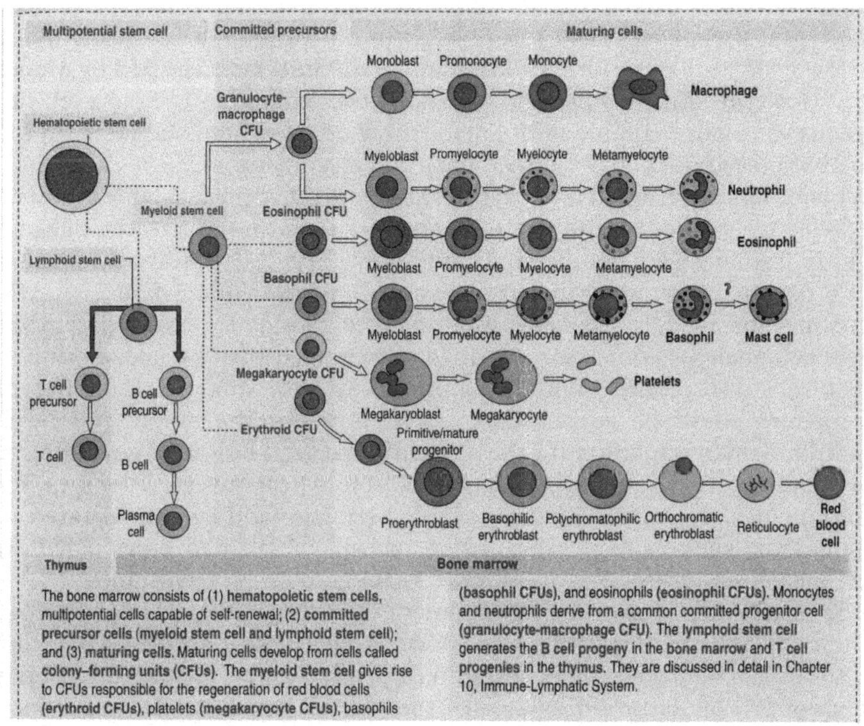

Figure 12.1 Schematic drawing of hierarchy in hematopoietic cells,[90] where stem cells produce different types of mature blood cells after a stepwise process of division and maturation. Essentially, few immature cells give rise to a progeny of many differentiated cells, which then leave the marrow cavity and enter the blood stream. Reproduced from ref. 90 with permission from Elsevier, Copyright 2007.

mice (see later). In addition, it was observed that such variations could be influenced by external illumination.

The availability of methods for labelling cells in DNA synthesis with ^{3}H-thymidine and for counting mitoses enabled sequence studies of proliferative activity in various tissues. In the bone marrow as well as in other organs of rodents, reproducible day–night variations were observed. In addition, the concept of stem cells in the bone marrow was verified experimentally by the introduction of the so-called *spleen colony technique* by Till and McCulloch in 1961.[5] The background information is that when normal bone marrow cells were injected into the tail vein of irradiated mice, a few haematopoietic stem cells would make colonies in the spleen, which could then be counted to estimate the number of stem cells. In the following two decades, a series of studies on rodents showed reproducible circadian variations in the bone marrow as well as in the peripheral blood cells.

Due to the high proliferative activity in the bone marrow, haematopoietic precursor cells were very susceptible to cytotoxic influences, such as cytostatic drugs and irradiation. Surprisingly, lethal effects on mice exhibited strong variations with day and night (see *e.g.*, Scheving *et al.*, 1974,[6] and Scheving and Halberg, 1980[7]). In 1962, diurnal variations in human bone marrow cells were described by Killmann *et al.*[8] and later verified by Mauer *et al.*[9] However, directly observed variations over complete 24-hour cycles in humans were not available until Smaaland *et al.* presented such data in the late 1980s (see later).

A major objection against this type of study was that the available methods for sequence studies on the rate of haematopoiesis throughout the day and night in humans were cumbersome, and only a few cells could be evaluated over a reasonable time period. This obstacle was overcome when flow cytometry became widely available around 1970. Suddenly, multi-parameter single cell analyses could be performed automatically with a speed of up to 10 000 cells per second.[10] With appropriate computer analysis, the different parameters could be correlated to each other at the single cell level. In addition, subpopulations of the cell samples could be sorted according to the measurements, and used for further biological investigations or for morphological controls. This also enabled analysis and sorting of rare cell populations, such as stem cells, a type of cells which had previously only been measured indirectly, for example, through cloning techniques.[10]

The next major step in the development of automated analytical cytology techniques was the construction of flow cytometric cell sorters, which operated under high pressure, leading to a speed of up to 100 000 single cells per second. This occurred as part of the *human genome project*, and such instruments became generally available by the late 1990s. A combination of ultra-speed cell sorting techniques with molecular biology, for example, expression analyses, later became an important basis for exploring the importance of clock genes.

A summary of the state of the art in the chronobiology of circulating blood leukocytes was given in 1994 by Haus[11] and for the human bone marrow by

Smaaland and Laerum[12] in the textbook "*Biologic rhythms in clinical and laboratory medicine*", edited by Touitou and Haus.[13] This represented in a way the end of the "pre-clock gene era" in chronobiology.

12.4 General Outline of Haematopoiesis, from Stem and Progenitor Cells to Mature Blood Cells

As reviewed a few years ago by Quesenberry *et al.*,[14] characterization of mouse stem cells by introduction of the spleen colony technique in 1961[5] was followed by *in vitro* cloning techniques introduced by Pluznik and Sachs in 1965[15] and by Bradley and Metcalf in 1966.[16] The principle was that a suspension of bone marrow cells was embedded into a semi-solid medium, either agar or methylcellulose. In addition, a source of growth factors was present in the medium, or by adding a separate feeder layer of producer cells, *e.g.*, mesenchymal cells. Progenitor cells would then produce local colonies consisting of one or more lineages of haematopoietic cells, primarily granulocytic or macrophage cells or both. Later, the system was expanded to erythropoietic cells, either as clusters from late erythropoiesis or large bursts from earlier progenitor cells. It could also be shown that both single type colonies and mixed colonies with several lineages could be produced.

These relatively simple culture procedures led to extensive research along two paths: one was centered on characterization of the resulting colony types and their progenitors, and the other involved isolating different types of growth factors, which were necessary for production of haematopoietic cultures. A multitude of such factors were purified and characterized, both chemically and biologically, thereby making it possible to obtain colonies from virtually all types of haematopoietic cells. It was also a common opinion that these growth factors, which were mainly protein and peptide structures, served as the actual regulators of haematopoietic lineages.

At the end of the last millennium, between 50 and 100 different, more or less defined growth factors and other regulators were considered as essential factors for maturation as well as cell proliferation in the haematopoietic system. As will be further discussed below, the term *regulator* was used too liberally for any type of biological substances, which could interfere with maturation and cell cycling in haematopoiesis.

The next step of development was the introduction of suspension cultures of bone marrow cells, in which all lineages of haematopoiesis as well as progenitor and stem cells were generated. The resulting cell population was then dependent on the added sources of more or less crude extracts of growth factors in addition to feeder layers.[17]

In parallel, flow cytometric analysis and cell sorting methods were further developed and refined. Thereby, it became possible for the first time to obtain purified preparations of even rare subpopulations of cells, such as multipotent stem cells, progenitor cells and different stages of precursor and fully differentiated mature cells. Therefore, from the 1990s, a series of different methodological approaches made it possible to study morphologically even very small subpopulations of cells in pure suspensions.

It is therefore not surprising that during the last two decades, the combination of easy accessibility to human and animal bone marrow and new methods has enabled a fruitful research wave. In the next part, earlier and current knowledge about circadian rhythms in haematopoietic and blood cells will be summarised.

12.5 Complexity of the Progeny

Some notes on the complexity of the resulting mature cell populations, leaving the bone marrow in large numbers, are necessary before we enter the field of their time dependent functionality.

Monocytes that have a close relationship with granulocytes further mature into macrophages when reaching their final destinations in the peripheral tissues. The macrophages are considered as the master regulators of local inflammation and fibrosis.[18] Through a feedback mechanism, they also serve as regulators of what we could call a steady-state and demand-adapted haematopoiesis, including cell renewal along the myelopoietic pathway as well as erythropoiesis (Figure 12.2).[19]

Although neutrophil granulocytes are the most numerous white blood cells, they only spend a short period in peripheral blood before they migrate into the tissues, with a half-life of about 6 hours. Basophil granulocytes, which were described in 1879, represent a rare and highly specialized cell

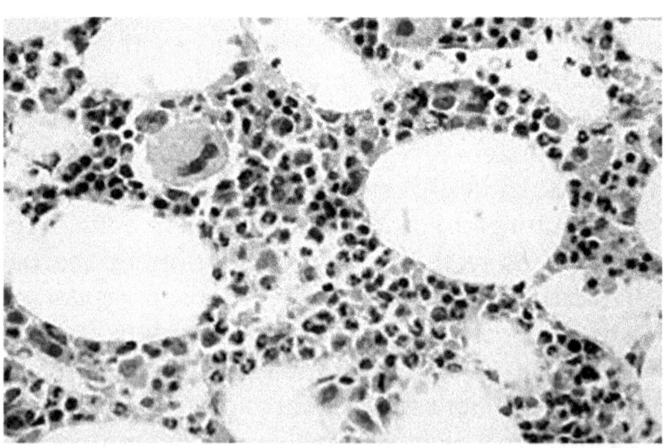

Figure 12.2 Microscopical picture of the human bone marrow showing maturing cells lying in clusters. The small cells with condensed nuclei are erythropoietic cells, while myelopoietic cells are larger and have round or lobulated nuclei. The cells with strongly lobulated nuclei are mature granulocytes ready to leave the bone marrow and move into the blood stream. The large cell is a megakaryocyte. Some lymphocytes are also present. Hematoxylin and eosin stain, initial magnification ×20. Picture taken by the author.

group closely related to mast cells residing in the peripheral tissues. They amount to about 1 percent of the circulating granulocytes.[20]

On the other hand, the large megakaryocytes undergo polyploidization during the maturation process, followed by extension of long, branched cytoplasmic extensions. These extensions undergo fission resulting in the formation of many platelets, which are released into the blood stream.[21] Both in peripheral blood and in the tissues, they participate in the formation of thrombi, a process that also experiences circadian variations. The non-nucleated thrombocytes in peripheral blood also show circadian variations in number with maximal values during the day.[11]

12.6 Chronobiology of Erythrocytes

It has been known for many years that the release of newly formed erythrocytes, which are anucleate, but contain RNA, *i.e.*, *reticulocytes*, undergo significant variations. Although mature erythrocytes lack a nucleus, it has been shown that their different cellular functions undergo significant circadian variations. This includes their reversible haemoglobin oxidation, temperature cycles and NADH/NADPH oscillations. Moreover, these rhythms are entrainable by environmental factors.[22,23]

12.7 Chronobiological Aspects of Peripheral Blood Cells

Circadian variations in peripheral blood cells are prominent in adult humans and have also been observed during fetal life.[24] In adults, roughly, 4.5 to 5.5 liters of blood circulate all over the body. In his review from 1992, Haus[11] summarised the available data from peripheral blood in clinically healthy humans, including his own data from many years of research.

Both in women and men, the total white blood cell counts exhibit reproducible circadian variations with maximal values in the late afternoon. This pattern applies to neutrophil granulocytes, monocytes and platelets. Eosinophil granulocytes, lymphocytes and reticulocytes reach their maximal values some hours later, occurring at the beginning of the dark phase. When studying peripheral blood from healthy, male volunteers, Haus[11] reported that the highest values for haemoglobin, haematocrit and red blood cell counts were around noon.[25]

It should not be forgotten that the concentration of several hormones coincides with the variations in peripheral blood cells, such as cortisol, testosterone and growth hormone, which may be associated with their regulation functions. Thus, their high values during late night and early morning may be partly responsible for a peak in both total blood cell counts and the numbers of neutrophils and lymphocytes (see Figure 12.3).

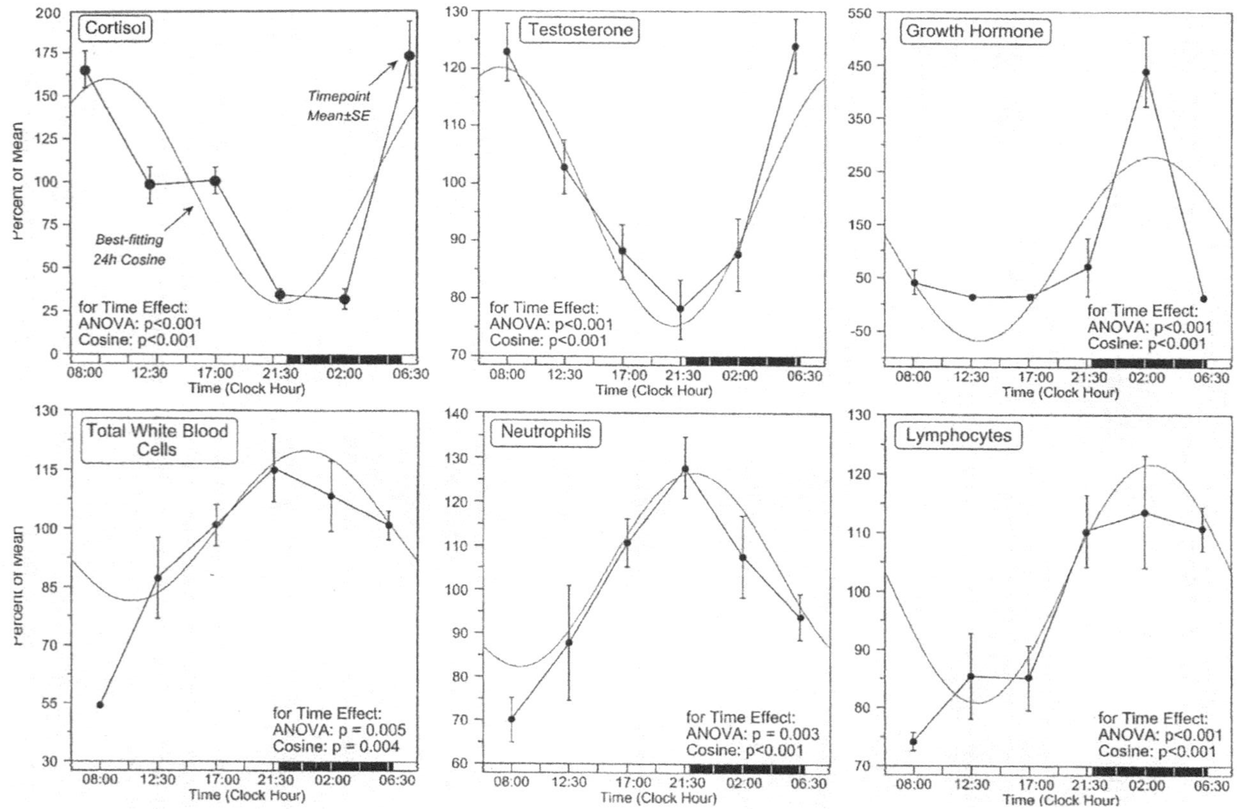

Figure 12.3 Lower panel: circadian variations in human peripheral blood leukocytes showing the highest values at the beginning of the dark phase. Upper panel: opposite variations of cortisol and testosterone in peripheral blood and synchronous variation of growth hormone. Each point is the mean value ± standard error of the mean of a group of healthy male volunteers. Reproduced from ref. 63 with permission from SAGE Publications, Copyright 2009.

12.8 Precursor Cells

12.8.1 Human Experiments and General Considerations

At any time, enormous numbers of cells belonging to various stages of different lineages are being produced. Principally, at each stage they undergo mitotic division, and their actual numbers are doubled.

After studies on time-dependent variations in the bone marrow of laboratory animals, systematic investigations on healthy volunteers were performed throughout the 1980s and early 1990s.

12.8.2 Proliferative Parameters

In the 1960s and up to the 1980s, experimental data on haematopoiesis in rodents, including mice, rats and rabbits were presented.[6] Studies using methods such as mitotic counts, DNA labelling with ^{3}H-thymidine and cell cycle distribution obtained by flow cytometry showed significant circadian variations in different lineages of haematopoiesis. In addition, there were seasonal variations.[26] A limitation was that the circadian patterns were unstable, and the location of the acrophase (phase of the predicted maximum) could vary from time to time, even when the animals were kept under strictly isolated conditions with alternating light and darkness.[27]

Mitotic counts and ^{3}H-labelling of bone marrow cells in DNA synthesis were also presented from bone marrow samples of human volunteers.[8,9] Although complete 24-hour cycles were not studied, the data could indicate high proliferative activity during the light phase.

These problems were partially overcome by Smaaland *et al.*,[28,29] who conducted complete 24-hour cycles of bone marrow sampling of healthy male volunteers. Using DNA-distribution curves obtained through flow cytometry, an all-over estimate of proliferative activity in the marrow could be obtained. Although some biological variability was present, it was convincingly shown that the number of cells in DNA synthesis peaked in the middle of the day.

These studies were further extended to two-parameter measurements of DNA and CD45 by Abrahamsen *et al.*[30] Thereby, the cell cycle distribution could be discriminated between myelopoietic cells, erythropoietic cells, and lymphocytes and monocytes together. Extensive control experiments were performed to ensure that the cell sampling techniques and accidental peripheral blood aspiration did not interfere with the results.[31] Marker rhythms such as blood cortisol were studied in parallel. Another potential source of error, *i.e.*, dilution of the cell samples through local bleeding could also be corrected for.[32,33]

Samples obtained at 4- to 5-hour intervals during 24-hour spans showed significant variations in the fractions of cells in the DNA synthetic phase in different lineages (*i.e.*, S-phase). Eighty percent of the volunteers showed the highest activity of myeloid cells in the DNA synthetic phase and 75 percent displayed the highest activity in erythroid cells during the day. Lowest activities were observed between midnight and 04:00 am. As expected in proliferating cells, there was also a temporal relationship between circadian variation of the S-phase and the G2 and M-phase. Highest variations over 24-hour periods were seen in myelopoiesis. The covariation between myelopoietic and erythropoietic cell cycling indicated that the pattern was due to a common cue.[30]

A limiting factor was that such analyses are based on relative measurements and only give an indication of the total activity in the whole haematopoietic system. For example, it is known that exercise has a rapid effect on the mobilization of mature blood cells from the marrow cavity to peripheral blood.[34] In contrast to rodent studies, where a whole femur can be flushed and quantitated, this objection is also relevant for all the other human studies, which will be mentioned below (see Figure 12.4).

12.9 Stem and Progenitor Cells

12.9.1 Some General Aspects

As already indicated, stem cells are capable of both extensive renewal and generating a specialized progeny through differentiation. During their early stage as pluripotent cells, stem cells can principally produce any type of cells and tissues in the body, and even during fetal life, these pluripotent cells undergo circadian variations. In their next step as multipotent or unipotent cells, stem cells also have a high regeneration capacity, but only for specific lineages or tissues[35] (see also Figure 12.1).

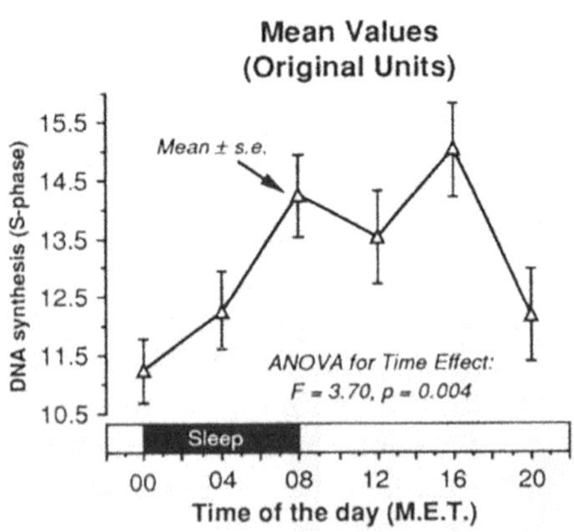

Figure 12.4 Circadian variations in proliferative activity, *i.e.*, fraction of cells in DNA synthesis of a group of healthy male volunteers with the highest values in the middle of the day (light phase), and lowest values at night. Reproduced from ref. 28 with permission from Elsevier, Copyright 1991.

When looking at time-dependent variations in the bone marrow, it should be remembered that stem cells are located in the so-called niches containing a specialized microenvironment.[36] The stem cells are thus residing in close proximity to and functionally dependent on *mesenchymal stem cells*, which contain *Nestin*, a component essential for stem cell renewal.[37] In addition, haematopoietic stem cells constitute a heterogeneous group of immature cells, which are closely related to progenitor cells, as their next step of commitment to the maturing cell lineages.[38] In these niches, early lymphoid progenitors are also present, indicating that a complex cellular micro-environment is necessary for normal haematopoiesis.[39] Use of many surface markers, such as CD 150, CD 48 and CD 244 can help distinguish between these subpopulations of stem and progenitor cells.[40,41]

Small numbers of haematopoietic stem cells (HSCs) are constantly circulating in the peripheral blood of humans as well as other species. In mice under strictly regulated light conditions, their release from the bone marrow cavity exhibits circadian variations with a maximum about 5 hours after initiation of the light and reaching a nadir 5 hours after the onset of darkness. The circadian variations in both HSCs and their progenitors fluctuate in an opposite phase of the chemokine CXCL12, indicating a regulatory role.[42]

Interestingly, there is evidence that during human fetal life, primitive macrophages control haematopoietic stem cell mobilization, leading to their colonization in peripheral tissues *via* the blood stream.[24,43] Therefore, cell interactions may be of critical importance for the development of cyclic behaviour in the bone marrow. In the adult bone marrow, cells also interact *via* extracellular vesicles and not only through direct cell-to-cell communication.[14] In addition, there are extensive nerve projections between the cell clusters in the marrow, aligning with several systems of cell-to-cell interactions and regulatory signals.

12.9.2 Circadian Variations of Stem and Progenitor Cells

Since Stoney *et al.*[11] reported that the clonability of murine haematopoietic progenitor cells for myelopoiesis showed circadian variations, numerous articles have been published, both for characterization of different cell types and for the study of regulatory aspects. Also the number of multipotent stem cells, as estimated using the spleen colony technique, shows a circadian variation (see, *e.g.*, ref. 45–47). For the precursor cells, maximal values were generally found during the active stage, followed by low values at rest.

Smaaland *et al.*[48] reported that the colony formation by human CD34$^+$ cells from the bone marrow of healthy male volunteers varied significantly over 24-hour periods, with maximal values during the day and low values during sleep. In addition, there was a covariation with the proliferative activity in the precursor cells in the bone marrow as estimated by the fraction of cells in DNA synthesis. This finding was further explored by Abrahamsen *et al.*,[49] who reported that there was a circadian variation in the yield of

CD34+ cells in the bone marrow and also a covariation with their fractions of cells in DNA synthesis.

In conclusion, there seems to be a general pattern where all cell types in haematopoiesis and their proliferative activity as well as clonability show covariation, from the different lineages of precursor cells to stem cells. Both in humans and in laboratory animals, maximal values were observed during the active phase and minimal values were observed at rest. This led to a general conclusion that *haematopoiesis occurs in synchronized rhythms.*[50,51] At least in laboratory animals, seasonal variations have also been observed,[26] but for obvious reasons, there were limitations for performing bone marrow punctures throughout the year in humans.

12.10 Clock Genes in Haematopoiesis

12.10.1 General Aspects

After the clock genes had been discovered in the brains of insects and later also in the brains of all species, including humans, they were considered to have central regulatory functions in the body. It remained an open question to what extent also cells in peripheral organs had their own clockworks with subsequent local regulatory functions. The existence of clock genes in haematopoietic cells was unknown. In addition, there was a widespread assumption that immature, resting stem cells did not express clock genes.

In 2002, Bourin *et al.*[52] hypothesized that there also existed endogenous oscillators in the bone marrow. They showed that murine bone marrow progenitor cells could form CFU-GM colonies with a circadian rhythmic pattern during the first 4 days of liquid culture after explantation. The authors concluded that bone marrow progenitors sustain circadian rhythmicity due to the presence of internal time-keeping mechanisms.

Tsinkalovsky *et al.*[53] later reported that clock genes were expressed in mouse haematopoietic stem cells. Following staining of bone marrow cells with Hoechst 33 342, the stem cells were enriched through high-speed flow cytometric cell sorting of the side population (SP cells). Test for colony formation was performed to ensure that the cells were clonogenic. By mRNA isolation and PCR, it was shown that the major clock genes, *Per1*, *Per2*, *Bmal1*, *Cry1*, *Clock* and *Rev-erbα* were expressed in stem cells. They could also show that, with the exception of *Per2*, the studied clock genes were not fully oscillating in a circadian manner. This pattern was similar to immature cells in rapidly proliferating organs like testis and thymus, indicating that circadian variations in clock gene expression in the bone marrow stem cells were developmentally regulated.[54]

Using sequence sampling of bone marrow every 4 hours for 24-hour periods in ten healthy male volunteers, the same group could show that clock genes were also expressed in human CD34+ cells obtained through high-speed flow sorting.[55] A circadian rhythm was observed for *PER1*, *PER2* and *CRY2* in CD34+ cells, and for *PER1* in the whole bone marrow with maximal values from the morning until noon. However, *BMAL1* was not showing rhythmic oscillations. These findings indicate clock gene expression in stem or progenitor cells in human bone marrow follow circadian patterns similar to those reported in other fully differentiating tissues (Figure 12.5).

Interestingly, it has later been shown that clock gene deficient C57BL mice lacked oscillations of immature haematopoietic cells as well as circulating leukocyte subsets.[56] Probably, circadian regulation of CXCL12 *via* clock genes is part of the normal regulatory mechanism (also see later).

12.10.2 A Universal System in Haematopoietic Cells

During the last few years, extensive research has given a more differentiated picture of the importance of clock genes for the various types of haematopoietic cells and their progeny of fully mature cells. Thus, it has been reported that the expression of *Per1* and *Per2* exhibit circadian oscillations in mouse leukocytes, while only *PER1* was active in the corresponding human cells. In contrast, *PER2* and *BMAL1* only showed a weak rhythm in their mRNA.[57] In addition, human peripheral blood mononuclear cells, *PER2*, *PER3*, and *DEC1* also exhibited a circadian pattern.[58]

Splenic macrophages, dendritic cells and B cells in mice revealed a similar and even more extensive circadian expression of a series of clock genes. In mouse peritoneal macrophages, however, *Bmal1* seemed to be of critical importance for regulating the circadian pattern.[59] Even human mast cells and eosinophils seem to be locally regulated by the whole series of clock genes.[60] In contrast, platelet activity seems to be mainly regulated by *Clock*.[61]

Bone marrow mesenchymal cells are of high importance for regulation of stem, progenitor and precursor cells in the microenvironment. Generally, fibroblastic cells have circadian rhythms of their own, which can be reinduced in cell culture through serum shock.[62] Using cultured human mesenchymal stem cells, the same could be induced by cyclic AMP, after which several rounds of oscillations alternating between stimulatory and inhibitory clock genes can be followed.[63] The clock genes are also able to modulate differentiation, migration and cell cycling of mesenchymal cells *in vitro* (Figure 12.6).[64,65]

12.11 Regulatory Aspects

In conjunction with the direct examination of the different clock genes and their importance for regulating haematopoiesis, several different mechanisms related to circadian rhythms have been unraveled. Thus, circadian variations in noradrenaline are associated with haematopoietic cell cycling in the mouse bone marrow.[66] Similarly, the circadian rhythm of leukocytes and some lymphocyte subsets has been related to the autonomic nervous system.[67] Rhythmic modulation of the haematopoietic niche is connected to the 24-hour cycles of release of neutrophil leukocytes and their migration back to the bone marrow, where they are eliminated.[68] Cholinergic signals

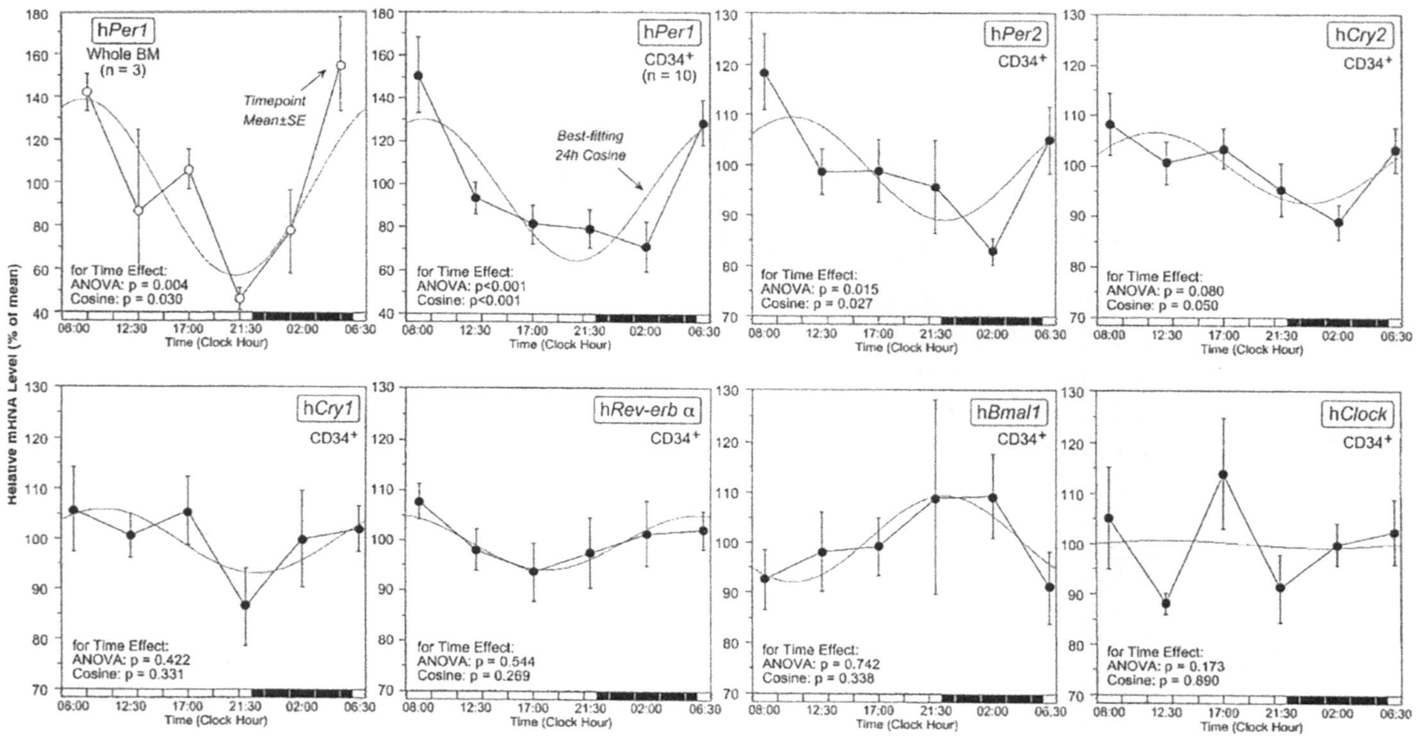

Figure 12.5 Example of circadian variations in the expression of the different clock genes in bone marrow $CD34^+$ stem/progenitor cells from male volunteers. *Per* and *Cry* seem to be synchronous to variations of leukocytes (Figure 12.3), and DNA synthesis (Figure 12.4). Variations in the other clock genes (lower panel) were not significant. Reproduced from ref. 55 with permission from SAGE Publications, Copyright 2007.

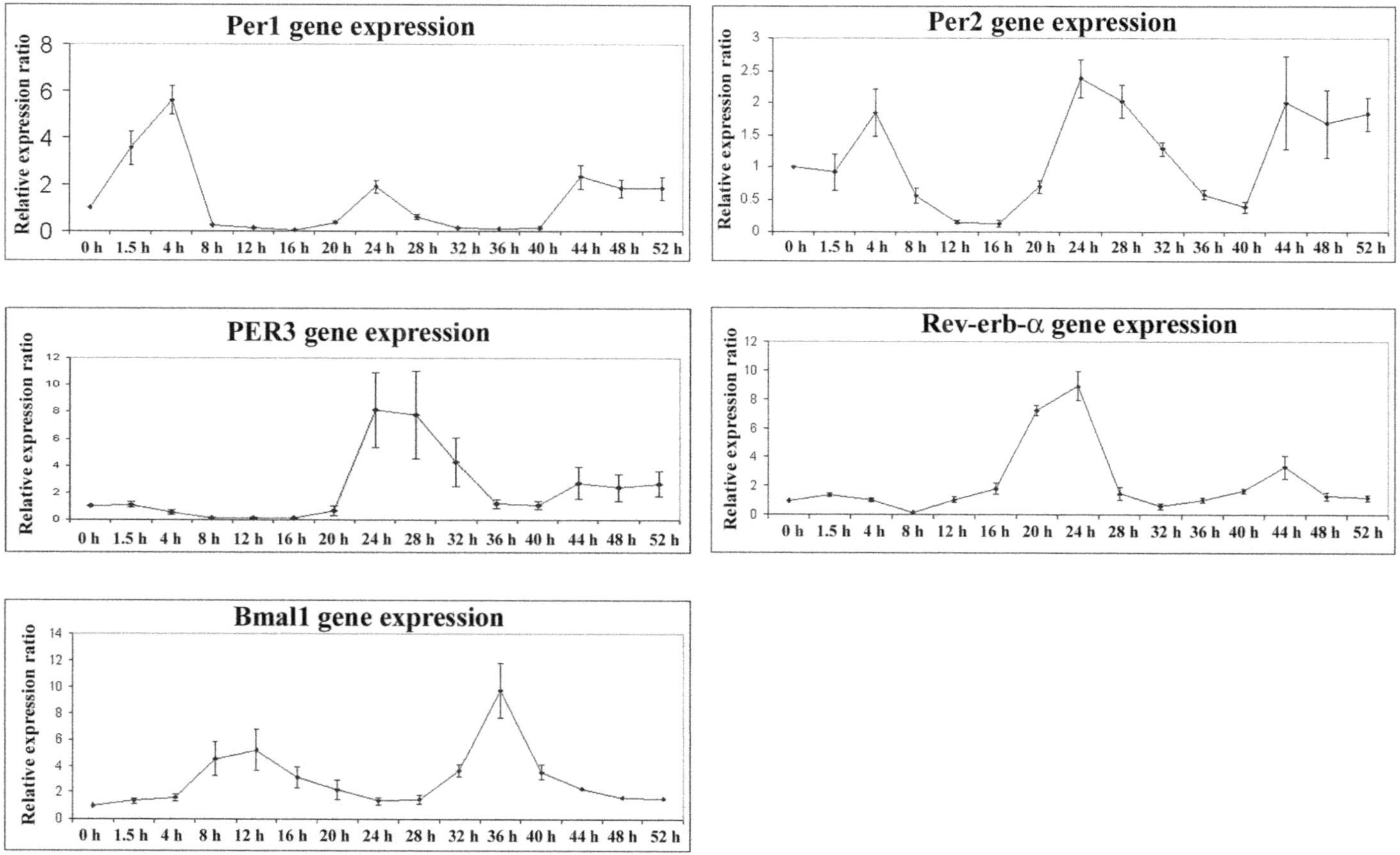

Figure 12.6 Rhythmic expression of clock genes *Per1*, *Per2*, *Per3*, *Rev-erbα* and *Bmal1*, induced by addition of a single dose of cyclic-AMP to the medium of cultured human mesenchymal stem cells, showing variations with different phasing, where *Bmal1* alternates with *Per* and *Rev-erbα*. Reproduced from ref. 63 with permission from Taylor & Francis, Copyright 2009.

also seem to regulate 24-hour migration of haematopoietic stem cells as well as leukocytes in mice.[69,70]

Circadian variations in haematopoietic stem cells seem to be controlled *via* TNF/ROS signals.[71] Not least, different chemokines and their receptors appear to be of critical importance.[72] Key factors are *CXCR4*,[73] *CXCL12*,[56] and *SDF-1*.[74,75] This also includes the cross-talk between clock genes and general regulatory mechanisms.[76] Thus, the release of haematopoietic stem cells from the bone marrow also appears to be regulated by such mechanisms. In contrast, *Clock* seems to be the main factor for differentiating mouse pluripotent stem cells.[77,78]

There is even evidence suggesting that sleep centers in the brain may modify the circadian clock.[79] The complexity of regulatory factors mediated through the various clock genes and their transcription has led to large projects with genome-wide RNAi screens for identifying modifiers of the circadian clock in humans.[80]

The complexity of regulatory functions related to the clockwork is illustrated by the role of cortisol. Its 24-hour oscillations regulate both hematopoietic stem and progenitor cell proliferation, and also the *CXCL12* expression by bone marrow stromal progenitors.[81,82]

On the other hand, *Rev-erbα* seems to regulate both stem cell proliferation and bone formation in mice.[83] In pluripotent stem cells, however, *Bmal1* has been reported to coordinate both energy metabolism and differentiation. This is in line with a report on macrophages, where a cross-talk between circadian rhythmicity, the dynamics of mitochondria and bactericidal activity has been observed.[84]

Finally, it should not be forgotten that the onset of the light phase of the day serves as a main regulator for setting the circadian phase, a process in which the clock machinery of every cell may, in principle, be involved.[85]

12.12 Implications for Chemotherapy of Malignant Tumours

Since chronotherapy is dealt with in another chapter, only a few aspects will be mentioned here. Studies on toxic influences related to time and light conditions were crucial for both the field of chronotherapy and for general chronobiological studies.[6,86] Not least, the timing of therapy became critical for avoiding serious side effects from cytotoxic drugs. The whole field of chronobiology, therefore, focuses on light and on timing.[87]

With the rapid development of chronobiology and clock gene research, however, the cellular systems look more and more complex. Therefore, it is not evident where the most rewarding prospects may be found. Large cellular networks both between same types of cells and between cells of different origin add to the complexity. Haematopoiesis is no exception (see *e.g.*, Pritchett and Reddy[88]).

12.13 General Conclusions

Circadian variations have been observed in all aspects of haematopoiesis, including all types of differentiation and stages of development, from stem and progenitor cells to mature leukocytes, macrophages and erythrocytes.

In all these cells, clock genes are present and actively transcribed. However, there are great variations between individual cell types with regard to the functions and importance of the different clock genes. Altogether, they contribute to an enormous complexity through their interactions with other regulatory mechanisms, influenced by the nervous system, by circulation and local cellular functions in peripheral tissues.[89] In the end, this complexity could lead to great opportunities for identifying specific interactions with disease in single cell types, opening a new avenue for therapy.

References

1. N. Mechanik, *Z. Anat.Entw. Gesch.*, 1926, **79**, 58.
2. O. D. Laerum, Chronobiological aspects of bone marrow and blood cells, in *Biological rhythms in structure and function*, ed. E. A. Vidrio, Alan R.Liss, inc., New York, 1981, pp. 87–97.
3. M. Weger, N. Diotel, A.-C. Dorsemans, T. Dickmeis and B. D. Weger, *Dev. Biol.*, 2017, 111–123.
4. F. Rud, *The eosinophil count in health and in mental disease. A biometrical study*, MD thesis, University of Oslo, 1947, Johan Grundt Tanum Forlag, Oslo, 1947.
5. J. E. Till and E. A. McCulloch, *Radiat. Res.*, 1961, **14**, 213.
6. *Chronobiology*, ed. L. E. Scheving, F. Halberg and J. E. Pauly, Igaku Shoin Ltd, Tokyo, 1974.
7. L. Scheving and F. Halberg, *Chronobiology: Principles and Applications to Shifts in Schedules*, Sijthoff & Noordhoff, Alphen aan den Rijn and Rockville, 1980.
8. S.-Å. Killmann, E. P. Cronkite, T. M. Fliedner and V. P. Bond, *Blood*, 1962, **70**, 1418.
9. A. M. Mauer, *Blood*, 1965, **26**, 1.
10. O. D. Laerum and R. Bjerknes, *Flow Cytometry in Hematology*, Academic Press, London, San Diego, New York, Boston, Sydney, Tokyo and Toronto, 1992.
11. E. Haus, Chronobiology of circulating blood cells and platelets, in *Biological Rhythms in Clinical and Laboratory Medicine*, ed. Y. Touitou and E. Haus, Springer-Verlag, Berlin, Heidelberg, New York, London, Paris, Tokyo, Hong Kong, Barcelona and Budapest, 1992, p. 504.
12. R. Smaaland and O. D.Laerum, Chronobiology of human bone marrow, in *Biological Rhythms in Clinical and Laboratory Medicine*, ed. Y. Touitou and E. Haus, Springer-Verlag, Berlin, Heidelberg, New York, London, Paris, Tokyo, Hong Kong, Barcelona and Budapest, 1992, p. 527.

13. *Biological Rhythms in Clinical and Laboratory Medicine*, ed. Y. Touitou and E. Haus, Springer-Verlag, Berlin, Heidelberg, New York, London, Paris, Tokyo, Hong Kong, Barcelona and Budapest, 1992.
14. P. J. Quesenberry, L. Goldberg, J. Aliotta and M. Dooner, Marrow hematopoietic stem cells revisited: they exist in a continuum and are not defined by standard purification approaches; then there are the micro-vesicles, *Front. Oncol.*, 2014, **4**, 1.
15. D. H. Pluznik and L. Sachs, *J. Cell. Comp. Physiol.*, 1965, **66**, 319.
16. R. Bradley and D. Metcalf, *Aust. J. Exp. Biol. Med. Sci.*, 1966, **44**, 287.
17. T. M. Dexter, T. D. Allen, L. G. Lajtha, R. Schofield and B. I. Lord, *J. Cell. Physiol.*, 1973, **82**, 461.
18. T. A. Wynn and L. Barron, *Semin. Liver Dis.*, 2010, **13**, 245.
19. A. McCabe and C. MacNamara, *Exp. Hematol.*, 2016, **44**, 213.
20. G. Marone, F. Borriello, G. Varrichi, A. Genovese and F. Granata, in *History of allergy*, ed. K.-C. Bergmann and J. Ring, Karger, Basel, 2014, p. 172.
21. K. R. Machlus and J. E. Italiano Jr, *J. Cell Biol.*, 2013, **201**, 785.
22. J. S. O'Neill and A. B. Reddy, *Nature*, 2011, **469**, 498.
23. R. H. Clark and D. R. Korst, *Science*, 1969, **166**, 236.
24. S. Servais, E. Baudoux and B. Brichard, *et al.*, *Haematologica*, 2015, **100**, 32.
25. M. Sani, H. Sebai, N. Ghanem-Boughanmi, N. A. Boughattas and M. Ben-Attia, Circadian (about 24 hours) variation in malondialdehyde content and catalase activity of mouse erythrocytes, *Redox Rep.*, 2015, **20**, 26.
26. N. P. Aardal, *Exp. Hematol.*, 1984, **12**, 61.
27. N. P. Aardal and O. D. Laerum, *Exp. Hematol.*, 1983, **11**, 792.
28. R. Smaaland, O. D. Laerum, K. Lote, O. Sletvold, R. B. Sothern and R. Bjerknes, *Blood*, 1991, **77**, 2603.
29. R. Smaaland, O. D. laerum, R. B. Sothern, O. Sletvold, R. Bjerknes and K. Lote, *Blood*, 1992, **79**, 2281.
30. J. F. Abrahamsen, R. Smaaland, R. B. Sothern and O. D. Laerum, *Eur. J. Haematol.*, 1997, **58**, 333.
31. J. F. Abrahamsen, R. B. Sothern, S. Sandberg, A. Aakvaag, O. D. Laerum and R. Smaaland, *Biol. Rhythm Res.*, 1999, **30**, 29.
32. J. F. Abrahamsen, R. Smaaland, R. Skjaerven and O. D. Laerum, *Eur. J. Haematol.*, 1996, **56**, 138.
33. J. F. Abrahamsen, F. Lund-Johansen, O. D. Laerum, B. C. Schem, O. Sletvold and R. Smaaland, *Cytometry*, 1995, **19**, 77.
34. G. G. Wardyn, S. I. Rennard, S. K. Brusnahan, T. R. McGuire, M. L. Carlson, L. M. Smith, S. McGranaghan and J. G. Sharp, *Exp. Hematol.*, 2008, **36**, 216.
35. A. De Los Angeles, F. Ferrari and R. Xi, *et al.*, *Nature*, 2015, **525**, 469.
36. G. M. Crane, E. Jeffery and S. Morrison, *Nat. Rev. Immunol.*, 2017, **17**, 573.
37. S. Méndez-Ferrer, T. V. Michurina and F. Ferraro, *et al.*, *Nature*, 2010, **466**, 829.
38. K. D. Kokkaliaris, D. Lucas, I. Beerman, D. G. Kent and L. Perié, *Exp. Hematol.*, 2016, **44**, 447.

39. L. Ding and S. J. Morrison, *Nature*, 2013, **495**, 231.
40. H. Oguro, L. Ding and S. J. Morrison, *Cell Stem Cell*, 2010, **13**, 102.
41. B. R. Susa, R. C. Perreira and E. A. Fonseca, *et al.*, *Cytometry*, 2014, **85A**, 43.
42. S. Méndes-Ferrer, D. Lucas, M. Battista and P. S. Frenette, *Nature*, 2008, **452**, 442.
43. J. Travnickova, V. T. Chau and E. Julien, *et al.*, *Nat. Commun.*, 2015, **6**, 6227.
44. P. J. Stoney, F. Halberg and H. W. Simpson, *Chronobiologia*, 1975, **2**, 319.
45. L. E. Scheving, E. R. Burns, J. E. Pauly and T.-N. Tsai, *Anat. Rec.*, 1978, **191**, 479.
46. O. Sletvold and O. D. Laerum, *Eur. J. Haematol.*, 1988, **41**, 230.
47. P. A. Wood, S. de la Pena and W. J. M. Hrushesky, *Annu. Rev. Chronopharmacol.*, 1990, **7**, 173.
48. R. Smaaland, O. D. Laerum, R. B. Sothern and O. Sletvold, *Blood*, 1991, **77**, 2603.
49. J. F. Abrahamsen, R. Smaaland, R. B. Sothern and O. D. Laerum, *Eur. J. Haematol.*, 1998, **60**, 7.
50. O. D. Laerum, *Exp. Hematol.*, 1995, **23**, 1145.
51. F. Levi, I. Blazcek and A. Ferle-Vidovic, *Exp. Hematol.*, 1988, **16**, 696.
52. P. Bourin, A. F. Ledain, J. Beau, D. Mille and F. Lévi, *Chronobiol. Int.*, 2002, **19**, 57.
53. O. Tsinkalovsky, B. Rosenlund, O. D. Laerum and H. G. Eiken, *Exp. Hematol.*, 2007, **13**, 100.
54. O. Tsinkalovsky, E. Filipski, B. Rosenlund, R. B. Sothern, H. G. Eiken, M. W. Wu, B. Claustrat, J. Bayer, F. Lévi and O. D. Laerum, *Exp. Hematol.*, 2006, **34**, 1248.
55. O. Tsinkalovsky, R. Smaaland, B. Rosenlund, R. B. Sothern, A. Hirt, S. Steine, A. Badiee, J. F. Abrahamsen, H. G. Eiken and O. D. Laerum, *J. Biol. Rhythms*, 2007, **22**, 140.
56. M. Stenzinger, D. KP, F. Cho-Park, H. Xiongarpova, C. Unterrainer, S. Harenkamp, E. Wiercinska, K. Hoerster, M. Pfeffer, E. Maronde and H. Bonig, *Cells*, 2019, **8**, 1033.
57. H. Fukuya, N. Emoto, H. Nonaka, K. Yagita, H. Okamura and M. Yokoyama, *Biochem. Biophys. Res. Commun.*, 2007, **354**, 924.
58. D. B. Boivin, F. O. James, A. Wu, P. F. Cho-Park, H. Xiong and Z. S. Sun, *Blood*, 2003, **102**, 4143.
59. M. Hayashi, S. Shimba and M. Tezuka, *Biol. Pharm. Bull.*, 2007, **30**, 621.
60. A. Baumann, K. Feilhauer, S. C. Bischoff, O. Froy and A. Lorentz, *Mol. Immunol.*, 2015, **64**, 76.
61. N. Ohkura, K. Oishi, T. Sudo, H. Hayashi, K. Shikata, N. Ishida, J. Matsuda and S. Horie, *J. Thrombosis. Res.*, 2009, **123**, 523.
62. A. Balsalobre, F. Damiola and E. Schibler, *Cell*, 1998, **93**, 929.
63. T. S. Huang, G. Grodeland, L. Sleire, M. Y. Wang, G. Kvalheim and O. D. Laerum, *Chronobiol. Int.*, 2009, **26**, 242.
64. H. Boucher, V. Vanneaux, T. Domet, A. Parouchev and J. Larghero, *PLoS One*, 2016, **11**, e0146674.
65. B. Anthony and D. C. Link, *Trends Immunol.*, 2014, **35**, 32.

66. G. J. Maestroni, M. Cosentino, F. Marino, M. Togni, A. Conti, S. Lecchini and G. Frigo, *Exp. Hematol.*, 1998, **26**, 1172.
67. S. Suzuki, S. Toyabe, T. Moroda, T. Tada, A. Tsukahara, T. Iiai, M. Minagawa, S. Maruyama, K. Hatakeyama, K. Endoh and T. Abo, *Clin. Exp. Immunol.*, 1997, **110**, 500.
68. M. Casanova-Acebes, C. Pitaval, L. Weiss and A. Hidalgo, *Cell*, 2013, **153**, 1025.
69. A. G. Garcia, C. Korn, M. Garcia-Fernandez, O. Dominguez, J. Villadiego, D. Martin-Pérez, J. Isern, J. A. Bejarano-Garcia, J. Zimmer, J. A. Pérez-Simón, J. J. Toledo-Aral, T. Michel, M. S. Airaksinen and S. Méndez-Ferrer, *Blood*, 2019, **133**, 224.
70. A. G. Garcia and S. Méndez-Ferrer, *Front. Immunol.*, 2020, **11**, 956.
71. J. S. Chavez and E. M. Pietras, *Cell Stem Cell*, 2018, **23**, 459.
72. H. E. Broxmeyer, *Curr. Opin. Hematol.*, 2008, **15**, 49.
73. Y. Nie, Y.-C. Han and Y.-R. Zou, *J. Exp. Med.*, 2008, **205**, 777.
74. A. Aiuti, I. J. Webb, C. Bleul, T. Springer and J. C. Gutierrez-Ramos, *J. Exp. Med.*, 1997, **185**, 111.
75. C. Fielding and S. Méndez-Ferrer, *F1000Res.*, 2020, **9**, F1000 Faculty Rev-614.
76. S. Méndez-Ferrer, D. Lucas, M. Battista and P. S. Frenette, *Nature*, 2008, **452**, 442.
77. C. Lu, Y. Yang, R. Zhao, B. Hua, C. Xu, Z. Yan, N. Sung and R. Quian, *Protein Cell*, 2016, **7**, 820.
78. A. Ieyasu, Y. Tajima, S. Shimba, H. Nakauchi and S. Yamazaki, *J. Negat. Results Biomed.*, 2014, **13**, 4.
79. C. S. Colwell and S. Michel, *Nat. Neurosci.*, 2003, **6**, 1005.
80. E. E. Zhang, A. C. Liu, T. Hirota, L. J. Miraglia, G. Welch, Y. P. Pongsawakul, X. Liu, A. Atwood, J. W. Huss, J. Janes, A. I. Su, J. B. Hogenesch and S. A. Kay, *Cell*, 2009, **139**, 199.
81. O. Kollet, V. Vagima, G. D'Uva, K. Golan, J. Canaani, T. Itkin, S. Gur-Cohen, A. Kalinkovich, G. Caglio, C. Medaglia, A. Ludin, K. Lapid, E. Shezen, A. Neufield-Cohen, D. Varol, A. Chen and T. Lapidot, *Leukemia*, 2013, **27**, 2006.
82. X. Wu, G. Yu, H. Parks, T. Hebert, B. C. Goh, M. A. Dietrich, G. Pelled, R. Izadpanah, D. Gazit, B. A. Bunnell and J. M. Gimble, *Bone*, 2008, **42**, 861.
83. Y. He, F. Lin, Y. Chen, Z. Tan, D. Bai and Q. Zhao, *Stem Cells Dev.*, 2015, **24**, 1194.
84. J. Oliva-Ramirez, M. M. Moreno-Altamirano, B. Pineda-Olvera, P. Cauich-Sanchez and F. J. Sanchez-Garcia, *Immunology*, 2014, **143**, 490.
85. K. Golan, A. Kumari and O. Kollet, *et al.*, *Cell Stem Cell*, 2018, **23**, 572.
86. F. Lévi and U. Schibler, *Ann. Rev. Pharmacol.*, 2007, **47**, 593.
87. H. O'Leary, *Blood*, 2019, **133**, 186.
88. D. Pritchett and A. B. Reddy, *J. Biol. Rhythms*, 2015, **30**, 374.
89. M. A. Rieger and T. Schroeder, *Cold Spring Harbor Perspect. Biol.*, 2012, **4**, a008250.
90. A. L. Kierszenbaum, *Histology and Cell Biology. An Introduction to Pathology*, Mosby Elsevier, Philadelphia, PA, 2007.

CHAPTER 13

How Do Skeletal Tissues Keep Time? Circadian Rhythms in Cartilage and Bone

CÁTIA F. GONÇALVES[a,b] AND QING-JUN MENG*[a,b]

[a] Wellcome Centre for Cell Matrix Research, University of Manchester, UK;
[b] Faculty of Biology, Medicine and Health, University of Manchester, UK
*Email: qing-jun.meng@manchester.ac.uk

13.1 Introduction

Life on Earth is attuned to daily fluctuations in light intensity, temperature and humidity generated by the planet's rotation around its axis. It is therefore not surprising that organisms have evolved strategies not only to adapt to these 24-hour oscillations in environmental parameters but also to anticipate them. This is illustrated by behavioural aspects, such as the sleep–wake, rest–activity and feeding–fasting cycles, and by the 24-hour molecular oscillations that have been woven into the fabric of almost every lifeform on Earth – the circadian clock.[1]

The skeletal system is composed of bones and cartilage that are connected to ligaments, forming a framework for the rest of the body. It provides protection to vital organs (*e.g.*, brain, heart and lungs), allows for the movement of an organism, encloses the medullary cavities containing bone marrow wherein haematopoiesis occurs and is involved in the metabolic regulation of mineral homeostasis, functioning as a calcium and phosphate reservoir. In mammals, the central and peripheral circadian clocks have emerged as significant modulators of nearly all aspects of skeletal physiology

Chemical Biology No. 23
Chronobiology and Chronomedicine: From Molecular and Cellular Mechanisms to Whole Body Interdigitating Networks
Edited by Germaine Cornelissen and Tsuyoshi Hirota

Published by the Royal Society of Chemistry, www.rsc.org

in health and disease. In addition to the central clock in the SCN, peripheral clocks are present in most of the cell types in bone and cartilage, regulating various physiological functions, such as chondrocyte proliferation and differentiation, longitudinal bone growth, bone remodelling, and matrix turnover.[2–4] Throughout the ageing process, the circadian clock undergoes changes that lead to reduced oscillation amplitude, impaired coordination of rhythms across multiple organs and misalignment with the environment. In the skeletal system, these alterations have been associated with cellular dysfunction and the development of skeletal disorders such as osteoarthritis (OA), rheumatoid arthritis (RA) and osteoporosis (OP).[5–7] Therefore, understanding the interactions between the circadian clock and the skeletal system could provide opportunities to lessen the prevalence of skeletal diseases and/or improve outcomes.

In this chapter, we describe the mechanisms whereby the circadian clock regulates several facets of bone and cartilage homeostasis by summarising evidence from both *in vivo* and *in vitro* studies of mouse models. We also review the implications of circadian clock disruptions for the development and progression of skeletal diseases in humans. Finally, we discuss chronotherapies and other potential avenues to exploit circadian timing mechanisms in order to promote skeletal health.

13.2 The Skeletal System (Cartilage and Bone): Development and Maturation

During skeletal development, mesenchymal stem cells undergo chondrogenesis, a dynamic cellular process, through which chondrocytes originate and form cartilage, a specialised connective tissue that consists of a dense extracellular matrix (ECM) surrounding sparse populations of chondrocytes (see Figure 13.1).[8] Hyaline cartilage, the most common type of cartilage, forms the embryonic skeleton, which is subsequently replaced by bone *via* the process of endochondral ossification.[8,9] In the embryo, mesenchymal stem cells produce cartilaginous templates of all long bones that are enlarged through chondrocyte proliferation and the secretion of an ECM mainly comprising type II collagen and aggrecan.[9] Over time, chondrocytes near the centre of each skeletal element stop proliferating and differentiate into hypertrophic chondrocytes that produce a matrix rich in type X collagen.[9] Invasion of the hypertrophic cartilage by blood vessels is accompanied by the apoptosis of terminally differentiated chondrocytes, resorption of the calcified cartilage matrix by osteoclasts and the deposition of mineralised bone matrix enriched in type I collagen by osteoblasts.[9] By the time the foetal skeleton is fully formed, cartilage only remains in the articular surfaces of diarthrodial joints and between the diaphysis and epiphysis as the growth plate (or epiphyseal plate), which is responsible for longitudinal bone growth.[9] In humans, after birth and until early adulthood, cartilage is formed on the epiphyseal side of the growth plate and is ossified on the

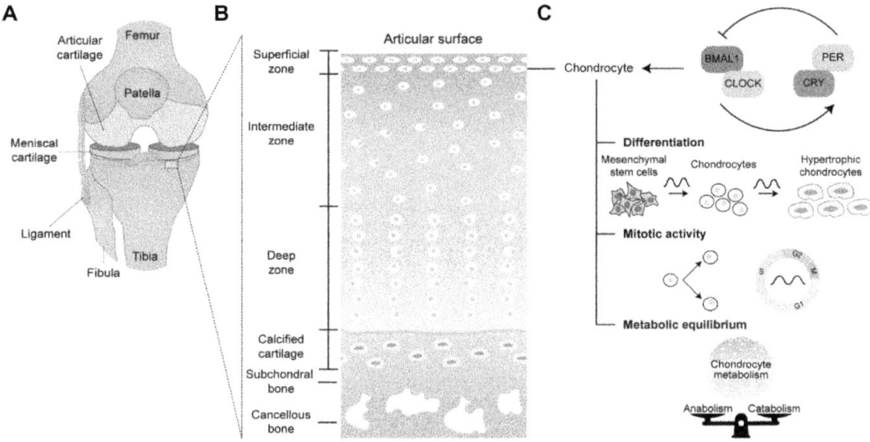

Figure 13.1 The circadian control of homeostasis in articular cartilage. (A) Schematic representation of the human knee joint showing the articular cartilage, a specialised type of connective tissue that covers the ends of the long bones where they articulate with each other within a joint. (B) Articular cartilage presents four distinct tissue histological zones that are correlated with chondrocyte morphology and orientation: the superficial (elongated chondrocytes aligned parallel to the articular surface), intermediate (obliquely distributed spherical chondrocytes), deep (chondrocytes organised in columns perpendicular to the joint surface) and calcified (hypertrophic chondrocytes that are embedded in a mineralised matrix) zones. (C) The circadian clock in chondrocytes is involved in the regulation of numerous biological processes crucial for chondrocyte function and cartilage homeostasis, including chondrocyte differentiation, and mitotic and metabolic activity (balance between anabolism and catabolism).

diaphyseal side. Bone lengthening stops in adults, the growth plate fuses and is replaced by bone, forming the epiphyseal line.[9] In mice, growth plate fusion does not occur during sexual maturation, albeit old mice present reduced longitudinal bone growth.[10]

In adulthood, chondrocytes rarely divide, meaning that the regenerative potential of articular cartilage is inherently limited.[11,12] In the case of bone, once formed, its structural integrity and mineral homeostasis is maintained through the processes of modelling and remodelling, which rely on the concerted action of three cell types: (1) osteoblasts, cells that produce the bone matrix and participate in its mineralisation; (2) osteocytes, derived from osteoblasts that have been incorporated into the ECM and possess extensive dendritic processes; and (3) osteoclasts, cells that degrade mineralised bone tissue (see Figure 13.2).[12–14] Bone modelling occurs mainly during childhood and continues at low rates in adulthood as an adaptive response to mechanical loading and physical exercise, and renal diseases.[15–19] In this process, the size and shape of a bone are modified by the uncoupled action of osteoblasts and osteoclasts on separate surfaces, thereby ensuring appropriate bone morphology and mass.[13,15] In contrast, bone remodelling

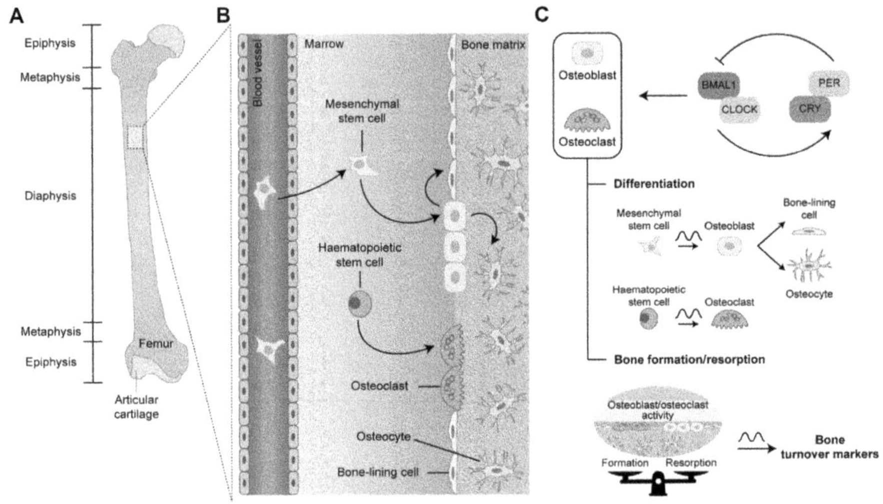

Figure 13.2 The circadian control of homeostasis in bone. (A) Schematic representation of a long bone, the human femur, showing the epiphysis, metaphysis and diaphysis as well as the articular cartilage. (B) Bone is a dynamic hard tissue that undergoes remodelling to maintain its function and integrity. In a finely tuned balance, haematopoietic stem cell-derived osteoclasts resorb the bony matrix and mesenchymal stem cell-derived osteoblasts generate newly synthesised matrix as they differentiate. Terminally differentiated osteoblasts that are incorporated into the mineralised bone matrix are designated osteocytes, and help sense mechanical strain. Bone-lining cells also belong to the osteoblast lineage and are thought to be involved in the coordination of bone resorption and bone formation. (C) In bone, the circadian clock not only coordinates bone cell differentiation but also coordinates the activity of these cells, leading to 24-hour rhythms in bone remodelling processes and turnover markers.

results from the coupled action of osteoclasts and osteoblasts on the same surface, leading to the replacement of bone matrix and the repair of small defects such as microfractures.[13,15]

13.3 The Mammalian Circadian System: A Multitude of Oscillators

The mammalian circadian system is organised in a hierarchy comprising multiple oscillators: the central clock, located in the suprachiasmatic nucleus (SCN) of the anterior hypothalamus, and peripheral oscillators scattered throughout the body.[20] The central clock is directly entrained by photic information (light–dark cycles), but it can also be reset by non-photic stimuli such as behavioural arousal, melatonin and serotonergic activation.[21–25] The SCN clock is not only involved in the regulation of 24-hour rhythms in behaviour (*e.g.*, locomotor activity and sleep–wake cycles) and physiology (*e.g.*, body temperature, pineal melatonin secretion and adrenal corticosterone release), but also acts as a pacemaker that relays temporal cues to peripheral oscillators *via* neural and humoral outputs.[26–33]

On a molecular level, the circadian clock consists of self-regulatory transcription–translation feedback loops, cycling with a period of approximately 24 hours, that regulate the rhythmic expression of clock-controlled genes.[34,35] In the primary feedback loop, the transcription factors CLOCK and BMAL1 heterodimerise in the cytoplasm before translocating into the nucleus, binding to E-box elements in the promoter region of target genes, including *Per* (*Per1*, *Per2* and *Per3*) and *Cry* (*Cry1* and Cry2).[36–40] Negative feedback is achieved by PER–CRY heterodimers that translocate back to the nucleus to repress their own transcription by inhibiting the activity of CLOCK–BMAL1 complexes.[36,41] The CLOCK–BMAL1 complex also activates the transcription of the genes encoding RORα and RORγ, which activate the transcription of *Bmal1*, and NR1D1 (or REV-ERBα), a nuclear receptor that competes with ROR proteins for their shared DNA binding elements and represses *Bmal1* expression.[42–44] A third loop involves the transcriptional activators DBP, TEF and HLF, whose expression is driven by CLOCK–BMAL1, and the repressor NFIL3 (or E4BP4), which is regulated by the REV–ERB–ROR loop; these proteins interact at sites containing D-boxes.[45,46] All in all, these interlocked feedback loops generate complex transcription–translation cycles with various phases of expression and govern multiple processes in cells.[47]

13.4 Circadian Rhythms in Skeletal Physiology

Circadian variations in the physiology of the skeletal system have been known since the sixties.[48] In rats, initial studies of endochondral ossification utilised isotopic labelling and demonstrated that the most intense window of matrix synthesis occurred during the photoperiod while mineralisation occurred at night.[49,50] This is supported by studies showing that the incorporation of [^{3}H]proline, [^{3}H]galactose, [^{14}C]proline or [^{3}H]thymidine peaked during the light phase in rat growth plates.[51,52] Circadian rhythms in the mitotic activity of growth plate chondrocytes in rats have also been described, with the highest mitotic index occurring in the early morning.[53,54] In humans, 24-hour variations in the serum levels of bone metabolism markers have been observed, including osteocalcin, alkaline phosphatase, calcium, sclerostin and parathyroid hormone (PTH).[55–57] In the context of skeletal disorders, 24-hour rhythms in the serum concentration of cartilage oligomeric matrix protein (COMP) have been described in patients with knee OA or RA.[58] Circadian variations in biochemical markers such as aggrecan, type II collagen, hyaluronic acid, keratan sulphate and transforming growth factor-β (TGF-β) have also been reported in patients with radiographic knee OA.[59] Furthermore, the symptoms of RA follow a 24-hour rhythm, with maximal pain and stiffness in the early morning.[60] In fact, this circadian variation in pain and stiffness is so important that in 1959 it became one of the diagnostic criteria of this disease.[61] Although the clinical relevance of circadian rhythms has been long appreciated, the molecular mechanisms behind these 24-hour rhythms in skeletal physiology were poorly understood until the discovery of the cell intrinsic molecular circadian clocks.

13.5 What Do Skeletal Tissues Use Circadian Timing For?

One way to infer the role of the circadian clock in cartilage and bone is to analyse their circadian rhythmic targets and compare them with those of other tissues. In the last decade, tissue-specific atlases of circadian gene expression have been compiled for primates and rodents, revealing that 5–20% of transcripts are circadian in any given tissue.[62–66] Strikingly, system-level analysis indicated that more than half of the genes encoding proteins exhibited 24-hour rhythms in their expression in at least one tissue.[62,64,65] These multi-organ studies have also demonstrated that circadian gene expression is tissue-specific, with little overlap between the genes under circadian control in each tissue.[62–66] These results highlight the need to investigate circadian timing mechanisms in different tissue systems and their niche environments. The molecular mechanisms driving this tissue specificity are not completely understood, but it is thought that it arises from cell type-specific differences in chromatin accessibility and the ability of circadian transcriptional activators such as BMAL1 and REV-ERBα to interact with tissue-specific transcription factors.[67,68]

Transcriptomic analysis of circadian gene expression in cartilage isolated from the xiphoid process or femoral head of mice, or from the ribs or femoral head of rats, has revealed that approximately 4% of the genes presented 24-hour rhythms.[69–71] Many of these genes participate in biological processes crucial for the maintenance of cartilage homeostasis, including genes encoding ECM structural components (*e.g.*, aggrecan and collagen), ECM remodelling enzymes (*e.g.*, *Adamts1*, *Adamts4* and *Adamts9*), calcium-dependent metalloproteinases (*e.g.*, *Mmp14*), inhibitors of metalloproteinases (*e.g.*, *Timp3* and *Timp4*) and growth factors/receptors (*e.g.*, *Tgfa*, *Tgfbr3* and *Egf*).[3,69–71] In calvarial bone isolated from mice, 26% of the genes represented in the microarray were rhythmic, including genes encoding proteins involved in BMP, WNT and FGF signalling, ECM structural components (*e.g.*, fibrillins and collagens) and remodelling enzymes (*e.g.*, *Adamts1*, *Adamts3* and *Adamts5*).[3,72] To the best of our knowledge, circadian transcriptomes in long bones or bone cells have not yet been reported. Nonetheless, circadian rhythms of PER2::Luc have been detected in *ex vivo* cultures of long (proximal femoral ends and radiuses) and flat (calvariae and scapulae) bones.[73] Moreover, the importance of the circadian regulation of bone physiology is further supported by mouse models, which will be discussed in the subsequent section. In summary, the available transcriptomic data suggest that circadian clocks in cartilage and bone are likely specialised in imparting temporal information to cell metabolism; their role is to co-ordinate and segregate the anabolic and catabolic pathways in time.

13.6 Roles of Circadian Clocks in the Skeletal System: Evidence from Mouse Studies

13.6.1 Cartilage

In mice, conditional deletion of *Bmal1* in *Col2a1*-expressing cells (*Col2a1-Bmal1*$^{-/-}$) led to early-onset cartilage degeneration and contributed to an imbalance between anabolic and catabolic factors by disrupting the time-dependent expression of many of these genes.[71] *Col2a1* is a marker of chondrocytes, the only cell type in articular cartilage. Remarkably, TGF-β signalling was dysregulated in articular cartilage from *Col2a1-Bmal1*$^{-/-}$ mice.[71] This cytokine not only plays a key role in cartilage development during embryogenesis, but is also an important regulator of cartilage homeostasis in adulthood.[74] For example, reduced TGF-β signalling in aged murine chondrocytes has been linked to a diminished repair capacity.[75,76] In human osteoarthritic cartilage, alterations in TGF-β signalling have been correlated with an increase in the expression of *Mmp13*, a matrix metalloproteinase.[77]

Targeted ablation of *Bmal1* in chondrocytes also led to a reduction in the expression of the gene encoding NFATc2 (also known as NFAT1 and NFATp), a transcription factor belonging to the NFAT family.[71] *In vitro*, the shRNA-mediated knockdown of *Nfatc2* in articular chondrocytes reduced the expression of chondrogenic markers (*e.g.*, *Acan* and *Col2a1*) but increased the expression of hypertrophic markers (*e.g.*, *Col10a1*), proinflammatory cytokines (*e.g.*, *Il1b* and *Il17a*) and matrix-degrading proteinases (*e.g.*, *Mmp13* and *Adamts5*).[78] In *Nfatc2*$^{-/-}$ mice, the ectopic formation of cartilage in extra-articular soft tissues was initiated at three months of age. Later in life, extensive cartilage degradation in appendicular joints was observed.[79,80] *Nfatc2*$^{-/-}$ mice also presented OA-like changes in the articular cartilage and synovium of the facet joints,[81] which were attributed to an imbalance in the expression of catabolic and anabolic genes favouring catabolism.[80,81] Indeed, ChIP assays demonstrated that NFATc2 regulates the promoter activities and expression of specific anabolic (*e.g.*, *Col2a1*, *Acan*, *Bmp7* and *Tgfb1*) and catabolic (*e.g.*, *Il1b*, *Tnf*, *Adamts5*, *Mmp13* and *Timp1*) genes in chondrocytes.[82]

The behaviourally arrhythmic *Clock*$^{\Delta 19}$ mice were reported to show accelerated cartilage degradation characterised by the loss of proteoglycans, cartilage and meniscus lesions, and an upregulation of genes encoding proinflammatory cytokines (*e.g.*, *Il1b* and *Il6*).[83] In cartilage from *Clock*$^{\Delta 19}$ mice, acetylation of NF-κB at Lys310 decreased and phosphorylation at Ser276 increased, thereby promoting the translocation of NF-κB into the nucleus.[83] Recently, it has been shown that CLOCK interacts with nuclear lamina components and heterochromatin-associated proteins to preserve the youth of human mesenchymal stem cells. Downregulation of CLOCK resulted in accelerated cellular senescence.[84] Strikingly, intra-articular administration of lentiviral vectors encoding CLOCK attenuated ageing-associated articular cartilage degeneration in mice, and promoted tissue regeneration.[84]

In human osteoarthritic cartilage, *CRY2* expression was significantly reduced.[85] At the protein level, CRY2 was expressed in all zones of articular cartilage in healthy tissue from mouse and human but reduced in osteoarthritic cartilage.[86] *Cry2*$^{-/-}$ mice with experimental OA (through surgical destabilisation of the medial meniscus) presented increased severity of histopathological changes in cartilage, subchondral bone and synovium.[86] Global transcriptome analysis supported a dysregulation of extracellular matrix remodelling in the absence of this core clock protein CRY2.[86]

Disruption of genes whose transcription is under the control of the circadian clock has also been associated with homeostatic imbalances in cartilage. For example, expression of *Adamts4* and *Mmp14*, which encode matrix-degrading enzymes, in mouse cartilage exhibited a 24-hour rhythm.[69,70] ADAMTS4 is involved in the degradation of proteoglycans, including aggrecan, one of the main constituents of cartilage.[87] In humans, this aggrecanase is upregulated in osteoarthritic cartilage and can be induced by proinflammatory cytokines.[88–90] MMP14, the only membrane-associated MMP, is involved in the regulation of TGF-β availability in cells and in the activation of other metalloproteinases, including MMP13 and MMP9, highlighting its significance in arthritic conditions.[91,92] Ablation of *Mmp14* in mice caused severe arthritis that resembled murine collagen-induced arthritis and human RA.[93]

Melatonin, a hormone primarily synthesised by the pineal gland during darkness, is a critical output of the circadian system. This hormone has been reported to enhance the chondrogenic differentiation of human mesenchymal stem cells as well as the synthesis of cartilage matrix by porcine chondrocytes.[94,95] Recently, it was shown that chondrocytes express the genes encoding melatonin-synthesising enzymes (AANAT and HIOMT) and melatonin receptors (MT1 and MT2) in the growth plate. Exogenous melatonin stimulated chondrocyte proliferation, upregulated the expression of cartilage-associated genes (*e.g.*, *Ihh*, *Col2a1*, *Acan* and *Sox9*) and modified the rhythmic expression of *Bmal1*, *Mt1*, *Mt2* and *Aanat*,[96] supporting a potential pharmacological use of melatonin.[96]

Another circadian hormone relevant for skeletal physiology is PTH, with peak levels occurring between 2 AM and 6 AM in humans.[97] In mouse femoral growth plate, bioluminescence imaging has demonstrated that PTH can reset PER2::Luc oscillations in a time and dose-dependent manner.[98] Furthermore, mice carrying a mutation in the gene encoding PTHrP (parathyroid hormone-related peptide) presented multiple abnormalities in the process of endochondral ossification resulting in a lethal skeletal dysplasia,[99] which was associated with a decrease in chondrocyte proliferation that was linked to premature chondrocyte maturation and accelerated bone formation, suggesting that PTHrP is a modulator of chondrocyte differentiation.[99]

13.6.2 Bone

A growing body of evidence using transgenic mouse models supports the notion that bone homeostasis relies on the temporal coordination of osteoclast-mediated bone resorption and osteoblast-mediated bone formation. In adult mice, global loss of BMAL1 resulted in a low bone mass phenotype characterised by significant reductions in bone mineral density and volume as well as in the number of active osteoblasts and osteocytes.[100–102] Bone marrow-derived mesenchymal stem cells isolated from *Bmal1*$^{-/-}$ mice demonstrated a reduced ability to differentiate into osteoblasts *in vitro*.[102] In addition, it has been demonstrated that BMAL1 modulates bone resorption and osteoclastogenesis by regulating the calcitriol-induced expression of *Rankl* in osteoblasts. The bone mass of vertebrae from osteocyte-specific *Bmal1*-deficient mice was comparable to that of *Bmal1* floxed mice, suggesting that osteocytes might contribute less to the bone mass phenotype presented by *Bmal1*$^{-/-}$ mice.[101] The low bone mass phenotype was recapitulated in mice with conditional *Bmal1* deletion in mesenchymal cells (*Bmal1*$^{f/f}$. *Prx1-cre*) by promoting osteoclast function.[103] These findings are in stark contrast to another study, which demonstrated that *Bmal1*$^{-/-}$ adolescent mice had a higher number of osteoblasts and exhibited increased rates of bone formation and mineral apposition.[104] The role of BMAL1 in bone resorption has been further investigated by specifically knocking out *Bmal1* in osteoclasts.[105] These mice presented a high bone mass phenotype due to a decrease in the differentiation of osteoclasts. *In vitro* studies revealed that the CLOCK–BMAL1 complex upregulated the transcription of *Nfatc1*, a master regulator of osteoclast differentiation, by binding to its promoter.[105,106] These conflicting results suggest that BMAL1 might have different roles according to the stage of skeletal development, and in different cell types within the bone. Moreover, some of the phenotypes observed in *Bmal1*$^{-/-}$ mice (*e.g.*, abnormal bone calcification) could not be recapitulated upon tamoxifen-induced ablation of *Bmal1* in adult mice, indicating that some of these might result from clock-independent BMAL1 functions.[107]

Moreover, BMAL1 expression was reduced in bone marrow cells isolated from a rat model of type 2 diabetes mellitus, a condition that is associated with bone loss. Subsequent *in vitro* studies indicated that this core clock protein regulates the differentiation of bone marrow mesenchymal stem cells through modulation of the NF-κB pathway, fine-tuning the balance between osteogenesis and osteoclastogenesis.[108] Despite these genetic mouse models, very little effort has been directed towards demonstrating cell intrinsic circadian rhythms in different bone cell types, which clearly warrants further investigation.

Mice lacking *Per1* and *Per2* (*Per1*$^{-/-}$*Per2*$^{-/-}$), *Per1* and the Per-Arnt-Sim (PAS) domain of *Per2* (*Per1*$^{-/-}$*Per2*$^{m/m}$) or *Cry1* and *Cry2* (*Cry1*$^{-/-}$*Cry2*$^{-/-}$) presented a high bone mass phenotype characterised by an increase in the number of osteoblasts, increased mineral apposition and higher bone formation rate.[104] A subsequent study utilising *Per2*$^{m/m}$*Cry2*$^{-/-}$ indicated that *Cry2* and *Per2* modulate osteoclast and osteoblast function, respectively.[109] All in all, the evidence provided by mouse models deficient in core circadian clock components strongly suggests that bone remodelling is under circadian control, with effects varying depending on the cell type and developmental stage.

Bone formation/resorption is also modulated by systemic signalling through hormones such as glucocorticoids and leptin. Stimulation of osteoblasts and osteoclasts with dexamethasone, a glucocorticoid hormone analogue, induced the rhythmic expression of circadian genes.[110,111] In bone tissue from adrenalectomised mice, the altered circadian rhythms of osteoclast-specific genes (*e.g.*, *Ctsk* and *Nfatc1*) were restored after intraperitoneal injection of dexamethasone.[111] *In vitro* assays have also demonstrated that overexpression of the transcriptional regulator *Gilz* (glucocorticoid-induced leucine zipper) blocked *Ppar*γ transcription, thereby enhancing osteogenic differentiation of bone marrow stromal cells.[112–114] Leptin, a hormone predominantly secreted by adipocytes, is involved in the modulation of bone remodelling through its action on both osteoblasts and osteoclasts.[104,115–117] In osteoblasts, leptin-mediated sympathetic signalling operated through β2-adrenergic receptors on the surface of the cell, activating two antagonistic pathways: (1) the upregulation of *Per1* and *Per2*, which in turn inhibited osteoblast proliferation by suppressing the expression of G1 cyclins; and (2) the upregulation of the AP-1 family of transcription factors, which stimulated osteoblast proliferation by upregulating G1 cyclins.[104,118]

In addition to its mechanical functions, bone is an important calcium and phosphate reservoir and regulates the metabolic pathways linked to mineral ion homeostasis.[119] In mice, hypophosphatemia led to abnormal growth plate maturation and bone mineralisation due to an impairment in the caspase-mediated apoptosis of hypertrophic chondrocytes.[120,121] In mice fed a low phosphate diet, expression of circadian clock genes was higher than in the normal diet group. A network analysis of genes differentially regulated by hypophosphatemia revealed *Ezh2* as one of the main regulators associated with the observed transcriptome changes.[122] EZH2 is a key epigenetic regulator of skeletal longitudinal growth that also binds to the promoter region of *Per1* and *Per2*.[123–125] Short hairpin RNA (shRNA)-mediated silencing of this histone methyltransferase disrupted circadian oscillations in mouse fibroblasts.[123] In early pubertal mice, deletion of *Ezh2* in nestin-expressing cells was associated with a depletion in the pool of mesenchymal progenitor cells, premature cellular senescence, impaired osteogenesis and a higher risk of osteoporosis later in adulthood.[126]

The skeleton has been proposed as an endocrine organ that is central to the regulation of energy metabolism.[5,127] Nocturnin (*Noc*) encodes an NADPH phosphatase that participates in the post-transcriptional regulation of genes necessary for nutrient uptake, metabolism and storage.[128,129] CLOCK–BMAL1 heterodimers bind to the E-box element in the promoter region of *Noc*, thereby driving its circadian transcription.[130] NOC is involved in the differentiation of bone marrow stromal cells by stimulating adipogenesis and suppressing osteoblastogenesis. Ablation of *Noc* in mice ($Noc^{-/-}$) was linked to a reduction in the number of adipocytes, indicating that in the absence of this protein, osteoblastogenesis is favoured.[131,132] NOC also modulates the activity of peroxisome proliferator-activated receptor γ (PPARγ) by promoting its nuclear translocation.[131] The activation of PPARγ favoured the differentiation of bone stromal cells into adipocytes and suppressed osteogenic signalling pathways (*e.g.*, BMP, TGF-β and WNT).[133] *In vivo* studies have also demonstrated that Pparγ silencing decreased the number of bone marrow adipocytes and increased trabecular bone formation.[134] In bone mesenchymal stem cells, overexpression of *Nr1d1*, a direct target of PPARγ and BMAL1, and itself a core clock gene, led to the inhibition of cell proliferation and osteogenesis.[135,136]

13.7 Circadian Rhythm Disruption and Skeletal Disease: Clocking in on Novel Therapies

In humans, circadian rhythm disruption has been increasingly recognised as a risk factor for the development of skeletal disease. Shift work (*e.g.*, night shifts, rotating shifts and irregular schedules) constitutes one of the main causes of circadian rhythm disruption. It leads to a misalignment between endogenous clock-controlled circadian rhythms and work, sleep and eating schedules.[137] Epidemiological studies have found a correlation between shift work and low mineral density in trabecular and cortical bones as well as a greater risk of hip and wrist fractures, and knee OA.[138–140] In addition, alterations in the normal function of the circadian clock have also been found in RA and OA patients, suggesting that circadian rhythm disruption may not only contribute to the development of pathophysiological changes but also contribute to disease progression. For example, patients with RA have disturbances in the hypothalamic–pituitary–adrenal axis that are reflected in altered circadian rhythms of serum cortisol, melatonin and IL-6.[141–143] In synovial fibroblasts from RA patients, the circadian expression of core clock genes is disturbed to the extent that *ARNTL* (or *BMAL1*) and *NPAS2* are expressed in phase with *PER1*.[144] This is accompanied by strong BMAL1 expression in the cytoplasm (which is normally expressed predominantly in the nucleus) and altered production of IL-6 and IL-1β even after a serum shock.[144,145] Genome-wide expression studies have revealed that circadian rhythm pathways are among the most disrupted in human osteoarthritic cartilage, a result supported by several studies showing that the expression of *BMAL1*/BMAL1 is reduced.[71,146–149] Moreover, knocking down of *BMAL1* in human chondrocytes led to an increase in the expression of catabolic genes such as *MMP1*, *MMP3*, *MMP13* and *ADAMTS5*, and abnormal TGF-β signalling.[146,148–150]

Although the misalignments between the circadian clock and external time cues (*e.g.*, light, food, temperature, and physical exercise) can lead to circadian disruption, these time cues also have the potential to promote skeletal homeostasis if provided at the right times (see Figure 13.3). For example, light therapy has been utilised to improve mood and restore sleep–wake cycles in patients with depression.[151] Regular physical exercise has been found to improve musculoskeletal function and lessen symptom

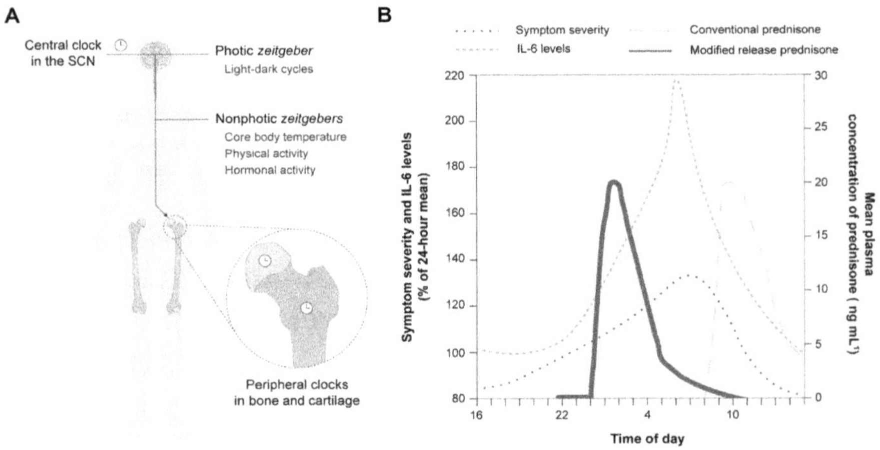

Figure 13.3 Harnessing the circadian timing mechanism to improve outcomes in rheumatic diseases. (A) The central clock located in the suprachiasmatic nucleus of the hypothalamus can be synchronised by photic Zeitgebers (time cues), such as the light–dark cycles. This clock modulates the circadian rhythms of numerous physiological processes, including 24-hour fluctuations in core body temperature, physical activity and hormonal signalling. These time cues can be sensed by the peripheral clocks in bone and cartilage and utilised to entrain local circadian rhythms. Disruptions of these Zeitgebers have been associated with pathophysiological changes in bone and cartilage. Restoration of this circadian timing (*e.g.*, scheduled physical exercise) has the potential to improve rhythmic processes in these tissues. (B) As an example, patients with active rheumatoid arthritis have circadian variations in symptoms (pain, joint stiffness and functional disability) that are correlated with the nocturnal increase in the levels of circulating proinflammatory cytokines such as IL-6, and reach a peak around 6 AM. Conventional prednisone is administrated in the morning, which misses the night-time initiation of pathophysiological processes. Administration of modified-release prednisone allows for convenient bedtime administration and drug concentration is maximal at the optimal time for suppression of the peak of inflammatory response.[169,170]

severity in patients with OA and RA.[152,153] In addition, evidence suggests that exercise is capable of phase shifting circadian rhythms, as observed through shifts in melatonin oscillations. These shifts depend on the duration and intensity of the activity and have the potential to accelerate adaptation to schedule readjustments.[154–157] Studies in mice also indicate that exercise can help counter ageing-related circadian rhythm dysregulation as aged mice with access to a wheel presented improved sleep–wake cycles.[158,159]

The marked circadian variation in skeletal physiology and its dysregulation in diseased tissue raise the intriguing possibility that targeting circadian rhythms could prompt new drug therapies or improve existing ones. Circadian medicine approaches that maximise treatment effectiveness and minimise adverse effects by timing drug intake/interventions and tailoring it to the patient's circadian rhythms were first described for OA and RA more than 40 years ago.[141,160,161] For instance, indomethacin evening dosing was more effective in OA patients with predominantly nocturnal or morning pain whereas patients with afternoon or evening pain benefitted more from a morning or noon dose.[161] In RA, coordinating glucocorticoid administration with the night-time rise in serum levels of IL-6 greatly reduced joint pain and stiffness in comparison to the standard administration of prednisolone early in the morning.[162] A modified-release prednisone tablet designed to release the drug approximately four hours after ingestion allowed for convenient bedtime administration while still markedly increasing treatment response rates and improving physical function in RA patients.[163–165] In OP patients, administration of teriparatide in the morning for 12 months resulted in a larger increase in the lumbar spine bone mineral density than the evening application.[166] In patients with postsurgical hypoparathyroidism, pump delivery of PTH1-34 to mimic normal circadian rhythms of PTH (parathyroid hormone) led to a remarkable decrease in both urine and calcium excretion in comparison to twice-daily administration. The use of pump delivery was also associated with the restoration of bone turnover markers (*e.g.*, osteocalcin and alkaline phosphatase) to normal levels.[167] These examples provide strong evidence in favour of evaluating time-of-administration in clinical trials when introducing new drugs or even to optimise old ones. In fact, a recent analysis of more than 100 human clinical trials has revealed that 75% of them showed a dosing-time-dependent efficacy or toxicity.[168]

13.8 Concluding Remarks

Over the last decade or so, emerging evidence has pointed towards the critical roles of circadian clocks in directing tissue physiology to not only adapt to 24-hour changes in the tissue microenvironment but also anticipate them. Indeed, many aspects of skeletal biology in cartilage and bone are under rhythmic control, such as cell division, metabolism, tissue remodelling and turnover. Cell intrinsic circadian clocks have been demonstrated in key cell types within the skeletal system and tissue-specific targets revealed by circadian time series "omics" approaches. Importantly, genetic mouse models have revealed key functional links between clock genes and tissue homeostasis in cartilage and bone, indicating that circadian biology is an important yet often overlooked critical regulatory mechanism in these tissues. Understanding how skeletal tissues perform in accordance with the internal circadian time could have relevance to sports medicine and athletic performance. In addition, symptoms of skeletal diseases tend to show exacerbations that are underpinned by rhythmic variations of pathophysiological factors. As such, further investigation of circadian rhythms and skeletal diseases has the potential to lead to the development of improved treatment strategies (a "chrono-medicine" approach) for skeletal disorders, many of which are highly prevalent, debilitating, painful, and currently lack disease-modifying drugs.

References

1. D. Bell-Pedersen, V. M. Cassone, D. J. Earnest, S. S. Golden, P. E. Hardin, T. L. Thomas and M. J. Zoran, *Nat. Rev. Genet.*, 2005, **6**, 544–556.
2. M. Dudek and Q.-J. Meng, *Biochem. J.*, 2014, **463**, 1–8.
3. N. Yang and Q.-J. Meng, *J. Biol. Rhythms*, 2016, **31**, 415–427.
4. C. F. Gonçalves and Q.-J. Meng, *J. Endocrinol.*, 2019, **243**, R29–R46.
5. M. Kawai and C. J. Rosen, *Nat. Rev. Endocrinol.*, 2010, **6**, 629–636.
6. J. E. Gibbs and D. W. Ray, *Arthritis Res. Ther.*, 2013, **15**, 205.
7. F. Berenbaum and Q.-J. Meng, *Nat. Rev. Rheumatol.*, 2016, **12**, 508–516.
8. A. L. Mescher, in *Junqueira's Basic Histology: Text and Atlas*, McGraw-Hill Education, New York, NY, 14th edn, 2016, pp. 129–137.
9. E. J. Mackie, Y. A. Ahmed, L. Tatarczuch, K.-S. Chen and M. Mirams, *Int. J. Biochem. Cell Biol.*, 2008, **40**, 46–62.
10. O. Vidal, M. Lindberg, L. Sävendahl, D. B. Lubahn, E. M. Ritzen, J. Å. Gustafsson and C. Ohlsson, *Biochem. Biophys. Res. Commun.*, 1999, **265**, 569–571.
11. C. W. Archer and P. Francis-West, *Int. J. Biochem. Cell Biol.*, 2003, **35**, 401–404.
12. M. Iwamoto, Y. Ohta, C. Larmour and M. Enomoto-Iwamoto, *Birth Defects Res., Part C*, 2013, **99**, 192–202.
13. M. R. Allen and D. B. Burr, in *Basic and Applied Bone Biology*, ed. D. B. Burr and M. R. Allen, Academic Press, San Diego, 2014, pp. 75–90.
14. R. Florencio-Silva, G. R. da S. Sasso, E. Sasso-Cerri, M. J. Simões and P. S. Cerri, *Biomed Res. Int.*, 2015, **2015**, 1–17.
15. R. Bartl and C. Bartl, in *Bone Disorders : Biology, Diagnosis, Prevention, Therapy*, ed. R. Bartl and C. Bartl, Springer International Publishing, Cham, 2017, pp. 21–30.
16. Y. Ubara, T. Tagami, S. Nakanishi, N. Sawa, J. Hoshino, T. Suwabe, H. Katori, F. Takemoto, S. Hara and K. Takaichi, *Kidney Int.*, 2005, **68**, 833–839.
17. Y. Ubara, T. Fushimi, T. Tagami, N. Sawa, J. Hoshino, M. Yokota, H. Katori, F. Takemoto and S. Hara, *Kidney Int.*, 2003, **63**, 1809–1816.

18. D. B. Burr, M. B. Schaffler, K. H. Yang, D. D. Wu, M. Lukoschek, D. Kandzari, N. Sivaneri, J. D. Blaha and E. L. Radin, *Bone*, 1989, **10**, 215–221.
19. H. Krahl, U. Michaelis, H. G. Pieper, G. Quack and M. Montag, *Am. J. Sports Med.*, 1994, **22**, 751–757.
20. S. Honma, *J. Physiol. Sci.*, 2018, **68**, 207–219.
21. S. Panda, T. K. Sato, A. M. Castrucci, M. D. Rollag, W. J. DeGrip, J. B. Hogenesch, I. Provencio and S. A. Kay, *Science*, 2002, **298**, 2213–2216.
22. C. Dibner, U. Schibler and U. Albrecht, *Annu. Rev. Physiol.*, 2010, **72**, 517–549.
23. E. Challet and P. Pévet, *Front. Biosci.*, 2003, **8**, s246–s257.
24. J. J. Gooley, J. Lu, T. C. Chou, T. E. Scammell and C. B. Saper, *Nat. Neurosci.*, 2001, **4**, 1165.
25. S. Hattar, H. W. Liao, M. Takao, D. M. Berson and K. W. Yau, *Science*, 2002, **295**, 1065–1070.
26. R. Y. Moore and V. B. Eichler, *Brain Res.*, 1972, **42**, 201–206.
27. R. Y. Moore and D. C. Klein, *Brain Res.*, 1974, **71**, 17–33.
28. C. I. Eastman, R. E. Mistlberger and A. Rechtschaffen, *Physiol. Behav.*, 1984, **32**, 357–368.
29. A. Kramer, F.-C. Yang, P. Snodgrass, X. Li, T. E. Scammell, F. C. Davis and C. J. Weitz, *Science*, 2001, **294**, 2511–2515.
30. F. A. J. L. Scheer, C. Pirovano, E. J. W. Van Someren and R. M. Buijs, *Neuroscience*, 2005, **132**, 465–477.
31. A. Balsalobre, S. A. Brown, L. Marcacci, F. Tronche, C. Kellendonk, H. M. Reichardt, G. Schütz and U. Schibler, *Science*, 2000, **289**, 2344–2347.
32. U. Albrecht, *Neuron*, 2012, **74**, 246–260.
33. F. N. Buijs, F. Cazarez, M. C. Basualdo, F. A. J. L. Scheer, M. Perusquía, D. Centurion and R. M. Buijs, *Neuroscience*, 2014, **266**, 197–207.
34. J. A. Mohawk, C. B. Green and J. S. Takahashi, *Annu. Rev. Neurosci.*, 2012, **35**, 445–462.
35. S. M. Reppert and D. R. Weaver, *Nature*, 2002, **418**, 935–941.
36. P. L. Lowrey and J. S. Takahashi, *Adv. Genet.*, 2011, **74**, 175–230.
37. D. P. King, Y. Zhao, A. M. Sangoram, L. D. Wilsbacher, M. Tanaka, M. P. Antoch, T. D. Steeves, M. H. Vitaterna, J. M. Kornhauser, P. L. Lowrey, F. W. Turek and J. S. Takahashi, *Cell*, 1997, **89**, 641–653.
38. N. Gekakis, D. Staknis, H. B. Nguyen, F. C. Davis, L. D. Wilsbacher, D. P. King, J. S. Takahashi and C. J. Weitz, *Science*, 1998, **280**, 1564–1569.
39. K. Kume, M. J. Zylka, S. Sriram, L. P. Shearman, D. R. Weaver, X. Jin, E. S. Maywood, M. H. Hastings and S. M. Reppert, *Cell*, 1999, **98**, 193–205.
40. L. P. Shearman, S. Sriram, D. R. Weaver, E. S. Maywood, I. Chaves, B. Zheng, K. Kume, C. C. Lee, G. T. van der Horst, M. H. Hastings and S. M. Reppert, *Science*, 2000, **288**, 1013–1019.
41. C. Lee, J. P. Etchegaray, F. R. Cagampang, A. S. Loudon and S. M. Reppert, *Cell*, 2001, **107**, 855–867.

42. N. Preitner, F. Damiola, L. Lopez-Molina, J. Zakany, D. Duboule, U. Albrecht and U. Schibler, *Cell*, 2002, **110**, 251–260.
43. Y. Zhang, B. Fang, M. J. Emmett, M. Damle, Z. Sun, D. Feng, S. M. Armour, J. R. Remsberg, J. Jager, R. E. Soccio, D. J. Steger and M. A. Lazar, *Science*, 2015, **348**, 1488–1492.
44. T. K. Sato, S. Panda, L. J. Miraglia, T. M. Reyes, R. D. Rudic, P. McNamara, K. A. Naik, G. A. FitzGerald, S. A. Kay and J. B. Hogenesch, *Neuron*, 2004, **43**, 527–537.
45. F. Gachon, P. Fonjallaz, F. Damiola, P. Gos, T. Kodama, J. Zakany, D. Duboule, B. Petit, M. Tafti and U. Schibler, *Genes Dev.*, 2004, **18**, 1397–1412.
46. S. Mitsui, S. Yamaguchi, T. Matsuo, Y. Ishida and H. Okamura, *Genes Dev.*, 2001, **15**, 995–1006.
47. H. R. Ueda, S. Hayashi, W. Chen, M. Sano, M. Machida, Y. Shigeyoshi, M. Iino and S. Hashimoto, *Nat. Genet.*, 2005, **37**, 187–192.
48. D. J. Simmons, *Nature*, 1964, **202**, 906–907.
49. J. E. Russell, B. Grazman and D. J. Simmons, *Proc. Soc. Exp. Biol. Med.*, 1984, **176**, 342–345.
50. D. J. Simmons, C. Arsenis, S. W. Whitson, S. E. Kahn, A. L. Boskey and N. Gollub, *Miner. Electrolyte Metab.*, 1983, **9**, 28–37.
51. J. E. Russell, D. J. Simmons, B. Huber and B. A. Roos, *Endocrinology*, 1983, **113**, 2035–2042.
52. K. Igarashi, S. Saeki and H. Shinoda, *Orthod. Waves*, 2013, **72**, 11–15.
53. K. V. Walker and N. F. Kember, *Cell Tissue Kinet.*, 1972, **5**, 401–408.
54. S. Stevenson, E. B. Hunziker, W. Herrmann and R. K. Schenk, *J. Orthop. Res.*, 1990, **8**, 132–135.
55. C. M. Gundberg, M. E. Markowitz, M. Mizruchi and J. F. Rosen, *J. Clin. Endocrinol. Metab.*, 1985, **60**, 736–739.
56. S. L. Greenspan, R. Dresner-Pollak, R. A. Parker, D. London and L. Ferguson, *Calcif. Tissue Int.*, 1997, **60**, 419–423.
57. C. Swanson, S. A. Shea, P. Wolfe, S. Markwardt, S. W. Cain, M. Munch, C. A. Czeisler, E. S. Orwoll and O. M. Buxton, *Osteoporos Int.*, 2017, **28**, 3205–3213.
58. M. L. E. Andersson, I. F. Petersson, K. E. Karlsson, E. N. Jonsson, B. Månsson, D. Heinegård and T. Saxne, *Ann. Rheum. Dis.*, 2006, **65**, 1490–1494.
59. S. Y. Kong, T. V. Stabler, L. G. Criscione, A. L. Elliott, J. M. Jordan and V. B. Kraus, *Arthritis Rheum.*, 2006, **54**, 2496–2504.
60. G. Labrecque, J.-P. Bureau and A. E. Reinberg, *Pharmacol. Ther.*, 1995, **66**, 285–300.
61. M. W. Ropes, G. A. Bennett, S. Cobb, R. Jacox and R. A. Jessar, *Arthritis Rheum.*, 1959, **2**, 16–20.
62. M. D. Ruben, G. Wu, D. F. Smith, R. E. Schmidt, L. J. Francey, Y. Y. Lee, R. C. Anafi and J. B. Hogenesch, *Sci. Transl. Med.*, 2018, **10**, eaat8806.
63. R. C. Anafi, L. J. Francey, J. B. Hogenesch and J. Kim, *Proc. Natl. Acad. Sci. U. S. A.*, 2017, **114**, 5312–5317.

64. R. Zhang, N. F. Lahens, H. I. Ballance, M. E. Hughes and J. B. Hogenesch, *Proc. Natl. Acad. Sci. U. S. A.*, 2014, **111**, 16219–16224.
65. L. S. Mure, H. D. Le, G. Benegiamo, M. W. Chang, L. Rios, N. Jillani, M. Ngotho, T. Kariuki, O. Dkhissi-Benyahya, H. M. Cooper and S. Panda, *Science*, 2018, **359**, eaao0318.
66. P. D. Mavroudis, D. C. DuBois, R. R. Almon and W. J. Jusko, *PLoS One*, 2018, **13**, e0197258.
67. J. R. Beytebiere, A. J. Trott, B. J. Greenwell, C. A. Osborne, H. Vitet, J. Spence, S.-H. Yoo, Z. Chen, J. S. Takahashi, N. Ghaffari and J. S. Menet, *Genes Dev.*, 2019, **33**, 294–309.
68. G. Caratti, M. Iqbal, L. Hunter, D. Kim, P. Wang, R. M. Vonslow, N. Begley, A. J. Tetley, J. L. Woodburn, M. Pariollaud, R. Maidstone, I. J. Donaldson, Z. Zhang, L. M. Ince, G. Kitchen, M. Baxter, T. M. Poolman, D. A. Daniels, D. R. Stirling, C. Brocker, F. Gonzalez, A. S. I. Loudon, D. A. Bechtold, M. Rattray, L. C. Matthews and D. W. Ray, *J. Clin. Invest.*, 2018, **128**, 4454–4471.
69. N. Gossan, L. Zeef, J. Hensman, A. Hughes, J. F. Bateman, L. Rowley, C. B. Little, H. D. Piggins, M. Rattray, R. P. Boot-Handford and Q.-J. Meng, *Arthritis Rheum.*, 2013, **65**, 2334–2345.
70. K. K. Honda, T. Kawamoto, H. R. Ueda, A. Nakashima, T. Ueshima, R. G. Yamada, M. Nishimura, R. Oda, S. Nakamura, T. Kojima, M. Noshiro, K. Fujimoto, S. Hashimoto and Y. Kato, *J. Biochem.*, 2013, **154**, 373–381.
71. M. Dudek, N. Gossan, N. Yang, H.-J. Im, J. P. D. Ruckshanthi, H. Yoshitane, X. Li, D. Jin, P. Wang, M. Boudiffa, I. Bellantuono, Y. Fukada, R. P. Boot-Handford and Q.-J. Meng, *J. Clin. Invest.*, 2016, **126**, 365–376.
72. S. Zvonic, A. A. Ptitsyn, G. Kilroy, X. Wu, S. A. Conrad, L. K. Scott, F. Guilak, G. Pelled, D. Gazit and J. M. Gimble, *J. Bone Miner. Res.*, 2007, **22**, 357–365.
73. N. Okubo, Y. Minami, H. Fujiwara, Y. Umemura, Y. Tsuchiya, T. Shirai, R. Oda, H. Inokawa, T. Kubo and K. Yagita, *PLoS One*, 2013, **8**, e78306.
74. W. Wang, D. Rigueur and K. M. Lyons, *Birth Defects Res., Part C*, 2014, **102**, 37–51.
75. A. Scharstuhl, H. M. van Beuningen, E. L. Vitters, P. M. van der Kraan and W. B. van den Berg, *Ann. Rheum. Dis.*, 2002, **61**, 1095–1098.
76. E. N. Blaney Davidson, A. Scharstuhl, E. L. Vitters, P. M. van der Kraan and W. B. van den Berg, *Arthritis Res. Ther.*, 2005, **7**, R1338–R1347.
77. E. N. Blaney Davidson, D. F. G. Remst, E. L. Vitters, H. M. van Beuningen, A. B. Blom, M.-J. Goumans, W. B. van den Berg and P. M. van der Kraan, *J. Immunol.*, 2009, **182**, 7937–7945.
78. M. Rodova, Q. Lu, Y. Li, B. G. Woodbury, J. D. Crist, B. M. Gardner, J. G. Yost, X. Zhong, H. C. Anderson and J. Wang, *J. Bone Miner. Res.*, 2011, **26**, 1974–1986.
79. A. M. Ranger, L. C. Gerstenfeld, J. Wang, T. Kon, H. Bae, E. M. Gravallese, M. J. Glimcher and L. H. Glimcher, *J. Exp. Med.*, 2000, **191**, 9–22.

80. J. Wang, B. M. Gardner, Q. Lu, M. Rodova, B. G. Woodbury, J. G. Yost, K. F. Roby, D. M. Pinson, O. Tawfik and H. C. Anderson, *J. Pathol.*, 2009, **219**, 163–172.
81. J. Wang, Q. Lu, Y. Feng and D. Burton, *Osteoarthritis Cartilage*, 2020, **28**, S192.
82. M. Zhang, Q. Lu, T. Budden and J. Wang, *Bone Jt. Res.*, 2019, **8**, 90–100.
83. G. Yuan, L. Xu, T. Cai, B. Hua, N. Sun, Z. Yan, C. Lu and R. Qian, *Osteoarthritis Cartilage*, 2019, **27**, 922–931.
84. C. Liang, Z. Liu, M. Song, W. Li, Z. Wu, Z. Wang, Q. Wang, S. Wang, K. Yan, L. Sun, T. Hishida, Y. Cai, J. C. I. Belmonte, P. Guillen, P. Chan, Q. Zhou, W. Zhang, J. Qu and G.-H. Liu, *Cell Res.*, 2020, 1–19.
85. K. M. Fisch, R. Gamini, O. Alvarez-Garcia, R. Akagi, M. Saito, Y. Muramatsu, T. Sasho, J. A. Koziol, A. I. Su and M. K. Lotz, *Osteoarthritis Cartilage*, 2018, **26**, 1531–1538.
86. H. Bekki, T. Duffy, N. Okubo, M. Olmer, O. Alvarez-Garcia, K. Lamia, S. Kay and M. Lotz, *Osteoarthritis Cartilage*, 2020, **28**, 966–976.
87. P. Verma and K. Dalal, *J. Cell. Biochem.*, 2011, **112**, 3507–3514.
88. H. I. Roach, N. Yamada, K. S. C. Cheung, S. Tilley, N. M. P. Clarke, R. O. C. Oreffo, S. Kokubun and F. Bronner, *Arthritis Rheum.*, 2005, **52**, 3110–3124.
89. S. Naito, T. Shiomi, A. Okada, T. Kimura, M. Chijiiwa, Y. Fujita, T. Yatabe, K. Komiya, H. Enomoto, K. Fujikawa and Y. Okada, *Pathol. Int.*, 2007, **57**, 703–711.
90. R.-H. Song, M. D. Tortorella, A.-M. Malfait, J. T. Alston, Z. Yang, E. C. Arner and D. W. Griggs, *Arthritis Rheum.*, 2007, **56**, 575–585.
91. V. Knäuper, L. Bailey, J. R. Worley, P. Soloway, M. L. Patterson and G. Murphy, *FEBS Lett.*, 2002, **532**, 127–130.
92. M. A. Chellaiah and T. Ma, *Biomed Res. Int.*, 2013, **2013**, 302392.
93. K. Holmbeck, P. Bianco, J. Caterina, S. Yamada, M. Kromer, S. A. Kuznetsov, M. Mankani, P. Gehron Robey, A. R. Poole, I. Pidoux, J. M. Ward and H. Birkedal-Hansen, *Cell*, 1999, **99**, 81–92.
94. M. Pei, F. He, L. Wei and A. Rawson, *J. Pineal Res.*, 2009, **46**, 181–187.
95. W. Gao, M. Lin, A. Liang, L. Zhang, C. Chen, G. Liang, C. Xu, Y. Peng, C. Chen, D. Huang and P. Su, *J. Pineal Res.*, 2014, **56**, 62–70.
96. S. Fu, M. Kuwahara, Y. Uchida, S. Kondo, D. Hayashi, Y. Shimomura, A. Takagaki, T. Nishida, Y. Maruyama, M. Ikegame, A. Hattori, S. Kubota and T. Hattori, *J. Endocrinol.*, 2019, **241**, 161–173.
97. W. D. Fraser, A. M. Ahmad and J. P. Vora, *Curr. Opin. Nephrol. Hypertens.*, 2004, **13**, 437–444.
98. N. Okubo, H. Fujiwara, Y. Minami, T. Kunimoto, T. Hosokawa, Y. Umemura, H. Inokawa, M. Asada, R. Oda, T. Kubo and K. Yagita, *Acta Orthop.*, 2015, **86**, 627–631.
99. A. C. Karaplis, A. Luz, J. Glowacki, R. T. Bronson, V. L. Tybulewicz, H. M. Kronenberg and R. C. Mulligan, *Genes Dev.*, 1994, **8**, 277–289.
100. R. V. Kondratov, A. A. Kondratova, V. Y. Gorbacheva, O. V. Vykhovanets and M. P. Antoch, *Genes Dev.*, 2006, **20**, 1868–1873.

101. T. Takarada, C. Xu, H. Ochi, R. Nakazato, D. Yamada, S. Nakamura, A. Kodama, S. Shimba, M. Mieda, K. Fukasawa, K. Ozaki, T. Iezaki, K. Fujikawa, Y. Yoneda, R. Numano, A. Hida, H. Tei, S. Takeda and E. Hinoi, *J. Bone Miner. Res.*, 2017, **32**, 872–881.
102. W. E. Samsa, A. Vasanji, R. J. Midura and R. V. Kondratov, *Bone*, 2016, **84**, 194–203.
103. K. Tsang, H. Liu, Y. Yang, J. F. Charles and J. Ermann, *Bone*, 2019, **121**, 172–180.
104. L. Fu, M. S. Patel, A. Bradley, E. F. Wagner and G. Karsenty, *Cell*, 2005, **122**, 803–815.
105. C. Xu, H. Ochi, T. Fukuda, S. Sato, S. Sunamura, T. Takarada, E. Hinoi, A. Okawa and S. Takeda, *J. Bone Miner. Res.*, 2016, **31**, 1344–1355.
106. J. H. Kim and N. Kim, *J. Bone Metab.*, 2014, **21**, 233–241.
107. G. Yang, L. Chen, G. R. Grant, G. Paschos, W.-L. Song, E. S. Musiek, V. Lee, S. C. McLoughlin, T. Grosser, G. Cotsarelis and G. A. FitzGerald, *Sci. Transl. Med.*, 2016, **8**, 324ra16.
108. X. Li, N. Liu, B. Gu, W. Hu, Y. Li, B. Guo and D. Zhang, *Mol. Biol. Rep.*, 2018, **45**, 1691–1704.
109. E. Maronde, A. F. Schilling, S. Seitz, T. Schinke, I. Schmutz, G. van der Horst, M. Amling and U. Albrecht, *PLoS One*, 2010, **5**, e11527.
110. S. Komoto, H. Kondo, O. Fukuta and A. Togari, *Chronobiol. Int.*, 2012, **29**, 66–74.
111. Y. Fujihara, H. Kondo, T. Noguchi and A. Togari, *Bone*, 2014, **61**, 1–9.
112. X. Shi, W. Shi, Q. Li, B. Song, M. Wan, S. Bai and X. Cao, *EMBO Rep.*, 2003, **4**, 374–380.
113. W. Zhang, N. Yang and X.-M. Shi, *J. Biol. Chem.*, 2008, **283**, 4723–4729.
114. G. Pan, J. Cao, N. Yang, K. Ding, C. Fan, W.-C. Xiong, M. Hamrick, C. M. Isales and X.-M. Shi, *J. Biol. Chem.*, 2014, **289**, 19373–19382.
115. P. Ducy, M. Amling, S. Takeda, M. Priemel, A. F. Schilling, F. T. Beil, J. Shen, C. Vinson, J. M. Rueger and G. Karsenty, *Cell*, 2000, **100**, 197–207.
116. S. Takeda, F. Elefteriou, R. Levasseur, X. Liu, L. Zhao, K. L. Parker, D. Armstrong, P. Ducy and G. Karsenty, *Cell*, 2002, **111**, 305–317.
117. F. Elefteriou, J. D. Ahn, S. Takeda, M. Starbuck, X. Yang, X. Liu, H. Kondo, W. G. Richards, T. W. Bannon, M. Noda, K. Clement, C. Vaisse and G. Karsenty, *Nature*, 2005, **434**, 514–520.
118. B. de Crombrugghe, *Cell*, 2005, **122**, 651–653.
119. D. H. Copp and S. S. Shim, *Oral Surg., Oral Med., Oral Pathol.*, 1963, **16**, 738–744.
120. Y. Sabbagh, T. O. Carpenter and M. B. Demay, *Proc. Natl. Acad. Sci. U. S. A.*, 2005, **102**, 9637–9642.
121. S. U. Miedlich, A. Zalutskaya, E. D. Zhu and M. B. Demay, *J. Biol. Chem.*, 2010, **285**, 18270–18275.
122. T. Noguchi, A. I. Hussein, N. Horowitz, D. Carroll, A. C. Gower, S. Demissie and L. C. Gerstenfeld, *Sci. Rep.*, 2018, **8**, 13756.
123. J.-P. Etchegaray, X. Yang, J. P. DeBruyne, A. H. F. M. Peters, D. R. Weaver, T. Jenuwein and S. M. Reppert, *J. Biol. Chem.*, 2006, **281**, 21209–21215.

124. A. Dudakovic, E. T. Camilleri, F. Xu, S. M. Riester, M. E. McGee-Lawrence, E. W. Bradley, C. R. Paradise, E. A. Lewallen, R. Thaler, D. R. Deyle, A. N. Larson, D. G. Lewallen, A. B. Dietz, G. S. Stein, M. A. Montecino, J. J. Westendorf and A. J. van Wijnen, *J. Biol. Chem.*, 2015, **290**, 27604–27617.
125. J. C. Lui, P. Garrison, Q. Nguyen, M. Ad, C. Keembiyehetty, W. Chen, Y. H. Jee, E. Landman, O. Nilsson, K. M. Barnes and J. Baron, *Nat. Commun.*, 2016, **7**, 13685.
126. C. Li, Y. Chai, L. Wang, B. Gao, H. Chen, P. Gao, F.-Q. Zhou, X. Luo, J. L. Crane, B. Yu, X. Cao and M. Wan, *Nat. Commun.*, 2017, **8**, 1312.
127. N. K. Lee, H. Sowa, E. Hinoi, M. Ferron, J. D. Ahn, C. Confavreux, R. Dacquin, P. J. Mee, M. D. McKee, D. Y. Jung, Z. Zhang, J. K. Kim, F. Mauvais-Jarvis, P. Ducy and G. Karsenty, *Cell*, 2007, **130**, 456–469.
128. M. A. Estrella, J. Du, L. Chen, S. Rath, E. Prangley, A. Chitrakar, T. Aoki, P. Schedl, J. Rabinowitz and A. Korennykh, *Nat. Commun.*, 2019, **10**, 2367.
129. I. Laothamatas, P. Gao, A. Wickramaratne, C. G. Quintanilla, A. Dino, C. A. Khan, J. Liou and C. B. Green, *Proc. Natl. Acad. Sci. U. S. A.*, 2020, **117**, 993–999.
130. R. Li, J. Yue, Y. Zhang, L. Zhou, W. Hao, J. Yuan, B. Qiang, J. M. Ding, X. Peng and J.-M. Cao, *Mol. Cell. Biochem.*, 2008, **317**, 169–177.
131. M. Kawai, C. B. Green, B. Lecka-Czernik, N. Douris, M. R. Gilbert, S. Kojima, C. Ackert-Bicknell, N. Garg, M. C. Horowitz, M. L. Adamo, D. R. Clemmons and C. J. Rosen, *Proc. Natl. Acad. Sci. U. S. A.*, 2010, **107**, 10508–10513.
132. A. R. Guntur, M. Kawai, P. Le, M. L. Bouxsein, S. Bornstein, C. B. Green and C. J. Rosen, *Ann. N. Y. Acad. Sci.*, 2011, **1237**, 58–63.
133. K. R. Shockley, O. P. Lazarenko, P. J. Czernik, C. J. Rosen, G. A. Churchill and B. Lecka-Czernik, *J. Cell. Biochem.*, 2009, **106**, 232–246.
134. A. W. James, J. Shen, K. Khadarian, S. Pang, G. Chung, R. Goyal, G. Asatrian, O. Velasco, J. Kim, X. Zhang, K. Ting and C. Soo, *Tissue Eng., Part A*, 2014, **20**, 2699–2710.
135. C. Fontaine, G. Dubois, Y. Duguay, T. Helledie, N. Vu-Dac, P. Gervois, F. Soncin, S. Mandrup, J.-C. Fruchart, J. Fruchart-Najib and B. Staels, *J. Biol. Chem.*, 2003, **278**, 37672–37680.
136. Y. He, F. Lin, Y. Chen, Z. Tan, D. Bai and Q. Zhao, *Stem Cells Dev.*, 2014, **24**, 1194–1204.
137. K. P. Wright, R. K. Bogan and J. K. Wyatt, *Sleep Med. Rev.*, 2013, **17**, 41–54.
138. I. Quevedo and A. M. Zuniga, *J. Clin. Densitom.*, 2010, **13**, 467–469.
139. D. Feskanich, S. E. Hankinson and E. S. Schernhammer, *Osteoporos Int.*, 2009, **20**, 537–542.
140. M. Zhou, S. Yang, Y. Guo, D. Wang, W. Qiu, B. Wang, G. Mu, L. Cao, Z. Zhang and W. Chen, *Scand. J. Work, Environ. Health*, 2020, **46**, 152–160.
141. I. C. Kowanko, R. Pownall, M. S. Knapp, A. J. Swannell and P. G. Mahoney, *Br. J. Clin. Pharmacol.*, 1981, **11**, 477–484.

142. M. Cutolo, B. Villaggio, K. Otsa, O. Aakre, A. Sulli and B. Seriolo, *Autoimmun. Rev.*, 2005, **4**, 497–502.
143. I. C. Chikanza, P. Petrou, G. Kingsley, G. Chrousos and G. S. Panayi, *Arthritis Rheum.*, 1992, **35**, 1281–1288.
144. V.-P. Kouri, J. Olkkonen, E. Kaivosoja, M. Ainola, J. Juhila, I. Hovatta, Y. T. Konttinen and J. Mandelin, *PLoS One*, 2013, **8**, e54049.
145. N. Yang, N. J. Smyllie, H. Morris, C. F. Gonçalves, M. Dudek, D. R. J. Pathiranage, J. E. Chesham, A. Adamson, D. G. Spiller, E. Zindy, J. Bagnall, N. Humphreys, J. Hoyland, A. S. I. Loudon, M. H. Hastings and Q.-J. Meng, *PLoS Genet.*, 2020, **16**, e1008729.
146. R. Akagi, Y. Akatsu, K. M. Fisch, O. Alvarez-Garcia, T. Teramura, Y. Muramatsu, M. Saito, T. Sasho, A. I. Su and M. K. Lotz, *Osteoarthritis Cartilage*, 2017, **25**, 943–951.
147. J. Soul, S. L. Dunn, S. Anand, F. Serracino-Inglott, J.-M. Schwartz, R. P. Boot-Handford and T. E. Hardingham, *Ann. Rheum. Dis.*, 2018, **77**, 423.
148. S. J. B. Snelling, A. Forster, S. Mukherjee, A. J. Price and R. C. Poulsen, *Chronobiol. Int.*, 2016, **33**, 574–579.
149. W. Yang, X. Kang, J. Liu, H. Li, Z. Ma, X. Jin, Z. Qian, T. Xie, N. Qin, D. Feng, W. Pan, Q. Chen, H. Sun and S. Wu, *Endocrinology*, 2016, **157**, 3096–3107.
150. S. Khurana, A. Bokkers, D. J. Geijs, S. Schivo, M. Karperien and J. N. Post, *Osteoarthritis Cartilage*, 2019, **27**, S193–S194.
151. A. Wirz-Justice, M. Terman, D. A. Oren, F. K. Goodwin, D. F. Kripke, P. C. Whybrow, K. L. Wisner, J. C. Wu, R. W. Lam, M. Berger, K. V. Danilenko, S. Kasper, E. Smeraldi, K. Takahashi, C. Thompson and R. H. van den Hoofdakker, *Science*, 2004, **303**, 467–469.
152. E. Roddy, W. Zhang and M. Doherty, *Ann. Rheum. Dis.*, 2005, **64**, 544–548.
153. J. K. Cooney, R.-J. Law, V. Matschke, A. B. Lemmey, J. P. Moore, Y. Ahmad, J. G. Jones, P. Maddison and J. M. Thom, *J. Aging Res.*, 2011, **2011**, 681640.
154. Y. Yamanaka, K. Honma, S. Hashimoto, N. Takasu, T. Miyazaki and S. Honma, *Sleep Biological. Rhythms*, 2006, **4**, 199–206.
155. O. M. Buxton, S. A. Frank, M. L'Hermite-Balériaux, R. Leproult, F. W. Turek and E. Van Cauter, *Am. J. Physiol.*, 1997, **273**, E536–E542.
156. C. I. Eastman, E. K. Hoese, S. D. Youngstedt and L. Liu, *Physiol. Behav.*, 1995, **58**, 1287–1291.
157. K. P. Schmidt, W. K. Koehler, G. Fleissner and B. Pflug, *J. Interdiscip. Cycle Res.*, 1990, **21**, 243–245.
158. T. L. Leise, M. E. Harrington, P. C. Molyneux, I. Song, H. Queenan, E. Zimmerman, G. S. Lall and S. M. Biello, *Age*, 2013, **35**, 2137–2152.
159. D. K. Welsh, G. S. Richardson and W. C. Dement, *Physiol. Behav.*, 1988, **43**, 771–777.
160. E. C. Huskisson, R. T. Taylor, D. Burston, P. J. Chuter and F. D. Hart, *Ann. Rheum. Dis.*, 1970, **29**, 393–396.

161. F. Levi, C. Le Louarn and A. Reinberg, *Clin. Pharmacol. Ther.*, 1985, **37**, 77–84.
162. N. G. Arvidson, B. Gudbjornsson, A. Larsson and R. Hallgren, *Ann. Rheum. Dis.*, 1997, **56**, 27–31.
163. F. Buttgereit, G. Doering, A. Schaeffler, S. Witte, S. Sierakowski, E. Gromnica-Ihle, S. Jeka, K. Krueger, J. Szechinski and R. Alten, *Lancet*, 2008, **371**, 205–214.
164. F. Buttgereit, G. Doering, A. Schaeffler, S. Witte, S. Sierakowski, E. Gromnica-Ihle, S. Jeka, K. Krueger, J. Szechinski and R. Alten, *Ann. Rheum. Dis.*, 2010, **69**, 1275–1280.
165. F. Buttgereit, D. Mehta, J. Kirwan, J. Szechinski, M. Boers, R. E. Alten, J. Supronik, I. Szombati, U. Romer, S. Witte and K. G. Saag, *Ann. Rheum. Dis.*, 2013, **72**, 204–210.
166. D. Michalska, M. Luchavova, V. Zikan, I. Raska, A. A. Kubena and J. J. Stepan, *Osteoporos Int.*, 2012, **23**, 2885–2891.
167. K. K. Winer, B. Zhang, J. A. Shrader, D. Peterson, M. Smith, P. S. Albert and G. B. Cutler, *J. Clin. Endocrinol. Metab.*, 2012, **97**, 391–399.
168. M. D. Ruben, D. F. Smith, G. A. FitzGerald and J. B. Hogenesch, *Science*, 2019, **365**, 547–549.
169. H. Derendorf, K. Ruebsamen, L. Clarke, A. Schaeffler and J. R. Kirwan, *J. Clin. Pharmacol.*, 2013, **53**, 326–333.
170. R. H. Straub and M. Cutolo, *Arthritis Rheum.*, 2007, **56**, 399–408.

CHAPTER 14

Circadian Aspects of Skeletal Muscle Biology

JULIEN DELEZIE AND KATJA LAMIA*

Department of Molecular Medicine, The Scripps Research Institute, 10550 North Torrey Pines Road, MB-21, La Jolla, CA 92037, USA
*Email: klamia@scripps.edu

14.1 Introduction

With nearly 650 individual parts, skeletal muscle is the most abundant tissue in the human body. Its functions include the maintenance of postural support, generation of force and power during voluntary movements, breathing, and thermoregulation. Skeletal muscle also plays a prominent role in lipid and glucose metabolism and contributes significantly to organismal oxidative metabolism and insulin-stimulated glucose disposal.[1] It is likewise a large reservoir for proteins, mitigating the decrease in circulating amino acids during prolonged fasting and supplying the liver with glucogenic amino acids to maintain homeostasis.[2] Perhaps, less appreciated, skeletal muscle has the ability to function in an endocrine-like manner, releasing cytokines and peptides, so-called *myokines*,[3] in addition to *exerkines*[4] (*i.e.*, exosomes and/or microvesicles containing peptides, nucleic acids, lipids, metabolites, or diverse RNA species), thereby influencing the metabolic activity of other tissues to ensure whole-body homeostasis.[4,5]

Chemical Biology No. 23
Chronobiology and Chronomedicine: From Molecular and Cellular Mechanisms to Whole Body Interdigitating Networks
Edited by Germaine Cornelissen and Tsuyoshi Hirota

Published by the Royal Society of Chemistry, www.rsc.org

Metabolic flexibility is critical to survival for all species and mammalian metabolism accordingly undergoes daily fluctuations. These variations are partly driven by the circadian[†] clock, an evolutionarily conserved time-keeping molecular mechanism that is present in virtually all cells of the body.[6] Increasing evidence indicates that circadian disruption contributes to the development of many common diseases and metabolic conditions, including cancer and type 2 diabetes.[7,8] During the past two decades, there has been a growing interest in the role played by the circadian clock in skeletal muscle biology. In this chapter, we first provide an overview of the muscle circadian clock and circadian regulation of the skeletal muscle transcriptome. Next, we present key studies demonstrating the involvement of the circadian clock in skeletal muscle growth, regeneration, and metabolic function. The chapter will conclude by examining the relationship between physical activity, food, and circadian aspects of muscle biology, along with some directions for future studies.

14.2 Molecular Orchestration of the Skeletal Muscle Circadian Symphony

14.2.1 The Molecular Circadian Clockwork

Virtually all cells of the human body harbour a functional circadian clock. The circadian timing system[‡] regulates daily cycles in behaviour and physiology with a near 24-hour period, ensuring that most aspects of our biology align with the external light–dark cycle. In mammals, circadian oscillations in sleep–wake cycles are driven by the suprachiasmatic nucleus (SCN) of the hypothalamus, a master pacemaker[§] that is entrained to the environmental light–dark cycle. The SCN orchestrates the coordination of other body clocks *via* neuronal, humoral, and other, as yet undefined, signals.[9] The mammalian cell-autonomous molecular clock is generated by interlocking transcription–translation feedback loops that produce robust 24-hour rhythms of core clock gene expression and, in turn, of their transcriptional outputs, the so-called clock-controlled genes (CCGs) (Figure 14.1).

At the core of this mechanism, the transcription factors brain and muscle Arnt-like 1 (BMAL1) and circadian locomotor output cycles kaput (CLOCK) form a complex that rhythmically transactivates genes containing specific DNA sequences (E-boxes) in their promoter regions, including the *Period* (*Per1-3*) and *Cryptochrome* (*Cry1-2*) genes. When

---

† *circa* meaning 'about' and *dies* meaning 'day'.

‡ In mammals, this system comprises several, hierarchically organized biological clocks in anatomically discrete locations (*e.g.*, multiple brain areas and peripheral organs).

§ Recently, it has become clear that the SCN acts more like a 'conductor' that enables guided synchrony of a system of autonomous clocks that are present throughout the body.

PER and CRY proteins reach a certain threshold, they form heterodimers that translocate into the nucleus to repress transcriptional activity induced by CLOCK:BMAL1, leading to their own repression, ultimately starting a new cycle. An additional loop involves the nuclear receptors *Rev-erbs* (α, β) and *Rors* (α, γ), which can translocate into the nucleus to modulate *Bmal1* transcription *via* opposite actions on a RORE sequence located in its promoter (Figure 14.1). This loop ensures the fine-tuning of circadian oscillations[¶]. In addition, post-translational mechanisms, *e.g.*, those involving the casein kinase 1 family (CK1), which affect stabilization, degradation, and subcellular localization of clock proteins, likewise contribute to circadian oscillations (for review see ref. 10).

14.2.2 Circadian Rhythms in Muscle Gene Expression

In 2007, McCarthy *et al.*[11] used gene expression profiling analysis with microarray technology to provide the first characterization of the circadian transcriptome in adult mouse skeletal muscle. In addition to core components of the circadian molecular clockwork (*e.g.*, *Bmal1*, *Clock*, *Pers*, and *Crys*), hundreds of genes were found to be rhythmically expressed among less than 9000 genes that were reliably detected, suggesting that 3–10% of expressed genes are under circadian control in muscle. These genes are linked to broad functional categories, including transcription, protein degradation, ion transport, vesicular trafficking, and lipid metabolism. In addition, these genes are involved in muscle structural and functional adaptations. Since then, other groups have complemented and expanded these findings by analyzing daily/circadian[‖] gene expression using next-generation sequencing in different mouse skeletal muscle types (that is, oxidative and glycolytic),[12] at greater time resolution,[13] and/or in different species, including humans.[14,15] The fraction of transcripts determined to be rhythmic in each study depends on the statistical thresholds employed, making direct comparisons difficult; however some consistent themes emerge from this growing body of literature. In studies that examine multiple mouse organs, the mouse skeletal muscle transcriptome seems to contain a smaller fraction of rhythmically expressed genes compared to other organs.[13,16] In studies that compare multiple muscle groups, a greater proportion of the muscle transcriptome is found to exhibit rhythmic expression in muscle groups that contain a greater proportion of oxidative fiber types.[12,14] Despite presumably greater heterogeneity among individuals, a greater proportion of detectably expressed transcripts are

---

¶ The molecular clockwork also includes other core clock genes, including *Ciart*, *Dbp*, *Dec1-2*, and *Npas2*.

‖ The term 'daily' is used when rhythmic gene expression analysis is performed from tissues/organs collected from mice exposed to a standard light–dark cycle, as opposed to 'circadian' when animals are exposed to constant darkness, a free-running circadian state in which the daily rhythms of an organism persist in the absence of external cues such as light.

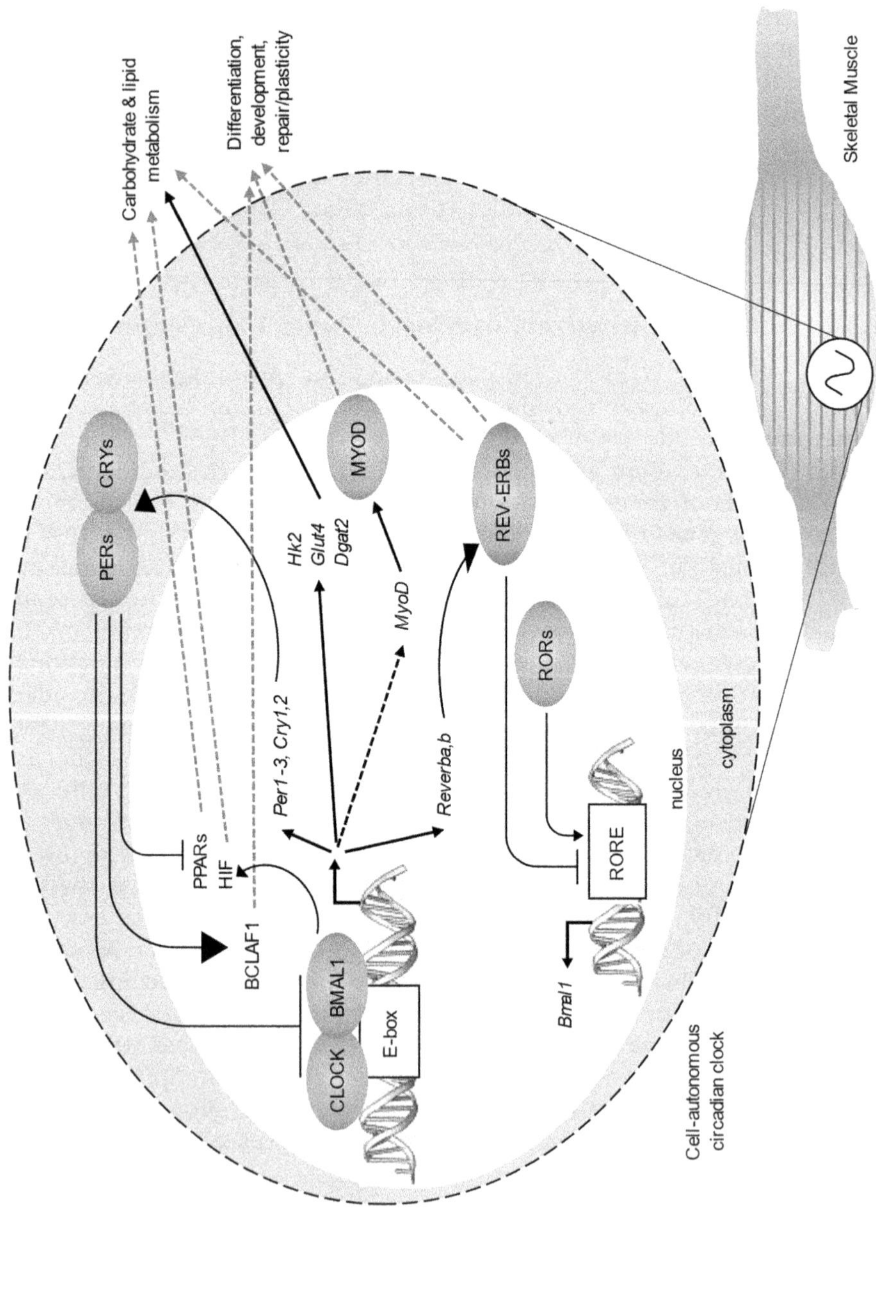
Carbohydrate & lipid metabolism
Differentiation, development, repair/plasticity
Skeletal Muscle
PERs
CRYs
MYOD
REV-ERBs
RORs
CLOCK
BMAL1
E-box
BCLAF1
PPARs
HIF
Per1-3, Cry1,2
Hk2
Glut4
Dgat2
MyoD
Reverba,b
RORE
Bmal1
nucleus
cytoplasm
Cell-autonomous circadian clock

reported to be rhythmically expressed in the muscles of primates (humans[15] or baboons[14]) compared to those of mice. Importantly, part of this rhythmicity depends on a functional circadian clock[15,17,18] and it is likewise influenced by daily cycles of fasting–feeding and rest–activity[12,19,20] and by age.[21] Together, these observations suggest that rhythmic gene expression in muscles may be heavily influenced by the environment and behaviour.

The rhythmicity of CCGs is highly tissue-specific[14,22,23] and given that most tissues/organs, including skeletal muscle, are composed of several different cell types, an open question is whether gene expression is rhythmic in all cell types in a coherent manner, or if only some cell types show rhythmicity. In addition to up to 10 different mononucleated cell types,[24] skeletal muscles are primarily made up of multinucleated myotubes, that is, muscle fibers containing hundreds of nuclei,[25] with each nucleus being potentially a circadian oscillator. It is currently unknown how circadian rhythms in gene expression within a single myofiber are coordinated among the multiple resident nuclei. Recent evidence suggests that nuclear subtypes at anatomically distinct locations (*e.g.*, neuromuscular junctions), unrelated to fiber type identity, exhibit distinct gene programs.[26,27] Transcriptional diversity between myonuclei and high cellular heterogeneity certainly add complexity to the generation and maintenance of circadian rhythmicity in the skeletal muscle matrix. Future approaches using single-cell gene-based analysis or fluorescently tagged clock proteins[28–30] at different time points of the light/dark cycle combined with spatial reconstruction may reveal spatiotemporal rhythmic patterns associated with specific groups of myonuclei, myofibers, and other muscle-resident cells.

In the next two sections of this chapter, we discuss the multiple connections between the muscle-intrinsic clock mechanism and muscle function such as growth and repair, lipid and glucose metabolism, mitochondrial bioenergetics and dynamics (Figure 14.1).

Figure 14.1 Circadian clock components regulate the key aspects of muscle development and function. Cell-autonomous molecular oscillators drive daily oscillations in gene expression through a network of transcriptional activators (*e.g.*, CLOCK, BMAL1, RORs) and repressors (*e.g.*, PERs, CRYs, and REV-ERBs). It is unclear how individual myonuclear clocks are organized and interact with each other within multi-nucleated muscle fibers, indicated by the dashed cell outline. The circadian repressors PER1, PER2, PER3, CRY1 and CRY2 not only contribute to rhythmic activity of the CLOCK–BMAL1 heterodimer but also rhythmically repress additional transcription networks in a tissue-specific manner. In addition to its role in the core clock activator as a partner for CLOCK, BMAL1 also heterodimerizes with HIF1a. Selected transcription networks and key transcripts within them that play critical roles in muscle differentiation, development, and metabolism and have been linked to core clock components are indicated. See text for details.

14.3 Circadian Aspects of Muscle Growth and Repair

14.3.1 A Functional Clock in Satellite Cells

Skeletal muscle is largely composed of postmitotic multinucleated elongated cells packed with a large number of arranged filaments or myofibrils. During early development, a population of mesoderm progenitors called myoblasts fuse, giving rise to myofibers that are dramatically growing in size in the neonatal/juvenile stages while maintaining a constant number.[31] Postnatal growth of muscle fibers involves a unique population of muscle-resident stem cells or satellite cells (SCs). SCs, like other stem cells, have the capacity for self-renewal and can either repair existing myofibers or give rise to new ones, thereby enabling maintenance and regeneration** of skeletal muscle throughout life.

The circadian clock has emerged as a strong regulator of tissue homeostasis and adult stem cells.[32,33] SCs, in particular, harbour a functional circadian clock and exhibit broad rhythmic gene expression programs in relation to, *e.g.*, cell proliferation, myotube differentiation, and insulin response. Interestingly, the daily transcriptome of SCs is extensively rewired during aging, switching from genes involved in tissue homeostasis to those implicated in stress response and autophagy, without changes in core clock gene oscillations.[34] Thus, despite the age-related decrease in both locomotor activity and amplitude of core clock gene oscillations in the master SCN clock, the SC clock remains robustly rhythmic during aging while a significant part of the rhythmic SC transcriptome is rewired to cope with new, age-related changes in the muscle environment. Although the mechanisms mediating this reprogramming remain unknown, activation of certain signalling pathways in response to specific environmental or pathological changes may influence several clock-controlled transcriptional outputs, for instance, by altering how CLOCK:BMAL1 regulate the chromatin landscape.[35–37] These results also show that the SC circadian clock can maintain robust rhythms during aging even in the presence of a dysfunctional SCN pacemaker. In response to injury, SCs are activated by the resulting transient hypoxic environment and they undergo rapid expansion followed by differentiation to repair the damaged tissue. It was recently demonstrated that the efficiency of SC activation and muscle tissue repair depends on the phase of the solar cycle at which the injury occurs.[38] Furthermore, inactivation of *Bmal1* in SCs of adult mice impairs muscle repair in response to injury.[38]

14.3.2 Circadian Clock Disruption and Muscle Development

Studies in genetically altered mouse and cell models have highlighted the contribution of circadian clock proteins to myogenic differentiation and

** Other muscle-resident fibro/adipogenic progenitors and immune cells also contribute to the regeneration of damaged myofibers.

muscle maintenance. For instance, healthy young adult mice in which REV-ERBα is deleted ubiquitously exhibit reduced cross-sectional area in skeletal muscle fiber types, together with altered expression of atrophy-related genes.[39] Modulation of REV-ERBα expression and activity in C2C12[††] cells was associated with altered proteasomal activity and fiber diameter.[39,40] Similarly, constitutive expression of the clock gene *Dec2*, a basic helix–loop–helix (bHLH) transcription factor, blocks myogenic differentiation of C2C12 myoblasts by interfering with MyoD, another member of the bHLH transcription factor family and master regulator of myogenesis.[41–43] Interestingly, MyoD is a target of the CLOCK:BMAL1 heterodimer and reciprocally regulates the clock,[44,45] though *Myod* expression is not affected by deletion of *Bmal1* in satellite cells.[38] Nevertheless, germline removal of *Clock* exon 19 or deletion of *Bmal1* leads to altered muscle structure and strength *in vivo*, which is associated with the loss of *Myod* rhythmicity,[45] as well as further signs of premature aging and sarcopenia[‡‡] in *Bmal1*-deficient mice.[46] Notably, depletion of *Bmal1 in vitro* disrupts the ability of myoblasts to differentiate into myotubes, whereas forced expression of *Bmal1* enhances differentiation. These cell-autonomous effects of BMAL1 are furthermore associated with impaired Wnt signalling.[47] Muscle-specific deletion of *Bmal1* induced in adulthood exacerbated the ageing-associated deterioration of musculoskeletal structure and function,[48] though these changes were observed at a much later age than similar alterations described in germline *Bmal1*-deficient animals. In stark contrast, skeletal muscle-specific inactivation of *Bmal1*, whether throughout development or induced in early adulthood, had no impact on adult mouse skeletal or muscle structure and function *in vivo*.[18] Several technical differences between these studies may underlie the differences between outcomes, including different Cre lines used for muscle-specific deletion, use of different control animals (Cre-positive without induction *vs.* Cre-negative treated with tamoxifen), assessment of function in different muscle groups, the ages at which *Bmal1* was inactivated, and the ages at which outcomes were measured.[49,50] Regardless, these studies (and another one in which *Bmal1* was inactivated ubiquitously in early adulthood[51]) indicate that the sarcopenia and arthropathy observed in *Bmal1*-deficient mice are not primarily caused by the loss of *Bmal1* in skeletal muscle, though muscle *Bmal1* may provide protection against progressive musculoskeletal deterioration with age. Finally, systemic *Bmal1* disruption, which incidentally abolishes core clockwork function, does not recapitulate any features of the age-associated reprogramming of the SC daily transcriptome.[34] Thus, even though global *Bmal1*-deficient animals clearly display premature aging-like symptoms, including sarcopenia,[46] these alterations could be linked to other noncircadian functions of *Bmal1* early in development, particularly in extra-muscle tissues. In any case, results obtained with different

[††] An immortalized myoblast cell line derived from the thigh muscle of a female mouse donor.
[‡‡] Sarcopenia is the loss of muscle mass and strength that occurs during normal aging.

mouse models based on germline clock gene inactivation should be interpreted with caution.[51,52]

14.3.3 Influence of the Circadian Clock on Muscle Regeneration

A bidirectional relation between the cell cycle and the circadian clock is well established.[53,54] Many studies have shown that cell cycle regulators are under circadian control, locking cell proliferation and DNA replication to specific phases of the light–dark cycle.[55] In muscle specifically, CRY2 regulates myogenesis by specifically interacting with the Bcl-2 associated factor 1 (BCLAF1), a gene under circadian regulation in muscle and essential for stabilizing mRNAs encoding *Cyclin D1*, a G1/S phase transition regulator, and *Tmem176b*, a transmembrane regulator for myogenic cell fusion.[56] In response to cardiotoxin-induced injury, *Bmal1* expression is robustly induced in skeletal muscle of control animals and its absence alters myogenic gene expression, and impairs satellite cell expansion and the muscle regenerative process.[38,57] Conversely, genetic or pharmacological disruption of REV-ERBα[§§] stimulates muscle repair in response to cardiotoxin *in vivo*, which is linked to a REV-ERB-dependent regulation of the activity of the nuclear factor-Y complex, a key regulator of cell cycle genes.[58] Consistent with this, ubiquitous deletion of REV-ERBα in the *mdx* mouse model of muscular dystrophy reduced muscle wasting and markers of muscle damage.[59] The timing of PER1 and PER2 protein expression appears to be essential to ensure appropriate myoblast differentiation *in vitro* and muscle regeneration *in vivo*. In particular, mouse muscle regeneration is more efficient when muscles are injured during the early active, feeding period, when PER proteins and energy intake are high, *versus* the early inactive, resting period, when PER proteins and energy levels are low.[60] Moreover, insulin-like growth factor 2 (*Igf2*), a gene essential for muscle growth and myoblast differentiation, shows a daily pattern of expression in mouse skeletal muscle with a peak in the early night, in phase with *Per1* and *Per2* expression. Finally, both PER proteins are required for the induction of *Igf2* in response to injury, suggesting that a PER-IGF2 axis could facilitate the activation of anabolic pathways at night, when the mouse is mostly active and feeding.[60]

The relation between solar cycles and muscle growth is further illustrated by a study performed on zebrafish, a diurnal species, which contrary to the mouse, is mostly active during the day.[61] Myotome volume in post-hatch larvae is greater under a standard 12 h light: 12 h dark cycle as compared to constant light or darkness, highlighting the importance of a normal day/night cycle for optimal muscle growth. Furthermore, the myotome of zebrafish grows more during the day than at night, irrespective of the developmental stage. Importantly, both the circadian clock and swimming

[§§] See Dierickx *et al.* 2018, however, for a note of caution on the specificity of the SR9009 REV-ERB ligand.

activity promoted muscle anabolism, as the abolition of all muscle activity in anesthetized fish did not fully prevent daily cycles of muscle growth and of muscle protein synthesis during the daytime. This result indicates that the regulation of myotome growth by the circadian clock is independent of muscle electrical and contractile activity; however, circadian- and activity-dependent growth converge, leading to a synergistic effect on volumetric growth. Given that food was not provided in this study and/or could not be assimilated by anesthetized and young larvae, whether circadian variations in internal energy availability, imposed by a central clock, limited anabolism and thus muscle growth remains an open question. Nevertheless, temporal variations in muscle TORC1 activity, a central regulator of cell growth and proliferation,[62] are observed, while TORC1 inhibition prevented the activity-mediated muscle growth during the day (but not at night), suggesting a time-dependent regulation of TORC1 activity by the circadian clock. Incidentally, a reciprocal influence between clock factors and the mTORC1 signalling pathway is also shown in the mouse liver,[63,64] and the extracellular signal-regulated kinase 1/2 (ERK1/2) pathway, a regulator of protein synthesis, likewise oscillates in rat muscles.[65] Furthermore, rats subjected to downhill running on a treadmill in the middle of the light phase exhibited greater activation of mTORC1 signalling and a greater increase in muscle mass than rats subjected to the same exercise during the dark phase.[66] Taken together, these data provide evidence that the circadian clock and anabolic pathways converge in skeletal muscle.

Given that obesity and diabetes are associated with circadian disruptions,[67,68] and impede skeletal muscle mass and regeneration,[69,70] pharmacological and/or exercise-based therapeutic interventions could be further optimized by taking into account circadian rhythms in (muscle) physiology.

14.4 Circadian Aspects of Muscle Metabolism

14.4.1 Glucose and Lipid Metabolism

Skeletal muscle fibers must quickly respond to deviations in cellular homeostasis, *e.g.*, in response to hormonal signals such as insulin or the rapid increase in contractile activity and energy demand, especially during the active phase. Clock proteins integrate multiple robust changes in their environment, notably through the feedback of AMPK and other energy sensors,[71–74] theoretically allowing time- and tissue-specific fine-tuning of energy homeostasis.[75] Similar to rhythms of gene expression, rhythmicity in energy substrates and metabolites is observed in skeletal muscle. For instance, the lipid composition of human skeletal muscle undergoes daily oscillation, which moreover persists in synchronized human myotubes *in vitro*, and thus does not require feeding rhythms.[76] Global metabolite profiling of mouse muscle tissue detected over 600 metabolites, a quarter of which exhibited fluctuations in concentration dependent on the phase of the light–dark cycle at which samples were collected.[77,78] Carbohydrates peak early in the resting (day) phase, whereas nucleotides peak in the late day, and amino acids peak at night. This temporal compartmentalization of metabolite accumulation was furthermore impacted when mice were fed a high-fat diet, with, *e.g.*, a large temporal displacement of amino acid accumulation to the end of the light (inactive) phase. This shift may underlie a lower anabolic response to nutrients similar to what is seen in obese patients, who alongside diverse metabolic complications, suffer from a deterioration of muscle mass and functionality.[79,80] Similarly, the circadian transcriptome of human primary myotubes is altered in response to palmitate, a cell model of lipid overload, suggesting that accumulation of intramuscular lipids seen in obesity may rewire the muscle circadian machinery *in vivo*.[81] Finally, simulated shift-work (circadian misalignment) in healthy men alters morning–evening variations in the skeletal muscle lipidome, which may contribute to the development of skeletal muscle insulin resistance.[82]

Genetic linkage analyses in humans uncovered an association of *Cry2* and *Per2* variants with blood glucose levels.[83–90] The circadian timing system accordingly contributes to rhythmic variations in plasma glucose and glucose tolerance, with higher tolerance in the early active phase both in humans and rodents,[91–93] which are somewhat attenuated in individuals with diabetes[94–96] and/or shifted in a mouse model of shift work.[97] Skeletal muscle is a potent regulator of glucose homeostasis, especially in the post-prandial state. However, it is unclear to what extent skeletal muscle contributes to daily variations in glycaemia and glucose tolerance. Measurements of basal and insulin-stimulated *ex vivo* glucose uptake in skeletal muscles collected at different phases of the light–dark cycle do not correlate with daily variations in whole-body insulin sensitivity,[98] but temporal differences in insulin sensitivity *in vivo* were observed only after a longer exposure to insulin than what was performed *ex vivo*. Regarding divergent reports on the role of muscle-intrinsic *Bmal1* in musculoskeletal health, selective inactivation of *Bmal1* in mouse skeletal muscles has been shown to either have no impact on blood glucose levels and insulin tolerance[18] or to cause dramatic hyperglycemia.[99] Notably, deletion of *Bmal1* exclusively in skeletal muscles consistently leads to nearly complete loss of *ex vivo* insulin-stimulated glucose uptake in isolated muscle groups, which is associated with reduced expression of hexokinase 2 (HK2) and the insulin-dependent glucose transporter GLUT4.[18,99] Furthermore, skeletal muscles display a rhythm of insulin-stimulated glucose uptake alongside mediators of the insulin signalling pathway and glucose transport during the early dark/active phase in the mouse.[98] Whether fluctuations in skeletal muscle insulin sensitivity across the day result from regulation by both the muscle intrinsic clock and physical activity, as for the aforementioned temporal regulation of muscle protein synthesis, remains to be clarified. Moreover, Dyar *et al.*[100] showed that a significant fraction of the metabolome oscillates in a BMAL1-dependent manner in mouse muscle. They likewise found that BMAL1 binds to the promoter of the *Dgat2* gene in a timely fashion, essential for the regulation of daily muscle triglyceride synthesis. Yet, with BMAL1 being the main driver of core clock gene oscillations, as its deletion abolishes core clock gene oscillations, it is difficult to delineate the exact contribution of BMAL1 *vs.* other clock factors to the temporal regulation of muscle metabolic function. Several studies have proposed a role for other clock components, downstream of the CLOCK:BMAL1 heterodimer, in muscle metabolism. For instance, muscle-specific expression of a truncated (dominant negative) form of the *Rorα* gene, a target and regulator of CLOCK: BMAL1, leads to glucose intolerance and impaired muscle-specific glucose uptake in the mouse, concomitant with altered AMPK and AKT signalling.[101] Nobiletin, a natural flavonoid targeting the ROR nuclear receptors, furthermore enhances day–night rhythms of gene expression and mitochondrial bioenergetics in skeletal muscle through a direct effect on circadian clock amplitude.[102,103] Circadian control of insulin release by pancreatic islets and sensitivity of extra-muscle tissues to insulin also likely contribute to day–night changes in systemic insulin tolerance.[98,104–107] Altogether, these results provide evidence that the muscle intrinsic clock regulates both glucose and lipid homeostasis in a timely fashion, and it might particularly prepare muscle cells for periods of feeding and for increased functional demand.

14.4.2 Mitochondrial Dynamics and Bioenergetics

The maintenance of mitochondrial structure and metabolism is essential to ensure adequate muscle mass, strength, and function.[108,109] Despite the clear physiological and clinical significance, our knowledge of the regulators of mitochondrial biology remains rudimentary, particularly concerning the biological programs that control the expression, transport, and activity of enzymes involved in, *e.g.*, cellular respiration and mitochondrial quality. This lack of understanding is partly due to the highly complex and dynamic proteome within muscle fibers and other muscle-resident cells, constantly adapting to external and internal stimuli.[110] Similarly, in addition to being major nodes for nutrient integration and ATP generation, mitochondria are highly plastic organelles, undergoing fusion and fission according to the cell nutritional status, and engaging in intimate cross-talk with other cellular compartments.[111] In this context, clock proteins have emerged as previously unappreciated modulators of mitochondrial dynamics and metabolism in the liver and in cultured fibroblasts.[112–114]

The circadian clock contributes to rhythms in oxidative capacity in human muscles,[115] which are affected in older and metabolically compromised individuals.[116] While rhythms of skeletal muscle respiration have yet to be demonstrated in the mouse, C2C12 cells clearly exhibit circadian cycles of respiration and NAD^+ accumulation, in phase with PER2 oscillation, upon synchronization.[117] Incidentally, the temporal expression pattern of *Pers* in skeletal muscle is intriguing, peaking right at the awakening phase both in humans and in the mouse.[15,118] In addition to its core clock function, PER2 is known to modulate cardiac and adipose tissue metabolism by regulating the activity of the transcription factors HIF-1α and PPARγ, respectively.[119,120] Moreover, PER2 (and PER1) drives rhythmic oscillations of rate-limiting mitochondrial enzymes that process lipids and carbohydrates in the mouse liver.[112] Whether this is similarly true in skeletal muscle remains to be investigated. The circadian repressors CRY1 and CRY2 play a similar central role in modulating nuclear receptor activities and/or the coordination of muscle mitochondrial oxidative metabolism. In particular, CRY-mediated repression of peroxisome proliferator-activated receptor delta (PPARδ) target genes influences muscle metabolism and exercise capacity.[121] CRYs also limit the transcriptional response to hypoxia *in vitro via* repression of the hypoxia-inducible factor 1α (HIF-1α)-BMAL1 heterodimer.[122,123] HIF-1α is a critical regulator of muscle function, influencing ATP production and energy substrate utilization, and likewise modulating the adaptation of skeletal muscle to endurance training (*e.g.*, mitochondrial plasticity and angiogenesis).[124,125] Stabilization of HIF-1α due to hypoxia in response to muscle injury also contributes to the activation of satellite cells and this is impaired in the absence of BMAL1.[38] It would thus be interesting to study the tissue-specific repressive function of CRYs in limiting HIF-1α activation *in vivo*, which may be beneficial for muscle oxidative capacity, training performance, and regeneration after injury.[126] This is especially important in light of the time-dependent induction of HIF-1α by exercise,[127] together with the reciprocal regulation between the circadian clock and hypoxia signalling in muscle cells and other tissues.[128–130] Finally, significant changes in muscle mitochondrial gene, protein, and structure are observed in global clock-deficient mice,[45,121] and the ROR nuclear receptors are required for the effect of nobiletin on mitochondrial respiration and biogenesis.[102] Additional studies are needed to reveal the specific contribution of circadian clock factors to mitochondrial function and to the subsequent functional and structural changes induced by aerobic-based exercise training in skeletal muscle.

14.5 Time, Exercise, and the Muscle-intrinsic Clock

14.5.1 Circadian Aspects of Muscle Performance

Alongside time-dependent variations in the skeletal muscle transcriptome, metabolome, and metabolic function, differences in skeletal muscle/exercise performance across the day are likewise observed. In healthy humans, a peak of performance is observed when aerobic exercise such as running or cycling is performed in the late afternoon/evening.[131–135] In resistance-based exercise, greater human strength is also observed in the evening hours.[132,134–136] Importantly, for those untrained individuals engaging in exercise training programs, similar improvements in strength, muscle mass, and/or endurance are detected whether training is performed in the morning or the evening.[137–140] Nonetheless, morning training attenuates the morning–evening variations in performance observed at baseline, whereas evening training accentuates them.[137,141] Finally, a circadian specificity of training (*i.e.*, a temporal relationship between time of training and the time of subsequent performance) is seen in most studies.[133,134,137,140,142] Together, despite potential confounding effects (*e.g.*, gender, genetic background, age, nutrition, health, sleep, chronotype[¶¶], mood, psychological stress, motivation, personal goals, and social interaction) inherent to studies involving humans, there is a good amount of evidence indicating that exercise capacity is affected by when the exercise is performed relative to the 24-hour day.

When tested under dynamic treadmill exercise conditions (*i.e.*, under a maximal running test), the running performance of untrained mice fluctuates with a peak occurring in the late part of the dark, active period.[118] Similar results are observed when mice are subjected to an endurance-type protocol (*i.e.*, using moderate exercise running intensities).[143] Hence, like in humans, mice show greater physical performance in their late active phase. Interestingly, mice chronically engaging over several weeks in low-to-moderate intensity, endurance-like wheel running activity primarily in the early night, only show a greater subsequent running performance when challenged with a treadmill test in the late night.[20] Notwithstanding the differences between these two running exercise modalities,[144,145] that is, wheel running and treadmill, there could be very little or no circadian specificity of training in the mouse, as opposed to humans. Whether this difference is purely linked to metabolic constraints (suggested in ref. 118) and/or to a lack of plasticity of the muscle intrinsic clock influencing muscle metabolic and contractile properties needs further investigation, the latter being particularly influenced by signalling pathways regulating calcium homeostasis and the sarcomere unit.[146] Future studies should likewise examine the influence of, *e.g.*, the cardiovascular and endocrine systems, neuronal inputs, and body temperature on the daily variations in maximal exercise/strength performance.

14.5.2 Zeitgeber Properties of Exercise

Light is the predominant time cue or Zeitgeber[‖‖] of the mammalian circadian timing system, synchronizing the activity of the SCN, which in turn distributes this temporal information to adjust other central and peripheral oscillators. Other 'non-photic' Zeitgebers, such as food and exercise, can also modulate circadian clock function in central and peripheral tissues.[147] This is particularly seen in, *e.g.*, human shift-workers, who are compelled to work at night during their sleep phase or to rotate shifts, which leads to misalignment between external signals (*e.g.*, light and food) and rhythmic physiological parameters (*e.g.*, sleep, metabolism and hormonal secretions).[148,149]

¶¶ Known in humans as the night owl and the morning lark.

‖‖ A Zeitgeber (from the German 'Zeit' = time and 'geber' = giver) is any environmental cue that is able to reset a biological clock.

Importantly, irregular timing of eating represents a major contributor to internal desynchronization in humans (reviewed in ref. 150 and 151), which is especially supported by studies of peripheral desynchronization in rodent models.[152–156]

Physical activity similarly exerts Zeitgeber properties on the circadian timing system (discussed in ref. 157). For instance, exercise performed in the morning or at night alters the rhythmic phase of plasma melatonin secretion in young[158–162] and similarly, in older adults.[163] In contrast, exercise at varied times of the day has a much smaller effect on the core temperature rhythm.[164] Importantly, the entraining effects of exercise on the human circadian timing system, as evaluated *via* the measure of hormonal and physiological outputs, may depend on the duration and intensity of an exercise bout, as well as its time-dependent repetition over time.[157,165] Conversely, a single bout of exercise acutely regulates core clock gene expression in skeletal muscles of untrained humans and mice.[118,166] In addition, the timing of an acute bout of exercise is a significant modifier of the subsequent oscillations of genes and metabolites in skeletal muscle of sedentary mice.[127] This result suggests that the muscle-intrinsic clock may be sensitive to repeated bouts of exercise at different times of the day. Accordingly, when enforced, daily scheduled wheel running or treadmill activities can phase-shift the rhythmic expression of PER2::luciferase as measured from mouse muscle explants *ex vivo*.[167] Strenuous exercise is well established to produce acute physiological stress, including hormonal and immunological responses similar to those produced by other physical stressors; the degree to which stress responses are activated by exercise depends on the intensity of the exercise stimulus.[168,169] Accordingly, enforced running in rodents induces a greater stress response than access to a wheel for voluntary running, evidenced by increased secretion of noradrenaline and corticosterone, the latter being a major internal Zeitgeber for multiple peripheral organs.[170] Indeed, it has been reported that forced exercise has a greater impact on peripheral clock time than voluntary exercise.[171] Consistent with this finding, the rhythmicity of core clock genes in skeletal muscles of mice chronically and spontaneously running in the wheel at night was not severely impacted by training, despite a significant remodelling of the daily muscle transcriptome.[20,172] Consistent with this observation, long-term spontaneous wheel-running exercise during the usual resting period does not lead to robust changes in the phase of core clock gene expression in mouse muscles.[118] Furthermore, denervation caused relatively minor changes in the daily expression of most core clock genes in muscle, indicating that nerve input and, by extension, locomotor activity, have only a slight role in regulating core clock gene expression.[12] Emblematic of the robustness of the muscle intrinsic clock to external perturbations, meal timing somewhat poorly influences the phase of core clock gene expression in mouse skeletal muscle,[118,173] in stark contrast to the liver clock.[174] Similar results have been observed in skeletal muscles of overweight/obese individuals challenged with time-restricted eating.[175] Thus, collectively, the muscle molecular clock appears resilient to external perturbations such as those induced by chronic exercise, even though training resulted in a clear remodelling of the skeletal muscle daily transcriptome.[20] Further research is needed to rule out a potential (phase-shifting) effect of training on core clock protein abundance and post-translational modification. Finally, it is possible that the induction of some of the core clock transcripts in response to an acute exercise bout observed in multiple studies[118,127,128,143,166] could be associated with their noncircadian function.

14.5.3 Therapeutic Potential of Timed Exercise

A quarter of the world's adult population is at risk of developing or exacerbating diseases from insufficient physical activity.[176–178] Conversely, health conditions, such as diabetes, obesity and cancer, are known risk factors for various disorders of the locomotor system, including the decline in mass and function of skeletal muscle.[179–181] Moreover, chronic circadian disruption resulting from irregular sleep–wake cycles, light exposure or meal timing (the modern lifestyle) at night, has been associated with obesity and other metabolic conditions, cancer, inflammation, and cognitive and mental disorders.[182–187] Physical activity represents one of the most effective strategies to reduce the prevalence and incidence of chronic illnesses and age-related disorders.[188–190] Endurance- and resistance-based training exert many positive influences throughout the body, in particular by modulating the cardiovascular, musculoskeletal, respiratory, metabolic, and neuroendocrine systems.[191,192]

The beneficial effects of physical activity are dependent on several parameters, including the type, duration, frequency, and intensity of a workout.[193] The timing of exercise has been put forward as a parameter for maximizing the positive outcomes of regular training.[194] For instance, afternoon exercise elicits superior beneficial effects compared to morning exercise on glycaemic control and peripheral insulin sensitivity in metabolically compromised individuals in most,[195–197] but not all studies.[198] Conversely, early morning aerobic exercise has more positive effects on blood pressure and sleep quality in healthy humans.[199–201] However, to date, there is no study comprehensively comparing the long-term physiological outcomes of morning *vs.* afternoon training in healthy individuals or patients. There is thus a need to further conduct randomized controlled trials to determine the effect of timed exercise on disease symptoms and progression in different populations, insofar as these studies can help 'control' for an individual's lifestyle over a prolonged period of time and/or recapitulate real-world situations that include circadian disturbances. It is also unclear whether the enhanced effects of timed exercise training on health parameters in both clinical and preclinical contexts depend on the Zeitgeber properties of exercise, as discussed above, and given that the volume and intensity of exercise needed to influence the circadian system may not be achievable by the majority of people.[157] Finally, experiments investigating the relationship between scheduled training and health improvements should likewise consider the timing of nutrition.[202–204] In any case, formulating exercise recommendation in patients is a challenge, and the success of a chronotherapeutic approach of exercise would likewise depend on whether training plans can be tailored for each individual, taking medical (*e.g.*, health status and prescription drug), societal (*e.g.*, work and family), lifestyle (*e.g.*, resting and dietary habits), and psychological aspects (*e.g.*, personal goals, compliance, and motivation) into consideration.

14.6 Conclusion

Due to the multicellular nature of skeletal muscle, the study of muscle clock (s) sparks numerous exciting questions and involves multiple challenges. First, the contribution of myofibers *vs.* other muscle-resident cells to the rhythmic transcriptome, metabolome or lipidome, is not delineated. Similarly, the organization of circadian symphony between hundreds of myonuclei and the different cell types of the skeletal muscle matrix remains unexplored. As for the liver clock,[205,206] the manipulation of SCN input(s), food, and/or physical activity could shed light on the inter-dependency of the different cell types that compose skeletal muscle tissue *vis-à-vis* its circadian function, and the cellular cross-talk influencing muscle cell-specific circadian processes. The identification of autocrine, paracrine and endocrine signalling pathways in skeletal muscle would ultimately yield insights into how the myofiber-intrinsic clock impacts the properties of other muscle-resident cells and interacts with other organs, and *vice versa*, in a timely manner.

While feedback from physical activity to the circadian timing system is apparent, how scheduled training *per se* affects the muscle clock (*e.g.*, at sub-transcriptional levels) needs further investigation, including in animal models and human studies accounting for sex/gender differences. The endocrine-like properties of skeletal muscle cells may play an important role in influencing the metabolic activities of other organs when training at specific times of the day. Yet, exercise is a physiological stressor that exerts countless and complex systemic effects; circadian feedback, if any, from the muscle clock to other tissues may only play a minor role. Future research exploring the additive entraining effects of physical activity and nutrition should, however, be encouraged. Regardless, the endogenous mechanisms driving circadian rhythms in, *e.g.*, muscle gene expression, growth, metabolism, and performance, as reviewed here, have many implications for fine-tuning timing of training sessions in athletes, other healthy individuals, and patients. Emerging evidence suggesting that circadian clocks modulate muscle development and regeneration indicates that pharmacological targeting of clock components holds promise for mitigating muscle injury or degeneration. Cardiac muscles are distinct from skeletal muscles but share some of the same properties and are also modulated by cell-intrinsic circadian clocks and several studies indicate an important role for circadian regulation of cardiovascular health and in the treatment of cardiovascular disease.[207–209] Finally, advancing our functional understanding of clock protein interactions with muscle-specific and ubiquitous transcription factors may provide the basis for developing new time-based therapeutic strategies (*e.g.*, pharmacological) for preventing and/or treating disease and injury.

Acknowledgements

This work was funded by NIH grant R01 DK112927 (to K.A.L.). We thank Y. Slivers and J. Valecko for administrative assistance.

References

1. R. A. DeFronzo, E. Jacot, E. Jequier, E. Maeder, J. Wahren and J. P. Felber, *Diabetes*, 1981, **30**, 1000–1007.
2. R. R. Wolfe, *Am. J. Clin. Nutr.*, 2006, **84**, 475–482.
3. B. K. Pedersen, A. Steensberg, C. Fischer, C. Keller, P. Keller, P. Plomgaard, M. Febbraio and B. Saltin, *J. Muscle Res. Cell Motil.*, 2003, **24**, 113–119.
4. A. Safdar, A. Saleem and M. A. Tarnopolsky, *Nat. Rev. Endocrinol.*, 2016, **12**, 504–517.
5. M. Whitham and M. A. Febbraio, *Nat. Rev. Drug Discovery*, 2016, **15**, 719–729.
6. E. D. Buhr and J. S. Takahashi, *Handb. Exp. Pharmacol.*, 2013, 3–27.
7. M. Pariollaud and K. A. Lamia, *Cancer Discovery*, 2020, **10**, 1455–1464.
8. I. C. Mason, J. Qian, G. K. Adler and F. Scheer, *Diabetologia*, 2020, **63**, 462–472.
9. A. Patke, M. W. Young and S. Axelrod, *Nat. Rev. Mol. Cell Biol.*, 2020, **21**, 67–84.
10. J. S. Takahashi, *Nat. Rev. Genet.*, 2017, **18**, 164–179.
11. J. J. McCarthy, J. L. Andrews, E. L. McDearmon, K. S. Campbell, B. K. Barber, B. H. Miller, J. R. Walker, J. B. Hogenesch, J. S. Takahashi and K. A. Esser, *Physiol. Genomics*, 2007, **31**, 86–95.
12. K. A. Dyar, S. Ciciliot, G. M. Tagliazucchi, G. Pallafacchina, J. Tothova, C. Argentini, L. Agatea, R. Abraham, M. Ahdesmäki, M. Forcato, S. Bicciato, S. Schiaffino and B. Blaauw, *Mol. Metab.*, 2015, **4**, 823–833.
13. R. Zhang, N. F. Lahens, H. I. Ballance, M. E. Hughes and J. B. Hogenesch, *Proc. Natl. Acad. Sci. U. S. A.*, 2014, **111**, 16219–16224.
14. L. S. Mure, H. D. Le, G. Benegiamo, M. W. Chang, L. Rios, N. Jillani, M. Ngotho, T. Kariuki, O. Dkhissi-Benyahya, H. M. Cooper and S. Panda, *Science*, 2018, **359**, eaao0318.
15. L. Perrin, U. Loizides-Mangold, S. Chanon, C. Gobet, N. Hulo, L. Isenegger, B. D. Weger, E. Migliavacca, A. Charpagne, J. A. Betts, J. P. Walhin, I. Templeman, K. Stokes, D. Thompson, K. Tsintzas, M. Robert, C. Howald, H. Riezman, J. N. Feige, L. G. Karagounis,

J. D. Johnston, E. T. Dermitzakis, F. Gachon, E. Lefai and C. Dibner, *eLife*, 2018, 7, e34114.
16. B. H. Miller, E. L. McDearmon, S. Panda, K. R. Hayes, J. Zhang, J. L. Andrews, M. P. Antoch, J. R. Walker, K. A. Esser, J. B. Hogenesch and J. S. Takahashi, *Proc. Natl. Acad. Sci. U. S. A.*, 2007, **104**, 3342–3347.
17. B. H. Miller, E. L. McDearmon, S. Panda, K. R. Hayes, J. Zhang, J. L. Andrews, M. P. Antoch, J. R. Walker, K. A. Esser, J. B. Hogenesch and J. S. Takahashi, *Proc. Natl. Acad. Sci. U. S. A.*, 2007, **104**, 3342.
18. K. A. Dyar, S. Ciciliot, L. E. Wright, R. S. Bienso, G. M. Tagliazucchi, V. R. Patel, M. Forcato, M. I. Paz, A. Gudiksen, F. Solagna, M. Albiero, I. Moretti, K. L. Eckel-Mahan, P. Baldi, P. Sassone-Corsi, R. Rizzuto, S. Bicciato, H. Pilegaard, B. Blaauw and S. Schiaffino, *Mol. Metab.*, 2014, **3**, 29–41.
19. T. Shavlakadze, T. Anwari, Z. Soffe, G. Cozens, P. J. Mark, C. Gondro and M. D. Grounds, *Am. J. Physiol.: Cell Physiol.*, 2013, **305**, C26–C35.
20. N. Casanova-Vallve, D. Duglan, M. E. Vaughan, M. Pariollaud, M. K. Handzlik, W. Fan, R. T. Yu, C. Liddle, M. Downes, J. Delezie, R. Mello, A. B. Chan, P. O. Westermark, C. M. Metallo, R. M. Evans and K. A. Lamia, *Mol. Metab.*, 2022, **61**, 101504.
21. C. A. Wolff, M. A. Gutierrez-Monreal, L. Meng, X. Zhang, L. G. Douma, H. M. Costello, C. M. Douglas, E. Ebrahimi, B. R. Alava, A. R. Morris, M. M. Endale, G. R. Crislip, K.-y. Cheng, E. A. Schroder, B. P. Delisle, A. J. Bryant, M. L. Gumz, Z. Huo, A. C. Liu and K. A. Esser, *Cell Rep.*, 2023, **42**, 111982.
22. S. M. Reppert and D. R. Weaver, *Nature*, 2002, **418**, 935–941.
23. K.-F. Storch, O. Lipan, I. Leykin, N. Viswanathan, F. C. Davis, W. H. Wong and C. J. Weitz, *Nature*, 2002, **417**, 78–83.
24. L. Giordani, G. J. He, E. Negroni, H. Sakai, J. Y. C. Law, M. M. Siu, R. Wan, A. Corneau, S. Tajbakhsh, T. H. Cheung and F. Le Grand, *Mol. Cell*, 2019, **74**, 609–621.e606.
25. J. C. Bruusgaard, K. Liestøl, M. Ekmark, K. Kollstad and K. Gundersen, *J. Physiol.*, 2003, **551**, 467–478.
26. F. Chemello, Z. Wang, H. Li, J. R. McAnally, N. Liu, R. Bassel-Duby and E. N. Olson, *Proc. Natl. Acad. Sci. U. S. A.*, 2020, **117**, 29691.
27. M. Kim, V. Franke, B. Brandt, E. D. Lowenstein, V. Schöwel, S. Spuler, A. Akalin and C. Birchmeier, *Nat. Commun.*, 2020, **11**, 6375.
28. N. Yang, N. J. Smyllie, H. Morris, C. F. Goncalves, M. Dudek, D. R. J. Pathiranage, J. E. Chesham, A. Adamson, D. G. Spiller, E. Zindy, J. Bagnall, N. Humphreys, J. Hoyland, A. S. I. Loudon, M. H. Hastings and Q. J. Meng, *PLoS Genet.*, 2020, **16**, e1008729.
29. N. J. Smyllie, V. Pilorz, J. Boyd, Q. J. Meng, B. Saer, J. E. Chesham, E. S. Maywood, T. P. Krogager, D. G. Spiller, R. Boot-Handford, M. R. White, M. H. Hastings and A. S. Loudon, *Curr. Biol.*, 2016, **26**, 1880–1886.
30. C. H. Gabriel, M. Del Olmo, A. Zehtabian, M. Jager, S. Reischl, H. van Dijk, C. Ulbricht, A. Rakhymzhan, T. Korte, B. Koller, A. Grudziecki,

B. Maier, A. Herrmann, R. Niesner, T. Zemojtel, H. Ewers, A. E. Granada, H. Herzel and A. Kramer, *Nat. Commun.*, 2021, **12**, 3796.
31. H. Yin, F. Price and M. A. Rudnicki, *Physiol. Rev.*, 2013, **93**, 23–67.
32. P. Dierickx, L. W. Van Laake and N. Geijsen, *EMBO Rep.*, 2018, **19**, 18–28.
33. P. Janich, Q. J. Meng and S. A. Benitah, *Curr. Opin. Cell Biol.*, 2014, **31**, 8–15.
34. G. Solanas, F. O. Peixoto, E. Perdiguero, M. Jardi, V. Ruiz-Bonilla, D. Datta, A. Symeonidi, A. Castellanos, P. S. Welz, J. M. Caballero, P. Sassone-Corsi, P. Munoz-Canoves and S. A. Benitah, *Cell*, 2017, **170**, 678–692.e620.
35. A. J. Trott and J. S. Menet, *PLoS Genet.*, 2018, **14**, e1007156.
36. K. L. Eckel-Mahan, V. R. Patel, S. de Mateo, R. Orozco-Solis, N. J. Ceglia, S. Sahar, S. A. Dilag-Penilla, K. A. Dyar, P. Baldi and P. Sassone-Corsi, *Cell*, 2013, **155**, 1464–1478.
37. S. Masri, T. Papagiannakopoulos, K. Kinouchi, Y. Liu, M. Cervantes, P. Baldi, T. Jacks and P. Sassone-Corsi, *Cell*, 2016, **165**, 896–909.
38. P. Zhu, N. X. Hamlish, A. V. Thakkar, A. W. T. Steffeck, E. J. Rendleman, N. H. Khan, N. J. Waldeck, A. W. DeVilbiss, M. S. Martin-Sandoval, T. P. Mathews, N. S. Chandel and C. B. Peek, *Genes Dev.*, 2022, **36**, 149–166.
39. A. Mayeuf-Louchart, Q. Thorel, S. Delhaye, J. Beauchamp, C. Duhem, A. Danckaert, S. Lancel, B. Pourcet, E. Woldt, A. Boulinguiez, L. Ferri, M. Zecchin, B. Staels, Y. Sebti and H. Duez, *Sci. Rep.*, 2017, **7**, 14383.
40. E. Woldt, Y. Sebti, L. A. Solt, C. Duhem, S. Lancel, J. Eeckhoute, M. K. Hesselink, C. Paquet, S. Delhaye, Y. Shin, T. M. Kamenecka, G. Schaart, P. Lefebvre, R. Neviere, T. P. Burris, P. Schrauwen, B. Staels and H. Duez, *Nat. Med.*, 2013, **19**, 1039–1046.
41. M. Buckingham and P. W. J. Rigby, *Dev. Cell*, 2014, **28**, 225–238.
42. S. Azmi, A. Ozog and R. Taneja, *J. Biol. Chem.*, 2004, **279**, 52643–52652.
43. B. M. Ling, S. Gopinadhan, W. K. Kok, S. R. Shankar, P. Gopal, N. Bharathy, Y. Wang and R. Taneja, *Mol. Biol. Cell*, 2012, **23**, 4778–4785.
44. B. A. Hodge, X. Zhang, M. A. Gutierrez-Monreal, Y. Cao, D. W. Hammers, Z. Yao, C. A. Wolff, P. Du, D. Kemler, A. R. Judge and K. A. Esser, *eLife*, 2019, **8**, e43017.
45. J. L. Andrews, X. Zhang, J. J. McCarthy, E. L. McDearmon, T. A. Hornberger, B. Russell, K. S. Campbell, S. Arbogast, M. B. Reid, J. R. Walker, J. B. Hogenesch, J. S. Takahashi and K. A. Esser, *Proc. Natl. Acad. Sci. U. S. A.*, 2010, **107**, 19090–19095.
46. R. V. Kondratov, A. A. Kondratova, V. Y. Gorbacheva, O. V. Vykhovanets and M. P. Antoch, *Genes Dev.*, 2006, **20**, 1868–1873.
47. S. Chatterjee, D. Nam, B. Guo, J. M. Kim, G. E. Winnier, J. Lee, R. Berdeaux, V. K. Yechoor and K. Ma, *J. Cell Sci.*, 2013, **126**, 2213–2224.
48. E. A. Schroder, B. D. Harfmann, X. Zhang, R. Srikuea, J. H. England, B. A. Hodge, Y. Wen, L. A. Riley, Q. Yu, A. Christie, J. D. Smith, T. Seward, E. M. Wolf Horrell, J. Mula, C. A. Peterson, T. A. Butterfield and K. A. Esser, *J. Physiol.*, 2015, **593**, 5387–5404.

49. K. A. Dyar, S. Schiaffino and B. Blaauw, *J. Physiol.*, 2016, **594**, 3161–3162.
50. E. Schroder, B. Hodge, L. Riley, X. Zhang and K. Esser, *J. Physiol.*, 2016, **594**, 3163–3164.
51. G. Yang, L. Chen, G. R. Grant, G. Paschos, W.-L. Song, E. S. Musiek, V. Lee, S. C. McLoughlin, T. Grosser, G. Cotsarelis and G. A. FitzGerald, *Sci. Transl. Med.*, 2016, **8**, 324ra316.
52. S. Schiaffino, B. Blaauw and K. A. Dyar, *Skeletal Muscle*, 2016, **6**, 33.
53. E. Farshadi, G. T. J. van der Horst and I. Chaves, *J. Mol. Biol.*, 2020, **432**, 3515–3524.
54. J. Gaucher, E. Montellier and P. Sassone-Corsi, *Trends Cell Biol.*, 2018, **28**, 368–379.
55. T. Matsuo, S. Yamaguchi, S. Mitsui, A. Emi, F. Shimoda and H. Okamura, *Science*, 2003, **302**, 255.
56. M. Lowe, J. Lage, E. Paatela, D. Munson, R. Hostager, C. Yuan, N. Katoku-Kikyo, M. Ruiz-Estevez, Y. Asakura, J. Staats, M. Qahar, M. Lohman, A. Asakura and N. Kikyo, *Cell Rep.*, 2018, **22**, 2118–2132.
57. S. Chatterjee, H. Yin, D. Nam, Y. Li and K. Ma, *Exp. Cell Res.*, 2015, **331**, 200–210.
58. R. D. Welch, C. Guo, M. Sengupta, K. J. Carpenter, N. A. Stephens, S. A. Arnett, M. J. Meyers, L. M. Sparks, S. R. Smith, J. Zhang, T. P. Burris and C. A. Flaveny, *Mol. Metab.*, 2017, **6**, 703–714.
59. X. Xiong, H. Gao, Y. Lin, V. Yechoor and K. Ma, *Exp. Cell Res.*, 2021, **406**, 112766.
60. N. Katoku-Kikyo, E. Paatela, D. L. Houtz, B. Lee, D. Munson, X. Wang, M. Hussein, J. Bhatia, S. Lim, C. Yuan, Y. Asakura, A. Asakura and N. Kikyo, *J. Cell Biol.*, 2021, **220**, e202101057.
61. J. J. Kelu, T. G. Pipalia and S. M. Hughes, *Proc. Natl. Acad. Sci. U. S. A.*, 2020, **117**, 31208.
62. N. Rion, P. Castets, S. Lin, L. Enderle, J. R. Reinhard, C. Eickhorst and M. A. Rüegg, *Development*, 2019, dev.172460.
63. R. Wu, F. Dang, P. Li, P. Wang, Q. Xu, Z. Liu, Y. Li, Y. Wu, Y. Chen and Y. Liu, *Cell Metab.*, 2019, **29**, 653–667.e656.
64. J. O. Lipton, E. D. Yuan, L. M. Boyle, D. Ebrahimi-Fakhari, E. Kwiatkowski, A. Nathan, T. Güttler, F. Davis, J. M. Asara and M. Sahin, *Cell*, 2015, **161**, 1138–1151.
65. S.-W. Chang, T. Yoshihara, S. Machida and H. Naito, *Biochem. Biophys. Rep.*, 2017, **9**, 153–158.
66. S. W. Chang, T. Yoshihara, T. Tsuzuki, T. Natsume, R. Kakigi, S. Machida and H. Naito, *PLoS One*, 2022, **17**, e0264171.
67. J. Delezie and E. Challet, *Ann. N. Y. Acad. Sci.*, 2011, **1243**, 30–46.
68. T. Roenneberg, K. V. Allebrandt, M. Merrow and C. Vetter, *Curr. Biol.*, 2012, **22**, 939–943.
69. D. M. D'Souza, D. Al-Sajee and T. J. Hawke, *Front. Physiol.*, 2013, **4**, 379–379.
70. X. Fu, M. Zhu, S. Zhang, M. Foretz, B. Viollet and M. Du, *Diabetes*, 2016, **65**, 188–200.

71. K. A. Lamia, U. M. Sachdeva, L. DiTacchio, E. C. Williams, J. G. Alvarez, D. F. Egan, D. S. Vasquez, H. Juguilon, S. Panda, R. J. Shaw, C. B. Thompson and R. M. Evans, *Science*, 2009, **326**, 437–440.
72. J. H. Um, J. S. Pendergast, D. A. Springer, M. Foretz, B. Viollet, A. Brown, M. K. Kim, S. Yamazaki and J. H. Chung, *PLoS One*, 2011, **6**, e18450.
73. J. H. Um, S. Yang, S. Yamazaki, H. Kang, B. Viollet, M. Foretz and J. H. Chung, *J. Biol. Chem.*, 2007, **282**, 20794–20798.
74. E. E. Zhang, Y. Liu, R. Dentin, P. Y. Pongsawakul, A. C. Liu, T. Hirota, D. A. Nusinow, X. Sun, S. Landais, Y. Kodama, D. A. Brenner, M. Montminy and S. A. Kay, *Nat. Med.*, 2010, **16**, 1152–1156.
75. V. D. Longo and S. Panda, *Cell Metab.*, 2016, **23**, 1048–1059.
76. U. Loizides-Mangold, L. Perrin, B. Vandereycken, J. A. Betts, J.-P. Walhin, I. Templeman, S. Chanon, B. D. Weger, C. Durand, M. Robert, J. Paz Montoya, M. Moniatte, L. G. Karagounis, J. D. Johnston, F. Gachon, E. Lefai, H. Riezman and C. Dibner, *Proc. Natl. Acad. Sci. U. S. A.*, 2017, **114**, E8565.
77. S. Sato, E. B. Parr, B. L. Devlin, J. A. Hawley and P. Sassone-Corsi, *Mol. Metab.*, 2018, **16**, 1–11.
78. K. A. Dyar, D. Lutter, A. Artati, N. J. Ceglia, Y. Liu, D. Armenta, M. Jastroch, S. Schneider, S. de Mateo, M. Cervantes, S. Abbondante, P. Tognini, R. Orozco-Solis, K. Kinouchi, C. Wang, R. Swerdloff, S. Nadeef, S. Masri, P. Magistretti, V. Orlando, E. Borrelli, N. H. Uhlenhaut, P. Baldi, J. Adamski, M. H. Tschop, K. Eckel-Mahan and P. Sassone-Corsi, *Cell*, 2018, **174**, 1571–1585.e1511.
79. A. Kalinkovich and G. Livshits, *Ageing Res. Rev.*, 2017, **35**, 200–221.
80. J. W. Beals, N. A. Burd, D. R. Moore and S. van Vliet, *Front. Nutr.*, 2019, **6**, 87.
81. L. S. Puig, A. Altõntaş, S. Casaní-Galdón, B. M. Gabriel, R. Barrès, A. Conesa, A. V. Chibalin, E. Näslund, A. Krook, N. J. Pillon and J. R. Zierath, bioRxiv, 2021, preprint, 2021.2002.2023.432336, DOI: 10.1101/2021.02.23.432336.
82. J. F. Harmsen, N. van Polanen, M. van Weeghel, J. Wefers, J. Hoeks, F. M. Vaz, M. L. Pras-Raves, A. H. C. van Kampen, G. Schaart, D. van Moorsel, J. Hansen, M. K. C. Hesselink, R. H. Houtkooper and P. Schrauwen, *FASEB J.*, 2021, **35**, e21611.
83. A. Englund, L. Kovanen, S. T. Saarikoski, J. Haukka, A. Reunanen, A. Aromaa, J. Lönnqvist and T. Partonen, *J. Circadian Rhythms*, 2009, **7**, 5.
84. J. Dupuis, C. Langenberg, I. Prokopenko, R. Saxena, N. Soranzo, A. U. Jackson, E. Wheeler, N. L. Glazer, N. Bouatia-Naji, A. L. Gloyn, C. M. Lindgren, R. Magi, A. P. Morris, J. Randall, T. Johnson, P. Elliott, D. Rybin, G. Thorleifsson, V. Steinthorsdottir, P. Henneman, H. Grallert, A. Dehghan, J. J. Hottenga, C. S. Franklin, P. Navarro, K. Song, A. Goel, J. R. Perry, J. M. Egan, T. Lajunen, N. Grarup, T. Sparso, A. Doney, B. F. Voight, H. M. Stringham, M. Li, S. Kanoni, P. Shrader, C. Cavalcanti-Proenca, M. Kumari, L. Qi, N. J. Timpson, C. Gieger, C. Zabena, G. Rocheleau, E. Ingelsson, P. An, J. O'Connell,

J. Luan, A. Elliott, S. A. McCarroll, F. Payne, R. M. Roccasecca, F. Pattou, P. Sethupathy, K. Ardlie, Y. Ariyurek, B. Balkau, P. Barter, J. P. Beilby, Y. Ben-Shlomo, R. Benediktsson, A. J. Bennett, S. Bergmann, M. Bochud, E. Boerwinkle, A. Bonnefond, L. L. Bonnycastle, K. Borch-Johnsen, Y. Bottcher, E. Brunner, S. J. Bumpstead, G. Charpentier, Y. D. Chen, P. Chines, R. Clarke, L. J. Coin, M. N. Cooper, M. Cornelis, G. Crawford, L. Crisponi, I. N. Day, E. J. de Geus, J. Delplanque, C. Dina, M. R. Erdos, A. C. Fedson, A. Fischer-Rosinsky, N. G. Forouhi, C. S. Fox, R. Frants, M. G. Franzosi, P. Galan, M. O. Goodarzi, J. Graessler, C. J. Groves, S. Grundy, R. Gwilliam, U. Gyllensten, S. Hadjadj, G. Hallmans, N. Hammond, X. Han, A. L. Hartikainen, N. Hassanali, C. Hayward, S. C. Heath, S. Hercberg, C. Herder, A. A. Hicks, D. R. Hillman, A. D. Hingorani, A. Hofman, J. Hui, J. Hung, B. Isomaa, P. R. Johnson, T. Jorgensen, A. Jula, M. Kaakinen, J. Kaprio, Y. A. Kesaniemi, M. Kivimaki, B. Knight, S. Koskinen, P. Kovacs, K. O. Kyvik, G. M. Lathrop, D. A. Lawlor, O. Le Bacquer, C. Lecoeur, Y. Li, V. Lyssenko, R. Mahley, M. Mangino, A. K. Manning, M. T. Martinez-Larrad, J. B. McAteer, L. J. McCulloch, R. McPherson, C. Meisinger, D. Melzer, D. Meyre, B. D. Mitchell, M. A. Morken, S. Mukherjee, S. Naitza, N. Narisu, M. J. Neville, B. A. Oostra, M. Orru, R. Pakyz, C. N. Palmer, G. Paolisso, C. Pattaro, D. Pearson, J. F. Peden, N. L. Pedersen, M. Perola, A. F. Pfeiffer, I. Pichler, O. Polasek, D. Posthuma, S. C. Potter, A. Pouta, M. A. Province, B. M. Psaty, W. Rathmann, N. W. Rayner, K. Rice, S. Ripatti, F. Rivadeneira, M. Roden, O. Rolandsson, A. Sandbaek, M. Sandhu, S. Sanna, A. A. Sayer, P. Scheet, L. J. Scott, U. Seedorf, S. J. Sharp, B. Shields, G. Sigurethsson, E. J. Sijbrands, A. Silveira, L. Simpson, A. Singleton, N. L. Smith, U. Sovio, A. Swift, H. Syddall, A. C. Syvanen, T. Tanaka, B. Thorand, J. Tichet, A. Tonjes, T. Tuomi, A. G. Uitterlinden, K. W. van Dijk, M. van Hoek, D. Varma, S. Visvikis-Siest, V. Vitart, N. Vogelzangs, G. Waeber, P. J. Wagner, A. Walley, G. B. Walters, K. L. Ward, H. Watkins, M. N. Weedon, S. H. Wild, G. Willemsen, J. C. Witteman, J. W. Yarnell, E. Zeggini, D. Zelenika, B. Zethelius, G. Zhai, J. H. Zhao, M. C. Zillikens, I. B. Borecki, R. J. Loos, P. Meneton, P. K. Magnusson, D. M. Nathan, G. H. Williams, A. T. Hattersley, K. Silander, V. Salomaa, G. D. Smith, S. R. Bornstein, P. Schwarz, J. Spranger, F. Karpe, A. R. Shuldiner, C. Cooper, G. V. Dedoussis, M. Serrano-Rios, A. D. Morris, L. Lind, L. J. Palmer, F. B. Hu, P. W. Franks, S. Ebrahim, M. Marmot, W. H. Kao, J. S. Pankow, M. J. Sampson, J. Kuusisto, M. Laakso, T. Hansen, O. Pedersen, P. P. Pramstaller, H. E. Wichmann, T. Illig, I. Rudan, A. F. Wright, M. Stumvoll, H. Campbell, J. F. Wilson, R. N. Bergman, T. A. Buchanan, F. S. Collins, K. L. Mohlke, J. Tuomilehto, T. T. Valle, D. Altshuler, J. I. Rotter, D. S. Siscovick, B. W. Penninx, D. I. Boomsma, P. Deloukas, T. D. Spector, T. M. Frayling, L. Ferrucci, A. Kong, U. Thorsteinsdottir, K. Stefansson, C. M. van Duijn, Y. S. Aulchenko, A. Cao, A. Scuteri, D. Schlessinger, M. Uda, A. Ruokonen, M. R. Jarvelin,

D. M. Waterworth, P. Vollenweider, L. Peltonen, V. Mooser, G. R. Abecasis, N. J. Wareham, R. Sladek, P. Froguel, R. M. Watanabe, J. B. Meigs, L. Groop, M. Boehnke, M. I. McCarthy, J. C. Florez and I. Barroso, *Nat. Genet.*, 2010, **42**, 105–116.

85. A. Barker, S. J. Sharp, N. J. Timpson, N. Bouatia-Naji, N. M. Warrington, S. Kanoni, L. J. Beilin, S. Brage, P. Deloukas, D. M. Evans, A. Grontved, N. Hassanali, D. A. Lawlor, C. Lecoeur, R. J. F. Loos, S. J. Lye, M. I. McCarthy, T. A. Mori, N. C. Ndiaye, J. P. Newnham, I. Ntalla, C. E. Pennell, B. St Pourcain, I. Prokopenko, S. M. Ring, N. Sattar, S. Visvikis-Siest, G. V. Dedoussis, L. J. Palmer, P. Froguel, G. D. Smith, U. Ekelund, N. J. Wareham and C. Langenberg, *Diabetes*, 2011, **60**, 1805.
86. H. Fujita, K. Hara, N. Shojima, M. Horikoshi, M. Iwata, Y. Hirota, K. Tobe, S. Seino and T. Kadowaki, *J. Hum. Genet.*, 2012, **57**, 776–779.
87. C. Hu, R. Zhang, C. Wang, J. Wang, X. Ma, X. Hou, J. Lu, W. Yu, F. Jiang, Y. Bao, K. Xiang and W. Jia, *PLoS One*, 2010, **5**, e15542.
88. T. W. Boesgaard, N. Grarup, T. Jørgensen, K. Borch-Johnsen, T. Hansen and O. Pedersen, *Diabetologia*, 2010, **53**, 1647–1655.
89. R. Wagner, K. Dudziak, S. A. Herzberg-Schäfer, F. Machicao, N. Stefan, H. Staiger, H.-U. Häring and A. Fritsche, *PLoS One*, 2011, **6**, e23639.
90. C. Liu, H. Li, L. Qi, R. J. Loos, Q. Qi, L. Lu, W. Gan and X. Lin, *PLoS One*, 2011, **6**, e21464.
91. N. W. Nowell, *Diabetologia*, 1970, **6**, 488–492.
92. E. Van Cauter, K. S. Polonsky and A. J. Scheen, *Endocr. Rev.*, 1997, **18**, 716–738.
93. K. F. Carroll and P. J. Nestel, *Diabetes*, 1973, **22**, 333.
94. J. Qian and F. A. J. L. Scheer, *Trends Endocrinol. Metab.*, 2016, **27**, 282–293.
95. C. H. Walsh and A. D. Wright, *Postgrad. Med. J.*, 1975, **51**, 169.
96. R. J. Jarrett and H. Keen, *Br. Med. J.*, 1969, **2**, 341.
97. L.-X. Zhong, X.-N. Li, G.-Y. Yang, X. Zhang, W.-X. Li, Q.-Q. Zhang, H.-X. Pan, H.-H. Zhang, M.-Y. Zhou, Y.-D. Wang, W.-W. Zhang, Q.-S. Hu, W. Zhu and B. Zhang, *PLoS One*, 2019, **14**, e0225813.
98. A. L. Basse, E. Dalbram, L. Larsson, Z. Gerhart-Hines, J. R. Zierath and J. T. Treebak, *Front. Physiol.*, 2018, **9**, 1198.
99. B. D. Harfmann, E. A. Schroder, M. T. Kachman, B. A. Hodge, X. Zhang and K. A. Esser, *Skeletal Muscle*, 2016, **6**, 12.
100. K. A. Dyar, M. J. Hubert, A. A. Mir, S. Ciciliot, D. Lutter, F. Greulich, F. Quagliarini, M. Kleinert, K. Fischer, T. O. Eichmann, L. E. Wright, M. I. Peña Paz, A. Casarin, V. Pertegato, V. Romanello, M. Albiero, S. Mazzucco, R. Rizzuto, L. Salviati, G. Biolo, B. Blaauw, S. Schiaffino and N. H. Uhlenhaut, *PLoS Biol.*, 2018, **16**, e2005886.
101. S. Raichur, R. L. Fitzsimmons, S. A. Myers, M. A. Pearen, P. Lau, N. Eriksson, S. M. Wang and G. E. O. Muscat, *Nucleic Acids Res.*, 2010, **38**, 4296–4312.
102. K. Nohara, V. Mallampalli, T. Nemkov, M. Wirianto, J. Yang, Y. Ye, Y. Sun, L. Han, K. A. Esser, E. Mileykovskaya, A. D'Alessandro, C. B. Green,

J. S. Takahashi, W. Dowhan, S.-H. Yoo and Z. Chen, *Nat. Commun.*, 2019, **10**, 3923.
103. B. He, K. Nohara, N. Park, Y.-S. Park, B. Guillory, Z. Zhao, J. M. Garcia, N. Koike, C. C. Lee, J. S. Takahashi, S.-H. Yoo and Z. Chen, *Cell Metab.*, 2016, **23**, 610–621.
104. B. Marcheva, K. M. Ramsey, E. D. Buhr, Y. Kobayashi, H. Su, C. H. Ko, G. Ivanova, C. Omura, S. Mo, M. H. Vitaterna, J. P. Lopez, L. H. Philipson, C. A. Bradfield, S. D. Crosby, L. JeBailey, X. Wang, J. S. Takahashi and J. Bass, *Nature*, 2010, **466**, 627–631.
105. V. Petrenko, N. R. Gandasi, D. Sage, A. Tengholm, S. Barg and C. Dibner, *Proc. Natl. Acad. Sci. U. S. A.*, 2020, **117**, 2484.
106. K. A. Lamia, K. F. Storch and C. J. Weitz, *Proc. Natl. Acad. Sci. U. S. A.*, 2008, **105**, 15172–15177.
107. L. A. Sadacca, K. A. Lamia, A. S. Delemos, B. Blum and C. J. Weitz, *Diabetologia*, 2011, **54**, 120–124.
108. J. Rahman and S. Rahman, *Lancet*, 2018, **391**, 2560–2574.
109. J. Y. Jang, A. Blum, J. Liu and T. Finkel, *J. Clin. Invest.*, 2018, **128**, 3662–3670.
110. K. Ohlendieck, *Skeletal Muscle*, 2011, **1**, 6.
111. M. Y. W. Ng, T. Wai and A. Simonsen, *Dev. Cell*, 2021, **56**, 881–905.
112. A. Neufeld-Cohen, M. S. Robles, R. Aviram, G. Manella, Y. Adamovich, B. Ladeuix, D. Nir, L. Rousso-Noori, Y. Kuperman, M. Golik, M. Mann and G. Asher, *Proc. Natl. Acad. Sci. U. S. A.*, 2016, **113**, E1673.
113. D. Jacobi, S. Liu, K. Burkewitz, N. Kory, N. H. Knudsen, R. K. Alexander, U. Unluturk, X. Li, X. Kong, A. L. Hyde, M. R. Gangl, W. B. Mair and C. H. Lee, *Cell Metab.*, 2015, **22**, 709–720.
114. K. Schmitt, A. Grimm, R. Dallmann, B. Oettinghaus, L. M. Restelli, M. Witzig, N. Ishihara, K. Mihara, J. A. Ripperger, U. Albrecht, S. Frank, S. A. Brown and A. Eckert, *Cell Metab.*, 2018, **27**, 657–666.e5.
115. D. van Moorsel, J. Hansen, B. Havekes, F. Scheer, J. A. Jorgensen, J. Hoeks, V. B. Schrauwen-Hinderling, H. Duez, P. Lefebvre, N. C. Schaper, M. K. C. Hesselink, B. Staels and P. Schrauwen, *Mol. Metab.*, 2016, **5**, 635–645.
116. J. Wefers, N. J. Connell, C. E. Fealy, C. Andriessen, V. de Wit, D. van Moorsel, E. Moonen-Kornips, J. A. Jörgensen, M. K. C. Hesselink, B. Havekes, J. Hoeks and P. Schrauwen, *Mol. Metab.*, 2020, **41**, 101050.
117. C. B. Peek, A. H. Affinati, K. M. Ramsey, H.-Y. Kuo, W. Yu, L. A. Sena, O. Ilkayeva, B. Marcheva, Y. Kobayashi, C. Omura, D. C. Levine, D. J. Bacsik, D. Gius, C. B. Newgard, E. Goetzman, N. S. Chandel, J. M. Denu, M. Mrksich and J. Bass, *Science*, 2013, **342**, 1243417.
118. G. Maier, J. Delezie, P. O. Westermark, G. Santos, D. Ritz and C. Handschin, *J. Physiol.*, 2021, **600**, 769–796.
119. T. Eckle, K. Hartmann, S. Bonney, S. Reithel, M. Mittelbronn, L. A. Walker, B. D. Lowes, J. Han, C. H. Borchers, P. M. Buttrick,

D. J. Kominsky, S. P. Colgan and H. K. Eltzschig, *Nat. Med.*, 2012, **18**, 774–782.
120. B. Grimaldi, M. M. Bellet, S. Katada, G. Astarita, J. Hirayama, R. H. Amin, J. G. Granneman, D. Piomelli, T. Leff and P. Sassone-Corsi, *Cell Metab.*, 2010, **12**, 509–520.
121. S. D. Jordan, A. Kriebs, M. Vaughan, D. Duglan, W. Fan, E. Henriksson, A. L. Huber, S. J. Papp, M. Nguyen, M. Afetian, M. Downes, R. T. Yu, A. Kralli, R. M. Evans and K. A. Lamia, *Cell Metab.*, 2017, **26**, 243–255.e6.
122. M. E. Vaughan, M. Wallace, M. K. Handzlik, A. B. Chan, C. M. Metallo and K. A. Lamia, *iScience*, 2020, **23**, 101338.
123. E. Y. Dimova, M. Jakupovic, K. Kubaichuk, D. Mennerich, T. F. Chi, F. Tamanini, M. Oklejewicz, J. Hänig, N. Byts, K. A. Mäkelä, K. H. Herzig, P. Koivunen, I. Chaves, G. van der Horst and T. Kietzmann, *iScience*, 2019, **13**, 284–304.
124. F. B. Favier, F. A. Britto, D. G. Freyssenet, X. A. Bigard and H. Benoit, *Cell. Mol. Life Sci.*, 2015, **72**, 4681–4696.
125. M. E. Lindholm and H. Rundqvist, *Exp. Physiol.*, 2016, **101**, 28–32.
126. S. D. Mason, R. A. Howlett, M. J. Kim, I. M. Olfert, M. C. Hogan, W. McNulty, R. P. Hickey, P. D. Wagner, C. R. Kahn, F. J. Giordano and R. S. Johnson, *PLoS Biol.*, 2004, **2**, e288.
127. S. Sato, A. L. Basse, M. Schonke, S. Chen, M. Samad, A. Altintas, R. C. Laker, E. Dalbram, R. Barres, P. Baldi, J. T. Treebak, J. R. Zierath and P. Sassone-Corsi, *Cell Metab.*, 2019, **30**, 92–110 e114.
128. C. B. Peek, D. C. Levine, J. Cedernaes, A. Taguchi, Y. Kobayashi, S. J. Tsai, N. A. Bonar, M. R. McNulty, K. M. Ramsey and J. Bass, *Cell Metab.*, 2017, **25**, 86–92.
129. Y. Adamovich, B. Ladeuix, M. Golik, M. P. Koeners and G. Asher, *Cell Metab.*, 2017, **25**, 93–101.
130. Y. Wu, D. Tang, N. Liu, W. Xiong, H. Huang, Y. Li, Z. Ma, H. Zhao, P. Chen, X. Qi and E. E. Zhang, *Cell Metab.*, 2017, **25**, 73–85.
131. E. Facer-Childs and R. Brandstaetter, *Curr. Biol.*, 2015, **25**, 518–522.
132. G. Atkinson and T. Reilly, *Sports Med.*, 1996, **21**, 292–312.
133. T. Bernard, M. Giacomoni, O. Gavarry, M. Seymat and G. Falgairette, *Eur. J. Appl. Physiol. Occup. Physiol.*, 1997, **77**, 133–138.
134. H. Chtourou and N. Souissi, *J. Strength Cond. Res.*, 2012, **26**, 1984–2005.
135. G. G. Mirizio, R. S. M. Nunes, D. A. Vargas, C. Foster and E. Vieira, *Sci. Rep.*, 2020, **10**, 9485.
136. D. Duglan and K. A. Lamia, *Trends Endocrinol. Metab.*, 2019, **30**, 347–356.
137. J. Grgic, B. Lazinica, A. Garofolini, B. J. Schoenfeld, N. J. Saner and P. Mikulic, *Chronobiol. Int.*, 2019, **36**, 449–460.
138. M. Sedliak, T. Finni, S. Cheng, M. Lind and K. Häkkinen, *J. Strength Cond. Res.*, 2009, **23**, 2451–2457.

139. M. Sedliak, M. Zeman, G. Buzgó, J. Cvecka, D. Hamar, E. Laczo, M. Okuliarova, M. Vanderka, T. Kampmiller, K. Häkkinen, J. P. Ahtiainen, J. J. Hulmi, T. S. Nilsen, H. Wiig and T. Raastad, *Chronobiol. Int.*, 2018, **35**, 450–464.
140. N. Souissi, A. Gauthier, B. Sesboüé, J. Larue and D. Davenne, *J. Sports Sci.*, 2002, **20**, 929–937.
141. M. Sedliak, T. Finni, J. Peltonen and K. Häkkinen, *J. Sports Sci.*, 2008, **26**, 1005–1014.
142. D. W. Hill, K. J. Cureton and M. A. Collins, *Ergonomics*, 1989, **32**, 79–92.
143. S. Ezagouri, Z. Zwighaft, J. Sobel, S. Baillieul, S. Doutreleau, B. Ladeuix, M. Golik, S. Verges and G. Asher, *Cell Metab.*, 2019, **30**, 78–91.e4.
144. S. Ghosh, S. Golbidi, I. Werner, B. C. Verchere and I. Laher, *Clin. Sci.*, 2010, **119**, 57–74.
145. Y. J. Kim, H. J. Kim, W. J. Lee and J. K. Seong, *Lab. Anim. Res.*, 2020, **36**, 3.
146. A. M. Gordon, E. Homsher and M. Regnier, *Physiol. Rev.*, 2000, **80**, 853–924.
147. E. Challet, *Nat. Rev. Endocrinol.*, 2019, **15**, 393–405.
148. S. L. Chellappa, N. Vujovic, J. S. Williams and F. Scheer, *Trends Endocrinol. Metab.*, 2019, **30**, 767–779.
149. J. Arendt, *Occup. Med.*, 2010, **60**, 10–20.
150. A. Chaix, E. N. C. Manoogian, G. C. Melkani and S. Panda, *Annu. Rev. Nutr.*, 2019, **39**, 291–315.
151. A. Lowden, C. Moreno, U. Holmback, M. Lennernas and P. Tucker, *Scand. J. Work, Environ. Health*, 2010, **36**, 150–162.
152. R. Salgado-Delgado, M. Angeles-Castellanos, M. R. Buijs and C. Escobar, *Neuroscience*, 2008, **154**, 922–931.
153. J. Delezie, S. Dumont, C. Sandu, S. Reibel, P. Pevet and E. Challet, *Sci. Rep.*, 2016, **6**, 29386.
154. J. Reznick, E. Preston, D. L. Wilks, S. M. Beale, N. Turner and G. J. Cooney, *Biochim. Biophys. Acta*, 2013, **1832**, 228–238.
155. J. L. Barclay, J. Husse, B. Bode, N. Naujokat, J. Meyer-Kovac, S. M. Schmid, H. Lehnert and H. Oster, *PLoS One*, 2012, **7**, e37150.
156. R. C. Salgado-Delgado, N. Saderi, C. Basualdo Mdel, N. N. Guerrero-Vargas, C. Escobar and R. M. Buijs, *PLoS One*, 2013, **8**, e60052.
157. G. Atkinson, B. Edwards, T. Reilly and J. Waterhouse, *Eur. J. Appl. Physiol.*, 2007, **99**, 331–341.
158. O. M. Buxton, S. A. Frank, M. L'Hermite-Balériaux, R. Leproult, F. W. Turek and E. Van Cauter, *Am. J. Physiol.*, 1997, **273**, E536–E542.
159. J. M. Thomas, P. A. Kern, H. M. Bush, K. J. McQuerry, W. S. Black, J. L. Clasey and J. S. Pendergast, *JCI Insight*, 2020, **5**, e134270.
160. T. Miyazaki, S. Hashimoto, S. Masubuchi, S. Honma and K. I. Honma, *Am. J. Physiol.: Regul., Integr. Comp. Physiol.*, 2001, **281**, R197–R205.
161. O. Van Reeth, J. Sturis, M. M. Byrne, J. D. Blackman, M. L'Hermite-Balériaux, R. Leproult, C. Oliner, S. Refetoff, F. W. Turek and E. Van Cauter, *Am. J. Physiol.*, 1994, **266**, E964–E974.

162. S. D. Youngstedt, J. A. Elliott and D. F. Kripke, *J. Physiol.*, 2019, **597**, 2253–2268.
163. E. K. Baehr, C. I. Eastman, W. Revelle, S. H. Olson, L. F. Wolfe and P. C. Zee, *Am. J. Physiol.: Regul., Integr. Comp. Physiol.*, 2003, **284**, R1542–R1550.
164. B. Edwards, J. Waterhouse, G. Atkinson and T. Reilly, *J. Sports Sci.*, 2002, **20**, 725–732.
165. O. M. Buxton, C. W. Lee, M. L'Hermite-Baleriaux, F. W. Turek and E. Van Cauter, *Am. J. Physiol.: Regul., Integr. Comp. Physiol.*, 2003, **284**, R714–R724.
166. A. C. Zambon, E. L. McDearmon, N. Salomonis, K. M. Vranizan, K. L. Johansen, D. Adey, J. S. Takahashi, M. Schambelan and B. R. Conklin, *Genome Biol.*, 2003, **4**, R61.
167. G. Wolff and K. A. Esser, *Med. Sci. Sports Exercise*, 2012, **44**, 1663–1670.
168. B. K. Pedersen and L. Hoffman-Goetz, *Physiol. Rev.*, 2000, **80**, 1055–1081.
169. D. Hansen, R. Meeusen, A. Mullens and P. Dendale, *Sports Med.*, 2012, **42**, 415–431.
170. H. Oster, E. Challet, V. Ott, E. Arvat, E. R. de Kloet, D. J. Dijk, S. Lightman, A. Vgontzas and E. Van Cauter, *Endocr. Rev.*, 2017, **38**, 3–45.
171. H. Sasaki, Y. Hattori, Y. Ikeda, M. Kamagata, S. Iwami, S. Yasuda, Y. Tahara and S. Shibata, *Sci. Rep.*, 2016, **6**, 27607.
172. Y. Yasumoto, R. Nakao and K. Oishi, *PLoS One*, 2015, **10**, e0116476.
173. P. De Goede, S. Sen, Y. Su, E. Foppen, V.-J. Poirel, E. Challet and A. Kalsbeek, *Int. J. Mol. Sci.*, 2018, **19**, 3171.
174. F. Damiola, N. Le Minh, N. Preitner, B. Kornmann, F. Fleury-Olela and U. Schibler, *Genes Dev.*, 2000, **14**, 2950–2961.
175. L. S. Lundell, E. B. Parr, B. L. Devlin, L. R. Ingerslev, A. Altõntaş, S. Sato, P. Sassone-Corsi, R. Barrès, J. R. Zierath and J. A. Hawley, *Nat. Commun.*, 2020, **11**, 4643.
176. R. Guthold, G. A. Stevens, L. M. Riley and F. C. Bull, *Lancet Global Health*, 2018, **6**, e1077–e1086.
177. P. C. Hallal, L. B. Andersen, F. C. Bull, R. Guthold, W. Haskell and U. Ekelund, *Lancet*, 2012, **380**, 247–257.
178. F. W. Booth, C. K. Roberts and M. J. Laye, *Compr. Physiol.*, 2012, **2**, 1143–1211.
179. A. Christ and E. Latz, *Nat. Rev. Immunol.*, 2019, **19**, 267–268.
180. H. Wu and C. M. Ballantyne, *J. Clin. Invest.*, 2017, **127**, 43–54.
181. S. Cohen, J. A. Nathan and A. L. Goldberg, *Nat. Rev. Drug Discovery*, 2015, **14**, 58–74.
182. S. Panda, *Science*, 2016, **354**, 1008–1015.
183. C. A. McClung, *Pharmacol. Ther.*, 2007, **114**, 222–232.
184. T. A. Bedrosian, L. K. Fonken and R. J. Nelson, *Annu. Rev. Physiol.*, 2016, **78**, 109–131.

185. C. M. Depner, E. R. Stothard and K. P. Wright Jr, *Curr. Diabetes Rep.*, 2014, **14**, 507.
186. K. Wulff, S. Gatti, J. G. Wettstein and R. G. Foster, *Nat. Rev. Neurosci.*, 2010, **11**, 589–599.
187. J. Bass and M. A. Lazar, *Science*, 2016, **354**, 994–999.
188. D. Ding, K. D. Lawson, T. L. Kolbe-Alexander, E. A. Finkelstein, P. T. Katzmarzyk, W. van Mechelen and M. Pratt, *Lancet*, 2016, **388**, 1311–1324.
189. F. Lobelo, M. Stoutenberg and A. Hutber, *Br. J. Sports Med.*, 2014, **48**, 1627–1633.
190. B. K. Pedersen and B. Saltin, *Scand. J. Med. Sci. Sports*, 2015, **25**(Suppl. 3), 1–72.
191. J. Delezie and C. Handschin, *Front. Neurol.*, 2018, **9**, 698.
192. J. A. Hawley, M. Hargreaves, M. J. Joyner and J. R. Zierath, *Cell*, 2014, **159**, 738–749.
193. WHO, in *Global Recommendations on Physical Activity for Health*, Geneva, 2010.
194. B. M. Gabriel and J. R. Zierath, *Nat. Rev. Endocrinol.*, 2019, **15**, 197–206.
195. M. Savikj, B. M. Gabriel, P. S. Alm, J. Smith, K. Caidahl, M. Bjornholm, T. Fritz, A. Krook, J. R. Zierath and H. Wallberg-Henriksson, *Diabetologia*, 2019, **62**, 233–237.
196. T. Moholdt, E. B. Parr, B. L. Devlin, J. Debik, G. Giskeødegård and J. A. Hawley, *Diabetologia*, 2021, **64**, 2061–2076.
197. R. Mancilla, B. Brouwers, V. B. Schrauwen-Hinderling, M. K. C. Hesselink, J. Hoeks and P. Schrauwen, *Physiol. Rep.*, 2021, **8**, e14669.
198. S. Y. M. Teo, J. A. Kanaley, K. J. Guelfi, K. J. Marston and T. J. Fairchild, *Med. Sci. Sports Exercise*, 2020, **52**, 323–334.
199. L. C. de Brito, R. A. Rezende, N. D. da Silva Junior, T. Tinucci, D. E. Casarini, J. Cipolla-Neto and C. L. Forjaz, *PLoS One*, 2015, **10**, e0132458.
200. K. Fairbrother, B. Cartner, J. R. Alley, C. D. Curry, D. L. Dickinson, D. M. Morris and S. R. Collier, *Vasc. Health Risk Manage.*, 2014, **10**, 691–698.
201. Y. Yamanaka, S. Hashimoto, N. N. Takasu, Y. Tanahashi, S. Y. Nishide, S. Honma and K. Honma, *Am. J. Physiol.: Regul., Integr. Comp. Physiol.*, 2015, **309**, R1112–R1121.
202. E. B. Parr, L. K. Heilbronn and J. A. Hawley, *Exercise Sport Sci. Rev.*, 2020, **48**, 4–10.
203. R. M. Edinburgh, H. E. Bradley, N.-F. Abdullah, S. L. Robinson, O. J. Chrzanowski-Smith, J.-P. Walhin, S. Joanisse, K. N. Manolopoulos, A. Philp, A. Hengist, A. Chabowski, F. M. Brodsky, F. Koumanov, J. A. Betts, D. Thompson, G. A. Wallis and J. T. Gonzalez, *J. Clin. Endocrinol. Metab.*, 2020, **105**, 660–676.
204. J. A. Hawley, L. M. Burke, S. M. Phillips and L. L. Spriet, *J. Appl. Physiol.*, 2010, **110**, 834–845.

205. D. Guan, Y. Xiong, T. M. Trinh, Y. Xiao, W. Hu, C. Jiang, P. Dierickx, C. Jang, J. D. Rabinowitz and M. A. Lazar, *Science*, 2020, **369**, 1388.
206. F. Sinturel, P. Gos, V. Petrenko, C. Hagedorn, F. Kreppel, K.-F. Storch, D. Knutti, A. Liani, C. Weitz, Y. Emmenegger, P. Franken, L. Bonacina, C. Dibner and U. Schibler, *Genes Dev.*, 2021, **35**, 329–334.
207. E. A. Schroder, M. Ono, S. R. Johnson, E. R. Rozmus, D. E. Burgess, K. A. Esser and B. P. Delisle, *J. Physiol.*, 2022, **600**, 2037–2048.
208. I. S. Aziz, A. M. McMahon, D. Friedman, I. Rabinovich-Nikitin, L. A. Kirshenbaum and T. A. Martino, *Curr. Opin. Pharmacol.*, 2021, **57**, 60–70.
209. P. Mistry, A. Duong, L. Kirshenbaum and T. A. Martino, *Heart Failure Clin.*, 2017, **13**, 657–672.

CHAPTER 15

Integumentary System

GERMAINE CORNELISSEN*[a] AND KUNIAKI OTSUKA*[a,b]

[a] Halberg Chronobiology Center, University of Minnesota, Minneapolis, Minnesota, USA; [b] Executive Medical Center, Totsuka Royal Clinic, Tokyo Women's Medical University, Tokyo, Japan
*Emails: corne001@umn.edu; frtotk99@ba2.so-net.ne.jp

15.1 Introduction

Skin is the largest organ in the body, covering its entire external surface. It consists of three major layers, the epidermis, dermis, and the hypodermis, which vary in their anatomy and function. The skin's structure consists of an intricate network serving as the body's initial protection against pathogens, chemicals, UV light, and mechanical injury. It also regulates temperature and the amount of water, urea and ammonia released into the environment, and permits the sensations of touch, heat, and cold.[1]

The uppermost layer of the epidermis is made up of keratin and horny scales produced by dead keratinocytes (anucleate squamous cells), which secrete defensins, serving as the first line of immune defense. Between the uppermost and deepest layers are three other layers, with the middle one containing glycolipids that are secreted to the surface of the cells to keep them stuck together. The deepest layer (*stratum basale*), separated from the dermis by the basement membrane, contains melanocytes and mitotically active stem cells that are constantly producing keratinocytes.[1]

Keratinocytes are responsible for the formation of the epidermal water barrier by making and secreting lipids. They also regulate calcium absorption by activating cholesterol precursors when exposed to UVB light, leading to the formation of vitamin D, which is important for normal bone metabolism.

Chemical Biology No. 23
Chronobiology and Chronomedicine: From Molecular and Cellular Mechanisms to Whole Body Interdigitating Networks
Edited by Germaine Cornelissen and Tsuyoshi Hirota

Published by the Royal Society of Chemistry, www.rsc.org

Melanocytes mostly produce melanin, which is responsible for the pigmentation of the skin. Melanin secretion is stimulated by UVB light as a protective response against UV radiation. The epidermis also contains Langerhans' cells, which play a significant role in antigen presentation, and Merkel cells, which serve a sensory function as mechanoreceptors for light touch.[1]

The dermis mainly contains extracellular matrix (ECM) composed of collagen, elastic fibers, proteoglycans, and glycoproteins. It has two layers, the upper papillary layer consisting of loose connective tissue and the deeper reticular layer containing dense connective tissue composed of tightly packed bundles of collagen fibers. Collagen strengthens skin by giving elasticity to its elastic fibers. The extracellular matrix is secreted by fibroblasts, which are the main cell type present in the dermis. Other cells residing in the dermis are T cells, mast cells, macrophages, dendritic cells, and dermal white adipocytes, which participate in immune responses and wound healing. The dermis houses sweat glands, sebaceous glands, hair, hair follicles, muscles, sensory neurons, and blood vessels. Sweat production is a common thermoregulatory mechanism controlled by sweat glands. Sensory receptors include pain sensing nociceptors, itch sensing prurireceptors, temperature sensitive thermoreceptors, and touch sensitive mechanoreceptors.[2]

The hypodermis or subcutaneous fascia contains adipose lobules along with some skin appendages like hair follicles, sensory neurons, and blood vessels. It serves to isolate, cushion and provide energy to skin. It also contains fibroblasts and macrophages. Although it is no longer considered as part of the skin,[1] dermal adipocytes are involved in a number of processes, such as hair growth,[3] wound healing,[4] skin aging,[5,6] temperature regulation,[7,8] and the immune response to injury.[9] They are also involved in the development and progression of several skin disorders.[10]

This chapter provides a brief review of the role of the molecular clock implicated in some of the skin's functions, which are also manifested at the organismic level. A better understanding of the bidirectional relationship between circadian disruption and skin diseases such as infections, inflammation, wound healing, sunburn, hair loss, cancer, and aging is likely to help identify new therapeutic targets. At the interface between the organism and its environment, the skin uses a complex network of defense mechanisms. Although much work remains to be carried out to fully understand the intricate interplay between the core clock network and fundamental physiological processes in skin, knowledge has increased significantly in the last decade. Findings have provided mechanistic insight to account for early observations made at the organismic level.

15.2 Some 24-hour Rhythms Characterizing Skin at the Organismic Level

As it exists at the interface between the organism and its environment, skin serves many functions, with most of them exhibiting circadian variations.

Many of these rhythms have been known and characterized for a long time in both humans and other species. Some of these early findings are briefly reviewed to illustrate the diversity of processes involved.

15.2.1 Animal Experiments

As early as 1956, mitotic activity in the pinna and interscapular epidermis of ZBC mice kept under a 12 hours of light alternating with 12 hours of darkness (LD12:12) schedule was reported to be statistically significantly lower during the night than during the day.[11] Both rhythms were roughly synchronized with each other and with the daily rhythm in the number of circulating tail blood eosinophils. A similar study in hamsters showed a significant decrease in mitoses around 20:30 as compared to around 12:30 in both pouch and pinna.[12] Concomitantly measured rectal temperature was significantly higher during the daily dark span (18:00–06:00) than during the daily light span (06:00–18:00).

On a population of ZBC mice previously found to exhibit a marked mitotic rhythm in the pinna epidermis, hydrocortisone and 9α-fluorocortisol were shown to have anti-mitotic activity at the anticipated time of day when mitotic activity is relatively high.[13] Checking for differences in proportions of mitotic stages between treated and untreated mice, the group of mice injected intraperitoneally with 100 γ of cortisol acetate had a higher proportion of prophases and a lower proportion of telophases than the untreated group.[13] Epinephrine (10 γ/20 g body weight) also inhibited mitoses in the pinnal epidermis of ZBC mice studied under similar conditions.[14] Interestingly, bilateral adrenalectomy obliterated the 24-hour rhythm in pinnal mitosis.[15]

Another study[16] investigated the relative synchronizing effects of the lighting regimen and the feeding schedule on the mitotic rhythm in murine corneal epithelium. CDF_1 mice kept in LD12:12 with lights on from 06:00 to 18:00 were placed on one of five different feeding schedules. One group was fed *ad lib*, and the other four groups had food access limited to 4 hours a day, either in the early or late light phase or in the early or late dark phase. A large-amplitude 24-hour rhythm in the corneal mitotic index characterized all five groups. While the mitotic rhythm is nearly reversed after a subsequent 180° inversion of the light–dark schedule, the timing of the daily meal only shifted the acrophase (phase of the predicted daily maximum) by a few hours. Compared to the *ad lib* group, which peaked at ZT09:08 (95% confidence limits: ZT08:32 and ZT09:44), food-restricted groups peaked slightly earlier; the group fed in the early light phase peaked at ZT05:32 and the group fed in the late dark phase peaked at ZT07:52.[16]

Skin tensile strength was also documented to vary depending on the time of day.[17] Male and female mice of two inbred strains were studied: C57-Black/1, traditionally free of malignant neoplasia, and Bagg-Albino, traditionally prone to develop tumors in females. The load required to rupture standardized 3-mm wide dorsal skin strips was determined in groups of mice sampled at 4-hour intervals over 24 hours. Skin tensile strength was highest around ZT08:40, and the 24-hour rhythm was demonstrated to be statistically significant for female mice of both strains.

More recently, susceptibility of the skin to DNA damage caused by UV radiation was shown to follow a 24-hour rhythm. More DNA damage occurred when the dorsal skin of mice was exposed to UVB radiation at ZT20 than at ZT08.[18] Exposing the dorsal skin of mice to moderate levels of radiation results in worse erythema at ZT21 than at ZT09, as well as higher levels of inflammatory cytokines and apoptosis.[19] Immune responses and outcomes of skin infections also depend on the time of infection and the host's circadian rhythms.[20–22]

15.2.2 Human Studies

Several investigators have shown a time-dependent variation in the *stratum corneum* barrier of human skin, the skin permeability barrier being higher during the night and evening than in the morning.[23–25] These studies also suggested that the timing of treatment during the day affected the resorption, efficacy, and side effects of many drugs applied to the skin. The skin blood flow rate has also been shown to be circadian periodic, peaking in the late afternoon and at night.[26,27] The rhythm in skin blood flow and barrier function was further shown to be maintained during treatment with corticosteroids in both healthy and irritated skin.[28]

The skin's response to allergens depends on the time of day. The erythematous response of clinically healthy adults to the intradermal injections of histamine is highest around 23:08 (21:44, 00:48). In allergic patients, the erythematous response to fixed doses of house dust extract also peaks around 23:00, while the response to penicillin is highest around 20:32 (18:48, 04:00). In allergic patients, the time of highest sensitivity to histamine coincides with the time of day when the excretion of adrenocortical metabolites is at its lowest.[29] Similar results are reported in another study of atopic and non-atopic men: The skin response to histamine was highest during the usual resting span, around 01:30, while the rhythms of total histamine release by leucocytes peaked early in the afternoon.[30]

Electrical skin resistance also follows a 24-hour variation, peaking around 14:30, as determined from analyses of published data and actual around-the-clock measurements on a clinically healthy man.[31]

A 24-hour rhythm in basal pain sensitivity has also been reported in healthy volunteers. Pressure pain threshold and tolerance as well as cold pain threshold and tolerance were assessed at 4-hour intervals for 24 hours from 30 volunteers, together with skin sympathetic nerve activity. Overall, all measured variables reached a minimum around 04:00.[32] Itch from psoriasis[33] and pruritus[34,35] is most pronounced at night. The time of day when a wound is incurred is reported to affect the length of recovery; a wound incurred during the daytime heals considerably faster than a wound occurring at night.[36]

The night, hence, appears to be a time of higher susceptibility to allergens, pain, and itch, and a time when wounds take longer to heal. These physiological manifestations of circadian rhythms in various processes in skin and the skin's response to different kinds of injury and treatment all point to fundamental mechanisms implicating the circadian clock machinery, which will be reviewed in the next section.

15.3 Some Aspects of the Cutaneous Circadian Clock System

Very briefly, organisms have an autonomous, evolutionarily conserved internal timing system. The circadian clock in the suprachiasmatic nuclei (SCN) of the brain's hypothalamus is synchronized by the 24-hour alternation between light and dark in the environment. Light information is transmitted to the SCN *via* intrinsically photosensitive retinal ganglion cells in the retina. While the SCN orchestrates circadian rhythms throughout the body, the molecular circuitry of the circadian clock is present in almost all cells, including those of peripheral organs, which are also affected by other external inputs such as temperature, food intake, and exercise. Each cell contains a biochemical oscillator consisting of interlocked transcription–translation feedback loops, including in the primary loop the core clock genes *Bmal1* and *Clock* on the positive arm and *Per* and *Cry* on the negative arm of the primary loop.[37,38] Directly or indirectly, the network of core clock genes modulates the expression of multiple genes and biological processes throughout the body.[18] Reportedly, over 1400 genes related to multiple functions show circadian variations in expression in mouse skin, indicating that various aspects of skin physiology depend on the circadian clock.[18]

Studies in mice show that the SCN synchronizes circadian clocks in skin. The expression of the clock and clock-controlled genes, which is circadian rhythmic under constant dark conditions, is lost in *Cry1*/*Cry2* knockout mice.[39] PER2 expression in the nuclei of keratinocytes in the epidermis and hair follicles peaks at CT16, lagging by 4 to 8 hours behind its mRNA expression; however it is no longer rhythmic after SCN ablation, even in LD12:12.[39] Allowing the expression of endogenous BMAL1 in the epidermis of conventional full-body *Bmal1* knockout mice, however, resulted in the circadian oscillation of the core circadian transcriptional machinery in the epidermis under light–dark conditions, indicating that light synchronizes the *Bmal1*-dependent circadian machinery in single tissues in the absence of *Bmal1* in all other tissues (including the SCN).[40] The presence of light-sensitive opsin proteins in skin cells such as keratinocytes, melanocytes, fibroblasts and hair follicle cells may contribute to this result.[41]

Duan *et al.* provided a comprehensive review of the cutaneous circadian system.[42] The role of the circadian clock in some of the important physiological processes in skin are briefly reviewed. As skin is a large and complex organ composed of multiple cell types, a useful framework is to consider multiple independent, but coordinated peripheral clocks functioning within distinct anatomic compartments of the skin instead of a single-entity "skin clock".[43] The circadian clock indeed likely affects the expression of distinct gene sets and physiological functions in the different cell types of skin. Human skin keratinocytes, melanocytes, and fibroblasts were all shown to contain distinct circadian clock machineries.[44]

15.3.1 Interfollicular Epidermal Clock

Twenty four-hour rhythms in proliferation[18] and the coordination between cell division and metabolism[45] depend on the intrinsic circadian clock in keratinocyte stem cells of the epidermis. The keratinocyte progeny in the *stratum basale* progressively differentiates as cells move upwards to the upper *stratum corneum* layer, where the dead keratin-filled cells are shed. As reviewed in more detail elsewhere,[43] epidermal progenitors display a robust 24-hour cycle in the S-phase, but the S-phase is basically constant in the *Bmal1* null epidermis; epidermal barrier functions such as hydration of the upper corneal layer become arrhythmic in *Clock* mutant mice; epidermis exposure to UV at night increases the skin cancer rate in mice, but UV-induced DNA damage is more or less constant in *Bmal1* mutant mice; wound healing is impaired in *Bmal1* mutant mice; and many immune-related genes in skin are regulated by the circadian clock.

The interfollicular epidermal pattern of clock activity and function extends to the upper epithelial compartments of the hair follicle, the infundibulum and the isthmus, which house stem cells that exhibit similar behavior to epidermal stem cells in terms of their proliferation activity. These cells contribute to the maintenance and repair of the nearby epidermis.[43]

The functions of the differentiated suprabasal epidermal compartments are also influenced by the circadian clock, as evidenced by 24-hour variations in skin pH, transepidermal water loss, and *stratum corneum* capacitance.[23,25,46] Daily changes in *stratum corneum* hydration and transepidermal water loss were shown to depend on aquaporin 3 (Aqp3), a membrane transporter for water and glycerol, and a putative epidermal clock output gene whose expression is controlled by the CLOCK-regulated transcription factor DBP.[47] In *Clock* mutant mice, *Aqp3* expression and *stratum corneum* hydration are significantly decreased.[43]

15.3.2 Hair Follicle Clock

As reviewed in more detail elsewhere,[48,49] hair follicles are miniature, anatomically complex ectodermal appendages located below the epidermis, extending over the dermal and adipose layers. Hair follicles have several concentric layers of epithelial cells sitting above and partially surrounding a cluster of specialized dermal papilla cells. Hair follicles follow a growth cycle as they go through consecutive phases of active regeneration (anagen), apoptotic involution (catagen), and relative proliferative quiescence (telogen) to grow new hair shafts. Hair regeneration is supported by several distinct stem cell populations, notably in the bulge. Bulge stem cells give rise, over the course of the hair cycle, to all epithelial cell types constituting the lower portion of the follicle. Dormant hair germ progenitors during telogen alternate with their actively dividing progeny at the beginning of anagen and hair matrix cells during full anagen. Circadian genes are prominently expressed in hair germ progenitors during telogen and at the telogen-to-anagen transition, while during anagen, clock genes and clock output genes are primarily expressed in the epithelial hair matrix cells in the hair bulb. Clock activity then also occurs in the specialized fibroblasts of the dermal papilla. This difference between telogen and anagen is reflected by the small proportion of clock-regulated genes overlapping between these two phases of the hair growth cycle.[18]

Germline deletion of *Bmal1* in mice led to a delayed expansion of the secondary hair germ and initiation of anagen.[50] Deletion of *Cry1*/*Cry2* altered the 24-hour mitotic rhythm of hair matrix cells.[51] Comparative gene expression profiling showed that WNT and TGFβ signaling pathways are implicated in bulge stem cell activation.[52] In the telogen bulge, the clock may help to identify which cells are ready to enter the cell cycle upon activation by low levels of Tgfβ and high Wnt activity, determining which of the bulge progenitors are destined to divide and contribute to hair follicle renewal and which cells will remain dormant for future cycles.[49,52] Telogen can thus be viewed as an interpreter of circadian output influencing the timing of anagen initiation and as the key stage in preparing for the subsequent organ regeneration (anagen) phase by suppressing molecular brakes on hair growth while activating pro-regenerative signals. The circadian clock hence seems to be involved in modulating cell proliferation activity at distinct stages of the hair cycle.[48,49]

15.3.2.1 Hair Follicle Cells for Assessing Circadian Phase Noninvasively

Knowing a person's internal circadian phase is important for diagnostic purposes and for guiding the timing of treatment. It usually requires multiple samples and data analysis. Hair follicle cells have recently gained attention due to the emergence of a new practical and accurate method to assess circadian phase based on these cells.[53] Hair follicles are indeed an easily accessed tissue and an ideal sample source for determining the circadian phase,[50,54,55] which can be accomplished from a single sample.[56,57] The method allowed the investigation of masked social jetlag and circadian-related sleep problems in shift workers[58] as well as in ordinary citizens[59,60] (see Figure 3 in the study by Matsumura *et al.*[59]). It can serve as a critical model for monitoring the circadian clock and help provide an understanding of the potential mechanisms of circadian-rhythm-related conditions.[61] From a chronotherapy perspective, this method could support personalized treatment scheduling based on the gene expression profile expressed in hair follicle cells[62,63] or in skin biopsy from the dorsal region[64] (see Figure 1 in the study by Plikus & Andersen[53]).

15.3.3 Dermal and Adipose Clocks

The dermis is a fibroblast-rich structure providing tensile strength to the skin and supporting hair follicles through an extensive collagen network. Below the dermis is a thin layer of white adipocytes. Circadian genes are expressed in at least a subset of cells in the skin dermis and adipose tissue.[43] Dermal white adipose tissue has been shown to be an important contributor to a multitude of processes, including immune response, wound healing and scarring, hair follicle growth and thermoregulation.[10] Moreover, it was shown to be decreased by caloric restriction, which also changed the structure of mouse skin by inducing faster hair regrowth after shaving, thicker epidermis, and greater vascular network.[42]

Primary fibroblasts exhibit well-characterized circadian rhythms in clock gene activity, which persist under constant conditions *in vitro*.[36] *In vitro* behavioral characterization studies of dorsal skin dermal fibroblasts in mice showed that *Npas2*$^{-/-}$ mutation was associated with increased proliferation, migration, and cell contraction measured through methods such as floating collagen gel contraction and single-cell force contraction assays.[65] The authors concluded that the induced *Npas2* expression in fibroblasts may interfere with skin homeostasis, wound healing, and dermal tissue reconstruction, in agreement with the observed upregulation of the clock gene *Npas2* along with fibroblast aging. Melatonin, which is present in the skin, has the ability to dose-dependently stimulate PER1 clock gene expression in normal human dermal fibroblasts and normal human epidermal keratinocytes.[66]

15.3.4 Immune, Vascular, and Neuronal Clocks

The skin contains a large number of immune cells that are, in part, influenced by factors expressed and secreted by epidermal keratinocytes, which play an important role in the skin's ability to respond to injuries and infections.[67] Langerhans cells and different types of T cells reside within the epidermis, while macrophages, mast cells, T cells, and dendritic cells abound within the dermis.[68] Evidence suggesting that the skin's immune system is influenced by the circadian clock includes the finding that several immune function-related genes show circadian regulation in mouse skin.[43] Studies specifically addressing how the circadian clock affects the skin's immune system implicated the role of clock genes in several skin conditions such as skin allergic reactions thought to model human allergic contact dermatitis.[43] The reciprocal relationship between the circadian clock and skin inflammation was reviewed by Greenberg *et al.*,[22] showing time-of-day-dependent activation of the interferon pathway, a key pathway involved in the host defense response.

Since the skin is highly vascularized and densely innervated, humoral and neuronal pathways receive inputs from the circadian rhythmic central nervous system.[43] Whether cells of the skin vasculature and neurons are themselves under circadian control deserves further investigation.

15.3.5 Circadian Clock in Melanocytes

Melanocytes determine the pigment patterning of the epidermis and hair. They synthesize and actively transfer UV-absorbing melanin granules to their neighboring keratinocytes, endowing the latter with natural sunscreen properties, also providing protection against UV radiation.[43] Melanocytes harbor an active circadian clock in mammals.[44,69] siRNA-mediated silencing of *BMAL1* or *PER1* results in increased hair pigmentation, an effect involving changes at different levels of melanocyte function.[43]

Current evidence suggests that melatonin and some of its metabolites inhibit melanogenesis by reducing tyrosinase activity, and melanocyte proliferation by stimulating melatonin membrane receptors MT1 and MT2. In melanocytes, melatonin and its metabolites protect against harmful agents, senescence, and oxidative stress by operating as free radical scavengers, stimulating the synthesis and activity of ROS scavenging enzymes and other antioxidants, promoting DNA repair, and enhancing mitochondrial function.[70] As photosensitive cells in the skin, melanocytes generate a photosensitive system by expressing several opsins responding to ambient light, thus playing an important role in the regulation of peripheral circadian rhythms.[71] Apart from its light receptor function, melanopsin mediates the heat activation of *Per1* and *Bmal1* in cultured murine Melan-a melanocytes and acts as a thermos-receptor that ultimately regulates the local circadian clock.[72]

15.3.6 Circadian Cell Cycle Gating in Skin

The proliferation of epidermal progenitor/stem cells occurs periodically over a 24-hour cycle. The highest proportion of epidermal progenitors is in the S-phase during the night in mice, while in humans, the highest proportion of keratinocytes is in the S-phase during the day.[73] This 24-hour variation was shown to depend on the presence of an intact circadian clock both at the systemic level and within keratinocytes.[18,51,74] In mice, the proportion of cells in the S-phase is constant and high in the absence of *Bmal1* in keratinocytes.[18] Moreover, the circadian expression of XPA (the nucleotide excision repair factor xeroderma pigmentosum group A) is in antiphase with S-phase progression.[74] When activated during the S-phase, XPA interferes with DNA replication.[43]

Compared to progenitor/stem cells in the basal epidermis, hair matrix cells proliferate more rapidly. Unlike the epidermis, the hair matrix shows a daily mitotic rhythm rather than a DNA replication rhythm, which is thought to depend on circadian gating of the G2/M cell cycle checkpoint.[51] As mitotic cells are vulnerable to double-stranded DNA breaks, exposure to the same dose of γ-radiation causes more severe hair loss during the mitotic peak in the morning than in the mid-afternoon during the mitotic decline. This variation along the 24-hour scale is not observed in arrhythmic *Cry1*/*Cry2* null mice, which experience significant hair loss irrespective of the time of day.[43]

The clock has been implicated in the regulation of cellular senescence pathways. Skin aging is accompanied with hair graying, hair loss, decreased cutaneous fat layer, and delay in skin wound healing.[43] Several signs of premature skin aging, such as delayed hair regrowth, thinning of the cutaneous fat layer,[50,75] and wound closure deficiency[76] characterize *Bmal1*-deficient mice.

Although somewhat counterintuitive, the programmed activation of senescence in wound myofibroblasts has been shown to be an integral part of the skin's injury repair response.[76,77] NONO, a multifunctional nuclear factor and a binding partner of Per proteins, has also been implicated in timing the replicative senescence in wound myofibroblasts.[76]

As briefly reviewed above, core clock genes play critical roles in each type of skin cells, as they relate to the different physiological functions of skin. Since skin cells differentiate rapidly, the cell cycle also plays an important role in these different processes, the cell cycle being itself gated by the circadian clock. Such involvement of circadian rhythms in all skin functions offers opportunities to improve the management and treatment of diseases of the skin, as reviewed below in a few examples.

15.4 Circadian Rhythms and Skin Diseases

As reviewed above, skin is a complex organ hosting diverse processes, many of them following predictable 24-hour variations. The determination of any association between circadian disruption and skin diseases is complicated by the role played by the cell cycle in relation to UV-induced DNA damage in conditions such as cancer, as well as by sleep, which is disturbed in several inflammatory conditions of the skin. Both sleep and the cell cycle have a circadian component of their own. One study suggesting the benefits of using a chronobiologic approach in treating skin disorders used intermittent circadian fasting.[78] In this multicenter, prospective observational study involving 72 patients suffering from different dermatoses, the circadian rhythm seemed to play a role in modulating the severity of dermatoses. A brief overview of some studies documenting the involvement of the circadian clock in a few skin disorders is provided below. A more detailed review is provided by Duan *et al.*[42]

15.4.1 Skin Cancer

Carcinoma, derived from keratinocytes, and melanoma, derived from melanocytes, are the two main forms of skin cancer. The major cause of these conditions stems from exposure to UV radiation associated with DNA mutations due to nucleotide changes or DNA damage caused by increased levels of reactive oxygen species.[42] Two possible mechanisms underlying the time-of-day-dependent susceptibility of the skin to UV radiation-induced DNA damage relate to the susceptibility of the S phase of the cell cycle to UV-induced DNA damage[18] and to the 24-hour changes in DNA repair.[74] Such 24-hour variation in DNA damage is no longer observed in *Bmal1*$^{-/-}$ mice, indicating the involvement of the circadian clock in susceptibility to UVB radiation.[42] Melanin produced by melanocytes protects the skin from UV radiation.[79] In the presence of BMAL1, the microphthalmia-associated transcription factor (MITF), a rate-limiting protein in melanin synthesis, is 24-hour rhythmic in human melanocytes and melanoma cells under synchronized clock conditions. BMAL1 binds to the promoter region of MITF and transcriptionally regulates its expression, which positively influences melanin synthesis. An increase in melanin concentrations due to BMAL1 overexpression protects human melanoma cells from UVB.[79]

The circadian clock is also involved in the progression of skin cancer, as the clock network is dampened in cancer cells.[80] Conversely, *in vivo* melanoma growth in mice is reduced by dexamethasone-induced clock gene activation in tumor cells, leading to cell cycle arrest.[9,81] Knocking down *Bmal1* in tumors prevents the growth-dampening effect of dexamethasone.[42] Agonists of nuclear hormone receptors REV-ERBα and REV-ERBβ stimulate apoptosis in cancer cells, showing that interventions targeting the core clock machinery impair melanoma proliferation.[82] Circadian effects on UV-induced damage and mutations have recently been reviewed.[83]

15.4.2 Skin Infection

The host's circadian rhythms and the time of infection are thought to influence immune responses and clinical outcomes of skin infections.[18,20,21,43] Bacteria such as *Staphylococcus aureus* survive longer on mouse skin when applied at ZT22 than at ZT16, perhaps due to 24-hour changes in the expression of antimicrobial peptides.[42] Likewise, the severity of parasite infections in skin is 24-hour rhythmic. Inflammation resulting from the injection of the *Leishmania major* parasite into mice at different times of the day was minimal when administered at ZT03, with minimal macrophage recruitment observed 6 hours after injection. The attachment of the parasite to mouse macrophages was shown to follow a circadian rhythm, which depended on the circadian machinery, also involved in the regulation of the time-dependent release of chemokines and cytokines.[21,42] The circadian clock control of antiviral factors expressed in the skin may account for 24-hour changes in sensitivity to the topical application of the viral mimic IMQ, which results in greater expression of antiviral defense response genes during the day compared to application at night, and in heightened levels of the same genes in the absence of *Bmal1*.[22]

15.4.3 Inflammatory Skin Diseases

Skin diseases occur when skin inflammation is triggered by self-peptides or persists following an infection.[42] Inflammatory responses are related to circadian rhythms and circadian disruption can lead to inflammatory diseases. In humans, psoriasis and atopic dermatitis are two of the most common autoimmune skin diseases. Since itch occurs at night, symptoms associated with atopic dermatitis and psoriasis can impair sleep quality, which in turn disrupts circadian rhythms and increases severity of the disease.[84]

Patients with psoriasis exhibit changes in the circadian oscillations of blood pressure and heart rate, supporting the notion of wider circadian disruption. Shift work has also been associated with psoriasis.[85] The expression of the core clock genes was assessed in human psoriatic lesional and non-lesional skin and in human adult low calcium temperature (HaCaT) keratinocytes after stimulation with pro-inflammatory cytokines, and clock proteins in skin biopsies from psoriatic patients were examined using immunohistochemistry. These studies showed that the clock gene transcripts were elevated in psoriatic lesions, especially in non-lesional psoriatic areas, except for *Reverba*, which was consistently downregulated in the psoriatic samples.[86] Moreover, the epidermal distribution of the REV-ERBα protein differed in non-lesional skin from that in healthy skin, and changes in the amplitude of the *Bmal1*, *Cry1*, *Reverba* and *Per1* mRNA oscillations were observed in cytokine-treated HaCaT cells.[86] Bulk RNA-seq studies on skin samples from human patients showed that *CRY2*, *PER3*, *REVERBA*, and *RORC* genes are downregulated in both the psoriasis lesion and the adjacent normal skin as compared to skin from non-psoriatic individuals.[22,87] In an animal model of psoriasis, the effects of maxacalcitol, a vitamin D analogue, were more pronounced when administered during the early to middle of the active span than those after dosing during the early to middle of the inactive span.[88]

Atopic dermatitis is thought to relate to barrier impairment, dysregulated immune system, and/or dysregulated commensal microbiota.[42] Symptoms tend to be more severe during the night than during the day, thus impairing sleep quality. Circadian disruption in mice was associated with loss of rhythmicity in circadian gene expressions and increased severity of symptoms from atopic dermatitis.[89] The expression of histamine and cytokines, including TNF-α and interleukins (IL-4, IL-10, and IL-13) was also increased.[42] At the molecular level, *Clock* deletion led to more severe delayed-type skin allergic reactions.[90] Reduced IFN-γ and IL-1β production is also found in *Per2*-knockout mice when exposed to lipopolysaccharide[91] and they show alleviated atopic dermatitis-like symptoms.[92] Studies in human keratinocytes showed that the transcription factor EGR1 interacted with the PER2 promoter and promoted IL-4-induced transcriptional activation of PER2.[93] Results indicated that IL-4 can use both MAPK and Jak signaling pathways to induce EGR1-mediated PER2 expression, and c-Jun N-terminal kinase 1/2 can augment IL-4-induced activation of the Jak-signal transducer and activator of the transcription 3 pathway.[93] Agerarin, a natural JAK1/2/3 inhibitor that has a protective effect on the epidermal skin barrier, reduced IL-4-induced PER2 mRNA expression in human keratinocytes by suppressing the JAK-STAT3 pathway, and restored the oscillatory rhythmicity of PER2 promoter activity altered by IL-4.[94]

15.4.4 Wound Healing

Complex repair mechanisms requiring regulation and coordination among cell types throughout the skin take place after skin injuries. Some of these elements interact with the core clock machinery.[42] The five steps involved in the process of wound healing are hemostasis, inflammation, growth, re-epithelialization, and remodeling. Hemostasis, the process of stopping bleeding, occurs periodically over a 24-hour cycle. As discussed above, the circadian clock is involved in relation to the inflammation step. The actin polymeric state involved in the growth step also follows a 24-hour rhythm and depends on the cells' internal circadian machinery.[36] Senescence of myofibroblasts through NONO, a binding partner of PER that regulates cell cycle inhibitors, is also affected by the circadian clock.[76] The circadian clock thus regulates both proliferation and migration of fibroblasts, which are the key functions required for wound healing.[42] Re-epithelization may also display circadian rhythms. Healing is faster in scratch wounds on human keratinocyte monolayers incurred at the peak rather than at the trough time of fibroblast motility.[36] Although more mouse epidermal keratinocytes are in the S phase at night than during the day,[18] cell proliferation is increased and no longer rhythmic in the presence of inflammation.[22]

Several converging experiments suggest that the invasion of fibroblasts into the wounded area, and thus possibly the recovery of the wound, depends on the time of wounding in mice.[36] A retrospective analysis of the Burn Injury Database also shows that burns incurred during the day healed faster than burns that occurred during the night.[36] Melatonin is reported to accelerate wound repair in full-thickness incisional wounds and to promote diabetic wound healing *in vitro* by regulating keratinocyte activity.[95] Circadian disruption delays skin wound healing,[96] but manipulation of the circadian machinery does not always impede wound healing.[42] Improved healing of dermal excisional wounds was reported in circadian clock *Npas2*-null mice.[65] High-throughput drug screening identified Dwn1 as a compound capable of downregulating *Npas2* activity.[97] Dwn1 was found to suppress circadian *Npas2* expression, increase murine dermal fibroblast cell migration, and decrease collagen synthesis *in vitro*. When topically applied to iatrogenic full-thickness dorsal cutaneous wounds in a murine model, the wounds healed faster with favorable mechanical strength.[97]

In chronic non-healing wounds (*e.g.*, diabetic foot ulcers, pressure sores, venous and arterial ulcers), several factors work together to hinder the process of wound repair. On the other hand, hypertrophic scars result from overactive repair mechanisms that fail to cease in a timely fashion, leading to excessive production of extracellular matrix components such as collagen.[98] Recently, there has been an increasing interest in the roles of microRNAs in wound repair with a growing recognition of their involvement in circadian processes, either as regulators or targets of clock activity or direct responders to external circadian stimuli. In their review, Fawcett *et al.*[98] explored the intersection of mechanisms through which circadian and microRNAs responses interact with each other in relation to wound repair in the skin.

15.5 Concluding Remarks

Skin blood flow and barrier function and responses of the skin to pathogens, chemicals, UV light and mechanical injury are all characterized by overt circadian rhythms. Multiple mechanisms are responsible for their manifestation, ranging from the regulation by the central clock in the SCN to the intermodulation of a network of core clock genes present in the different cell types in all skin layers. The rapid differentiation of skin cells also implicates the role of the cell cycle, which is itself regulated by the circadian clock.

As briefly reviewed herein, multiple independent, but coordinated clocks function within distinct anatomic compartments, likely affecting the expression of distinct gene sets and physiological functions in the different cell types in skin. The rhythmic expression of core clock genes in the different cell types in skin and their interaction with other molecular pathways contribute to the 24-hour changes observed in physiological functions of skin, from skin hydration and thermogenesis to responses to injury from pathogens and UV light, as well as tissue repair and remodeling.

Despite the significant progress made recently in understanding how the molecular clock of the skin is affected by environmental stimuli such as UV and infrared radiation, visible light and temperature, and exposure to chemicals and pathogens, much work still remains to be done. Much of the molecular work carried out on nocturnal rodents awaits translation to diurnal humans. Humans also have a higher degree of heterogeneity than animals studied in the laboratory.

The important role that the circadian system plays in skin health is a fundamental concept. Assessing variations in skin function and characteristics throughout the 24-hour day should help better understand patient symptoms and maximize therapeutic benefits. Circadian medicine is receiving increasing attention as part of prevention and patient care. It can be implemented in practice through several approaches. As originally envisioned, chronotherapy and chronopharmacology aim at delivering a given treatment at the circadian stage of maximal efficacy while minimizing side effects.[99–104] Clinical trials are needed to assess optimal times for treating skin diseases. Studies in mice indeed indicated that maxacalcitol, a vitamin D3 analogue used for psoriasis symptom relief, was more effective when applied during the active phase rather than during the resting phase.[101]

With the current understanding that circadian disruption is associated with an increased risk of poor health in diverse disease conditions,[105] another approach of circadian medicine involves strengthening the circadian system by means of lifestyle adjustments for health maintenance and optimization. Exposure to light and dark, exercise, and meal schedules are features amenable to manipulation to strengthen the circadian system. As feeding can influence skin circadian rhythms in mice, adjusting feeding behavior may be a way to support circadian function in the skin.[106]

In contrast to altering the genome by means of gene editing, the idea of targeting the clock machinery at the molecular level is yet another approach in circadian medicine. For instance, cell proliferation and tumor growth in B16 melanoma tumors is reduced by dexamethasone, which activates the circadian clock.[94] As another example, melanoma proliferation is impaired by using agonists of REV-ERBs.[95]

Combining these different approaches in circadian medicine offers great new opportunities to prevent and manage skin diseases.

References

1. E. McLafferty, C. Hendry and F. Alistair, The integumentary system: anatomy, physiology and function of skin, *Nurs. Stand.*, 2012, **27**(3), 35–42.
2. W. D. James, Skin: Basic Structure and Function, in *Andrews' Diseases of the Skin: Clinical Dermatology*, ed. W. D.James, D. M.Elston, J. R. Treat, M. A.Rosenbach and I. M.Neuhaus, Elsevier, 13th edn, 2020, pp. 1–10.
3. E. Festa, J. Fretz, R. Berry, B. Schmidt, M. Rodeheffer, M. Horowitz and V. Horsley, Adipocyte lineage cells contribute to the skin stem cell niche to drive hair cycling, *Cell*, 2011, **146**(5), 761–771.
4. V. I. Alexaki, D. Simantiraki, M. Panayiotopoulou, O. Rasouli, M. Venihaki, O. Castana, D. Alexakis, M. Kampa, E. N. Stathopoulos and E. Castanas, Adipose tissue-derived mesenchymal cells support skin reepithelialization through secretion of KGF-1 and PDGF-BB: comparison with dermal fibroblasts, *Cell Transplant.*, 2012, **21**(11), 2441–2454.
5. Y. Luo, M. Toyoda, M. Nakamura and M. Morohashi, Morphological analysis of skin in senescence-accelerated mouse P10, *Med. Mol. Morphol.*, 2002, **35**(1), 31–45.
6. L. Q. Sun, D. W. Lee, Q. Zhang, W. Xiao, E. H. Raabe, A. Meeker, D. Miao, D. L. Huso and R. J. Arceci, Growth retardation and premature aging phenotypes in mice with disruption of the SNF2-like gene, PASG, *Genes Dev.*, 2004, **18**(9), 1035–1046.
7. E. L. Gregory, Thermoregulatory aspects of adipose tissue, *Clin. Dermatol.*, 1989, **7**(4), 78–92.
8. E. A. Tansey and C. D. Johnson, Recent advances in thermoregulation, *Adv. Physiol. Educ.*, 2015, **39**(3), 139–148.

9. L. V. M. de Assis, M. N. Moraes and A. M. L. Castrucci, The molecular clock in the skin, its functionality, and how it is disrupted in cutaneous melanoma: a new pharmacological target?, *Cell. Mol. Life Sci.*, 2019, **76**(19), 3801–3826.
10. I. L. Kruglikov and P. E. Scherer, Dermal adipocytes: from irrelevance to metabolic targets?, *Trends Endocrinol. Metab.*, 2016, **27**(1), 1–10.
11. A. P. Chaudhry, F. Halberg and J. J. Bittner, Mitoses in pinna and interscapular epidermis of mice in relation to physiologic 24 hours periodicity, *Fed. Proc.*, 1956, **15**(1), 34.
12. A. P. Chaudhry, F. Halberg, C. E. Keenan, R. N. Harner and J. J. Bittner, Daily rhythms in rectal temperature and in epithelial mitoses of hamster pinna and pouch, *J. Appl. Physiol.*, 1958, **12**, 221–224.
13. A. P. Chaudhry, F. Halberg and J. J. Bittner, Reduction of mitotic activity in pinna epidermis of mice given cortisol or 9α-fluorocortisol, *Proc. Soc. Exp. Biol. Med.*, 1956, **91**, 602–604.
14. A. P. Chaudhry, F. Halberg and J. J. Bittner, Epinephrine and mitotic activity in pinnal epidermis of the mouse, *J. Appl. Physiol.*, 1956, **9**, 265–267.
15. A. P. Chaudhry and F. Halberg, Adrenalectomy effect on temperature and epidermal mitotic rhythms in hamsters. Abstract, 39th General Meeting, International Association for Dental Research, Boston, March 23-26, 1961, *J. Dent. Res.*, 1961, **40**, 760–761.
16. L. E. Scheving, J. E. Pauly, E. R. Burns, F. Halberg, S. Tsai and H. O. Betterton, Lighting regimen dominates interacting meal schedules and synchronizes mitotic rhythm in mouse corneal epithelium, *Anat. Rec.*, 1974, **180**, 47–52.
17. D. E. Lofstrom, W. J. L. Felts and F. Halberg, Circadian variation in skin tensile strength of two inbred strains of mice, *Int. J. Chronobiol.*, 1973, **1**, 259–267.
18. M. Geyfman, V. Kumar, Q. Liu, R. Ruiz, W. Gordon, F. Espitia, E. Cam, S. E. Millar, P. Smyth, A. Ihler, J. S. Takahashi and B. Andersen, Brain and muscle Arnt-like protein-1 (BMAL1) controls circadian cell proliferation and susceptibility to UVB-induced DNA damage in the epidermis, *Proc. Natl. Acad. Sci.*, 2012, **109**, 11758–11763.
19. S. Gaddameedhi, C. P. Selby, M. G. Kemp, R. Ye and A. Sancar, The circadian clock controls sunburn apoptosis and erythema in mouse skin, *J. Invest. Dermatol.*, 2015, **135**, 1119–1127.
20. B. Bilska, A. Zegar, A. T. Slominski, K. Kleszczynski, J. Cichy and E. Pyza, Expression of antimicrobial peptide genes oscillates along day/night rhythm protecting mice skin from bacteria, *Exp. Dermatol.*, 2021, **30**, 1418–1427.
21. S. Kiessling, G. Dubeau-Laramee, H. Ohm, N. Labrecque, M. Olivier and N. Cermakian, The circadian clock in immune cells controls the magnitude of leishmania parasite infection, *Sci. Rep.*, 2017, **7**, 10892.
22. E. N. Greenberg, M. E. Marshall, S. Jin, S. Venkatesh, M. Dragan, L. C. Tsoi, J. E. Gudjonsson, Q. Nie, J. S. Takahashi and B. Andersen,

Circadian control of interferon-sensitive gene expression in murine skin, *Proc. Natl. Acad. Sci. U. S. A.*, 2020, **117**, 5761–5771.

23. G. Yosipovitch, G. L. Xiong, E. Haus, L. Sackett-Lundeen, I. Ashkenazi and H. I. Maibach, Time dependent variations of the skin barrier function in humans: Transepidermal water loss, stratum corneum hydration, skin surface pH and skin temperature, *J. Invest. Dermatol.*, 1998, **110**, 20–23.
24. M. Denda and T. Tsuchiya, Barrier recovery rate varies time dependently in human skin, *Br. J. Dermatol.*, 2000, **142**, 881–884.
25. I. Le Fur, A. Reinberg, S. Lopez, F. Morizot, M. Mechkouri and E. Tschachler, Analysis of circadian and ultradian rhythms of skin surface properties of face and forearm of healthy women, *J. Invest. Dermatol.*, 2001, **117**, 718–724.
26. J. Smolander, M. Harma, A. Lindqvist, P. Kolari and L. A. Laitinen, Circadian variation in peripheral blood flow in relation to core temperature at rest, *Eur. J. Appl. Physiol.*, 1993, **67**, 192–196.
27. L. Caspary, S. Nordbruch, R. Lange and A. Creutzig, Circadian variation of skin perfusion in arterial occlusive disease, *Vasa*, 1997, **26**, 194–198.
28. G. Yosipovitch, L. Sackett-Lundeen, A. Goon, C. Yiong Huak, C. Leok Goh and E. Haus, Circadian and ultradian (12 h) variations of skin blood flow and barrier function in non-irritated and irritated skin – effect of topical corticosteroids, *J. Invest. Dermatol.*, 2004, **122**, 824–829.
29. A. Reinberg, Z. W. Zagula-Mally, J. Ghata and F. Halberg, Circadian reactivity rhythm of human skin to house dust, penicillin and histamine, *J. Allergy*, 1969, **44**, 292–306.
30. B. Roitman-Johnson, R. B. Sothern, F. Halberg and M. N. Blumenthal, Circadian immunologic rhythms and their implications in the diagnosis and treatment of atopic disorders, in *Recent Advances in the Chronobiology of Allergy and Immunology*, ed. M. H. Smolensky, A. Reinberg and J. P. McGovern, Pergamon Press, Oxford/New York, 1980, pp. 65–72.
31. J. Harvey, F. Halberg, J. Wu and G. Cornelissen, Circadian electrical skin resistance (ESR) & cardiovascular rhythms, *Chronobiologia*, 1987, **14**, 183–184.
32. Y. Zhu, A. Yan, B. Shu, X. Chen, Y. Chen, G. Duan and H. Huang, The diurnal orofile of human basal pain sensitivity and skin sympathetic nerve activity: a healthy volunteer study, *Front. Neurosci.*, 2022, **16**, 810166.
33. F. J. Ferguson, G. Lada, H. J. A. Hunter, C. Bundy, A. L. Henry, C. E. M. Griffiths and C. E. Kleyn, Diurnal and seasonal variation in psoriasis symptoms, *J. Eur. Acad. Dermatol. Venereol.*, 2021, **35**, e45–e47.
34. T. Patel, Y. Ishiuji and G. Yosipovitch, Nocturnal itch: why do we itch at night?, *Acta Derm.-Venereol.*, 2007, **87**(4), 295–298.
35. C. Jeon, D. Yan, M. Nakamura, S. Sekhon, T. Bhutani, T. Berger and W. Liao, Frequency and management of sleep disturbance in adults with atopic dermatitis: a systematic review, *Dermatol. Ther.*, 2017, 7(3), 349–364.

36. N. P. Hoyle, E. Seinkmane, M. Putker, K. A. Feeney, T. P. Krogager, J. E. Chesham, L. K. Bray, J. M. Thomas, K. Dunn, J. Blaikley, J. Blakley and J. S. O'Neill, Circadian actin dynamics drive rhythmic fibroblast mobilization during wound healing, *Sci. Transl. Med.*, 2017, **9**, eaal2774.
37. P. L. Lowrey and J. S. Takahashi, Mammalian circadian biology: elucidating genome-wide levels of temporal organization, *Annu. Rev. Genomics Hum. Genet.*, 2004, **5**, 407–441.
38. K. H. Cox and J. S. Takahashi, Circadian clock genes and the transcriptional architecture of the clock mechanism, *J. Mol. Endocrinol.*, 2019, **63**, R93–R102.
39. M. Tanioka, H. Yamada, M. Doi, H. Bando, Y. Yamaguchi, C. Nishigori and H. Okamura, Molecular clocks in mouse skin, *J. Invest. Dermatol.*, 2009, **129**, 1225–1231.
40. P-S Welz, V. M. Zinna, A. Symeonidi, K. B. Koronowski, K. Kinouchi, J. G. Smith, I. M. Guillen, A. Castellanos, S. Furrow, F. Aragon, G. Crainiciuc, N. Prats, J. M. Caballero, A. Hidalgo and P. Sassone-Corsi, BMAL1-driven tissue clocks respond independently to light to maintain homeostasis, *Cell*, 2019, **177**, 1436–1447.e12.
41. S. Suh, E. H. Choi and N. A. Mesinkovska, The expression of opsins in the human skin and its implications for photobiomodulation: A systematic review, *Photodermatol., Photoimmunol. Photomed.*, 2020, **36**, 329–338.
42. J. Duan, E. N. Greenberg, S. S. Karri and B. Andersen, The circadian clock and diseases of the skin, *FEBS Lett.*, 2021, **595**(19), 2413–2436.
43. M. V. Plikus, E. N. Van Spyk, K. Pham, M. Geyfman, V. Kumar, J. S. Takahashi and B. Andersen, The circadian clock in skin: implications for adult stem cells, tissue regeneration, cancer, aging, and immunity, *J. Biol. Rhythms*, 2015, **30**(3), 163–182.
44. C. Sandu, M. Dumas, A. Malan, D. Sambakhe, C. Marteau, C. Nizard, S. Schnebert, E. Perrier, E. Challet, P. Pevet and M. P. Felder-Schmittbuhl, Human skin keratinocytes, melanocytes, and fibroblasts contain distinct circadian clock machineries, *Cell. Mol. Life Sci.*, 2012, **69**(19), 3329–3339.
45. C. Stringari, H. Wang, M. Geyfman, V. Crosignani, V. Kumar, J. S. Takahashi, B. Andersen and E. Gratton, In vivo single-cell detection of metabolic oscillations in stem cells, *Cell Rep.*, 2015, **10**, 1–7.
46. A. E. Reinberg, Y. Touitou, E. Soudant, D. Bernard, R. Bazin and M. Mechkouri, Oral contraceptives alter circadian rhythm parameters of cortisol, melatonin, blood pressure, heart rate, skin blood flow, transepidermal water loss, and skin amino acids of healthy young women, *Chronobiol. Int.*, 1996, **13**, 199–211.
47. N. Matsunaga, K. Itcho, K. Hamamura, E. Ikeda, H. Ikeyama, Y. Furuichi, M. Watanabe, S. Koyanagi and S. Ohdo, 24 hours rhythm of aquaporin-3 function in the epidermis is regulated by molecular clocks, *J. Invest. Dermatol.*, 2014, **134**, 1636–1644.

48. R. Paus, S. Muller-Rover, C. Van Der Veen, M. Maurer, S. Eichmuller, G. Ling, U. Hofmann, K. Foitzik, L. Mecklenburg and B. Handjiski, A comprehensive guide for the recognition and classification of distinct stages of hair follicle morphogenesis, *J. Invest. Dermatol.*, 1999, **113**, 523–532.
49. M. Geyfman, M. V. Plikus, E. Treffeisen, B. Andersen and R. Paus, Resting no more: re-defining telogen, the maintenance stage of the hair growth cycle, *Biol. Rev. Cambridge Philos. Soc.*, 2015, **90**(4), 1179–1196.
50. K. K. Lin, V. Kumar, M. Geyfman, D. Chudova, A. T. Ihler, P. Smyth, R. Paus, J. S. Takahashi and B. Andersen, Circadian clock genes contribute to the regulation of hair follicle cycling, *PLoS Genet.*, 2009, **5**, e1000573.
51. M. V. Plikus, C. Vollmers, D. de la Cruz, A. Chaix, R. Ramos, S. Panda and C. M. Chuong, Local circadian clock gates cell cycle progression of transient amplifying cells during regenerative hair cycling, *Proc. Natl. Acad. Sci. U. S. A.*, 2013, **110**, E2106–E2115.
52. P. Janich, G. Pascual, A. Merlos-Suarez, E. Batlle, J. Ripperger, U. Albrecht, H. Y. Cheng, K. Obrietan, L. Di Croce and S. A. Benitah, The circadian molecular clock creates epidermal stem cell heterogeneity, *Nature*, 2011, **480**, 209–214.
53. M. V. Plikus and B. Andersen, Skin as a window to body-clock time, *Proc. Natl. Acad. Sci. U. S. A.*, 2018, **115**(48), 12095–12097.
54. M. Akashi, H. Soma, T. Yamamoto, A. Tsugitomi, S. Yamashita, T. Yamamoto, E. Nishida, A. Yasuda, J. K. Liao and K. Node, Non-invasive method for assessing the human circadian clock using hair follicle cells, *Proc. Natl. Acad. Sci. U. S. A.*, 2010, **107**(35), 15643–15648.
55. M. Watanabe, A. Hida, S. Kitamura, M. Enomoto, Y. Ohsawa, Y. Katayose, K. Nozaki, Y. Moriguchi, S. Aritake, S. Higuchi, M. Tamura, M. Kato and K. Mishima, Rhythmic expression of circadian clock genes in human leukocytes and beard hair follicle cells, *Biochem. Biophys. Res. Commun.*, 2012, **425**(4), 902–907.
56. G. Wu, M. D. Ruben, R. E. Schmidt, L. J. Francey, D. F. Smith, R. C. Anafi, J. J. Hughey, R. Tasseff, J. D. Sherrill, J. E. Oblong, K. J. Mills and J. B. Hogenesch, Population-level rhythms in human skin with implications for circadian medicine, *Proc. Natl. Acad. Sci. U. S. A.*, 2018, **115**(48), 12313–12318.
57. G. Wu, M. D. Ruben, L. J. Francey, D. F. Smith, J. D. Sherrill, J. E. Oblong, K. J. Mills and J. B. Hogenesch, A population-based gene expression signature of molecular clock phase from a single epidermal sample, *Genome Med.*, 2020, **12**(1), 73.
58. M. Hattammaru, Y. Tahara, T. Kikuchi, K. Okajima, K. Konishi, S. Nakajima, K. Sato, K. Otsuka, H. Sakura, S. Shibata and T. Nakaoka, The effect of night shift work on the expression of clock genes in beard hair follicle cells, *Sleep Med.*, 2019, **56**, 164–170.
59. R. Matsumura, K. Node and M. Akashi, Estimation methods for human circadian phase by use of peripheral tissues, *Hypertens. Res.*, 2016, **39**(9), 623–627.

60. M. Akashi, R. Sogawa, R. Matsumura, A. Nishida, R. Nakamura, I. T. Tokuda and K. Node, A detection method for latent circadian rhythm sleep-wake disorder, *EBioMedicine*, 2020, **62**, 103080.
61. L. P. Liu, M. H. Li and Y. W. Zheng, Hair follicles as a critical model for monitoring the circadian clock, *Int. J. Mol. Sci.*, 2023, **24**(3), 2407.
62. J. E. Lubov, W. Cvammen and M. G. Kemp, The impact of the circadian clock on skin physiology and cancer development, *Int. J. Mol. Sci.*, 2021, **22**(11), 6112.
63. S. Stenger, H. Grasshoff, J. E. Hundt and T. Lange, Potential effects of shift work on skin autoimmune diseases, *Front. Immunol.*, 2023, **13**, 1000951.
64. A. Hida, S. Kitamura, Y. Ohsawa, M. Enomoto, Y. Katayose, Y. Motomura, Y. Moriguchi, K. Nozaki, M. Watanabe, S. Aritake, S. Higuchi, M. Kato, Y. Kamei, S. Yamazaki, Y. Goto, M. Ikeda and K. Mishima, *In vitro* circadian period is associated with circadian/sleep preference, *Sci. Rep.*, 2013, **3**, 2074.
65. H. Sasaki, A. Hokugo, L. Wang, K. Morinaga, J. T. Ngo, H. Okawa and I. Nishimura, *Neuronal PAS domain 2* (*Npas2*)-deficient fibroblasts accelerate skin wound healing and dermal collagen reconstruction, *Anat. Rec.*, 2020, **303**(6), 1630–1641.
66. K. Dong, E. Goyarts, A. Rella, E. Pelle, Y. H. Wong and N. Pernodet, Age associated decrease of MT-1 melatonin receptor in human dermal skin fibroblasts impairs protection against UV-induced DNA damage, *Int. J. Mol. Sci.*, 2020, **21**(1), 326.
67. P. Di Meglio, G. K. Perera and F. O. Nestle, The multitasking organ: recent insights into skin immune function, *Immunity*, 2011, **35**, 857–869.
68. M. Pasparakis, I. Haase and F. O. Nestle, Mechanisms regulating skin immunity and inflammation, *Nat. Rev. Immunol.*, 2014, **14**, 289–301.
69. Z. Lengyel, C. Lovig, S. Kommedal, R. Keszthelyi, G. Szekeres, Z. Battyani, V. Csernus and A. D. Nagy, Altered expression patterns of clock gene mRNAs and clock proteins in human skin tumors, *Tumour Biol.*, 2013, **34**, 811–819.
70. A. Sevilla, J. Cheret, R. M. Slominski, A. T. Slominski and R. Paus, Revisiting the role of melatonin in human melanocyte physiology: A skin context perspective, *J. Pineal Res.*, 2022, **72**(3), e12790.
71. Y. Y. Chen, L. P. Liu, H. Zhou, Y. W. Zheng and Y. M. Li, Recognition of melanocytes in immuno-neuroendocrinology and circadian rhythms: beyond the conventional melanin synthesis, *Cells*, 2022, **11**(13), 2082.
72. M. N. Moraes, L. V. M. de Assis, K. K. Magalhaes-Marques, M. O. Poletini, L. de Lima and A. M. L. Castrucci, Melanopsin, a canonical light receptor, mediates thermal activation of clock genes, *Sci. Rep.*, 2017, **7**, 13977.
73. W. R. Brown, A review and mathematical analysis of circadian rhythms in cell proliferation in mouse, rat, and human epidermis, *J. Invest. Dermatol.*, 1991, **97**, 273–280.

74. S. Gaddameedhi, C. P. Selby, W. K. Kaufmann, R. C. Smart and A. Sancar, Control of skin cancer by the circadian rhythm, *Proc. Natl. Acad. Sci. U. S. A.*, 2011, **108**, 18790–18795.
75. R. V. Kondratov, A. A. Kondratova, V. Y. Gorbacheva, O. V. Vykhovanets and M. P. Antoch, Early aging and age-related pathologies in mice deficient in BMAL1, the core component of the circadian clock, *Genes Dev.*, 2006, **20**, 1868–1873.
76. E. Kowalska, J. A. Ripperger, D. C. Hoegger, P. Bruegger, T. Buch, T. Birchler, A. Mueller, U. Albrecht, C. Contaldo and S. A. Brown, NONO couples the circadian clock to the cell cycle, *Proc. Natl. Acad. Sci. U. S. A.*, 2013, **110**, 1592–1599.
77. J. I. Jun and L. F. Lau, The matricellular protein CCN1 induces fibroblast senescence and restricts fibrosis in cutaneous wound healing, *Nat. Cell Biol.*, 2010, **12**, 676–685.
78. N. L. Bragazzi, K. Trabelsi, S. Garbarino, A. Ammar, H. Chtourou, A. Pacifico, P. Malagoli, H. Kocic, R. R. Z. Conic and K. Kridin, Young Dermatologists Italian Network, Kridin K, Pigatto PDM, Damiani G. Can intermittent, time-restricted circadian fasting modulate cutaneous severity of dermatological disorders? Insights from a multicenter, observational, prospective study, *Dermatol. Ther.*, 2021, **34**, 949.
79. S. Sarkar, K. I. Porter, P. P. Dakup, R. P. Gajula, B. S. C. Koritala, R. Hylton, M. G. Kemp, K. Wakamatsu and S. Gaddameedhi, Circadian clock protein BMAL1 regulates melanogenesis through MITF in melanoma cells, *Pigm. Cell Melanoma Res.*, 2021, **34**(5), 955–965.
80. S. Masri, K. Kinouchi and P. Sassone-Corsi, Circadian clocks, epigenetics, and cancer, *Curr. Opin. Oncol.*, 2015, **27**, 5056.
81. S. Kiessling, L. Beaulieu-Laroche, I. D. Blum, D. Landgraf, D. K. Welsh, K-F Storch, N. Labrecque and N. Cermakian, Enhancing circadian clock function in cancer cells inhibits tumor growth, *BMC Biol.*, 2017, **15**, 13.
82. G. Sulli, A. Rommel, X. Wang, M. J. Kolar, F. Puca, A. Saghatelian, M. V. Plikus, I. M. Verma and S. Panda, Pharmacological activation of REV-ERBs is lethal in cancer and oncogene-induced senescence, *Nature*, 2018, **553**, 351–355.
83. D. Goodenow, A. J. Greer, S. J. Cone and S. Gaddameedhi, Circadian effects on UV-induced damage and mutations, *Mutat. Res., Rev. Mutat. Res.*, 2022, **789**, 108413.
84. B. Tas and V. Kabeloglu, Prevalence of metabolic syndrome and its parameters and their correlations with psoriasis duration, severity, and sleep quality in psoriasis patients: A cross-sectional study, *Dermatol. Pract. Concept.*, 2021, **11**(3), e2021049.
85. A. Luengas-Martinez, R. Paus, M. Iqbal, L. Bailey, D. W. Ray and H. S. Young, Circadian rhythms in psoriasis and the potential of chronotherapy in psoriasis management, *Exp. Dermatol.*, 2022, **31**(11), 1800–1809.
86. V. Nemeth, S. Horvath, A. Kinyo, R. Gyulai and Z. Lengyel, Expression patterns of clock gene mRNAs and clock proteins in human psoriatic skin samples, *Int. J. Mol. Sci.*, 2021, **23**(1), 121.

87. Z. Yu, Y. Gong, L. Cui, Y. Hu, Q. Zhou, Z. Chen, Y. Yu, Y. Chen, P. Xu, X. Zhang, C. Guo and Y. Shi, High-throughput transcriptome and pathogenesis analysis of clinical psoriasis, *J. Dermatol. Sci.*, 2020, **98**, 109–118.
88. D. Yoshioka, H. Ando, K. Ushijima, M. Kumazaki and A. Fujimura, Chronotherapy of maxacalcitol on skin inflammation induced by topical 12-otetradecanoylphorbol-13-acetate in mice, *Chronobiol. Int.*, 2018, **35**, 1269–1280.
89. K. Hiramoto, K. Orita, Y. Yamate, E. Kasahara, S. Yokoyama and E. F. Sato, The clock genes are involved in the deterioration of atopic dermatitis after day-and-night reversed physical stress in NC/nga mice, *Open Biochem. J.*, 2018, **12**, 87–102.
90. E. Takita, S. Yokota, Y. Tahara, A. Hirao, N. Aoki, Y. Nakamura, A. Nakao and S. Shibata, Biological clock dysfunction exacerbates contact hypersensitivity in mice, *Br. J. Dermatol.*, 2013, **168**, 39–46.
91. J. Liu, G. Malkani, X. Shi, M. Meyer, S. Cunningham-Runddles, X. Ma and Z. S. Sun, The circadian clock Period 2 gene regulates gamma interferon production of NK cells in host response to lipopolysaccharide-induced endotoxic shock, *Infect. Immun.*, 2007, **75**(8), 4186.
92. T. J. Kang, A. V. V. Prakash and J. Seong, The role of circadian clock period 2 (per 2) gene and the expression of per2 related gene in atopic dermatitis animal model, *J. Immunol.*, 2020, **204**(1_Supplement), 147.13.
93. H. Yeo, S. S. Ahn, E. Jung, Y. Lim, Y. H. Lee and S. Y. Shin, Transcription factor EGR1 regulates the expression of the clock gene PER2 under IL-4 stimulation in human keratinocytes, *J. Invest. Dermatol.*, 2022, **142**(10), 2677–2686e9.
94. J. Y. Lee, E. Jung, H. Yeo, S. S. Ahn, Y. Lim and Y. H. Lee, The natural Janus kinase inhibitor Agerarin downregulates interleukin-4-induced PER2 expression in HaCaT keratinocytes, *Molecules*, 2022, **27**(13), 4205.
95. R. Song, L. Ren, H. Ma, R. Hu, H. Gao, L. Wang, X. Chen, Z. Zhao and J. Liu, Melatonin promotes diabetic wound healing *in vitro* by regulating keratinocyte activity, *Am. J. Transl. Res.*, 2016, **8**(11), 4682–4693.
96. H. Walker William, O. H. Melendez-Fernandez and R. J. Nelson, Prior exposure to dim light at night impairs dermal wound healing in female C57BL/6 mice, *Arch. Dermatol. Res.*, 2019, **311**, 573–576.
97. Y. Shibuya, A. Hokugo, H. Okawa, T. Kondo, D. Khalil, L. Wang, Y. Roca, A. Clements, H. Sasaki, E. Berry, I. Nishimura and R. Jarrahy, Therapeutic downregulation of *neuronal PAS domain 2* (*Npas2*) promotes surgical skin wound healing, *eLife*, 2022, **11**, e71074.
98. S. Fawcett, R. Al Kassas, I. M. Dykes, A. T. Hughes, F. Ghali and K. Ross, A time to heal: microRNA and circadian dynamics in cutaneous wound repair, *Clin. Sci.*, 2022, **136**(8), 579–597.
99. A. Reinberg and F. Halberg, Circadian chronopharmacology. Annual Review of Pharmacology, 1971, **11**, 455–492.
100. F. Halberg, E. Haus, S. S. Cardoso, L. E. Scheving, J. F. Kuhl, R. Shiotsuka, G. Rosene, J. E. Pauly, W. Runge, J. F. Spalding, J. K. Lee

and R. A. Good, Toward a chronotherapy of neoplasia: tolerance of treatment depends upon host rhythms, *Experientia*, 1973, **29**(8), 909–934.
101. F. Halberg, W. Nelson, G. Cornelissen, E. Haus, L. E. Scheving and R. A. Good, On methods for testing and achieving cancer chronotherapy, *Cancer Treat. Rep.*, 1979, **63**(8), 1428–1430.
102. F. Halberg, H. F. Kabat and P. Klein, Chronopharmacology: a therapeutic frontier, *Am. J. Hosp. Pharm.*, 1980, **37**(1), 101–106.
103. E. Halberg, G. Cornelissen and F. Halberg, Optimization of the chronotherapeutic index in the experimental animal laboratory, *In Vivo*, 1992, **6**(4), 371–385.
104. G. Cornelissen and F. Halberg, The chronobiologic pilot study with special reference to cancer research: Is chronobiology or, rather, its neglect wasteful?, in *Cancer Growth and Progression*, ed. A. L. Goldson and H. Kaiser, Kluwer Academic Publ., Dordrecht, 1989, vol. 9, ch. 9, pp. 103–133.
105. A. B. Fishbein, K. L. Knutson and P. C. Zee, Circadian disruption and human health, *J. Clin. Invest.*, 2021, **131**(19), e148286.
106. N. L. Bragazzi, M. Sellami, I. Salem, R. Conic, M. Kimak, P. D. M. Pigatto and G. Damiani, Fasting and its impact on skin anatomy, physiology, and physiopathology: A comprehensive review of the literature, *Nutrients*, 2019, **11**(2), 249.

CHAPTER 16

The Interplay Between Female Reproductive Function and the Circadian System

TAKAHIRO J. NAKAMURA,[a] NANA N. TAKASU,[b] GERMAINE CORNELISSEN,[c] KUNIAKI OTSUKA[c,d] AND WATARU NAKAMURA*[b]

[a] Laboratory of Animal Physiology, School of Agriculture, Meiji University, Kawasaki, Kanagawa, 214-8571, Japan; [b] Department of Oral Chrono-Physiology, Graduate School of Biomedical Sciences, Nagasaki University, Nagasaki, 852-8588, Japan; [c] Halberg Chronobiology Center, University of Minnesota, Minneapolis, Minnesota, USA; [d] Executive Medical Center, Totsuka Royal Clinic, Tokyo Women's Medical University, Tokyo, Japan
*Email: wataru_nakamura@nagasaki-u.ac.jp

16.1 Introduction

The human menstrual cycle usually lasts approximately 28 days (24–32 days). In humans, the ovarian and uterine endometrial cycles are known as the menstrual cycle. However, it is called the oestrous cycle in mammals other than primates, including humans, as no significant menstrual bleeding is observed for them. Rodents such as mice, rats, and hamsters, often used in experiments, exhibit an oestrous cycle lasting 4 to 5 days called the "incomplete oestrous cycle", which does not include a long luteal phase, as observed in humans.[1,2] Female reproductive function and circadian rhythm influence each other in a bidirectional manner. For example, in adult

Chemical Biology No. 23
Chronobiology and Chronomedicine: From Molecular and Cellular Mechanisms to Whole Body Interdigitating Networks
Edited by Germaine Cornelissen and Tsuyoshi Hirota

Published by the Royal Society of Chemistry, www.rsc.org

women, oestrogen and progesterone are ovarian steroid hormones affecting the circadian rhythm, which undergoes large and rapid changes within the menstrual cycle, possibly causing sleep/wake rhythm problems. Insomnia and hypersomnia reportedly have been associated with premenstrual tension and dysphoria that appear physically and mentally during the menstrual cycle. These sleep disturbances are thought to be due to variations in the secretion of ovarian steroid hormones.[3] Women contemplating pregnancy may follow chronobiological principles to change their lifestyle to make it easier for them to conceive naturally. An overview of the relationship between circadian rhythms and the menstrual cycle, from the molecular to the organismic level, represents an important biological and medical topic in countries with declining birth rates and delayed first childbirths.

This chapter describes the relationship between the circadian rhythm and the oestrous cycle, which represents the phenotype of reproductive function, from two aspects. The first aspect pertains to "effects of the oestrous cycle on the circadian rhythm", and the second focuses on "effects of the circadian rhythm on the reproductive function." Recent findings, including our data on these two aspects, are discussed.

16.2 Circadian Timekeeping Mechanism in Mammals

In mammals, circadian rhythms are physiological and behavioural cycles with a periodicity of approximately 24 hours, which are generated by the endogenous biological pacemaker, the suprachiasmatic nucleus (SCN), located in the anterior hypothalamus. The SCN exists in pairs on the left and right sides of the third ventricle in the midline directly above the optic chiasm, where the optic nerves intersect and have a characteristic structure in which approximately 10 000 cells are quite tightly packed on one side. Experimental lesions of the rodent SCN result in the loss of free-running and photic entrainment of circadian rhythms. Additionally, when the SCN of an adult rodent is lesioned and is then replaced with a foetal SCN that has a genetically different free-running period, its activity rhythm aligns with the period from its donor SCN,[4] indicating that the SCN is the central circadian clock.

Circadian clocks orchestrate physiological rhythms *via* the temporal regulation of specific gene expression to control core "clock genes" and the proteins they encode. The molecular oscillator involves negative autoregulatory feedback loops of transcription and translation. In these loops, transcription factors, Bmal1 and Clock, positively regulate the expression of the *Period* (*Per1*, *Per2*, and *Per3*) and *Cryptochrome* (*Cry1* and *Cry2*) genes at the beginning of the cycle. *Per* and *Cry* gene products accumulate, dimerise, and form a complex which, in turn, suppresses the activity of *Bmal1* and *Clock*.[4,5] Circadian clocks have been found in virtually every peripheral tissue.

Although the SCN acts as a master clock that orchestrates circadian rhythms through various timing cues such as humoral and neural signals, feeding cycles, and temperature, peripheral clocks do not respond identically to cues from the SCN. Circadian rhythms play a critical role in regulating physiological processes, and disruption of circadian rhythms can lead to metabolic syndrome, diabetes, and cardiovascular diseases. Clock gene-deficient or mutant animals exhibit shortened, lengthened, or arrhythmic circadian activity rhythms, and also show symptoms of lifestyle-related diseases such as diabetes and hypertension, along with disruptions in the circadian rhythm.[6,7]

16.3 Effects of the Oestrous Cycle on Circadian Rhythms

16.3.1 Variations in Locomotor Activity Throughout the Oestrous Cycle

The oestrous cycle in experimental rodents, which lasts 4 to 5 days, consists of four stages: metestrus, diestrus, proestrus, and oestrus. Fluctuations in blood levels of ovarian steroid hormones during the oestrous cycle in animal models have been used to gain knowledge that can be applied to humans[8] (see Figure 16.1A).

It is known that the fluctuations in ovarian steroid hormones associated with the oestrous cycle affect the locomotor activity rhythm. Since rodents prefer wheel running, mice run in the dark phase and rest in the light phase when a wheel is placed in a cage. When wheel running activities were recorded over a long period and double-plotted on an actogram, their rest–activity patterns became readily apparent, as shown in Figure 16.1B. When monitoring the activity rhythm of adult female mice with a regular estrous cycle, it is evident that the level of activity during the dark phase is higher during proestrus and the onset of the active phase advances (see Figure 16.1B, right). As the rise in activity and advance in activity onset are not seen in other oestrous stages, the active phase appears to display a ripple pattern in the actogram, a phenomenon called "scalloping".[9,10] Although scalloping is not commonly seen in healthy women, some women reportedly show sleep problems such as insomnia or hypersomnia associated with premenstrual syndrome (PMS) and premenstrual dysphoric disorder (PMDD).[3,11,12] Moreover, since the phase angle difference between the activity and the light–dark (LD) cycle depends on the free-running period, the scalloping seen in rodents indicates that the free-running period is shortened in proestrus and oestrus. To prove this experimentally, when ovariectomised rodents are subcutaneously implanted with oestrogen, the amount of activity is increased and the free-running period is shortened.[13]

16.3.2 Effects of the Oestrous Cycle on the Expression Rhythm of Clock Genes

We focused on the regulation of the circadian rhythm and investigated whether peripheral clocks, including those related to reproductive organs, and the SCN are affected by the oestrous cycle (see Figure 16.2). The analysis of the expression rhythms of clock genes (*Per1*, *Per2*, and *Bmal1*) in rats conducted through *in situ* hybridization and quantitative real-time PCR showed that the SCN was rarely affected by the oestrous cycle. However, peripheral tissues (liver, uterus, and ovary) were significantly affected by the oestrous cycle.[14] The effect was not constant depending on the type of gene or tissue. In the uterus and ovary, the phase of the expression of the *Per2* mRNA rhythm during proestrus was advanced and the amplitude was also

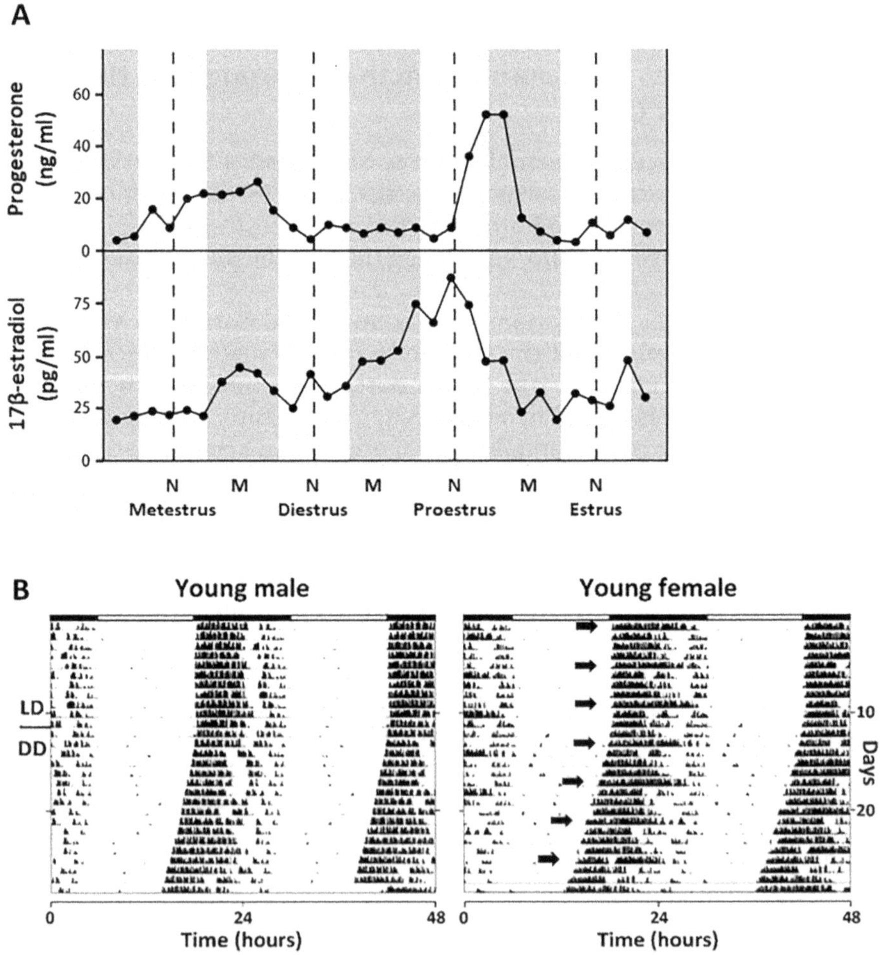

greater compared to other stages (see Figure 16.2). Additionally, our behavioural experiment indicated that phase responses to light did not differ throughout the oestrous cycle, suggesting that the SCN rhythm was not influenced by the oestrous cycle.[15] Since Amir's group also reported that the expression rhythm of PER2 protein in the SCN was not affected by the oestrous cycle,[16] it is concluded that the central clock in the SCN is not affected by the oestrous cycle.

Next, we examined the effect of the oestrous cycle on cultured tissues using PER2::LUC knock-in mice,[17] which can measure bioluminescence by linking the luciferase gene to *Per2*. While the SCN rhythm was not affected by the oestrous cycle in the culture system, the phase of the PER2::LUC rhythm was advanced during proestrus as compared to metestrus, and the amplitude was also larger in the uterus (see Figure 16.2). This result is consistent with the aforementioned *in vivo* experiments. The oestrous cycle thus influenced the expression rhythms of clock genes in the uterus, but not in the SCN.[13]

We built a pseudo-hormonal environment similar to that of *in vivo* cultures and examined the effect of combined oestrogen and progesterone on the PER2::LUC rhythm in the uterus. When the hormonal environment of oestrogen and progesterone was prepared to resemble proestrus conditions, the amplitude of the PER2::LUC rhythm increased compared to the control (vehicle) environment. In the subsequent pseudo-oestrus stage, the large amplitude of the PER2::LUC rhythm was maintained, and the phase of the rhythm was delayed. This result is similar to the *Per2* mRNA rhythm in the uterus observed *in vivo*, suggesting that changes in oestrogen and progesterone concentrations associated with the oestrous cycle regulate the clock gene expression rhythm.[14]

Figure 16.1 Changes in ovarian steroid hormones associated with the (incomplete) oestrous cycle of rodents and sex differences in wheel-running rhythm of rodents. (A) Fluctuations in plasma oestrogen (17β-oestradiol) and progesterone concentrations during the oestrous cycle in rats. Oestrogen increases from proestrus, and progesterone increases from proestrus to oestrus. N: noon and M: midnight. Reproduced from ref. 8 with permission from Oxford University Press, Copyright 1974. (B) Representative double-plot actogram of adult male (left) and female (right) mice. The black part inside the figure shows the activity, and the white and black bars at the top show the time of the light phase (white) and the dark (black) phase. The animals were housed under light/dark (LD) conditions (12 h: 12 h) for the first 11 days and then kept under constant dark (DD) conditions. The activity of the mouse is synchronized with the LD conditions, and both males and females rest in the light phase and are active in the dark phase. Under DD conditions, animals cannot obtain timing information from the environment. Therefore, they maintain a rhythm in the cycle of their internal clock. This is called a free-running rhythm. In female animals (right), increased activity and advanced active phase are observed on the day corresponding to proestrus, indicated by the arrowheads.

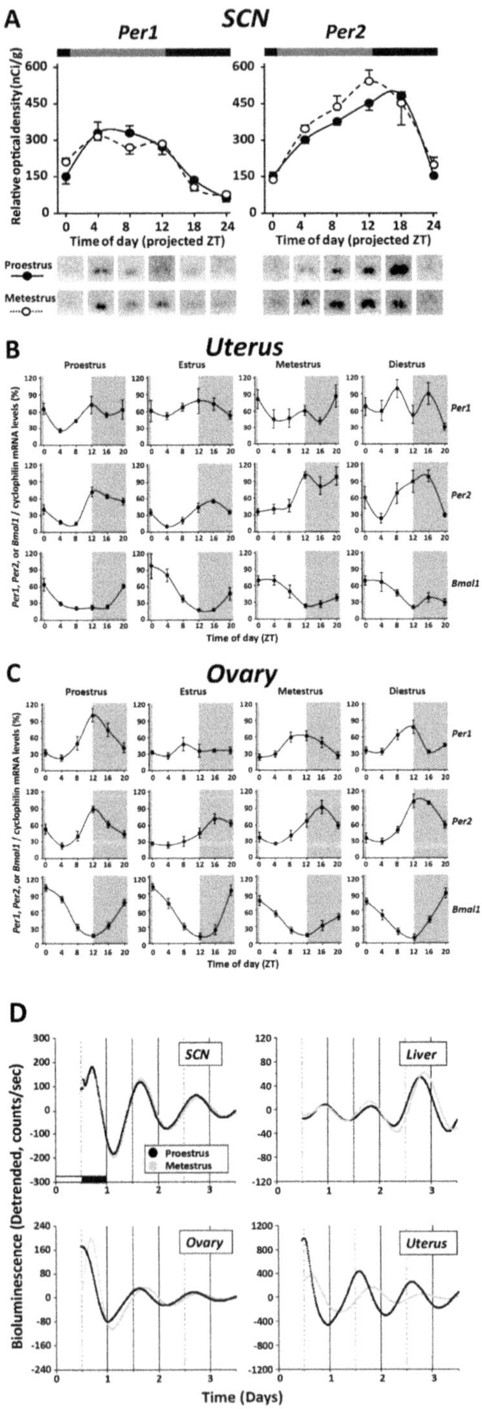
A
SCN
Per1
Per2
Relative optical density (nCi/g)
Time of day (projected ZT)
Proestrus
Metestrus
B
Uterus
Proestrus
Estrus
Metestrus
Diestrus
Per1, Per2, or Bmal1 / cyclophilin mRNA levels (%)
Per1
Per2
Bmal1
Time of day (ZT)
C
Ovary
D
SCN
Liver
Ovary
Uterus
Bioluminescence (Detrended, counts/sec)
Time (Days)

Furthermore, we investigated the effects of each hormone separately. We found that oestrogen shortens the free-running period of the PER2::LUC rhythm in the uterus. Additionally, raloxifene, which acts as an oestrogen antagonist, attenuated this effect. A similar experiment performed on the SCN did not affect the period or amplitude of the PER2::LUC rhythm.[18]

Collectively, the results show that ovarian steroid hormones influence the timing of peripheral clocks but do not affect the central clock. Applying these results to humans, it is possible that the circadian clock system may become unbalanced due to significant hormonal fluctuations that occur during pre-ovulation and the pre-menstrual period throughout the menstrual cycle.

16.4 Effects of the Circadian Rhythm on Female Reproductive Functions

16.4.1 Circadian Regulation in Oestrous Cycles: A Concept of Daily "Timing Signals"

The timing of ovulation is controlled by the hypothalamus–pituitary–gonadal (HPG) axis.[19] Oestrogen produced in the ovaries functions as negative feedback in the HPG axis. The oestrogen level gradually increases in the late stage of the oestrous cycle. When it exceeds the threshold level due to follicle growth, oestrogen increasingly acts on the HPG axis, resulting in a surge in luteinizing hormone (LH), which causes ovulation. The pre-ovulatory LH surge is induced by a positive feedback effect of oestrogen.

Figure 16.2 Effects of the oestrous cycle on clock gene expression rhythms. (A) The expression rhythm of *Period* (*Per*)1 and *Per*2 mRNA in the suprachiasmatic nucleus (SCN) was measured by *in situ* hybridization. Rats were examined during proestrus and metestrus, as determined using vaginal smears, and sampling was performed every 4 hours for 24 hours after 3 days of DD conditions to eliminate the influence of light. *Per1* and *Per2* mRNA expression rhythms in the SCN are almost unaffected by the oestrous cycle stage. (B and C) The expression rhythms of *Per1*, *Per2*, and *Bmal*1 mRNA in the uterus and ovaries of rats were assessed by quantitative real-time PCR. Rats were examined during proestrus, oestrus, metestrus, and diestrus, as determined through vaginal smears, and sampling was performed every 4 hours for 24 hours under LD conditions. In the uterus and ovary, each gene has an altered form of expression rhythm during each oestrous cycle stage. (D) Representative examples of PER2::LUC rhythms in the SCN, liver, ovary, and uterus of adult female PER2:: LUC knock-in mice sampled in proestrus (dark curve) and metestrus (light curve). The stage of the oestrous cycle was determined through vaginal smears, and tissue cultures of the SCN, liver, ovary, and uterus were performed. In the uterus, the phase of the PER2::LUC rhythm in proestrus is advanced, and the amplitude is larger as compared to metestrus. (B–D) Reproduced from ref. 14 with permission from Elsevier, Copyright 2010.

At the beginning of the 1950s, there were discussions about the neuroendocrine mechanisms responsible for triggering a cascade of HPG events for establishing a reproductive cycle, which controls ovulation. The importance of daily "timing signals"[20] and of the neuronal input that affects the pituitary gland from the hypothalamus was recognized. Everett and Sawyer reported that female rats reared under 24-hour LD cycles (14 hours of light and 10 hours of darkness, light onset at 5:00 am) had a 4-day oestrus cycle and ovulated between 1:00 am and 2:30 am. In this study, the administration of a short-acting barbiturate sedative at 2:00 pm delayed ovulation by one day (24 hours). In contrast, ovulation occurred at the regular time when the sedative was administered at 4:00 pm. This result indicates the existence of "a time window" set at a particular time of the day for neural signals to induce the LH surge successfully (see Figure 16.3), thus revealing the involvement of the circadian system in regulating the HPG axis.

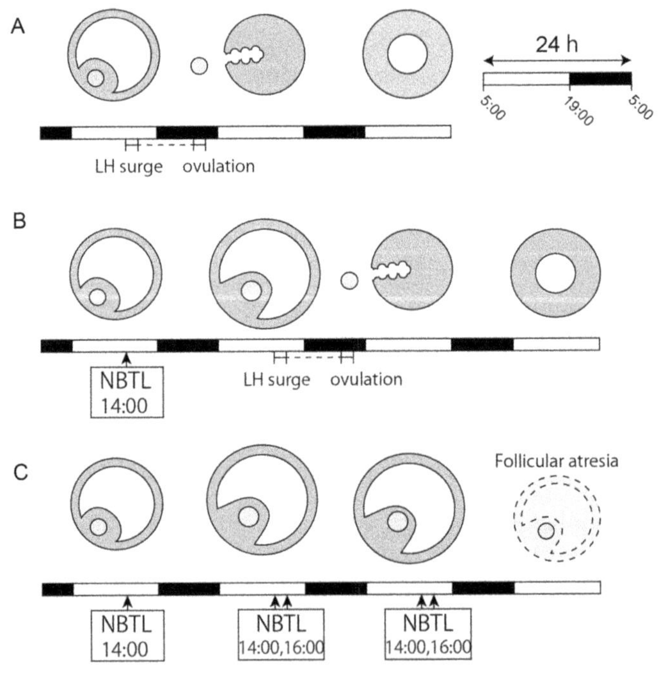

Figure 16.3 Schematic representations of the critical "time window" for normal ovulation. (A) Under 24-hour LD cycles, indicated by the LD bar, ovulation occurs with regularity between 1:00 AM and 2:30 AM. (B) A single injection of Nembutal at 14:00 h causes a delay in ovulation by 24 hours. "NBTL" indicates intraperitoneal injection of Nembutal (30 mg kg^{-1}). (C) A single injection during proestrus and double injections on the next 2 days causes consecutive skipping of ovulations and early atresia. Reproduced from ref. 20 with permission from Oxford University Press, Copyright 1950.

16.4.2 Maintenance of Oestrous Cycles by the Master Circadian Clock

The fact that the 4- to 5-day period of oestrous cycles is an integral multiple of the 24-hour cycle also indicates an association of circadian rhythms with oestrous cycles.[21] The measuring system for wheel-running behaviours, which records the activity rhythms of female mice of reproductive age (2- to 6-month-old) when oestrous cycles normally occur, shows an increase in wheel-running activity every four days (see Figure 16.4). Here, let us compare the oestrous cycles in mice that are deficient in *Cry1*, a clock gene that shortens the period of the activity rhythm (*Cry1* KO, circadian period: 22.5 hours) with those deficient in *Cry2* causing prolongation of the period (*Cry2* KO, circadian period: 24.5 hours). Under constant dark (DD) conditions, the length of the oestrous cycle in *Cry1* KO mice is 22.5 hours $\times$ 4 cycles = 90 hours. In contrast, the oestrous cycle in *Cry2* KO mice is 24.5 hours $\times$ 4 cycles = 98 hours. These results indicate that the oestrous cycle in mice is not an integer multiple of 24 hours (4 days or 96 hours), but rather an integer multiple of the circadian period. Ablation of the SCN resulted not only in the loss of circadian rhythms but also in the loss of oestrous cycles in female rats.[22] In contrast to the circadian rhythm, the loss of reproductive function resulting from SCN lesions cannot be restored by transplantation of the SCN,[23] indicating a strong association between the central neural control of the circadian rhythm, the SCN, and the oestrous cycle.

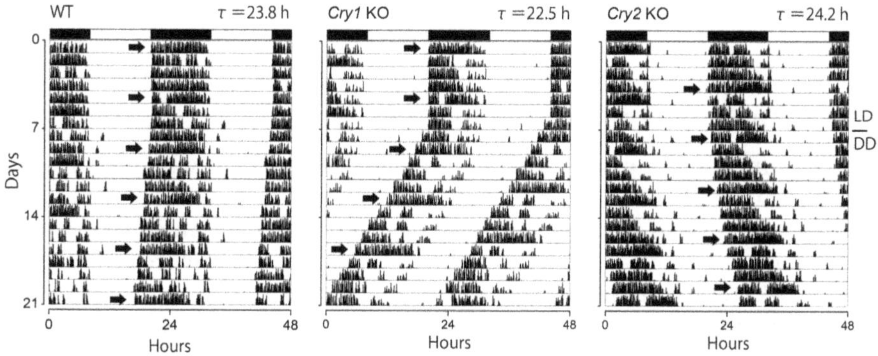

Figure 16.4 Circadian behavioural phenotypes in *Cry*-deficient female mice. All animals were maintained under LD conditions for the first 7 days, as indicated by the bar above the record. Then, mice were released in constant darkness, DD. Under LD and DD conditions, females of each genotype showed increasing wheel-running activity every four cycles of daily and circadian rhythms (indicated by arrowheads). The day of peak activity coincides with the proestrus stage.[15] Reproduced from ref. 24 with permission from Elsevier, Copyright 2015 The Authors.

16.4.3 Dysfunction of the Female Reproductive System Caused by Dissociation of the Circadian Clock from the External Environment

How does the circadian rhythm affect female reproductive functions, including the oestrous cycle? In 2- to 6-month-old mice, which is a suitable age for reproduction, the pregnancy success rate is 100% when female mice are mated with male mice in the dark phase during proestrus, when wheel-running activity is increased (see Figure 16.5). A normal oestrous cycle is apparent in wheel running activity in most 2- to 6-month-old female mice, but it can no longer be observed in older mice after 24 months of age. We observed a normal oestrous cycle in all 2- to 6-month-old female *Cry1* and *Cry2* KO mice, but a clear oestrous cycle was not observed in 8- to 12-month-old *Cry1* or *Cry2* KO mice, an early ageing stage for mice. Their pregnancy success rate when mated with male wild-type mice was approximately 10% (*Cry1* KO: 10.0%, *Cry2* KO: 12.5%, see Figure 16.5). The mice were usually housed under 24-hour LD conditions (12 hours of light alternating with 12 hours of darkness). When the light–dark cycle matches the 22.5- or 24.5-hour free-running period of *Cry1* or *Cry2* KO mice, respectively, the pregnancy success rate in the early ageing stage is partly restored to reach approximately 70% (*Cry1* KO: 70.0%, *Cry2* KO: 66.7%, see Figure 16.5). Since the experiment was only performed using a light–dark schedule matching the intrinsic cycle of individual mice, it cannot be concluded that the *Cry1* and *Cry2* genes themselves are directly involved in the regulation of the reproductive system. Therefore, it can only be strongly suggested that the dissociation of the circadian clock from the external environment causes dysfunction in the female reproductive system.[19]

16.4.4 "Social Jet Lag" Causes Female Reproductive Dysfunction

Jet lag occurs when the body's internal clock is out of sync with cues from a new environment such as when travelling across time zones. It may also occur without travelling abroad, when there is a discrepancy between an individual's circadian rhythm and the social clock, usually reflected by a marked difference in sleep timing between work or school days and free days. Returning to the weekday schedule after the weekend may then be associated with fatigue and depression, symptoms that are similar to those experienced after jet lag. This condition is called "social jet lag".[25] Mice on a schedule that induces social jet lag desynchronize their activity rhythm from the light–dark cycle starting on the first day of the week. Additionally, when social jet lag is applied to female mice at a suitable age for pregnancy, the oestrous cycle can be maintained, but it is disrupted during early ageing[24] (see Figure 16.6). Translating these results to humans, social jet lag can be viewed as a factor that promotes female reproductive dysfunction, such as irregular menstruation.

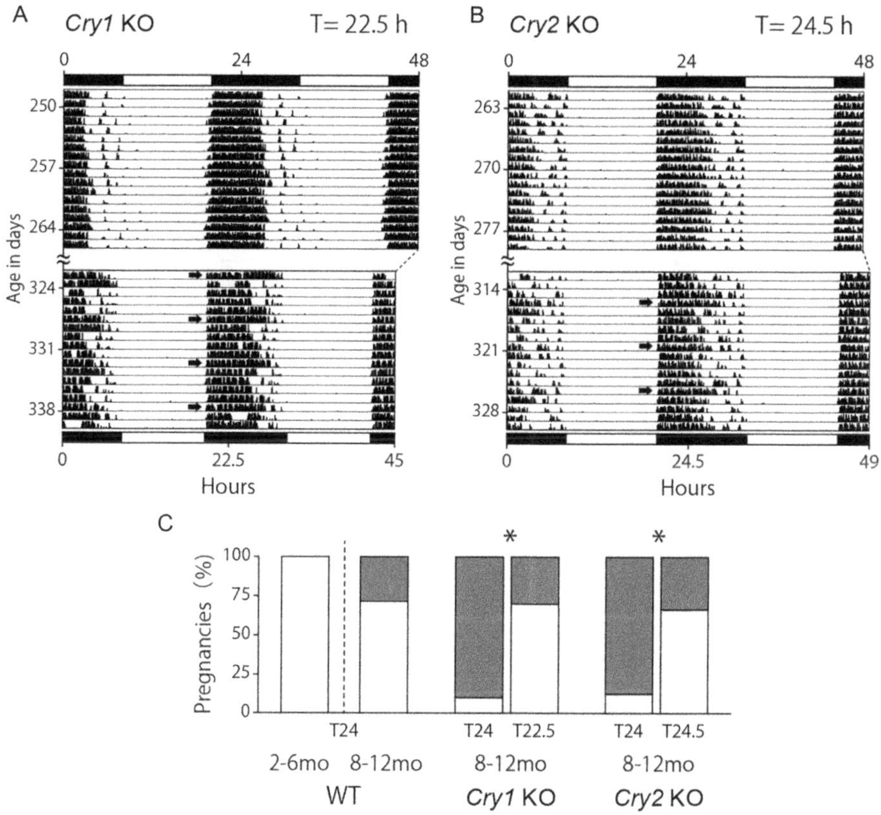

Figure 16.5 Improvement of the oestrous cycle and counteracting irregularities and infertility by adjusting the external environmental cycle to the intrinsic circadian cycle. (A and B) Representative double plot actograms showing wheel-running activity rhythms of 8- to 12-month-old female *Cryptochrome* (*Cry*)*1* KO mice (free-running period of 22.5 hours) and *Cry2* KO mice (free-running period of 24.5 hours). In the upper panels, mice were housed in a 24-hour LD cycle (12 hours of light and 12 hours of darkness). In the lower panel, mice were transferred to an LD cycle matching their own circadian periods (*Cry1* KO: $T = 22.5$ hours, *Cry2* KO: $T = 24.5$ hours), according to the free-running period of each *Cry1* or *Cry2* KO mouse. By matching the intrinsic and external environmental cycles, the "scalloping" characteristic of a normal oestrous cycle appears. (C) Comparison of the pregnancy success rate in 2- to 6-month-old and 8- to 12-month old (early ageing) wild-type mice, and in similar-age *Cry1* KO and *Cr 2* KO mice kept under 24-hour LD conditions or in LD cycles matching their intrinsic cycle. The pregnancy success rate increases in *Cry1,2* KO mice when housed in an LD cycle that matches their own free-running period. Reproduced from ref. 24 with permission from Elsevier, Copyright 2015 The Authors. *$p < 0.05$ (Fisher's exact test).

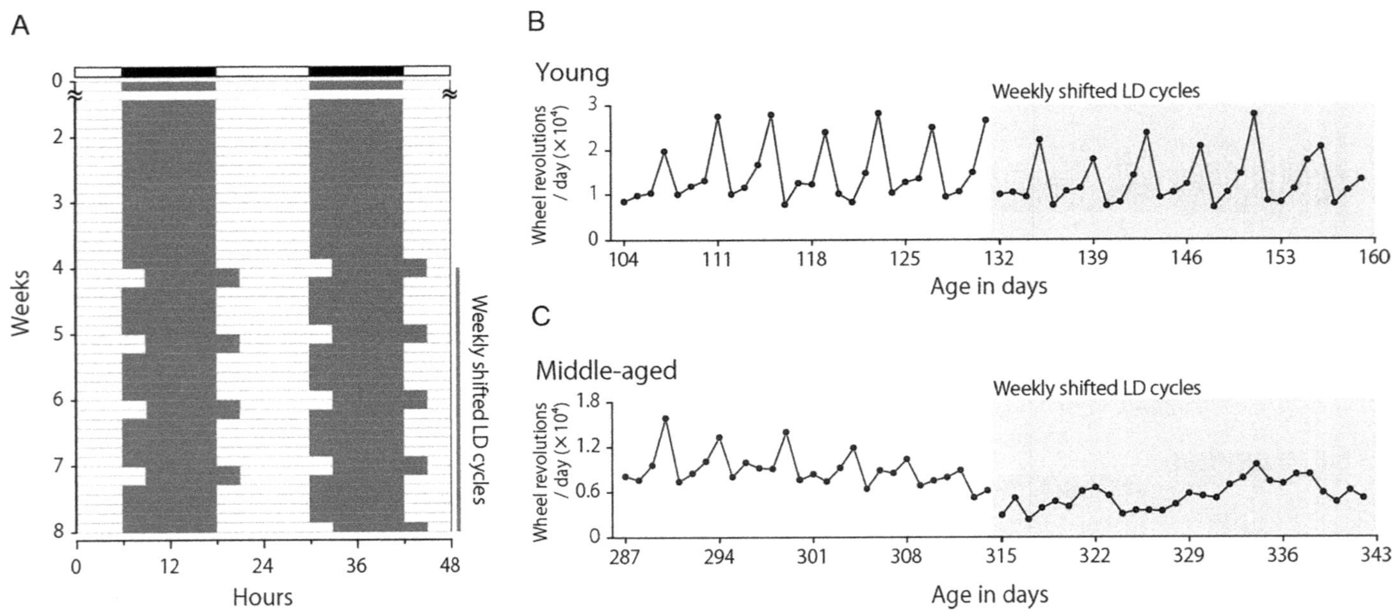

Figure 16.6 Irregular oestrus cycle caused by "social jet lag". (A) A light–dark schedule that causes "social jet lag" in mice. White represents the light phase, and grey represents the dark phase. From the 4th week onwards, the lighting conditions were delayed by 3 hours for 2 days every 7 days (simulating Saturdays and Sundays). (B and C) Wheel-running revolutions in young (pregnancy-appropriate) (B) and middle-aged (C) female mice, which were exposed to "social jet lag", as shown in A. Mice were housed in usual light/dark condition for the first 28 days and then transferred to a schedule simulating "social jet lag" for the next 28 days. Young animals showed a stable sexual cycle of 4 to 5 days, even under "social jet lag" conditions, whereas early-ageing animals had irregular sexual cycles due to "social jet lag". Reproduced from ref. 24 with permission from Elsevier, Copyright 2015 The Authors.

16.4.5 Mechanism of Female Reproductive Dysfunction Caused by Dissociation of the Circadian Clock from the External Environment

What is the mechanism accounting for the role of the circadian rhythm in the age-related decline in female reproductive function? During a normal oestrous cycle, elevated oestrogen levels trigger a surge in the secretion of LH (LH surge) released from the anterior pituitary during the late afternoon/early night of proestrus. As a result, mature eggs are released from the ovary.[26] In rodents, LH surges are known to occur only in the late light phase of proestrus, which requires timing signals from the central circadian clock in the SCN.[26] Day information from the elevated oestrogen that signifies "proestrus" and timing information from the SCN that indicates "late afternoon" work in combination to trigger the LH surge once during each oestrous cycle. This hypothesis is supported by the following experiments: when a silicone tube containing oestrogen is implanted in a female mouse, the animal exhibits an LH surge every day,[27] and an animal in which the SCN has been destroyed does not experience an LH surge.[28] We reported that the tens of thousands of neurons comprising the SCN lose their ability to synchronize their rhythms with advancing age,[29] and that the timing outputs from the SCN weaken.[30] Together, these reports suggest that the timing output from the SCN is required to drive the oestrous cycle accurately, and that the output weakens with age, similar to the decrease in synchronization with the environmental cycle. Consequently, the LH surge is not generally triggered, and the oestrous cycle cannot follow its usual course;[31] this phenomenon can also be shown using our mathematical model.[32]

16.5 Conclusions

Fluctuations in ovarian steroid hormones associated with the oestrous cycle affect the biological clock in terms of the effects of the oestrous cycle on the circadian rhythm. Our results suggest that the causes of sleep problems such as insomnia and hypersomnia which are part of the symptoms of PMS and PMDD are primarily related to the "misalignment" between the central and peripheral clocks. This misalignment is generally caused by an irregular lifestyle. However, fluctuations in ovarian steroid hormones can also cause it, even under a regular lifestyle.

Additionally, an inappropriate light environment and social jet lag (irregular lifestyle) disturb female reproductive function, implicating effects on the circadian rhythm. Disturbances in reproductive function tend to occur with advancing age. Considering the tendency toward a later age of first childbirth, it is critical to physiologically comprehend the basis of the onset of menstrual cycle irregularity that can lead to infertility. A new therapy is anticipated to be developed to alleviate problems of infertility through lifestyle modification that considers changes in the body clock with ageing, such as increasing the regularity of daily schedules.

16.6 Appendix

16.6.1 Circadian Disruption and Reproductive Function

Sleep and sleep disturbances are increasingly recognized as determinants of women's health and well-being, particularly in the context of the menstrual cycle, pregnancy, and menopause,[33] related, at least in part, to the influence of oestrogens.[34] Chronotype has been associated with reproductive function in women.[35] A cross-sectional study found that infertile women had poorer sleep quality, and a higher likelihood of having an evening chronotype compared to fertile women.[36] In addition, irregular sleep schedules may be problematic since more than one hour of social jetlag was associated with severe menstrual symptoms, independently of sleep duration and late chronotype.[37] Epidemiological evidence also showed that shift-working women are at high risk of irregular menstrual cycles, spontaneous abortions, and low-birth-weight babies.[38]

These results are corroborated by animal studies: while exposure of female mice to a single phase-shift had minor reproductive effects, exposure to chronically disrupted light/dark cycles markedly impaired the occurrence of the pre-ovulatory luteinizing hormone (LH) surge, leading to reduced fertility.[39] Exposure of female rats to constant light disrupted the oestrous cycle, which was partially restored by time-restricted feeding.[40] Long-term exposure to constant dark conditions caused polycystic ovarian syndrome (PCOS)-like reproductive abnormalities in female rats, including oestrous cycle disorder, polycystic ovaries, LH elevation, hyper-androgenism, and glucose intolerance.[41] It was associated with decreased expression of melatonin receptor 1A in ovarian granulosa cells. Melatonin receptor-agonist application and exposure to LD12:12 relieved hyper-androgenism in these animals.[41] In post-adolescent female rats, feeding restricted to the rest span instead of the active span or feeding *ad lib* was associated with decreased frequency and number of ovulations.[42]

These observations suggest the involvement of the circadian system in reproduction. Despite recent research work on the intricate relation between core clock genes in all cells participating in the reproductive function, the coordination between central and peripheral reproductive clocks in females is still poorly understood.[43] As such, much of the current research is still focusing on discovering, cataloguing, and characterizing circadian rhythms and clock mechanisms and their effects on reproduction, rather than on the evaluation or optimization of treatments. For this reason, emphasis will be placed herein on female rodent studies first, followed by a few notes on how impaired clock function affects human reproduction in a negative way, and ending with some suggestions for future investigations of ways in which circadian medicine could help resolve issues related to human reproduction.

16.6.2 Circadian Rhythm Involvement in Reproduction

Few challenges are more critical to the survival of a species than reproduction. Reproductive success depends on tightly orchestrated physiological processes. Boden *et al.*[44] reviewed the involvement of circadian rhythms in various aspects of reproductive physiology, including regulation of the oestrus cycle, the LH surge and ovulation, the production and maturation of sperm and the timing of insemination and fertilization. The authors also describe the role of circadian rhythms in supporting the preimplantation embryo in the oviduct, implantation and placentation, and in controlling parturition and early postnatal life.[44]

Briefly, the environmental alternation between light and darkness synchronizes the central clock in the suprachiasmatic nucleus (SCN), which has afferent connections with various hypothalamic regions, including interneurons of the medial hypothalamus, neuroendocrine neurons in the paraventricular nucleus (PVN) and arcuate nucleus (ARC), and preautonomic neurons in the PVN.[45] Peripheral tissues, although orchestrated by the SCN, also express circadian rhythms of gene expression and function, and tissue-specific rhythmicity can regulate the responsiveness of these peripheral targets to various stimuli, including incoming temporal information from the SCN.[44]

Initiation of pregnancy depends on a series of tightly coordinated events. The primordial follicles in the ovary need to develop and mature; ovulation needs to occur; and mating behaviour needs to coincide with the release of the mature oocyte for fertilization to occur. Plasma estradiol concentrations increase in the late follicular phase, along with an increase in the pulsatile frequency of LH release. As the dominant follicle reaches the mature preovulatory stage, the influence of estradiol on gonadotropin-releasing hormone (GnRH) secretion changes from inhibitory to stimulatory, resulting in a concerted release of GnRH from the medial peri-optic area (POMA) and a subsequent sustained increase in LH release from the anterior pituitary, leading to the release of the oocyte from the ovary.[44]

Both the SCN and intact neuronal signalling are needed for the integrity of ovarian rhythms and circadian regulation of LH release.[46] The neurotransmitters vasoactive intestinal peptide (VIP), gamma-aminobutyric acid (GABA), arginine vasopressin (AVP), and gastrin releasing peptide (GRP) play important roles in facilitating communication between the SCN's ventrolateral core and dorsomedial shell. The ventrolateral SCN uses VIP to project to POMA's GnRH-containing neurons in order to gate ovulation to a particular time of day. The dorsomedial SCN contains receptors for sampling plasma oestrogenic concentration; it uses AVP to communicate with the rostral periventricular preoptic area (RP3V), including the periventricular preoptic nucleus (PeN) and the anteroventral periventricular nuclei (AVPv). The neurotransmitter kisspeptin plays a role in projecting this region to POMA to stimulate GnRH release. It also enables communication between the dorsomedial SCN and the ARC, which projects to POMA. The SCN also projects to the gonadotropin inhibitory hormone (GnIH or RFRP3) positive cells of the dorsomedial nucleus, which project to the POMA to inhibit GnRH release.[47] As the follicles develop, the increase in estradiol concentrations leads to a selective increase in synaptogenesis between SCN projections and GnRH positive cells, and the induction of kisspeptin expression in the RP3V.[48] The effect of estradiol shifts towards promoting GnRH release, increasing the GnRH concentration in the portal veins and leading to the subsequent release of LH and ovulation. Key differences exist, however, with respect to the source of kisspeptin signalling to POMA between rodents and humans.[44]

Considering that kisspeptin neurons may mediate VIP's effects on the reproductive axis, Mansano *et al.*[49] investigated the effects of VIP on hypothalamic kisspeptin neurons in female mice during oestrogen negative feedback. These authors found that VIP can recruit different signalling pathways to modulate AVPv/PeN or ARH kisspeptin neurons, resulting in distinct cellular responses, possibly to influence pulsatile LH secretion.

The sensitivity of the ovary to LH also changes across 24 hours, with maximal responsiveness consolidated to the middle of the night of proestrus in rats.[50] Rhythmic expression of core clock genes has been reported in the granulosa and thecal cells of the follicle (early, pre-antral and antral), glandular tissue and corporal luteal cells, but evidence is less clear in other cell types (oocytes and stromal fibroblasts).[44] Implantation depends upon the synchronized development of a competent blastocyst and a uterine environment that is receptive to attachment. Clock genes and their proteins are rhythmically expressed within the luminal epithelium, stroma and myometrium of the uterus.[44] Ovarian steroid hormones alter clock gene expression; progesterone increases *Npas2*, *Clock*, *Cry1* and *Per1* and decreases *Rev-erbβ* and *RORγ* mRNA expression *via* progesterone receptor binding.[51] Clock genes may also play a role in the development and functions of the placenta.[44]

16.6.3 Role of Oestrogens

Oestrogens interact with the circadian system to regulate biological processes. At the molecular level, oestrogens and circadian genes interact to regulate gene expression and cell biology. Oestrogens also regulate circadian behaviour across the oestrous cycle. The timing of ovulation during the oestrous cycle requires coincident oestrogen and SCN signals.[52] Oestrogens are steroid hormones mostly produced by ovaries, with contributions from the brain, adipose tissue, stomach, bone, and skin in female rodents.[53] Oestrogen signalling is mediated by several types of oestrogen receptors that are expressed throughout the body.[54] Ovulation and the induction of oestrus are caused by a surge in LH released from the anterior pituitary during the late afternoon/early night of proestrus.[52]

A series of classical studies showed how the regulation of the oestrous cycle and ovulation in female rodents depends on the SCN.[52] Molecular oestrogen signalling mechanisms are complex. Oestrogens can directly regulate the expression of some circadian genes (such as *Per2* and *Clock*) by means of the oestrogen response element (ERE), but effects vary depending on the tissue and species.[52] The circadian system could also directly regulate oestrogen signalling if oestrogen receptors contain E-boxes in their regulatory regions. Moreover, the reciprocal regulation between circadian rhythms and oestrogen signalling involves direct binding of clock and ER proteins and post-translational modifications of clock proteins.[52] Evidence suggests that oestrogens do not have marked, direct effects on SCN timekeeping.[52] Estradiol, however, directly alters neuronal activity in the SCN. It increases SCN cell excitability that is mediated by oestrogen receptors, and regulates expression of c-Fos in SCN neurons.[52] In sexually differentiated female rodents, estradiol treatment shortens the free-running period of the locomotor activity rhythm, whereas ovariectomy lengthens it.[52]

The hypothalamic–pituitary–gonadal (HPG) axis in female rodents governs the oestrous cycle and reproduction. GnRH is secreted into the hypophyseal portal circulation, stimulating the release of LH and follicle stimulating hormone (FSH) from gonadotropes in the anterior pituitary into systemic circulation, thereby promoting follicle development and production of oestrogens and progesterone. As noted above, oestrogens exert negative feedback to inhibit the release of LH and the development of more follicles until the afternoon of proestrus. Then, at their highest concentration, their feedback to GnRH neurons shifts from negative to positive, causing rapid firing of GnRH neurons, the release of a GnRH surge into portal circulation, and an LH surge just before dark. The LH surge stimulates ovulation about 12 hours later during oestrus. The precise timing of the GnRH and LH surges is critical to align fertility with sexual receptivity.[52,55] In many species, there is a circadian component to the LH surge, restricting its occurrence to specific times of day. The precise cell targets, neural signalling factors and receptors, hormonal pathways, and molecular mechanisms through which ovarian-derived estradiol indirectly stimulates GnRH surge secretion remain incompletely known. How the circadian clock interacts with endocrine signals to ultimately time LH surge generation also remains incompletely understood.

Elevated circulating oestrogens and gating of the time window by the SCN are necessary for GnRH to be released in the afternoon of proestrus. ERα is necessary and sufficient for positive oestrogen feedback on the GnRH surge in female mice; oestrogen-induced LH surges and c-Fos expression in GnRH neurons are absent in ERα KO female mice and restored in ovariectomized mice treated with an ERα agonist.[56–58] The SCN is also needed to signal to GnRH neurons, rendering them sensitive to activation by high concentrations of oestrogens. Signalling occurs every day, as an LH surge takes place every afternoon when ovariectomized rodents are treated with constitutively high estradiol. Evidence indicates that it is mediated by neural rather than paracrine mechanisms.[52] It is thought that the integration of circadian and oestrogenic signals regulating GnRH neurons depends on AVPv kisspeptin neurons, which detect the increase in the concentration of estradiol and also the circadian signals generated at the SCN.[59]

The role of progesterone in biological clock gene expression is illustrated by differences between late follicular phase and mid luteal phase in the expression patterns of *Bmal1*, *Cry1* and *Per2* mRNA in peripheral blood mononuclear cells from female monkeys, when there is no difference in oestrogen levels.[60] Using PER2::luciferase (PER2::LUC) circadian reporter mice, Yaw *et al.*[61] determined the time of day of the PER2::LUC peak in the SCN, pituitary, uterus, and ovary. The relationships between reproductive tissue PER2::LUC phases were found to change throughout the oestrous cycle and late pregnancy and were accompanied by changes in the PER2::LUC period in the SCN, uterus, and ovary. Furthermore, progesterone was found to increase *Per2*-luciferase expression in immortalized SCN (SCN2.2) and arcuate nucleus (KTAR) cells, and to shorten the PER2::LUC period in *ex vivo* uterine tissue recordings collected during pregnancy.[61]

Since oestrogens are potent regulators of metabolism,[53] it is not surprising that restricting food intake to the inactive phase impaired ovarian function in post-adolescent female rats[42] and female pre-ovulatory LH surges in mice.[62] Conversely, time-restricted feeding cycles partially restored the oestrous cycle of female rats kept in constant light.[40]

16.6.3.1 Sex Differences in Circadian Timing Systems

Evidence indicates that sex differences in circadian timing systems are endocrinologically and genetically mediated. They are seen at many levels of the neuroendocrine and endocrine systems, from cells, glands and organs, to whole behaviour, including rest-activity cycles, cognitive functions, responses to external stimuli and to medicines.[63–67] Steroid hormones modulate every circadian characteristic, including the period, amplitude and phase. As such, there are compelling reasons to study the role of steroids and sex differences in the circadian timing system.[67]

Sleep and sleep disorders tend to affect women more than men.[68,69] Aging-related conditions, including sleep disturbances and memory impairments, are also more common in women than in men. Harrington *et al.*[70] attribute these sex differences partially to fluctuations in levels of female sex hormones (*i.e.*, oestrogen and progesterone). Women indeed often experience sleep and memory disturbances across the menstrual cycle, during pregnancy and during the peri-menopausal period. These authors reviewed the inter-relationships among female sex hormones, sleep, and memory across the female lifespan, and proposed possible mediating and moderating mechanisms linking these variables. They also considered implications for the risk of developing Alzheimer's disease later in life.[70,71]

16.6.4 Role of Core Clock Genes

16.6.4.1 Animal Studies

Studies on genetic circadian clock mutants have shown that disabling circadian clocks disrupts the oestrous cycle and the timing of the LH surge, thereby compromising fertility.[72–74] Female mice deficient in the *Bmal1* gene do not exhibit the proestrous LH surge. However, spontaneous ovulation on the day of oestrus was unaffected in these animals. *Bmal1*$^{-/-}$ females were also deficient in the proestrous FSH surge, which, like the LH surge, is GnRH-dependent. *Bmal1*$^{-/-}$ females continued to cycle in constant darkness, albeit with increased cycle length and time spent in oestrus. Female pituitaries rhythmically expressed clock components throughout all cycle stages. Mice that specifically lack BMAL1 in gonadotropes exhibited a modest elevation in both proestrous and baseline LH levels across all oestrous stages, and showed increased variability in oestrous cycle length, but were otherwise reproductively normal. The intrinsic clock in gonadotropes is dispensable for LH surge regulation but contributes to oestrous cycle robustness.[74]

As stated earlier, mice lacking the core circadian clock gene *Bmal1* are infertile and show evidence of neuroendocrine disruption, including the absence of the preovulatory LH surge,[75] as well as enhanced responsiveness to exogenous kisspeptin. Conditional deletion of *Bmal1* in arginine vasopressin (*Bmal1*$^{fl/fl:Avpcre}$), vasoactive intestinal peptide (*Bmal1*$^{fl/fl:Vipcre}$), both AVP and VIP (*Bmal1*$^{fl/fl:Avpcre+Vipcre}$), and neuromedin-*s* (*Bmal1*$^{fl/fl:Nmscre}$) neurons was found to have substantial effects on home-cage circadian activity and temperature rhythms. Nonetheless, female mice from these lines demonstrated normal oestrus cycles, fecundity, kisspeptin responsiveness, and inducible LH surge. While conditional *Bmal1* knockout in AVP, VIP, or NMS neurons was sufficient to disrupt locomotor activity, this disruption was insufficient to recapitulate the neuroendocrine reproductive effects of the whole-body *Bmal1* knockout.[75] Another study also suggests that circadian function in multiple cell types may be necessary for appropriate timing of the LH surge, since the timing of the LH surge was altered to varying degrees by the conditional deletion of *Bmal1* in AVP^{Cre}, Kiss^{CreBAC}, and GnRH^{CreBAC} mice.[76]

Rhythms of molecular clock gene expression are detected in the uterus, oviduct, and ovary.[77] Both oestrogen and progesterone can influence the timing and amplitude of clock gene expression in the uterine endometrium and myometrium.[78] *Clock* and *Bmal1* affect reproductive function not only at the hypothalamic–pituitary level but also in the ovary.[79] Theca cell-specific disruption of *Bmal1* leads to changes in the circadian expression of the LH receptor (*Lhcgr*), reduced sensitivity of the ovaries to LH, and subfertility, suggesting that BMAL1 acts as a pacemaker of LH sensitivity.[80] Disruption of *Bmal1* in mice expressing *Sf1-Cre* inhibits the expression of Star, progesterone production in corpora lutea, and implantation, indicating that BMAL1 also affects steroidogenesis.[80] Thus, disruption of this circadian regulatory pathway in theca cells might contribute to PCOS.[81]

Novel roles have been identified for the homeodomain transcription factors SIX3 (sine oculis homeobox 3) and VAX1 (ventral anterior homeobox 1) in neurons, where they are required for appropriate molecular circadian clock function, SCN rhythmic output, and female fertility.[82] The deletion of the core clock gene *Bmal1* significantly decreased the oocyte fertilization rate, early embryo development, and implantation potential in female mice, possibly due to excess reactive oxygen species (ROS) levels generated in the ovary and oviduct.[83] The loss of BMAL1 in ovarian steroidogenic cells resulted in implantation failure in female mice, which was restored by progesterone supplementation.[84] Mice with a mutation in the core clock gene *Clock* had abnormal oestrous cycles and did not exhibit an LH surge in the afternoon of proestrus due to a defect at the hypothalamic level. Vasopressin can act as a circadian signal to regulate the proestrous release of LH, a signal that is deficient in the *Clock* mutant.[85]

As clock genes are expressed in the ovary, several studies showed a link between the clock and follicular development.[77] Clock genes are expressed in both cell types of the follicle (granulosa cells and theca cells), the oocyte, and interstitial or stromal cells.[78] In the ovary, molecular clock function has been linked to steroidogenesis, folliculogenesis, cellular differentiation, responsiveness to gonadotropins, and ovulation.[86] The clock in the ovarian theca cells contributes to the timing of ovulation, perhaps through its control of phasic sensitivity to gonadotropins. Furthermore, it has been suggested that, although linked to the development of the follicle, molecular clock function in the theca cells is not required for folliculogenesis.[77]

The exact mechanisms through which mutations in clock genes disrupt reproductive function, however, are still only poorly understood. The challenge is that tissues at all levels of the HPG axis contain circadian clocks. Circadian clocks in GnRH and kisspeptin neurons are required for appropriate timing of the LH surge since the latter is disrupted in mice lacking circadian clocks in GnRH and kisspeptin neurons.[76] In contrast, the pituitary circadian clock in gonadotropes is not required for the appropriate timing of the LH surge since gonadotrope-specific *Bmal1* KO mice have normal LH surges.[74] The circadian clock in ovarian thecal cells is critical for properly timed LH induction of ovulation. Indeed, while female mice with *Bmal1* KO in either ovarian thecal or granulosa cells have normal oestrous cycles and LH surges, the timing of ovulation is disrupted in thecal cell-*Bmal1* KO females, likely due to the loss of phasic responsiveness of the follicle to LH.[87]

Experimental evidence also exists for circadian control of neuroendocrine events indispensable for ovulation that occur prior to proestrus.[59] Regulatory processes other than the pre-ovulatory surge of gonadotropins also affect ovulation. Injection of pentobarbital during the afternoon of oestrus, metestrus and diestrus inhibits the pre-ovulatory surge of LH and ovulation, thus indicating the importance of "critical windows" at cycle stages other than proestrus. Additional evidence is derived from studies blocking different neurotransmitter systems at different times of each stage of the oestrous cycle, as reviewed by Silva and Dominguez.[59]

Novel small molecules that alter the timing and activity of select clock gene transcription factors, including CRY, REV-ERBα, and retinoic acid receptor-related orphan receptor-α have been developed.[77] Melatonin can also modulate the timing of the clock and novel agonists with greater potency and selectivity for the melatonin 2 receptor have been developed. These compounds can modulate downstream activity of the clock in relation to immune function, metabolism, and reproductive function.[77]

VIP-deficient females exhibit a disrupted oestrous cycle: ovulation occurs less frequently and results in the release of fewer oocytes compared with controls.[88] VIP-deficient SCN tissue exhibits lower amplitude oscillations with altered phase relationships between the SCN and peripheral oscillators as measured by PER2-driven bioluminescence. Loss of VIP thus resulted in a weakened SCN oscillator, which reduced the synchronization of the female circadian system.[88]

Per1/*Per2* disruption reduced testosterone synthesis and sperm motility by affecting the PKA-StAR pathway, leading to decreased fertility.[89] Mice homozygous for a dominant-negative allele of the *Clock* gene ($Clock^{\Delta 19/\Delta 19}$) exhibited slightly but significantly decreased male fertility.[90] Specifically knocking down the *Clock* gene expression in the testes of male mice led to smaller litter size, a lower *in vitro* fertility rate, lower blastula formation rate, and lower acrosin activity of the knockdown sperm. These results indicate that *Clock* gene expression in round spermatids is essential for maintaining male reproductivity, and that acrosin may be a novel regulatory target of the *Clock* gene that could regulate the fertilization and early embryonic development to the blastula stage.[90]

16.6.4.2 Circadian Clock Regulation in Oocytes

As reviewed in other chapters, circadian rhythms are not detected in early embryonic stages and only develop gradually during ontogeny.[91] Similarly, circadian genes do not seem to be involved in circadian clock regulation in mouse oocytes and preimplantation embryos. In addition to the central role of the SCN, most tissues, cells, and cultured fibroblast cell lines contain an intrinsic circadian oscillator that controls cellular physiology in a temporal manner. The lack of 24-hour oscillations in mouse oocytes and preimplantation embryos, despite the presence of transcripts of these genes was an unanticipated discovery.[92] Using a bioluminescence imaging system to monitor clock gene expression, Yagita *et al.*[93] showed that the circadian bioluminescence rhythm is not detected in mouse embryonic stem (ES) cells, suggesting that the ES cells likely lack the transcriptional–translational feedback loop regulation for clock gene expression. Studies by Umemura *et al.*[91,94,95] indicated that the CLOCK protein is required for molecular oscillations to be discernible in differentiated cells; the post-transcriptional regulation of *Clock* plays a role in setting the timing for the emergence of the circadian clock oscillation during mammalian development.

Umemura *et al.*[91] further showed that human induced pluripotent stem cells (iPSCs) do not exhibit discernible circadian molecular oscillations; in addition, *in vitro* differentiation culture of iPSCs required a longer duration for the emergence of circadian oscillations than that required in mice. These results suggest that CLOCK protein expression could be suppressed post-transcriptionally in early developmental stages not only in mice but also in humans.[91] Yagita's team[96] recently reported similar findings of post-transcriptional suppression of CLOCK protein in some human epigenetic cancers, such as Wilms tumours and malignant rhabdoid tumours, suggesting that the post-transcriptional suppression of CLOCK protein could offer a novel perspective for understanding the pathophysiological nature of cancers by exploring the correlation between cellular differentiation and the circadian clock, which warrants further investigation.

16.6.4.3 Human Studies

PCOS affects 5% to 10% of women of childbearing age and is characterised by features including a polycystic ovary, anovulation, and elevated serum androgen. Approximately 40% of these women are affected by metabolic syndrome. Developmental exposure to excess androgen is thought to contribute to the aetiology of fertility disorders, including PCOS.[97] Analysis of plasma exosome miRNAs sequenced from PCOS patients and healthy controls identified differentially expressed miRNAs, which target functions including axon guidance, mitogen-activated protein kinase signalling, endocytosis, circadian rhythms, and cancer pathways. The expression of these miRNAs correlated with the menstrual cycle, antral follicle count, and hormone level.[98]

Recurrent miscarriage is characterized by two or more consecutive losses of a clinically established intrauterine pregnancy at early gestation. Impaired decidualization is believed to predispose women to subsequent pregnancy failure.[99] Lv *et al.*[99] found that the expression of *BMAL1* was reduced in decidual tissues from recurrent miscarriage cases, compared to normal pregnant women. Genetic variability in the *CLOCK* gene has also been associated with idiopathic recurrent spontaneous abortion in a case-control study involving 268 women with a history of three or more idiopathic recurrent spontaneous abortions and 284 women with at least two live births and no history of pathologic pregnancies.[100] Decidual macrophages play crucial roles in maintaining immune balance at the maternal–foetal interface; abnormal macrophage polarization has been related to adverse pregnancy outcomes, such as infertility, recurrent spontaneous abortion, and preterm labor.[101] Cui *et al.*[101] reported a type-1 macrophage-like dominance in decidual macrophages from human miscarriages, with a decreased expression of *Rev-erbα* compared to that of normal pregnancies. Zhang *et al.*[102] showed that the expression of the clock gene *PER1* increased immediately and remained elevated during human endometrial decidualization. In their exploration of the functional role and the mechanisms of *CRY2* in regulating trophoblast migration and invasion, Wu *et al.*[103] showed that *CRY2* was highly expressed in human villous specimens from cases of recurrent spontaneous abortion.

Low concentrations of circulating melatonin have been associated with low implantation rates and difficulties in maintaining pregnancy.[104] Since melatonin is involved in the maintenance of uterine homeostasis through regulation of numerous pathways associated with uterine receptivity and gestation, Chuffa *et al.*[104] suggested that melatonin replacement may offer an opportunity to restore normal physiological function of the affected tissues.

16.6.5 Concluding Remarks

These results point to the intricate involvement of the circadian system in the multiple physiological processes participating in reproduction. Synchronizing the circadian system through light exposure and meal timing, in addition to maintaining a regular schedule, helps prevent circadian disruption and associated negative effects from occurring. As a circadian-regulating hormone, melatonin supplementation may deserve further consideration as a potential intervention, as already suggested in relation to implantation issues.[104] In the laboratory, melatonin receptors have also been shown to play a role in relieving hyper-androgenism, which could be of potential interest in the context of PCOS.[41]

Treatment optimization by timing (chronotherapy) also deserves consideration. The effectiveness of a 5-day course of oral letrozole, a non-steroidal aromatase inhibitor, was compared between two groups of PCOS patients with fertility requirements who received treatment for ovulation induction, starting either from the third or fifth day of the menstrual cycle. As compared to the former, the latter regimen was associated with higher ovulation and conception rates and shorter time-to-pregnancy.[105] The treatment of infertile women with PCOS using clomiphene citrate similarly showed higher rates of pregnancy when initiated early (days 1–5) rather than later (days 5–9) in the menstrual cycle.[106] Exogenous gonadotrophins, given to either the donor or the recipient, reportedly could increase the number of developing follicles, the effect's magnitude being dependent on the timing of the injections relative to the time of grafting.[107] The optimal timing of administration and daily dose of recombinant luteinizing hormone (rLH) supplementation were investigated in infertile patients considered to be poor responders to *in vitro* fertilization and undergoing GnRH-ant treatment. Ovarian responses were reportedly affected by the timing of administration more than the total-dose of rLH. The optimal window for administering rLH appeared to be the mid-to-late follicular phase, despite the fact that rLH-supplementation in the early-follicular phase seemed to increase endometrial thickness and enhance its morphology.[108] Further studies are warranted to optimize treatment timing, as well as studies aimed at directly targeting underlying molecular mechanisms. Small molecules have already been identified that can alter the timing and activity of select clock gene transcription factors.[77]

Acknowledgements

This work was supported by JSPS KAKENHI (grant number 17K07592 and 21K06363; to N. N. T. and 19K10046; to W. N.), and the Naito Science and Engineering Foundation (to N. N. T.), Takeda Science Foundation (to W. N.), and Terumo Foundation for Life Sciences and Arts (to W. N.).

References

1. H. Naito and Y. Suzuki, The Rat Estrous Cycle Associated with Functional Corpora Lutea and the Mechanism of Its Recurrence, *Endocrinol. Jpn.*, 1978, **25**, 367–376.
2. J. Cohen, Reproduction in mammals, *Reproduction*, 1977, ch. 16, pp. 263–291.
3. M. L. Moline, L. Broch, R. Zak and V. Gross, Sleep in women across the life cycle from adulthood through menopause, *Sleep Med. Rev.*, 2003, **7**, 155–177.
4. M. R. Ralph, R. G. Foster, F. C. Davis and M. Menaker, Transplanted suprachiasmatic nucleus determines circadian period, *Science*, 1990, **247**, 975–978.
5. J. S. Takahashi, Transcriptional architecture of the mammalian circadian clock, *Nat. Rev. Genet.*, 2017, **18**, 164–179.
6. M. Doi, Y. Takahashi, R. Komatsu, F. Yamazaki, H. Yamada, S. Haraguchi, N. Emoto, Y. Okuno, G. Tsujimoto, A. Kanematsu and O. Ogawa, Salt-sensitive hypertension in circadian clock-deficient Cry-null mice involves dysregulated adrenal Hsd3b6, *Nat. Med.*, 2010, **16**, 67–74.
7. B. Marcheva, K. M. Ramsey, E. D. Buhr, Y. Kobayashi, H. Su, C. H. Ko, G. Ivanova, C. Omura, S. Mo, M. H. Vitaterna and J. P. Lopez, Disruption of the clock components CLOCK and BMAL1 leads to hypoinsulinaemia and diabetes, *Nature*, 2010, **466**, 627–631.
8. R. L. Butcher, W. E. Collins and N. W. Fugo, Plasma concentration of LH, FSH, prolactin, progesterone and estradiol-17beta throughout the 4 days estrous cycle of the rat, *Endocrinology*, 1974, **94**, 1704–1708.
9. F. Wollnik and F. W. Turek, Estrous correlated modulations of circadian and ultradian wheel-running activity rhythms in LEW/Ztm rats, *Physiol. Behav.*, 1988, **43**, 389–396.
10. C. Kopp, V. Ressel, E. Wigger and I. Tobler, Influence of estrus cycle and ageing on activity patterns in two inbred mouse strains, *Behav. Brain Res.*, 2006, **167**, 165–174.
11. H. Suzuki, M. Uchiyama, K. Shibui, K. Kim, H. Tagaya and K. Shinohara, Long-term rectal temperature measurements in a patient with menstrual-associated sleep disorder, *Phychiatry Clin. Neurosci.*, 2002, **56**, 475–478.
12. K. Shinohara, M. Uchiyama, M. Okawa, K. Saito, M. Kawaguchi, T. Funabashi and F. Kimura, Menstrual changes in sleep, rectal temperature and melatonin rhythms in a subject with premenstrual syndrome, *Neurosci. Lett.*, 2000, **281**, 159–162.
13. L. P. Morin, Effect of ovarian hormones on synchrony of hamster circadian rhythms, *Physiol. Behav.*, 1980, **24**, 741–749.
14. T. J. Nakamura, M. T. Sellix, T. Kudo, N. Nakao, T. Yoshimura, S. Ebihara, C. S. Colwell and G. D. Block, Influence of the estrous cycle on clock gene expression in reproductive tissues: Effects of fluctuating ovarian steroid hormone levels, *Steroids*, 2010, **75**, 203–212.

15. S. Mizuta, M. Sugiyama, I. T. Tokuda, W. Nakamura and T. J. Nakamura, Photic phase-response curves for cycling female mice, *Horm. Behav.*, 2018, **105**, 41–46.
16. J. S. Perrin, L. A. Segall, V. L. Harbour, B. Woodside and S. Amir, The expression of the clock protein PER2 in the limbic forebrain is modulated by the estrous cycle, *Proc. Natl. Acad. Sci. U. S. A.*, 2006, **103**, 5591–5596.
17. S. H. Yoo, S. Yamazaki, P. L. Lowrey, K. Shimomura, C. H. Ko, E. D. Buhr, S. M. Siepka, H. K. Hong, W. J. Oh, O. J. Yoo and M. Menaker, PERIOD2::LUCIFERASE real-time reporting of circadian dynamics reveals persistent circadian oscillations in mouse peripheral tissues, *Proc. Natl. Acad. Sci. U. S. A.*, 2004, **101**, 5339–5346.
18. T. J. Nakamura, M. T. Sellix, M. Menaker and G. D. Block, Estrogen directly modulates circadian rhythms of PER2 expression in the uterus, *Am. J. Physiol.: Endocrinol. Metab.*, 2008, **295**, E1025–E1031.
19. G. W. Harris, Neural control of the pituitary gland, *Physiol. Rev.*, 1948, **28**, 139–179.
20. J. W. Everett and C. H. Sawyer, A 24-hour periodicity in the "LH-release apparatus" of female rats, disclosed by barbiturate sedation, *Endocrinology*, 1950, **47**, 198–218.
21. M. S. Carmichael, R. J. Nelson and I. Zucker, Hamster activity and estrous cycles: control by a single *versus* multiple circadian oscillator(s), *Proc. Natl. Acad. Sci. U. S. A.*, 1981, **78**, 7830–7834.
22. K. Brown-Grant and G. Raisman, Abnormalities in reproductive function associated with the destruction of the suprachiasmatic nuclei in female rats, *Proc. R. Soc. London, Ser. B*, 1977, **198**, 279–296.
23. E. L. Meyer-Bernstein, A. E. Jetton, S. I. Matsumoto, J. F. Markuns, M. N. Lehman and E. L. Bittman, Effects of suprachiasmatic transplants on circadian rhythms of neuroendocrine function in golden hamsters, *Endocrinology*, 1999, **140**, 207–218.
24. N. N. Takasu, T. J. Nakamura, I. T. Tokuda, T. Todo, G. D. Block and W. Nakamura, Recovery from age-related infertility under environmental light-dark cycles adjusted to the intrinsic circadian period, *Cell Rep.*, 2015, **12**, 1407–1413.
25. T. Roenneberg, K. V. Allebrandt, M. Merrow and C. Vetter, Social jetlag and obesity, *Curr. Biol.*, 2012, **22**, 939–943.
26. W. P. Williams 3rd and L. J. Kriegsfeld, Circadian control of neuroendocrine circuits regulating female reproductive function, *Front. Endocrinol.*, 2012, **3**, 60.
27. H. O. de la Iglesia and W. J. Schwartz, Mini review: timely ovulation: circadian regulation of the female hypothalamo-pituitary-gonadal axis, *Endocrinology*, 2006, **147**, 1148–1153.
28. S. J. Wiegand, E. Terasawa and W. E. Bridson, Persistent estrus and blockade of progesterone-induced LH release follows lesions which do not damage the suprachiasmatic nucleus, *Endocrinol*, 1978, **102**, 1645–1648.
29. T. J. Nakamura, W. Nakamura, I. T. Tokuda, T. Ishikawa, T. Kudo, C. S. Colwell and G. D. Block, Age-related changes in the circadian system

unmasked by constant conditions, *eNeuro*, 2015, **2**, ENEURO.0064-15. 2015.

30. T. J. Nakamura, W. Nakamura, S. Yamazaki, T. Kudo, T. Cutler, C. S. Colwell and G. D. Block, Age-related decline in circadian output, *J. Neurosci.*, 2011, **31**, 10201–10205.
31. T. J. Nakamura, N. N. Takasu and W. Nakamura, The suprachiasmatic nucleus: age-related decline in biological rhythms, *J. Physiol. Sci.*, 2016, **66**, 367–374.
32. T. Ohara, T. J. Nakamura, W. Nakamura and I. T. Tokuda, Modeling circadian regulation of ovulation timing: age-related disruption of estrous cyclicity, *Sci. Rep.*, 2020, **10**, 16767.
33. J. D. Kloss, M. L. Perlis, J. A. Zamzow, E. J. Culnan and C. R. Gracia, Sleep, sleep disturbance, and fertility in women, *Sleep Med. Rev.*, 2015, **22**, 78–87.
34. A. Dzaja, S. Arber, J. Hislop, M. Kerkhofs, C. Kopp, T. Pollmacher, P. Polo-Kantola, D. J. Skene, P. Stenuit, I. Tobler and T. Porkka-Heiskanen, Women's sleep in health and disease, *J. Psychiatr. Res.*, 2005, **39**(1), 55–76.
35. E. Toffol, I. Merikanto, T. Lahti, R. Luoto, O. Heikinheimo and T. Partonen, Evidence for a relationship between chronotype and reproductive function in women, *Chronobiol. Int.*, 2013, **30**(6), 756–765.
36. C. Ozcelik, B. Varli, A. Gokce, T. Takmaz, C. Cetin and P. Ozcan, Evaluation of chronotype and sleep quality in infertile population and comparison with fertile population: a cross-sectional study, *J. Psychosom. Obstet. Gynecol.*, 2023, **44**(1), 2148523.
37. Y. Komada, Y. Ikeda, M. Sato, A. Kami, C. Masuda and S. Shibata, Social jetlag and menstrual symptoms among female university students, *Chronobiol. Int.*, 2019, **36**(2), 258–264.
38. J. Park, M. H. Kim, D. H. Han, J. Y. Noh, E. S. Ji, S. H. Lee, C. J. Kim and S. Cho, Mealtime shift delays conception in mice, *Reproduction*, 2022, **163**(5), 323–331.
39. T. Bahougne, M. Kretz, E. Angelopoulou, N. Jeandidier and V. Simonneaux, Impact of circadian disruption on female mice reproductive function, *Endocrinology*, 2020, **161**(4), bqaa028.
40. N. N. Guerrero-Vargas, E. Espitia-Bautista, R. Escalona, H. Lugo-Martinez, M. Gutierrez-Perez, R. Navarro-Espindola, M. F. Setien, S. Boy-Waxman, E. A. Retana-Flores, B. Ortega, R. M. Buijs and C. Escobar, Timed restricted feeding cycles drive daily rhythms in female rats maintained in constant light but only partially restore the estrous cycle, *Front. Nutr.*, 2022, **9**, 999156.
41. W. Chu, S. Li, X. Geng, D. Wang, J. Zhai, G. Lu, W. Y. Chan, Z. J. Chen and Y. Du, Long-term environmental exposure of darkness induces hyperandrogenism in PCOS *via* melatonin receptor 1A and aromatase reduction, *Front. Cell Dev. Biol.*, 2022, **10**, 954186.
42. T. Fujiwara, R. Nakata, M. Ono, M. Mieda, H. Ando, T. Daikoku and H. Fujiwara, Time restriction of food intake during the circadian cycle

is a possible regulator of reproductive function in postadolescent female rats, *Curr. Dev. Nutr.*, 2019, **3**(4), nzy093.
43. M. Das, S. R. Mohanty, T. Minocha, N. K. Mishra, S. K. Yadav and C. Haldar, Circadian desynchronization in pregnancy of Golden hamster following long time light exposure: Involvement of Akt/FoxO1 pathway, *J. Photochem. Photobiol., B*, 2022, **234**, 112508.
44. M. J. Boden, T. J. Varcoe and D. J. Kennaway, Circadian regulation of reproduction: from gamete to offspring, *Prog. Biophys. Mol. Biol.*, 2013, **113**(3), 387–397.
45. A. Kalsbeek, F. A. Scheer, S. Perreau-Lenz, S. E. La Fleur, C. X. Yi, E. Fliers and R. M. Buijs, Circadian disruption and SCN control of energy metabolism, *FEBS Lett.*, 2011, **585**, 1412–1426.
46. R. Silver, J. LeSauter, P. A. Tresco and M. N. Lehman, A diffusible coupling signal from the transplanted suprachiasmatic nucleus controlling circadian locomotor rhythms, *Nature*, 1996, **382**, 810–813.
47. L. J. Kriegsfeld, E. M. Gibson, W. P. Williams 3rd, S. Zhao, A. O. Mason, G. E. Bentley and K. Tsutsui, The roles of RFamide-related peptide-3 in mammalian reproductive function and behaviour, *J. Neuroendocrinol.*, 2010, **22**, 692–700.
48. B. Vida, L. Deli, E. Hrabovszky, T. Kalamatianos, A. Caraty, C. W. Coen, Z. Liposits and I. Kallo, Evidence for suprachiasmatic vasopressin neurones innervating kisspeptin neurones in the rostral periventricular area of the mouse brain: regulation by oestrogen, *J. Neuroendocrinol.*, 2010, **22**, 1032–1039.
49. N. D. S. Mansano, R. S. Paradela, T. M. Bohlen, I. M. Zanardi, F. M. Chaves, M. A. Silveira, M. R. Tavares, J. Donato Jr and R. Frazao, Vasoactive intestinal peptide exerts an excitatory effect on hypothalamic kisspeptin neurons during estrogen negative feedback, *Mol. Cell. Endocrinol.*, 2022, **542**, 111532.
50. M. T. Sellix, T. Yoshikawa and M. Menaker, A circadian egg timer gates ovulation, *Curr. Biol.*, 2010, **20**, R266–R267.
51. C. A. Rubel, R. B. Lanz, R. Kommagani, H. L. Franco, J. P. Lydon and F. J. DeMayo, Research resource: genome-wide profiling of progesterone receptor binding in the mouse uterus, *Mol. Endocrinol.*, 2012, **26**, 1428–1442.
52. V. M. Alvord, E. J. Kantra and J. S. Pendergast, Estrogens and the circadian system, *Semin. Cell Dev. Biol.*, 2022, **126**, 56–65.
53. F. Mauvais-Jarvis, D. J. Clegg and A. L. Hevener, The role of estrogens in control of energy balance and glucose homeostasis, *Endocr. Rev.*, 2013, **34**(3), 309–338.
54. S. C. Hewitt and K. S. Korach, Estrogen receptors: new directions in the new millennium, *Endocr. Rev.*, 2018, **39**(5), 664–675.
55. A. S. Kauffman, Neuroendocrine mechanisms underlying estrogen positive feedback and the LH surge, *Front. Neurosci.*, 2022, **16**, 953252.
56. G. E. Hoffman, M. S. Smith and J. G. Verbalis, c-Fos and related immediate early gene products as markers of activity in neuroendocrine systems, *Front. Neuroendocrinol.*, 1993, **14**(3), 173–213.

57. J. F. Couse, M. M. Yates, V. R. Walker and K. S. Korach, Characterization of the hypothalamic-pituitary-gonadal axis in estrogen receptor (ER) Null mice reveals hypergonadism and endocrine sex reversal in females lacking ERalpha but not ERbeta, *Mol. Endocrinol.*, 2003, **17**(6), 1039–1053.
58. M. Wintermantel, R. E. Campbell, R. Porteous, D. Bock, H. J. Grone, M. G. Todman, K. S. Korach, E. Greiner, C. A. Perez, G. Schutz and A. E. Herbison, Definition of estrogen receptor pathway critical for estrogen positive feedback to gonadotropin-releasing hormone neurons and fertility, *Neuron*, 2006, **52**(2), 271–280.
59. C. C. Silva and R. Dominguez, Clock control of mammalian reproductive cycles: Looking beyond the pre-ovulatory surge of gonadotropins, *Rev. Endocr. Metab. Disord.*, 2020, **21**(1), 149–163.
60. J. Xu, Y. Xu, B. Miao, M. Deng, Y. Wang, P. Xiang and C. Zhou, Influence of menstrual cycle on the expression of clock genes in peripheral blood mononuclear cells in Macaca fascicularis, *Eur. J. Obstet. Gynecol. Reprod. Biol.*, 2015, **186**, 54–58.
61. A. M. Yaw, T. V. Duong, D. Nguyen and H. M. Hoffmann, Circadian rhythms in the mouse reproductive axis during the estrous cycle and pregnancy, *J. Neurosci. Res.*, 2021, **99**(1), 294–308.
62. A. Kukino, T. J. Walbeek, L. J. Sun, A. T. Watt, J. H. Park, A. S. Kauffman and M. P. Butler, Mistimed restricted feeding disrupts circadian rhythms of male mating behavior and female preovulatory LH surges in mice, *Horm. Behav.*, 2022, **145**, 105242.
63. R. A. Wever, Sex differences in human circadian rhythms: intrinsic periods and sleep fractions, *Experientia*, 1984, **40**(11), 1226–1234.
64. E. Iwahana, I. Karatsoreos, S. Shibata and R. Silver, Gonadectomy reveals sex differences in circadian rhythms and suprachiasmatic nucleus androgen receptors in mice, *Horm. Behav.*, 2008, **53**(3), 422–430.
65. J. A. Mong, F. C. Baker, M. M. Mahoney, K. N. Paul, M. D. Schwartz, K. Semba and R. Silver, Sleep, rhythms, and the endocrine brain: influence of sex and gonadal hormones, *J. Neurosci.*, 2011, **31**(45), 16107–16116.
66. M. Bailey and R. Silver, Sex differences in circadian timing systems: implications for disease, *Front. Neuroendocrinol.*, 2014, **35**(1), 111–139.
67. L. Yan and R. Silver, Neuroendocrine underpinnings of sex differences in circadian timing systems, *J. Steroid Biochem. Mol. Biol.*, 2016, **160**, 118–126.
68. J. A. Mong and D. M. Cusmano, Sex differences in sleep: impact of biological sex and sex steroids, *Philos. Trans. R. Soc., B*, 2016, **371**, 20150110.
69. S. Suh, N. Cho and J. Zhang, Sex differences in insomnia: from epidemiology and etiology to intervention, *Curr. Psychiatry Rep.*, 2018, **20**(9), 69.
70. Y. A. Harrington, J. M. Parisi, D. Duan, D. M. Rojo-Wissar, C. Holingue and A. P. Spira, Sex hormones, sleep, and memory: interrelationships across the adult female lifespan, *Front. Aging Neurosci.*, 2022, **14**, 800278.

71. T. H. Hicks, T. N. C. Magalhães, H. K. Ballard, T. B. Jackson, S. J. Cox and J. A. Bernard, Network segregation in aging females and evaluation of the impact of sex steroid hormones, *Front. Hum. Neurosci.*, 2023, **17**, 1059091.
72. B. H. Miller, S. L. Olson, F. W. Turek, J. E. Levine, T. H. Horton and J. S. Takahashi, Circadian clock mutation disrupts estrous cyclicity and maintenance of pregnancy, *Curr. Biol.*, 2004, **14**(15), 1367–1373.
73. C. K. Ratajczak, K. L. Boehle and L. J. Muglia, Impaired steroidogenesis and implantation failure in Bmal1$^{-/-}$ mice, *Endocrinology*, 2009, **150**(4), 1879–1885.
74. A. Chu, L. Zhu, I. D. Blum, O. Mai, A. Leliavski, J. Fahrenkrug, H. Oster, U. Boehm and K. F. Storch, Global but not gonadotrope-specific disruption of Bmal1 abolishes the luteinizing hormone surge without affecting ovulation, *Endocrinology*, 2013, **154**(8), 2924–2935.
75. K. J. Tonsfeldt, E. L. Schoeller, L. E. Brusman, L. J. Cui, J. Lee and P. L. Mellon, The contribution of the circadian gene Bmal1 to female fertility and the generation of the preovulatory luteinizing hormone surge, *J. Endocr. Soc.*, 2019, **3**(4), 716–733.
76. E. L. Bittman, Circadian function in multiple cell types is necessary for proper timing of the preovulatory LH surge, *J. Biol. Rhythms*, 2019, **34**(6), 622–633.
77. A. Sen and M. T. Sellix, The circadian timing system and environmental circadian disruption: from follicles to fertility, *Endocrinology*, 2016, **157**(9), 3366–3373.
78. K. Isayama, H. Chen, N. Yamauchi and MA. Hattori, REV-ERB inhibits the PTGS2 expression in bovine uterus endometrium stromal and epithelial cells exposed to ovarian steroids, *J. Reprod. Dev.*, 2014, **60**(5), 362–370.
79. R. Li, S. Cheng and Z. Wang, Circadian clock gene plays a key role on ovarian cycle and spontaneous abortion, *Cell. Physiol. Biochem.*, 2015, **37**, 911–920.
80. M. M. Mahoney, Shift work, jet lag, and female reproduction, *Int. J. Endocrinol.*, 2010, **2010**, 813764.
81. M. Menaker, Z. C. Murphy and M. T. Sellix, Central control of peripheral circadian oscillators, *Curr. Opin. Neurobiol.*, 2013, **23**, 741–746.
82. H. M. Hoffmann, J. D. Meadows, J. A. Breuer, A. M. Yaw, D. Nguyen, K. J. Tonsfeldt, A. Y. Chin, B. M. Devries, C. Trang, H. J. Oosterhouse, J. S. Lee, J. W. Doser, M. R. Gorman, D. K. Welsh and P. L. Mellon, The transcription factors SIX3 and VAX1 are required for suprachiasmatic nucleus circadian output and fertility in female mice, *J. Neurosci. Res.*, 2021, **99**(10), 2625–2645.
83. J. Xu, Y. Li, Y. Wang, Y. Xu and C. Zhou, Loss of Bmal1 decreases oocyte fertilization, early embryo development and implantation potential in female mice, *Zygote*, 2016, **24**(5), 760–767.
84. Y. Liu, B. P. Johnson, A. L. Shen, J. A. Wallisser, K. J. Krentz, S. M. Moran, R. Sullivan, E. Glover, A. F. Parlow, N. R. Drinkwater, L. A. Schuler and C. A. Bradfield, Loss of BMAL1 in ovarian steroidogenic

cells results in implantation failure in female mice, *Proc. Natl. Acad. Sci. U. S. A.*, 2014, **111**(39), 14295–14300.
85. B. H. Miller, S. L. Olson, J. E. Levine, F. W. Turek, T. H. Horton and J. S. Takahashi, Vasopressin regulation of the proestrous luteinizing hormone surge in wild-type and Clock mutant mice, *Biol. Reprod.*, 2006, **75**(5), 778–784.
86. M. T. Sellix, Circadian clock function in the mammalian ovary, *J. Biol. Rhythms*, 2015, **30**, 7–19.
87. A. L. Mereness, Z. C. Murphy, A. C. Forrestel, S. Butler, C. Ko, J. S. Richards and M. T. Sellix, Conditional deletion of bmal1 in ovarian theca cells disrupts ovulation in female mice, *Endocrinology*, 2016, **157**(2), 913–927.
88. D. H. Loh, D. A. Kuljis, L. Azuma, Y. Wu, D. Truong, H. B. Wang and C. S. Colwell, Disrupted reproduction, estrous cycle, and circadian rhythms in female mice deficient in vasoactive intestinal peptide, *J. Biol. Rhythms*, 2014, **29**(5), 355–369.
89. Q. Liu, H. Wang, H. Wang, N. Li, R. He and Z. Liu, Per1/Per2 disruption reduces testosterone synthesis and impairs fertility in elderly male mice, *Int. J. Mol. Sci.*, 2022, **23**(13), 7399.
90. X. Liang, S. Cheng, X. Jiang, X. He, Y. Wang, Z. Jiang, W. Hou, S. Li, Y. Liu and Z. Wang, The noncircadian function of the circadian Clock gene in the regulation of male fertility, *J. Biol. Rhythms*, 2013, **28**(3), 208–217.
91. Y. Umemura, I. Maki, Y. Tsuchiya, N. Koike and K. Yagita, Human circadian molecular oscillation development using induced pluripotent stem cells, *J. Biol. Rhythms*, 2019, **34**(5), 525–532.
92. T. Amano, A. Matsushita, Y. Hatanaka, T. Watanabe, K. Oishi, N. Ishida, M. Anzai, T. Mitani, H. Kato, S. Kishigami, K. Saeki, Y. Hosoi, A. Iritani and K. Matsumoto, Expression and functional analyses of circadian genes in mouse oocytes and preimplantation embryos: Cry1 is involved in the meiotic process independently of circadian clock regulation, *Biol. Reprod.*, 2009, **80**(3), 473–483.
93. K. Yagita, K. Horie, S. Koinuma, W. Nakamura, I. Yamanaka, A. Urasaki, Y. Shigeyoshi, K. Kawakami, S. Shimada, J. Takeda and Y. Uchiyama, Development of the circadian oscillator during differentiation of mouse embryonic stem cells *in vitro*, *Proc. Natl. Acad. Sci. U. S. A.*, 2010, **107**(8), 3846–3851.
94. Y. Umemura, N. Koike, M. Ohashi, Y. Tsuchiya, Q. J. Meng, Y. Minami, M. Hara, M. Hisatomi and K. Yagita, Involvement of posttranscriptional regulation of Clock in the emergence of circadian clock oscillation during mouse development, *Proc. Natl. Acad. Sci. U. S. A.*, 2017, **114**(36), E7479–E7488.
95. Y. Umemura, N. Koike, Y. Tsuchiya, H. Watanabe, G. Kondoh, R. Kageyama and K. Yagita, Circadian key component CLOCK/BMAL1 interferes with segmentation clock in mouse embryonic organoids, *Proc. Natl. Acad. Sci. U. S. A.*, 2022, **119**(1), e2114083119.
96. M. Ohashi, Y. Umemura, N. Koike, Y. Tsuchiya, Y. Inada, H. Watanabe, T. Tanaka, Y. Minami, O. Ukimura, T. Miki, T. Tajiri, G. Kondoh,

Y. Yamada and K. Yagita, Disruption of circadian clockwork in *in vivo* reprogramming-induced mouse kidney tumors, *Genes Cells*, 2018, **23**(2), 60–69.

97. A. L. Mereness, Z. C. Murphy and M. T. Sellix, Developmental programming by androgen affects the circadian timing system in female mice, *Biol. Reprod.*, 2015, **92**(4), 88.
98. X. Jiang, J. Li, B. Zhang, J. Hu, J. Ma, L. Cui and Z. J. Chen, Differential expression profile of plasma exosomal microRNAs in women with polycystic ovary syndrome, *Fertil. Steril.*, 2021, **115**(3), 782–792.
99. S. Lv, N. Wang, J. Ma, W. P. Li, Z. J. Chen and C. Zhang, Impaired decidualization caused by downregulation of circadian clock gene BMAL1 contributes to human recurrent miscarriage, *Biol. Reprod.*, 2019, **101**(1), 138–147.
100. A. Hodzic, P. Lavtar, M. Ristanovic, I. Novakovic, J. Dotlic and B. Peterlin, Genetic variation in the CLOCK gene is associated with idiopathic recurrent spontaneous abortion, *PLoS One*, 2018, **13**(5), e0196345.
101. L. Cui, X. Jin, F. Xu, S. Wang, L. Liu, X. Li, H. Lin and M. Du, Circadian rhythm-associated Rev-erbalpha modulates polarization of decidual macrophage *via* the PI3K/Akt signaling pathway, *Am. J. Reprod. Immunol.*, 2021, **86**(3), e13436.
102. Y. Zhang, N. Meng, H. Bao, Y. Jiang, N. Yang, K. Wu, J. Wu, H. Wang, S. Kong and Y. Zhang, Circadian gene PER1 senses progesterone signal during human endometrial decidualization, *J. Endocrinol.*, 2019, JOE-19-0284.R2.
103. L. Wu, B. Cheng, Q. Liu, P. Jiang and J. Yang, CRY2 suppresses trophoblast migration and invasion in recurrent spontaneous abortion, *J. Biochem.*, 2020, **167**(1), 79–87.
104. L. G. A. Chuffa, L. A. Lupi, M. S. Cucielo, H. S. Silveira, R. J. Reiter and F. R. F. Seiva, Melatonin promotes uterine and placental health: potential molecular mechanisms, *Int. J. Mol. Sci.*, 2019, **21**(1), 300.
105. L. Shi, S. Ye, M. Gao, Y. Chen, X. Jin and Z. Zhang, Effect of different timing of letrozole initiation on pregnancy outcome in polycystic ovary syndrome, *Front. Endocrinol.*, 2022, **13**, 1059609.
106. S. Dehbashi, H. Vafaei, M. D. Parsanezhad and S. Alborzi, Time of initiation of clomiphene citrate and pregnancy rate in polycystic ovarian syndrome, *Int. J. Gynaecol. Obstetr.*, 2006, **93**(1), 44–48.
107. B. Imthurn, S. L. Cox, G. Jenkin, A. O. Trounson and J. M. Shaw, Gonadotrophin administration can benefit ovarian tissue grafted to the body wall: implications for human ovarian grafting, *Mol. Cell. Endocrinol.*, 2000, **163**(1-2), 141–146.
108. S. Gizzo, A. Andrisani, M. Noventa, S. Manfe, A. Oliva, M. Gangemi, G. B. Nardelli and G. Ambrosini, Recombinant LH supplementation during IVF cycles with a GnRH-antagonist in estimated poor responders: A cross-matched pilot investigation of the optimal daily dose and timing, *Mol. Med. Rep.*, 2015, **12**(3), 4219–4229.

Part III: Chronopharmaceutics and Chronotheranostics Relevant to Biological Processes

CHAPTER 17

Growth: Emergence of the Circadian Clock

YOSHIKI TSUCHIYA AND KAZUHIRO YAGITA*

Department of Physiology and Systems Bioscience, Kyoto Prefectural University of Medicine, Kawaramachi-Hirokoji, Kamigyo-ku, Kyoto, 602-8566, Japan
*Email: kyagita@koto.kpu-m.ac.jp

17.1 Introduction

It is widely known that a broad range of our behavioural and physiological activities are orchestrated by the internal circadian clock. Considering the potential influence of the circadian clock on human physiology, it is important to understand how the circadian clock is established during early development. Since the 1950s, observational studies have revealed that sleep–wake behaviour shows fragmented patterns in the first several weeks of the postnatal period and gradually develops into 24-hour cycles.[1,2] These observations indicate that the human circadian system is not completely established at birth and that a 24-hour pattern in sleep–wake activity is observed as early as 3–4 weeks after birth.[3] However, subsequent studies have demonstrated that foetal body movement, breathing and heart rate exhibit rhythmic patterns even in the prenatal period.[4–9] Animal studies have further strengthened the notion that circadian rhythms are present before birth.[10–12] Until now, a number of studies have demonstrated the presence of prenatal circadian clock in foetal tissues, and the spontaneous development of the circadian clock during cellular differentiation.[13]

Chemical Biology No. 23
Chronobiology and Chronomedicine: From Molecular and Cellular Mechanisms to Whole Body Interdigitating Networks
Edited by Germaine Cornelissen and Tsuyoshi Hirota

Published by the Royal Society of Chemistry, www.rsc.org

This chapter summarizes the current understanding of the emergence of the circadian clock during prenatal development and cellular differentiation. In this chapter, the words "embryo" and "foetus" are used to refer to the stages of approximately the first third and the last two thirds of the prenatal human development, respectively. As a fundamental property of the cell, the circadian clock is closely linked with cellular differentiation and coordinates functional rhythms of cellular physiology. The latter part of this chapter also describes the association between the circadian clock and cancer in terms of cellular differentiation. Accumulating evidence indicates that cell stemness, dedifferentiation, and misdifferentiation are associated with tumour development in some types of cancer. The nature of differentiation-coupled development of the circadian clock may provide important insights into the characteristics of cancer and potentially pave the way for a circadian clock-targeted therapeutic approach and cancer chronotherapy.

17.2 Emergence of the Circadian Clock

17.2.1 Maternal Influence on Foetal Circadian Rhythms

Historically, earlier studies on newborn infants showed that sleep–wake cycles are unstable at birth and then gradually become established after birth. However, circadian rhythms were also observed during the prenatal period in various foetal activities such as body movement and heart rate changes. Thus, there has been a keen interest in determining whether these prenatal circadian variations are under control of the intrinsic foetal circadian clock or simply reflect maternal circadian systemic signals.

In this context, it is important to understand two aspects of the development of foetal circadian rhythms – the establishment of the intrinsic circadian oscillator and its entrainment to maternal and/or environmental time cues. From observational studies on prematurely born infants, it became clear that preterm infants have an intrinsic circadian system that regulates rest–activity cycles and body temperature rhythms.[14] Animal studies since the 1980s have revealed that an entrainable circadian clock is present in the foetal suprachiasmatic nucleus (SCN) before birth and that the maternal clock plays a crucial role in the entrainment of the foetal circadian phase.[10–12,15] It is worth noting that in human studies, blood pressure and heart rate peak shortly after midnight during the first 2–3 days of life, almost in antiphase with maternal rhythms.[16] Although the mechanisms of maternal–foetal entrainment have not been fully elucidated, body temperature cycles and several humoral factors, including melatonin, dopamine, and glucocorticoid have been postulated as candidate time cues for maternal–foetal entrainment.[17,18]

As for the foetal circadian clock, spontaneous development of the intrinsic circadian clock was suggested by studies demonstrating that the circadian locomotor activity rhythms of mouse pups are observed even when maternal rhythms are disrupted by SCN lesion on day 7 of gestation.[19,20]

Furthermore, it has been reported that mouse pups born to either the *Per1*^{Brdm1}/*Per2*^{Brdm1} or the *Per2*^{Brdm1}/*Cry1*$^{-/-}$ double-mutant arrhythmic females mated with the wild-type males, where Brdm1 mutants produce a truncated form of PER proteins and thus have impaired PER function, show intact postnatal activity rhythms without any external synchronizer, although the synchrony among pups is lost.[21] These studies collectively indicate that the development of foetal circadian rhythms does not depend on the maternal clock. Instead, the synchronization of circadian clocks within litters is the result of maternal entrainment of individual foetal clocks, while mutual synchronization among litters has not yet been observed.

Of note, several reports on animal studies have suggested that disturbance of maternal circadian rhythms could lead to changes in neurobehavioural functions such as increased anxiety-like behaviours and in metabolic homeostasis, including impaired glucose tolerance in the adult progeny.[22,23] However, it is not clear whether these effects are mediated by disturbance of the foetal circadian clock and the circadian regulatory system. Human studies investigating long-term effects of circadian disruption during pregnancy on the progeny are also lacking. Thus, the overall role of maternal influence on the foetal circadian clock remains to be elucidated.

It is also important to note that our global gene expression analysis has identified very few genes that are rhythmically expressed in mouse foetal hearts on embryonic days 10–12 (E10–12).[24] This result suggests that foetal tissues at early developmental stages neither have an established intrinsic circadian clock nor are controlled by systemic signals from the maternal circadian clock. Thus, it is plausible that foetal circadian physiology is regulated by both the intrinsic circadian clock and maternal systemic time signals especially during the late gestation period, although further studies are needed to fully understand the maternal influence during prenatal development (Figure 17.1).

17.2.2 Emergence of the Circadian Clock in Prenatal Development

It has long been a challenge to elucidate the exact timing of the emergence of the circadian clock during foetal development because of the difficulties in monitoring foetal rhythms directly *in vivo*. In humans, circadian rhythms of foetal heart rate, body movement and respiratory movement can be observed as early as 20–22 weeks after conception, although it remains unclear whether these rhythms are under the control of the foetal intrinsic clock or are simply masked by maternal rhythms.[7] Earlier studies have revealed that the foetal SCN exhibits clear day–night variations in its glucose utilization at E19 in rats and during very late (about 90% of term) gestation in squirrel monkeys.[11,12] These observations indicate the existence of circadian physiology in prenatal development.

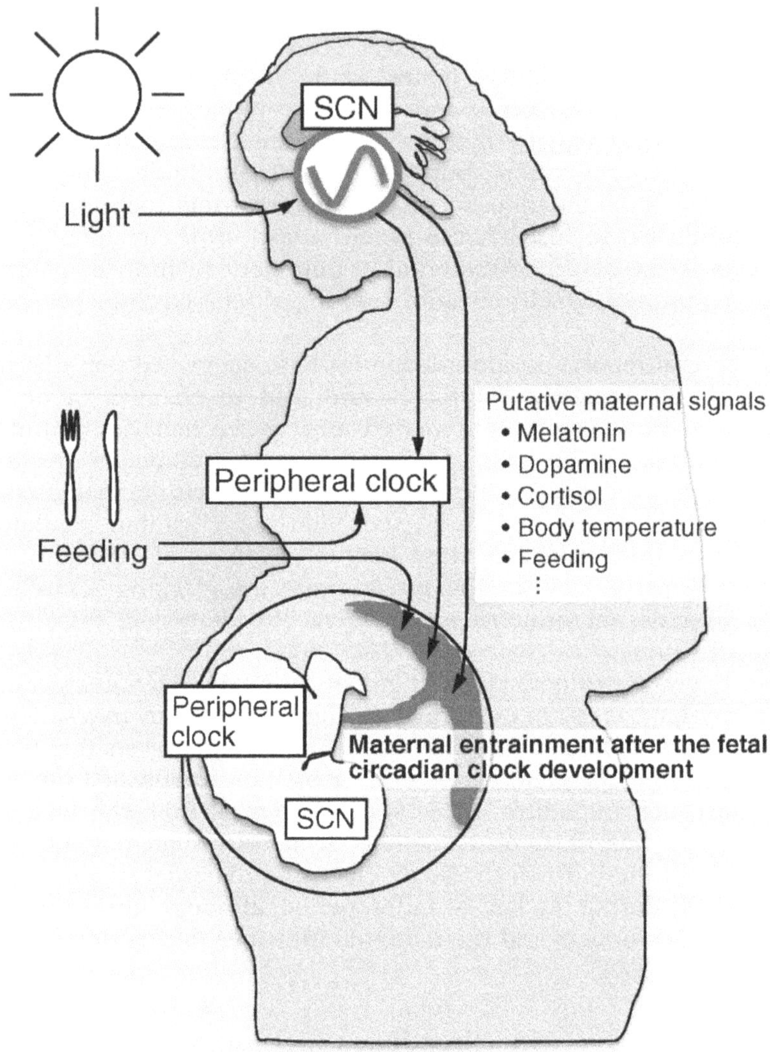

Figure 17.1 Circadian environment surrounding the foetus. The foetal circadian clock develops even without the maternal clock. The foetus is exposed to various maternal influences and entrained by maternal signals, especially in late gestation, whereas in earlier developmental stages, the foetal clock seems to be shielded from maternal time cues by yet unidentified mechanisms.

The expression rhythms of core clock genes during the prenatal period have been examined extensively in animal models. In an earlier study, *Per1* mRNA expression rhythms were detected at E17 in the mouse SCN.[25] Expression levels of *Per1* and *Per2* were also shown to exhibit 24-hour variations at E20 in the rat SCN.[26] Another report demonstrated the formation of *Per1* expression rhythms at E20 in the rat SCN.[27] In capuchin monkeys, *Per2* and *Bmal1* were found to be expressed rhythmically at 90% of gestation in the SCN.[28] More recent studies have demonstrated that *Per2::luciferase* (*Per2^{Luc}*) reporter expression exhibits circadian rhythms in *ex vivo* cultures of the mouse foetal SCN that is explanted at E14.5 or later.[29–31] Given that the SCN neurogenesis is completed at around E15.5 in mice,[32,33] neuronal differentiation should be associated with circadian clock development in the SCN. In addition, the postnatal maturation of neural networks would further enhance the robustness of circadian oscillations of the SCN neuronal network activity.

As for peripheral tissues, circadian expression rhythms of clock genes have also been observed during the prenatal period. In mice, circadian expression of clock proteins, including PER1, PER2, CRY1, and CRY2, was detected at E18 in pars tuberalis.[34] In rats, *Per2* and *Bmal1* mRNA expression rhythms were observed at E18 in the adrenal gland.[35] Expression rhythms of *Per2* and *Bmal1* were also observed during late gestation in adrenal glands of capuchin monkeys.[28] Strikingly, our global expression analysis of mouse foetal hearts prepared on E10–12 and E17–19 demonstrated that most genes, including core clock genes, do not show circadian expression rhythms at E10–12, while as many as 4.0% of genes (483 genes), including *Per2*, *Cry1*, and *Bmal1* exhibit rhythms at E17–19.[24] These findings also indicate that foetal tissues at E10–12 have not yet been entrained to the maternal circadian rhythms, as described above. As for Per2^{Luc} reporter bioluminescence analyses, it has been reported that Per2^{Luc} bioluminescence rhythms are observed in *ex vivo* cultures of the mouse liver, heart, and kidney prepared at E18.[36] We have demonstrated that *ex vivo* cultures of the mouse submandibular gland (SMG) prepared at E15.5 or later exhibit Per2^{Luc} circadian bioluminescence rhythms.[37] We have also reported that mouse embryonic fibroblasts (MEFs) prepared at E15.5 but not at E10.5 exhibit circadian rhythms of Per2^{Luc} bioluminescence. Importantly, both SMG cultured at E13.5 and MEFs prepared at E10.5 do not show Per2^{Luc} bioluminescence rhythms but develop rhythms after several days in culture.[37] These studies provide direct evidence that the foetal circadian clock is formed during the gestation period and that foetal tissues and cells have an ability to develop the circadian clock in the absence of maternal influences.

In contrast to these peripheral oscillators, the SCN explanted at E13.5 or earlier does not develop circadian rhythms in *ex vivo* culture.[31] The SCN may require some maternal factors and/or an *in vivo* environmental cue to develop the appropriate oscillatory function, while other studies have shown the spontaneous development of SCN rhythms in culture.[29,30] Taken together, the foetal circadian clock gradually develops during the prenatal period and mid-to-late gestation would be critical for the establishment of both central and peripheral oscillators.[38–40]

17.2.3 Differentiation-coupled Development of the Circadian Molecular Clock

While the circadian clock resides in most cells in the body, its absence in cells of early stage embryos has been reported.[37,41–43] Considering that the circadian clock is a fundamental property of the cell, it is also intriguing to know why the circadian clock does not function in embryos during early developmental stages, and how cells acquire the circadian clock as ontogenesis progresses.

In 2010, we reported that mouse embryonic stem (ES) cells do not show circadian expression rhythms of clock genes, while *in vitro* differentiation of ES cells leads to spontaneous emergence of circadian gene expression rhythms[13] (Figure 17.2).

Furthermore, the reprogramming of differentiated cells into induced pluripotent stem (iPS) cells eliminates previously established circadian oscillations in gene expression. This study has set a conceptual frame of circadian clock development by (1) clearly demonstrating the absence of the circadian transcriptional circuit in ES cells, (2) indicating that the circadian clock emerges *in vitro* in the complete absence of maternal influence, and (3) suggesting that development of the circadian clock is tightly coupled with cellular differentiation. Another study has also demonstrated the absence of circadian gene expression rhythms in ES cells.[44] The established recapitulation of circadian clock development in differentiating ES cells has provided an epoch-making approach to investigate the mechanisms underlying the emergence of the circadian clock.[13,45] Since then, our understanding of the emergence of the circadian clock has been growing rapidly through related research works conducted by our group and others. This section summarizes mechanistic insights into the differentiation-coupled emergence of the circadian clock by introducing related studies.

The important first step in differentiation is the exit from the pluripotent state, which is associated with dynamic changes in the epigenetic landscape

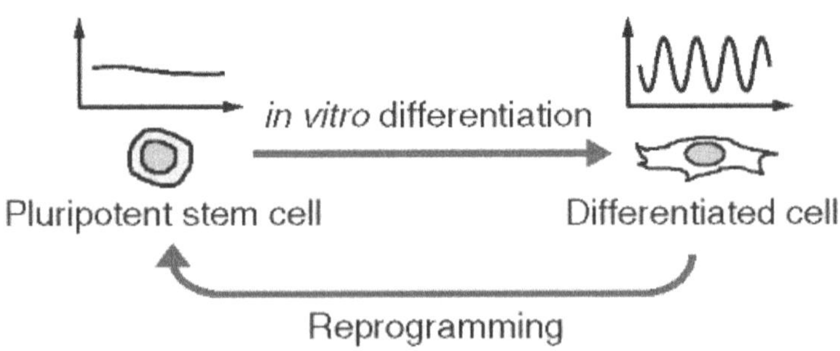

Figure 17.2 Differentiation-coupled formation of the circadian clock. Circadian gene expression rhythms spontaneously emerge during the course of *in vitro* differentiation of pluripotent stem cells. Reprogramming of differentiated cells into iPS cells abrogates the previously established circadian oscillation of gene expression.

and associated transcriptome. In this context, formation of the circadian transcriptional circuit would include appropriate epigenetic regulation as an essential prerequisite. Interestingly, the ectopic expression of c-Myc impairs the differentiation-coupled emergence of circadian rhythms in mouse ES cells.[46] Given the role of c-Myc as a universal regulator of gene expression in both cancerous cells and ES cells,[47,48] the appropriate regulation of gene expression programs during differentiation likely contributes to the establishment of the circadian transcriptional circuit. Also, the deficiency of an epigenetic regulator DNA methyltransferase 1 (DNMT1), the most abundant DNA methyltransferase that predominantly methylates cytosines at hemimethylated CpG dinucleotides, abrogates the differentiation-coupled emergence of circadian gene expression rhythms.[46] The result suggests that DNMT1-mediated epigenetic modifications play a critical role in the development of the circadian clock during cellular differentiation.[46] In a recent study, epigenetic marks by DNMT1 and DMNT3a have also been suggested to be involved in inheritable period regulation of the circadian clock in mouse cultured fibroblasts.[49] Cellular differentiation triggers drastic changes in the epigenetic landscape that define the cell type-specific transcriptional network. It has also been demonstrated that histone modifications regulate the circadian clock. Considering that the circadian clock regulates multiple cellular functions and physiology by controlling a set of output genes in a tissue-specific manner, differentiation-coupled development of the circadian clock would be advantageous in forming the circadian transcriptome, as observed in mouse foetal hearts.[24] Thus, appropriate epigenetic modifications during cellular differentiation should play a key role in the development of the circadian transcriptional circuit as well as tissue-specific circadian output functions.

In the ES cell differentiation assay, it takes about 14 days from the onset of differentiation to develop discernible circadian gene expression rhythms, although expression of pluripotency factors such as *Oct3/4* and *Nanog* is terminated as early as day 7 of differentiation.[37,46] This observation indicates that the loss of pluripotency alone is not sufficient for the generation of circadian transcriptional rhythms. Although underlying molecular mechanisms that contribute to the emergence of the circadian clock have not been fully elucidated, several important mechanistic insights have been provided. Importantly, the expression patterns of some clock proteins are differently regulated between undifferentiated ES cells and differentiated ES cells, while the mRNAs of core clock genes, including *Per1*, *Per2*, *Cry1*, *Cry2*, *Bmal1*, and *Clock* are expressed both in undifferentiated and differentiated ES cells as well as 3T3 fibroblasts.[13,44,46,50,51]

For example, PER proteins are excluded from the nucleus in undifferentiated pluripotent stem cells, whereas they are predominantly localized in the nucleus in clock-oscillating differentiated cells.[46] From a global mRNA expression analysis, a set of genes have been identified as being differentially expressed between differentiated ES cells that show normal circadian rhythms and undifferentiated or misdifferentiated ES cells that lack apparent rhythms. Among them, *Kpna2* that encodes importin-α2 was identified as a downregulated gene in clock-oscillating cells. KPNA2 has previously been shown to differentially regulate the nuclear import of transcription factors, including OCT3/4 and OCT6.[52] Ectopic expression of *Kpna2* impaired differentiation-coupled formation of circadian rhythms.[46] Because physical interaction between KPNA2 and PER2 is quite weak or undetectable, the differentiation-associated decrease in *Kpna2* expression should contribute to the progression of normal differentiation.[46] Of note, regulation of appropriate localization of PER proteins is associated with intact cellular differentiation, which is required for a functional circadian clock.

We have also reported that post-transcriptional control of *Clock* expression is a key mechanism for completing the circadian transcriptional circuit during cellular differentiation.[24] In pluripotent stem cells such as ES cells, multipotent germline stem (mGS) cells, and iPS cells, the expression of CLOCK proteins is strongly suppressed, while the level of *Clock* mRNA expression is comparable to that in differentiated cells. Further analysis has revealed the significant nuclear retention of *Clock* mRNA in undifferentiated cells, which may prevent translation of CLOCK proteins. As DICER- and DGCR8-deficient ES cells express CLOCK proteins, DICER/DGCR8-mediated post-transcriptional regulation likely contributes to the repression of CLOCK protein expression in undifferentiated ES cells.[24] An analysis of the untranslated region (UTR) of *Clock* suggests the existence of an important *cis*-regulatory region in *Clock* 3'UTR and the potential involvement of several miRNA species, which have been previously proposed to target *Clock* 3'UTR.[53,54] It has also been reported that post-transcriptional regulation of *Clock via* its 3'UTR region plays a key role in the development and maintenance of the *Drosophila* circadian clock,[55] suggesting a conserved role of *Clock* 3'UTR.

In mouse ES cells, CLOCK proteins can be detected on day 14 of differentiation or later and its amount gradually increases during differentiation.[24] The temporal relationship between the increase in CLOCK proteins and the development of circadian gene expression rhythms further strengthens the notion that post-transcriptional regulation of *Clock* is crucial for initiating cycling of the circadian clock. This notion is also supported by the fact that in mouse foetal hearts, CLOCK protein is absent at E10 but is clearly detected at E18, while *Clock* mRNA shows comparable expression levels at both stages. In addition, the absence of CLOCK protein is also observed in human iPS cells.[56] Therefore, it is plausible that the suppression of CLOCK protein expression is a universal mechanism in the embryonic development of the circadian clock. Intriguingly, human iPS cells require a 3- to 4-fold longer differentiation period than mouse ES/iPS cells to initiate circadian oscillations of gene expression.[56] This observation might reflect the difference in gestation periods between mice and humans, although further studies are needed to verify this notion.

Collectively, accumulating data have provided important mechanistic insights into the emergence of circadian gene expression rhythms, in which exit from pluripotency, initiation of cellular differentiation, and associated epigenetic modifications followed by appropriate spatiotemporal regulation of core clock proteins, including PER2 and CLOCK all have essential functions (Figure 17.3).

It is also possible that a primitive metabolic oscillator exhibits circadian variations at an earlier developmental stage before maturation of the circadian transcriptional circuit.[50] Elucidating the complete picture of the molecular basis for emergence of the circadian clock is critical for understanding cellular differentiation as well as embryonic and foetal development in mammals.

17.2.4 Physiological Function of the Circadian Clock in Stem Cells

Given that the circadian clock emerges during ontogenesis, it is important to define the physiological role of the circadian clock in developing embryos and foetuses. Although clock gene-deficient arrhythmic mice are born after the normal gestation period, they show multiple features of metabolic abnormalities and premature aging.[57–59] Intriguingly, aging phenotypes, including shortened life span and impaired glucose homeostasis, which are observed in *Bmal1*-deficient mice, are not replicated in an inducible *Bmal1* knockout after birth.[60] This observation suggests that the circadian clock during prenatal and/or postnatal development plays a critical role in maintaining normal physiology in adults, although it is also possible that non-clock function of *Bmal1* is more relevant for appropriate development.

A long-term effect of maternal circadian disruption on adult progeny has been reported as described previously,[22,23] which also suggests the importance of prenatal circadian rhythms. A more recent study has demonstrated that foetal clock disruption by global or ureteric bud-specific deficiency of *Bmal1* leads to impaired circadian expression of genes involved in kidney development and structural abnormalities in the kidneys, emphasizing the critical role of circadian clock components in organ development.[61] As in this study, *Bmal1* heterozygous parents with intact circadian clocks were used, the observed abnormality should be due to *Bmal1* deficiency in pups. Taken together, it is plausible that circadian disruption during the prenatal period can impact the development and physiology of the progeny, although the essential role of the foetal circadian clock still remains to be elucidated.

As for cellular differentiation, accumulating evidence has demonstrated that, in contrast to the case of pluripotent stem cells, a circadian clock is present in various types of adult tissue stem cells that plays a role in regulating stem cell homeostasis and function.[62] For instance, it was reported that epidermal stem cells possess a functional circadian clock to respond to homeostatic and differentiation time cues at the appropriate circadian phase.[63] The clock in epidermal stem cells controls the expression of genes

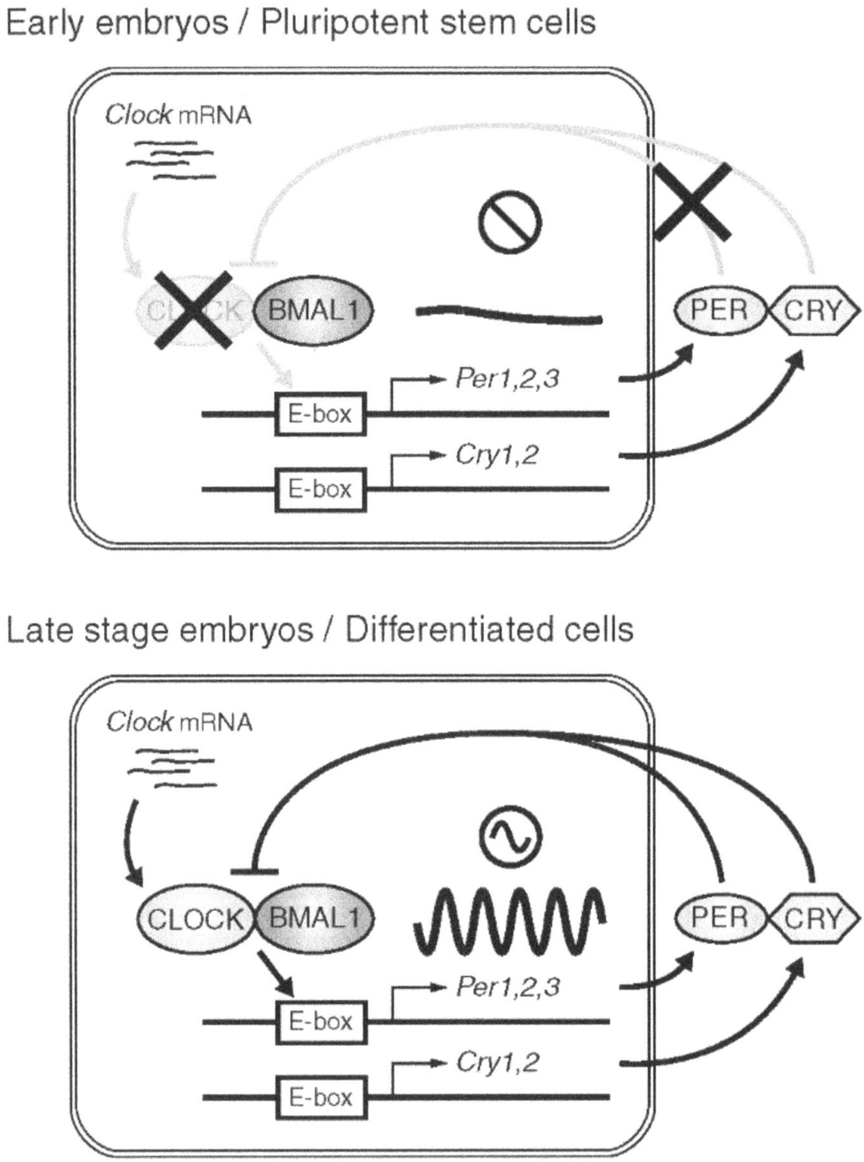

Figure 17.3 Molecular mechanisms underlying the absence and presence of the circadian transcriptional/translational feedback loop. In ES/iPS cells, the circadian transcriptional /translational feedback loop is abrogated by multiple mechanisms. For instance, the expression of CLOCK proteins is suppressed by a DICER/DGCR8-dependent mechanism. Nuclear translocation of PER2 proteins is also inhibited in pluripotent stem cells. In late-stage embryos and differentiated cells, appropriate post-transcriptional regulations of these clock genes enable the cell-autonomous circadian oscillation of gene expression. Appropriate differentiation processes, including global epigenetic changes, are required for the emergence of the circadian clock (see text).

involved in stem cell homeostasis such as metabolism and differentiation in young adult mice, whereas, in aged mice, the circadian transcriptome is drastically rewired to genes that regulate stress responses.[64] As for mammary epithelial cells, a previous study has shown that mice carrying *Clock* gene mutations have impaired ability for stem cell renewal.[65] It has also been reported that, in mice, decreased postnatal pup survival is associated with impaired mammary development in *Clock* mutant dams.[66] Another study has demonstrated that the circadian clock controls the timing of the cell cycle in intestinal stem cells by regulating WNT production and secretion from Paneth cells.[67] These studies signify the importance of the circadian clock in adult tissue stem cell maintenance and function that should be regulated properly in a circadian time-dependent manner.

With a considerable body of evidence suggesting the potential role of the circadian clock in cell cycle control, the circadian clock in adult stem cells would play a pivotal role in the regulation of 24-hour variation of cell proliferation.[68,69] The *in vivo* evidence supporting the role of the circadian clock in regulating the timing of the cell-cycle entry and exit has also been provided by studies on hepatocytes in regenerating liver[70] and neural progenitor cells in adult hippocampal neurogenesis.[71] During regenerative hair cycling, analysis in *Clock* and *Bmal1* mutant mice revealed delay in hair growth progression,[72] and hair follicle cells show daily mitotic rhythms that are abrogated in *Cry1*/*Cry2* double-knockout mice.[73] Thus, the circadian clock in regenerating tissues including adult stem cells and/or progenitor cells plays a key role in maintaining tissue homeostasis through appropriate and strict time-dependent regulation of cell proliferation and differentiation in response to changes in both external and internal environments. In this context, the functional significance of the foetal circadian clock observed in late gestation may be represented by coordination of foetal physiology for better adaptation of the timing of postnatal development and diverse physiological processes to the environmental cycles occurring after birth.

In contrast to the late developmental period, the circadian clock apparently does not oscillate in the early developmental stages. As described previously, the circadian clock is not established at E12.5 or earlier in mice, and it takes a considerable period to develop circadian gene expression rhythms in *in vitro* differentiation cultures of ES cells. Given these observations, it is intriguing to examine the possibility that the circadian clock could be actively suppressed in early development. In this context, our recent observations have provided a new perspective on this possibility. Early developmental events include important morphogenic processes exemplified by the genetic control of the body plan by homeobox genes and sequential morphological changes, such as rapid cell division, proliferation, and migration as well as non-circadian oscillatory processes such as somitogenesis. We have demonstrated that the expression of CLOCK/BMAL1 can affect the expression of *Hes7*, which encodes a transcription factor essential for the somite segmentation clock.[74] Furthermore, the expression of CLOCK/BMAL1 abrogated both the oscillation of *Hes7* promoter-driven luciferase expression and the formation of segment-like structures in gastruloids, the ES cell-derived embryonic organoids.[74] Of note, the expression of *Hes7* and another segmentation clock gene *Lfng* shows circadian oscillation in mouse foetal hearts at E17–19, in which foetal tissue-autonomous circadian clocks have been established. These observations suggest that the circadian clock can interfere with the somite segmentation clock. Intriguingly, *Per1* oscillates in the same phase as *Hes7* in *in vitro*-derived presomitic mesoderm cells,[75] suggesting that the segmentation clock can also affect circadian clock gene expression in early-stage embryos during somitogenesis. Thus, it is reasonable to suggest that these two biological clocks, each with a distinct frequency, may switch their function at a specific developmental window to avoid interfering with each other. The circadian clock oscillation would be strictly suppressed by multiple mechanisms as described above[13,24,46] when the mouse somitogenesis occurs from E7.5 to E13.5. Given the fact that, in humans, somitogenesis occurs at 20–35 days post fertilization, the species-specific timing of somitogenesis and other early developmental events could be a reason for the later emergence of the circadian clock in human cells compared to that in mouse cells.[56] In this context, it is conceivable that the apparent absence of the circadian clock might be more beneficial for early development.

Intriguingly, as demonstrated in a transcriptome analysis of the foetal heart, very few genes demonstrated circadian variations in mouse foetal hearts at E10–12,[24] raising the possibility that the embryos in early gestation are protected, even from maternal circadian systemic signals, which actually affect the foetal clock in later gestation. Considering the unusual situation in early embryonic development, it is possible that the circadian clock function is actively suppressed at early developmental stages so that essential developmental programs are not disrupted by undesired signals from intrinsic and environmental/maternal circadian time cues.

Although further investigation is needed to fully elucidate the physiological impact of circadian clock development, the emergence of the circadian oscillation would establish the daily temporal order in late gestation foetuses, leading to the development of an optimal coordination of physiological functions (Figure 17.4).

Especially, it would be therapeutically beneficial to understand how circadian transcriptional networks are wired to control output cellular circadian physiology and what is the consequence of circadian disruption during prenatal development. Recent progress in technical innovations, including *in vivo* imaging, single cell analysis, as well as improved *in vitro* differentiation systems such as 3D-organoid models would serve as useful tools for in-depth observation of rhythms and further elucidation of mechanisms underlying differentiation-coupled development of the circadian clock.

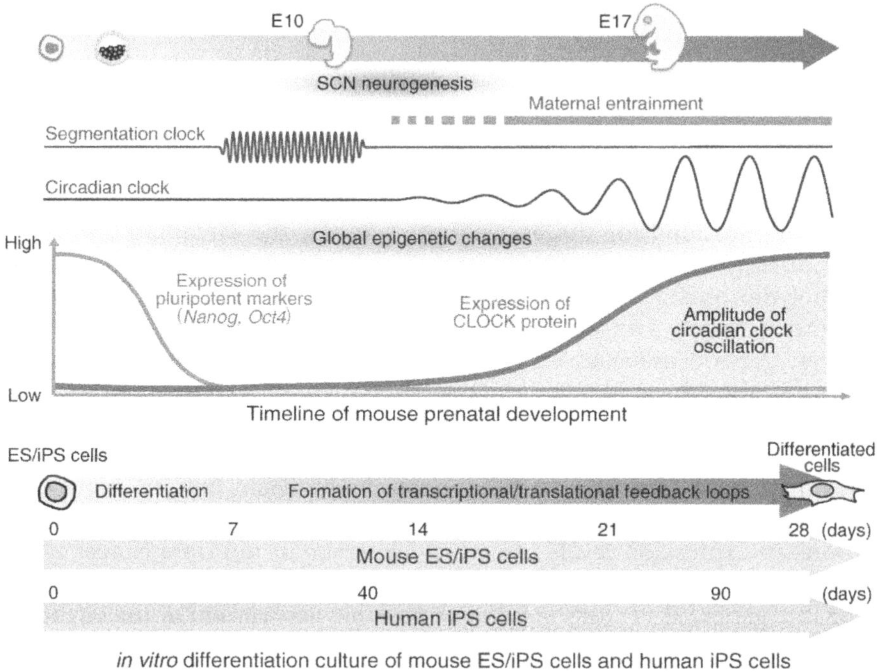

Figure 17.4 Timeline of the emergence and development of the circadian clock and associated developmental features. In early developmental stages, the circadian clock is suppressed for successful implementation of developmental programs, including somitogenesis. During the mid-to-late gestation period, the circadian clock emerges to coordinate circadian homeostasis of the developing foetus, which is thereafter under the control of maternal entrainment cues. The emergence of a cell-autonomous circadian clock can be recapitulated using *in vitro* differentiation systems applied to both mouse and human pluripotent stem cells, making them valuable tools for further investigation into the mechanism underlying circadian clock emergence.

17.3 The Circadian Clock and Cancer Cells

17.3.1 Dysregulation of the Circadian Clock in Cancer

The association between the circadian clock and cancer was proposed as early as the 1960s. It has been reported that disruption of circadian rhythms through constant light exposure could increase the incidence of carcinogen-induced mammary gland tumours in rats.[76] Since then, a number of animal studies have revealed that circadian disruption contributes to tumour incidence and development.[77–81] Epidemiological studies on shift workers have also suggested the association between carcinogenesis and disturbance of circadian rhythms.[82–88] In 2007, the International Agency for Research on Cancer (IARC) declared that shift work that involves circadian disruption is probably carcinogenic to humans (group 2A), based on evidence from experimental animals and human breast cancer patients.[89–91]

A large number of studies have implicated the deregulation of core circadian clock genes in various types of human cancers, including breast, prostate, gastric, colorectal, pancreatic, renal, bladder, oral, and non-small cell lung cancers as well as in glioma, hepatocellular carcinoma, head and neck squamous cell carcinoma, and myeloid leukaemia.[92–107] There are also genetic implications for the association between the circadian clock and cancer. It has been reported that genetic variations including single nucleotide polymorphism (SNP) in the core clock genes, *Per1*, *Per2*, *Per3*, *Cry1*, *Cry2*, *Clock*, *Npas2*, and *Bmal1*, are associated with specific cancers.[108–111] However, genetic associations are not always consistent across different studies and thus the significance of deregulation and/or genetic variations of clock genes in cancer remains unknown. A recent study by Ye *et al.*[112] has comprehensively characterized genetic mutations and altered mRNA expression of clock and clock-related genes in multiple human cancers by analysing multiple cancer genome databases. The result demonstrates alterations in expression levels of core clock genes in multiple cancer types, and the correlation between the transcription of clock genes and clinically actionable genes. This study suggests a possible association of the circadian clock with cancer, although further studies are needed to elucidate their reciprocal relationship.

A growing body of evidence indicating that dysfunction of clock genes contributes to the alteration of cancer susceptibility and tumour development has been obtained from animal models with clock gene deficiency.[78,80,113–118] Genetic alterations are implicated in mechanisms underlying the deregulation of clock gene expression that may play functional roles in cancer pathophysiology. It has been reported that the patients with core clock gene mutations show reduced overall survival, although the mutation frequency is relatively low.[112] Altered epigenetic modifications in clock genes are also prevalent in various cancers. It has been reported that DNA methylation levels in promoter regions of core clock genes are negatively associated with their expression levels in multiple cancers, suggesting that clock genes are epigenetically regulated in cancer.[107,112] Collectively, circadian disruption and/or deregulation of clock genes are correlated with cancer development, although their causal relationship is still enigmatic and further studies are required to understand the whole picture of their interaction.

17.3.2 Dedifferentiation- and Misdifferentiation-related Circadian Disruption in Cancer Development

Cancer epigenetics has aroused growing interest in recent decades. Dynamic changes of epigenetic modifications are prevalent in most forms of carcinogenesis, which are closely associated with cellular dedifferentiation and abnormal differentiation processes.[119–122] For example, epigenetic changes in carcinogenesis include DNA hypermethylation of the promoter region of key tumour suppressor genes such as *CDKN2A*, *VHL*, *P53*, and *BRCA1*.[123,124]

Importantly, a genome-wide remodelling of the epigenetic landscape is often observed in cell fate determination during the course of cellular differentiation of pluripotent stem cells. It is well known that some types of cancer cells share common characteristics with pluripotent stem cells, such as similar epigenetic marks and metabolic requirements for rapid and unlimited proliferation.[125–129] The similarity between pluripotent cells and cancer stem cells is also exemplified by studies demonstrating that forced expression of reprogramming factors, *Oct3/4*, *Sox2*, *Klf4*, and *c-Myc* (OSKM), leads to dedifferentiation-mediated formation of teratomas and kidney tumours.[130,131] The developed kidney tumours resemble human Wilms tumour, a common paediatric nephroblastoma, and, importantly, have no cancer-related gene mutations.[131] These studies consistently suggest that dedifferentiation and epigenetic reprogramming are involved in specific types of carcinogenesis.

Given that OSKM-mediated reprogramming of differentiated cells into iPS cells leads to the disruption of circadian gene expression rhythms,[13] it is intriguing to investigate whether the circadian clock is also impaired by dedifferentiation-triggered cancer development. Our previous study clearly demonstrated that circadian gene expression rhythms are attenuated in kidney tumours developed by *in vivo* reprogramming in chimeric mice from tetracycline-inducible OSKM ES cells with *Per2-Luc* reporter.[132] Importantly, the morphology and structure as well as Per2-Luc rhythms were normal in the kidney before the induction of OSKM reprogramming factors. However, after the induction of reprogramming factors, the kidney became markedly enlarged and developed human Wilms-like tumours composed of undifferentiated cells. More strikingly, Per2-Luc rhythms in developed tumours were significantly attenuated and essentially no detectable circadian oscillation was observed in any small regions of tumours, indicating circadian disruption in every tumour cell rather than desynchronization of individual cells.[132] This observation provides strong evidence for the disruption of the circadian clock during dedifferentiation-associated cancer development.

Another circadian feature showing similarity between ES cells and *in vivo* reprogramming-mediated kidney tumours is the expression level of CLOCK proteins. As observed in pluripotent stem cells,[24] CLOCK protein is not detected in the developed kidney tumours, while *CLOCK* mRNA is evidently expressed.[132] Furthermore, CLOCK proteins are also undetectable in human Wilms tumour and another paediatric tumour, malignant rhabdoid tumour (MRT), which is caused by the loss of SMARCB1, a subunit of the SWI/SNF chromatin remodelling complex.[132–134] These observations clearly indicate that CLOCK expression is repressed post-transcriptionally in both ES cells and dedifferentiation-induced tumours, suggesting a common mechanism underlying the disruption of the circadian clock. Of note, repression of CLOCK protein expression has also been observed in different cancer types. Bu *et al.*[135] demonstrated that the unfolded protein response (UPR) can be observed in the mouse liver upon a reversal of light–dark cycles and that UPR induces DICER- and miRNA-mediated downregulation of CLOCK and BMAL1 expression. Although CLOCK and BMAL1 proteins were undetectable in Burkitt's lymphoma cell lines, the inhibition of UPR signalling by a PERK inhibitor or miRNA inhibitor restored not only the expression of CLOCK and BMAL1 proteins but also the circadian oscillation of *CLOCK* and *BMAL1* mRNA expression. This study highlights the significance of post-transcriptional regulation of *CLOCK* and *BMAL1* in cancer pathophysiology and suggests that UPR could be a possible mechanism for clock disruption in cancer cells.

While evidence is increasing for circadian clock disruption in multiple cancers, there are also many cancer types that maintain a functional circadian clock. For instance, it has been reported that an intact circadian clock exists in acute myeloid leukaemia (AML) cells, in which *Clock* and *Bmal1* play a crucial role in AML cell proliferation.[136] Thus, the relationship between circadian properties and cancer characteristics still remains elusive.

An earlier study has shown that patients with fast-growing, poorly differentiated tumours exhibit a short period of tumour temperature rhythm, suggesting a correlation between tumour malignancy and circadian disruption.[137] A comparative study using multiple cultured colon cancer cell lines has revealed a considerable variation in the amplitude of circadian gene expression rhythms among these cell lines.[138] This study has shown that the circadian phase, amplitude, and period can be changed by the activation of Ras/MAPK signalling in tumour cells. Consistently, constitutive activation of Ras signalling impaired the circadian oscillation of gene expression in colorectal cancer cells and human keratinocytes, suggesting that the upregulation of Ras signalling is a potential mechanism for circadian clock disruption. Another study comparing breast cancer cell lines has shown that circadian bioluminescence rhythms of *Bmal1-Luc* and *Per2-Luc* reporters are observed in low-grade luminal MCF7 breast cancer cells but not in high-grade basal MDA-MB-231 breast cancer cells.[139] This result may indicate the association between circadian clock disruption and aggressiveness of some types of cancer. These studies collectively suggest that disruption of the circadian clock is correlated to the metastatic characteristics of tumours, such as poorly differentiated and highly proliferative states.

17.3.3 Impact of Circadian Clock Disruption on Cancer Development

As described above, evidence is still accumulating that suggests reciprocal regulation between the disruption of the circadian clock and cancer development. The circadian clock controls diverse cellular processes, including the cell cycle, DNA repair, apoptosis, and energy metabolism by regulating a number of genes such as *p16*, *p21*, *Cyclin D1*, *Wee1*, *p53*, and *Pdk1*.[118,140–143] Dysregulation of these essential cellular processes will lead to multiple cancer phenotypes such as rapid proliferation, impaired DNA repair, resistance to apoptosis, and elevated glycolysis. Thus, these molecular links further support the notion that the disruption of circadian regulation should play a critical role in cancer pathophysiology. Considering the significance of dedifferentiation processes in some forms of carcinogenesis, it is important to understand the effects of dedifferentiation-associated global epigenetic changes on the circadian clock and mechanisms through which the circadian clock is regulated in developing cancer cells.[144]

17.4 Future Perspectives

Recent progress in the field of circadian clock ontogeny has revealed the differentiation-coupled development of the circadian clock during embryogenesis as well as the dedifferentiation-coupled disruption of the circadian clock during carcinogenesis. The presence or absence of a functional circadian clock may be indicative of the specific cancer cell type or stem cell type, and could serve as an indicator of cell stemness or malignancy. Thus, an important aspect of studying circadian disruption is to determine whether dysregulation of the circadian clock, including the loss of CLOCK protein expression, serves as a marker of poorly differentiated malignant tumour cells. If so, we could make better use of the circadian properties of tumour cells for diagnosis and prognosis of a wide range of cancer types.[145–148]

For a sustainable development of a 24-hour society, there will be a continuous demand for shift work in various industries, including transportation and health care. Thus, as the global burden of circadian rhythm disorders and their associated comorbidities increases, it will become more important to provide further mechanistic insight into circadian disruption and cancer development. Progress in gaining such insights is a great challenge, leading to the identification of circadian clock-related prognostic markers and therapeutic targets for specific cancers.

Acknowledgements

The authors are grateful to members of the Yagita lab for their valuable comments and discussions.

References

1. N. Kleitman and T. G. Engelmann, *J. Appl. Physiol.*, 1953, **6**, 269–282.
2. A. H. Parmelee Jr, W. H. Wenner and H. R. Schulz, *J. Pediatr.*, 1964, **65**, 576–582.
3. T. Hellbrügge, *Cold Spring Harbor Symp. Quant. Biol.*, 1960, **25**, 311–323.
4. J. Patrick, K. Campbell, L. Carmichael, R. Natale and B. Richardson, *Obstet. Gynecol.*, 1980, **56**, 24–30.
5. J. Patrick, K. Campbell, L. Carmichael, R. Natale and B. Richardson, *Am. J. Obstet. Gynecol.*, 1982, **142**, 363–371.

6. G. H. Visser, J. D. Goodman, D. H. Levine and G. S. Dawes, *Am. J. Obstet. Gynecol.*, 1982, **142**, 535–544.
7. J. I. de Vries, G. H. Visser, E. J. Mulder and H. F. Prechtl, *Early Hum. Dev.*, 1987, **15**, 333–348.
8. J. Sletten, G. Cornelissen, J. Assmus, T. Kiserud, S. Albrechtsen and J. Kessler, *Acta Physiol.*, 2018, **224**, e13093.
9. J. Sletten, A. Lund, C. Ebbing, G. Cornelissen, J. Aßmus, T. Kiserud, S. Albrechtsen and J. Kessler, *Chronobiol. Int.*, 2019, **36**, 481–492.
10. S. M. Reppert and W. J. Schwartz, *Science*, 1983, **220**, 969–971.
11. S. M. Reppert and W. J. Schwartz, *J. Neurosci.*, 1984, **4**, 1677–1682.
12. S. M. Reppert and W. J. Schwartz, *Neurosci. Lett.*, 1984, **46**, 145–149.
13. K. Yagita, K. Horie, S. Koinuma, W. Nakamura, I. Yamanaka, A. Urasaki, Y. Shigeyoshi, K. Kawakami, S. Shimada, J. Takeda and Y. Uchiyama, *Proc. Natl. Acad. Sci. U. S. A.*, 2010, **107**, 3846–3851.
14. M. Mirmiran, J. H. Kok, K. Boer and H. Wolf, *Neurosci. Biobehav. Rev.*, 1992, **16**, 371–378.
15. S. Honma, K. I. Honma, T. Shirakawa and T. Hiroshige, *Endocrinology*, 1984, **114**, 1791–1796.
16. G. Cornelissen, J. Siegelova and F. Halberg, *Proceedings, Symposium: The Importance of Chronobiology in Diagnosing and Therapy of Internal Diseases, Faculty of Medicine*, Masaryk University, Brno, Czech Republic, 2002, pp. 58–96.
17. M. Seron-Ferre, G. J. Valenzuela and C. Torres-Farfan, *Birth Defects Res., Part C*, 2007, **81**, 204–214.
18. K. Bates and E. D. Herzog, *Front. Endocrinol.*, 2020, **11**, 198.
19. S. M. Reppert and W. J. Schwartz, *J. Neurosci.*, 1986, **6**, 2724–2729.
20. F. C. Davis and R. A. Gorski, *J. Comp. Physiol., A*, 1988, **162**, 601–610.
21. C. Jud and U. Albrecht, *J. Biol. Rhythms*, 2006, **21**, 149–154.
22. F. G. Amaral, A. M. Castrucci, J. Cipolla-Neto, M. O. Poletini, N. Mendez, H. G. Richter and M. T. Sellix, *J. Neuroendocrinol.*, 2014, **26**, 603–612.
23. T. J. Varcoe, K. L. Gatford and D. J. Kennaway, *Am. J. Physiol.: Regul., Integr. Comp. Physiol.*, 2018, **314**, R231–R241.
24. Y. Umemura, N. Koike, M. Ohashi, Y. Tsuchiya, Q. J. Meng, Y. Minami, M. Hara, M. Hisatomi and K. Yagita, *Proc. Natl. Acad. Sci. U. S. A.*, 2017, **114**, E7479–E7488.
25. H. Shimomura, T. Moriya, M. Sudo, H. Wakamatsu, M. Akiyama, Y. Miyake and S. Shibata, *Eur. J. Neurosci.*, 2001, **13**, 687–693.
26. H. Ohta, S. Honma, H. Abe and K. Honma, *Eur. J. Neurosci.*, 2003, **17**, 1628–1634.
27. Z. Kovacikova, M. Sladek, Z. Bendova, H. Illnerova and A. Sumova, *J. Biol. Rhythms*, 2006, **21**, 140–148.
28. C. Torres-Farfan, V. Rocco, C. Monso, F. J. Valenzuela, C. Campino, A. Germain, F. Torrealba, G. J. Valenzuela and M. Seron-Ferre, *Endocrinology*, 2006, **147**, 4618–4626.
29. D. Wreschnig, H. Dolatshad and F. C. Davis, *J. Biol. Rhythms*, 2014, **29**, 299–310.

30. D. Landgraf, C. Achten, F. Dallmann and H. Oster, *Chronobiol. Int.*, 2015, **32**, 416–427.
31. V. Carmona-Alcocer, J. H. Abel, T. C. Sun, L. R. Petzold, F. J. Doyle 3rd, C. L. Simms and E. D. Herzog, *J. Neurosci.*, 2018, **38**, 1326–1334.
32. J. L. Bedont and S. Blackshaw, *Front. Syst. Neurosci.*, 2015, **9**, 74.
33. V. Carmona-Alcocer, K. E. Rohr, D. A. M. Joye and J. A. Evans, *Eur. J. Neurosci.*, 2020, **51**, 82–108.
34. N. Ansari, M. Agathagelidis, C. Lee, H. W. Korf and C. von Gall, *Eur. J. Neurosci.*, 2009, **29**, 477–489.
35. C. Torres-Farfan, N. Mendez, L. Abarzua-Catalan, N. Vilches, G. J. Valenzuela and M. Seron-Ferre, *Endocrinology*, 2011, **152**, 1891–1900.
36. H. Dolatshad, A. J. Cary and F. C. Davis, *PLoS One*, 2010, **5**, e9855.
37. Y. Inada, H. Uchida, Y. Umemura, W. Nakamura, T. Sakai, N. Koike and K. Yagita, *FEBS Lett.*, 2014, **588**, 459–465.
38. M. Seron-Ferre, N. Mendez, L. Abarzua-Catalan, N. Vilches, F. J. Valenzuela, H. E. Reynolds, A. J. Llanos, A. Rojas, G. J. Valenzuela and C. Torres-Farfan, *Mol. Cell. Endocrinol.*, 2012, **349**, 68–75.
39. D. Landgraf, C. E. Koch and H. Oster, *Front. Neuroanat.*, 2014, **8**, 143.
40. A. Sumova and V. Cecmanova, *Eur. J. Neurosci.*, 2020, **51**, 300–309.
41. D. Morse, N. Cermakian, S. Brancorsini, M. Parvinen and P. Sassone-Corsi, *Mol. Endocrinol.*, 2003, **17**, 141–151.
42. J. D. Alvarez, D. Chen, E. Storer and A. Sehgal, *Biol. Reprod.*, 2003, **69**, 81–91.
43. T. Amano, A. Matsushita, Y. Hatanaka, T. Watanabe, K. Oishi, N. Ishida, M. Anzai, T. Mitani, H. Kato, S. Kishigami, K. Saeki, Y. Hosoi, A. Iritani and K. Matsumoto, *Biol. Reprod.*, 2009, **80**, 473–483.
44. E. Kowalska, E. Moriggi, C. Bauer, C. Dibner and S. A. Brown, *J. Biol. Rhythms*, 2010, **25**, 442–449.
45. Y. Umemura, J. Yoshida, M. Wada, Y. Tsuchiya, Y. Minami, H. Watanabe, G. Kondoh, J. Takeda, H. Inokawa, K. Horie and K. Yagita, *PLoS One*, 2013, **8**, e67241.
46. Y. Umemura, N. Koike, T. Matsumoto, S. H. Yoo, Z. Chen, N. Yasuhara, J. S. Takahashi and K. Yagita, *Proc. Natl. Acad. Sci. U. S. A.*, 2014, **111**, E5039–5048.
47. J. Kim, A. J. Woo, J. Chu, J. W. Snow, Y. Fujiwara, C. G. Kim, A. B. Cantor and S. H. Orkin, *Cell*, 2010, **143**, 313–324.
48. Z. Nie, G. Hu, G. Wei, K. Cui, A. Yamane, W. Resch, R. Wang, D. R. Green, L. Tessarollo, R. Casellas, K. Zhao and D. Levens, *Cell*, 2012, **151**, 68–79.
49. Y. Li, Y. Shan, G. K. Kilaru, S. Berto, G. Z. Wang, K. H. Cox, S. H. Yoo, S. Yang, G. Konopka and J. S. Takahashi, *eLife*, 2020, **9**, e54186.
50. J. K. Paulose, E. B. Rucker 3rd and V. M. Cassone, *PLoS One*, 2012, **7**, e49555.
51. P. Dierickx, M. W. Vermunt, M. J. Muraro, M. P. Creyghton, P. A. Doevendans, A. van Oudenaarden, N. Geijsen and L. W. Van Laake, *EMBO Rep.*, 2017, **18**, 1199–1212.

52. N. Yasuhara, R. Yamagishi, Y. Arai, R. Mehmood, C. Kimoto, T. Fujita, K. Touma, A. Kaneko, Y. Kamikawa, T. Moriyama, T. Yanagida, H. Kaneko and Y. Yoneda, *Dev. Cell*, 2013, **26**, 123–135.
53. Q. Gao, L. Zhou, S. Y. Yang and J. M. Cao, *Sci. Rep.*, 2016, **6**, 30070.
54. A. J. Gruber, W. A. Grandy, P. J. Balwierz, Y. A. Dimitrova, M. Pachkov, C. Ciaudo, E. Nimwegen and M. Zavolan, *Nucleic Acids Res.*, 2014, **42**, 9313–9326.
55. I. Lerner, O. Bartok, V. Wolfson, J. S. Menet, U. Weissbein, S. Afik, D. Haimovich, C. Gafni, N. Friedman, M. Rosbash and S. Kadener, *Nat. Commun.*, 2015, **6**, 7056.
56. Y. Umemura, I. Maki, Y. Tsuchiya, N. Koike and K. Yagita, *J. Biol. Rhythms*, 2019, **34**, 525–532.
57. E. A. Yu and D. R. Weaver, *Aging*, 2011, **3**, 479–493.
58. E. Maury, H. K. Hong and J. Bass, *Diabetes Metab.*, 2014, **40**, 338–346.
59. R. V. Kondratov, A. A. Kondratova, V. Y. Gorbacheva, O. V. Vykhovanets and M. P. Antoch, *Genes Dev.*, 2006, **20**, 1868–1873.
60. G. Yang, L. Chen, G. R. Grant, G. Paschos, W. L. Song, E. S. Musiek, V. Lee, S. C. McLoughlin, T. Grosser, G. Cotsarelis and G. A. FitzGerald, *Sci. Transl. Med.*, 2016, **8**, 324ra316.
61. H. Dan, T. Ruan and R. V. Sampogna, *Cell Rep.*, 2020, **31**, 107661.
62. M. Weger, N. Diotel, A. C. Dorsemans, T. Dickmeis and B. D. Weger, *Dev. Biol.*, 2017, **431**, 111–123.
63. P. Janich, G. Pascual, A. Merlos-Suarez, E. Batlle, J. Ripperger, U. Albrecht, H. Y. Cheng, K. Obrietan, L. Di Croce and S. A. Benitah, *Nature*, 2011, **480**, 209–214.
64. G. Solanas, F. O. Peixoto, E. Perdiguero, M. Jardi, V. Ruiz-Bonilla, D. Datta, A. Symeonidi, A. Castellanos, P. S. Welz, J. M. Caballero, P. Sassone-Corsi, P. Munoz-Canoves and S. A. Benitah, *Cell*, 2017, **170**, 678–692.e620.
65. N. Yang, J. Williams, V. Pekovic-Vaughan, P. Wang, S. Olabi, J. McConnell, N. Gossan, A. Hughes, J. Cheung, C. H. Streuli and Q. J. Meng, *Nat. Commun.*, 2017, **8**, 14287.
66. T. Casey, J. Crodian, A. Suárez-Trujillo, E. Erickson, B. Weldon, K. Crow, S. Cummings, Y. Chen, A. Shamay, S. J. Mabjeesh and K. Plaut, *Am. J. Physiol. Regul. Integr. Comp. Physiol.*, 2016, **311**, R1125–R1134.
67. T. Matsu-Ura, A. Dovzhenok, E. Aihara, J. Rood, H. Le, Y. Ren, A. E. Rosselot, T. Zhang, C. Lee, K. Obrietan, M. H. Montrose, S. Lim, S. R. Moore and C. I. Hong, *Mol. Cell*, 2016, **64**, 900–912.
68. M. V. Plikus, E. N. Van Spyk, K. Pham, M. Geyfman, V. Kumar, J. S. Takahashi and B. Andersen, *J. Biol. Rhythms*, 2015, **30**, 163–182.
69. T. Matsu-Ura, S. R. Moore and C. I. Hong, *J. Biol. Rhythms*, 2018, **33**, 5–14.
70. T. Matsuo, S. Yamaguchi, S. Mitsui, A. Emi, F. Shimoda and H. Okamura, *Science*, 2003, **302**, 255–259.
71. P. Bouchard-Cannon, L. Mendoza-Viveros, A. Yuen, M. Kaern and H. Y. Cheng, *Cell Rep.*, 2013, **5**, 961–973.

72. K. K. Lin, V. Kumar, M. Geyfman, D. Chudova, A. T. Ihler, P. Smyth, R. Paus, J. S. Takahashi and B. Andersen, *PLoS Genet.*, 2009, **5**, e1000573.
73. M. V. Plikus, C. Vollmers, D. de la Cruz, A. Chaix, R. Ramos, S. Panda and C. M. Chuong, *Proc. Natl. Acad. Sci. U. S. A.*, 2013, **110**, E2106–2115.
74. Y. Umemura, N. Koike, Y. Tsuchiya, H. Watanabe, G. Kondoh, R. Kageyama and K. Yagita, *Proc. Natl. Acad. Sci. U. S. A.*, 2022, **119**, e2114083119.
75. M. Matsuda, Y. Yamanaka, M. Uemura, M. Osawa, M. K. Saito, A. Nagahashi, M. Nishio, L. Guo, S. Ikegawa, S. Sakurai, S. Kihara, T. L. Maurissen, M. Nakamura, T. Matsumoto, H. Yoshitomi, M. Ikeya, N. Kawakami, T. Yamamoto, K. Woltjen, M. Ebisuya, J. Toguchida and C. Alev, *Nature*, 2020, **580**, 124–129.
76. T. Hamilton, *Br. J. Surg.*, 1969, **56**, 764–766.
77. E. Filipski and F. Levi, *Integr. Cancer Ther.*, 2009, **8**, 298–302.
78. S. Lee, L. A. Donehower, A. J. Herron, D. D. Moore and L. Fu, *PLoS One*, 2010, **5**, e10995.
79. M. W. Greene, *Cancer Lett.*, 2012, **318**, 115–123.
80. N. M. Kettner, H. Voicu, M. J. Finegold, C. Coarfa, A. Sreekumar, N. Putluri, C. A. Katchy, C. Lee, D. D. Moore and L. Fu, *Cancer Cell*, 2016, **30**, 909–924.
81. T. Papagiannakopoulos, M. R. Bauer, S. M. Davidson, M. Heimann, L. Subbaraj, A. Bhutkar, J. Bartlebaugh, M. G. Vander Heiden and T. Jacks, *Cell Metab.*, 2016, **24**, 324–331.
82. S. Davis, D. K. Mirick and R. G. Stevens, *J. Natl. Cancer Inst.*, 2001, **93**, 1557–1562.
83. T. Kubo, K. Ozasa, K. Mikami, K. Wakai, Y. Fujino, Y. Watanabe, T. Miki, M. Nakao, K. Hayashi, K. Suzuki, M. Mori, M. Washio, F. Sakauchi, Y. Ito, T. Yoshimura and A. Tamakoshi, *Am. J. Epidemiol.*, 2006, **164**, 549–555.
84. A. N. Viswanathan, S. E. Hankinson and E. S. Schernhammer, *Cancer Res.*, 2007, **67**, 10618–10622.
85. M. Conlon, N. Lightfoot and N. Kreiger, *Epidemiology*, 2007, **18**, 182–183.
86. L. G. Sigurdardottir, U. A. Valdimarsdottir, K. Fall, J. R. Rider, S. W. Lockley, E. Schernhammer and L. A. Mucci, *Cancer Epidemiol., Biomarkers Prev.*, 2012, **21**, 1002–1011.
87. M. G. Wendeu-Foyet and F. Menegaux, *Cancer Epidemiol., Biomarkers Prev.*, 2017, **26**, 985–991.
88. K. Papantoniou, E. E. Devore, J. Massa, S. Strohmaier, C. Vetter, L. Yang, Y. Shi, E. Giovannucci, F. Speizer and E. S. Schernhammer, *Int. J. Cancer*, 2018, **143**, 2709–2717.
89. K. Straif, R. Baan, Y. Grosse, B. Secretan, F. El Ghissassi, V. Bouvard, A. Altieri, L. Benbrahim-Tallaa and V. Cogliano, *Lancet Oncol.*, 2007, **8**, 1065–1066.
90. I. W. Humans, *IARC Monogr. Eval. Carcinog. Risks Hum.*, 2010, **98**, 9–764.
91. I. M. V. group, *Lancet Oncol.*, 2019, **20**, 1058–1059.

92. S. T. Chen, K. B. Choo, M. F. Hou, K. T. Yeh, S. J. Kuo and J. G. Chang, *Carcinogenesis*, 2005, **26**, 1241–1246.
93. K. L. Pogue-Geile, J. Lyons-Weiler and D. C. Whitcomb, *Cancer Lett.*, 2006, **243**, 55–57.
94. M. Y. Yang, J. G. Chang, P. M. Lin, K. P. Tang, Y. H. Chen, H. Y. Lin, T. C. Liu, H. H. Hsiao, Y. C. Liu and S. F. Lin, *Cancer Sci.*, 2006, **97**, 1298–1307.
95. S. Gery, N. Komatsu, N. Kawamata, C. W. Miller, J. Desmond, R. K. Virk, A. Marchevsky, R. McKenna, H. Taguchi and H. P. Koeffler, *Clin. Cancer Res.*, 2007, **13**, 1399–1404.
96. S. L. Winter, L. Bosnoyan-Collins, D. Pinnaduwage and I. L. Andrulis, *Neoplasia*, 2007, **9**, 797–800.
97. Y. M. Lin, J. H. Chang, K. T. Yeh, M. Y. Yang, T. C. Liu, S. F. Lin, W. W. Su and J. G. Chang, *Mol. Carcinog.*, 2008, **47**, 925–933.
98. N. Mostafaie, E. Kallay, E. Sauerzapf, E. Bonner, S. Kriwanek, H. S. Cross, K. R. Huber and W. Krugluger, *Mol. Carcinog.*, 2009, **48**, 642–647.
99. Q. Cao, S. Gery, A. Dashti, D. Yin, Y. Zhou, J. Gu and H. P. Koeffler, *Cancer Res.*, 2009, **69**, 7619–7625.
100. B. Jung-Hynes, W. Huang, R. J. Reiter and N. Ahmad, *J. Pineal Res.*, 2010, **49**, 60–68.
101. H. C. Xia, Z. F. Niu, H. Ma, S. Z. Cao, S. C. Hao, Z. T. Liu and F. Wang, *Can. J. Neurol. Sci.*, 2010, **37**, 365–370.
102. M. Y. Yang, W. C. Yang, P. M. Lin, J. F. Hsu, H. H. Hsiao, Y. C. Liu, H. J. Tsai, C. S. Chang and S. F. Lin, *J. Biol. Rhythms*, 2011, **26**, 136–148.
103. C. M. Hsu, S. F. Lin, C. T. Lu, P. M. Lin and M. Y. Yang, *Tumour Biol.*, 2012, **33**, 149–155.
104. G. Mazzoccoli, A. Piepoli, M. Carella, A. Panza, V. Pazienza, G. Benegiamo, O. Palumbo and E. Ranieri, *Biomed. Pharmacother.*, 2012, **66**, 175–179.
105. J. Litlekalsoy, K. Rostad, K. H. Kalland, J. G. Hostmark and O. D. Laerum, *BMC Cancer*, 2016, **16**, 549.
106. H. Xiong, Y. Yang, K. Yang, D. Zhao, H. Tang and X. Ran, *Int. J. Oncol.*, 2018, **52**, 279–287.
107. Z. Liu, K. Yu, J. Zheng, H. Lin, Q. Zhao, X. Zhang, W. Feng, L. Wang, J. Xu, D. Xie, Z. X. Zuo, Z. X. Liu and Q. Zheng, *Cancer Med.*, 2019, **8**, 1710–1720.
108. L. Zhang, L. J. Ptacek and Y. H. Fu, *Prog. Mol. Biol. Transl. Sci.*, 2013, **119**, 51–81.
109. F. J. Valenzuela, J. Vera, C. Venegas, S. Munoz, S. Oyarce, K. Munoz and C. Lagunas, *Int. J. Endocrinol.*, 2016, **2016**, 2746909.
110. E. Reszka, M. Przybek, O. Muurlink and B. Peplonska, *Cancer Lett.*, 2017, **390**, 137–145.
111. S. Morales-Santana, S. Morell, J. Leon, A. Carazo-Gallego, J. C. Jimenez-Lopez and M. Morell, *Front. Endocrinol.*, 2019, **10**, 104.
112. Y. Ye, Y. Xiang, F. M. Ozguc, Y. Kim, C. J. Liu, P. K. Park, Q. Hu, L. Diao, Y. Lou, C. Lin, A. Y. Guo, B. Zhou, L. Wang, Z. Chen, J. S. Takahashi, G. B. Mills, S. H. Yoo and L. Han, *Cell Syst.*, 2018, **6**, 314328.e312.

113. L. Fu, H. Pelicano, J. Liu, P. Huang and C. Lee, *Cell*, 2002, **111**, 41–50.
114. A. Mteyrek, E. Filipski, C. Guettier, A. Okyar and F. Levi, *Oncotarget*, 2016, **7**, 85832–85847.
115. P. A. Wood, X. Yang, A. Taber, E. Y. Oh, C. Ansell, S. E. Ayers, Z. Al-Assaad, K. Carnevale, F. G. Berger, M. M. Pena and W. J. Hrushesky, *Mol. Cancer Res.*, 2008, **6**, 1786–1793.
116. L. Fu, M. S. Patel, A. Bradley, E. F. Wagner and G. Karsenty, *Cell*, 2005, **122**, 803–815.
117. S. Gery, N. Komatsu, L. Baldjyan, A. Yu, D. Koo and H. P. Koeffler, *Mol. Cell*, 2006, **22**, 375–382.
118. L. Fu and N. M. Kettner, *Prog. Mol. Biol. Transl. Sci.*, 2013, **119**, 221–282.
119. G. Q. Daley, *Cold Spring Harbor Symp. Quant. Biol.*, 2008, **73**, 171–174.
120. D. Friedmann-Morvinski and I. M. Verma, *EMBO Rep.*, 2014, **15**, 244–253.
121. A. P. Feinberg, M. A. Koldobskiy and A. Gondor, *Nat. Rev. Genet.*, 2016, **17**, 284–299.
122. G. L. Brien, D. G. Valerio and S. A. Armstrong, *Cancer Cell*, 2016, **29**, 464–476.
123. S. B. Baylin and P. A. Jones, *Nat. Rev. Cancer*, 2011, **11**, 726–734.
124. Y. Cheng, C. He, M. Wang, X. Ma, F. Mo, S. Yang, J. Han and X. Wei, *Signal Transduction Targeted Ther.*, 2019, **4**, 62.
125. M. Widschwendter, H. Fiegl, D. Egle, E. Mueller-Holzner, G. Spizzo, C. Marth, D. J. Weisenberger, M. Campan, J. Young, I. Jacobs and P. W. Laird, *Nat. Genet.*, 2007, **39**, 157–158.
126. I. Ben-Porath, M. W. Thomson, V. J. Carey, R. Ge, G. W. Bell, A. Regev and R. A. Weinberg, *Nat. Genet.*, 2008, **40**, 499–507.
127. J. Kim and S. H. Orkin, *Genome Med.*, 2011, **3**, 75.
128. C. Hadjimichael, K. Chanoumidou, N. Papadopoulou, P. Arampatzi, J. Papamatheakis and A. Kretsovali, *World J. stem cells*, 2015, **7**, 1150–1184.
129. A. M. Intlekofer and L. W. S. Finley, *Nat. Metab.*, 2019, **1**, 177–188.
130. M. Abad, L. Mosteiro, C. Pantoja, M. Canamero, T. Rayon, I. Ors, O. Grana, D. Megias, O. Dominguez, D. Martinez, M. Manzanares, S. Ortega and M. Serrano, *Nature*, 2013, **502**, 340–345.
131. K. Ohnishi, K. Semi, T. Yamamoto, M. Shimizu, A. Tanaka, K. Mitsunaga, K. Okita, K. Osafune, Y. Arioka, T. Maeda, H. Soejima, H. Moriwaki, S. Yamanaka, K. Woltjen and Y. Yamada, *Cell*, 2014, **156**, 663–677.
132. M. Ohashi, Y. Umemura, N. Koike, Y. Tsuchiya, Y. Inada, H. Watanabe, T. Tanaka, Y. Minami, O. Ukimura, T. Miki, T. Tajiri, G. Kondoh, Y. Yamada and K. Yagita, *Genes Cells*, 2018, **23**, 60–69.
133. I. Versteege, N. Sevenet, J. Lange, M. F. Rousseau-Merck, P. Ambros, R. Handgretinger, A. Aurias and O. Delattre, *Nature*, 1998, **394**, 203–206.
134. R. S. Lee, C. Stewart, S. L. Carter, L. Ambrogio, K. Cibulskis, C. Sougnez, M. S. Lawrence, D. Auclair, J. Mora, T. R. Golub, J. A. Biegel, G. Getz and C. W. Roberts, *J. Clin. Invest.*, 2012, **122**, 2983–2988.

135. Y. Bu, A. Yoshida, N. Chitnis, B. J. Altman, F. Tameire, A. Oran, V. Gennaro, K. E. Armeson, S. B. McMahon, G. B. Wertheim, C. V. Dang, D. Ruggero, C. Koumenis, S. Y. Fuchs and J. A. Diehl, *Nat. Cell Biol.*, 2018, **20**, 104–115.
136. R. V. Puram, M. S. Kowalczyk, C. G. de Boer, R. K. Schneider, P. G. Miller, M. McConkey, Z. Tothova, H. Tejero, D. Heckl, M. Jaras, M. C. Chen, H. Li, A. Tamayo, G. S. Cowley, O. Rozenblatt-Rosen, F. Al-Shahrour, A. Regev and B. L. Ebert, *Cell*, 2016, **165**, 303–316.
137. M. Gautherie and C. Gros, *Chronobiologia*, 1977, **4**, 1–17.
138. A. Relogio, P. Thomas, P. Medina-Perez, S. Reischl, S. Bervoets, E. Gloc, P. Riemer, S. Mang-Fatehi, B. Maier, R. Schafer, U. Leser, H. Herzel, A. Kramer and C. Sers, *PLoS Genet.*, 2014, **10**, e1004338.
139. S. S. Lellupitiyage Don, H. H. Lin, J. J. Furtado, M. Qraitem, S. R. Taylor and M. E. Farkas, *Cell Cycle*, 2019, **18**, 2447–2453.
140. A. Shostak, *Int. J. Mol. Sci.*, 2017, **18**, 873.
141. J. Gaucher, E. Montellier and P. Sassone-Corsi, *Trends Cell Biol.*, 2018, **28**, 368–379.
142. A. A. Shafi and K. E. Knudsen, *Cancer Res.*, 2019, **79**, 3806–3814.
143. K. Kinouchi and P. Sassone-Corsi, *Nat. Rev. Cancer*, 2020, **20**, 645–661.
144. F. Hernandez-Rosas, C. A. Lopez-Rosas and M. V. Saavedra-Velez, *Biochem. Genet.*, 2020, **58**, 189–209.
145. H. Tokunaga, Y. Takebayashi, H. Utsunomiya, J. Akahira, M. Higashimoto, M. Mashiko, K. Ito, H. Niikura, S. Takenoshita and N. Yaegashi, *Acta Obstet. Gynecol. Scand.*, 2008, **87**, 1060–1070.
146. E. J. Lewintre, C. R. Martin, C. G. Ballesteros, D. Montaner, R. F. Rivera, J. R. Mayans and J. Garcia-Conde, *Haematologica*, 2009, **94**, 280–284.
147. L. Eisele, R. Prinz, L. Klein-Hitpass, H. Nuckel, K. Lowinski, J. Thomale, L. C. Moeller, U. Duhrsen and J. Durig, *Eur. J. Haematol.*, 2009, **83**, 320–327.
148. C. C. Yu, L. C. Chen, C. Y. Chiou, Y. J. Chang, V. C. Lin, C. Y. Huang, I. L. Lin, T. Y. Chang, T. L. Lu, C. H. Lee, S. P. Huang and B. Y. Bao, *Cancer Cell Int.*, 2019, **19**, 87.

CHAPTER 18

Disease Conditions

GERMAINE CORNELISSEN,*[a] DENIS GUBIN[b,c] AND KUNIAKI OTSUKA[a,d]

[a] Halberg Chronobiology Center, University of Minnesota, Minneapolis, Minnesota, USA; [b] Tyumen State Medical University, Tyumen, Russia; [c] Tyumen Cardiology Research Center, Tomsk National Research Medical Center, Russian Academy of Science, Tomsk, Russia; [d] Executive Medical Center, Totsuka Royal Clinic, Tokyo Women's Medical University, Tokyo, Japan

*Email: corne001@umn.edu

18.1 Introduction

Reviewing the role played by clock genes in pathology clearly reveals the extent to which disease conditions result from disturbances in an interdigitating network of different physiological systems, as illustrated in this chapter. The bidirectional relationship between circadian disruption and an increased disease risk appears to be a common feature, spanning from neurodegenerative diseases to metabolic, endocrine, immune, respiratory and cardiovascular conditions, as well as various types of cancer. The fact that circadian disruption is almost inseparable from disturbed sleep adds a layer of complexity to the problem as sleep disturbances also involve alterations in sleep stages, including the ultradian rapid eye movement (REM)-NREM cycle. For health maintenance and also in some cases as part of interventions to restore health, strategies aimed at reinforcing the circadian system have been implemented with some success. Maintaining a regular daily routine and relying on external synchronizers such as light, meal timing, and exercise to strengthen the circadian system can be beneficial, as

Chemical Biology No. 23
Chronobiology and Chronomedicine: From Molecular and Cellular Mechanisms to Whole Body Interdigitating Networks
Edited by Germaine Cornelissen and Tsuyoshi Hirota

Published by the Royal Society of Chemistry, www.rsc.org

reported in relation to metabolic disease, skin conditions, and cancer, among others. Given that hormones also drive the internal clock, some benefits may be derived from melatonin supplementation in the evening, as it is now used for certain sleep disorders in older adults. Novel treatments currently under development and in testing use small molecules to target the circadian clock directly, offering a complementary approach to chronotherapy, which is conventionally practiced by timing treatments at the optimal circadian phase to maximize efficacy while minimizing side effects.

Detailing the role played by clock genes in relation to the numerous disease conditions is beyond the scope of this chapter. Instead, based on a few examples, it highlights the pervasive role played by the circadian system in pathology, just as it does in all physiological systems, as described in Part II of this book. Doing so also brings to the fore how circadian disruption goes hand-in-hand with sleep disturbance and how the failure of one physiological system often implicates changes in other physiological systems as well. In particular, the immune system is involved in many other aspects of pathophysiology. Inflammation has indeed emerged as a key factor in serious diseases beyond infectious diseases, such as obesity and diabetes, cardiovascular disease, Alzheimer's disease, and cancer. Autoimmune conditions can be related to the skeletal system (rheumatoid arthritis), the central nervous system (multiple sclerosis), and the endocrine system (insulin-dependent diabetes mellitus). Immune reactions can trigger inflammation in the skin (as seen in psoriasis) and other organs (systemic lupus erythematosus can affect the joints, skin, brain, lungs, kidneys, and blood vessels).

Bio-energetic and metabolic features also play an essential role in the pathophysiology characterization of many diseases as metabolism influences and/or is influenced by all cellular processes. Studies of the intracellular metabolic pathways in immune cells have contributed to a better understanding of the mechanisms and functions of the immune system in both health and disease and led to the consideration of metabolism as a potential therapeutic target for the treatment of chronic inflammatory disorders.[1] Since mitochondria play a central role in the regulation of cellular metabolism, their dysfunction has been considered a major pathogenic mechanism underlying cardiovascular diseases and metabolic disorders.[2] Below, the role of the circadian system in a few selected disease conditions is briefly reviewed to illustrate the combined involvement of several physiological systems.

18.2 Sleep Disorders

Sleep is important for an organism to function properly. Disturbed sleep patterns can markedly influence essential functions, including memory consolidation and repair mechanisms, and ultimately result in a variety of disorders. The circadian system plays a key role in the sleep–wake cycle. A mismatch between the behavioral timing of sleep and the circadian timing of sleepiness/alertness can contribute to insomnia.[3] Together with bright light therapy, scheduled dim light, and blue-blocking glasses, melatonin supplementation is commonly used for treating sleep disorders in older adults.[4]

Some circadian sleep disorders have been linked to the circadian clock, using a computational approach.[5] A decrease in PER phosphorylation is accompanied by shortening of the endogenous period and a phase advance of the circadian clock, as seen in familial, advanced sleep phase syndrome. Likewise, an increase in PER phosphorylation is associated with the delayed sleep phase syndrome, characterized by lengthening of the autonomous period. In the model, the inability to entrain to the light–dark cycle corresponds to the non-24-hour sleep–wake cycle syndrome, in which the phase of the circadian clock drifts as a function of time.[5] Since the primary sleep disorders are partly heritable, defining heritable pathophysiologic mechanisms could delineate diagnoses and suggest treatment. Such a study was conducted in participants of European ancestry, which showed that several single-nucleotide polymorphisms of clock genes associated with several markers of obstructive sleep apnea, decreased polysomnographic sleep duration, periodic limb movements in sleep, and perhaps also with sleep latency and time of falling asleep.[6]

Insomnia is a condition characterized by problems with sleep onset, sleep maintenance, and/or early morning awakenings. Chronic insomnia is highly prevalent and associated with significant morbidity, as it confers risk for multiple psychiatric and medical disorders. It has been suggested that ultradian pulses in cortisol are involved in the maintenance of wakefulness during the day and their relative absence at night may allow for the consolidation of sleep and/or shorter nocturnal awakenings.[7] If so, a better understanding of the role played by ultradian rhythms and their coordination with the circadian system may help develop countermeasures to alleviate such a pervasive disorder.

Ultradian rhythms also play an important role in relation to narcolepsy.[8] Narcolepsy type 1 is caused by hypocretin/orexin deficiency and is associated with an inability to consolidate sleep. The main functions of this neuropeptide in regulating the organization of sleep and wake in humans are thought to include maintenance of wakefulness in the presence of moderate sleep deprivation; passive promotion of wakefulness, notably in the evening (driven by the circadian clock); and inhibition of REM sleep, with possible differential modulating effects on various subcomponents of the sleep stage.[9]

Obstructive sleep apnea is a prevalent form of sleep-disordered breathing, which results in sleep fragmentation and deprivation. It is associated with various cardiovascular and metabolic complications, including type 2 diabetes mellitus and obesity. Recent studies have indicated that circadian clock genes were dysregulated in obstructive sleep apnea, and that the impairment of circadian clocks drives the progression of the metabolic syndrome. A recent study[10] found that *Bmal1* and *Dec1* were upregulated, while *Per1* was reduced in patients with obstructive sleep apnea, and that these changes correlated with the apnea–hypopnea index and the presence of components of the metabolic syndrome. *Dec1* expression also had a relationship with oxidative indicators and inflammatory biomarkers in these patients, suggesting that *Dec1* can be used as a specific biomarker and a therapeutic target in patients with obstructive sleep apnea and the metabolic syndrome. Several studies have also reported the role of hypoxia-inducible factor 1α (HIF1α, a key regulator of oxygen metabolism) in obstructive sleep apnea, notably in relation to the development of circadian rhythm disorders.[11] Considering that current treatments aimed at correcting the respiratory disturbances are not universally successful and a non-negligible proportion of patients who are correctly using available therapies do not experience symptomatic relief, complementary interventions involving exposure to bright light in the morning and exercise aimed at strengthening the circadian rhythm have been proposed.[12]

18.3 Mental Disorders

A misalignment between the endogenous circadian system and the sleep–wake cycle is thought to contribute to the clinical status of patients suffering from a variety of psychiatric disorders. Sleep disturbances in major depressive disorder, for instance, are characterized by increased sleep latency, poorer sleep efficiency, reduced latency to the first REM sleep episode, and early-morning awakening. In patients with bipolar disorder, the circadian rhythm is advanced and REM sleep latency is altered.[13] Genotyping six genetic polymorphisms in BMAL1, PER2 and CRY2 in 566 patients with schizophrenia and 926 controls showed that the rs2290036-C variant of BMAL1 was over-represented in psychosis patients, and that the variants rs934945-G and rs10462023-G of PER2 were associated with a more severe psychotic disorder.[14] Genetic deficiencies in the neuronal PAS domain proteins NPAS1 and NPAS3 have also been linked to a variety of human psychiatric disorders, including bipolar disease, autism spectrum disorders, and schizophrenia.[15] Given the capability of multi-ligand binding of the wider mammalian bHLH-PAS family, it was proposed as an ideal class of transcription factors for direct targeting by small-molecule drugs.[15]

Rhythmic expression of CRY1 and PER2 was reportedly lost in fibroblasts from skin samples obtained from patients with chronic schizophrenia, in comparison to cells from healthy controls. Decreased expression of CLOCK, PER2 and CRY1 was also found in mononuclear blood cells from patients with schizophrenia experiencing their first episode of psychosis, as compared to blood cells from healthy controls.[16] Genome-wide association studies (GWASs) of chronotype confirmed the prominent role of circadian genes in psychiatric disorders. In GWAS on psychiatric traits, only BMAL1 was identified as one of the few loci differentiating schizophrenia from bipolar disorder. In support of the role of BMAL1 in schizophrenia, macaque monkeys in which BMAL1 was knocked out display symptoms similar to schizophrenia.[17]

Mood disorders such as depression, anxiety, and bipolar disorder are often associated with disrupted daily activity rhythms, observed, for instance, in shift workers. Several genetic variations in clock-related genes have also been reported in patients with mood disorders compared to healthy controls. The REV-ERBα protein was recently proposed as a key circadian nuclear receptor linking behavioral rhythms to mood regulation.[18] A study in mice showed that knocking down *Rev-erbα* disrupts the serotonergic system, which plays a central role in stress-induced mood behaviors. The expression of a gene that encodes the rate-limiting enzyme of serotonin synthesis is decreased in the raphe nuclei of these mice, along with 5-HT concentrations in the prefrontal cortex, which receives strong inputs from the raphe nuclei and controls stress-related behaviors.[18] Administration of SR8278, a REV-ERBα antagonist, to 6-hydroxydopamine-lesioned mice produced antidepressant and anxiolytic effects in a circadian time-dependent manner in this murine model of Parkinson's disease, suggesting that REV-ERBα is a potential therapeutic target for mood disorders related to circadian disturbances, such as for sundowning syndrome in Parkinson's disease.[19] However, the mechanism through which circadian rhythms contribute to mood disorders remains unknown. There is a growing body of evidence showing how the interaction between circadian and neurotransmitter systems influences the mood and behavior, including the role of current antidepressants and mood stabilizers. Treatments designed specifically to target circadian rhythm regulators, such as sleep, light, and melatonin, are being developed to produce an antidepressant response.[20]

PER3 was among the top six genes differentially expressed in blood between patients with major depressive disorder who died by suicide or by other means.[21] Major depressive disorder is the most common diagnosis among suicides, which have increased to over 48 000 deaths yearly in the United States. PER3 was increased both in the blood and in the brains of those who died by suicide.[21] Clock genes, including *Per3* have been associated with major depressive disorder and can affect the patient's response to selective serotonin-reuptake-inhibitor antidepressants.[22] Three single nucleotide polymorphisms of *Per3* were reported to predict adverse effects of these medications and drug compliance in patients with major depressive disorder.[22] A 2018 review that summarized findings regarding the role of clock genes in the etiology of major depressive disorder found that the most consistent result reported the association between SNP rs2287161 of CRY1 and the development of major depressive disorder.[23]

18.4 Neurodegenerative Conditions

Neurodegenerative diseases result in the progressive deterioration of the nervous system, with motor and cognitive impairments being the two most observable problems. Disrupted sleep and circadian rhythms have long been shown to be associated with reduced alertness and cognitive performance, even in young, clinically healthy individuals.[24] Impaired sleep and alertness are common symptoms of neurodegenerative disorders, and circadian dysfunction might exacerbate the disease process. Clinical studies and experiments in animal models of neurodegenerative disorders have documented the progressive nature of circadian dysfunction throughout the course of neurodegeneration, suggesting strategies for the restoration of circadian rhythmicity by involving behavioral and pharmacological interventions that target the sleep–wake cycle.[25] While circadian disturbances have generally been considered consequences of the neurodegeneration, circadian disruption may also contribute to the neurodegenerative process.[26] The circadian system has thus become an attractive target for research and clinical care innovations.

Alzheimer's disease is an age-related neurodegenerative disorder and the most common cause of dementia. Relative expression levels of PER1, PER2, and BMAL1 were detected in the bed nucleus of the stria terminalis and cingulate cortex, areas known to be involved in decision making and motivated behaviors, and in the pineal gland, in the brains of Alzheimer's disease patients and aged controls. The phase of these clock gene rhythms and phase relationships between genes and regions were found to differ in the brains of Alzheimer's disease patients compared to those of controls.[27] The expression of vasopressin mRNA in the human suprachiasmatic nucleus was also considerably reduced in Alzheimer's disease patients compared to age- and time-of-death-matched controls, and its circadian variation observed in controls was not detected in patients.[28] In Alzheimer's disease patients, cerebrospinal fluid orexin concentrations are elevated and correlate with total tau protein levels and with decreased sleep efficiency and REM sleep, which are associated with cognitive decline.[29] It has been proposed that orexin receptor antagonists, which can consolidate sleep, may reduce Alzheimer's disease pathology.[30] Expression of inducible nitric oxide synthase, which produces nitric oxide, important in REM sleep regulation is also increased during REM sleep in patients with Alzheimer's disease.[31] Since energy metabolism has been linked to Alzheimer's disease, promoting metabolic health in disease models and healthy aging may be desirable. Studies in a mouse model of Alzheimer's disease showed that the natural compound Nobiletin can alter the expression of several core clock genes (*Bmal1*, *Npas2*, and *Rorα*) in the cerebral cortex and activate various clock-controlled metabolic genes involved in insulin signaling and mitochondrial function. The expression of several genes related to Alzheimer's disease was also attenuated, and Aβ pathology in the cortex was improved.[32]

Sleep disorder is one of the most common and early non-motor symptoms in Parkinson's disease, a progressive neurodegenerative disease characterized by the degeneration of the dopaminergic neurons of the substantia nigra, leading to impaired motor control and movement. Nicotinamide adenine dinucleotide (NAD^+), one of the most abundant metabolites in the human body, is a critical cofactor in numerous reactions, notably those related to energy metabolism. NAD^+ also functions as a substrate for certain NAD^+-including sirtuins, poly-(ADP-ribose) polymerases (Parps), CD38/157, and sterile α and TIR motif-containing protein 1 (SARM1), used for DNA repair, the circadian clock, and protein deacetylation, rendering the maintenance of intracellular NAD^+ levels crucial for cellular functions and survival.[33] The therapeutic potential of NAD^+ precursors in Parkinson's disease is currently under investigation. Patients who received daily niacin treatment for 12 months showed improved motor function assessment scores and lower levels of neuro-inflammatory markers. Other trials have examined the effect of nicotinamide riboside on Parkinson's disease progression.[34]

Amyotrophic lateral sclerosis (ALS) is a progressive neurological disorder characterized by the degeneration of motor neurons in the spinal cord, brain stem, and motor cortex, limb weakness, paralysis, and death.[33] Mitochondrial dysfunction, oxidative stress, and neuro-inflammation, along with other impairments affect ALS patients, who show indications of NAD^+ homeostatic disruption, as they have reduced concentrations of serum and cerebrospinal fluid nicotinamide.[35] Studies in animal models indicate that increasing nicotinamide phosphoribosyltransferase activity can enhance motor neuron survival in ALS.[36] A pilot study in ALS patients showed that compared to placebo, 4 months of treatment with nicotinamide riboside and pterostilbene slowed the progression of symptoms significantly.[37] The expression of clock and clock-controlled genes was investigated in multiple tissues (SCN, spinal cord, gastrocnemius muscle, and liver) from mutant hSOD1-linked ALS mouse models; these genes underwent tissue-specific changes, leading to altered daily patterns, which may contribute to metabolic and redox impairment in ALS.[38] In the spinal cord of symptomatic mice expressing mutant hSOD1, Rev-erbα protein and mRNA expression were significantly downregulated.[39] Downregulation of Rev-erbα in primary astrocyte cultures further induced a pro-inflammatory phenotype and decreased the survival of co-cultured motor neurons, suggesting that Rev-erbα could be a potential therapeutic target to prevent astrocyte-mediated motor neuron toxicity in ALS.[39] It has also been reported that PPARγ, which contributes to the regulation of several key circadian genes, is dysregulated in neurodegenerative diseases.[40]

Glaucoma is a neurodegenerative disease in which retinal ganglion cells (RGCs) and the optic nerve are primarily affected.[41,42] Glaucoma patients share common genes with other neurodegenerative diseases such as Alzheimer's disease, Parkinson's disease, and ALS.[42] Due to RGCs damage, photic and non-photic signal transduction is compromised in glaucoma.[43] In advanced glaucoma, loss of intrinsically photo-receptive retinal ganglion cells occurs,[44,45] resulting in altered light perception, which provokes disruptions of systemic[46,47] and local[48] circadian rhythms, deterioration of sleep[49] and mood.[50–52] Similar to other neurodegenerative processes, potent anti-oxidants and substances that protect against neuro-inflammation such as NAD^+/NADH[53,54] and melatonin[43,47] can serve as biomarkers and therapeutic targets for glaucoma. Importantly, both NAD^+/NADH and melatonin have profound circadian rhythmicity; when their bioavailability is targeted, their administration must be properly timed.

18.5 Cancer

The circadian system plays an important role in the entire process of tumorigenesis and has the potential to be applied in tumor prevention, diagnosis, and treatment.

Disrupted expression of core clock genes has been associated with poor prognosis of cancers in many studies, but other investigations found no such association and some even reported that overexpression of some of the same genes was associated with unfavorable prognosis in cancer patients. A *meta*-analysis undertaken to clarify the influence of low expression of seven circadian clock genes suggested that low expression of *Per1*, *Per2* and *Npas2* played a role in cancer progression, and that low expressions of *Per1* and *Per2* could be unfavorable indicators for cancer prognosis, notably in the case of gastrointestinal cancers.[55] Efforts continue to find clinical indicators to predict patient prognosis and responsiveness to treatment, and to identify potential clock-drugs that could be used for adjuvant cancer treatment.[56]

The relation between cancer and the circadian system is bidirectional. Indeed, genes regulated by the circadian clock include various oncogenes and tumor suppressor genes, and the core genes of the circadian clock are also regulated by oncogenes and tumor suppressor genes, which are involved in tumor onset and malignancy. The circadian clock also regulates genes associated with metabolic and endocrine functions involved in tumor development.[57] Moreover, circadian disruption promotes the malignant progression of tumors by weakening the body's immune system. Tumor cells can develop multiple mechanisms to escape from immune system defenses. Cancer cells are surrounded by various stromal cells and by factors in the tumor microenvironment, which affects tumor progression and treatment efficacy. Angiogenesis, tumor-promoting inflammation, and immune evasion are mechanisms through which clock components in cancer cells regulate the biology of the tumor microenvironment.[58]

The circadian clock participates in cancer progression through several mechanisms. Tumor cell growth and senescence, cell proliferation and apoptosis, and DNA damage repair and various metabolic processes depend on biochemical reactions that are partially regulated by the circadian clock.[59] The circadian clock also plays an important role in the stemness of self-renewal cancer cell subsets as cancer stem cells undergo tumorigenesis, development, and metastasis.[60] Moreover, clock components regulate the expression of angiogenic factors such as hypoxia-inducible factor 1α, aryl hydrocarbon receptor nuclear translocator, and vascular endothelial growth factor in cancer cells. The circadian clock further regulates the inflammation mediated by myeloid cells, a crucial cancer marker.[57] Immune escape is another integral part of cancer progression.[61] The roles of CLOCK/BMAL1 have been reviewed by Huang *et al.*[57] for several types of cancer, notably hepatocellular carcinoma, colorectal cancer, breast cancer, and glioma. Specific circadian clock components in cancer cells or in cells of the tumor microenvironment, which affect their cross-talk with cancer cells, are examined by Xuan *et al.*,[58] who also discuss potential treatments.

The different mechanisms involving the circadian clock can be used to guide treatment. *Per2*'s role in sensitizing cancer cells to treatment has been documented in relation to doxorubicin-induced cell death in the case of breast cancer cells,[62] and in relation to cisplatin sensitivity in ovarian cancer[63] and esophageal cancer cells.[64] Bevacizumab, a monoclonal antibody that interferes with the ability of cancer cells to grow and spread, inhibits angiogenesis by preventing the binding of vascular endothelial growth factor A to its receptor, thereby achieving antitumor therapeutic effects in some patients with colorectal cancer.[57,65] The BMAL1–CLOCK–PER1–HK2 axis, which regulates glycolysis can be disrupted with metformin-based chronotherapy to overcome resistance to trastuzumab (a monoclonal antibody against human epidermal growth factor receptor 2) in trastuzumab-resistant HER2-positive gastric cancer.[66] Overexpression of both *Clock* and *Bmal1* suppressed cell growth in human colon cancer cells by preventing the cells from entering S from the G1 phase of the cell cycle by means of suppressing *CyclinD1* expression, thereby acquiring resistance to paclitaxel.[67] The latter results illustrate the complex role of clock genes in cancer. In another example, BMAL1 can increase the sensitivity of cells to anticancer drugs, but at the same time, enhance the invasiveness of tumors and lead to distant metastasis.[57]

Components associated with mechanisms through which the circadian system affects cancer progression have the potential to be used as targets for treatment. Some drugs may exert anti-tumor effects by regulating circadian rhythms, as reviewed by Huang *et al.*[57] Circadian-mediated immune responses can also be the focus of therapeutic approaches.[58] A correlation analysis of tumor mutation burden and BMAL1 gene expression in different cancers showed that BMAL1 has different or even opposite effects on the function of the immune system.[57] Although it is well established that circadian clock disruption is linked to an increased risk of cancer, details of the molecular mechanisms underlying this association are limited. Xuan *et al.*[58] highlighted some of the molecular circuitry that underlies the clock component-regulated cross-talk between cancer cells and cells of the tumor microenvironment, including innate myeloid cells, lymphocytes, and endothelial cells.

Owing to the rhythmic expression of circadian clock genes, administration times for clock-targeted drugs and chemoradiotherapy could be optimized to maximize anti-tumor effects and minimize toxic and adverse effects. Treating cancer by administering mTOR inhibitors at the peak of mTOR expression, for instance, was shown to improve the survival rate in mice.[68,69] The checkpoint inhibitor nivolumab, which targets T(CD8) cells, was found to be nearly four times as effective when administered in the following morning as compared to afternoon dosing in a cohort of patients with metastatic non-small-cell cancer.[70] Numerous studies have examined the main circadian hormone melatonin as a promising substance for reducing cancer risk or combat cancer.[71–75] Depending on its timing and dosing, melatonin not only has variable efficacy in preventing circadian disruption, but it also acts as an antioxidant, immune-modulator, and anti-inflammatory agent, and it also impacts apoptosis.[73] From an epigenetic viewpoint, melatonin affects the status of DNA methylation in cancer cells and models such as malignant glioma and breast carcinoma.[76] Another putative mechanism of action of melatonin in cancer was recently reported, namely, targeting the selective expression of different types of non-coding RNAs such as microRNAs, long ncRNAs, and circular RNAs that may time-specifically affect the expression of mRNAs, thereby differentially modulating functions of normal *versus* cancerous cells.[77,78]

To summarize, the involvement of clock genes has been documented in different kinds of cancers as well as in different stages of development, from tumor initiation and tumor promotion to malignant conversion and tumor progression. Liquid biopsies are being developed to predict patient prognosis and responsiveness to treatment. The exact nature of how circadian-controlled physiological processes interact at the host and tumor level, however, is very complex and remains poorly understood. Contradictions remain in the current state of knowledge regarding the underlying mechanisms. Inter-individual variability among patients also led to the consideration of personalized cancer chronotherapy.[79]

18.6 Role of the 12-hour Component in Various Disease Conditions

Biological rhythms expand beyond the circadian system.[80,81] In the early stage of Alzheimer's disease, the circadian variation of blood pressure and heart rate exhibited a reduced harmonic content.[82,83] The 12-hour and 8-hour components, which usually contribute to the circadian waveform of these variables, were not detectable for patients with Alzheimer's disease, whereas the 24-hour rhythm was maintained. Loss of ultradian variation in plasma 17-OHCS was also reported for patients with Down syndrome, compared to clinical health.[84]

Recent observations suggest that the 12-hour (circasemidian) component could help consolidate the circadian system. It could provide a ray of hope for human health in our current 24-hour society, which leads to internal desynchrony and plays a major role in the development of adverse health consequences from circadian disruption. Clear 12-hour transcriptional rhythms are detected in the liver, also in the lung, kidney, adrenal gland, heart, and even in the hypothalamus of the central nervous system.[85] A series of studies from Pittsburgh's group[86–93] identified a 12-hour oscillator independent from the 24-hour circadian clock. It is thought to be regulated by the unfolded protein response (UPR) transcription factor, the spliced form of X-Box Binding Protein 1 (XBP1s). A large protein called Son is essential for appropriate subnuclear organization of pre-mRNA splicing factors and for promoting normal cell cycle progression.[94,95] An evolutionarily conserved XBP1s-SON axis coordinates a rapidly functioning feedforward loop connecting nuclear speckle liquid–liquid phase separation and proteostasis, resulting in a highly efficient genetic information transfer that operates at multiple temporal scales. At the molecular level, it modulates the rates of chemical reactions; at the mesoscale, it organizes large structures within cells; and at the cellular level, it facilitates the localization of cellular materials and homeostatic responses.[91,95,96]

Emerging from an XBP1s–SON axis, recent studies of the 12-hour rhythm indicate its involvement in immune regulation[92,94] and in several disease conditions, including cardiovascular dysfunction and morning hypertension,[97–99] and nonalcoholic fatty liver disease.[86,88–91] In addition, because the XBP1s–SON axis amplifies p53-mediated gene expression, which is involved in the DNA damage response as *p53* is one of the tumor suppressor genes,[100] it may enhance overall anti-aging effects and longevity.[88,91] Otsuka *et al.*[98] reported that together with enhanced 8-hour and/or 6-hour components, an amplified 12-hour rhythm of heart rate variability could help consolidate the circadian system, suggesting the additional involvement of the immune function by means of NF-kB signaling[85,92] and/or of the hypothalamic–pituitary–adrenal (HPA) axis[101,102] in anti-aging effects in space.

The 12-hour rhythm may provide an evolutionary advantage, as it may have evolved earlier than the circadian clock: 12-hour clock genes are conserved in even more divergent species.[80,81,95,98,103–108] As shown by Otsuka *et al.*,[99] optimizing the circadian–circasemidian coupling, *i.e.*, dynamically balancing the interactions between circadian and ultradian oscillations could play an important role in improving human health. It could be part of a precision medicine based on chronomodecine, including consideration of several rhythmic components in a broad time structure to harmonize their constellation.

18.7 Concluding Remarks

As clock genes are involved in all physiological systems, they also play a role in most disease conditions. Sleep is often disturbed in the presence of disease, from psychological to cardio-metabolic and neurodegenerative conditions. Sleep disruption also modifies the immune system. Studying the complex interactions between clock genes, metabolic, inflammatory and other pathways as constellations rather than isolated conditions is giving rise to novel biomarkers for predicting disease activities and therapeutic efficacies. At the organismic level, marker rhythms provide information beyond differences in mean values, changes in their amplitude, phase, and/or period, potentially serving as additional guides for determining the kind and timing of needed interventions.

Maintaining a robust, synchronized circadian system remains a hallmark of health. In the presence of an elevated disease risk or overt disease, circadian disruption can be counteracted by reinforcing the circadian system by means of lifestyle changes and optimally timed medications to help prevent disease development or even restore health. Directly targeting the circadian clock has shown promise in the laboratory. While circadian medicine has made progress since the first realization that a broad time structure is endogenous, the complexity of the interactions of clock genes with a number of important pathways requires additional research to gain a better understanding of how a chronobiologic approach can best help prevent, diagnose, and treat disease conditions. In this respect, recent evidence suggests how the 12-hour component could also be targeted to help consolidate the circadian system, as reviewed above.

Bridging knowledge gained at the cellular and molecular level with results from outcome studies complemented by information from physiological monitoring and accurate diaries of daily activities, including the timing of medications, may pave the way toward a more personalized approach to treatment. Accounting for both pharmacokinetics and pharmacodynamics of the drugs used and the fact that they can also undergo circadian changes can further improve outcomes. On the other hand, finding general rules regarding effects of gene expression in different disease conditions and at different stages of a given disease will bring further clarity regarding the safety and efficacy of treatments, notably those targeting the circadian system more directly. Meta-analyses such as those undertaken within the scope of cancer[55,60,61,109] are likely to help in this respect.

References

1. M. C. Maiuri and P. Maffia, Cellular metabolism and diseases, *Br. J. Pharmacol.*, 2021, **178**, 2031–2033.
2. R. Tian, W. S. Colucci, Z. Arany, M. M. Bachschmid, S. W. Ballinger, S. Boudina, J. E. Bruce, D. W. Busija, S. Dikalov, G. W. I. I. Dorn, Z. S. Galis, R. A. Gottlieb, D. P. Kelly, R. N. Kitsis, M. J. Kohr, D. Levy, E. D. Lewandowski, J. M. McClung, D. Mochly-Rosen, K. D. O'Brien, B. O'Rourke, J. Y. Park, P. Ping, M. N. Sack, S. S. Sheu, Y. Shi, S. Shiva, D. C. Wallace, R. G. Weiss, H. J. Vernon, R. Wong and L. S. Longacre, Unlocking the secrets of mitochondria in the cardiovascular system: Path to a cure in heart failure — A report from the 2018 National Heart, Lung, and Blood Institute Workshop, *Circulation*, 2019, **140**(14), 1205–1216.
3. L. M. Swanson and G. B. Raglan, Circadian interventions as adjunctive therapies to cognitive-behavioral therapy for insomnia, *Sleep Med. Clin.*, 2023, **18**(1), 21–30.
4. C. Tuft, E. Matar, Z. Menczel Schrire, R. R. Grunstein, B. J. Yee and C. M. Hoyos, Current insights into the risks of using melatonin as a treatment for sleep disorders in older adults, *Clin. Interventions Aging*, 2023, **18**, 49–59.

5. A. Goldbeter and J. C. Leloup, From circadian clock mechanism to sleep disorders and jet lag: Insights from a computational approach, *Biochem. Pharmacol*, 2021, **191**, 114482.
6. D. F. Kripke, L. E. Kline, C. M. Nievergelt, S. S. Murray, F. F. Shadan, A. Dawson, J. S. Poceta, J. Cronin, S. M. Jamil, G. J. Tranah, R. T. Loving, A. P. Grizas and E. K. Hahn, Genetic variants associated with sleep disorders, *Sleep Med.*, 2015, **16**(2), 217–224.
7. I. Vargas, A. N. Vgontzas, J. L. Abelson, R. T. Faghih, K. H. Morales and M. L. Perlis, Altered ultradian cortisol rhythmicity as a potential neurobiologic substrate for chronic insomnia, *Sleep Med. Rev.*, 2018, **41**, 234–243.
8. P. Passouant, F. Halberg, R. Genicot, L. Popoviciu and M. Baldy-Moulinier, La périodicité des accès narcoleptiques et le rythme ultradien du sommeil rapide., *Rev. Neurol.*, 1969, **121**, 155–164.
9. E. Mignot, J. Zeitzer, F. Pizza and G. Plazzi, Sleep problems in narcolepsy and the role of hypocretin/orexin deficiency, *Front. Neurol. Neurosci.*, 2021, **45**, 103–116.
10. X. Li, X. Liu, Q. Meng, X. Wu, X. Bing, N. Guo, X. Zhao, X. Hou, B. Wang, M. Xia and H. Li, Circadian clock disruptions link oxidative stress and systemic inflammation to metabolic syndrome in obstructive sleep apnea patients, *Front. Physiol.*, 2022, **13**, 932596.
11. T. Xie, D. Guo, J. Luo, Z. Guo, S. Zhang, A. Wang, X. Wang, X. Wang, W. Cao, L. Su, J. Guo, R. Huang and Y. Xiao, The relationship between HIF1alpha and clock gene expression in patients with obstructive sleep apnea, *Nat. Sci. Sleep*, 2022, **14**, 381–392.
12. I. Soreca, The role of circadian rhythms in Obstructive Sleep Apnea symptoms and novel targets for treatment, *Chronobiol. Int.*, 2021, **38**(9), 1274–1282.
13. E. W. Lamont, D. Legault-Coutu, N. Cermakian and D. B. Boivin, The role of circadian clock genes in mental disorders, *Dialogues Clin. Neurosci.*, 2007, **9**(3), 333–342.
14. J. J. Liu, D. Sudic Hukic, Y. Forsell, M. Schalling, U. Osby and C. Lavebratt, Depression-associated ARNTL and PER2 genetic variants in psychotic disorders, *Chronobiol. Int.*, 2015, **32**(4), 579–584.
15. D. Wu, X. Su, N. Potluri, Y. Kim and F. Rastinejad, NPAS1-ARNT and NPAS3-ARNT crystal structures implicate the bHLH-PAS family as multi-ligand binding transcription factors, *eLife*, 2016, **5**, e18790.
16. A. S. Johansson, B. Owe-Larsson, J. Hetta and G. B. Lundkvist, Altered circadian clock gene expression in patients with schizophrenia, *Schizophrenia Research*, 2016, **174**(1-3), 17–23.
17. M. von Schantz, M. A. Leocadio-Miguel, M. J. McCarthy, S. Papiol and D. Landgraf, Genomic perspectives on the circadian clock hypothesis of psychiatric disorders, *Adv. Genet.*, 2021, **107**, 153–191.
18. T. Otsuka, H. T. Le, Z. L. Thein, H. Ihara, F. Sato, T. Nakao and A. Kohsaka, Deficiency of the circadian clock gene Rev-erbalpha induces mood disorder-like behaviours and dysregulation of the serotonergic system in mice, *Physiol. Behav.*, 2022, **256**, 113960.

19. J. Kim, I. Park, S. Jang, M. Choi, D. Kim, W. Sun, Y. Choe, J. W. Choi, C. Moon, S. H. Park, H. K. Choe and K. Kim, Pharmacological rescue with SR8278, a circadian nuclear receptor REV-ERBalpha antagonist as a therapy for mood disorders in Parkinson's disease, *Neurotherapeutics*, 2022, **19**(2), 592–607.
20. M. R. Scott and C. A. McClung, Circadian rhythms in mood disorders, *Adv. Exp. Med. Biol.*, 2021, **1344**, 153–168.
21. F. Mamdani, M. D. Weber, B. Bunney, K. Burke, P. Cartagena, D. Walsh, F. S. Lee, J. Barchas, A. F. Schatzberg, R. M. Myers, S. J. Watson, H. Akil, M. P. Vawter, W. E. Bunney and A. Sequeira, Identification of potential blood biomarkers associated with suicide in major depressive disorder, *Transl. Psychiatry Psychiatry*, 2022, **12**(1), 159.
22. Y. Xu, H. Ma, T. Zhao, D. Wen, Y. Wen, D. Qiao and Z. Liu, Association between *Period 3* gene polymorphisms and adverse effects of antidepressants for major depressive disorder, *Genet. Test. Mol. Biomarkers*, 2019, **23**(12), 843–849.
23. M. Buoli, M. Serati, S. Grassi, L. Pergoli, L. Cantone, A. C. Altamura and V. Bollati, The role of clock genes in the etiology of Major Depressive Disorder: Special Section on "Translational and Neuroscience Studies in Affective Disorders". Section Editor, Maria Nobile MD, PhD. This Section of JAD focuses on the relevance of translational and neuroscience studies in providing a better understanding of the neural basis of affective disorders. The main aim is to briefly summaries relevant research findings in clinical neuroscience with particular regards to specific innovative topics in mood and anxiety disorders, *J. Affective Disord.*, 2018, **234**, 351–357.
24. D. J. Dijk, J. F. Duffy and C. A. Czeisler, Circadian and sleep/wake dependent aspects of subjective alertness and cognitive performance, *J. Sleep Res.*, 1992, **1**(2), 112–117.
25. A. Videnovic, A. S. Lazar, R. A. Barker and S. Overeem, The clocks that time us'–circadian rhythms in neurodegenerative disorders, *Nat. Rev. Neurol.*, 2014, **10**(12), 683–693.
26. M. Nassan and A. Videnovic, Circadian rhythms in neurodegenerative disorders, *Nat. Rev. Neurol.*, 2022, **18**(1), 7–24.
27. N. Cermakian, E. W. Lamont, P. Boudreau and D. B. Boivin, Circadian clock gene expression in brain regions of Alzheimer 's disease patients and control subjects, *J. Biol. Rhythms*, 2011, **26**(2), 160–170.
28. R. Y. Liu, J. N. Zhou, W. J. Hoogendijk, J. van Heerikhuize, W. Kamphorst, U. A. Unmehopa, M. A. Hofman and D. F. Swaab, Decreased vasopressin gene expression in the biological clock of Alzheimer disease patients with and without depression, *J. Neuropathol. Exp. Neurol.*, 2000, **59**(4), 314–322.
29. C. Liguori, A. Romigi, M. Nuccetelli, S. Zannino, G. Sancesario, A. Martorana, M. Albanese, N. B. Mercuri, F. Izzi, S. Bernardini, A. Nitti, G. M. Sancesario, F. Sica, M. G. Marciani and F. Placidi, Orexinergic system dysregulation, sleep impairment, and cognitive decline in Alzheimer disease, *JAMA Neurol.*, 2014, **71**(12), 1498–1505.

30. F. Gao, T. Liu, M. Tuo and S. Chi, The role of orexin in Alzheimer disease: From sleep-wake disturbance to therapeutic target, *Neurosci. Lett.*, 2021, **765**, 136247.
31. I. J. Tseng, H. C. Liu, R. Y. Yuan, J. J. Sheu, J. M. Yu and C. J. Hu, Expression of inducible nitric oxide synthase (iNOS) and period 1 (PER1) clock gene products in different sleep stages of patients with cognitive impairment, *J. Clin. Neurosci.*, 2010, **17**(9), 1140–1143.
32. E. Kim, K. Nohara, M. Wirianto, G. Escobedo Jr, J. Y. Lim, R. Morales, S. H. Yoo and Z. Chen, Effects of the clock modulator Nobiletin on circadian rhythms and pathophysiology in female mice of an Alzheimer's disease model, *Biomolecules*, 2021, **11**(7), 1004.
33. S. Lautrup, D. A. Sinclair, M. P. Mattson and E. F. Fang, NAD^+ in brain aging and neurodegenerative disorders, *Cell Metab.*, 2019, **30**, 630–655.
34. S. Lundt and S. Ding, NAD^+ metabolism and diseases with motor dysfunction, *Genes*, 2021, **12**, 1776.
35. E. Blacher, S. Bashiardes, H. Shapiro, D. Rothschild, U. Mor, M. Dori-Bachash, C. Kleimeyer, C. Moresi, Y. Harnik, M. Zur, M. Zabari, R. B. Brik, D. Kviatcovsky, N. Zmora, Y. Cohen, N. Bar, I. Levi, N. Amar, T. Mehlman, A. Brandis, I. Biton, Y. Kuperman, M. Tsoory, L. Alfahel, A. Harmelin, M. Schwartz, A. Israelson, L. Arike, M. E. V. Johansson, G. C. Hansson, M. Gotkine, E. Segal and E. Elinav, Potential roles of gut microbiome and metabolites in modulating ALS in mice, *Nature*, 2019, **572**, 474–480.
36. R. Tesla, H. P. Wolf, P. Xu, J. Drawbridge, S. J. Estill, P. Huntington, L. McDaniel, W. Knobbe, A. Burket, S. Tran, R. Starwalt, L. Morlock, J. Naidoo, N. S. Williams, J. M. Ready, S. L. McKnight and A. A. Pieper, Neuroprotective efficacy of aminopropyl carbazoles in a mouse model of amyotrophic lateral sclerosis, *Proc. Natl. Acad. Sci. U. S. A.*, 2012, **109**(42), 17016–17021.
37. J. E. de la Rubia, E. Drehmer, J. L. Platero, M. Benlloch, J. Caplliure-Llopis, C. Villaron-Casales, N. de Bernardo, J. AlarcÓn, C. Fuente, S. Carrera, D. Sancho, P. GarcÍa-Pardo, R. Pascual, M. JuÁrez, M. Cuerda-Ballester, A. Forner, S. Sancho-Castillo, C. Barrios, E. Obrador, P. Marchio, R. Salvador, H. E. Holmes, R. W. Dellinger, L. Guarente and J. M. Estrela, Efficacy and tolerability of EH301 for amyotrophic lateral sclerosis: a randomized, double-blind, placebo-controlled human pilot study, *Amyotrophic Lateral Scler. Frontotemporal Degener.*, 2019, **20**(1-2), 115–122.
38. K. M. Killoy, M. Pehar, B. A. Harlan and M. R. Vargas, Altered expression of clock and clock-controlled genes in a hSOD1-linked amyotrophic lateral sclerosis mouse model, *FASEB J.*, 2021, **35**(2), e21343.
39. K. M. Killoy, B. A. Harlan, M. Pehar and M. R. Vargas, NR1D1 down-regulation in astrocytes induces a phenotype that is detrimental to cocultured motor neurons, *FASEB J.*, 2022, **36**(4), e22262.
40. A. Vallee, Y. Lecarpentier, R. Guillevin and J. N. Vallee, Thermodynamics in neurodegenerative diseases: interplay between canonical

WNT/beta-catenin pathway-PPARgamma, energy metabolism and circadian rhythms, *NeuroMol. Med.*, 2018, **20**(2), 174–204.
41. N. Gupta and Y. H. Yücel, Glaucoma as a neurodegenerative disease, *Curr. Opin. Ophthalmol.*, 2007, **18**(2), 110–114.
42. A. Artero-Castro, F. J. Rodriguez-Jimenez, P. Jendelova, K. B. VanderWall, J. S. Meyer and S. Erceg, Glaucoma as a neurodegenerative disease caused by intrinsic vulnerability factors, *Prog. Neurobiol.*, 2020, **193**, 101817.
43. D. Gubin and D. Weinert, Melatonin, circadian rhythms and glaucoma: current perspective, *Neural Regener. Res.*, 2022, **17**(8), 1759–1760.
44. B. Feigl, D. Mattes, R. Thomas and A. J. Zele, Intrinsically photosensitive (melanopsin) retinal ganglion cell function in glaucoma, *Invest. Ophthalmol. Visual Sci.*, 2011, **52**, 4362–4367.
45. E. A. Obara, J. Hannibal, S. Heegaard and J. Fahrenkrug, Loss of melanopsin-expressing retinal ganglion cells in severely staged glaucoma patients, *Invest. Ophthalmol. Visual Sci.*, 2016, **57**, 4661–4667.
46. D. G. Gubin, T. N. Malishevskaya, Y. S. Astakhov, S. Y. Astakhov, G. Cornelissen, V. A. Kuznetsov and D. Weinert, Progressive retinal ganglion cell loss in primary open-angle glaucoma is associated with temperature circadian rhythm phase delay and compromised sleep, *Chronobiol. Int.*, 2019, **36**, 564–577.
47. D. Gubin, V. Neroev, T. Malishevskaya, G. Cornelissen, S. Y. Astakhov, S. Kolomeichuk, N. Yuzhakova, Y. Kabitskaya and D. Weinert, Melatonin mitigates disrupted circadian rhythms, lowers intraocular pressure, and improves retinal ganglion cells function in glaucoma, *J. Pineal Res.*, 2021, **70**, e12730.
48. V. Neroev, T. Malishevskaya, D. Weinert, S. Astakhov, S. Kolomeichuk, G. Cornelissen, Y. Kabitskaya, E. Boiko, I. Nemtsova and D. Gubin, Disruption of 24 hours rhythm in intraocular pressure correlates with retinal ganglion cell loss in glaucoma, *Int. J. Mol. Sci.*, 2020, **22**, 359.
49. C. P. Gracitelli, G. L. Duque-Chica, M. Roizenblatt, A. L. Moura, B. V. Nagy, G. Ragot de Melo, P. D. Borba, S. H. Teixeira, S. Tufik, D. F. Ventura and A. Paranhos Jr., Intrinsically photosensitive retinal ganglion cell activity is associated with decreased sleep quality in patients with glaucoma, *Ophthalmology*, 2015, **122**, 1139–1148.
50. T. Yoshikawa, K. Obayashi, K. Miyata, T. Ueda, N. Kurumatani, K. Saeki and N. Ogata, Association between glaucomatous optic disc and depressive symptoms independent of light exposure profiles: a cross-sectional study of the HEIJO-KYO cohort, *Br. J. Ophthalmol.*, 2019, **103**(8), 1119–1122.
51. T. Yoshikawa, K. Obayashi, K. Miyata, K. Saeki and N. Ogata, Lower cognitive function in patients with functionally and structurally severe glaucoma: The LIGHT Study, *J. Glaucoma*, 2021, **30**(10), 882–886.
52. D. Gubin, V. Neroev, T. Malishevskaya, S. Kolomeichuk, G. Cornelissen, N. Yuzhakova, A. Vlasova and D. Weinert, Depression scores are associated with retinal ganglion cells loss, *J. Affective Disord.*, 2023, **333**, 290–296.

53. G. Cimaglia, M. Votruba, J. E. Morgan, H. André and P. A. Williams, Potential therapeutic benefit of NAD^+ supplementation for glaucoma and age-related macular degeneration, *Nutrients*, 2020, **12**(9), 2871.
54. B. Petriti, P. A. Williams, G. Lascaratos, K. Y. Chau and D. F. Garway-Heath, Neuroprotection in glaucoma: NAD^+/NADH redox state as a potential biomarker and therapeutic target, *Cells*, 2021, **10**(6), 1402.
55. J. Zhang, H. Lv, M. Ji, Z. Wang and W. Wu, Low circadian clock genes expression in cancers: A meta-analysis of its association with clinicopathological features and prognosis, *PLoS One*, 2020, **15**(5), e0233508.
56. Y. Liu, S. Guo, Y. Sun, C. Zhang, J. Gan, S. Ning and J. Wang, CRS: a circadian rhythm score model for predicting prognosis and treatment response in cancer patients, *J. Transl. Med.*, 2023, **21**(1), 185.
57. C. Huang, C. Zhang, Y. Cao, J. Li and F. Bi, Major roles of the circadian clock in cancer, *Cancer Biol. Med.*, 2023, **20**(1), 1–24.
58. W. Xuan, F. Khan, C. D. James, A. B. Heimberger, M. S. Lesniak and P. Chen, Circadian regulation of cancer cell and tumor microenvironment crosstalk, *Trends Cell Biol.*, 2021, **31**(11), 940–950.
59. A. Sancar and R. N. van Gelder, Clocks, cancer, and chronochemotherapy, *Science*, 2021, **371**, eabb0738.
60. Y. Lee and A. S. Tanggono, Potential role of the circadian clock in the regulation of cancer stem cells and cancer Ttherapy, *Int. J. Mol. Sci.*, 2022, **23**(22), 14181.
61. S. Pesce, S. Trabanelli, C. Di Vito, M. Greppi, V. Obino, F. Guolo, P. Minetto, M. Bozzo, M. Calvi, E. Zaghi, S. Candiani, R. M. Lemoli, C. Jandus, D. Mavilio and E. Marcenaro, Cancer immunotherapy by blocking immune checkpoints on innate lymphocytes, *Cancers*, 2020, **12**(12), 3504.
62. M. I. Mitchell and A. M. Engelbrecht, Circadian rhythms and breast cancer: the role of *Per2* in doxorubicin-induced cell death, *J. Toxicol.*, 2015, **2015**, 392360.
63. Z. Wang, F. Li, M. Wei, S. Zhang and T. Wang, Circadian clock protein PERIOD2 suppresses the PI3K/Akt pathway and promotes cisplatin sensitivity in ovarian cancer, *Cancer Manage. Res.*, 2020, **12**, 11897–11908.
64. J. A. Redondo, R. Bibes, A. Vercauteren Drubbel, B. Dassy, X. Bisteau, E. Maury and B. Beck, PER2 Circadian oscillation sensitizes esophageal cancer cells to chemotherapy, *Biology*, 2021, **10**(4), 266.
65. E. Burgermeister, F. Battaglin, F. Eladly, W. Wu, F. Herweck, N. Schulte, J. Betge, N. Hartel, J. N. Kather, C. A. Weis, T. Gaiser, A. Marx, C. Weiss, R. Hofheinz, I. S. Miller, F. Loupakis, H. J. Lenz, A. T. Byrne and M. P. Ebert, Aryl hydrocarbon receptor nuclear translocator-like (ARNTL/BMAL1) is associated with bevacizumab resistance in colorectal cancer *via* regulation of vascular endothelial growth factor A, *EBioMedicine*, 2019, **45**, 139–154.
66. J. Wang, Q. Huang, X. Hu, S. Zhang, Y. Jiang, G. Yao, K. Hu, X. Xu, B. Liang, Q. Wu, Z. Ma, Y. Wang, C. Wang, Z. Wu, X. Rong, W. Liao and

M. Shi, Disrupting circadian rhythm *via* the PER1-HK2 axis reverses trastuzumab resistance in gastric cancer, *Cancer Res.*, 2022, **82**(8), 1503–1517.

67. W. Sakamoto and S. Takenoshita, Overexpression of both *Clock* and *Bmal1* inhibits entry to S phase in human colon cancer cells, *Fukushima J. Med. Sci.*, 2015, **61**(2), 111–124.
68. D. Mullins, D. Proulx, A. Saoudi and C. E. Ng, Chronomodulation of topotecan or X-radiation treatment increases treatment efficacy without enhancing acute toxicity, *Int. J. Radiat. Oncol., Biol., Phys.*, 2005, **62**, 230–237.
69. N. Ozturk, D. Ozturk, Z. Pala-Kara, E. Kaptan, S. Sancar-Bas, N. Ozsoy, S. Cinar, G. Deniz, X. M. Li, S. Giacchetti, F. Levi and A. Okyar, The immune system as a chronotoxicity target of the anticancer mTOR inhibitor everolimus, *Chronobiol. Int.*, 2018, **35**(5), 705–718.
70. A. Karaboue, T. Collon, I. Pavese, V. Bodiguel, J. Cucherousset, E. Zakine, P. F. Innominato, M. Bouchahda, R. Adam and F. Levi, Time-dependent efficacy of checkpoint inhibitor nivolumab: results from a pilot study in patients with metastatic non-small-cell lung cancer, *Cancers*, 2022, **14**(4), 896.
71. G. Cornelissen, F. Halberg, F. Perfetto, R. Tarquini, C. Maggioni and L. Wetterberg, Melatonin involvement in cancer: methodological considerations, in *The Pineal Gland and Cancer: Neuroimmunoendocrine Mechanisms in Malignancy*, ed. C. Bartsch, H. Bartsch, D. E. Blask, D. P. Cardinali, W. J. M. Hrushesky and W. Mecke, Springer, Heidelberg, 2001, pp. 117–149.
72. P. Lissoni, M. Chilelli, S. Villa, L. Cerizza and G. Tancini, Five years survival in metastatic non-small cell lung cancer patients treated with chemotherapy alone or chemotherapy and melatonin: a randomized trial, *J. Pineal Res.*, 2003, **35**(1), 12–15.
73. R. J. Reiter, S. A. Rosales-Corral, D. X. Tan, D. Acuna-Castroviejo, L. Qin, S. F. Yang and K. Xu, Melatonin, a full service anti-cancer agent: inhibition of initiation, progression and metastasis, *Int. J. Mol. Sci.*, 2017, **18**(4), 843.
74. G. M. Zharinov, O. A. Bogomolov, I. V. Chepurnaya, N. Y. Neklasova and V. N. Anisimov, Melatonin increases overall survival of prostate cancer patients with poor prognosis after combined hormone radiation treatment, *Oncotarget*, 2020, **11**(41), 3723–3729.
75. L. Wang, C. Wang and W. S. Choi, Use of melatonin in cancer treatment: where are we?, *Int. J. Mol. Sci.*, 2022, **23**(7), 3779.
76. A. Davoodvandi, B. Nikfar, R. J. Reiter and Z. Asemi, Melatonin and cancer suppression: insights into its effects on DNA methylation, *Cell. Mol. Biol. Lett.*, 2022, **27**(1), 73.
77. J. Lu, Y. Luo, S. Mei, Y. Fang, J. Zhang and S. Chen, The effect of melatonin modulation of non-coding RNAs on central nervous system disorders: an updated review, *Curr. Neuropharmacol.*, 2021, **19**(1), 3–23.
78. A. Mafi, A. Keshavarzmotamed, N. Hedayati, Z. Y. Boroujeni, R. J. Reiter, R. M. Dehmordi, M. H. Aarabi, M. Rezaee and Z. Asemi,

Melatonin targeting non-coding RNAs in cancer: Focus on mechanisms and potential therapeutic targets, *Eur. J. Pharmacol.*, 2023, 175755.

79. R. J. W. Hill, P. F. Innominato, F. Levi and A. Ballesta, Optimizing circadian drug infusion schedules towards personalized cancer chronotherapy, *PLoS Comput. Biol.*, 2020, **16**(1), e1007218.
80. F. Halberg, The 24-hour scale: a time dimension of adaptive functional organization, *Perspect. Biol. Med.*, 1960, **3**, 491–527.
81. F. Halberg, G. Cornelissen, R. B. Sothern, G. S. Katinas, O. Schwartzkopff and K. Otsuka, Cycles tipping the scale between death and survival (="Life"), *Prog. Theor. Phys.*, 2008, (Supp.)**173**, 153–181.
82. P. Cugini, M. C. Gori, C. M. Petrangeli, P. Tisei and F. Giubilei, Maintained blood pressure and heart rate circadian rhythm in early stage Alzheimer's disease, *J. Gerontol. Med. Sci.*, 1999, **54A**, M304–M308.
83. F. Halberg and G. Cornelissen, Commentary: Chronomes: Time structures within the physiological range identify early disease risk aiming at primary prevention, *J. Gerontol., Ser. A*, 1999, **54**(6), M309–M311.
84. F. Halberg, Circadian temporal organization and experimental pathology, in *Proceedings of VII Conferenza Internazionale della Societa per lo Studio dei Ritmi Biologici*, Edizioni Panminerva Medica, Siena, 1960, pp. 1–20.
85. M. E. Hughes, L. DiTacchio, K. R. Hayes, C. Vollmers, S. Pulivarthy, J. E. Baggs, S. Panda and J. B. Hogenesch, Harmonics of circadian gene transcription in mammals, *PLoS Genet.*, 2009, **5**(4), e1000442.
86. G. Cretenet, M. Le Clech and F. Gachon, Circadian clock-coordinated 12 hr period rhythmic activation of the IRE1alpha pathway controls lipid metabolism in mouse liver, *Cell Metab.*, 2010, **11**, 47–57.
87. B. Zhu, Q. Zhang, Y. Pan, E. M. Mace, B. York, A. C. Antoulas, C. C. Dacso and B. W. O'Malley, A cell-autonomous mammalian 12 hr clock coordinates metabolic and stress rhythms, *Cell Metab.*, 2017, **25**(6), 1305–1319.e9.
88. B. Zhu, C. C. Dacso and B. W. O'Malley, Unveiling "Musica Universalis" of the cell: a brief history of biological 12 hours rhythms, *J. Endocr. Soc.*, 2018, **2**, 727–752.
89. H. Meng, N. M. Gonzales, D. M. Lonard, N. Putluri, B. Zhu, C. C. Dacso, B. York and B. W. O'Malley, XBP1 links the 12 hours clock to NAFLD and regulation of membrane fluidity and lipid homeostasis, *Nat. Commun.*, 2020, **11**, 6215.
90. H. Meng, N. M. Gonzales, S. Y. Jung, Y. Lu, N. Putluri, B. Zhu, C. C. Dacso, D. M. Lonard and B. W. O'Malley, Defining the mammalian coactivation of hepatic 12-h clock and lipid metabolism, *Cell Rep.*, 2022, **38**(10), 110491.
91. W. Dion, H. Ballance, J. Lee, Y. Pan, S. Irfan, C. Edwards, M. Sun, J. Zhang, X. Zhang, S. Liu and B. Zhu, Four-dimensional nuclear speckle phase separation dynamics regulate proteostasis, *Sci. Adv.*, 2022, **8**(1), eabl4150.

92. Y. Pan, H. Ballance, H. Meng, N. Gonzalez, S. M. Kim, L. Abdurehman, B. York, X. Chen, Y. Schnytzer, O. Levy, C. C. Dacso, C. A. McClung, B. W. O'Malley, S. Liu and B. Zhu, 12-h clock regulation of genetic information flow by XBP1s, *PLoS Biol.*, 2020, **18**(1), e3000580.
93. B. Zhu, Decoding the function and regulation of the mammalian 12-h clock, *J. Mol. Cell Biol.*, 2020, **12**, 752–758.
94. H. Ballance and B. Zhu, Revealing the hidden reality of the mammalian 12-h ultradian rhythms, *Cell. Mol. Life Sci.*, 2021, **78**(7), 3127–3140.
95. A. Sharma, H. Takata, K. Shibahara, A. Bubulya and P. A. Bubulya, Son is essential for nuclear speckle organization and cell cycle progression, *Mol. Biol. Cell*, 2010, **21**(4), 650–663.
96. X. Lu, H. H. Ng and P. A. Bubulya, The role of SON in splicing, development, and disease, *Wiley Interdiscip. Rev.: RNA*, 2014, **5**(5), 637–646.
97. K. Otsuka, G. Cornelissen and F. Halberg, Circadian rhythmic fractal scaling of heart rate variability in health and coronary artery disease, *Clin. Cardiol.*, 1997, **20**, 631–638.
98. K. Otsuka, G. Cornelissen, S. Furukawa, K. Shibata, Y. Kubo, K. Mizuno, T. Aiba, H. Ohshima and C. Mukai, Unconscious mind activates central cardiovascular network and promotes adaptation to microgravity possibly anti-aging during 1-year-long spaceflight, *Sci. Rep.*, 2022, **12**(1), 11862.
99. K. Otsuka, S. Murakami, K. Okajima, K. Shibata, Y. Kubo, D. Gubin, L. A. Beaty and G. Cornelissen, Appropriate circadian-circasemidian coupling protects blood pressure from morning surge and promotes human resilience and wellbeing, *Clin. Invest. Aging*, 2023, **18**, 755–769.
100. K. A. Alexander, A. Coté, S. C. Nguyen, L. Zhang, O. Gholamalamdari, P. Agudelo-Garcia, E. Lin-Shiao, K. M. A. Tanim, J. Lim, N. Biddle, M. C. Dunagin, C. R. Good, M. R. Mendoza, S. C. Little, A. Belmont, E. F. Joyce, A. Raj and S. L. Berger, p53 mediates target gene association with nuclear speckles for amplified RNA expression, *Mol. Cell*, 2021, **81**(8), 1666–1681.e6.
101. D. Ono, K. I. Honma and S. Honma, Circadian and ultradian rhythms of clock gene expression in the suprachiasmatic nucleus of freely moving mice, *Sci. Rep.*, 2015, **5**, 12310.
102. Y. E. Wu, R. Enoki, Y. Oda, Z. L. Huang, K. I. Honma and S. Honma, Ultradian calcium rhythms in the paraventricular nucleus and subparaventricular zone in the hypothalamus, *Proc. Natl. Acad. Sci. U. S. A.*, 2018, **115**(40), E9469–E9478.
103. M. Fossel, *J. Anti-Aging Med.*, 1998, **1**, 239.
104. D. Lloyd and D. B. Murray, The temporal architecture of eukaryotic growth, *FEBS Lett.*, 2006, **580**(12), 2830–2835.
105. A. S. Loudon, Circadian biology: a 2.5 billion year old clock, *Curr. Biol.*, 2012, **22**, R570–R571.
106. R. Aviram, Y. Adamovich and G. Asher, Circadian organelles: rhythms at all scales, *Cells*, 2021, **10**, 2447.

107. M. Mofatteh, F. Echegaray-Iturra, A. Alamban, F. Dalla Ricca, A. Bakshi and M. G. Aydogan, Autonomous clocks that regulate organelle biogenesis, cytoskeletal organization, and intracellular dynamics, *eLife*, 2021, **10**, e72104.
108. A. S. Lyon, W. B. Peeples and M. K. Rosen, A framework for understanding the functions of biomolecular condensates across scales, *Nat. Rev. Mol. Cell Biol.*, 2021, **22**(3), 215–235.
109. E. Cash, S. Sephton, C. Woolley, A. M. Elbehi, R. I. Anu, B. Ekine-Afolabi and V. C. Kok, The role of the circadian clock in cancer hallmark acquisition and immune-based cancer therapeutics, *J. Exp. Clin. Cancer Res.*, 2021, **40**(1), 119.

CHAPTER 19

Circadian Clocks and Metabolism

ETIENNE CHALLET* AND PATRICK VUILLEZ

Institute of Cellular and Integrative Neurosciences, Centre National de la Recherche Scientifique, Université de Strasbourg, 8 allée du général Rouvilllois, 67000 Strasbourg, France
*Email: challet@inci-cnrs.unistra.fr

19.1 The Circadian Timing System

At cellular, tissue and organism levels, metabolism is organized in a rhythmic manner by the circadian multi-oscillatory system on a daily scale. For instance, resting energy expenditure, in humans, displays circadian variations with a peak during the biological evening. Furthermore, fasting carbohydrate oxidation follows a rhythmic pattern and is highest during the biological morning, while fasting lipid oxidation also displays a rhythmic pattern, but it is in antiphase to carbohydrate utilisation.[1] The circadian timing system comprises a master hypothalamic clock and numerous secondary clocks in the brain and peripheral organs. Thus, the circadian system influences metabolism throughout the body, including temporal cues from the master clock *via* the autonomic nervous system and neuroendocrine pathways, messages from local clocks in a given tissue, and the synchronisation of external or internal factors, such as light, ambient temperature and feeding time.[2]

The clock machinery involves interconnected molecular feedback loops that generate self-sustained oscillations with a nearly 24-hour (*i.e.*, circadian) periodicity. A handful of specific genes, called clock genes, constitute the

Chemical Biology No. 23
Chronobiology and Chronomedicine: From Molecular and Cellular Mechanisms to Whole Body Interdigitating Networks
Edited by Germaine Cornelissen and Tsuyoshi Hirota

Published by the Royal Society of Chemistry, www.rsc.org

molecular clockwork that delivers temporal cues through the rhythmic transcription of clock-controlled genes. Two transcription factors, CLOCK (for circadian locomotor output cycles kaput) and BMAL1 (for brain and muscle aryl-hydrocarbon receptor nuclear translocator-like protein 1), dimerise to activate the transcription of genes containing E-box sequences in their promoter, including other clock genes such as *Period* (*Per 1* and *Per2*) and *Cryptochrome* (*Cry1* and *Cry2*), and clock-controlled genes. In turn, PER and CRY proteins form multimeric complexes that are translocated into the nucleus to suppress their own transcription *via* interactions with CLOCK/BMAL1 heterodimers (Figure 19.1).[3] The inactivation of PER genes and CRY genes through ubiquitination and proteasome degradation by β-transducin repeat-containing protein (β-TrCP) and F-box proteins (FBXL3 and 21) alleviates their inhibiting effect on CLOCK/BMAL1 transactivation, allowing a new circadian cycle to restart.[4–6] Moreover, phosphorylation of clock proteins such as CRY2, PER2, CLOCK and BMAL1 by glycogen synthase kinase (GSK) 3β modulates their stability, eventually triggering their rhythmic

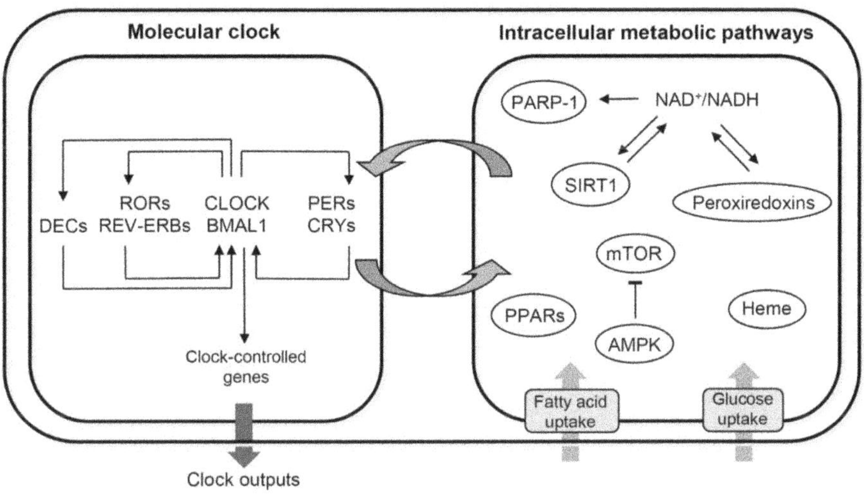

Figure 19.1 Schematic representation of the cross-talk between the circadian clocks and intracellular metabolic factors at the cellular level. These interactions lead to many possible pathways through which signals related to food intake, exercise and energy availability may affect circadian clocks. Conversely, these interactions may partly underlie the deleterious effects of circadian disturbances on metabolic health. For details, see the text. AMPK, AMP-activated protein kinase; BMAL, brain and muscle Aryl-hydrocarbon receptor nuclear translocator-like protein; CLOCK, circadian locomotor output cycles kaput; CRY, cryptochrome; mTOR, mammalian/mechanistic target of rapamycin; DEC, differentially expressed in chondrocytes protein; NAD, nicotinamide adenine dinucleotide; PARP, poly(ADP-ribose) polymerase; PER, period; PPAR, peroxisome proliferator-activated receptor; REV-ERB, reverse viral erythroblastis oncogene product; ROR, retinoic acid receptor-related orphan receptor; and SIRT, sirtuin.

degradation.[7–10] Isoforms of casein kinase I can also phosphorylate core clock proteins, thus controlling their cytoplasmic retention and stability.[11] The nuclear receptors REV-ERBs (for reverse viral erythroblastis oncogene products) and RORs (for retinoic acid receptor-related orphan receptors) are additional components of the core clock factors that modulate the transcription of BMAL1 and CLOCK through RORE sequences in their promoter.[12] An additional feedback loop includes DEC1 (also called STRA13, SHARP2 or BHLHE40) and DEC2 (SHARP1, or BHLHE41), whose transcription is activated by CLOCK/BMAL1 and they repress their own expression *via* interactions with CLOCK/BMAL1 (Figure 19.1).[13] What is more, CLOCK/BMAL1 heterodimers also activate the transcription of albumin site D-binding protein (DBP), which positively regulates targeted genes through a D-box sequence, while E4 promoter-binding protein 4 (E4BP4) suppresses their transcription through the same sequence.[14,15]

In addition to the genetic basis of circadian oscillations mentioned above, there exists an epigenetic regulation of the clockwork. Besides its role as a circadian transcription factor, CLOCK has the ability to modify the chromatin structure *via* histone acetyltransferase activity. Notably, CLOCK leads to acetylation of not only histone proteins (H)3 and H4,[16] but also of non-histone proteins, such as its circadian partner BMAL1 and the glucocorticoid receptor.[17,18] Another link bridging epigenetics with circadian clocks is mediated by sirtuin (SIRT)1, a nicotinamide adenine dinucleotide (NAD^+)-dependent histone deacetylase that serves as an intracellular metabolic detector and a redox sensor. As detailed below, SIRT1 exhibits rhythmic activity and directly interacts with clock proteins to amplify circadian oscillations.[19,20]

In contrast to most biochemical reactions in which an increase in temperature speeds up their rate, the period of the circadian clocks (*i.e.*, the speed of their oscillations) is resistant to fluctuations of temperature. This process, known as temperature compensation, is a fundamental property of circadian clocks even at a cellular level, ensuring them with a constant period close to 24 hours. Both temperature-dependent phosphoswitch between two phosphorylation sites of the clock protein PER2 and the time-dependent rate of PER2 degradation are the molecular mechanisms critical for temperature compensation.[21]

In mammals, the master circadian clock is located in the suprachiasmatic nuclei of the hypothalamus (SCN). Although the intracellular clock machinery described above is quite similar between the master clock and secondary clocks, the intercellular clock coupling within the master clock appears to be much higher, leading to robust and long-term self-sustained oscillations.[22] The SCN clock is involved in controlling the timing of most behavioural rhythms, such as the sleep/wake cycle and feeding/fasting rhythm, and physiological rhythms, such as body temperature and hormonal rhythms. The regulation of internal timing by the SCN clock is mediated through autonomic neuronal, endocrine, and eventually behavioural temporal signals to synchronise circadian clocks in the brain and peripheral organs. This process leads to an internal synchronisation within the body. The respective importance of each internal synchronising factor may depend on each organ. For instance, blood-borne signals seem to be more important for synchronising the liver and kidney clocks, compared to heart or skeletal muscle clocks.[23] Among hormonal rhythms, at least pineal melatonin and adrenal glucocorticoids have time-giving properties because they are closely synchronised with the SCN clock. Melatonin is always secreted by the pineal gland during the night in both diurnal and nocturnal mammals.[24] In addition to inducing phase-shifts of the master clock (see below), light exposure during nighttime immediately inhibits melatonin synthesis and secretion, thus altering temporal messages transmitted by the daily rhythm of plasma melatonin.[25] Moreover, the daily peak of glucocorticoids from the adrenal glands is in phase with the onset of the active span, that is, around dawn and dusk in diurnal and nocturnal animals, respectively.[26] Stress, by blunting the endogenous rhythm at high levels of circulating glucocorticoids, can impair the time-giving properties of circulating glucocorticoids.[27] Because the SCN clock also controls the sleep/wake cycle, additional internal time-givers are provided by the SCN-driven alternation of vigilance states. Lesions of the SCN lead, for instance, to a loss of the feeding/fasting rhythm.[28] This observation does not necessarily demonstrate that the master clock is the only driver of the feeding/fasting rhythm because feeding arrhythmicity in SCN-lesioned animals may be an indirect consequence of totally disrupted timing of sleep and wakefulness.[29]

The molecular clock of the SCN oscillates at comparable astronomical phases in diurnal and nocturnal mammals. For instance, *Per1* expression in the SCN usually peaks close to the middle of the light phase for a given photoperiod.[30,31] Similarly, the master clock is always more active during daytime (*e.g.*, in terms of metabolic activity or firing rate) in both categories of animals.[32] In contrast, there are lags (8–12 hours) in the temporal pattern of clock gene expression of secondary clocks that follow the sleep–wake cycle of each species.[33] Experimental evidence highlights functional differences downstream of the SCN clock, albeit the mechanisms underlying the transition between diurnality and nocturnality remain to be fully unraveled.[34,35] Nevertheless, for the subject of the present chapter, one should keep in mind that 24-hour variations in carbohydrate and lipid metabolism match the sleep–wake cycle and, accordingly, are in opposite astronomical phases between diurnal and nocturnal species.[36]

External synchronisation of the circadian timing system is mostly achieved through ambient light, which serves as the main synchroniser of the SCN clock. Indeed, photic cues activate photoreceptors in the retina, particularly melanopsin expressed by intrinsically photosensitive ganglion cells.[30] Next, photic cues are transmitted along the retinohypothalamic tract to reset the SCN. Light-induced phase shifts of the master clock in the SCN depend on light wavelength, intensity, duration, and importantly, on the time of the day when light is perceived. Blue-enriched light is the most efficient at activating melanopsin. Photic cues in the late day and early night lead to subsequent phase delays of the SCN, while the same light cues in the late night and morning result in phase advances of the SCN.[37] Light cues reaching the SCN trigger glutamate release and increase in intracellular calcium, leading to the up-regulation of *Per1* and *Per2* expression *via* binding of phosphorylated Ca^{2+}/cAMP-responsive element binding protein (CREB) to CRE sequences.[38,39] Moreover, light resetting of the SCN is also associated with dynamic remodelling of chromatin.[40]

Besides ambient light, several other factors can shift circadian clocks. These factors are classified as "non-photic" in contrast to photic cues. They include feeding-related cues, temperature and exercise, among others. As detailed below, scheduled meals have synchronising effects on the circadian clocks. But unlike light, meal time induces phase-shifts in the majority of secondary clocks in the brain and peripheral organs, without affecting the master clock.[41–44] Timed physical exercise, voluntary or forced, is a non-photic synchroniser that can phase-shift muscle clocks and the SCN, depending on its intensity and time of execution.[45,46] Finally, another possible non-photic synchroniser is temperature. In endothermic animals like mammals, episodes of hyperthermia or hypothermia can reset the phase of the secondary clocks.[47] High temperature triggers interactions between heat-shock protein factor 1 (HSF1) and CLOCK/BMAL1 heterodimers, leading to the activation of the transcription of the clock gene, *Per2*.[48] As a complement, low temperature triggers expression of cold-inducible RNA-binding protein (CIRP), which, in turn, modulates post-transcriptionally the expression of core clock components such as *Clock* and *Rorα*.[49] In most mammalian species, temperature cycles do not reset or entrain the master clock in the SCN.[47] Nevertheless, in two mammals adapted to desert conditions, namely, the camel and the Moroccan goat, daily variations in external temperature can entrain the master clock.[50,51]

19.2 Clock Genes and Nutrient-sensing Pathways

Multiple, sometimes reciprocal, interactions have been demonstrated between clock genes and essential factors of intracellular metabolism. These interactions create many possible pathways through which signals related to food intake, exercise and energy availability can impact circadian clocks.

A number of transcriptional interactions are mediated by PPARs (for peroxisome proliferator-activated receptors) that are transcription factors involved in lipid and glucose metabolism. PPARs are activated through the binding of fatty acids, whose concentrations mainly increase during fasting or a high-fat diet. Accordingly, they may serve as nutrient sensors in metabolic tissues.[52] PPARα, which regulates genes involved in lipid metabolism and inflammation, can also activate the transcription of the core clock gene, *Bmal1*.[53] Of note, PPARα is considered an important transducer in the synchronisation of secondary clocks in peripheral organs through timed-restricted feeding.[54] Furthermore, PPARα regulates the daily expression of intestinal nutrient transporters under the influence of feeding-induced secretion of bile acids, but through clock-independent mechanisms.[55,56] Although PPARα is not expressed at sizeable levels in the SCN,[57] PPARβ (also called PPARδ) signalling modulates the synchronisation to light of the master clock.[58] PPARγ is a likely mediator of molecular changes within the secondary circadian clocks in response to high-fat feeding.[59] Furthermore, PER2 can directly repress PPARγ, thus affecting adipogenesis and inflammatory responses.[60]

Many cellular functions involve redox reactions. This is also the case for circadian oscillations. In particular, DNA binding of CLOCK/BMAL1 is increased by the reduced form NADH of nicotinamide adenine dinucleotide (NAD), while DNA binding is decreased by the oxidized form, NAD^+.[61] Further interactions involve the rhythmic transcription by CLOCK/BMAL1 heterodimers of nicotinamide phosphoribosyl-transferase (*Nampt*), the enzyme controlling the synthesis of NAD^+.[19,62] Given that the NAD^+/NADH ratio is modified according to the energetic state of the cell, the redox state may thus provide a metabolic feedback to the clockwork. Circadian oscillations in redox pathways also implicate rhythmic oxidation of peroxiredoxins *via* their cyclic S-sulfinylation. Strikingly, these metabolic oscillations can occur independently of the transcriptional–translational feedback loops of the circadian clocks.[63]

SIRT1 is an $NAD(^+)$-dependent histone deacetylase, which has rhythmic activity. In addition, SIRT1 interacts directly with CLOCK/BMAL1 heterodimers, promotes deacetylation and degradation of PER2, and increases the magnitude of circadian oscillations. SIRT1 also contributes to the clock-controlled expression of *Nampt* reported above.[19,20] SIRT1 amplifies circadian oscillations in the SCN, as shown by dampened and increased rhythmic transcription of several clock genes in the SCN of mice with brain deletion or overexpression of *Sirt1*, respectively.[64]

Another intracellular metabolic detector is AMP-activated protein kinase (AMPK). Its activity in cells is enhanced under conditions of low energy, as seen during sleep or fasting. It is noteworthy that AMPK in the brain, more precisely in the hypothalamus and hindbrain, is considered to play a central role in the regulation of energy balance and metabolism.[65] The kinase activity of AMPK is intertwined with the molecular clockwork because it destabilizes the clock proteins, CRY1 and PER2, leading to their degradation.[66,67] AMPK activity thus provides another direct metabolic modulation of cellular clocks.

The mammalian/mechanistic target of rapamycin (mTOR) is an essential intracellular factor that also modulates the speed of the circadian clocks in the SCN and peripheral organs.[68] In turn, clock proteins such as PER2 and REV-ERBα modify mTOR activity, thus defining an additional loop of regulation between intracellular metabolism and the molecular clockwork.[69,70] Furthermore, mTOR signalling participates in photic resetting of the master clock, and differentially modifies light-induced phase-advances and delays.[71] In the liver clock, tight junction protein 1 (TJP1) interacts with PER1, thereby inhibiting its nuclear translocation. mTOR can phosphorylate TJP1, thus promoting the nuclear translocation of PER1. Because this effect is associated with feeding, mTOR and TJP1 interactions may contribute to the synchronising effects of timed meals on secondary clocks.[72]

The circadian molecular cogs also respond to changes in intracellular pH. In response to hypoxia, a low pH dampens or suppresses oscillations of the clockwork, *via* silencing of mTOR signalling.[73] Moreover, an alkaline pH shock can reset secondary clocks *via* the transforming growth factor β (TGF-β)-dependent pathway that activates *Dec1* transcription.[74]

Poly(ADP-ribose) polymerase (PARP)-1 is an NAD^+-dependent enzyme ADP-ribosyl-transferase. Besides its role in DNA damage repair, PARP-1 modulates the transcriptional activity of CLOCK *via* poly(ADP)-ribosylation. Considering that the cyclic activity of PARP-1 is controlled by feeding cues, the activity of PARP-1 may serve as a signal of the fed state of the cells.[75]

Among other critical biological functions, haem can bind or release oxygen transported within haemoglobin. In addition, haem availability is connected to the molecular clock. On the one hand, haem modulates transcription of *Per1* and *Per2* expression.[76] On the other hand, the clock proteins REV-ERBα and REV-ERBβ can bind to haem and as such, they are considered as haem sensors.[77,78] Haem also interacts with human CLOCK protein to inhibit its binding to E-box sequences, thus altering circadian regulation of transcription.[79] These examples highlight that haem bioavailability constitutes another functional link between metabolic pathways and circadian clocks (Figure 19.1).

19.3 Circadian Function of Mitochondria

Mitochondria are specific organelles that produce ATP within the cell during aerobic respiration. Mitochondrial activity and ATP production are not constant; instead, they undergo daily fluctuations that may be controlled by circadian clocks. In parallel, depending on various factors, such as exercise, mitochondria display large changes in mass and number, a process called mitochondrial biogenesis. PPARγ coactivator (PGC)-1α is a critical regulator of mitochondrial biogenesis. Activation of PGC-1α is induced by SIRT1 and AMPK, through respective deacetylation and phosphorylation.[80] Thus, SIRT1 and AMPK may indirectly provide a clock-controlled modulation of mitochondrial biogenesis.

It has been long known that the mitochondrial structure changes over time, being remodelled by both fusion and fission processes according to nutrient availability. Fusion during fasting increases bioenergetic efficiency, while fission during feeding promotes uncoupled respiration. Thus, mitochondrial fission may be a protective mechanism against nutrient overload, such as during obesity and type 2 diabetes.[81] Mitochondrial fusion and fission involve several proteins such as mitofusins (MFN1 and MFN2), two key factors for fusion, dynamin-related protein (DRP1), a key factor for fission, and parkin, an E3 ubiquitin ligase involved in mitochondrial dynamics in a broad sense. It is only recently that the involvement of clocks has been demonstrated in the daily alternation of mitochondrial fission and fusion.[82,83]

Furthermore, the molecular clockwork generates oscillations in oxidative capacity of the mitochondria, *via* $NAD(^+)$-dependent activity of SIRT3 that modulates mitochondrial protein acetylation. Accordingly, mitochondrial respiration is altered in cultured cells sampled from mice knockout for clock genes, as in the liver of mice with targeted hepatic deletion of *Bmal1*.[84,85] The circadian control of mitochondrial respiration in turn is modulated by DRP1 and parkin.[82,83]

Together, the numerous molecular interactions discussed in the preceding sections provide mechanistic insights into why changes in circadian rhythmicity can lead to detrimental consequences on metabolic health (see below).

19.4 Microbiome

In recent decades, there has been a growing interest in the gut microbiota. Essential descriptive studies provide information about the large number of microorganisms and their species diversity,[86] their gradual establishment during growth,[87] and the importance of environmental parameters on microbiotic populations.[88]

The adult human gut is mainly colonised by two phylotypes of bacteria, each comprising a large number of genera: Firmicutes (*Clostridium*, *Enterococcus*, *Lactobacillus*, and *Ruminococcus*) and Bacteroidetes (*Bacteroides* and *Prevotella*). What we eat serves as microbial substrates; thus gut microbiota composition is influenced by dietary habits.[89] A high-fibre diet seems to be positively associated with bacterial diversity. Long-term diets rich in either protein and animal fat or carbohydrates are linked to the prevalence of *Bacteroides* and *Prevotella*, respectively.[90] Moreover, alterations in the composition of gut bacteria (*i.e.*, dysbiosis) are associated with various disorders, mainly metabolic ones such as obesity and type 2 diabetes.[91] However, the precise characteristics of the microbiota and their correlation to certain pathologies are still controversial.[92,93]

Nevertheless, an excess of fat in the regular diet clearly increases the proportion of gram-negative bacteria such as *Bacteroides*. As a consequence, high-fat diet increases the presence of locally inflammatory lipopolysaccharides (LPSs), which pass into the bloodstream and tissues. The low-noise inflammation that increases in these tissues on a chronic basis promotes insulin resistance prior to type 2 diabetes and obesity (so-called metabolic endotoxaemia). Furthermore, small extracellular vesicles containing LPS, probably originated from these bacteria, are increased in patients suffering from metabolic syndrome.[94]

Not only the nature of the food, but also the timing of food intake, seems important for gut bacterial populations. The composition and function of mice microbiota show robust 24-hour rhythmicity that mainly responds to the feeding/fasting cycle of the host. Whether there are functional clocks in microbiota is still a subject of debate. Nevertheless, the daily rhythmicity in gut microbiota is almost completely lost in *Per1*/*Per2* mutant mice, as well as in *Bmal1* knock-out mice, suggesting a tight control by the host circadian system.[95,96] Intestinal dysbiosis caused by a high-fat and high-sugar diet is increased if the mice also undergo chronic jetlag due to weekly phase-shifts of the light–dark cycle.[97]

Reciprocally, the microbiota can influence the circadian organisation of the host (Figure 19.2). Fermentation of plant carbohydrates by the microbiota leads to the production of short-chain saturated fatty acid metabolites such as butyrate, and the latter can modulate the expression of the liver clock genes, *Per2* and *Bmal1*.[98,99] In antibiotic-treated or germ-free mice, the absence of the microbiota eliminates the daily rhythmicity of serum metabolites (amino acids and polyamines). Moreover, daily oscillations of microbiota, together with the host circadian clock, regulate the gut and hepatic epigenome and transcriptome.[100] In addition, mutual interactions between the quality and duration of sleep, and gut microbiota and its end products are also documented.[101]

All together, these data raise many questions, but they also suggest possible solutions as simple as probiotic supplementation in several medical domains, including chronomedicine. It is already well known that a regular sleep pattern and a healthy diet eaten during daytime are important for

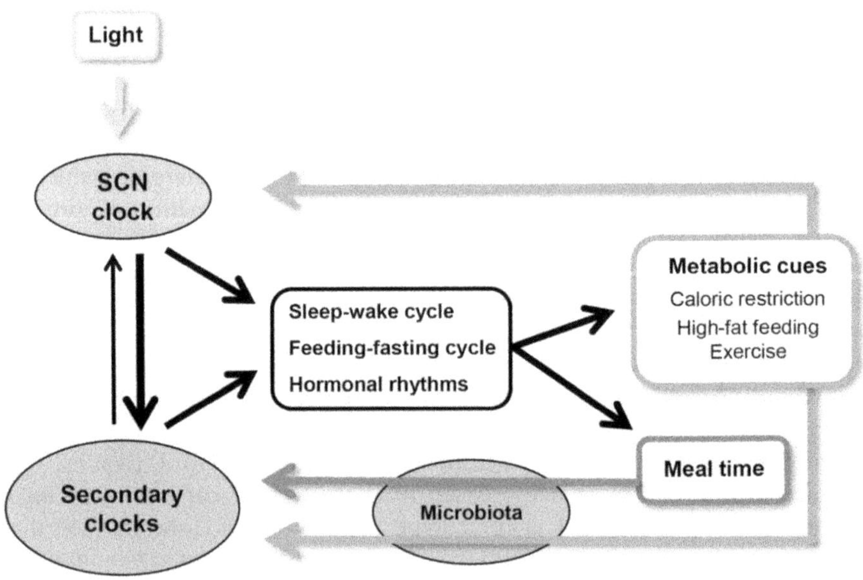

Figure 19.2 Schematic representation of the cross-talk between the circadian clocks and intracellular metabolic factors at the systemic level. The master clock in the suprachiasmatic nuclei (SCN) of the hypothalamus is mainly synchronised by ambient light, without being phase-shifted by feeding at atypical times (*e.g.*, at night in humans). In contrast, meal timing is a potent synchroniser of the secondary clocks, leading to phase-shifts if meals are taken at unusual times (*e.g.*, at night in humans). The synchronising effect of meal timing on secondary clocks may involve in part gut microbiota. Furthermore, metabolic cues associated either with a high-fat diet, caloric restriction or exercise, differentially modulate the function of the SCN and secondary clocks.

metabolic health. Part of these benefits may be achieved through maintaining the microbial balance in the gut. Accordingly, controlling daily rhythmicity of gut microbes and their metabolites could therefore be a potential field of chrononutrition.[102]

19.5 Circadian Disruption and Metabolic Disorders

19.5.1 Circadian Alterations Affect Metabolism

19.5.1.1 Circadian Disruption Due to Genetic Mutations or Variations

The functional links between circadian alterations and metabolism will now be discussed at the levels of organs and organisms. One of the first experimental arguments revealing the negative impact of circadian alterations on metabolic health emerged from the metabolic phenotyping of mice with mutations or deletions of clock genes. Following the pioneering study describing a metabolic syndrome in mice with mutations in the *Clock* gene,[103] subsequent studies have confirmed various metabolic disorders including, obesity, diabetes, atherosclerosis and/or hypertension in rodents with genetically defective circadian clocks due to global knock-out or mutations in clock genes.[104–107] Metabolic disorders are even observed when the circadian clock of a single organ, for example, liver or pancreas, is defective due to targeted deletion of clock genes.[105,108,109]

Furthermore, genetic variations in various human clock genes, such as *Bmal1*, *Clock* and its paralog *Npas2*, *Per2*, *Cry1-2*, and *Rev-erbα* have been associated with higher risks of developing metabolic syndrome, including obesity, type 2 diabetes and hypertension.[110–115] This non-exhaustive list indicates that polymorphisms in a large number of core clock elements lead to comparable effects on metabolic health, thus suggesting that the common denominator may be an impaired circadian clockwork, rather than potentially specific roles in metabolism for all these different clock genes.

19.5.1.2 Circadian Disruption Due to an Altered Rhythmic Environment

With the advent of artificial lighting and with a society that tends to become active 24 hours a day, 7 days a week, modern life favours conditions of circadian desynchronisation. These unfavourable conditions include very irregular or shifted lifestyle rhythms, meals eaten during the usual phase of sleep (*i.e.*, at night in humans), recurrent exposure to bright light at night and daytime sleep, often correlated with sleep debt. Below we will give some illustrations of the detrimental effects on metabolic health of meals taken at atypical times. These meals are likely to (de)synchronise the secondary clocks, while the main clock will remain in phase with the light–dark cycle.

Such uncoupling between secondary clocks and the SCN may lead to internal desynchronisation. Another possible negative aspect is the uncoupling between circadian clocks in different peripheral organs. In nocturnal rodents, meals ingested during the resting phase correspond to daytime food intake. In diurnal humans, this scheme would be equivalent to eating at night. Mice fed exclusively during the day on a standard (*i.e.*, balanced) diet display spontaneous hyperphagia and body mass gain as compared to control mice fed with the same diet available only at night.[116] In comparison to control animals fed exclusively at night, Wistar rats fed only during the day ingest spontaneously fewer calories than controls but nevertheless gain more body mass.[117] When an unbalanced (*i.e.*, high-fat) diet is given to mice, food access limited to daytime makes them fatter than when food is only available at night.[118] In humans, heavy and late dinners, as well as late-night meals are correlated with high adiposity and a high body mass index.[119] In addition, overweight individuals ingest on average more calories in the evening than normal-weight controls.[120]

If the daily conditions of exposure to ambient light are changed such as during shift work or night work, this will directly impact the photic synchronisation of the main clock, as well as the nocturnal secretion of melatonin, which is immediately inhibited by light. A 10-hour chronic phase shift of the light–dark cycle twice a week in rats results in, from the first month, overweight compared to control animals kept in the regular cycle of light and dark.[121] Impaired secretion of insulin has also been reported in chronically jet-lagged rats.[122] The grass rat *Arvicanthis* is an African rodent with daytime activity and nighttime sleep, similar to the human species. Therefore, it is an appropriate animal model in terms of biomedical applications to study the pathophysiological consequences of circadian disorders. In this diurnal rodent, circadian desynchronisation resulting from chronic phase shifts of the light–dark cycle has diabetogenic consequences (sharp reduction in glucose tolerance) and also causes an acceleration of cellular aging (shortening of telomeres).[123] Of interest, SIRT1 levels are reduced in various animal models of circadian desynchronisation.[123,124] In turn, reduced SIRT1 leads to dysfunction of both the circadian clocks and telomere homeostasis.[125] Using a mathematical model, Sadria and Layton predicted that a long-term reduction in bioavailability in SIRT1 and NAD^+ may be detrimental to clock function, thus mimicking the effects of aging on circadian clocks.[126]

In humans, circadian misalignment under controlled laboratory conditions causes major alterations of metabolic hormones, such as reduced leptin, increased insulin, and reversed cortisol secretion.[127] Furthermore, acute circadian desynchronisation has diabetogenic (glucose intolerance and insulin resistance) and inflammatory effects, which are additive to the same effects caused to a lower degree by sleep curtailment.[128] A number of epidemiological studies involving staff exposed to shift work or night work indicate the same trend, insofar as these patterns of professional activity at atypical hours promote the prevalence of metabolic risk factors (diabetes, obesity, atherosclerosis, and hypertension).[129–131]

The duration and phase of the eating interval can be used to define the eating pattern of every person in a chronobiological perspective. Using this parameter has been proven to be useful for assessing changes in meal timing between working- and free-days, a concept called "eating jetlag" or "metabolic jetlag".[132,133] A cross-sectional study reveals that eating jetlag, mainly due to delayed breakfast timing, is positively correlated with body mass index.[132]

19.5.2 Metabolic Disorders Affect Circadian Rhythms

Let us now consider the functional links between metabolic disturbances and circadian rhythms. To begin with, we will assess the possible impact of metabolic disturbances on circadian clocks and the rhythms they control.

Animal models of genetic diabetes and obesity show alterations in circadian rhythms at various levels. For example, rodents with a mutation in the leptin receptor, whether Zucker rats (*fa*/*fa*) or *db*/*db* mice, are hyperphagic and become obese and severely diabetic. The overeating in these animals, to a large extent, involves a very large increase in food intake during the usual rest span.[134,135] The SCN clock of *db*/*db* mice oscillates more slowly than in *db*/+ control animals, as evidenced by the longer endogenous period of circadian rhythms of locomotor activity and body temperature.[136] At the molecular level, clock gene oscillations are reduced in amplitude in the liver of Zucker rats.[137] In obese *ob*/*ob* mice that cannot synthesise leptin, alterations in secondary molecular clocks appear before the first signs of metabolic disturbances.[138] In addition, the liver clock and other secondary clocks in peripheral organs of rodents can be differentially phase-shifted with streptozotocin-induced type 1 diabetes.[139,140] Diet-induced obesity, often associated with type 2 diabetes, also causes circadian changes. They will be considered in more detail in Section 19.8.

Alterations in circadian rhythms have also been reported in obese and/or diabetic human subjects. Among other findings, diabetes and obesity differentially affect the amplitude and phase of circulating metabolites.[141] Furthermore, daily rhythms in fasting blood glucose, body temperature and heart rate are progressively disrupted according to metabolic dysfunction and the development of prediabetes and type 2 diabetes in overweight middle-aged humans.[142] Regarding hormonal profiles, plasma rhythms of melatonin and leptin exhibit similar phases between healthy subjects and diabetics. In contrast, nighttime melatonin secretion is higher in obese non-diabetic subjects, possibly related to increased sympathetic tone.[143] The rhythmic expression of clock genes in white adipose tissue exhibits the same oscillatory characteristics, regardless of individual adiposity or diabetes.[144] Another study, however, highlighted a reduction in the daily rhythmicity of metabolic and clock gene expression in the white adipose tissue of patients with type 2 diabetes.[145]

19.6 Time-restricted Feeding

19.6.1 Synchronising Effects of Time-restricted Feeding on Circadian Clocks Outside the SCN

Meal timing is a potent synchroniser of secondary clocks such as the liver, kidney, heart, and gastrointestinal tract (Figure 19.2).[41,42,44] Food synchronisation occurs independently of the SCN because the master clock is not phase-shifted by meal timing and remains entrained by light. These findings have been initially demonstrated by providing food access to nocturnal rodents only during atypical times (*i.e.*, during daytime, which corresponds to their usual resting span).[41,42] The synchronising effects of restricted feeding are still observed in mice with SCN lesions.[146] Ultradian multiple-meal schedules (*i.e.*, providing food access every 3 or 4 hours) can be used to eliminate the daily rhythm in feeding behaviour. Of interest, ultradian multiple-meal schedules disturb metabolic rhythms in various peripheral organs, without major changes in their circadian clocks. These findings highlight that a robust daily periodicity of feeding serves as a key driver in the daily rhythmicity of peripheral metabolism.[147,148] Most cerebral clocks outside the SCN can also be synchronised by timed meals during the resting phase, possibly leading to a state of brain desynchronisation because of the master clock still being entrained by the light–dark cycle.[43,149]

In humans under controlled laboratory conditions, a switch between a single carbohydrate-rich meal taken in the morning or evening produces phase-shifts in peripheral rhythms, while not affecting the dim-light melatonin onset used as a clock hand of the master clock.[150] A 6-hour phase-advance in timed three meals leads to a 3-hour phase-advance in the daily rhythm in cardiac activity.[151] Moreover, a 5-hour phase-delay in timed three meals leads to delays in the plasma glucose rhythm and adipose clock. The same nutritional intervention does not modify the phase of plasma melatonin and cortisol rhythms, used as indirect phase-markers of the SCN clock.[152] Therefore, these studies indicate that as found in animal models, secondary clocks in humans can be phase-shifted by meals during atypical times. In obese individuals, time-restricted feeding by reducing the eating window from 15 to 8 hours per day during the usual active span increases the amplitude of gene expression in peripheral tissues, as expected, without changing the phase of the circadian clocks.[153] Shortening the eating span from ≥14 to 8 hours per day in individuals with metabolic syndrome leads to body mass loss and decreased adiposity, as well as reduction in blood pressure and atherogenic lipids.[154] In men with prediabetes, a more pronounced reduction in the eating span from 12 to 6 hours per day improves insulin sensitivity and reduces blood pressure without body mass loss.[155]

19.6.2 Physiological Transducers of Synchronising Effects of Time-restricted Feeding

Timing cues associated with food ingestion such as meal-induced hyperglycaemia and the resulting increase in plasma insulin and gluco-incretins may define post-feeding timers. Supply of glucose to cultured fibroblasts enhances clock gene oscillations following down-regulation of both *Per1* and *Per2* expression. Neither non-metabolisable mannitol, change in the NADH/NAD^+ ratio, nor the metabolism of pyruvate produce the same effects.[156] In immortalised hypothalamic neurons, changes in glucose concentration affect the clockwork *via Per2* expression.[157] The results reveal that increased availability of glucose, such as during meal-induced hyperglycaemia, could be a direct resetting signal for secondary clocks. Among the hormonal signals capable of resetting secondary clocks by meal time, insulin, a hypoglycaemic hormone secreted from β cells of the pancreas, has a prominent place. First, insulin induces phase-shifts in the liver clock, *via* up-regulation and down-regulation of *Per2* and *Rev-erbα* expression, respectively.[139,158] Insulin also promotes phosphorylation of BMAL1.[159] The synchronising effect of insulin on secondary clocks may further involve mTOR signalling and induction of E4BP4.[160,161] Circadian oscillations in SCN cells are not markedly reset by insulin cues, in accordance with the lack of food-mediated resetting of the master clock.[161] Insulin is important, but not necessary, for food entrainment of the liver clock. In fact, food synchronisation of secondary clocks is maintained in mice that no longer synthesise insulin due to streptozotocin-induced damage to β cells.[162] Insulin-like growth factor 1 released from the liver is regulated by CRY proteins, and may also participate in resetting effects of meal time.[161,163,164] Gluco-incretins, like glucagon-like peptide 1 and oxyntomodulin, are intestinal hormones released from the gut in response to food ingestion. These hormones can reset the liver clock, *via* up-regulation of *Per1* and *Per 2* expression.[165,166]

Timing cues associated with food deprivation and sleep, including fasting-induced increases in fatty acids and ketone bodies and hypergluconemia, may define pre-feeding timers. The effects of fatty acids on the clockwork depend on their degree of saturation. Palmitate, a saturated fatty acid, increases the amplitude of *Bmal1* expression and down-regulates the expression of both *Per2* and *Rev-erbα* in immortalised hypothalamic neurons, while docosahexaenoic acid, a polyunsaturated fatty acid, leads to phase-advanced expression of *Bmal1* in the same cell line.[167,168] As aforementioned, PPARα, which is activated by fatty acids, may play a role in the synchronisation of secondary clocks by timed restricted feeding.[54] Plasma ketone bodies rise in response to food deprivation, including overnight fast. Ketone bodies, such as β-hydroxybutyrate, are synthesised in the liver under the control of the clock protein, PER2.[169] Feeding mice with a ketogenic diet triggers a robust daily rhythm in plasma levels of β-hydroxyl-butyrate, and increased accumulation of PPARα in both the liver and intestine that may reset these circadian clocks.[170] Furthermore, besides providing energy to neurons, elevated levels of β-hydroxybutyrate have been shown to influence the brain, modulating food-seeking behaviour in anticipation of scheduled food access.[169] Glucagon, a hyperglycaemic hormone secreted from α cells of the pancreas, notably during fasting, has been involved in the synchronisation of the liver clock to timed meals.[164] The main target of glucagon signalling within the clockwork is *Bmal1*, whose transcription is activated by CREB and CREB-regulated transcription co-activator (CRTC2).[54,171]

To recapitulate, a temporal combination of physiological responses to food intake and low energy availability contributes to the synchronising effects of meal time on circadian clocks outside the SCN. As a clinical application, time-restricted eating can be useful in improving metabolic health in patients with diabetes, obesity or metabolic syndrome.[133,155,172]

19.7 Caloric Restriction

The master clock is not phase-shifted by timed-restricted feeding. This fact, however, does not mean that the SCN cells do not receive cues relative to peripheral metabolic status. First of all, when the light synchroniser is absent (*i.e.*, in constant darkness or constant light), time-restricted feeding may, although not systematically, drive the SCN clock.[54,173,174] In addition, *in vivo* parenteral nutrition or *in vitro* infusion of glucose is also likely to shift the master clock.[175,176]

Next, even if the light synchroniser is present, timed caloric restriction disrupts the activity/rest rhythm, leading nocturnal animals to display phase-advanced rhythms of rest/activity, body temperature and melatonin secretion controlled by the SCN. Thus, caloric-restricted nocturnal rodents fed during daytime become partially diurnal.[177,178] Phase-advances in the rest/activity rhythm in mice were observed regardless of the time of day when a calorie-restricted diet was given, although their magnitude was time-dependent.[178] Conversely, diurnal rodents that are fed a calorie-restricted diet in the middle of the night become partially nocturnal.[179] These changes in timing of the active phase during caloric restriction are associated with altered clockwork within the SCN.[178,180] As mentioned above, ultradian multiple-meal schedules (*e.g.*, providing food access every 4 hours) allow the suppression of the daily periodicity of feeding. When an ultradian 6-meal schedule is coupled to caloric restriction, expression of both clock and clock-controlled proteins is modified in the SCN.[181] Furthermore, the behavioural responses to caloric restriction are, in part, linked to increased synchronisation of the SCN to light cues.[180,182,183]

Metabolic cues may reach SCN cells *via* circulating metabolites such as glucose, fatty acids, and β-hydroxybutyrate. Reduced glucose availability, like glucoprivation induced by 2-deoxy-glucose, reduces the synchronising effects of light.[184] Glucose-sensitive neurons have been identified in dorsomedial neurons of the SCN. Of note, glucoprivation more specifically inhibits vasopressin-containing SCN neurons *via* opening K_{ATP} channels.[185] Considering that the concentration of circulating fatty acids increases with caloric restriction, PPARs in the SCN may ultimately be involved in the circadian phase changes induced by caloric restriction. As noted above, PPARβ/δ are potential modulators of light-induced glutamatergic activation within the master clock of the SCN.[58] Accordingly, treatment with pan-PPAR activator bezafibrate in nocturnal mice under a light–dark cycle leads to phase-advances of feeding/fasting and rest/activity rhythms,[186,187] somehow mimicking the behavioural response to caloric restriction.

Circulating peptides, such as fibroblast growth factor (FGF) 21 are synthesised in, and released from, the liver during fasting. Day–night variations in plasma FGF21 exhibit an increased amplitude and are phase-shifted by caloric restriction.[188] Because FGF21 can directly target the SCN, *via* FGF receptors and β-klotho complexes,[189] FGF21 may convey metabolic signals to the SCN. Unlike insulin,[161] other metabolic hormones, like leptin and ghrelin, can affect the phase of the SCN or its synchronisation with light. These effects are likely indirect, that is, hormonal cues would be first integrated by arcuate, ventromedial and dorsomedial hypothalamic nuclei that, in turn, send modulatory cues to the SCN.[190–192]

Caloric restriction is known as an efficient dietary intervention for extending life span in various animals, including mammals. The molecular mechanisms promoting the anti-aging effects of caloric restriction may involve upregulation of SIRT1 activity, combined with downregulation of mTOR.[193] It is possible that SIRT1 activity in the SCN is involved in the circadian effects of caloric restriction.[64] Along this line, SIRT1 activation through dietary supplementation with resveratrol shortens the free-running period of the SCN in a non-human primate,[194] again highlighting a metabolic feedback to the master clock.

19.8 High-fat Feeding

High-fat feeding is well known for inducing diet-induced obesity, atherosclerosis and, eventually, type 2 diabetes. High-fat feeding in mice triggers circadian disturbances, often marked by a diminished amplitude of daily variations, including rest/activity, body temperature, feeding/fasting and hormonal rhythms.[195,196] In nocturnal C57BL6J mice fed *ad libitum*, one of the first rhythmic parameters modified by high-fat feeding is the pattern of spontaneous food intake, which increases markedly during the day, that is, during the usual rest span of the mice.[196,197] Notably, despite its unbalanced nutritional content, high-fat feeding limited to the usual active period (*i.e.*, at night for nocturnal mice) prevents or delays most metabolic disturbances, including diet-induced obesity.[198]

High-fat feeding in mice may attenuate the amplitude or induce phase-shifts of clock gene oscillations in secondary clocks.[196,199] Moreover, high-fat diet has a major impact not only on intracellular metabolic pathways, but also on their temporal organisation across the 24-hour cycle. In addition, a drastic increase in *de novo* oscillating transcripts is detected in the liver. This effect is mediated by reprogramming the hepatic clockwork that involves modified recruitment of CLOCK/BMAL1 heterodimers to chromatin in target promoters. The transcriptional factors contributing to clock reprogramming are mainly PPARγ and, to a lesser extent, sterol-regulatory element binding protein (SREBP) 1, an intracellular cholesterol sensor.[59] The 24-hour profiles of metabolites are drastically modified by high-fat diet in circadian clocks of peripheral organs and the brain, such as the prefrontal cortex.[200]

High-fat feeding also impacts the functioning of the SCN clock, therefore providing another example of metabolic feedback on the master clock, in addition to caloric restriction. In particular, compared to a chow diet, many more 24-hour oscillations of lipids are detected in the SCN of mice fed a high-fat diet.[200] High-fat feeding in mice also lengthens the period of the SCN clock.[196] Furthermore, synchronisation to light is slower in mice made obese by a high-fat diet. Notably, after a phase shift in the light–dark cycle (mimicking a 6-hour transmeridian journey), obese mice take longer to re-synchronise to the new light cycle than normal-weight mice.[195] Although the underlying molecular mechanisms are still unclear, the downregulation of SIRT1 likely plays a role in the negative effects of a high-fat diet on lipid metabolism.[201] In the same line, the downregulation of SIRT1 in the SCN may be involved in the circadian effects of high-fat feeding.[64] The dopaminergic projection to the SCN plays a role in mediating the disturbing effects of a high-fat diet on circadian rhythmicity.[202]

19.9 Exercise

Physical activity is an important component of energy balance, as it has stimulating effects on energy expenditure. Besides its metabolic impact, exercise has a significant effect on the physiology of skeletal muscles. As demonstrated by muscle-specific deletion of *Bmal1*, the circadian clock in skeletal muscles participates not only in glucose metabolism,[203] but also in oxidative capacity.[204] Spontaneous physical activity is a rhythmic behaviour strongly influenced by circadian clocks in the brain, including the SCN. In turn, exercise is a non-photic synchroniser that can provide feedback to the circadian clocks.[45] Timed physical activity in nocturnal rodents induces phase-shifts in the circadian clocks in skeletal muscles and other secondary clocks in peripheral organs.[205,206] In humans, the robustness of circadian oscillations of clock genes, *Nampt,* and *Sirt1* is higher in primary myotubes sampled from trained athletes than in those of sedentary controls.[207] Using exercise to reset the circadian clock in skeletal muscles may be beneficial for muscle function and help counteract sarcopenia.[208] However, the molecular mechanisms through which physical activity can reset the muscle clocks are not fully identified yet. On the one hand, exercise activates AMPK that may provide feedback to the muscle clockwork.[209] On the other hand, acidification of the extracellular medium can induce circadian phase-shifts *via* the induction of E4BP4. It is thus possible that this mechanism also plays a role in phase-shifts of the skeletal muscle clock by accumulation of lactic acid in response to sustained exercise.[160] The muscle resetting may also implicate

Dec1 transcription in response to the exercise-induced increase in TGF-β levels.[74] Moreover, the synchronising effects of forced exercise on secondary clocks involve stress mediators, such as glucocorticoids and noradrenaline.[206]

In addition to its effects on secondary clocks, forced or voluntary exercise also causes phase-advances in the locomotor activity rhythm controlled by the SCN clock, but only if increased physical activity occurs during the subjective day, corresponding to the usual resting span of nocturnal animals.[210] In nocturnal rodents, the resetting effects of light are diminished by enhanced physical activity or arousal prior to light exposure at night.[35] In contrast, in humans, exercise in late night/morning and afternoon/evening leads to phase-advances and delays, respectively.[211,212] Furthermore, exercise in humans can potentiate the phase-shifting effects of ambient light.[213] Neurochemical pathways identified for transmitting exercise- and arousal-related information to the SCN include afferents releasing serotonin, neuropeptide Y, γ-aminobutyric acid (GABA), enkephalin, neurotensin, acetylcholine and orexin.[35,210] Within the SCN of nocturnal rodents, these non-photic signals lead to the transient downregulated expression of the clock genes *Per1* and *Per2*.[35,210,214]

In conclusion, numerous reciprocal interactions occur between metabolic intracellular pathways and the molecular clockwork. These interactions are notably affected by the meal timing, quantity and nature of the diet. Circadian misalignment has deleterious consequences for metabolic health, highlighting the importance of circadian hygiene for health and well-being. Personalised chronomedicine should be developed using nutritional and chronopharmacological interventions, together with appropriately phased exercise and meal timing to reduce or prevent circadian disruption and metabolic disorders.

Abbreviations

AMPK	AMP-activated protein kinase
BMAL	Brain and muscle aryl-hydrocarbon receptor nuclear translocator-like protein
CIRP	Cold-inducible RNA-binding protein
CLOCK	Circadian locomotor output cycles kaput
CREB	Ca^{2+}/cAMP-responsive element binding protein
CRTC2	CREB-regulated transcription co-activator
CRY	Cryptochrome
DBP	Albumin site D-binding protein
DEC	Differentially expressed in chondrocytes protein
DRP	Dynamin-related protein
E4BP	E4 promoter-binding protein
FBXL	F-box proteins
FGF	Fibroblast growth factor
GABA	γ-Aminobutyric acid
GSK	Glycogen synthase kinase

H	Histone protein
HSF	Heat-shock protein factor
LPS	Lipopolysaccharide
MFN	Mitofusin
mTOR	Mammalian/mechanistic target of rapamycin
NAD	Nicotinamide adenine dinucleotide
NAMPT	Nicotinamide phosphoribosyl-transferase
PARP	Poly(ADP-ribose) polymerase
PER	Period
PGC	PPARγ coactivator
PPAR	Peroxisome proliferator-activated receptor
REV-ERB	Reverse viral erythroblastis oncogene product
ROR	Retinoic acid receptor-related orphan receptor
SCN	Suprachiasmatic nuclei of the hypothalamus
SIRT	Sirtuin
SREBP	Sterol-regulatory element binding protein
TGF	Transforming growth factor
TJP	Tight junction protein
TrCP	Transducin repeat-containing protein

Acknowledgements

The authors acknowledge the support of the Centre National de la Recherche Scientifique (CNRS) and University of Strasbourg in their studies. The authors have no conflict of interest to declare.

References

1. K. M. Zitting, N. Vujovic, R. K. Yuan, C. M. Isherwood, J. E. Medina, W. Wang, O. M. Buxton, J. S. Williams, C. A. Czeisler and J. F. Duffy, *Curr. Biol.*, 2018, **28**, 3685–3690 e3683.
2. J. Delezie and E. Challet, *Ann. N. Y. Acad. Sci.*, 2011, **1243**, 30–46.
3. K. H. Cox and J. S. Takahashi, *J. Mol. Endocrinol.*, 2019, **63**, R93–R102.
4. E. J. Eide, M. F. Woolf, H. Kang, P. Woolf, W. Hurst, F. Camacho, E. L. Vielhaber, A. Giovanni and D. M. Virshup, *Mol. Cell. Biol.*, 2005, **25**, 2795–2807.
5. S. M. Siepka, S. H. Yoo, J. Park, W. Song, V. Kumar, Y. Hu, C. Lee and J. S. Takahashi, *Cell*, 2007, **129**, 1011–1023.
6. H. Dardente, J. Mendoza, J. M. Fustin, E. Challet and D. G. Hazlerigg, *PLoS One*, 2008, **3**, e3530.
7. Y. Harada, M. Sakai, N. Kurabayashi, T. Hirota and Y. Fukada, *J. Biol. Chem.*, 2005, **280**, 31714–31721.
8. M. L. Spengler, K. K. Kuropatwinski, M. Schumer and M. P. Antoch, *Cell Cycle*, 2009, **8**, 4138–4146.
9. S. Sahar, L. Zocchi, C. Kinoshita, E. Borrelli and P. Sassone-Corsi, *PLoS One*, 2010, **5**, e8561.

10. C. Iitaka, K. Miyazaki, T. Akaike and N. Ishida, *J. Biol. Chem.*, 2005, **280**, 29397–29402.
11. D. M. Virshup, E. J. Eide, D. B. Forger, M. Gallego and E. V. Harnish, *Cold Spring Harbor Symp. Quant. Biol.*, 2007, **72**, 413–420.
12. L. J. Everett and M. A. Lazar, *Trends Endocrinol. Metab.*, 2014, **25**, 586–592.
13. Y. Kato, T. Kawamoto, K. Fujimoto and M. Noshiro, *Curr. Top. Dev. Biol.*, 2014, **110**, 339–372.
14. M. Stratmann, D. M. Suter, N. Molina, F. Naef and U. Schibler, *Mol. Cell*, 2012, **48**, 277–287.
15. D. Yamajuku, Y. Shibata, M. Kitazawa, T. Katakura, H. Urata, T. Kojima, S. Takayasu, O. Nakata and S. Hashimoto, *FEBS Lett.*, 2011, **585**, 2217–2222.
16. M. Doi, J. Hirayama and P. Sassone-Corsi, *Cell*, 2006, **125**, 497–508.
17. J. Hirayama, S. Sahar, B. Grimaldi, T. Tamaru, K. Takamatsu, Y. Nakahata and P. Sassone-Corsi, *Nature*, 2007, **450**, 1086–1090.
18. N. Nader, G. P. Chrousos and T. Kino, *FASEB J.*, 2009, **23**, 1572–1583.
19. Y. Nakahata, S. Sahar, G. Astarita, M. Kaluzova and P. Sassone-Corsi, *Science*, 2009, **324**, 654–657.
20. G. Asher, D. Gatfield, M. Stratmann, H. Reinke, C. Dibner, F. Kreppel, R. Mostoslavsky, F. W. Alt and U. Schibler, *Cell*, 2008, **134**, 317–328.
21. R. Narasimamurthy and D. M. Virshup, *Front. Neurol.*, 2017, **8**, 161.
22. A. B. Webb, N. Angelo, J. E. Huettner and E. D. Herzog, *Proc. Natl. Acad. Sci. U. S. A.*, 2009, **106**, 16493–16498.
23. H. Guo, J. M. Brewer, A. Champhekar, R. B. Harris and E. L. Bittman, *Proc. Natl. Acad. Sci. U. S. A.*, 2005, **102**, 3111–3116.
24. P. Pevet and E. Challet, *J. Physiol.*, 2011, **105**, 170–182.
25. U. Redlin, *Chronobiol. Int.*, 2001, **18**, 737–758.
26. H. Oster, E. Challet, V. Ott, E. Arvat, E. R. de Kloet, D. J. Dijk, S. Lightman, A. Vgontzas and E. Van Cauter, *Endocr. Rev.*, 2017, **38**, 3–45.
27. T. Kino and G. P. Chrousos, *Endocr. Dev.*, 2011, **20**, 116–126.
28. K. Nagai, T. Nishio, H. Nakagawa, S. Nakamura and Y. Fukuda, *Brain Res.*, 1978, **142**, 384–389.
29. E. Challet, *Nat. Rev. Endocrinol.*, 2019, **15**, 393–405.
30. E. Challet, *Endocrinology*, 2007, **148**, 5648–5655.
31. B. B. Tournier, J. S. Menet, H. Dardente, V. J. Poirel, A. Malan, M. Masson-Pevet, P. Pevet and P. Vuillez, *Neuroscience*, 2003, **118**, 317–322.
32. W. J. Schwartz, S. M. Reppert, S. M. Eagan and M. C. Moore-Ede, *Brain Res.*, 1983, **274**, 184–187.
33. L. S. Mure, H. D. Le, G. Benegiamo, M. W. Chang, L. Rios, N. Jillani, M. Ngotho, T. Kariuki, O. Dkhissi-Benyahya, H. M. Cooper and S. Panda, *Science*, 2018, **359**, eaao0318.
34. A. Kalsbeek, L. A. Verhagen, I. Schalij, E. Foppen, M. Saboureau, B. Bothorel, R. M. Buijs and P. Pevet, *Eur. J. Neurosci.*, 2008, **27**, 818–827.

35. P. K. Jha, H. Bouaouda, A. Kalsbeek and E. Challet, *Neurosci. Biobehav. Rev.*, 2021, **123**, 48–60.
36. P. Kumar Jha, E. Challet and A. Kalsbeek, *Mol. Cell. Endocrinol.*, 2015, **418**(Pt 1), 74–88.
37. D. A. Golombek and R. E. Rosenstein, *Physiol. Rev.*, 2010, **90**, 1063–1102.
38. S. A. Tischkau, J. W. Mitchell, S. H. Tyan, G. F. Buchanan and M. U. Gillette, *J. Biol. Chem.*, 2003, **278**, 718–723.
39. Z. Travnickova-Bendova, N. Cermakian, S. M. Reppert and P. Sassone-Corsi, *Proc. Natl. Acad. Sci. U. S. A.*, 2002, **99**, 7728–7733.
40. C. Crosio, N. Cermakian, C. D. Allis and P. Sassone-Corsi, *Nat. Neurosci.*, 2000, **3**, 1241–1247.
41. K. A. Stokkan, S. Yamazaki, H. Tei, Y. Sakaki and M. Menaker, *Science*, 2001, **291**, 490–493.
42. F. Damiola, N. Le Minh, N. Preitner, B. Kornmann, F. Fleury-Olela and U. Schibler, *Genes Dev.*, 2000, **14**, 2950–2961.
43. H. Wakamatsu, Y. Yoshinobu, R. Aida, T. Moriya, M. Akiyama and S. Shibata, *Eur. J. Neurosci.*, 2001, **13**, 1190–1196.
44. W. A. Hoogerwerf, H. L. Hellmich, G. Cornelissen, F. Halberg, V. B. Shahinian, J. Bostwick, T. C. Savidge and V. M. Cassone, *Gastroenterology*, 2007, **133**, 1250–1260.
45. F. W. Turek, *J. Biol. Rhythms*, 1989, **4**, 135–147.
46. B. D. Harfmann, E. A. Schroder and K. A. Esser, *J. Biol. Rhythms*, 2015, **30**, 84–94.
47. E. D. Buhr, S. H. Yoo and J. S. Takahashi, *Science*, 2010, **330**, 379–385.
48. T. Tamaru, M. Hattori, K. Honda, I. Benjamin, T. Ozawa and K. Takamatsu, *PLoS One*, 2011, **6**, e24521.
49. J. Morf, G. Rey, K. Schneider, M. Stratmann, J. Fujita, F. Naef and U. Schibler, *Science*, 2012, **338**, 379–383.
50. H. Farsi, M. R. Achaaban, M. Piro, B. Bothorel, M. Ouassat, E. Challet, P. Pevet and K. El Allali, *Sci. Rep.*, 2020, **10**, 19515.
51. H. Farsi, D. Harti, M. R. Achaaban, M. Piro, V. Raverot, B. Bothorel, M. Ouassat, E. Challet, P. Pevet and K. E. Allali, *J. Pineal Res.*, 2020, e12634.
52. B. Desvergne and W. Wahli, *Endocr. Rev.*, 1999, **20**, 649–688.
53. L. Canaple, J. Rambaud, O. Dkhissi-Benyahya, B. Rayet, N. S. Tan, L. Michalik, F. Delaunay, W. Wahli and V. Laudet, *Mol. Endocrinol.*, 2006, **20**, 1715–1727.
54. A. Mukherji, A. Kobiita and P. Chambon, *Proc. Natl. Acad. Sci. U. S. A.*, 2015, **112**, E6683–E6690.
55. E. Wada, S. Koyanagi, N. Kusunose, T. Akamine, H. Masui, H. Hashimoto, N. Matsunaga and S. Ohdo, *Mol. Pharmacol.*, 2015, **87**, 314–322.
56. A. Okamura, S. Koyanagi, A. Dilxiat, N. Kusunose, J. J. Chen, N. Matsunaga, S. Shibata and S. Ohdo, *J. Biol. Chem.*, 2014, **289**, 25296–25305.
57. A. Mukherji, A. Kobiita, M. Damara, N. Misra, H. Meziane, M. F. Champy and P. Chambon, *Proc. Natl. Acad. Sci. U. S. A.*, 2015, **112**, E6691–E6698.

58. E. Challet, I. Denis, V. Rochet, J. Aioun, S. Gourmelen, H. Lacroix, B. Goustard-Langelier, C. Papillon, J. M. Alessandri and M. Lavialle, *Cell. Mol. Life Sci.*, 2013, **70**, 2003–2014.
59. K. L. Eckel-Mahan, V. R. Patel, S. de Mateo, R. Orozco-Solis, N. J. Ceglia, S. Sahar, S. A. Dilag-Penilla, K. A. Dyar, P. Baldi and P. Sassone-Corsi, *Cell*, 2013, **155**, 1464–1478.
60. B. Grimaldi, M. M. Bellet, S. Katada, G. Astarita, J. Hirayama, R. H. Amin, J. G. Granneman, D. Piomelli, T. Leff and P. Sassone-Corsi, *Cell Metab.*, 2010, **12**, 509–520.
61. J. Rutter, M. Reick, L. C. Wu and S. L. McKnight, *Science*, 2001, **293**, 510–514.
62. K. M. Ramsey, J. Yoshino, C. S. Brace, D. Abrassart, Y. Kobayashi, B. Marcheva, H. K. Hong, J. L. Chong, E. D. Buhr, C. Lee, J. S. Takahashi, S. Imai and J. Bass, *Science*, 2009, **324**, 651–654.
63. N. B. Milev, S. G. Rhee and A. B. Reddy, *Cold Spring Harbor Perspect. Biol.*, 2018, **10**, a027689.
64. H. C. Chang and L. Guarente, *Cell*, 2013, **153**, 1448–1460.
65. M. Lopez, R. Nogueiras, M. Tena-Sempere and C. Dieguez, *Nat. Rev. Endocrinol.*, 2016, **12**, 421–432.
66. K. A. Lamia, U. M. Sachdeva, L. DiTacchio, E. C. Williams, J. G. Alvarez, D. F. Egan, D. S. Vasquez, H. Juguilon, S. Panda, R. J. Shaw, C. B. Thompson and R. M. Evans, *Science*, 2009, **326**, 437–440.
67. J. H. Um, S. Yang, S. Yamazaki, H. Kang, B. Viollet, M. Foretz and J. H. Chung, *J. Biol. Chem.*, 2007, **282**, 20794–20798.
68. C. Ramanathan, N. D. Kathale, D. Liu, C. Lee, D. A. Freeman, J. B. Hogenesch, R. Cao and A. C. Liu, *PLoS Genet.*, 2018, **14**, e1007369.
69. R. Wu, F. Dang, P. Li, P. Wang, Q. Xu, Z. Liu, Y. Li, Y. Wu, Y. Chen and Y. Liu, *Cell Metab.*, 2019, **29**, 653–667 e656.
70. M. Dadon-Freiberg, N. Chapnik and O. Froy, *Mol. Cell. Endocrinol.*, 2021, **521**, 111108.
71. R. Cao, A. Li, H. Y. Cho, B. Lee and K. Obrietan, *J. Neurosci.*, 2010, **30**, 6302–6314.
72. Y. Liu, Y. Zhang, T. Li, J. Han and Y. Wang, *Nat. Commun.*, 2020, **11**, 589.
73. Z. E. Walton, C. H. Patel, R. C. Brooks, Y. Yu, A. Ibrahim-Hashim, M. Riddle, A. Porcu, T. Jiang, B. L. Ecker, F. Tameire, C. Koumenis, A. T. Weeraratna, D. K. Welsh, R. Gillies, J. C. Alwine, L. Zhang, J. D. Powell and C. V. Dang, *Cell*, 2018, **174**, 72–87 e32.
74. N. Kon, T. Hirota, T. Kawamoto, Y. Kato, T. Tsubota and Y. Fukada, *Nat. Cell Biol.*, 2008, **10**, 1463–1469.
75. G. Asher, H. Reinke, M. Altmeyer, M. Gutierrez-Arcelus, M. O. Hottiger and U. Schibler, *Cell*, 2010, **142**, 943–953.
76. K. Kaasik and C. C. Lee, *Nature*, 2004, **430**, 467–471.
77. S. Raghuram, K. R. Stayrook, P. Huang, P. M. Rogers, A. K. Nosie, D. B. McClure, L. L. Burris, S. Khorasanizadeh, T. P. Burris and F. Rastinejad, *Nat. Struct. Mol. Biol.*, 2007, **14**, 1207–1213.

78. L. Yin, N. Wu, J. C. Curtin, M. Qatanani, N. R. Szwergold, R. A. Reid, G. M. Waitt, D. J. Parks, K. H. Pearce, G. B. Wisely and M. A. Lazar, *Science*, 2007, **318**, 1786–1789.
79. S. L. Freeman, H. Kwon, N. Portolano, G. Parkin, U. Venkatraman Girija, J. Basran, A. J. Fielding, L. Fairall, D. A. Svistunenko, P. C. E. Moody, J. W. R. Schwabe, C. P. Kyriacou and E. L. Raven, *Proc. Natl. Acad. Sci. U. S. A.*, 2019, **116**, 19911–19916.
80. B. L. Tang, *Mol. Cells*, 2016, **39**, 87–95.
81. M. Liesa and O. S. Shirihai, *Cell Metab.*, 2013, **17**, 491–506.
82. K. Schmitt, A. Grimm, R. Dallmann, B. Oettinghaus, L. M. Restelli, M. Witzig, N. Ishihara, K. Mihara, J. A. Ripperger, U. Albrecht, S. Frank, S. A. Brown and A. Eckert, *Cell Metab.*, 2018, **27**, 657–666 e655.
83. C. Pacelli, G. Rotundo, L. Lecce, M. Menga, E. Bidollari, R. Scrima, O. Cela, C. Piccoli, T. Cocco, A. L. Vescovi, G. Mazzoccoli, J. Rosati and N. Capitanio, *Int. J. Mol. Sci.*, 2019, **20**, 2772.
84. C. B. Peek, A. H. Affinati, K. M. Ramsey, H. Y. Kuo, W. Yu, L. A. Sena, O. Ilkayeva, B. Marcheva, Y. Kobayashi, C. Omura, D. C. Levine, D. J. Bacsik, D. Gius, C. B. Newgard, E. Goetzman, N. S. Chandel, J. M. Denu, M. Mrksich and J. Bass, *Science*, 2013, **342**, 1243417.
85. D. Jacobi, S. Liu, K. Burkewitz, N. Kory, N. H. Knudsen, R. K. Alexander, U. Unluturk, X. Li, X. Kong, A. L. Hyde, M. R. Gangl, W. B. Mair and C. H. Lee, *Cell Metab.*, 2015, **22**, 709–720.
86. P. B. Eckburg, E. M. Bik, C. N. Bernstein, E. Purdom, L. Dethlefsen, M. Sargent, S. R. Gill, K. E. Nelson and D. A. Relman, *Science*, 2005, **308**, 1635–1638.
87. M. Derrien, A. S. Alvarez and W. M. de Vos, *Trends Microbiol.*, 2019, **27**, 997–1010.
88. T. S. Dong and A. Gupta, *Clin. Gastroenterol. Hepatol.*, 2019, **17**, 231–242.
89. S. Bibbo, G. Ianiro, V. Giorgio, F. Scaldaferri, L. Masucci, A. Gasbarrini and G. Cammarota, *Eur. Rev. Med. Pharmacol. Sci.*, 2016, **20**, 4742–4749.
90. G. D. Wu, J. Chen, C. Hoffmann, K. Bittinger, Y. Y. Chen, S. A. Keilbaugh, M. Bewtra, D. Knights, W. A. Walters, R. Knight, R. Sinha, E. Gilroy, K. Gupta, R. Baldassano, L. Nessel, H. Li, F. D. Bushman and J. D. Lewis, *Science*, 2011, **334**, 105–108.
91. Y. Fan and O. Pedersen, *Nat. Rev. Microbiol.*, 2021, **19**, 55–71.
92. A. Gorvitovskaia, S. P. Holmes and S. M. Huse, *Microbiome*, 2016, **4**, 15.
93. F. Magne, M. Gotteland, L. Gauthier, A. Zazueta, S. Pesoa, P. Navarrete and R. Balamurugan, *Nutrients*, 2020, **12**, 1474.
94. S. Ali, M. Malloci, Z. Safiedeen, R. Soleti, L. Vergori, X. Vidal-Gomez, C. Besnard, S. Dubois, S. Le Lay, J. Boursier, A. Chevrollier, F. Gagnadoux, G. Simard, R. Andriantsitohaina, M. C. Martinez and G. Metabol study, *Metabolism*, 2021, **118**, 154727.
95. C. A. Thaiss, D. Zeevi, M. Levy, G. Zilberman-Schapira, J. Suez, A. C. Tengeler, L. Abramson, M. N. Katz, T. Korem, N. Zmora, Y. Kuperman, I. Biton, S. Gilad, A. Harmelin, H. Shapiro, Z. Halpern, E. Segal and E. Elinav, *Cell*, 2014, **159**, 514–529.

96. X. Liang, F. D. Bushman and G. A. FitzGerald, *Proc. Natl. Acad. Sci. U. S. A.*, 2015, **112**, 10479–10484.
97. R. M. Voigt, C. B. Forsyth, S. J. Green, E. Mutlu, P. Engen, M. H. Vitaterna, F. W. Turek and A. Keshavarzian, *PLoS One*, 2014, **9**, e97500.
98. V. Leone, S. M. Gibbons, K. Martinez, A. L. Hutchison, E. Y. Huang, C. M. Cham, J. F. Pierre, A. F. Heneghan, A. Nadimpalli, N. Hubert, E. Zale, Y. Wang, Y. Huang, B. Theriault, A. R. Dinner, M. W. Musch, K. A. Kudsk, B. J. Prendergast, J. A. Gilbert and E. B. Chang, *Cell Host Microbe*, 2015, **17**, 681–689.
99. Y. Tahara, M. Yamazaki, H. Sukigara, H. Motohashi, H. Sasaki, H. Miyakawa, A. Haraguchi, Y. Ikeda, S. Fukuda and S. Shibata, *Sci. Rep.*, 2018, **8**, 1395.
100. C. A. Thaiss, M. Levy, T. Korem, L. Dohnalova, H. Shapiro, D. A. Jaitin, E. David, D. R. Winter, M. Gury-BenAri, E. Tatirovsky, T. Tuganbaev, S. Federici, N. Zmora, D. Zeevi, M. Dori-Bachash, M. Pevsner-Fischer, E. Kartvelishvily, A. Brandis, A. Harmelin, O. Shibolet, Z. Halpern, K. Honda, I. Amit, E. Segal and E. Elinav, *Cell*, 2016, **167**, 1495–1510 e1412.
101. B. A. Matenchuk, P. J. Mandhane and A. L. Kozyrskyj, *Sleep Med. Rev.*, 2020, **53**, 101340.
102. S. G. Parkar, A. Kalsbeek and J. F. Cheeseman, *Microorganisms*, 2019, **7**, 41.
103. F. W. Turek, C. Joshu, A. Kohsaka, E. Lin, G. Ivanova, E. McDearmon, A. Laposky, S. Losee-Olson, A. Easton, D. R. Jensen, R. H. Eckel, J. S. Takahashi and J. Bass, *Science*, 2005, **308**, 1043–1045.
104. A. Chaix, T. Lin, H. D. Le, M. W. Chang and S. Panda, *Cell Metab.*, 2018, **29**, 303–319 e4.
105. X. Pan, C. A. Bradfield and M. M. Hussain, *Nat. Commun.*, 2016, **7**, 13011.
106. J. Delezie, S. Dumont, H. Dardente, H. Oudart, A. Grechez-Cassiau, P. Klosen, M. Teboul, F. Delaunay, P. Pevet and E. Challet, *FASEB J.*, 2012, **26**, 3321–3335.
107. T. A. Martino, G. Y. Oudit, A. M. Herzenberg, N. Tata, M. M. Koletar, G. M. Kabir, D. D. Belsham, P. H. Backx, M. R. Ralph and M. J. Sole, *Am. J. Physiol.: Regul., Integr. Comp. Physiol.*, 2008, **294**, R1675–R1683.
108. K. A. Lamia, K. F. Storch and C. J. Weitz, *Proc. Natl. Acad. Sci. U. S. A.*, 2008, **105**, 15172–15177.
109. B. Marcheva, K. M. Ramsey, E. D. Buhr, Y. Kobayashi, H. Su, C. H. Ko, G. Ivanova, C. Omura, S. Mo, M. H. Vitaterna, J. P. Lopez, L. H. Philipson, C. A. Bradfield, S. D. Crosby, L. JeBailey, X. Wang, J. S. Takahashi and J. Bass, *Nature*, 2010, **466**, 627–631.
110. M. Garaulet, Y. C. Lee, J. Shen, L. D. Parnell, D. K. Arnett, M. Y. Tsai, C. Q. Lai and J. M. Ordovas, *Am. J. Clin. Nutr.*, 2009, **90**, 1466–1475.
111. M. Garaulet, M. D. Corbalan-Tutau, J. A. Madrid, J. C. Baraza, L. D. Parnell, Y. C. Lee and J. M. Ordovas, *J. Am. Diet. Assoc.*, 2010, **110**, 917–921.

112. A. Englund, L. Kovanen, S. T. Saarikoski, J. Haukka, A. Reunanen, A. Aromaa, J. Lonnqvist and T. Partonen, *J. Circadian Rhythms*, 2009, **7**, 5.
113. E. G. Ruano, S. Canivell and E. Vieira, *PLoS One*, 2014, **9**, e104065.
114. L. Kovanen, K. Donner, M. Kaunisto and T. Partonen, *Hypertens. Res.*, 2015, **38**, 186–192.
115. P. Y. Woon, P. J. Kaisaki, J. Braganca, M. T. Bihoreau, J. C. Levy, M. Farrall and D. Gauguier, *Proc. Natl. Acad. Sci. U. S. A.*, 2007, **104**, 14412–14417.
116. M. S. Bray, W. F. Ratcliffe, M. H. Grenett, R. A. Brewer, K. L. Gamble and M. E. Young, *Int. J. Obes.*, 2013, **37**, 843–852.
117. O. D. Ramirez-Plascencia, N. Saderi, C. Escobar and R. C. Salgado-Delgado, *Eur. J. Neurosci.*, 2017, **45**, 1325–1332.
118. D. M. Arble, J. Bass, A. D. Laposky, M. H. Vitaterna and F. W. Turek, *Obesity*, 2009, **17**, 2100–2102.
119. A. W. McHill, A. J. Phillips, C. A. Czeisler, L. Keating, K. Yee, L. K. Barger, M. Garaulet, F. A. Scheer and E. B. Klerman, *Am. J. Clin. Nutr.*, 2017, **106**, 1213–1219.
120. A. W. McHill, C. A. Czeisler, A. J. K. Phillips, L. Keating, L. K. Barger, M. Garaulet, F. Scheer and E. B. Klerman, *Nutrients*, 2019, **11**, 587.
121. L. L. Tsai, Y. C. Tsai, K. Hwang, Y. W. Huang and J. E. Tzeng, *Am. J. Physiol.: Endocrinol. Metab.*, 2005, **289**, E212–E217.
122. I. Bartol-Munier, S. Gourmelen, P. Pevet and E. Challet, *Int. J. Obes.*, 2006, **30**, 60–67.
123. E. Grosbellet, S. Zahn, M. Arrive, S. Dumont, S. Gourmelen, P. Pevet, E. Challet and F. Criscuolo, *FASEB J.*, 2015, **29**, 4794–4803.
124. R. C. Salgado-Delgado, N. Saderi, C. Basualdo Mdel, N. N. Guerrero-Vargas, C. Escobar and R. M. Buijs, *PLoS One*, 2013, **8**, e60052.
125. M. Osum and N. Serakinci, *DNA Repair*, 2020, **96**, 102993.
126. M. Sadria and A. T. Layton, *iScience*, 2021, **24**, 102245.
127. F. A. Scheer, M. F. Hilton, C. S. Mantzoros and S. A. Shea, *Proc. Natl. Acad. Sci. U. S. A.*, 2009, **106**, 4453–4458.
128. R. Leproult, U. Holmback and E. Van Cauter, *Diabetes*, 2014, **63**, 1860–1869.
129. A. Knutsson and A. Kempe, *Chronobiol. Int.*, 2014, **31**, 1146–1151.
130. F. Dutheil, J. S. Baker, M. Mermillod, M. De Cesare, A. Vidal, F. Moustafa, B. Pereira and V. Navel, *Atherosclerosis*, 2020, **313**, 156–169.
131. M. Khosravipour, P. Khanlari, S. Khazaie, H. Khosravipour and H. Khazaie, *Sleep Med. Rev.*, 2021, **57**, 101427.
132. M. F. Zeron-Rugerio, A. Hernaez, A. P. Porras-Loaiza, T. Cambras and M. Izquierdo-Pulido, *Nutrients*, 2019, **11**, 2980.
133. S. Gill and S. Panda, *Cell Metab.*, 2015, **22**, 789–798.
134. E. Grosbellet, S. Dumont, C. Schuster-Klein, B. Guardiola-Lemaitre, P. Pevet, F. Criscuolo and E. Challet, *Chronobiol. Int.*, 2015, **32**, 637–649.
135. R. E. Mistlberger, H. Lukman and B. G. Nadeau, *Appetite*, 1998, **30**, 255–267.
136. E. Grosbellet, S. Dumont, C. Schuster-Klein, B. Guardiola-Lemaitre, P. Pevet, F. Criscuolo and E. Challet, *Biochimie*, 2016, **124**, 198–206.

137. Y. Motosugi, H. Ando, K. Ushijima, T. Maekawa, E. Ishikawa, M. Kumazaki and A. Fujimura, *Chronobiol. Int.*, 2011, **28**, 968–972.
138. H. Ando, M. Kumazaki, Y. Motosugi, K. Ushijima, T. Maekawa, E. Ishikawa and A. Fujimura, *Endocrinology*, 2011, **152**, 1347–1354.
139. D. Yamajuku, T. Inagaki, T. Haruma, S. Okubo, Y. Kataoka, S. Kobayashi, K. Ikegami, T. Laurent, T. Kojima, K. Noutomi, S. Hashimoto and H. Oda, *Sci. Rep.*, 2012, **2**, 439.
140. J. Bostwick, D. Nguyen, G. Cornelissen, F. Halberg and W. A. Hoogerwerf, *Mol. Cell. Biochem.*, 2010, **338**, 203–213.
141. C. M. Isherwood, D. R. Van der Veen, J. D. Johnston and D. J. Skene, *FASEB J.*, 2017, **31**, 5557–5567.
142. D. G. Gubin, A. A. Nelaeva, A. E. Uzhakova, Y. V. Hasanova, G. Cornelissen and D. Weinert, *Chronobiol. Int.*, 2017, **34**, 1136–1148.
143. S. Mantele, D. T. Otway, B. Middleton, S. Bretschneider, J. Wright, M. D. Robertson, D. J. Skene and J. D. Johnston, *PLoS One*, 2012, **7**, e37123.
144. H. Ando, T. Takamura, N. Matsuzawa-Nagata, K. R. Shima, T. Eto, H. Misu, M. Shiramoto, T. Tsuru, S. Irie, A. Fujimura and S. Kaneko, *Diabetologia*, 2009, **52**, 329–335.
145. D. J. Stenvers, A. Jongejan, S. Atiqi, J. P. Vreijling, E. J. Limonard, E. Endert, F. Baas, P. D. Moerland, E. Fliers, A. Kalsbeek and P. H. Bisschop, *Diabetologia*, 2019, **62**, 704–716.
146. R. Hara, K. Wan, H. Wakamatsu, R. Aida, T. Moriya, M. Akiyama and S. Shibata, *Genes Cells*, 2001, **6**, 269–278.
147. P. de Goede, S. Sen, Y. Su, E. Foppen, V. J. Poirel, E. Challet and A. Kalsbeek, *Int. J. Mol. Sci.*, 2018, **19**, 3171.
148. B. J. Greenwell, A. J. Trott, J. R. Beytebiere, S. Pao, A. Bosley, E. Beach, P. Finegan, C. Hernandez and J. S. Menet, *Cell Rep.*, 2019, **27**, 649–657 e645.
149. M. Verwey and S. Amir, *Eur. J. Neurosci.*, 2009, **30**, 1650–1657.
150. K. Krauchi, C. Cajochen, E. Werth and A. Wirz-Justice, *J. Biol. Rhythms*, 2002, **17**, 364–376.
151. T. Yoshizaki, Y. Tada, A. Hida, A. Sunami, Y. Yokoyama, J. Yasuda, A. Nakai, F. Togo and Y. Kawano, *Eur. J. Appl. Physiol.*, 2013, **113**, 2603–2611.
152. S. M. T. Wehrens, S. Christou, C. Isherwood, B. Middleton, M. A. Gibbs, S. N. Archer, D. J. Skene and J. D. Johnston, *Curr. Biol.*, 2017, **27**, 1768–1775 e1763.
153. L. S. Lundell, E. B. Parr, B. L. Devlin, L. R. Ingerslev, A. Altintas, S. Sato, P. Sassone-Corsi, R. Barres, J. R. Zierath and J. A. Hawley, *Nat. Commun.*, 2020, **11**, 4643.
154. M. J. Wilkinson, E. N. C. Manoogian, A. Zadourian, H. Lo, S. Fakhouri, A. Shoghi, X. Wang, J. G. Fleischer, S. Navlakha, S. Panda and P. R. Taub, *Cell Metab.*, 2019, **31**, 92–104 e5.
155. E. F. Sutton, R. Beyl, K. S. Early, W. T. Cefalu, E. Ravussin and C. M. Peterson, *Cell Metab.*, 2018, **27**, 1212–1221 e1213.

156. T. Hirota, T. Okano, K. Kokame, H. Shirotani-Ikejima, T. Miyata and Y. Fukada, *J. Biol. Chem.*, 2002, **277**, 44244–44251.
157. J. E. Oosterman and D. D. Belsham, *PLoS One*, 2016, **11**, e0146969.
158. Y. Tahara, M. Otsuka, Y. Fuse, A. Hirao and S. Shibata, *J. Biol. Rhythms*, 2011, **26**, 230–240.
159. F. Dang, X. Sun, X. Ma, R. Wu, D. Zhang, Y. Chen, Q. Xu, Y. Wu and Y. Liu, *Nat. Commun.*, 2016, **7**, 12696.
160. H. Yoshitane, Y. Asano, A. Sagami, S. Sakai, Y. Suzuki, H. Okamura, W. Iwasaki, H. Ozaki and Y. Fukada, *Commun. Biol.*, 2019, **2**, 300.
161. P. Crosby, R. Hamnett, M. Putker, N. P. Hoyle, M. Reed, C. J. Karam, E. S. Maywood, A. Stangherlin, J. E. Chesham, E. A. Hayter, L. Rosenbrier-Ribeiro, P. Newham, H. Clevers, D. A. Bechtold and J. S. O'Neill, *Cell*, 2019, **177**, 896–909 e820.
162. K. Oishi, M. Kasamatsu and N. Ishida, *Biochem. Biophys. Res. Commun.*, 2004, **317**, 330–334.
163. A. Chaudhari, R. Gupta, S. Patel, N. Velingkaar and R. Kondratov, *Mol. Biol. Cell*, 2017, **28**, 834–842.
164. Y. Ikeda, M. Kamagata, M. Hirao, S. Yasuda, S. Iwami, H. Sasaki, M. Tsubosaka, Y. Hattori, A. Todoh, K. Tamura, K. Shiga, T. Ohtsu and S. Shibata, *EBioMedicine*, 2018, **28**, 210–224.
165. H. Ando, K. Ushijima and A. Fujimura, *PLoS One*, 2013, **8**, e81119.
166. D. Landgraf, A. H. Tsang, A. Leliavski, C. E. Koch, J. L. Barclay, D. J. Drucker and H. Oster, *eLife*, 2015, **4**, e06253.
167. L. J. Fick, G. H. Fick and D. D. Belsham, *Biochem. Biophys. Res. Commun.*, 2011, **413**, 414–419.
168. J. A. Greco, J. E. Oosterman and D. D. Belsham, *Am. J. Physiol.: Regul., Integr. Comp. Physiol.*, 2014, **307**, R1049–R1060.
169. R. Chavan, C. Feillet, S. S. Costa, J. E. Delorme, T. Okabe, J. A. Ripperger and U. Albrecht, *Nat. Commun.*, 2016, **7**, 10580.
170. P. Tognini, M. Murakami, Y. Liu, K. L. Eckel-Mahan, J. C. Newman, E. Verdin, P. Baldi and P. Sassone-Corsi, *Cell Metab.*, 2017, **26**, 523–538 e525.
171. X. Sun, F. Dang, D. Zhang, Y. Yuan, C. Zhang, Y. Wu, Y. Wang and Y. Liu, *J. Biol. Chem.*, 2015, **290**, 2189–2197.
172. M. J. Wilkinson, E. N. C. Manoogian, A. Zadourian, H. Lo, S. Fakhouri, A. Shoghi, X. Wang, J. G. Fleischer, S. Navlakha, S. Panda and P. R. Taub, *Cell Metab.*, 2020, **31**, 92–104 e105.
173. M. R. Castillo, K. J. Hochstetler, R. J. Tavernier Jr., D. M. Greene and A. Bult-Ito, *Am. J. Physiol.: Regul., Integr. Comp. Physiol.*, 2004, **287**, R551–R555.
174. I. Caldelas, C. A. Feillet, H. Dardente, F. Eclancher, A. Malan, S. Gourmelen, P. Pevet and E. Challet, *Eur. J. Neurosci.*, 2005, **22**, 921–929.
175. H. Miki, M. Yano, H. Iwanaga, T. Tsujinaka, M. Nakayama, M. Kobayashi, K. Oishi, H. Shiozaki, N. Ishida, K. Nagai and M. Monden, *NeuroReport*, 2003, **14**, 1457–1461.
176. A. C. Hall, R. M. Hoffmaster, E. L. Stern, M. E. Harrington and D. Bickar, *J. Biol. Rhythms*, 1997, **12**, 388–400.

177. E. Challet, P. Pevet, B. Vivien-Roels and A. Malan, *J. Biol. Rhythms*, 1997, **12**, 65–79.
178. E. Challet, L. C. Solberg and F. W. Turek, *Am. J. Physiol.: Regul., Integr. Comp. Physiol.*, 1998, **274**, R1751–R1761.
179. J. Mendoza, S. Gourmelen, S. Dumont, D. Sage-Ciocca, P. Pevet and E. Challet, *J. Physiol.*, 2012, **590**, 3155–3168.
180. J. Mendoza, P. Pevet and E. Challet, *Eur. J. Neurosci.*, 2007, **25**, 3691–3701.
181. S. Sen, H. Raingard, S. Dumont, A. Kalsbeek, P. Vuillez and E. Challet, *Chronobiol. Int.*, 2017, **34**, 17–36.
182. E. Challet, *J. Comp. Physiol., B*, 2010, **180**, 631–644.
183. D. Resuehr and J. Olcese, *Brain Res.*, 2005, **1048**, 146–152.
184. E. Challet, S. Losee-Olson and F. W. Turek, *Am. J. Physiol.: Regul., Integr. Comp. Physiol.*, 1999, **276**, R1063–R1070.
185. J. J. Yang, R. C. Cheng, P. C. Cheng, Y. C. Wang and R. C. Huang, *Sci. Rep.*, 2017, **7**, 640.
186. R. Gutman, M. Barnea, L. Haviv, N. Chapnik and O. Froy, *Int. J. Obes.*, 2012, **36**, 1131–1134.
187. K. Oishi, H. Shirai and N. Ishida, *NeuroReport*, 2008, **19**, 487–489.
188. A. C. Thompson, M. D. Bruss, N. Nag, A. Kharitonenkov, A. C. Adams and M. K. Hellerstein, *PLoS One*, 2014, **9**, e111418.
189. A. L. Bookout, M. H. de Groot, B. M. Owen, S. Lee, L. Gautron, H. L. Lawrence, X. Ding, J. K. Elmquist, J. S. Takahashi, D. J. Mangelsdorf and S. A. Kliewer, *Nat. Med.*, 2013, **19**, 1147–1152.
190. C. X. Yi, J. van der Vliet, J. Dai, G. Yin, L. Ru and R. M. Buijs, *Endocrinology*, 2006, **147**, 283–294.
191. C. X. Yi, E. Challet, P. Pevet, A. Kalsbeek, C. Escobar and R. M. Buijs, *Eur. J. Neurosci.*, 2008, **27**, 1965–1972.
192. E. Grosbellet, S. Gourmelen, P. Pevet, F. Criscuolo and E. Challet, *Endocrinology*, 2015, **156**, 1080–1090.
193. L. Guarente, *Genes Dev.*, 2013, **27**, 2072–2085.
194. F. Pifferi, A. Dal-Pan, M. Menaker and F. Aujard, *J. Biol. Rhythms*, 2011, **26**, 271–275.
195. J. Mendoza, P. Pevet and E. Challet, *J. Physiol.*, 2008, **586**, 5901–5910.
196. A. Kohsaka, A. D. Laposky, K. M. Ramsey, C. Estrada, C. Joshu, Y. Kobayashi, F. W. Turek and J. Bass, *Cell Metab.*, 2007, **6**, 414–421.
197. J. S. Pendergast, K. L. Branecky, W. Yang, K. L. Ellacott, K. D. Niswender and S. Yamazaki, *Eur. J. Neurosci.*, 2013, **37**, 1350–1356.
198. M. Hatori, C. Vollmers, A. Zarrinpar, L. DiTacchio, E. A. Bushong, S. Gill, M. Leblanc, A. Chaix, M. Joens, J. A. Fitzpatrick, M. H. Ellisman and S. Panda, *Cell Metab.*, 2012, **15**, 848–860.
199. H. Yanagihara, H. Ando, Y. Hayashi, Y. Obi and A. Fujimura, *Chronobiol. Int.*, 2006, **23**, 905–914.
200. K. A. Dyar, D. Lutter, A. Artati, N. J. Ceglia, Y. Liu, D. Armenta, M. Jastroch, S. Schneider, S. de Mateo, M. Cervantes, S. Abbondante, P. Tognini, R. Orozco-Solis, K. Kinouchi, C. Wang, R. Swerdloff, S. Nadeef, S. Masri, P. Magistretti, V. Orlando, E. Borrelli, N. H. Uhlenhaut, P. Baldi,

J. Adamski, M. H. Tschop, K. Eckel-Mahan and P. Sassone-Corsi, *Cell*, 2018, **174**, 1571–1585 e1511.
201. D. J. Lomb, G. Laurent and M. C. Haigis, *Biochim. Biophys. Acta, Bioenerg.*, 2010, **1804**, 1652–1657.
202. R. M. Grippo, Q. Tang, Q. Zhang, S. R. Chadwick, Y. Gao, E. B. Altherr, L. Sipe, A. M. Purohit, N. M. Purohit, M. D. Sunkara, K. J. Cios, M. Sidikpramana, A. J. Spano, J. N. Campbell, A. D. Steele, J. Hirsh, C. D. Deppmann, M. Wu, M. M. Scott and A. D. Guler, *Curr. Biol.*, 2020, **30**, 1352–1355.
203. K. A. Dyar, S. Ciciliot, L. E. Wright, R. S. Bienso, G. M. Tagliazucchi, V. R. Patel, M. Forcato, M. I. Paz, A. Gudiksen, F. Solagna, M. Albiero, I. Moretti, K. L. Eckel-Mahan, P. Baldi, P. Sassone-Corsi, R. Rizzuto, S. Bicciato, H. Pilegaard, B. Blaauw and S. Schiaffino, *Mol. Metab.*, 2014, **3**, 29–41.
204. T. Wada, Y. Ichihashi, E. Suzuki, Y. Kosuge, K. Ishige, T. Uchiyama, M. Makishima, R. Nakao, K. Oishi and S. Shimba, *Int. J. Mol. Sci.*, 2018, **19**, 2813.
205. G. Wolff and K. A. Esser, *Med. Sci. Sports Exercise*, 2012, **44**, 1663–1670.
206. H. Sasaki, Y. Hattori, Y. Ikeda, M. Kamagata, S. Iwami, S. Yasuda, Y. Tahara and S. Shibata, *Sci. Rep.*, 2016, **6**, 27607.
207. J. Hansen, S. Timmers, E. Moonen-Kornips, H. Duez, B. Staels, M. K. Hesselink and P. Schrauwen, *Sci. Rep.*, 2016, **6**, 35047.
208. Y. Choi, J. Cho, M. H. No, J. W. Heo, E. J. Cho, E. Chang, D. H. Park, J. H. Kang and H. B. Kwak, *Int. J. Mol. Sci.*, 2020, **21**, 3106.
209. D. G. Lassiter, R. J. O. Sjogren, B. M. Gabriel, A. Krook and J. R. Zierath, *Mol. Metab.*, 2018, **16**, 12–23.
210. A. T. Hughes and H. D. Piggins, *Prog. Brain Res.*, 2012, **199**, 305–336.
211. O. Van Reeth, J. Sturis, M. M. Byrne, J. D. Blackman, M. L'Hermite-Baleriaux, R. Leproult, C. Oliner, S. Refetoff, F. W. Turek and E. Van Cauter, *Am. J. Physiol.*, 1994, **266**, E964–E974.
212. S. D. Youngstedt, J. A. Elliott and D. F. Kripke, *J. Physiol.*, 2019, **597**, 2253–2268.
213. S. D. Youngstedt, C. E. Kline, J. A. Elliott, M. R. Zielinski, T. M. Devlin and T. A. Moore, *J. Circadian Rhythms*, 2016, **14**, 2.
214. T. Hamada, M. C. Antle and R. Silver, *Neurosci. Lett.*, 2004, **362**, 87–90.

CHAPTER 20

Circadian Clocks in Ageing

PEI ZHU[a,b] AND CLARA B. PEEK*[a,b]

[a] Department of Biochemistry and Molecular Genetics, Northwestern University Feinberg School of Medicine, Chicago, IL USA; [b] Department of Medicine-Endocrinology, Northwestern University Feinberg School of Medicine, Chicago, IL USA
*Email: c-peek@northwestern.edu

20.1 Introduction

Molecular circadian clocks are conserved throughout life and consist of genetically-encoded factors that work to align behavioural and physiological processes with the changing demands of the 24-hour day. In mammals, the molecular clock is a negative feedback loop that consists of basic helix–loop–helix/Per-ARNT-SIM (bHLH-PAS) transcriptional activators, CLOCK and BMAL1, which regulate rhythmic expression of genes, including their own repressors, period (PER1/2/3) and cryptochrome (CRY1/2). PER and CRY proteins undergo post-translational regulation and ultimately return to the nucleus to inhibit CLOCK/BMAL1 heterodimer activity. An additional mechanism, involving the nuclear receptors REVERBα/β and RORα/β, serves to refine this feedback loop through controlled repression and activation of *Bmal1* transcription.[1]

Mammalian molecular clocks are expressed throughout the body and function autonomously within individual cells. However, clock function within the neurons of the hypothalamic suprachiasmatic nucleus (SCN) serves as the central pacemaker of the body by directly sensing external light–dark signals from the retina.[2] The SCN then acts to 'reset' other clocks in surrounding neuronal and peripheral tissues *via* direct and indirect

Chemical Biology No. 23
Chronobiology and Chronomedicine: From Molecular and Cellular Mechanisms to Whole Body Interdigitating Networks
Edited by Germaine Cornelissen and Tsuyoshi Hirota

Published by the Royal Society of Chemistry, www.rsc.org

mechanisms. Individuals exposed to circadian disturbances such as shift work or jet lag undergo molecular clock resetting by the SCN, but chronic exposure to these conditions leads to a multitude of pathologies, including propensity for diabetes, obesity and cancer.[3,4] Furthermore, similar pathologies are also observed in animals and humans exposed to mistimed feeding, which has been shown to uncouple SCN and peripheral clocks, such that they are no longer aligned. Overall, each of these conditions of circadian disruption is widespread in modern society and coincides with age-associated dysfunction. Therefore, understanding the reciprocal effects of ageing and circadian rhythms is vital for unlocking new therapeutic strategies to minimize the adverse consequences of ageing.

20.2 Circadian Function During Ageing in Humans

A prominent effect of advanced age on circadian behaviour is the transition towards a 'morningness' chronotype, which means that individuals tend to display progressively earlier onset/offset phases of rest and activity, hormone secretion, body temperature, and other physiological processes.[5–16] Similar advanced peak phase and reduced amplitude of oscillations can be observed at the molecular level as well. For example, seniors over age 60 display changes in core molecular clock gene expression in the prefrontal cortex of the brain and in blood.[17,18] Finally, in addition to alterations in phase and amplitude, older age is associated with a reduction in the capacity to adapt to changes in the light–dark cycle (*i.e.*, jet lag)[13,19,20] and shift work.[21–23]

Although the above-described changes in chronotype and circadian clock resetting are well-characterized in ageing humans, the molecular mechanisms underlying these changes are unclear. Furthermore, it is not well understood what the impact of age-related circadian changes might be on overall health and life expectancy, or whether circadian changes in older adults are a cause or effect of living longer (for more comprehensive reviews of this topic, see Hood and Amir, *JCI* 2017 and Cornelissen and Otsuka, *Gerontol.* 2017[7,24]).

20.3 Molecular Circadian Clock Function During Ageing

20.3.1 Central Clocks

Similar to humans, animals such as non-human primates, rodents, and flies also display altered circadian function during ageing, including earlier peak phases and reduced amplitudes of behavioural (*i.e.*, sleep/wake, feeding/fasting) and other physiological and metabolic rhythms, and impaired clock resetting following phase shifts.[25–29] In addition, mice with genetic loss of molecular clock function[30–33] or those exposed to conditions of circadian disruption, mimicking jet lag or shift work,[31,34,35] display accelerated development of pathologies such as cancer, obesity, diabetes, and early mortality. A study assessing the effects of ageing on circadian function in mice demonstrated that older mice (22 months) placed in constant darkness display a longer free-running period and a longer time to re-entrain to a phase shift than younger mice.[36] Notably, another study demonstrated opposite effects on the free-running period in old mice,[37] which may be due to differences in strain background and the age of the mice (*i.e.*, early old-age *versus* very old).

Early studies aimed at establishing the importance of the SCN within the hypothalamus as the central circadian pacemaker utilized SCN lesioning and transplantation approaches in rodents.[38] Using these techniques in mice, rats, and hamsters, it was found that the youthful circadian amplitude, period and responsiveness to phase shifts in aged SCN-ablated animals could be restored by transplanting fetal SCN tissue, demonstrating that the SCN incurs age-related changes, leading to altered circadian function.[39–42] Furthermore, rodents and primates with a significant deviation in their endogenous free-running period length (or '*tau*') from 24 hours exhibit negative effects on lifespan, indicating that life-long circadian misalignment comes at a cost.[43] Interestingly, SCN-lesioned chipmunks display a significantly reduced survival rate compared with the pooled surgical and intact controls following a 14-month tracking period in an environmentally-controlled woodland field.[44] However, the authors noted that the difference in mortality could have been influenced by a reduced capacity to avoid death from predation and thus it does not necessarily pinpoint the importance of SCN-derived circadian entrainment in ageing.[40]

In addition to SCN-intrinsic changes during ageing, it is possible that extrinsic factors originating from peripheral tissues might contribute to changes in SCN function. For example, light serves a first-order external time cue ('Zeitgeber') that entrains the SCN.[45] It is received by photoreceptors in the retina, including the intrinsically-photosensitive retinal ganglion cells (iPRGs), which then relay signals to the SCN.[46] It is possible that the decline in photoentrainment during ageing is impacted by structural changes and a functional decline in this relay system. Indeed, more recent studies have identified age-related effects on the SCN in mouse models, including reduced light-derived signals from iPRGs and reduced coupling between individual neurons.[47–49] Therefore, weakening of input Zeitgeber signals as well as their transmission to and processing within the SCN could contribute to core clock changes in the SCN.

At the molecular level, clock gene expression in the SCN is altered during ageing in rodent models. For example, oscillations of *Clock* and *Bmal1* gene expression are altered in aged hamsters[50] and mice.[51] A reduced amplitude of *Per2* oscillations has also been observed in aged mice,[52] but not in rats[53] and hamsters.[50] This inconsistency might reflect differences in the techniques used to measure expression (*e.g.*, *in situ* hybridization *versus* quantitative real-time RT-PCR). In addition to the core clock genes in the primary regulatory feedback loop, transcription factors in additional clock-regulatory pathways, including *Reverb*, *Dbp,* and *Dec1* have been reported to lose their transcriptional oscillations in the SCN of aged mice.[51] In contrast, *PER1* and *CRY1* circadian rhythms in SCN have been shown to be preserved at an advanced age in mice,[51,52] rats,[53] and hamsters.[50] However, this does not rule out potential age-associated changes at the post-transcriptional and/or translational levels, as the circadian clock is known to regulate the proteome and phosphoproteome in the SCN and hippocampus of mice.[54–56] Finally, it is worth noting that animal-based studies on the impact of ageing on molecular clock rhythms in the SCN might be masked by the constant daily exposure to light–dark cycles. Indeed, mice exposed to constant darkness displayed a more profound age-dependent effect on circadian period and amplitude of locomotor behaviour.[57]

The mechanisms underlying changes in core clock function with age in the SCN are not well understood. One study found that the mammalian homolog of yeast Sir2, SIRT1, might regulate age-dependent changes in core clock expression and rhythms of locomotor behaviour.[58] SIRT1 is an NAD^+-dependent protein deacetylase that has been implicated in lifespan extension and age-related metabolic decline.[59] Another study demonstrated that young, brain-specific, *Sirt1*-deficient mice display circadian alterations similar to those observed in old mice, whereas ectopic SIRT1 expression counteracts the effects of ageing.[58] The authors proposed that SIRT1 and its target PPARγ coactivator 1 α (PGC-1α) bind directly to the *Bmal1* promoter to activate transcription. Furthermore, in fibroblasts, liver and adipose tissue, the deacetylase activity of SIRT1 is directly controlled by the circadian clock through the rhythmic synthesis of the NAD^+ biosynthetic pathway rate-limiting enzyme, nicotinamide phosphoribosyltransferase (Nampt).[60–63] NAD^+ levels in various cell types change with age and NAD^+ precursor supplementation can reduce age-related metabolic dysfunction in circadian mutants.[59] However, the exact role of NAD^+ and sirtuins in mediating clock changes with age, and the impact of clock/NAD^+ cross-talk in contributing to age-related pathologies and lifespan are unclear.

20.3.2 Peripheral Clocks

In addition to the SCN, age-associated changes in molecular clock function also occur in other tissue types throughout the body. In humans, a large cohort study demonstrated alterations in the expression patterns of core molecular clock factors, and over 1000 rhythmic genes in the prefrontal cortex as individuals age. These alterations included phase shifts, amplitude changes, and loss or gain of gene rhythmicity.[17] In another study, real-time quantitative PCR followed by stepwise multiple regression analysis in a cohort of healthy women ranging in age from 20 to 79 years demonstrated an age-dependent decrease in *Bmal1* and *Per3* mRNAs in peripheral blood cells.[18]

Mechanisms governing changes in peripheral clock gene expression during ageing are incompletely understood. In mammals, the SCN functions as a central circadian pacemaker and entrains peripheral circadian clocks to the external light–dark cycle. Consequently, disruptions in SCN function (*i.e.*, amplitude and phase of molecular clock oscillations) in advanced age could indirectly weaken peripheral clock function and expression of clock-controlled genes. These effects could be mediated through alterations in SCN-driven rhythms of food intake, thermoregulation, physical activity, and hormone signaling (*e.g.* glucocorticoids).[64–67] As an example, time-restricted feeding in mice can override rhythms of feeding behaviour and alter circadian clock rhythms in peripheral tissues with little or no influence on the entrainment of the circadian clock in the SCN.[68,69] Interestingly, recent studies have demonstrated that restricting food intake to the active/wake period can ameliorate age-related pathologies and increase lifespan, particularly when combined with calorie restriction.[70]

It is still unclear to what extent the loss of peripheral clock function with age is due to the loss of intrinsic cell-autonomous clock function. Some insights into this issue have been revealed in several studies using clock gene reporters to monitor circadian clock function in young and old animal tissues *ex vivo*.[71,72] In one study, the authors observed differences in a luciferase reporter controlled by the *Per1* gene in 10 different cultured tissues removed from young (2 months) and aged (24–26 months) rats.[71] Interestingly, the authors observed striking differences in the impact of age on *Per1*-luciferase rhythms in different tissue types. Furthermore, in some tissues such as the rat lung, loss of *Per1*-luciferase rhythmicity could be restored by adding forskolin, a known synchronization agent,[73,74] suggesting that clock uncoupling may contribute to the decline in peripheral clock function with age.[71] In mice, a recent study demonstrated similar effects on *Per2*-luciferase rhythmicity in *ex vivo* cultured skeletal muscle tissue.[72] Additionally, exogenous addition of the NAD^+ precursor nicotinamide riboside restored the rhythmicity of aged mouse muscle tissue, providing evidence for age-related regulation of circadian clocks through NAD^+ signaling.

In the fruit fly *Drosophila melanogaster*, several studies have also reported a decline in the amplitude of peripheral circadian clock gene expression throughout the body with age.[75–79] However, similar to what has been found in rodents, different tissue types exhibit different degrees and rates of change in molecular clock gene expression. For instance, *per* and *tim* genes display reduced oscillations in the head but not in the body of aged flies.[25] Finally, it is still not well understood how ageing might impact circasemidian rhythms (*e.g.*, 12-hour rhythms). Such rhythms have been observed in several physiological processes including body temperature, melatonin and slow-wave sleep.[80,81] Additionally, the mouse liver displays 12-hour rhythms in genes enriched in pathways important for the maintenance of metabolic homeostasis and stress adaptation, such as endoplasmic reticulum (ER)-stress- and mitochondria metabolism-related genes.[82] It would be interesting to explore further whether dysregulation of circasemidian rhythms causes pathologies such as loss of proteostasis and/or mitochondrial dysfunction, which are hallmarks of ageing.[83]

20.4 Circadian Control of Metabolism During Ageing

Metabolic diseases such as diabetes, obesity, cancer, and muscle wasting all become more prevalent in older individuals. In recent decades, much work has been done to characterize the molecular hallmarks of ageing and age-related metabolic dysfunction. For example, an analysis of population-based studies identified a set of 11 serum metabolic signatures that are related to oxidative stress, alterations in cell morphology, fatty acid beta-oxidation capacity, and vascular function during ageing in females.[84] In mice, gene expression and metabolomic analyses of the liver and muscle tissues in young (13-week old) and aged (93-week old) mice revealed changes related to glucose and fatty acid metabolism as well as redox homeostasis pathways.[85] Finally, a recent study involving 6421 people ranging in age from 8 days to 95 years reported that human total body energy expenditure is relatively stable from 20 to 60, followed by an annual decrease of 0.7% after the age of 60.[86] Yet, these reports do not elucidate the specific function of circadian timing in age-related metabolic decline. Specifically, it is still unclear whether disrupted central and peripheral clock function is a consequence of metabolic changes with age, or whether age-associated clock disruption causes metabolic dysfunction.

As mentioned in Section 20.3.1, a promising molecular connection between circadian clocks and age-related metabolic decline is the regulation of NAD^+ signaling.[87] NAD^+ biosynthesis and redox cycling are indispensable for central carbon metabolic pathways, including glycolysis, the citric acid cycle, oxidative phosphorylation, and fatty acid oxidation.[88] Several studies have measured an age-dependent reduction in NAD^+ levels in plasma and multiple solid organ tissues, a phenomenon that is highly conserved from *C. elegans* to rodents to humans.[89–91] Furthermore, boosting NAD^+ levels through delivery of NAD^+ biosynthetic precursors can combat age-related pathologies.[92–94] NAD^+ is a cofactor for the sirtuin proteins (SIRT1-7 in mammals), a family of enzymes possessing broad functions, including protein deacetylase, desuccinylase, demaloynylase, deglutarylase, long-chain deacylase, lipoamidase, and ADP-ribosyltransferase activity.[95] Loss of the yeast sirtuin Sir2 leads to a reduced replicative lifespan,[96] but the effect of sirtuins on lifespan in animals is controversial. However, sirtuins have been shown to play a role in the prevention of age-associated pathologies such as cancer susceptibility,[97–99] neurodegeneration,[100,101] metabolic syndrome and cardiovascular disease.[102–105] The cellular content of NAD^+ is determined through both the *de novo* biosynthesis from tryptophan and NAD^+ salvage pathway. The latter pathway is controlled by the circadian clock through the regulation of the enzyme nicotinamide phosphoribosyltransferase (NAMPT).[61,62] A recent study demonstrated that oral supplementation of the NAD^+ biosynthetic precursor, nicotinamide riboside (NR), to mice could alleviate the age-related loss of circadian clock function as well as downstream oscillatory transcripts in the mouse liver.[72] More work is needed to fully understand the role of clock-regulated NAD^+

synthesis and its downstream effectors such as sirtuins in the regulation of whole-animal life- and health-span.

Other known ageing-related factors interact with circadian clocks, including AMP-activated protein kinase (AMPK), mammalian/mechanistic target of rapamycin (mTOR), and nuclear factor κB (NFκB).[106–111] AMPK is a key energy-sensing enzyme that is activated under metabolic stress conditions such as nutrient limitation, and several studies have implicated AMPK in the communication of nutrient signals to peripheral circadian clocks. Specifically, it has been shown that AMPK can induce phosphorylation of CRY1 at Ser-71, which facilitates its ubiquitin-mediated degradation.[110] AMPK can also advance the degradation of PER2 through the phosphorylation of casein kinases I epsilon (CKIε) at Ser-389.[112] Further studies are required to determine whether ageing-associated loss of AMPK signaling results in altered circadian clock function through stabilization of the repressor arm of the circadian clock loop.

The mTOR complex is another intracellular sensor integrating nutrient availability and energy status, and modulation of mTORC1 activity can impact age-related metabolic changes.[106] mTOR-mediated translational control plays an important role in light-triggered clock entrainment in the SCN. Specifically, light induces mTOR activation, which facilitates the translation of mRNAs with 5′-terminal oligopyrimidine tracts.[113,114] Inhibition of mTOR activation attenuates light-induced PERIOD protein expression, leading to phase-shifted clock gene oscillations.[114] More recently, mTOR regulation of the translation regulator eIF4E-binding protein 1 (4E-BP1) was shown to repress the synthesis of vasoactive intestinal peptide (Vip), a neuropeptide responsible for neuronal synchronization in the SCN.[115,116] Finally, a recent study has uncovered mTOR regulation of peripheral clock period length and amplitude, including hepatocyte and adipocyte clocks.[117]

NFκB is well known for its role in regulating age-associated changes in inflammatory responses and immune function.[108,118] It was found that NFκB activation leads to the genomic relocalization of BMAL1 and CLOCK to sites convergent with NFkB binding, thus reprogramming circadian transcription.[111] More recently, NFκB was shown to directly alter BMAL1 function by regulating CRY1–BMAL1 binding and coactivator recruitment.[119] Overall, while both clocks and NFκB function are impacted by ageing, the degree to which these pathway interactions contribute to age-dependent decline in immune responses and/or circadian function remain unclear.

Finally, in addition to the above-described pathways, other factors downstream of circadian clocks may mediate their effects on ageing and interactions with other ageing-associated regulatory pathways. For example, the oxygen-sensing master regulator hypoxia-inducible factor (HIF) is known to control these pathways as well, and also displays bidirectional interactions with the circadian clock.[120–126] It would be intriguing to investigate whether clock-HIF cross-talk is required for the decline in circadian function during ageing, and whether age-related pathologies associated with clock-disruption occur through impaired HIF signaling. In support of this, one study has demonstrated clock regulation by HIF through the control of mTORC1, but the role of this axis during ageing was not tested.[127]

20.5 Circadian Regulation of Stem Cells During Ageing

Adult stem cells (also called somatic stem cells) are rare cell populations residing in most postnatal tissues. They are multipotent cells that possess the capacity to differentiate into limited types of corresponding lineage cells. In addition, adult stem cells must undergo continual self-renewal to prevent exhaustion of stem cell pools. The functions of adult stem cells decline with ageing, namely biological ageing, and this phenomenon contributes significantly to chronological ageing-related disorders (see Table 20.1 for studies regarding ageing of adult stem cells).

Mechanisms that underlie stem cell ageing occur at multiple levels, including transcriptional, translational, epigenetic and metabolomic changes.[128] For example, hematopoietic stem cells from old (22–24-month old) mice display increased expression of genes involved in protein aggregation response, stress response, and inflammatory response as compared to young (2–3-month old) mice.[129] Conversely, genes that are down-regulated in old mouse haematopoetic stem cells (HSCs) compared to young mice are associated with functions such as genomic integrity, chromatin remodeling, DNA repair, DNA replication and cell cycle. Similarly, aged hair follicle stem cells also demonstrate up-regulated expression of genes related to immune system function and lineage differentiation as well as down-regulated expression of genes associated with fatty acid biosynthesis, response to peptide hormones, extracellular matrix organization and mitotic nuclear division.[130] In addition to gene expression changes in aged stem cells, chromatin-associated epigenetic changes have also been observed. Indeed, epigenetic chromatin modifications were reported as determinants of chronological ageing of muscle stem cells, as evidenced by an observation of the genome-wide increase of H3K27me3 in aged *versus* young mouse muscle stem cells.[131]

Circadian clock rhythms are indeed present in postnatal adult stem cells and play important roles in their physiology and behaviours (for more comprehensive reviews on this topic, see Weger *et al.*, *Dev. Bio.* 2017 and Benitah and Welz, *Cell Stem Cell*, 2020[132,133]). A central feature of most adult stem cells is their low abundance and their location within niches surrounded by an excess of other cell types. This feature allows for potentially dynamic cross-talk between stem cells and their surrounding differentiated cells and extracellular matrix components. For example, sympathetic nerves in the bone marrow secrete noradrenaline in a circadian manner, which is transmitted to stromal cells by the β3-adrenergic receptor, leading to a reduction in the nuclear content of the SP1 transcription factor and the rapid downregulation of the chemokine CXCL12.[134] CXCL12 regulates the retention of HSCs in the bone marrow, and therefore HSCs are released in the

Table 20.1 Overview of stem cells in ageing and circadian clock disruption.

	Age-related phenotype	Clock-deficiency related phenotype	Ref.
Neural stem cells (NSCs)	• Increased dormancy • Decreased self-renewal, neuronal fate commitment and death	*Per2* knockout: • loss of cell cycle entrance of quiescent neural progenitor cells (NPCs) *Bmal1* knockout: • constitutively high levels of proliferation and delayed cell cycle exit • Increased cell death and neurospherical differentiation toward glia rather than neurons *Cry1* and *Cry2* double knockout: • Reduced growth of neurospheres and increased astrocyte proliferation during differentiation *Per2* mutation or deletion: • Higher density of dividing NPCs and increase in neuronal cell death	PMID: 21549330[168] PMID: 15385618[169] PMID: 14973255[170] PMID: 16957738[146] PMID: 24268780[171] PMID: 26439128[172] PMID: 19327139[173] PMID: 17368741[174]
Epidermal stem cells (epSCs)	• Decreased colony-forming ability • Do not become arrhythmic, but reprogram their oscillatory transcriptome to genes associated with DNA damage and stress responses	Over-expression of PER1 and PER2 in keratinocytes: • Spontaneous differentiation *Bmal1* deletion: • Progressive accumulation of dormant hair-follicle stem cells *Per1/2* deletion: • Progressive depletion of dormant hair-follicle stem cells Cultured hair follicle keratinocytes from shift workers *vs.* non-shift workers: • Reduced secretion of oxytocin, cortisol, β-endorphin, neurotensin, α-melanocyte-stimulating hormone, and melatonin	PMID: 2436229[175] PMID: 24120744[176] PMID: 22080954[177] PMID: 30396867[137] PMID: 28802040[136]
Muscle stem cells (MuSCs)	• Decrease in frequency • Decline in *in vitro* proliferation rate	*Bmal1*-deficient primary mouse myoblasts: • Impaired myogenic differentiation • Blunted expression of key myogenic regulatory factors	PMID: 16219688[178] PMID: 17218401[179] PMID: 6646160[180] PMID: 24531379[181]

Table 20.1 (*Continued*)

	Age-related phenotype	Clock-deficiency related phenotype	Ref.
	• Decline in *in vivo* engraftment and regeneration potential upon transplantation • Aged MuSCs convert from myogenic to fibrogenic lineage	• Reduced growth and proliferation *ex vivo* *Bmal1* induction: • Enhanced differentiation of C2C12 myoblasts. *Cry2*-deficient myoblasts: • premature cell cycle exits and form short myotubes due to inefficient cell fusion *Bmal1-deficient satellite cells:* • Reduced proliferation • Premature differentiation • Attenuated anaerobic glycolysis and cytosolic NAD^+ production	PMID: 14637002[182] PMID: 24531378[183] PMID: 24522534[184] PMID: 17690295[185] PMID: 23525013[186] PMID: 25218946[187] PMID: 29466738[188] PMID: 35115380[189] PMID: 35115380
Mesenchymal stem cells (MSCs)	• Bone MSCs exhibit age-dependent reduction in osteogenesis and increased adipogenesis • Decline in frequency, population doubling rate, colony forming efficiency, and differentiation capacity	Knockdown of *Clock* or *Per2*: • Impaired adipocyte differentiation but not osteoblastic differentiation capacity *Per2* deletion: • Decreased cell migration • Altered cell cycle phase distribution *Clock* deficiency: • Accelerated human MSC senescence CLOCK over-expression: • Rejuvenated physiologically and pathologically aged human MSCs (note: CLOCK transcriptional activity not required) Rev-erbα over-expression: • Inhibition of cell proliferation and osteogenesis	PMID: 26741371[190] PMID: 32737416[191] PMID: 25539035[192] PMID: 15342932[193] PMID: 10583212[194] PMID: 18248663[195] PMID: 18241911[196]
Intestinal stem cells (ISCs)	• Increased numbers of immunophenotypic ISCs due to enhanced proliferation and/or an accumulation of functionally differentiated cells that retain stem cell markers	Loss of *Per*: • Impaired G1 to S phase transition in ISCs following damage Loss of *Bmal1*:	PMID: 18940735[197] PMID: 18284659[198] PMID: 20157545[199] PMID: 22055505[200] PMID: 23583176[201]

	• Activation of environmental stress response pathways • Reduced mitochondrial function	• Increased self-renewal and production of tumor-initiating clones	PMID: 34534703[202]
Mammary stem cells (MaSCs)	• Enhanced basal stem cell number • Reduced differentiation capacity • Enhanced neoplastic transformation potential • Reduced luminal-restricted progenitors in the common marmoset monkey	Loss of *Clock* in MaSCs: • Reduced primary and secondary mammosphere-forming capacity indicating compromised stem cell self-renewal *Clock* knockdown in mammary epithelial cells: • Enhanced cell growth rate • Reduced expression of factors associated with mammary differentiation and milk synthesis *Bmal1* knockout in mammary epithelial: • Slower growth • Higher levels of reactive oxygen species and lower expression of superoxide dismutase 3 Ectopic expression of CLOCK in breast cancer stem cells: • Reduced stemness properties such as tumorigenicity and invasive potential *Per2* loss: • Increased basal-like cell populations in the mouse mammary gland	PMID: 28134247[203] PMID: 27707717[204] PMID: 34415905[205] PMID: 33890571[206] PMID: 29490985[207] PMID: 27852980[208] PMID: 27558284[209]
Haematopoietic stem cells (HSCs)	• Increased self-renewal and differential capacity to generate committed myeloid and lymphoid progenitors as well as diminished lymphoid potential • Increased numbers of immunophenotypic human HSCs that are less quiescent and exhibit myeloid-biased differentiation potential	Circadian expression of *Cxcl12* is lost in *Bmal1* knockout mice, as are the HSC release rhythms Disruption of canonical circadian pathway components produces anti-leukemic effects, including impaired proliferation, enhanced myeloid differentiation, and depletion of leukemia stem cells	PMID: 15967997[210] PMID: 22123971[211] PMID: 22560076[212] PMID: 18256599[134] PMID: 27058663[213]

circulation in a rhythmic manner. Additionally, CXCR4, which is a receptor for CXCL12 on the surface of HSCs, is also controlled by the core circadian clock machinery.[135] Such cross-talk of circadian networks highlights the complex nature of stem cell extracellular signaling and regulation. Yet, the role of stem cell clocks and the timing of their activation and maintenance throughout the lifespan has been relatively unexplored.

Similar to HSCs, epidermal stem cells (epSCs) and skeletal muscle stem cells (MuSCs) continuously proliferate and renew throughout life. In one study, the authors profiled gene expression of isolated epSCs and MuSCs from young (8-week-old) and old (102 to 116-week-old) mice across the circadian cycle. Interestingly, they did not observe significant differences in the rhythmic expression patterns of core circadian clock genes. Instead, they found that the oscillating transcriptome was affected by age. In young SCs, rhythmic genes included those involved in basal metabolism and homeostasis, whereas rhythmic genes in older SCs were enriched in pathways involved in oxidative stress responses, DNA damage and proteostasis.[136] Another study demonstrated that human hair follicle keratinocytes, interfollicular epidermal keratinocytes and hair follicle dermal papilla stem cells isolated from women shift workers exhibit not only dysregulated expression of core clock genes but also functional declines, including decreases in clonogenic potential as well as secretion of circadian neuropeptides and hormones such as oxytocin, β-endorphin, neurotensin, αMSH and melatonin, compared to those from regular diurnal workers.[137] These observations align with the age-dependent loss of epithelial and follicular stem/progenitor cell functions in humans[138–140] and mice.[141]

Genome-wide changes in epigenetic modifications and epigenetic drift are characteristic features of ageing, and age-associated epigenetic changes affect stem cell properties (for a more comprehensive review on this topic, see Zhang *et al.*, *Nat Rev Mol Cell Biol.* 2020[142]). For example, MuSCs from elderly subjects exhibit increased DNA methylation within gene bodies as compared to young subjects. This increase in methylation suppresses the expression of *SPRY1*, a gene that encodes a regulator of MuSC quiesence, thus impairing their self-renewal capacity.[143] In addition, aged HSCs display hypermethylated genomic regions associated with lymphoid and erythroid cell lineages, leading to skewed differentiation toward myeloid and lymphoid progenitors.[144] In addition to DNA methylation, stem cells also undergo age-associated changes in histone modifications. For example, aged neural stem cells (NSCs) display a loss of the repressive histone mark, H3K27me3, due to the down-regulation of the polycomb group gene *BMI1*.[145] This occurs at the *p16* gene locus, which encodes p16INK4a, a potent inhibitor of the G1/S-phase transition, cell cycle exit, and senescence. As a result, aged NSCs display a decline in self-renewal potential.[146] An intriguing possibility is that the decline in circadian function might be linked to these age-dependent epigenetic alterations, since the molecular clock is an important regulator of the dynamic regulation of histone modifications and DNA methylation.[147–150] Interestingly, sites of rhythmic DNA cytosine methylation overlap significantly with those sites that are altered during ageing in the mouse liver and lung[151] and human neutrophils.[152] However, this has not been investigated in stem cell models.

As summarized above and in Table 20.1, there is substantial overlap between the effects of ageing and clock disruption on adult stem cell function. This suggests potential connections between clock function and stem cell ageing; however, a causal relationship between these two processes remains unclear and deserves more comprehensive study.

20.6 Perspective

As described in the preceding sections, circadian rhythm characteristics (free-running period, amplitude, and phase) of many clock-controlled physiological and behavioural processes change along with ageing. While these changes could be consequences of ageing, it is evident that appropriate circadian reprogramming helps counter them. To fully understand this, it will be important to investigate the impact of lifestyle interventions that are known to combat age-related pathologies, such as caloric restriction and exercise, and to assess the extent to which they combat ageing through restoration of circadian clock function. Indeed, several studies have demonstrated altered circadian amplitude and phase in response to these interventions.[136,153–159] Furthermore, studies have shown powerful effects of both timed food delivery (both high and low caloric diets) and scheduled exercise on overall metabolic health and the promotion of healthy ageing and lifespan,[70,160–163] again highlighting the strong connection between circadian reprogramming and the prevention of age-associated decline.

In addition to what we have outlined in the preceding sections, there are other considerations for the connections between circadian rhythms and ageing. For example, with the development of aerospace technologies, the scope of human activities is no longer restricted to the earth, but extends into outer space. While still in its infancy, there is a growing area of research investigating the impact of spaceflight on circadian and ageing-related pathways. One study found that an astronaut exhibited lengthened telomeres as compared to his genetically matched twin after a 1-year long-duration spaceflight, including changes in telomere length.[164] Similar effects have also been observed in participants in a high-fidelity 520-day ground simulation experiment, revealing significantly decreased blood DNA methylation – a robust predictor of mortality risk correlated with chronological ageing.[165] Interestingly, circadian output genes were also significantly changed in fruit flies carried on the spaceship after only 13 days of spaceflight,[166] indicating that the circadian transcriptome may be reprogrammed during spaceflight. Similarly, microgravity – a condition mimicking weightlessness in space – amplifies oscillations of *Bmal1/Rev-erbα* in human keratinocytes.[167]

In conclusion, growing evidence supports a causal role for altered circadian clock function in the progression of ageing, and points towards the potential therapeutic benefits of circadian realignment to combat age-related disease and to promote healthy ageing.

References

1. J. S. Takahashi, Molecular components of the circadian clock in mammals, *Diabetes, Obes. Metab.*, 2015, **17**, 6–11.
2. M. H. Hastings, A. B. Reddy and E. S. Maywood, A clockwork web: circadian timing in brain and periphery, in health and disease, *Nat. Rev. Neurosci.*, 2003, **4**, 649–661.
3. J. Bass and M. A. Lazar, Circadian time signatures of fitness and disease, *Science*, 2016, **354**, 994–999.
4. J. Bass and J. S. Takahashi, Circadian Integration of Metabolism and Energetics, *Science*, 2010, **330**, 1349–1354.
5. J. F. Duffy, K.-M. Zitting and E. D. Chinoy, Aging and Circadian Rhythms, *Sleep Med. Clin.*, 2015, **10**, 423–434.
6. J. A. Horne and O. Ostberg, A self-assessment questionnaire to determine morningness-eveningness in human circadian rhythms, *Int. J. Chronobiol.*, 1976, **4**, 97–110.
7. S. Hood and S. Amir, The aging clock: circadian rhythms and later life, *J. Clin. Invest.*, 2017, **127**, 437–446.
8. J. Carrier, T. H. Monk, D. J. Buysse and D. J. Kupfer, Sleep and morningness-eveningness in the 'middle' years of life (20–59y), *J. Sleep Res.*, 1997, **6**, 230–237.
9. Y. Touitou and E. Haus, Alterations with Aging of the Endocrine and Neuroendocrine Circadian System in Humans, *Chronobiol. Int.*, 2009, **17**, 369–390.
10. D. Weinert and J. Waterhouse, The circadian rhythm of core temperature: Effects of physical activity and aging, *Physiol. Behav.*, 2007, **90**, 246–256.
11. J. F. Duffy, D.-J. Dijk, E. B. Klerman and C. A. Czeisler, Later endogenous circadian temperature nadir relative to an earlier wake time in older people, *Am. J. Physiol.: Regul., Integr. Comp. Physiol.*, 1998, **275**, R1478–R1487.
12. E. V. Cauter, R. Leproult and D. J. Kupfer, Effects of gender and age on the levels and circadian rhythmicity of plasma cortisol, *J. Clin. Endocrinol. Metab.*, 1996, **81**, 2468–2473.
13. J. Carrier, T. H. Monk, D. J. Buysse and D. J. Kupfer, Amplitude Reduction of the Circadian Temperature and Sleep Rhythms in the Elderly, *Chronobiol. Int.*, 2009, **13**, 373–386.
14. J. H. Kim and J. F. Duffy, Circadian Rhythm Sleep-Wake Disorders in Older Adults, *Sleep Med. Clin.*, 2018, **13**, 39–50.
15. E. A. Antoniadis, C. H. Ko, M. R. Ralph and R. J. McDonald, Circadian rhythms, aging and memory, *Behav. Brain Res.*, 2000, **114**, 221–233.
16. R. Singh, S. Sharma, R. K. Singh and G. Cornelissen, Circadian Time Structure of Circulating Plasma Lipid Components in Healthy Indians of Different Age Groups, *Indian J. Clin. Biochem.*, 2016, **31**, 215–223.
17. C.-Y. Chen, R. W. Logan, T. Ma, D. A. Lewis, G. C. Tseng, E. Sibille and C. A. McClung, Effects of aging on circadian patterns of gene

expression in the human prefrontal cortex, *Proc. Natl. Acad. Sci. U. S. A.*, 2016, **113**, 206–211.
18. H. Ando, K. Ushijima, M. Kumazaki, T. Takamura, N. Yokota, T. Saito, S. Irie, S. Kaneko and A. Fujimura, Influence of Age on Clock Gene Expression in Peripheral Blood Cells of Healthy Women, *J. Gerontol., Ser. A*, 2010, **65A**, 9–13.
19. T. H. Monk, D. J. Buysse, J. Carrier and D. J. Kupfer, Inducing jet-lag in older people: Directional asymmetry, *J. Sleep Res.*, 2000, **9**, 101–116.
20. T. H. Monk, D. J. Buysse, C. F. Reynolds and D. J. Kupfer, Inducing jet lag in older people: Adjusting to a 6-hour phase advance in routine, *Exp. Gerontol.*, 1993, **28**, 119–133.
21. P. Gander and L. Signal, Who Is Too Old for Shift Work? Developing Better Criteria, *Chronobiol. Int.*, 2009, **25**, 199–213.
22. S. Folkard, Shift Work, Safety, and Aging, *Chronobiol. Int.*, 2009, **25**, 183–198.
23. M. M. Blok and M. P. de Looze, What is the evidence for less shift work tolerance in older workers?, *Ergonomics*, 2011, **54**, 221–232.
24. G. Cornelissen and K. Otsuka, Chronobiology of Aging: A Mini-Review, *Gerontology*, 2017, **63**, 118–128.
25. K. Rakshit, N. Krishnan, E. M. Guzik, E. Pyza and J. M. Giebultowicz, Effects of Aging on the Molecular Circadian Oscillations in Drosophila, *Chronobiol. Int.*, 2012, **29**, 5–14.
26. T. J. Nakamura, W. Nakamura, S. Yamazaki, T. Kudo, T. Cutler, C. S. Colwell and G. D. Block, Age-Related Decline in Circadian Output, *J. Neurosci.*, 2011, **31**, 10201–10205.
27. I. V. Zhdanova, K. Masuda, C. Quasarano-Kourkoulis, D. L. Rosene, R. J. Killiany and S. Wang, Aging of Intrinsic Circadian Rhythms and Sleep in a Diurnal Nonhuman Primate, Macaca mulatta, *J. Biol. Rhythms*, 2011, **26**, 149–159.
28. K. Koh, J. M. Evans, J. C. Hendricks and A. Sehgal, A Drosophila model for age-associated changes in sleep:wake cycles, *Proc. Natl. Acad. Sci. U. S. A.*, 2006, **103**, 13843–13847.
29. S. M. Biello, Circadian clock resetting in the mouse changes with age, *Age*, 2009, **31**, 293–303.
30. R. V. Kondratov, A. A. Kondratova, V. Y. Gorbacheva, O. V. Vykhovanets and M. P. Antoch, Early aging and age-related pathologies in mice deficient in BMAL1, the core componentof the circadian clock, *Genes Dev.*, 2006, **20**, 1868–1873.
31. T. Papagiannakopoulos, M. R. Bauer, S. M. Davidson, M. Heimann, L. Subbaraj, A. Bhutkar, J. Bartlebaugh, M. G. Vander Heiden and T. Jacks, Circadian Rhythm Disruption Promotes Lung Tumorigenesis, *Cell Metab.*, 2016, **24**, 324–331.
32. B. Marcheva, K. M. Ramsey, E. D. Buhr, Y. Kobayashi, H. Su, C. H. Ko, G. Ivanova, C. Omura, S. Mo, M. H. Vitaterna, J. P. Lopez, L. H. Philipson, C. A. Bradfield, S. D. Crosby, L. JeBailey, X. Wang, J. S. Takahashi and J. Bass, Disruption of the clock components CLOCK

and BMAL1 leads to hypoinsulinaemia and diabetes, *Nature*, 2010, **466**, 627.
33. F. W. Turek, C. Joshu, A. Kohsaka, E. Lin, G. Ivanova, E. McDearmon, A. Laposky, S. Losee-Olson, A. Easton, D. R. Jensen, R. H. Eckel, J. S. Takahashi and J. Bass, Obesity and Metabolic Syndrome in Circadian *Clock* Mutant Mice, *Science*, 2005, **308**, 1043–1045.
34. M. P. Antoch, V. Y. Gorbacheva, O. Vykhovanets, I. A. Toshkov, R. V. Kondratov, A. A. Kondratova, C. Lee and A. Y. Nikitin, Disruption of the circadian clock due to the Clock mutation has discrete effects on aging and carcinogenesis, *Cell Cycle*, 2008, **7**, 1197–1204.
35. C. A. Thaiss, D. Zeevi, M. Levy, G. Zilberman-Schapira, J. Suez, A. C. Tengeler, L. Abramson, M. N. Katz, T. Korem, N. Zmora, Y. Kuperman, I. Biton, S. Gilad, A. Harmelin, H. Shapiro, Z. Halpern, E. Segal and E. Elinav, Transkingdom Control of Microbiota Diurnal Oscillations Promotes Metabolic Homeostasis, *Cell*, 2014, **159**, 514–529.
36. V. S. Valentinuzzi, K. Scarbrough, J. S. Takahashi and F. W. Turek, Effects of aging on the circadian rhythm of wheel-running activity in C57BL/6 mice, *Am. J. Physiol.: Regul., Integr. Comp. Physiol.*, 1997, **273**, R1957–R1964.
37. C. S. Pittendrigh and S. Daan, Circadian Oscillations in Rodents: A Systematic Increase of Their Frequency with Age, *Science*, 1974, **186**, 548–550.
38. K. M. Claire and M. Michael, Effect of Transplanting Suprachiasmatic Nuclei from Donors of Different Ages into Completely SCN Lesioned Hamsters, *Neural Plast.*, 1993, **4**, 257–265.
39. H. Guo, H. Guo, J. M. Brewer, M. N. Lehman and E. L. Bittman, Suprachiasmatic Regulation of Circadian Rhythms of Gene Expression in Hamster Peripheral Organs: Effects of Transplanting the Pacemaker, *J. Neurosci.*, 2006, **26**, 6406–6412.
40. M. W. Hurd and M. R. Ralph, The Significance of Circadian Organization for Longevity in the Golden Hamster, *J. Biol. Rhythms*, 1998, **13**, 430–436.
41. O. V. Reeth, Y. Zhang, P. C. Zee and F. W. Turek, Grafting fetal suprachiasmatic nuclei in the hypothalamus of old hamsters restores responsiveness of the circadian clock to a phase shifting stimulus, *Brain Res.*, 1994, **643**, 338–342.
42. H. Li and E. Satinoff, Fetal tissue containing the suprachiasmatic nucleus restores multiple circadian rhythms in old rats, *Am. J. Physiol.: Regul., Integr. Comp. Physiol.*, 1998, **275**, R1735–R1744.
43. C. A. Wyse, A. N. Coogan, C. Selman, D. G. Hazlerigg and J. R. Speakman, Association between mammalian lifespan and circadian free-running period: the circadian resonance hypothesis revisited, *Biol. Lett.*, 2010, **6**, 696–698.
44. P. J. DeCoursey and J. R. Krulas, Behavior of SCN-Lesioned Chipmunks in Natural Habitat: A Pilot Study, *J. Biol. Rhythms*, 1998, **13**, 229–244.

45. A. Ashton, R. G. Foster and A. Jagannath, Photic Entrainment of the Circadian System, *Int. J. Mol. Sci.*, 2022, **23**, 729.
46. S.-K. Chen, T. C. Badea and S. Hattar, Photoentrainment and pupillary light reflex are mediated by distinct populations of ipRGCs, *Nature*, 2011, **476**, 92–95.
47. D. Lupi, M. Semo and R. G. Foster, Impact of age and retinal degeneration on the light input to circadian brain structures, *Neurobiol. Aging*, 2012, **33**, 383–392.
48. P.-S. Welz and S. A. Benitah, Molecular Connections Between Circadian Clocks and Aging, *J. Mol. Biol.*, 2020, **432**, 3661–3679.
49. S. Farajnia, S. Michel, T. Deboer, H. T. vanderLeest, T. Houben, J. H. T. Rohling, A. Ramkisoensing, R. Yasenkov and J. H. Meijer, Evidence for Neuronal Desynchrony in the Aged Suprachiasmatic Nucleus Clock, *J. Neurosci.*, 2012, **32**, 5891–5899.
50. D. E. Kolker, H. Fukuyama, D. S. Huang, J. S. Takahashi, T. H. Horton and F. W. Turek, Aging Alters Circadian and Light-Induced Expression of Clock Genes in Golden Hamsters, *J. Biol. Rhythms*, 2003, **18**, 159–169.
51. M. Bonaconsa, G. Malpeli, A. Montaruli, F. Carandente, G. Grassi-Zucconi and M. Bentivoglio, Differential modulation of clock gene expression in the suprachiasmatic nucleus, liver and heart of aged mice, *Exp. Gerontol.*, 2014, **55**, 70–79.
52. H. Weinert, D. Weinert, I. Schurov, E. S. Maywood and M. H. Hastings, Impaired expression of the mPer2 CIRCADIAN clock gene in the suprachiasmatic nuclei of aging mice, *Chronobiol. Int.*, 2009, **18**, 559–565.
53. M. Asai, Y. Yoshinobu, S. Kaneko, A. Mori, T. Nikaido, T. Moriya, M. Akiyama and S. Shibata, Circadian profile of Per gene mRNA expression in the suprachiasmatic nucleus, paraventricular nucleus, and pineal body of aged rats, *J. Neurosci. Res.*, 2001, **66**, 1133–1139.
54. C.-K. Chiang, N. Mehta, A. Patel, P. Zhang, Z. Ning, J. Mayne, W. Y. L. Sun, H.-Y. M. Cheng and D. Figeys, The Proteomic Landscape of the Suprachiasmatic Nucleus Clock Reveals Large-Scale Coordination of Key Biological Processes, *PLoS Genet.*, 2014, **10**, e1004695.
55. C.-K. Chiang, B. Xu, N. Mehta, J. Mayne, W. Y. L. Sun, K. Cheng, Z. Ning, J. Dong, H. Zou, H.-Y. M. Cheng and D. Figeys, Phosphoproteome Profiling Reveals Circadian Clock Regulation of Posttranslational Modifications in the Murine Hippocampus, *Front. Neurol.*, 2017, **8**, 110.
56. P. Adler, C.-K. Chiang, J. Mayne, Z. Ning, X. Zhang, B. Xu, H.-Y. M. Cheng and D. Figeys, Aging Disrupts the Circadian Patterns of Protein Expression in the Murine Hippocampus, *Front. Aging Neurosci.*, 2020, **11**, 368.
57. T. J. Nakamura, W. Nakamura, I. T. Tokuda, T. Ishikawa, T. Kudo, C. S. Colwell and G. D. Block, Age-Related Changes in the Circadian System Unmasked by Constant Conditions1,2,3, *Eneuro*, 2015, **2**, ENEURO.0064-15.2015.

58. H.-C. Chang and L. Guarente, SIRT1 Mediates Central Circadian Control in the SCN by a Mechanism that Decays with Aging, *Cell*, 2013, **153**, 1448–1460.
59. L. Guarente, Sirtuins in Aging and Disease, *Cold Spring Harbor Symp. Quant. Biol.*, 2007, **72**, 483–488.
60. C. B. Peek, A. H. Affinati, K. M. Ramsey, H.-Y. Kuo, W. Yu, L. A. Sena, O. Ilkayeva, B. Marcheva, Y. Kobayashi, C. Omura, D. C. Levine, D. J. Bacsik, D. Gius, C. B. Newgard, E. Goetzman, N. S. Chandel, J. M. Denu, M. Mrksich and J. Bass, Circadian Clock NAD^+ Cycle Drives Mitochondrial Oxidative Metabolism in Mice, *Science*, 2013, **342**, 1243417.
61. K. M. Ramsey, J. Yoshino, C. S. Brace, D. Abrassart, Y. Kobayashi, B. Marcheva, H.-K. Hong, J. L. Chong, E. D. Buhr, C. Lee, J. S. Takahashi, S. Imai and J. Bass, Circadian Clock Feedback Cycle Through NAMPT-Mediated NAD^+ Biosynthesis, *Science*, 2009, **324**, 651–654.
62. Y. Nakahata, S. Sahar, G. Astarita, M. Kaluzova and P. Sassone-Corsi, Circadian Control of the NAD^+ Salvage Pathway by CLOCK-SIRT1, *Science*, 2009, **324**, 654–657.
63. G. Asher, D. Gatfield, M. Stratmann, H. Reinke, C. Dibner, F. Kreppel, R. Mostoslavsky, F. W. Alt and U. Schibler, SIRT1 Regulates Circadian Clock Gene Expression through PER2 Deacetylation, *Cell*, 2008, **134**, 317–328.
64. U. Schibler, I. Gotic, C. Saini, P. Gos, T. Curie, Y. Emmenegger, F. Sinturel, P. Gosselin, A. Gerber, F. Fleury-Olela, G. Rando, M. Demarque and P. Franken, Clock-Talk: Interactions between Central and Peripheral Circadian Oscillators in Mammals, *Cold Spring Harbor Symp. Quant. Biol.*, 2015, **80**, 223–232.
65. A. Balsalobre, S. A. Brown, L. Marcacci, F. Tronche, C. Kellendonk, H. M. Reichardt, G. Schütz and U. Schibler, Resetting of Circadian Time in Peripheral Tissues by Glucocorticoid Signaling, *Science*, 2000, **289**, 2344–2347.
66. D. R. Bruns, M. Yusifova, N. A. Marcello, C. J. Green, W. J. Walker and E. E. Schmitt, The Peripheral Circadian Clock and Exercise: Lessons from Young and Old Mice, *J. Circadian Rhythms*, 2020, **18**, 7.
67. S. A. Brown, G. Zumbrunn, F. Fleury-Olela, N. Preitner and U. Schibler, Rhythms of Mammalian Body Temperature Can Sustain Peripheral Circadian Clocks, *Curr. Biol.*, 2002, **12**, 1574–1583.
68. F. Damiola, N. L. Minh, N. Preitner, B. Kornmann, F. Fleury-Olela and U. Schibler, Restricted feeding uncouples circadian oscillators in peripheral tissues from the central pacemaker in the suprachiasmatic nucleus, *Genes Dev.*, 2000, **14**, 2950–2961.
69. K.-A. Stokkan, S. Yamazaki, H. Tei, Y. Sakaki and M. Menaker, Entrainment of the Circadian Clock in the Liver by Feeding, *Science*, 2001, **291**, 490–493.
70. V. Acosta-Rodríguez, F. Rijo-Ferreira, M. Izumo, P. Xu, M. Wight-Carter, C. B. Green and J. S. Takahashi, Circadian alignment of early onset

caloric restriction promotes longevity in male C57BL/6J mice, *Science*, 2022, **376**, 6598.

71. S. Yamazaki, M. Straume, H. Tei, Y. Sakaki, M. Menaker and G. D. Block, Effects of aging on central and peripheral mammalian clocks, *Proc. Natl. Acad. Sci. U. S. A.*, 2002, **99**, 10801–10806.
72. D. C. Levine, H. Hong, B. J. Weidemann, K. M. Ramsey, A. H. Affinati, M. S. Schmidt, J. Cedernaes, C. Omura, R. Braun, C. Lee, C. Brenner, C. B. Peek and J. Bass, NAD^+ Controls Circadian Reprogramming through PER2 Nuclear Translocation to Counter Aging, *Mol. Cell*, 2020, **78**, 835–849.e7.
73. A. Balsalobre, L. Marcacci and U. Schibler, Multiple signaling pathways elicit circadian gene expression in cultured Rat-1 fibroblasts, *Curr. Biol.*, 2000, **10**, 1291–1294.
74. K. Yagita and H. Okamura, Forskolin induces circadian gene expression of rPer1, rPer2 and dbp in mammalian rat-1 fibroblasts, *Febs Lett.*, 2000, **465**, 79–82.
75. C. Driver, The circadian clock in old Drosophila melanogaster, *Biogerontology*, 2000, **1**, 157–162.
76. W. Luo, W. Chen, Z. Yue, D. Chen, M. Sowcik, A. Sehgal and X. Zheng, Old flies have a robust central oscillator but weaker behavioral rhythms that can be improved by genetic and environmental manipulations, *Aging Cell*, 2012, **11**, 428–438.
77. J. Zhao, G. R. Warman and J. F. Cheeseman, Clock gene expression and locomotor activity predict death in the last days of life in Drosophila melanogaster, *Sci. Rep.*, 2018, **8**, 11923.
78. K. Rakshit and J. M. Giebultowicz, Cryptochrome restores dampened circadian rhythms and promotes healthspan in aging Drosophila, *Aging Cell*, 2013, **12**, 752–762.
79. R. C. Kuintzle, E. S. Chow, T. N. Westby, B. O. Gvakharia, J. M. Giebultowicz and D. A. Hendrix, Circadian deep sequencing reveals stress-response genes that adopt robust rhythmic expression during aging, *Nat. Commun.*, 2017, **8**, 14529.
80. M. Hayashi, T. Morikawa and T. Hori, Circasemidian 12 h cycle of slow wave sleep under constant darkness, *Clin. Neurophysiol.*, 2002, **113**, 1505–1516.
81. H. Ballance and B. Zhu, Revealing the hidden reality of the mammalian 12-h ultradian rhythms, *Cell. Mol. Life Sci.*, 2021, **78**, 3127–3140.
82. B. Zhu, Q. Zhang, Y. Pan, E. M. Mace, B. York, A. C. Antoulas, C. C. Dacso and B. W. O'Malley, A Cell-Autonomous Mammalian 12 hr Clock Coordinates Metabolic and Stress Rhythms, *Cell Metab.*, 2017, **25**, 1305–1319.e9.
83. C. López-Otín, L. Galluzzi, J. M. P. Freije, F. Madeo and G. Kroemer, Metabolic Control of Longevity, *Cell*, 2016, **166**, 802–821.
84. Z. Yu, G. Zhai, P. Singmann, Y. He, T. Xu, C. Prehn, W. Römisch-Margl, E. Lattka, C. Gieger, N. Soranzo, J. Heinrich, M. Standl, E. Thiering, K. Mittelstraß, H. Wichmann, A. Peters, K. Suhre, Y. Li, J. Adamski,

T. D. Spector, T. Illig and R. Wang-Sattler, Human serum metabolic profiles are age dependent, *Aging Cell*, 2012, **11**, 960–967.

85. R. H. Houtkooper, C. Argmann, S. M. Houten, C. Cantó, E. H. Jeninga, P. A. Andreux, C. Thomas, R. Doenlen, K. Schoonjans and J. Auwerx, The metabolic footprint of aging in mice, *Sci. Rep.*, 2011, **1**, 134.
86. H. Pontzer, Y. Yamada, H. Sagayama, P. N. Ainslie, L. F. Andersen, L. J. Anderson, L. Arab, I. Baddou, K. Bedu-Addo, E. E. Blaak, S. Blanc, A. G. Bonomi, C. V. C. Bouten, P. Bovet, M. S. Buchowski, N. F. Butte, S. G. Camps, G. L. Close, J. A. Cooper, R. Cooper, S. K. Das, L. R. Dugas, U. Ekelund, S. Entringer, T. Forrester, B. W. Fudge, A. H. Goris, M. Gurven, C. Hambly, A. E. Hamdouchi, M. B. Hoos, S. Hu, N. Joonas, A. M. Joosen, P. Katzmarzyk, K. P. Kempen, M. Kimura, W. E. Kraus, R. F. Kushner, E. V. Lambert, W. R. Leonard, N. Lessan, C. Martin, A. C. Medin, E. P. Meijer, J. C. Morehen, J. P. Morton, M. L. Neuhouser, T. A. Nicklas, R. M. Ojiambo, K. H. Pietiläinen, Y. P. Pitsiladis, J. Plange-Rhule, G. Plasqui, R. L. Prentice, R. A. Rabinovich, S. B. Racette, D. A. Raichlen, E. Ravussin, R. M. Reynolds, S. B. Roberts, A. J. Schuit, A. M. Sjödin, E. Stice, S. S. Urlacher, G. Valenti, L. M. V. Etten, E. A. V. Mil, J. C. K. Wells, G. Wilson, B. M. Wood, J. Yanovski, T. Yoshida, X. Zhang, A. J. Murphy-Alford, C. Loechl, A. H. Luke, J. Rood, D. A. Schoeller, K. R. Westerterp, W. W. Wong, J. R. Speakman and I. D. D. Consortium, Daily energy expenditure through the human life course, *Science*, 2021, **373**, 808–812.
87. D. C. Levine, K. M. Ramsey and J. Bass, Circadian NAD(P)(H) cycles in cell metabolism, *Semin. Cell Dev. Biol.*, 2022, **126**, 15–26.
88. N. Xie, L. Zhang, W. Gao, C. Huang, P. E. Huber, X. Zhou, C. Li, G. Shen and B. Zou, NAD^+ metabolism: pathophysiologic mechanisms and therapeutic potential, *Signal Transduction Targeted Ther.*, 2020, **5**, 227.
89. J. Clement, M. Wong, A. Poljak, P. Sachdev and N. Braidy, The Plasma NAD^+ Metabolome Is Dysregulated in "Normal" Aging, *Rejuvenation Res.*, 2019, **22**, 121–130.
90. R. Grant, J. Berg, R. Mestayer, N. Braidy, J. Bennett, S. Broom and J. Watson, A Pilot Study Investigating Changes in the Human Plasma and Urine NAD^+ Metabolome During a 6 Hour Intravenous Infusion of NAD^+, *Front. Aging Neurosci.*, 2019, **11**, 257.
91. M. R. McReynolds, K. Chellappa, E. Chiles, C. Jankowski, Y. Shen, L. Chen, H. C. Descamps, S. Mukherjee, Y. R. Bhat, S. R. Lingala, Q. Chu, P. Botolin, F. Hayat, T. Doke, K. Susztak, C. A. Thaiss, W. Lu, M. E. Migaud, X. Su, J. D. Rabinowitz and J. A. Baur, NAD^+ flux is maintained in aged mice despite lower tissue concentrations, *Cell Syst.*, 2021, **12**, 1160–1172.
92. K. F. Mills, S. Yoshida, L. R. Stein, A. Grozio, S. Kubota, Y. Sasaki, P. Redpath, M. E. Migaud, R. S. Apte, K. Uchida, J. Yoshino and S. Imai, Long-Term Administration of Nicotinamide Mononucleotide Mitigates Age-Associated Physiological Decline in Mice, *Cell Metab.*, 2016, **24**, 795–806.

93. E. F. Fang, H. Kassahun, D. L. Croteau, M. Scheibye-Knudsen, K. Marosi, H. Lu, R. A. Shamanna, S. Kalyanasundaram, R. C. Bollineni, M. A. Wilson, W. B. Iser, B. N. Wollman, M. Morevati, J. Li, J. S. Kerr, Q. Lu, T. B. Waltz, J. Tian, D. A. Sinclair, M. P. Mattson, H. Nilsen and V. A. Bohr, NAD^+ Replenishment Improves Lifespan and Healthspan in Ataxia Telangiectasia Models *via* Mitophagy and DNA Repair, *Cell Metab.*, 2016, **24**, 566–581.
94. H. Zhang, D. Ryu, Y. Wu, K. Gariani, X. Wang, P. Luan, D. D'Amico, E. R. Ropelle, M. P. Lutolf, R. Aebersold, K. Schoonjans, K. J. Menzies and J. Auwerx, NAD^+ repletion improves mitochondrial and stem cell function and enhances life span in mice, *Science*, 2016, **352**, 1436–1443.
95. C. S. S. Teixeira, N. M. F. S. A. Cerqueira, P. Gomes and S. F. Sousa, A Molecular Perspective on Sirtuin Activity, *Int. J. Mol. Sci.*, 2020, **21**, 8609.
96. S.-J. Lin, P.-A. Defossez and L. Guarente, Requirement of NAD and SIR2 for Life-Span Extension by Calorie Restriction in Saccharomyces cerevisiae, *Science*, 2000, **289**, 2126–2128.
97. R.-H. Wang, Y. Zheng, H.-S. Kim, X. Xu, L. Cao, T. Lahusen, M.-H. Lee, C. Xiao, A. Vassilopoulos, W. Chen, K. Gardner, Y.-G. Man, M.-C. Hung, T. Finkel and C.-X. Deng, Interplay among BRCA1, SIRT1, and Survivin during BRCA1-Associated Tumorigenesis, *Mol. Cell*, 2008, **32**, 11–20.
98. D. Herranz, M. Muñoz-Martin, M. Cañamero, F. Mulero, B. Martinez-Pastor, O. Fernandez-Capetillo and M. Serrano, Sirt1 improves healthy ageing and protects from metabolic syndrome-associated cancer, *Nat. Commun.*, 2010, **1**, 3.
99. H.-S. Kim, A. Vassilopoulos, R.-H. Wang, T. Lahusen, Z. Xiao, X. Xu, C. Li, T. D. Veenstra, B. Li, H. Yu, J. Ji, X. W. Wang, S.-H. Park, Y. I. Cha, D. Gius and C.-X. Deng, SIRT2 Maintains Genome Integrity and Suppresses Tumorigenesis through Regulating APC/C Activity, *Cancer Cell*, 2011, **20**, 487–499.
100. J. Pallos, L. Bodai, T. Lukacsovich, J. M. Purcell, J. S. Steffan, L. M. Thompson and J. L. Marsh, Inhibition of specific HDACs and sirtuins suppresses pathogenesis in a Drosophila model of Huntington's disease, *Hum. Mol. Genet.*, 2008, **17**, 3767–3775.
101. R. Luthi-Carter, D. M. Taylor, J. Pallos, E. Lambert, A. Amore, A. Parker, H. Moffitt, D. L. Smith, H. Runne, O. Gokce, A. Kuhn, Z. Xiang, M. M. Maxwell, S. A. Reeves, G. P. Bates, C. Neri, L. M. Thompson, J. L. Marsh and A. G. Kazantsev, SIRT2 inhibition achieves neuroprotection by decreasing sterol biosynthesis, *Proc. Natl. Acad. Sci. U. S. A.*, 2010, **107**, 7927–7932.
102. Q. Zhang, Z. Wang, H. Chen, S. Zhou, W. Zheng, G. Liu, Y. Wei, H. Cai, D. Liu and C. Liang, Endothelium-specific overexpression of class III deacetylase SIRT1 decreases atherosclerosis in apolipoprotein E-deficient mice, *Cardiovasc. Res.*, 2008, **80**, 191–199.
103. S. Stein, C. Lohmann, N. Schäfer, J. Hofmann, L. Rohrer, C. Besler, K. M. Rothgiesser, B. Becher, M. O. Hottiger, J. Borén, M. W. McBurney, U. Landmesser, T. F. Lüscher and C. M. Matter, SIRT1 decreases

Lox-1-mediated foam cell formation in atherogenesis, *Eur. Heart J.*, 2010, **31**, 2301–2309.
104. J. Krishnan, C. Danzer, T. Simka, J. Ukropec, K. M. Walter, S. Kumpf, P. Mirtschink, B. Ukropcova, D. Gasperikova, T. Pedrazzini and W. Krek, Dietary obesity-associated Hif1α activation in adipocytes restricts fatty acid oxidation and energy expenditure *via* suppression of the Sirt2-NAD^+ system, *Genes Dev.*, 2012, **26**, 259–270.
105. H.-C. Chang and L. Guarente, SIRT1 and other sirtuins in metabolism, *Trends Endocrinol. Metab.*, 2014, **25**, 138–145.
106. G. Y. Liu and D. M. Sabatini, mTOR at the nexus of nutrition, growth, ageing and disease, *Nat. Rev. Mol. Cell Biol.*, 2020, **21**, 183–203.
107. K. Burkewitz, Y. Zhang and W. B. Mair, AMPK at the Nexus of Energetics and Aging, *Cell Metab.*, 2014, **20**, 10–25.
108. S. Tabibzadeh, Signaling pathways and effectors of aging, *Front. Biosci.*, 2021, **26**, 50–96.
109. R. Wu, F. Dang, P. Li, P. Wang, Q. Xu, Z. Liu, Y. Li, Y. Wu, Y. Chen and Y. Liu, The Circadian Protein Period2 Suppresses mTORC1 Activity *via* Recruiting Tsc1 to mTORC1 Complex, *Cell Metab.*, 2019, **29**, 653–667.e6.
110. K. A. Lamia, U. M. Sachdeva, L. DiTacchio, E. C. Williams, J. G. Alvarez, D. F. Egan, D. S. Vasquez, H. Juguilon, S. Panda, R. J. Shaw, C. B. Thompson and R. M. Evans, AMPK Regulates the Circadian Clock by Cryptochrome Phosphorylation and Degradation, *Science*, 2009, **326**, 437–440.
111. H.-K. Hong, E. Maury, K. M. Ramsey, M. Perelis, B. Marcheva, C. Omura, Y. Kobayashi, D. C. Guttridge, G. D. Barish and J. Bass, Requirement for NF-κB in maintenance of molecular and behavioral circadian rhythms in mice, *Genes Dev.*, 2018, **32**, 1367–1379.
112. J. H. Um, S. Yang, S. Yamazaki, H. Kang, B. Viollet, M. Foretz and J. H. Chung, Activation of 5′-AMP-activated Kinase with Diabetes Drug Metformin Induces Casein Kinase Iε (CKIε)-dependent Degradation of Clock Protein mPer2*, *J. Biol. Chem.*, 2007, **282**, 20794–20798.
113. R. Cao, B. Lee, H. Cho, S. Saklayen and K. Obrietan, Photic regulation of the mTOR signaling pathway in the suprachiasmatic circadian clock, *Mol. Cell. Neurosci.*, 2008, **38**, 312–324.
114. R. Cao, A. Li, H. Y Cho, B. Lee and K. Obrietan, Mammalian Target of Rapamycin Signaling Modulates Photic Entrainment of the Suprachiasmatic Circadian Clock, *J. Neurosci.*, 2010, **30**, 6302–6314.
115. R. Cao, B. Robinson, H. Xu, C. Gkogkas, A. Khoutorsky, T. Alain, A. Yanagiya, T. Nevarko, A. C. Liu, S. Amir and N. Sonenberg, Translational Control of Entrainment and Synchrony of the Suprachiasmatic Circadian Clock by mTOR/4E-BP1 Signaling, *Neuron*, 2013, **79**, 712–724.
116. D. Liu, A. Stowie, N. de Zavalia, T. Leise, S. S. Pathak, L. R. Drewes, A. J. Davidson, S. Amir, N. Sonenberg and R. Cao, mTOR signaling in VIP neurons regulates circadian clock synchrony and olfaction, *Proc. Natl. Acad. Sci. U. S. A.*, 2018, **115**, E3296–E3304.

117. C. Ramanathan, N. D. Kathale, D. Liu, C. Lee, D. A. Freeman, J. B. Hogenesch, R. Cao and A. C. Liu, mTOR signaling regulates central and peripheral circadian clock function, *PLoS Genet.*, 2018, **14**, e1007369.
118. A. Oeckinghaus and S. Ghosh, The NF-κB Family of Transcription Factors and Its Regulation, *Cold Spring Harbor Perspect. Biol.*, 2009, **1**, a000034.
119. Y. Shen, M. Endale, W. Wang, A. R. Morris, L. J. Francey, R. L. Harold, D. W. Hammers, Z. Huo, C. L. Partch, J. B. Hogenesch, Z.-H. Wu and A. C. Liu, NF-κB modifies the mammalian circadian clock through interaction with the core clock protein BMAL1, *PLoS Genet.*, 2021, **17**, e1009933.
120. C. B. Peek, D. C. Levine, J. Cedernaes, A. Taguchi, Y. Kobayashi, S. J. Tsai, N. A. Bonar, M. R. McNulty, K. M. Ramsey and J. Bass, Circadian Clock Interaction with HIF1α Mediates Oxygenic Metabolism and Anaerobic Glycolysis in Skeletal Muscle, *Cell Metab.*, 2017, **25**, 86–92.
121. Y. Adamovich, B. Ladeuix, M. Golik, M. P. Koeners and G. Asher, Rhythmic Oxygen Levels Reset Circadian Clocks through HIF1α, *Cell Metab.*, 2017, **25**, 93–101.
122. Y. Wu, D. Tang, N. Liu, W. Xiong, H. Huang, Y. Li, Z. Ma, H. Zhao, P. Chen, X. Qi and E. E. Zhang, Reciprocal Regulation between the Circadian Clock and Hypoxia Signaling at the Genome Level in Mammals, *Cell Metab.*, 2017, **25**, 73–85.
123. M. Kobayashi, A. Morinibu, S. Koyasu, Y. Goto, M. Hiraoka and H. Harada, A circadian clock gene, PER2, activates HIF-1 as an effector molecule for recruitment of HIF-1α to promoter regions of its downstream genes, *Febs J*, 2017, **284**, 3804–3816.
124. M. T. Ghorbel, J. M. Coulson and D. Murphy, Cross-talk between hypoxic and circadian pathways: cooperative roles for hypoxia-inducible factor 1α and CLOCK in transcriptional activation of the vasopressin gene, *Mol. Cell. Neurosci.*, 2003, **22**, 396–404.
125. D. Chilov, T. Hofer, C. Uer, R. H. Wenger and M. Gasann, Hypoxia affects expression of circadian genes PER1 and CLOCK in mouse brain, *FASEB J.*, 2001, **15**, 2613–2622.
126. C. B. Peek, Metabolic Implications of Circadian–HIF Crosstalk, *Trends Endocrinol. Metab.*, 2020, **31**, 459–468.
127. Z. E. Walton, C. H. Patel, R. C. Brooks, Y. Yu, A. Ibrahim-Hashim, M. Riddle, A. Porcu, T. Jiang, B. L. Ecker, F. Tameire, C. Koumenis, A. T. Weeraratna, D. K. Welsh, R. Gillies, J. C. Alwine, L. Zhang, J. D. Powell and C. V. Dang, Acid Suspends the Circadian Clock in Hypoxia through Inhibition of mTOR, *Cell*, 2018, **174**, 72–87.e32.
128. M. B. Schultz and D. A. Sinclair, When stem cells grow old: phenotypes and mechanisms of stem cell aging, *Development*, 2016, **143**, 3–14.
129. S. M. Chambers, C. A. Shaw, C. Gatza, C. J. Fisk, L. A. Donehower and M. A. Goodell, Aging Hematopoietic Stem Cells Decline in Function and Exhibit Epigenetic Dysregulation, *PLoS Biol.*, 2007, **5**, e201.

130. B. E. Keyes, J. P. Segal, E. Heller, W.-H. Lien, C.-Y. Chang, X. Guo, D. S. Oristian, D. Zheng and E. Fuchs, Nfatc1 orchestrates aging in hair follicle stem cells, *Proc. Natl. Acad. Sci. U. S. A.*, 2013, **110**, E4950–E4959.
131. L. Liu, T. H. Cheung, G. W. Charville, B. M. C. Hurgo, T. Leavitt, J. Shih, A. Brunet and T. A. Rando, Chromatin Modifications as Determinants of Muscle Stem Cell Quiescence and Chronological Aging, *Cell Rep.*, 2013, **4**, 189–204.
132. M. Weger, N. Diotel, A.-C. Dorsemans, T. Dickmeis and B. D. Weger, Stem cells and the circadian clock, *Dev. Biol.*, 2017, **431**, 111–123.
133. S. A. Benitah and P.-S. Welz, Circadian Regulation of Adult Stem Cell Homeostasis and Aging, *Cell Stem Cell*, 2020, **26**, 817–831.
134. S. Méndez-Ferrer, D. Lucas, M. Battista and P. S. Frenette, Haematopoietic stem cell release is regulated by circadian oscillations, *Nature*, 2008, **452**, 442–447.
135. D. Lucas, M. Battista, P. A. Shi, L. Isola and P. S. Frenette, Mobilized Hematopoietic Stem Cell Yield Depends on Species-Specific Circadian Timing, *Cell Stem Cell*, 2008, **3**, 364–366.
136. G. Solanas, F. O. Peixoto, E. Perdiguero, M. Jardí, V. Ruiz-Bonilla, D. Datta, A. Symeonidi, A. Castellanos, P.-S. Welz, J. M. Caballero, P. Sassone-Corsi, P. Muñoz-Cánoves and S. A. Benitah, Aged Stem Cells Reprogram Their Daily Rhythmic Functions to Adapt to Stress, *Cell*, 2017, **170**, 678–692.e20.
137. N. Deshayes, G. Genty, F. Berthelot and M. Paris, Human long-term deregulated circadian rhythm alters regenerative properties of skin and hair precursor cells, *Eur. J. Dermatol.*, 2018, **28**, 467–475.
138. J. Lecardonnel, N. Deshayes, G. Genty, N. Parent, B. A. Bernard, M. Rathman-Josserand and M. Paris, Ageing and colony-forming efficiency of human hair follicle keratinocytes, *Exp. Dermatol.*, 2013, **22**, 604–606.
139. S. Pain, C. Dezutter, C. Reymermier, B. Vogelgesang, E. Delay and V. André, Age-related changes in pro-opiomelanocortin (POMC) and related receptors in human epidermis, *Int. J. Cosmet. Sci.*, 2010, **32**, 266–275.
140. N. Gago, V. Pérez-López, J. P. Sanz-Jaka, P. Cormenzana, I. Eizaguirre, A. Bernad and A. Izeta, Age-Dependent Depletion of Human Skin-Derived Progenitor Cells, *Stem Cells*, 2009, **27**, 1164–1172.
141. X. Su, M. Paris, Y. J. Gi, K. Y. Tsai, M. S. Cho, Y.-L. Lin, J. A. Biernaskie, S. Sinha, C. Prives, L. H. Pevny, F. D. Miller and E. R. Flores, TAp63 Prevents Premature Aging by Promoting Adult Stem Cell Maintenance, *Cell Stem Cell*, 2009, **5**, 64–75.
142. W. Zhang, J. Qu, G.-H. Liu and J. C. I. Belmonte, The ageing epigenome and its rejuvenation, *Nat. Rev. Mol. Cell Biol.*, 2020, **21**, 137–150.
143. A. Bigot, W. J. Duddy, Z. G. Ouandaogo, E. Negroni, V. Mariot, S. Ghimbovschi, B. Harmon, A. Wielgosik, C. Loiseau, J. Devaney, J. Dumonceaux, G. Butler-Browne, V. Mouly and S. Duguez, Age-Associated Methylation Suppresses SPRY1, Leading to a Failure of

Re-quiescence and Loss of the Reserve Stem Cell Pool in Elderly Muscle, *Cell Rep.*, 2015, **13**, 1172–1182.
144. I. Beerman, C. Bock, B. S. Garrison, Z. D. Smith, H. Gu, A. Meissner and D. J. Rossi, Proliferation-Dependent Alterations of the DNA Methylation Landscape Underlie Hematopoietic Stem Cell Aging, *Cell Stem Cell*, 2013, **12**, 413–425.
145. M. Abdouh, W. Chatoo, J. E. Hajjar, J. David, J. Ferreira and G. Bernier, Bmi1 Is Down-Regulated in the Aging Brain and Displays Antioxidant and Protective Activities in Neurons, *PLoS One*, 2012, **7**, e31870.
146. A. V. Molofsky, S. G. Slutsky, N. M. Joseph, S. He, R. Pardal, J. Krishnamurthy, N. E. Sharpless and S. J. Morrison, Increasing p16INK4a expression decreases forebrain progenitors and neurogenesis during ageing, *Nature*, 2006, **443**, 448–452.
147. T. M. Joska, R. Zaman and W. J. Belden, Regulated DNA Methylation and the Circadian Clock: Implications in Cancer, *Biology*, 2014, **3**, 560–577.
148. S. Katada and P. Sassone-Corsi, The histone methyltransferase MLL1 permits the oscillation of circadian gene expression, *Nat. Struct. Mol. Biol.*, 2010, **17**, 1414–1421.
149. H. J. Nam, K. Boo, D. Kim, D.-H. Han, H. K. Choe, C. R. Kim, W. Sun, H. Kim, K. Kim, H. Lee, E. Metzger, R. Schuele, S.-H. Yoo, J. S. Takahashi, S. Cho, G. H. Son and S. H. Baek, Phosphorylation of LSD1 by PKCα Is Crucial for Circadian Rhythmicity and Phase Resetting, *Mol. Cell*, 2014, **53**, 791–805.
150. N. Koike, S.-H. Yoo, H.-C. Huang, V. Kumar, C. Lee, T.-K. Kim and J. S. Takahashi, Transcriptional Architecture and Chromatin Landscape of the Core Circadian Clock in Mammals, *Science*, 2012, **338**, 349–354.
151. G. Oh, S. Ebrahimi, M. Carlucci, A. Zhang, A. Nair, D. E. Groot, V. Labrie, P. Jia, E. S. Oh, R. H. Jeremian, M. Susic, T. C. Shrestha, M. R. Ralph, J. Gordevičius, K. Koncevičius and A. Petronis, Cytosine modifications exhibit circadian oscillations that are involved in epigenetic diversity and aging, *Nat. Commun.*, 2018, **9**, 644.
152. G. Oh, K. Koncevičius, S. Ebrahimi, M. Carlucci, D. E. Groot, A. Nair, A. Zhang, A. Kriščiūnas, E. S. Oh, V. Labrie, A. H. C. Wong, J. Gordevičius, P. Jia, M. Susic and A. Petronis, Circadian oscillations of cytosine modification in humans contribute to epigenetic variability, aging, and complex disease, *Genome Biol.*, 2019, **20**, 2.
153. S. Sato, G. Solanas, F. O. Peixoto, L. Bee, A. Symeonidi, M. S. Schmidt, C. Brenner, S. Masri, S. A. Benitah and P. Sassone-Corsi, Circadian Reprogramming in the Liver Identifies Metabolic Pathways of Aging, *Cell*, 2017, **170**, 664–677.e11.
154. S. A. Patel, N. Velingkaar, K. Makwana, A. Chaudhari and R. Kondratov, Calorie restriction regulates circadian clock gene expression through BMAL1 dependent and independent mechanisms, *Sci. Rep.*, 2016, **6**, 25970.
155. M. Takahashi, A. Haraguchi, Y. Tahara, N. Aoki, M. Fukazawa, K. Tanisawa, T. Ito, T. Nakaoka, M. Higuchi and S. Shibata, Positive

association between physical activity and PER3 expression in older adults, *Sci. Rep.*, 2017, **7**, 39771.

156. A. T. L. Hughes, R. E. Samuels, B. Baño-Otálora, M. D. C. Belle, S. Wegner, C. Guilding, R. C. Northeast, A. S. I. Loudon, J. Gigg and H. D. Piggins, Timed daily exercise remodels circadian rhythms in mice, *Commun. Biol.*, 2021, **4**, 761.
157. D. Kemler, C. A. Wolff and K. A. Esser, Time-of-day dependent effects of contractile activity on the phase of the skeletal muscle clock, *J. Physiol.*, 2020, **598**, 3631–3644.
158. G. WOLFF and K. A. ESSER, Scheduled Exercise Phase Shifts the Circadian Clock in Skeletal Muscle, *Med. Sci. Sports Exercise*, 2012, **44**, 1663–1670.
159. E. Challet, L. C. Solberg and F. W. Turek, Entrainment in calorie-restricted mice: conflicting zeitgebers and free-running conditions, *Am. J. Physiol.: Regul., Integr. Comp. Physiol.*, 1998, **274**, R1751–R1761.
160. D. M. Arble, J. Bass, A. D. Laposky, M. H. Vitaterna and F. W. Turek, Circadian Timing of Food Intake Contributes to Weight Gain, *Obesity*, 2009, **17**, 2100–2102.
161. H. Sasaki, T. Ohtsu, Y. Ikeda, M. Tsubosaka and S. Shibata, Combination of meal and exercise timing with a high-fat diet influences energy expenditure and obesity in mice, *Chronobiol. Int.*, 2014, **31**, 959–975.
162. M. Hatori, C. Vollmers, A. Zarrinpar, L. DiTacchio, E. A. Bushong, S. Gill, M. Leblanc, A. Chaix, M. Joens, J. A. J. Fitzpatrick, M. H. Ellisman and S. Panda, Time-Restricted Feeding without Reducing Caloric Intake Prevents Metabolic Diseases in Mice Fed a High-Fat Diet, *Cell Metab.*, 2012, **15**, 848–860.
163. V. A. Acosta-Rodríguez, F. Rijo-Ferreira, C. B. Green and J. S. Takahashi, Importance of circadian timing for aging and longevity, *Nat. Commun.*, 2021, **12**, 2862.
164. F. E. Garrett-Bakelman, M. Darshi, S. J. Green, R. C. Gur, L. Lin, B. R. Macias, M. J. McKenna, C. Meydan, T. Mishra, J. Nasrini, B. D. Piening, L. F. Rizzardi, K. Sharma, J. H. Siamwala, L. Taylor, M. H. Vitaterna, M. Afkarian, E. Afshinnekoo, S. Ahadi, A. Ambati, M. Arya, D. Bezdan, C. M. Callahan, S. Chen, A. M. K. Choi, G. E. Chlipala, K. Contrepois, M. Covington, B. E. Crucian, I. D. Vivo, D. F. Dinges, D. J. Ebert, J. I. Feinberg, J. A. Gandara, K. A. George, J. Goutsias, G. S. Grills, A. R. Hargens, M. Heer, R. P. Hillary, A. N. Hoofnagle, V. Y. H. Hook, G. Jenkinson, P. Jiang, A. Keshavarzian, S. S. Laurie, B. Lee-McMullen, S. B. Lumpkins, M. MacKay, M. G. Maienschein-Cline, A. M. Melnick, T. M. Moore, K. Nakahira, H. H. Patel, R. Pietrzyk, V. Rao, R. Saito, D. N. Salins, J. M. Schilling, D. D. Sears, C. K. Sheridan, M. B. Stenger, R. Tryggvadottir, A. E. Urban, T. Vaisar, B. V. Espen, J. Zhang, M. G. Ziegler, S. R. Zwart, J. B. Charles, C. E. Kundrot, G. B. I. Scott, S. M. Bailey, M. Basner, A. P. Feinberg, S. M. C. Lee, C. E. Mason, E. Mignot, B. K. Rana, S. M. Smith, M. P. Snyder and F. W. Turek, The NASA Twins

Study: A multidimensional analysis of a year-long human spaceflight, *Science*, 2019, **364**(6436), eaau8650.
165. J. C. Nwanaji-Enwerem, U. Nwanaji-Enwerem, L. V. D. Laan, J. M. Galazka, N. S. Redeker and A. Cardenas, A Longitudinal Epigenetic Aging and Leukocyte Analysis of Simulated Space Travel: The Mars-500 Mission, *Cell Rep.*, 2020, **33**, 108406–108406.
166. L. Ma, J. Ma and K. Xu, Effect of Spaceflight on the Circadian Rhythm, Lifespan and Gene Expression of Drosophila melanogaster, *PLoS One*, 2015, **10**, e0121600.
167. D. Ranieri, A. Cucina, M. Bizzarri, M. Alimandi and M. R. Torrisi, Microgravity influences circadian clock oscillation in human keratinocytes, *Febs Open Bio*, 2015, **5**, 717–723.
168. J. M. Encinas, T. V. Michurina, N. Peunova, J.-H. Park, J. Tordo, D. A. Peterson, G. Fishell, A. Koulakov and G. Enikolopov, Division-Coupled Astrocytic Differentiation and Age-Related Depletion of Neural Stem Cells in the Adult Hippocampus, *Cell Stem Cell*, 2011, **8**, 566–579.
169. E. Enwere, T. Shingo, C. Gregg, H. Fujikawa, S. Ohta and S. Weiss, Aging Results in Reduced Epidermal Growth Factor Receptor Signaling, Diminished Olfactory Neurogenesis, and Deficits in Fine Olfactory Discrimination, *J. Neurosci.*, 2004, **24**, 8354–8365.
170. A. Y. Maslov, T. A. Barone, R. J. Plunkett and S. C. Pruitt, Neural Stem Cell Detection, Characterization, and Age-Related Changes in the Subventricular Zone of Mice, *J. Neurosci.*, 2004, **24**, 1726–1733.
171. P. Bouchard-Cannon, L. Mendoza-Viveros, A. Yuen, M. Kærn and H.-Y. M. Cheng, The Circadian Molecular Clock Regulates Adult Hippocampal Neurogenesis by Controlling the Timing of Cell-Cycle Entry and Exit, *Cell Rep.*, 2013, **5**, 961–973.
172. A. Malik, R. V. Kondratov, R. J. Jamasbi and M. E. Geusz, Circadian Clock Genes Are Essential for Normal Adult Neurogenesis, Differentiation, and Fate Determination, *PLoS One*, 2015, **10**, e0139655.
173. L. Borgs, P. Beukelaers, R. Vandenbosch, L. Nguyen, G. Moonen, P. Maquet, U. Albrecht, S. Belachew and B. Malgrange, Period 2 regulates neural stem/progenitor cell proliferation in the adult hippocampus, *BMC Neurosci.*, 2009, **10**, 30.
174. T. Moriya, K. Hiraishi, N. Horie, M. Mitome and K. Shinohara, Correlative association between circadian expression of mousePer2 gene and the proliferation of the neural stem cells, *Neuroscience*, 2007, **146**, 494–498.
175. Y. Barrandon and H. Green, Three clonal types of keratinocyte with different capacities for multiplication, *Proc. Natl. Acad. Sci. U. S. A.*, 1987, **84**, 2302–2306.
176. P. Janich, K. Toufighi, G. Solanas, N. M. Luis, S. Minkwitz, L. Serrano, B. Lehner and S. A. Benitah, Human Epidermal Stem Cell Function Is Regulated by Circadian Oscillations, *Cell Stem Cell*, 2013, **13**, 745–753.
177. P. Janich, G. Pascual, A. Merlos-Suárez, E. Batlle, J. Ripperger, U. Albrecht, H.-Y. M. Cheng, K. Obrietan, L. D. Croce and S. A. Benitah,

The circadian molecular clock creates epidermal stem cell heterogeneity, *Nature*, 2011, **480**, 209–214.

178. A. S. Brack, H. Bildsoe and S. M. Hughes, Evidence that satellite cell decrement contributes to preferential decline in nuclear number from large fibres during murine age-related muscle atrophy, *J. Cell Sci.*, 2005, **118**, 4813–4821.
179. C. A. Collins, P. S. Zammit, A. P. Ruiz, J. E. Morgan and T. A. Partridge, A Population of Myogenic Stem Cells That Survives Skeletal Muscle Aging, *Stem Cells*, 2007, **25**, 885–894.
180. M. C. Gibson and E. Schultz, Age-related differences in absolute numbers of skeletal muscle satellite cells, *Muscle Nerve*, 1983, **6**, 574–580.
181. J. D. Bernet, J. D. Doles, J. K. Hall, K. K. Tanaka, T. A. Carter and B. B. Olwin, p38 MAPK signaling underlies a cell-autonomous loss of stem cell self-renewal in skeletal muscle of aged mice, *Nat. Med.*, 2014, **20**, 265–271.
182. S. Bortoli, V. Renault, E. Eveno, C. Auffray, G. Butler-Browne and G. Piétu, Gene expression profiling of human satellite cells during muscular aging using cDNA arrays, *Gene*, 2003, **321**, 145–154.
183. B. D. Cosgrove, P. M. Gilbert, E. Porpiglia, F. Mourkioti, S. P. Lee, S. Y. Corbel, M. E. Llewellyn, S. L. Delp and H. M. Blau, Rejuvenation of the muscle stem cell population restores strength to injured aged muscles, *Nat. Med.*, 2014, **20**, 255–264.
184. P. Sousa-Victor, S. Gutarra, L. García-Prat, J. Rodriguez-Ubreva, L. Ortet, V. Ruiz-Bonilla, M. Jardí, E. Ballestar, S. González, A. L. Serrano, E. Perdiguero and P. Muñoz-Cánoves, Geriatric muscle stem cells switch reversible quiescence into senescence, *Nature*, 2014, **506**, 316–321.
185. A. S. Brack, M. J. Conboy, S. Roy, M. Lee, C. J. Kuo, C. Keller and T. A. Rando, Increased Wnt Signaling During Aging Alters Muscle Stem Cell Fate and Increases Fibrosis, *Science*, 2007, **317**, 807–810.
186. S. Chatterjee, D. Nam, B. Guo, J. M. Kim, G. E. Winnier, J. Lee, R. Berdeaux, V. K. Yechoor and K. Ma, Brain and muscle Arnt-like 1 is a key regulator of myogenesis, *J. Cell Sci.*, 2013, **126**, 2213–2224.
187. S. Chatterjee, H. Yin, D. Nam, Y. Li and K. Ma, Brain and muscle Arnt-like 1 promotes skeletal muscle regeneration through satellite cell expansion, *Exp. Cell Res.*, 2015, **331**, 200–210.
188. M. Lowe, J. Lage, E. Paatela, D. Munson, R. Hostager, C. Yuan, N. Katoku-Kikyo, M. Ruiz-Estevez, Y. Asakura, J. Staats, M. Qahar, M. Lohman, A. Asakura and N. Kikyo, Cry2 Is Critical for Circadian Regulation of Myogenic Differentiation by Bclaf1-Mediated mRNA Stabilization of Cyclin D1 and Tmem176b, *Cell Rep.*, 2018, **22**, 2118–2132.
189. P. Zhu, N. X. Hamlish, A. V. Thakkar, A. W. T. Steffeck, E. J. Rendleman, N. H. Khan, N. J. Waldeck, A. W. DeVilbiss, M. S. Martin-Sandoval, T. P. Mathews, N. S. Chandel and C. B. Peek, BMAL1 drives muscle repair

through control of hypoxic NAD$^+$ regeneration in satellite cells, *Genes Dev.*, 2022, **36**, 149–166.
190. H. Boucher, V. Vanneaux, T. Domet, A. Parouchev and J. Larghero, Circadian Clock Genes Modulate Human Bone Marrow Mesenchymal Stem Cell Differentiation, Migration and Cell Cycle, *PLoS One*, 2016, **11**, e0146674.
191. C. Liang, Z. Liu, M. Song, W. Li, Z. Wu, Z. Wang, Q. Wang, S. Wang, K. Yan, L. Sun, T. Hishida, Y. Cai, J. C. I. Belmonte, P. Guillen, P. Chan, Q. Zhou, W. Zhang, J. Qu and G.-H. Liu, Stabilization of heterochromatin by CLOCK promotes stem cell rejuvenation and cartilage regeneration, *Cell Res.*, 2021, **31**, 187–205.
192. Y. He, F. Lin, Y. Chen, Z. Tan, D. Bai and Q. Zhao, Overexpression of the Circadian Clock Gene Rev-erbα Affects Murine Bone Mesenchymal Stem Cell Proliferation and Osteogenesis, *Stem Cells Dev.*, 2015, **24**, 1194–1204.
193. M. A. Baxter, R. F. Wynn, S. N. Jowitt, J. E. Wraith, L. J. Fairbairn and I. Bellantuono, Study of Telomere Length Reveals Rapid Aging of Human Marrow Stromal Cells following In Vitro Expansion, *Stem Cells*, 2004, **22**, 675–682.
194. C. M. DiGirolamo, D. Stokes, D. Colter, D. G. Phinney, R. Class and D. J. Prockop, Propagation and senescence of human marrow stromal cells in culture: a simple colony-forming assay identifies samples with the greatest potential to propagate and differentiate, *Br. J. Haematol.*, 1999, **107**, 275–281.
195. S. Zhou, J. S. Greenberger, M. W. Epperly, J. P. Goff, C. Adler, M. S. LeBoff and J. Glowacki, Age-related intrinsic changes in human bone-marrow-derived mesenchymal stem cells and their differentiation to osteoblasts, *Aging Cell*, 2008, 7, 335–343.
196. A. Stolzing, E. Jones, D. McGonagle and A. Scutt, Age-related changes in human bone marrow-derived mesenchymal stem cells: Consequences for cell therapies, *Mech. Ageing Dev.*, 2008, **129**, 163–173.
197. B. Biteau, C. E. Hochmuth and H. Jasper, JNK Activity in Somatic Stem Cells Causes Loss of Tissue Homeostasis in the Aging Drosophila Gut, *Cell Stem Cell*, 2008, **3**, 442–455.
198. N.-H. Choi, J.-G. Kim, D.-J. Yang, Y.-S. Kim and M.-A. Yoo, Age-related changes in Drosophila midgut are associated with PVF2, a PDGF/VEGF-like growth factor, *Aging Cell*, 2008, **7**, 318–334.
199. J.-S. Park, Y.-S. Kim and M.-A. Yoo, The role of p38b MAPK in age-related modulation of intestinal stem cell proliferation and differentiation in Drosophila, *Aging*, 2009, **1**, 637–651.
200. M. Rera, S. Bahadorani, J. Cho, C. L. Koehler, M. Ulgherait, J. H. Hur, W. S. Ansari, T. Lo, D. L. Jones and D. W. Walker, Modulation of Longevity and Tissue Homeostasis by the Drosophila PGC-1 Homolog, *Cell Metab.*, 2011, **14**, 623–634.
201. P. Karpowicz, Y. Zhang, J. B. Hogenesch, P. Emery and N. Perrimon, The Circadian Clock Gates the Intestinal Stem Cell Regenerative State, *Cell Rep.*, 2013, **3**, 996–1004.

202. K. Stokes, M. Nunes, C. Trombley, D. E. F. L. Flôres, G. Wu, Z. Taleb, A. Alkhateeb, S. Banskota, C. Harris, O. P. Love, W. I. Khan, L. Rueda, J. B. Hogenesch and P. Karpowicz, The Circadian Clock Gene, Bmal1, Regulates Intestinal Stem Cell Signaling and Represses Tumor Initiation, *Cell. Mol. Gastroenterol. Hepatol.*, 2021, **12**, 1847–1872.e0.
203. N. Yang, J. Williams, V. Pekovic-Vaughan, P. Wang, S. Olabi, J. McConnell, N. Gossan, A. Hughes, J. Cheung, C. H. Streuli and Q.-J. Meng, Cellular mechano-environment regulates the mammary circadian clock, *Nat. Commun.*, 2017, **8**, 14287.
204. T. Casey, J. Crodian, A. Suárez-Trujillo, E. Erickson, B. Weldon, K. Crow, S. Cummings, Y. Chen, A. Shamay, S. J. Mabjeesh and K. Plaut, CLOCK regulates mammary epithelial cell growth and differentiation, *Am. J. Physiol.: Regul., Integr. Comp. Physiol.*, 2016, **311**, R1125–R1134.
205. T. Casey, A. Suarez-Trujillo, S. Cummings, K. Huff, J. Crodian, K. Bhide, C. Aduwari, K. Teeple, A. Shamay, S. J. Mabjeesh, P. S. Miguel, J. Thimmapuram and K. Plaut, Core circadian clock transcription factor BMAL1 regulates mammary epithelial cell growth, differentiation, and milk component synthesis, *PLoS One*, 2021, **16**, e0248199.
206. T. Ogino, N. Matsunaga, T. Tanaka, T. Tanihara, H. Terajima, H. Yoshitane, Y. Fukada, A. Tsuruta, S. Koyanagi and S. Ohdo, Post-transcriptional repression of circadian component CLOCK regulates cancer-stemness in murine breast cancer cells, *eLife*, 2021, **10**, e66155.
207. C. M. McQueen, E. E. Schmitt, T. R. Sarkar, J. Elswood, R. P. Metz, D. Earnest, M. Rijnkels and W. W. Porter, PER2 regulation of mammary gland development, *Development*, 2018, **145**, dev157966.
208. Q. Dong, H. Gao, Y. Shi, F. Zhang, X. Gu, A. Wu, D. Wang, Y. Chen, A. Bandyopadhyay, I.-T. Yeh, B. J. Daniel, Y. Chen, Y. Zou, V. L. Rebel, C. A. Walter, J. Lu, C. Huang and L.-Z. Sun, Aging is associated with an expansion of CD49fhi mammary stem cells that show a decline in function and increased transformation potential, *Aging*, 2016, **8**, 2754–2770.
209. A. Wu, Q. Dong, H. Gao, Y. Shi, Y. Chen, F. Zhang, A. Bandyopadhyay, D. Wang, K. M. Gorena, C. Huang, S. Tardif, P. W. Nathanielsz and L.-Z. Sun, Characterization of mammary epithelial stem/progenitor cells and their changes with aging in common marmosets, *Sci. Rep.*, 2016, **6**, 32190.
210. D. J. Rossi, D. Bryder, J. M. Zahn, H. Ahlenius, R. Sonu, A. J. Wagers and I. L. Weissman, Cell intrinsic alterations underlie hematopoietic stem cell aging, *Proc. Natl. Acad. Sci. U. S. A.*, 2005, **102**, 9194–9199.
211. W. W. Pang, E. A. Price, D. Sahoo, I. Beerman, W. J. Maloney, D. J. Rossi, S. L. Schrier and I. L. Weissman, Human bone marrow hematopoietic stem cells are increased in frequency and myeloid-biased with age, *Proc. Natl. Acad. Sci. U. S. A.*, 2011, **108**, 20012–20017.
212. M. C. Florian, K. Dörr, A. Niebel, D. Daria, H. Schrezenmeier, M. Rojewski, M.-D. Filippi, A. Hasenberg, M. Gunzer, K. Scharffetter-Kochanek,

Y. Zheng and H. Geiger, Cdc42 Activity Regulates Hematopoietic Stem Cell Aging and Rejuvenation, *Cell Stem Cell*, 2012, **10**, 520–530.

213. R. V. Puram, M. S. Kowalczyk, C. G. de Boer, R. K. Schneider, P. G. Miller, M. McConkey, Z. Tothova, H. Tejero, D. Heckl, M. Järås, M. C. Chen, H. Li, A. Tamayo, G. S. Cowley, O. Rozenblatt-Rosen, F. Al-Shahrour, A. Regev and B. L. Ebert, Core Circadian Clock Genes Regulate Leukemia Stem Cells in AML, *Cell*, 2016, **165**, 303–316.

CHAPTER 21

Chronotherapeutic Approaches

DENIS GUBIN[a,b]

[a] Tyumen State Medical University, Tyumen, Russia; [b] Tyumen Cardiology Research Center, Tomsk National Research Medical Center, Russian Academy of Science, Tomsk, Russia
Email: dgubin@mail.ru

21.1 Introduction

Diagnosis and treatment without considering the time factor become an anachronism. Many functional genes exhibit circadian rhythms in expression[1–5] with more than 80% of protein-coding genes in primates.[2] The pronounced 24-hour variations in most clinical indicators mean that ignoring such dynamics will inevitably lead to an incorrect interpretation of clinical data and cause systemic diagnostic errors. Translation of the latest achievements of chronobiology into health care practice is viewed as a requisite for a breakthrough in personalized medicine.[6,7] Applying knowledge in the practical field of medicine requires a more comprehensive understanding of molecular, genetic, and systemic factors underlying the circadian phenotype (CPT) of the person.[7,8] Knowledge of biological rhythmicity must not only improve the diagnosis but should also enhance the therapeutic index and treatment efficacy to help avoid misdiagnosis and over-treatment. We consider the administration of drugs along with other therapeutic or preventive measures and even the scheduling of surgery. Not only can the time of intervention or medication affect its result, but the frequency of intervention and managed floating dosing also matters. Criteria to identify individuals who will benefit most with fewer side-effects should be developed. The above considerations substantiate approaches for

Chemical Biology No. 23
Chronobiology and Chronomedicine: From Molecular and Cellular Mechanisms to Whole Body Interdigitating Networks
Edited by Germaine Cornelissen and Tsuyoshi Hirota

Published by the Royal Society of Chemistry, www.rsc.org

chronotherapy (CT). The criticism of "quantified self" tracking technologies mainly stems from the lack of gold standards for archiving, analysing, and interpreting streaming health data.[9] The incorporation of chronobiological principles into the core "self-tracking" philosophy is a sine qua non to overcome the stagnation of this concept.

21.1.1 Chronodiagnosis from Human Samples

State-of-the-art CT should extend beyond a choice between morning and evening intervention, and should rather tailor treatment based on individual rhythm parameters such as the circadian phase and amplitude to achieve personalized CT (P-CT). Then it becomes increasingly important to develop novel technologies of data sampling and analysis suitable to meet the demands of CT.

The rhythmic nature of the physiological/clinical variables implies that variability itself can be optimal, suboptimal, or abnormal. Therefore, amplitude, phase, and period, in addition to mean and/or time-specified values should be incorporated into screening/diagnosis because these endpoints can be affected by both the disease and the treatment. Deviations in amplitude and/or phase within the physiological range may occur before the disease becomes overt, making prophylactic intervention possible. Since knowledge of amplitude and phase can guide diagnostics and treatment timing, establishing reference values for the amplitude and phase is of paramount importance.

Suggesting a fixed clock hour as the optimal treatment time for everyone is grossly inadequate and should at least rely on marker rhythms before it can be substituted with a truly personalized CT, where intervention accounts for all aspects of the chronodiagnosis. Marker variables include sleep time, melatonin, and temperature phase that usually represent central body clocks, or other physiological variables that represent target peripheral clocks.

21.1.2 Toward Personalized Chronotherapy

Despite the evidence of benefits from timed treatment, most medications are still prescribed without considering circadian variations in drug efficacy or patient response. Studies seeking optimal treatment times generally aim at finding an optimal treatment time for a given intervention administered to patients with a specified disease condition. Although overall benefits can be achieved in comparison to conventional treatment, certain disease conditions do not manifest equally in all patients. When this is the case, determining the optimal treatment time for each individual patient becomes necessary. This situation has long been recognized for treating conditions such as type1 diabetes, where patients can personalize their treatment by programming the delivery pattern of their insulin pump.

Standard treatment may remain justified in cases of emergency care or on rare occasions when the relevant circadian rhythms involved in the drug's

ADME (absorption, distribution, metabolism, and excretion) or the targeted physiological variable have small amplitudes or highly labile phases. Even when the target physiological variable has a stable large-amplitude circadian rhythm with a similar phase across individuals, a population approach advocating an optimal fixed treatment time may not serve all patients equally well, potentially causing harm to some patients. As a minimum, the optimal fixed treatment time should be guided by a marker rhythm. Such an approach may be valid for the purpose of "training the clock" with chronobiotics. While personalized chronotherapy can be implemented through the physiological monitoring of marker rhythms, notably when chronobiological reference values in health are available,[10] it would also benefit from an understanding of each patient's clock phase or target tissue-specific phase.

Knowledge gained from personalized medicine could also be considered, including the determination of key genes involved in chronopharmacokinetcs/chronopharmacodynamics (CPK/CPD) of each given patient's medications. While most of the commonly prescribed drugs have cross-talk with the expression of clock genes (CGs) or clock-controlled genes (CCGs),[5] their circadian rhythms (CPK, CPD, and chronotoxicity) have not been investigated, and are still awaiting clinical trials.[6,11,12] These genes may influence the circadian phase position and variability. Inter-individual phase variability should also be considered.[8,13–15] Susceptibility to disease can be related to certain polymorphisms in CGs/CCGs, just as the disease can alter the chemical environment affecting the expression of CGs/CCGs.

Certain conditions necessitate the consideration of ultradian rhythms.[16] New pumps can now deliver corticoids for pulsatile delivery to patients with adrenal insufficiency.[17] In this case, treatment delivery can also be optimized through the appropriate programming of ultradian schedules.[17,18] The 12-hour component, which becomes prominent with increasing age[19,20] and at higher latitudes,[21–24] could also be a target for chronotherapy since its disruption has been linked to metabolic pathology.[25] Recently, genetic and molecular mechanisms underlying ultradian[25–28] and circatrigintan[29] (about-monthly) rhythms have been proposed.

Infradian modulation of treatment administration can further improve outcomes, as already documented in the laboratory.[30] Infradian schedules may be valuable when the toxicity of the drug prevents its daily administration, or when it may have physiological relevance (as in the case of alternate day feeding schedules). Floating regimens may be considered to prevent compensatory loss of drug function or counteract serious side-effects that prevent everyday consumption (*e.g.*, rapamycin associated with immunosuppression). Intermittent schemes for drug administration have been suggested.[31–33]

Personalized chronotherapy is the more urgently needed, given that "most drugs prescribed in the U.S. today are effective in fewer than 60% of treated patients".[34] Currently, the dosing of many medications is the same for everybody, irrespective of age, gender, or body size/mass, even though all these factors influence treatment efficacy.[35–37] Some drugs can exhibit drastic gender-related differences.[38] *REV-ERBα* (*NR1D1*) rs2071427 and rs2071570 are associated with sleep duration and body mass index in boys but not in girls.[39] Gender also affects the intrinsic circadian period,[40,41] blood pressure,[42,43] and sensitivity to exogenously administered melatonin.[44] Gender can also modify circadian-related mechanisms of the microbiome.[45] Different frequencies of certain genes and their combinations in different populations may cause considerable differences in their respective response to a given drug, as exemplified by the higher efficacy of the anti-hypertensive drug combination BiDil in African Americans.[46,47]

Despite the demonstration of the effectiveness of chronotherapy in over 40 clinical trials,[6,48,49] studies comparing outcomes obtained with CT *vs.* standard treatment (ST) are currently lacking. Examples of population-based chronotherapy include choosing the optimal time for immunomodulatory and cytotoxic therapy,[50–53] vaccination,[54,55] and anti-hypertensive treatment.[56,57] Circadian rhythms of immunity and host-viral interaction may severely affect the precision in the diagnosis of viral infection, especially for COVID-19 (SARS-CoV-2).[58,59] One approach toward personalizing chronotherapy involves timing the treatment based on chronodiagnosis.[10,60] The timing of treatment can be guided by a marker rhythm that can easily be monitored longitudinally, such as temperature or activity. The heart rate and blood pressure are particularly relevant in the context of cardiovascular conditions, where they can be used to gauge the response to treatment. Several studies have varied the treatment time from the time of awakening to bedtime in increments of about 3 hours to cover 6 different treatment times.[61,62] Such more demanding protocols are particularly suited when the circadian variation can present several kinds of abnormalities, each associated with an increased disease risk.[63,64]

Furthermore, infradian components[65,66] can modulate processes of virus-hiding exosomes, leading to subsequent reappearance episodes. This "Trojan horse" strategy for episodes of the viral RNA hiding and reappearing was noticed in the recovered COVID-19 patients within 7–14 days since discharge.[67] About 7-day (circaseptan) components are of particular interest since they are ubiquitous in relation to growth and immunity.[68]

A general principle underlying chronotherapy involves strengthening the circadian system by rhythm amplification and synchronization to achieve phase stability. Exceptions exist when the circadian amplitude may be excessive, perhaps as a compensatory mechanism. This is the case with blood pressure, which was extensively monitored in clinical health and in treated or untreated hypertensive patients, who had not experienced any prior cardiovascular morbid event.[64,69] These multiple outcome studies all show that abnormal features of blood pressure and heart rate variability (*e.g.*, circadian amplitude and phase) are invariably associated with an increased risk of cardiovascular disease. Treating hypertensive patients with an excessively large circadian amplitude of blood pressure (and thus an excessive nocturnal fall) in the evening (as the preferred dosing time) can thus be harmful.[70,71]

These patients represent a non-negligible fraction of the population, notably since an excessive circadian amplitude of blood pressure is a risk factor even for individuals who are otherwise normotensive.[64,69]

Understanding certain gene polymorphisms underlying individual differences in response to treatment may offer further guidelines for practicing personalized chronotherapy. The investigation of candidate genes[72] involved in altered circadian variation of blood pressure, as observed in spontaneously hypertensive stroke prone (SHRSP) rats[73] can be helpful, especially since one of the candidate genes, soluble epoxide hydrolase (Ephx2) appeared in the TimeSignature gene list for robust detection of the circadian state from gene expression.[74] Advancing toward personalized chronotherapy will benefit from the implemention of longitudinal monitoring of marker variables so that the timing of treatment can be adjusted to each individual patient's needs, while also mechanistically searching for the root cause of the abnormal circadian pattern. In addition to studies investigating polymorphisms in CGs and CCGs, the role of coupling strength between paired central SCN oscillators, and communication between cells within organs should be scrutinized in future research. Attention should also be given to comparing different times of intervention with both pharmacological and non-pharmacological agents in standardized clinical trials, and constructing phase response curves whenever possible. Comparison between patients with different chronotypes and known clock gene polymorphisms is also warranted.

The concept of "precision medicine" is closely related to a personalized approach. It is based on the consideration of individualized variations in safety, efficacy, and efficiency of administered treatment.[75] Gene polymorphisms, gut microbiota, and environmental factors lie at the core of the personalized differences in response that also include dependency on timing. For example, the gut microbiota produce chemical substances, some of which are produced *de novo* in a circadian manner.[76–79] On the other hand, administered drugs can be metabolized by gut microbiota in a circadian manner. Therefore, precision medicine should consider both factors.[80] Furthermore, the rhythmicity in gut microbiota depends on the host's circadian-system genetic landscape and gender,[43] can be modulated by the host's melatonin[77,81] and temperature,[77,82] and interacts with the host's feeding behaviour.[79]

21.1.3 Chronodiagnostics as the Foundation for Chronotherapy

The interpretation of single measurements should be based on time-specified reference values ("chronodesms") to avoid misdiagnoses. The larger the circadian amplitude of the given variable, the higher the likelihood of a misdiagnosis if circadian variability is disregarded. It is thus important to adjust single measurements for individual characteristics of the CPT, amplitude, and phase. For practical reasons, one should also consider the time of measurement in relation to the phase of the circadian rhythm of the variable being measured: if it is close to the time of the acrophase or batyphase, the error will be much less than if it is near the times of mid-line crossings.

Whenever possible, collecting data around the clock and estimating circadian rhythm characteristics is advocated, notably when gender- and age-specific reference values are available for the circadian amplitude and phase of variables such as certain hormones, endpoints of heart rate variability, and blood pressure. Since disease can alter circadian rhythm parameters beyond the mean value, this approach offers additional diagnostic criteria. The development of chronobiological reference values should be a priority for clinical applications.

N-of-1 studies that combine rigorous in-laboratory surveys with ambulatory monitoring and longitudinal self-surveillance data could be the first bricks in constructing tools to quantify reference values for clock and CPT.

21.2 Circadian Clock Drug Targets

To implement P-CT, it is important to determine whether basal clock properties are intact or compromised. Both the disease condition and the treatment modalities contribute to CPT alterations, apart from any irregular behaviour or shift work that may impose transient misalignment to an otherwise intact circadian clock. Phase stability, amplitude, and intrinsic alignment are best determined in clinically healthy individuals of known chronotype first before being compared to patients diagnosed with a given disease. Compromised circadian functioning decreases the survival rate in cancer patients[83] and diminishes the potential benefits of CT. For patients with a reduced circadian amplitude, restoring or retraining the clock should be considered as an additional treatment target. Clock preservation, the restoration of intrinsic circadian synchrony and amplitude enhancement can all be aligned strategies aimed at enhancing the CPT robustness or circadian clock training.[3,49]

Clock-training strategies can be achieved pharmacologically by influencing the cell clock machinery or by treatments aimed at enhancing CPT amplitude through better coupling among cells. A non-pharmacological approach involves synchronization of systemic CPT through timed light exposure, feeding, and exercise, consolidating a strong rest–activity schedule.

Clock enhancement can be different depending on whether the purpose of CT is the central clock or peripheral CPT, Table 21.1. Clock preservation, modulation, and enhancement strategies can be targeted on the molecular machinery within a single cell, cell-to-cell communication and coupling, or systemic alignment. Mechanisms available to manipulate circadian characteristics and drug treatments targeting the circadian clock all have their optimal phases. Such cellular/molecular treatment modalities should also be administered at optimal circadian stages with personalized phasing. Phase response curves of central or peripheral circadian processes to many substances may exist, similar to those described for melatonin.[84,85]

Table 21.1 Some factors that can be targeted to modulate the amplitude-phase of circadian rhythms.[a]

Circadian parameter	Factor (main role = gain of function unless otherwise specified)	Mechanism
Clock speed acceleration (decreasing/shortening τ, Φ advance)	Mutations or SNPs of *Clock*$^{-/-}$, *Cry1*$^{-/-}$, *Per1*$^{-/-}$, *Dbp*$^{-/-}$, *Rev-erbα*$^{-/-}$, *Rorα*$^{-/-}$	Modulation of expression CGs and CCGs
	CK1δ/ε	Phosphorylation rate of PER2
	Metformin	↑ CK1ε-dependent phosphorylation of PER2
	Ageing	Shortened or lowered melatonin production, changes in the expression of CGs
	RGS16	↑ *Per1* expression, facilitation of circadian oscillations of cyclic adenosine-monophosphate, cAMP, provides inter-neural coupling within the SCN
Clock speed slowdown (increasing/lengthening τ, Φ delay)	Mutations or SNPs of *Clock*$^{\Delta 19/\Delta 19}$, *Cry2*$^{-/-}$, *CK1δ*$^{+/-}$, *CK1ε*$^{-/-}$, *Pgc1a*$^{-/-}$	Modulation of expression of CGs and CCGs
	miR-17-5p	↓ expression of *Clock/Npas2*
	miR-24	↓ expression of *Per2*
	miR-132	Modulation of *Per2* expression
	miR-192/194	↓ PER production (liver)
	TOR, target of rapamycin	Inhibition of GSK-3β/SGG in the SCN
	FBXL3	Modulation of the PER/CRY dimerization rate by CRY1 stability and degradation
	↑ calories	↑ mTOR
	GSK-3β (τ decrease when inhibited)	Modulation of REV-ERBα and CRY2 degradation rates
	Lithium	↑ expression and amplitude of *Per2*, ↑ GSK-3β phosphorylation

Enhancing circadian amplitude	Bright daylight (morning)	Circadian synchronization, τ and phase alignment
	Melatonin (evening)	CGs synchronization (*Bmal1* и *Clock*), temperature circadian rhythm synchronization, coupling of peripheral clocks (?)
	SIRT1	↑ circadian amplitude of BMAL1, PER2
	Physical activity, exercise	Circadian synchronization
	Chrononutrition, timed fasting, ↓ calories	Circadian synchronization
	miR-182/196	↑ retinal melatonin
Decreasing circadian amplitude	CGs mutations	↓ intrinsic synchronization, central circadian clock, or tissue specific circadian misalignment
	REV-ERBα (NR1D1)	↓ circadian amplitude of *Bmal1* by downregulation of its transcription
	LAN	↓ melatonin, ↓ circadian alignment
	Sedentary lifestyle	↓ circadian alignment
	Toxic factors	↓ circadian alignment
	Ageing	↓ SCN neuronal coupling, ↓ melatonin, ↓ tissue-specific peripheral clocks coupling (?)
	Target organ damage	↓ tissue-specific intrinsic circadian alignment
	Feeding ad libitum, ↑ calories	↓ circadian alignment

[a] τ: tau (period); Φ: phase; CGs: clock genes; CCGs: clock-controlled genes; and LAN: light at night.

As was shown more than 50 years ago, rhythmicity in resistance and sensitivity seems to constitute a ubiquitous phenomenon.[86–89]

21.2.1 Molecular and Cellular Clock Machinery

21.2.1.1 Tau Genetics

Primary factors that determine clock speed and intrinsic tau (period) are polymorphic variants of the core clock genes (*PER1, 2, 3*; *CRY1, 2*; *BMAL1* and *CLOCK*). Clock speed and tau precision are influenced by proteins of the PER family. For example, the mouse tau depends linearly on the PER1/PER2 ratio. At the lowest PER1/PER2 ratio (knockout genotype $Per1^{-/-}$), tau extends up to the extreme values of the circadian range of about 28 hours in constant light, while in constant dark, tau increases only slightly.[90] The duration of existence of PER protein products is crucial for clock speed, which, in turn, is determined by the dynamics of their stabilization and phosphorylation. The activity of casein kinases (primarily CK1ε) in PER2 phosphorylation determines the Mendelian nature of tau inheritance in Syrian hamsters, exhibiting an autosomal dominant type with incomplete expression in SCN neurons.[91] A recent study conducted under controlled laboratory conditions (forced desynchrony protocol) discovered an association of the T-allele of the single nucleotide polymorphism (SNP) rs35333999 polymorphic marker of the human *PER2* gene with a tau elongation of 11–12 minutes.[92] Expression of another *Period* gene, *Per3*, is probably less strongly associated with light information than that of *Per1/Per2*.[93] This gene may be more strongly related to the quality and duration of sleep,[94–96] or to individual characteristics of activity and chronotype.[97–99] The literature on the role of *PER3* polymorphisms (*e.g.*, *PER3 VNTR*) in the regulation of tau is contradictory.[100–103]

CRY, the PER partner in the main loop of core clock genes, contributes to clock speed and its response to light signalling. The dominant allele of the *CRY1* gene (*CRY1* Δ11) was identified as an inherited factor that leads to the chronotype's extreme evening variant[104] with a frequency of 0.6%. Individuals who are heterozygous for this allele may have an intermediate alternative phenotype associated with sleep fragmentation. The time frames of PER and CRY presence and their bioavailability to each other and to their feedback partners are regulated by degradation, thereby regulating the clock phase.[104] For example, casein-kinases CK1δ and CK1ε phosphorylate the PER2 protein. Loss or reduction in CK1δ/ε activity results in the lengthening of intrinsic tau due to the high and prolonged PER presence.[106,107] A *CRY2* missense, which leads to an increase in the binding affinity of FBXL3 to CRY2, with subsequent degradation, results in advanced sleep syndrome in humans.[108] F-box and leucine-rich repeat protein 3, FBXL3, knock-out human cell lines, which were edited using CRISPR/Cas9 resulted in FBXL3 loss-of-function and a small-amplitude, long-tau phenotype.[109]

Since the 24-hour expression cycle of CRY and PER proteins determines the clock speed, factors affecting the rate of stabilization and phosphorylation of these proteins at the post-translational level should also affect tau.[110–113] AMP-activated protein kinase can serve as a link between energy exchange, nutrition, and clock speed. Thus, metabolic factors that can affect AMP-activated protein kinase[85] may also affect tau. The presence of glucose in the extracellular environment inhibits the expression of *Per1* and *Per2* genes.[114] Accordingly, the rate of its metabolism and entry into the cell, as well as sensitivity to insulin can modulate clock speed,[115] at least in peripheral tissues. Since external desynchronization factors (such as jet lag) are interrelated with dysbiosis, and since the microbiota participate in the 24-hour dynamics of glucose metabolism,[116] biochemical factors of the intestinal microbiome can also participate in the modulation of the tau and clock phase.

Conflicting results have been obtained regarding the role of the *CLOCK* gene, and one of its well-studied polymorphisms, *CLOCK* 3111T/C (rs1801260). Several studies have shown the relationship of the minor *CLOCK* 3111T/C C-allele with the evening chronotype, a longer tau or sleep phase delay,[117–120] while other authors have not confirmed such reports.[121–123] However, the results obtained recently in mice showed that the *CLOCK* 3111T/C C-allele increases the expression of CLOCK protein, as well as the expression of the *PER2* gene, thus offering a putative mechanism through which the C-allele is theoretically able to lengthen tau.[124]

The principal secondary loop repressor of the main BMAL1/CLOCK ↔ PER/CRY CG feedback loop, REV-ERBα, was recognized as a universal core clock component for targeting tissue-specific epigenomic modifications when metabolic processes are concerned.[125–127] It may also constitute a key element that modulates the circadian amplitude in gene expression at the cellular level.[128,129]

Several microRNAs are also able to extend tau by modulating the expression of *Per* (miR-24, miR-132) or *Clock* (miR-17-5p), particularly in peripheral tissues. Factors such as FBXL3, which regulates the rate of PER/CRY dimer formation, and GSK-3β, which reduces the rate of REV-ERB protein metabolism, also have a similar effect.[1]

21.2.1.2 Drugs that Target Circadian Clock Machinery

A dampened circadian amplitude is a common CPT endpoint, resulting from multi-layered circadian misalignment. It can serve as a useful test for clock synchronization and for health and well-being.[130–138] Therefore, the "training clock approach"[49] is promising.

As shown in Table 21.1, there are several ways to manipulate the circadian clock machinery by influencing the intrinsic circadian tau and phase to support challenged or compromised circadian rhythms. For example, lithium effects in bipolar patients depend on the intrinsic circadian period of a patient; circadian effects of lithium are muted in patients with longer intrinsic tau of fibroblasts.[139] Several recent works reviewed molecular mechanisms of "chronomodulation".[105,140] For example, ischaemia and reperfusion injury, which is associated with common pathologies such as myocardial infarction, stroke, and liver and kidney injuries,[140] exhibits daytime variation in its severity, influenced by circadian genetics and the rhythmicity of hypoxic resistance.[141] While defects in *Per2* are linked to a lower resistance to ischaemia and reperfusion and larger myocardial infarction size in an animal model,[140,142] light therapy promotes higher PER2 cardiac levels and increases resistance to cardiac hypoxia. As such, personalized light therapy according to the individual phase may become an important step of the perioperative strategy for preparing patients for an elective surgery.

Recently, efforts have been made to identify small molecules that can specifically affect the basic clock properties.[50,105,143–146] Pharmacological modulation of the circadian clock is appealing to the pharmaceutical industry since it has the potential of designing therapeutic drugs. Some of the existing drugs and their availability for CT have been reviewed recently.[147,148]

Sirtuins are evolutionarily conserved NAD^+-dependent deacetylases, localized in the nucleus (SIRT1, SIRT6, and SIRT7), cytoplasm (SIRT2), or mitochondria (SIRT3-5).[149] SIRT1 is engaged in fine-tuning circadian transcription and modulating the cross-talk between epigenetic and metabolic processes.[150–153] SIRT1, a NAD^+-consuming nuclear sirtuin, pivotal in regulating the gene expression, has emerged as another key therapeutic target. Physiological dynamics of cellular NAD^+ availability account for rhythmicity in SIRT1 activity since the Michaelis constant for NAD^+ in terms of SIRT1 lies within 24-hour fluctuations of NAD^+.[154] As such, the NAD^+ precursor, nicotinamide riboside, is a more promising counteragent to balance the age-related decline of metabolic and stress-related pathways. For example, in aged mice, the process of BMAL1 chromatin binding, the amplitude of BMAL1 oscillations, mitochondrial respiration, and activity rhythms are dampened, but they can be restored by nicotinamide riboside, which modulates core circadian clock cellular mechanisms, facilitating binding of BMAL1 to PER2.[154] As mechanisms of action of sirtuins and nicotinamide are tightly entwined with circadian clock machinery, knowledge of the circadian phase is crucial to ensure the best efficacy of pharmaceutical actions with these substances.

There are only few other examples of how interaction with cellular clock machinery can be used for CT. Enhancing cellular clock function in intact cells *vs.* cancer cells might represent a novel approach in controlling cancer progression.[155] These mechanisms are intricately linked to the circadian modulation of immunity and extrinsic factors driving immunomodulation, including those of the human microbiome.[13,156,157]

21.2.2 Cell-to-cell Communication

Robust oscillations depend fundamentally on oscillator coupling.[158,159] Large circadian amplitude relies on intrinsic synchronization or aligned background physiological processes. This alignment incorporates diverse processes such as inter-cellular coupling, day-to-day phase stability, and resilience to extrinsic stimuli. To ensure high-amplitude rhythms, the circadian system needs to resonate with both the interior milieu and the environment, but it should also ensure enough flexibility/plasticity to prevent frailty/rigidity. The theory of coupled oscillators posits that stable oscillations can rely on a tau somewhat different from exactly 24 hours. Strong oscillators in higher vertebrates, particularly within the SCN, allow wide phase distributions in overt circadian rhythms as a response to the minor mismatch between the intrinsic period and the environmental synchronizing signal. Recent simulation studies have demonstrated that period and phase distributions are narrower when the coupling strength is high, a situation that also allows maintaining large-amplitude rhythms due to resonance effects.[160]

Individual SCN neurons may have different phases and intrinsic tau values, which may be due to different PER1/PER2 expression ratios.[161] A strong inter-neural interaction among SCN neurons might provide both tau stability and a large circadian rhythm (CR) amplitude, but at the risk of rhythm rigidity in relation to external factors.[162] The amount of RGS16, a protein with a clear circadian rhythm of its production that inactivates the α-subunit of the G protein, increases at strictly defined hours ("circadian window"), which results in time-dependent activation of cyclic AMP in SCN neurons.[163] The RGS16-mediated activation of cAMP is necessary to accelerate the expression of *Per1* in dorsomedial cells of the SCN. *Rgs16*$^{-/-}$ knockout mice lack a circadian rhythm of cAMP and have a longer tau.[163] The distinct relationship between *RGS16* (*i.e.*, its SNP rs1144566) and the chronotype has been confirmed by several GWAS studies.[164,165]

21.2.3 Laboratory *vs.* Field (Daily Routine)

Many studies that provided references for the circadian parameters (mainly the circadian phase and intrinsic period) were conducted under strict laboratory conditions with controlled light, activity, and feeding regimens. These results are currently considered the gold standard for circadian clock assessment. It was established that the phase of the central biological clock in the SCN is usually more stable and differs from specific phases of peripheral rhythms.[2,5,6,166] Personal central clock phases, determined by the expression of SCN core CGs in experimental research, or by melatonin and temperature phases in humans, and circadian phenotypic phases of an amalgam of other physiological/biochemical variables and expression of CCGs differ greatly among individuals. The differences depend on genetic factors and sensitivity to synchronization by light, temperature and food (*e.g.*, human preferences for earlier or later sleep regimens; chronotypes). Knowledge of the local circadian phase can thus be important for organ-targeted CT. Information about tissue-specific circadian oscillators can often be assessed only *via* indirect biomarkers since the tissue itself is inaccessible.[8,167]

Clinical variables, interpreted in the light of presumably normal ranges, however, are generally measured under usual prevailing living conditions, and not under standardized laboratory conditions.

Comparing the estimation of phase and amplitude is best done between laboratory and constant routine conditions to identify the factors underlying consistent discrepancies. Modern monitoring of activity, temperature, heart rate, and light exposure and minimally invasive techniques provide novel opportunities to evaluate circadian rhythm parameters of numerous physiological variables. Biochemical variables of clinical interest can be determined in saliva and blood samples for hormones, lipids, and carbohydrates, and in mucosa fibroblasts and epidermis for gene expression. Since these real-life parameters are obtained under less standardized conditions, the estimation of their circadian parameters can be expected to be more variable.

21.2.4 Animal Models *vs.* Humans

The knowledge of circadian rhythms and their mechanisms is primarily based on animal studies. The most widely used mammalian models of laboratory animals are nocturnal. A fundamental discrepancy in adaptational mechanisms to light and temperature exists between these species and humans. For example, effects of preclinical drugs tested are in anti-phase from those in humans. Moreover, tests conducted during the daytime correspond to the rest span of animals, thereby exerting an inherent artificial stressful factor.[6]

Specific differences between nocturnal and diurnal animals lie downstream of the SCN. Melatonin represents a rare exception with phases tightly bound to darkness, irrespective of activity.[168–170] Peripheral clocks differ in phase in the relation to the central clock, often with specific phases in nocturnal *vs.* diurnal species.[2,5,166] Many hormones *e.g.*, glucocorticoids, leptin, and ghrelin, and metabolites such as glucose and free fatty acids show opposite phases in diurnal and nocturnal rodent species.[171] Chronotype-dependent phase differences for these chemicals are hence expected. Since diurnal and nocturnal species differ in the specific alignment of the clocks in peripheral organs,[166] individuals with different chronotypes can also differ in specific tissue phases. If so, the best timing to administer certain drugs depends on their site of action (organ/tissue-specific). Circadian drug delivery schedules can therefore be adjusted between different species.[166] Timing also needs to be adjusted individually according to the chronotype. Overall, the human "chronobiome"[172] should be assessed and mapped in correlation with circadian phenotypes (given that the central clock and clocks in peripheral tissues may differ in phase with specific phase lags).

One approach to CT could be to first determine the optimal circadian stage of administering a given drug in an animal model, while considering specific phase relations between the target organ and the central clock.

The information could then be translated for human use,[172] further correcting for the chronotype.

The phase determined by fitting cosine models to serial data is widely used in the analysis of marker variables such as activity, temperature, heart rates, blood pressure, or glucose. These variables have the advantage that they can be monitored over long spans. Day-to-day variability in all circadian characteristics can thus be assessed, together with any differences between working days and weekends.[173–175] Masking factors such as physical activity and ambient light[176,177] can now be considered. Devices that allow for the concomitant monitoring of blood pressure, accelerometer, and ambient light, along with the development of software for data purification (similar to what has been introduced for temperature) are also available.[178,179]

Among the methods aimed at determining the circadian phase, DLMO (dim light melatonin onset, the beginning of melatonin production in conditional darkness, at light intensity <10 Lux) is regarded as the most accurate method for estimating the internal phase[180,181] and remains the reference standard for verifying the accuracy of other techniques. It is, however, complex, and costly, and requires the interruption of sleep.

The simplest and most popular method for estimating the circadian phase is the assessment of chronotype using the Horne–Ostberg morningness–eveningness questionnaire[182] or several modifications thereof. Currently, the Munich ChronoType questionnaire (MCTQ)[183] is widely used. The chronotype determines individual preferences in activity and sleep. It is graded, depending on the individual score, into three main variants, morning type (larks), intermediate type (pigeons), and evening type (owls), also paying attention to specific phases on workdays and free weekdays to assess social jet lag.[184]

The estimation of the chronotype also gives an approximate idea of the internal circadian period.[185–188] The correlation between the intrinsic tau, chronotype, circadian phase, and time of awakening was considered a fundamental property of the circadian pacemaker,[185] a later phase being usually associated with a longer tau. Experimentally, the circadian time-dependent phase advancing effect of light is greater in owls than in larks, with a higher ratio of phase advance to phase delay,[186] which was interpreted as a longer circadian period in evening types. Chronotypes can be reflected in the SCN and in phases of peripheral clocks, as shown for the circadian phases of *PER1*, *PER2*, and *REV-ERBα* in the oral mucosa.[189]

SM-ART (single-measurement-analysis-reports-timing) technologies based on single sampling are another approach for defining the circadian phase. With the development of technologies for the study of genetic factors, the CGs, Table 21.2, the prospect of introducing methods for analysing the CR phase based on the direct assessment of the particular CGs is becoming more attractive.[190] From a practical point of view, genes in which the expression of mRNA is differently influenced by external light are interesting. Among the members of the *Per* family (*Per1*, *2*, and *3*), *Per1* and *Per2* mRNAs likely are more influenced by light,[93] whereas *Per3* seems to be more related

Table 21.2 Currently existing templates for determining the circadian rhythm phase based on genetic expression.

Biological material	Technology brand name	Ref.
Blood (monocytes)	*TimeSignature*	Braun *et al.*, 2018[74]
	BioClock	Agostinelli *et al.*, 2016[194]
	ZeitZeiger	Hughey *et al.*, 2017[196]
	PLSR	Laing *et al.*, 2017[197]
	BodyTime	Wittenbrink *et al.*, 2018[198]
Epidermis	*CYCLOPS*	Wu *et al.*, 2018, 2020[200,201]
Organ tissue samples	*CYCLOPS*	Anafi *et al.*, 2017[199]

to the rhythm of sleep and activity.[191] Therefore, the polymorphic variants of *PER3* were considered to be strongly associated with individual features of sleep and chronotype.[99] On the other hand, numerous CG-affecting mutations can be linked to the intrinsic circadian tau. For example, *PER2* (rs35333999) polymorphism, which, according to a genome-wide association study (GWAS),[192] represents the strongest among CG associations with the chronotype, also showed an association with the intrinsic circadian tau.[92] The dominant gain-of-function *CRY1* allelic variant is linked to a marginal CPT, delayed sleep phase disorder (DSPD).[104] The GWAS, which was aimed at searching for associations between genetic polymorphisms and chronotype in the UK Biobank database, confirmed the role of the SNPs in *PER2* and *PER3* genes.[164,165,192,193]

Technologies based on innovations that combine machine learning and modern genomic technologies can become a revolutionary method for evaluating the CR phase, Table 21.2. The novelty of this approach is that the evaluation of the CR phase does not require longitudinal sampling (as there is no need for multiple measurements), but rather uses a once-created database. Several methods have been proposed for evaluating the CR phase based on a single sample (a drop of blood or a skin sample). In the first case, the DNA of neutrophils is analysed, and in the second case, fibroblasts are analysed. Determining the rhythm phase through a single measurement is possible when analysing a sample of genes from a predetermined genes list (which is set using machine learning and genomic analysis).[194] Such CR phase estimation technologies can be defined as SM-ART technologies. Thus, reliable information about the CR phase can now be obtained using only one additional analysis for the activity of a specific set of genes in a blood sample,[74,195–198] tissues,[199] or skin,[200,201] Table 21.2. Non-invasive techniques based on the analysis of hair follicles,[202] or buccal scraping,[203] or even samples of saliva and sweat, which today appear to be promising material for routine transcriptome-proteome monitoring,[204,205] may appear in the future.[204,205] The cyclic ordering by periodic structure (CYCLOPS) algorithm, first developed for mapping time structures in the expression of normal *versus* cancerous lung and liver biopsy samples,[199] later proved to be a promising non-invasive technology. The molecular clock phase in the epidermal skin was shown to surpass the dermal clock in terms of precision.[200,201] This technology was used to create the SkinPhaser application for ongoing research and community datasets.[201]

Irrespective of the method used to estimate the circadian phase, phases can vary greatly among different individuals,[13,206] caused in part by large inter-individual differences in light sensitivity[207] and by effects of disease conditions and/or medications. Phases can vary from one day to another in the same individual, from one person to another, from different cells within a sample, or from different systemic functions that drive overt CPT, Figure 21.1. Synchronization is thus an important aspect of study designs. The unequal stacking in gene expression depends on the number of cells in a sample and the degree to which they are synchronized. The lack of synchronization can lead to a spurious small amplitude population rhythm, even when individual cells have maintained a robust rhythm. Rhythmic expression can thus be compromised due to uncoupling between cells. The greater the phase differences among cells, the more dampened the amplitude will be, unless single cells can be studied.[13] To overcome this technical issue, high-quality circadian transcriptomes with RNA-sequencing (RNA-Seq)

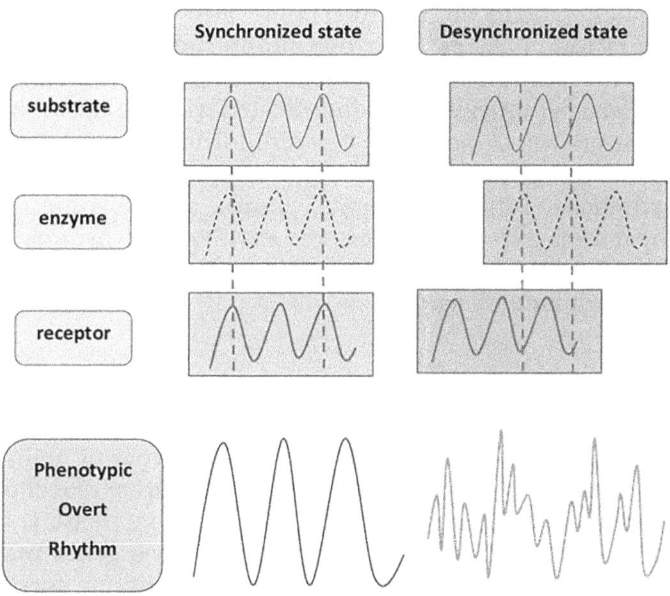

Figure 21.1 Simplified model of the interaction between the substrate, enzyme and receptor. The phenotypic overt rhythm (CPT) synchronized state relies on the synchrony of the circadian rhythms of its background elements – rhythms in substrate production and clearance rates, amount and activity of the substrate converting enzyme, and the density of the receptors for substrate's action. If one of the background elements shifts phase or loses its consistency (desynchronized state), the amplitude of an output CPT decreases, the variability pattern becomes less regular, and extra-circadian variability increases.

reads of the order of 20–40 million per sample are needed for the advancement of research.[13,190] Large inter-individual differences in the mean value can be handled by expressing individual data as a percentage of the mean value of each individual record.

21.3 Pharmacological Approaches

The effects of drugs vary across different timescales, depending on circadian (and other) rhythmic mechanisms. The sensitivity of tissues and organs to a given chemical depends on its pharmacokinetic/pharmacodynamic properties and on circadian mechanisms of the signal transduction within cells. These mechanisms can be specific to a given drug and follow universal principles of rhythmicity in ADME. The internal phase, specific to a given organism and predetermined by genetic background, also plays an important role. These genetic factors set personalized time frames for modulating basic ADME.

21.3.1 Chronopharmacokinetics and Chronopharmacodynamics

ADME dynamics depend on the metabolic rate, which is influenced by personalized factors such as gene polymorphisms, age, gender, activity, nutrition, and meal timing. They also depend on the microbiome, known to exhibit circadian rhythmicity and is interrelated with the host. Thus, ADME rhythmicity is subject to substantial inter-individual variability. Proteins needed for the absorption and metabolism of drugs also exhibit their own 24-hour rhythmicity. Different systems, tissues, and cells are not equally driven by the master circadian clock in the SCN. Robust rhythms or variable fluctuations may thus emerge as output of circadian coordination with a plethora of environmental signals.

21.3.1.1 Absorption

Drugs can be administered orally or parenterally. The rate of absorption varies across time scales and impacts the drug efficiency of orally administered drugs. Twenty-four-hour variations in the absorption rate of drugs were first described several decades ago.[208–210] The morning phase is typical for the absorption rate of many drugs.[211] Circadian clock genes may also influence the availability of the specific transporter chemical factors, which are mainly proteins. Mutations in clock genes can affect rhythms in ADME.[212]

Specific interactions between the host's circadian clock, circadian phenotypes, and microbiota rhythms are receiving increasing attention. Circadian dynamics in the absorption of chemicals, including drugs, are influenced not only by gastric and intestinal factors of the host organism,[211] but also by gut microbiota. For example, microbiota affect the circadian amplitude of the Nfil3 factor in intestinal epithelial cells, which governs lipid absorption and export in intestinal epithelial cells.[213]

21.3.1.2 Distribution

The distribution rate is influenced by the drug's chemical properties, such as its lipophilicity, by the abundance of plasma proteins (mainly albumin, which is circadian rhythmic), and by its binding affinity.[211]

Transport across cell membranes and mechanisms that determine it are key factors governing the distribution of administered substances. For comprehension of the dynamical principles of the trans-membrane transport of chemicals during 24 hours, studies of the circadian rhythms in receptor density[214] are crucial. For example, the abundance of opioid receptors, and their serotonergic circadian and ultradian modulation[208,215] lead to time-dependent therapeutic effects.

Some chemicals can be transferred to the brain and may directly affect the properties of the central clock. Sleep is entwined with the blood–brain barrier, which functions as a cycling regulatory interface with dynamically changing permeability. Therefore, xenobiotic metabolism and effects of chemicals, including drugs, are modified by circadian rhythmicity in the blood/brain barrier.[215,216] Circadian disruption and sleep deficit alter the pharmacokinetics and blood–brain barrier permeation, affecting the delivery and distribution of drugs targeted for the central nervous system.[215]

21.3.1.3 Metabolism

Drug metabolism in the liver includes two phases of enzymatic processes: (1) oxidation, reduction, and hydrolysis by monooxygenases; and (2) conjugation by transferases, resulting in the production of conjugates that can be readily excreted.[211] Both phases are basically circadian periodic.[217,218] Oscillations in nuclear receptors play a major role in the metabolism of chemicals entering peripheral cells within and outside the liver. Some nuclear receptors such as REV-ERBα and REV-ERBβ (encoded by *NR1D1* and *NR1D2*) and peroxisome proliferator-activated receptors play a key role in feedback with cellular core clock genes.[219] Since plenty of cycling enzymes participate in the metabolism of drugs and other chemicals, their background genetic factors can modify circadian dynamics.[219–221] Ultradians (*e.g.*, 12-hour components) may also be important for metabolic processes, particularly in the liver.[25]

21.3.1.4 Excretion

Kidney function and its tissue-specific circadian clocks strongly modulate the excretion rate of metabolized chemicals and drugs. Glomerular function, xenobiotic-metabolizing enzymes, and detoxification pathways, filtration rate, renal blood flow, tubular reabsorption, transporters, electrolytes, and urinary pH were all shown to exhibit circadian rhythmicity.[6,83,208,211,214,222–226] The circadian rhythm in physiological proteinuria even requires a 24-hour urine albumin excretion rate assessment that is regarded as the gold standard for measuring albuminuria. The same issue concerns urine extracellular vesicles.[227] Changes in circadian parameters derived from excretion patterns may themselves provide valuable diagnostic information and chronotherapeutic guidelines.

Taken together, the CPK and CPD depend on the drug's interaction with binding factors during ADME processes. Drug ADME tells how precise the clock phase needs to be. From a chronotherapeutic point of view, a shorter drug half-life requires a higher phase precision. For example, a 3-hour phase precision determined using the CYCLOPS algorithm for epidermal samples may be sufficient for drugs with a half-life longer than 6 hours.[201]

Since different drugs prescribed against the same disease were shown to have different CPK and CPD, as reported for different antihypertensive drugs,[228] the best timing for their administration can differ. This difference holds true whether timing is optimized across a population (treating-the-disease), or for a particular person (treating-the-patient). Over-generalization of conclusions for drugs with apparently different mechanisms of action and ADME has been criticized.[229] Different drugs having different formulae may vary in CPK/CPD, thereby affecting circadian parameters in a different manner. For instance, anti-hypertensive drugs unequally affect the circadian amplitude of blood pressure.[60,230] Anti-hypertensive drugs may also have different beneficial effects on heart rate variability.[231,232]

21.3.2 Chronotherapeutic Approaches

21.3.2.1 Increasing Efficacy and Minimizing Side Effects

The problem of "too much medicine" caused by the institutionalization of overdiagnosis and overtreatment have expanded exponentially,[233] raising increasing concerns.[234–237] Chronodiagnostics should become an effective tool in strategies against overdiagnosis and unnecessary treatment (treat only when treatment is really needed). This endeavour should help in minimizing "false alarms" and help reach correct interpretations of results from medical observations by reducing dosing and by prescribing appropriate timed treatment to help prevent the loss of drug effectiveness.[238,239]

21.3.2.2 Fighting Acquired and Compensatory Drug Resistance

Inter- and intra-patient variability in compensatory pathways impact the clinical effect of chronically administered drugs. For example, immunomodulatory therapy targeted against the tumour necrosis factor (TNF) or modulation of interleukin-6 (IL-6) in cancer or inflammatory diseases can face obstacles such as primary lack of response, or secondary loss of response.[238] Overcoming both situations can be managed by CT. Lack of personalization of delivery regimens of the medication should be overcome. Ilan *et al.*[239] stated that "differences in compensatory mechanisms may have a greater impact than pharmacogenomics on the pharmacodynamics of drugs". One size does not fit all: age, gender, genetics, chronotypes should all be considered for a personalized approach to CT, irrespective of its modality.[8,38,172]

Taking medication dosing into consideration as a variable should be beneficial. Introducing variability in chronic drug administration should help overcome compensatory effects and improve long-term efficacy. It was hypothesized that irregularity of therapeutic regimens may have a beneficial effect on drug.[238,239] From a chronotherapeutic point of view, it means that further therapeutic benefits could be derived from optimal ultradian (adjusting multiple interventions applied within 24-hour intervals) and infradian (administration on certain days, varying dosing, or programming the duration of "drug holidays") modulations of drug administration. Specific drug-delivery platforms[240–244] can be updated for such purposes.

21.3.2.3 Strategies of Chronotherapy

CT approaches can be designed to enhance treatment efficacy, reduce risks during periods of highest susceptibility, or maintain a robust intrinsic clock. The efficiency of targeted chronotherapy relies on achieving a reasonable balance between the rhythm stage of optimal effectiveness (sensitivity rhythm) and, when possible, the rhythm stage when side effects are minimal (toxicity and tolerability rhythms, linked to the drug's ADME). CT aims to administer treatment at a rhythm stage where it reaches its peak efficacy at the time when the chronorisk is highest, while considering the CPK/CPD of the drug and the chronophysiology of the disease condition. Methods of strengthening the intrinsic clock are reviewed in Section 21.2 above. Drugs can be developed or repurposed to target the expression of core clock genes or their products.[49,245,246] One may try to either "clock the drug" by choosing the best time for its administration, or "drug the clock" by developing drugs targeting core clock components.[49] Sulli *et al.*[49] also refer to "training-the-clock" by means of interventions aimed at combating compromised circadian rhythms having reduced amplitudes and/or phase stability. This strategy manipulates the timing of light exposure, meals, and sleep–activity and the use of melatonin. It applies equally to the prevention of age-related circadian disruption[136–138] and that of circadian disruption from other origins.

21.4 Non-pharmacological Approaches

21.4.1 Light

Light is the principal synchronizer of the central circadian clock that can drive circadian phase delays or advances according to established phase-response curves (PRCs).[247–250] Sensitivity to light and sensitivity thresholds differ markedly between experimental animals and humans and among human individuals.[207,251] Substantial individual variability in response to light causes high discordance in responsiveness to photic signalling[252] and considerable variability in overt rhythms (circadian phenotypes).

The clear demarcation of day and night, which existed during most of ancestral development, has changed remarkably in industrialized societies. Electric lighting has enabled modern life in a 24-hour/7-day society, where more than 90% of people have light in the evening and night hours. Sitting in front of screens and smartphones before falling asleep has become a common routine. Many people spend more than 22 hours a day indoors without being exposed to natural daylight. As a result, many people experience too little daylight during the day and consume a relatively large amount of artificial light in the evening or at night, with consequences on physiology, sleep, performance, and well-being.[176,253,254] Human-centric lighting (HCL) technologies for the optimization of circadian light hygiene and scheduling of the light regimen in offices and homes have become increasingly popular.[253–258] Solid scientific evidence is required, along with further personalization.

Human circadian responses to light in real-life depend on physical properties of light (intensity and wavelength), on exposure (circadian phase and duration), and on individual characteristics (gene polymorphisms, including CGs/CCGs, age, gender, and probable comorbid states). Individuals with a longer intrinsic clock and a later chronotype are thought to be more susceptible to evening light, causing even further delay in the circadian cycle.[188,259,260] The wide distribution of chronotypes in modern society can be linked to individual differences in light sensitivity (evening owls can be more sensitive to deviations in the social schedule from the natural lighting regimen than morning larks). Sensitivity to light[261–264] as well as melatonin production changes with age.[136–138,265,266] Non-pharmacological CT based on smart light approaches also benefits the aging process.[136–138] In adolescents, sensitivity to evening light can be higher.[261] It is possible to phase advance the circadian clock of adolescents who have late bedtimes and short school-night sleep within one weekend using light that begins shortly after their sleep midpoint.[200] In healthy aging, optimal non-visual sensitivity to light can be compensated for; an increase in light filtration through the ocular lens does not necessarily lead to a decrease in non-visual sensitivity (melatonin suppression) at these wavelengths, and the peak of non-visual sensitivity can be shifted to longer wavelengths.[267] Principles of light hygiene and human-centric lighting were recently proposed.[255]

Reduced exposure to daytime bright light and decreased sensitivity of receptors of the circadian system to bright light are among the causes of disrupted circadian rhythms. Likewise, exogenous melatonin has numerous positive effects during the nocturnal phase, particularly when it is lacking. Bright light therapy exerts beneficial effects on circadian rhythms and sleep quality *via* compensating deficits of its circadian effects during the daylight phase. Optimal timing, duration, and intensity of bright light therapy for the treatment and prevention of metabolic diseases, depression, and age-related circadian disruption should be established on a personalized basis. The timing of light exposure, number of sessions, and intensity can all be adjusted to yield optimal results to correct sleep disorders and help aged individuals. The efficiency of bright light therapy depends on the correct diagnosis of the major disorder that underpins circadian disruption, the mechanisms that underlie sleep disorders, and the individual's chronotype. The optimization of light therapy, including its timing, duration, and appropriate physical properties of light, is best personalized since its effects may vary depending on certain disease states, such as diabetes.[268]

Many are affected by vitamin D deficiency due to poor circadian hygiene related to light by night,[269] lack of sun exposure,[270,271] or both factors combined.[272] The uncertainty about the effectiveness of vitamin D supplementation against non-skeletal pathologies,[273–275] as defined by The Third International Conference on Controversies in Vitamin D,[276] lies within the chronobiological sphere of interest. Chronobiological aspects include the determination of daily doses of vitamin D needed to maintain acceptable concentrations in the general population, cutaneous production of vitamin D by UVB exposure, hepatic regulation of 25OHD metabolites, definition of vitamin D excess, vitamin D deficiency in acute illness, and vitamin D requirements during reproduction. Chronotherapeutic approaches can thus address the questions whether the lack of efficiency of supplements is related to a lack of personalization in the timing of their administration, and whether corrected light hygiene is needed for supplementation to be effective.

21.4.2 Melatonin

Melatonin is the best studied substance with natural chronobiotic properties that is capable of alleviating common manifestations of circadian misalignment.[135,277–279] Physiological effects of exogenous melatonin administration depend critically on timing, dosing, and duration of treatment. The phase response curve to melatonin[84,85] accounts for the differences in its effect on the circadian clock depending on the circadian stage, changing from a phase advance to a phase delay. When melatonin is administered at its habitual mean phase to modulate its average natural circadian pattern, the circadian amplitude increases, the phase stabilizes, and the robustness of the circadian clock can be maintained, as seen in the elderly.[135–138,278,279] Low doses of melatonin can be justified for their chronobiotic effects,[280,281] whereas antioxidant properties require higher doses.[282,283] Apart from being a pivotal component of the circadian system, melatonin has numerous other physiologic effects.[282,283] Melatonin improves sleep characteristics,[135,268–279,284–286] reduces blood pressure,[20,135,287–290] compensates for insufficient nocturnal decline in the heart rate,[20,135,291] provides a gentler morning rise of the heart rate,[135] serves as an anti-depressant,[292] improves mood,[279,293] counteracts inflammaging, metabolic syndrome, and age-associated neurodegeneration,[277,294] and combats viral infections.[295,296] Melatonin is a known effector modulating the immune system. Melatonin influences both the innate and specific immune system responses by means of combined mechanisms that mainly involve the modulation of cytokines and the production of oxidative stress.[297]

The physiological effects of melatonin on the immune system critically depend on timing and dosing, and on genetics, lifestyle, and environmental factors. Melatonin can turn from an immunostimulant to an immunosuppressor depending on any of these factors. It acts as immunostimulant under suppressed conditions, prompting a more rapid and efficient immune response against external factors such as viruses and bacteria. In the presence of a transient or chronic exacerbated immune response (*i.e.*, septic shock), melatonin could be considered as an anti-inflammatory molecule. The term "immunological buffer" was coined to define such pleiotropic effects of melatonin on the immune system.[297] Inflammatory reactions can be successfully modulated or buffered through the synergistic action of melatonin and vitamin D.[298]

A pronounced nocturnal reduction in blood pressure has been achieved with low-dose melatonin taken for two weeks,[20,135] perhaps as a compensation for an age-dependent melatonin deficit. Effects of melatonin can be greater in age-related neurodegenerative pathologies.[279,299–302] Specific effects of exogenous melatonin may critically depend on timing and be modified by specific receptors' density. Circadian dynamics of melatonin receptor (MT1, or MTNT1a, and MT2, or MTNR1b) concentrations in the SCN, and plasma concentrations of melatonin may generate a feedback loop with each other: pre-treatment with melatonin causes a decrease in melatonin receptors in a time- and concentration-dependent manner.[303,304] Individuals with a relatively reduced endogenous production of melatonin might thus benefit more from lower doses of exogenous melatonin, higher doses being less effective.[280,281] *MTNR1b* receptor gene polymorphism (*e.g.*, SNP rs10830963) is associated with an altered fasting glucose and increased risk of diabetes mellitus,[305] higher systolic blood pressure in hypertensive individuals,[306] and delayed offset of salivary melatonin and temperature rhythms in advanced glaucoma.[279,302] Melatonin was effective in combating metabolic dysfunction in obese patients, improving their lipid profile,[307–309] and in improving retinal ganglion function in patients with glaucoma.[279,302] In addition to its effects on the central and peripheral clocks, melatonin may impact lipid metabolism by orchestrating rhythmic metabolic processes related to gut microbiota.[310,311]

Melatonin can also facilitate the maintenance of the circadian rhythm in body temperature. A robust circadian temperature rhythm is considered as the "third signalling pathway" in the circadian coordination of sleep and circadian rhythmicity, in addition to synaptic and neurohormonal pathways. Mammals capable of maintaining a low metabolic rate and body temperature exhibit a large-amplitude circadian rhythm of core temperature and tend to live longer.[312] Most typical age-dependent changes such as diminished rhythm stability and synchronization with the 24-hour regimen can be partially ameliorated with low-dose exogenous melatonin taken by night.[135,136,278,279,302] A single daily melatonin dose stabilized and synchronized the body temperature rhythm by means of hypothermic and sleep-improving effects.

21.4.3 Sleep Scheduling

Optimal sleep duration can vary depending on the age,[313,314] gender,[188,315,316] and genetic background.[165,313,317,318] Both too short and too long sleep durations are associated with health problems and higher all-cause mortality,[319,320] notably related to cardiovascular conditions.[321] However, too long sleep duration is often secondary to primary health issues,[322] depression and low socioeconomic status,[323] while too short sleep duration could be a primary issue caused by social pressure.[324] The relationship between optimal sleep duration and morbidity is U-shaped, with an increased risk associated with sleep durations less than 6 hours and more than 9 hours.[325,326] Gender differences are also reported.[325] GWAS results[165] indicate that genes responsible for the sleep phase and sleep duration are significantly different. The relationship between sleep duration and health status is likely bidirectional. A causal relationship between sleep duration and the presence or absence of an elevated disease risk has not yet been established and could be performed along with defined gene polymorphisms.

Sleep consistency,[327,328] sleep phase,[329] and alignment of sleep[328] with the internal clock are also linked to health status and well-being. The sleep phase is related to the chronotype, which is influenced by genetic, epigenetic, and environmental factors with heritability that differs substantially among populations.[164,188] The chronotype and mean sleep phase depend on gender,[315,330–332] age,[331,332] season,[329,333] and social schedules.[324,334]

The evening chronotype is most susceptible to external circadian disruption (social jet lag).[183] Accumulated sleep debt and conflict between the biological clock and social demands is greater in late chronotypes. Severity of social jet lag correlates with complex metabolic disorders, including age-associated chronic pathology (decreased high-density lipoproteins, increased triglycerides, fasting insulin, insulin resistance, risk of metabolic syndrome, predisposition to type 2 diabetes and atherosclerosis),[335] and with a higher prevalence of sleep apnoea and higher concentrations of stress hormones.[336] Sleep disorders/sleep deficits have been associated with changes in lipid metabolism[337,338] and carbohydrate metabolism.[339,340] Circadian and sleep disruptions and corresponding alterations in hormonal variability deserve closer investigation in problems of infertility.[341] The association between certain chronotypes and the intrinsic tau deserves further attention.[188]

21.4.4 Time-restricted Feeding

The SCN phase coordinates phases in peripheral tissues by means of intrinsically aligned rhythms in temperature, hormones, neuropeptides,[342] and feeding.[343,344] The feeding schedule may also influence resetting of the SCN phase independently of certain peripheral circadian rhythms.[345,346] Specifically, insulin and insulin-like growth factor 1 (IGF-1) can induce PER proteins, and thereby reset peripheral clocks *in vivo* and *in vitro*;[347] these mechanisms also involve activation of targets of rapamycin (mTOR), increased phosphoinositide signalling, and down-regulation of microRNAs.

Timed feeding may facilitate intrinsic synchronization and help maintain large-amplitude circadian rhythms assumed to be associated with longevity.[348,349] Restricted caloric intake, the most reliable way of increaing life expectancy, is accompanied by an increase in the circadian amplitude of telemetered core temperature.[312] Longevity may be achieved only when a fortunate genotype meets good circadian hygiene.

The feeding schedule can serve as a powerful secondary synchronizer of the circadian system.[350–353] Mathematical modelling of circadian patterns in metabolism confirms benefits derived from time-restricted feeding.[354] Considering that glucose-stimulated insulin secretion rather than production is under circadian modulation, models showed that the timing and size of meals affect the insulin secretion rate and the amplitude of its circadian oscillation. Certain gene polymorphisms may predispose to a "natural born-healthy constitution" in terms of food intake regimens that are more resistant to challenges. For example, α-MUPA mice, which spontaneously consume 20–30% fewer calories than wild type mice, are long-lived and exhibit large-amplitude circadian rhythms in the expression of clock genes in the liver, coinciding with large-amplitude rhythms in food intake and body temperature.[355] Calorie restriction can modulate the expression of circadian core clock genes (*i.e.*, *Bmal1*), affecting their mRNA and protein levels by means of transcriptional and post-transcriptional mechanisms.[356] Targeted use of specific nutrients based on individual chronotype has the potential for immense clinical utility. Certain nutrients may even act as synchronizers, regulating CGs and CCGs in different tissues and organs. Such circadian clock manipulation by distinct nutrients can be tissue specific. Optimization of nutritional regimens for the treatment of metabolic disorders and the mitigation of age-related circadian disruption should be a strategy in personalized medicine, considering genetic factors.[357]

Feeding[358–361] and melatonin signalling[136,278,279,302,362] may be more effective synchronizers of peripheral rhythms when synchronization of the central clock by light is compromised, as after SCN ablation,[361] uncoupled time cue periodicities,[360] ageing,[136,218,219] or age-associated ocular pathology.[279,302] Time-restricted feeding can also reinforce circadian rhythms of compromised clock gene expression in the SCN[363,364] and may thus be used in theory to strengthen the circadian system. Timed feeding can also synchronize peripheral clocks without affecting the SCN.[345,346,365] As such, it can be used to correct intrinsic misalignment between the central and peripheral clocks when the latter are compromised by a poor circadian diet or light hygiene.

21.4.5 Exercise

Regular activity, timed exercise and maintaining social contacts[366] may all be beneficial for keeping circadian clocks robust and aligned, or for implementing training-the-clock strategies. Aged people sometimes do this rather intuitively by adopting a regular lifestyle.[367] Individuals who maintain an established daily schedule have a reduced incidence of insomnia.[368] Scheduled physical activity is another strategy to maintain circadian rhythms that contributes to healthy ageing.[136–138,369] Regular physical activity upholds feedback coupling with the central brain clock and helps preserve synchronized circadian rhythms.[370,371] Actimetry is useful for assessing the circadian rhythm in activity and its robustness.[372] Rodents with voluntary access to a running wheel have attenuated age-related changes in their circadian system, large-amplitude circadian rhythms in peripheral tissues and the SCN, and reduced resynchronization time after induced experimental jet lag.[373–375] Higher levels of physical activity and oxygen capacity in older individuals correlate with a larger circadian amplitude of *PER3* (but not *REV-ERBα*) expression[376] and of oral temperature[377] (Figure 21.2).

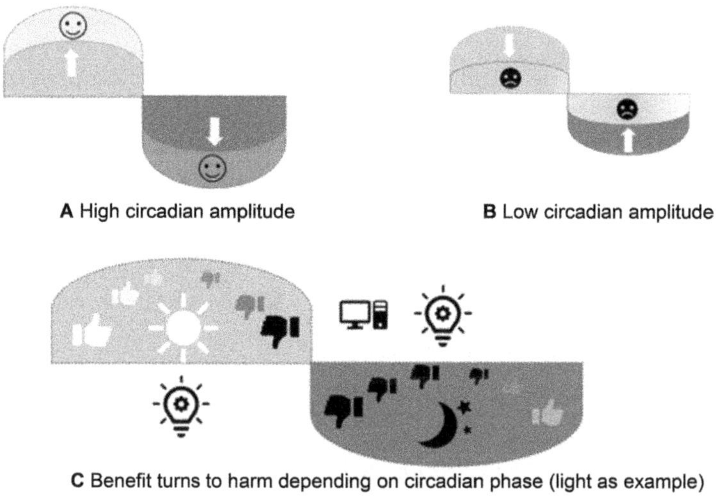

Figure 21.2 Clock-friendly non-pharmacological approaches: light, diet, work schedule, activity, and exercise. (A) Factors that increase the circadian amplitude usually have a positive effect on health. Accordingly, the lifestyle, daily routine, and nutrition that increase daytime variables (having a maximum during the waking period, which for a human usually corresponds to the light phase of the day) or reduce night-type variables (having a maximum during sleep, which for a human corresponds to the dark phase of the day) are chronobiologically justified and beneficial. (B) In contrast, factors that reduce the circadian amplitude in most cases should be considered as harmful to health. (C) Factors that affect the phase of the circadian rhythms have a multidirectional effect on health depending on the circadian time, in accordance with the phase response curves. In most cases, for the preservation and promotion of health, the conditional benefit of the "daytime" factor (for example, bright blue light) gradually decreases during the light phase. In the second part of the cycle, the factor passes into the category of harmful (for example, light pollution in the evening and night hours). A similar pattern applies to nocturnal factors (for example, melatonin).

21.5 Conclusion

It becomes increasingly obvious that circadian rhythms, and chronobiology more generally, should be taken into consideration in clinical practice, for screening, diagnosis, and timed treatment. Chronotherapy and pharmacogenomics represent two important avenues to make personalized treatment and primary prevention a reality. Conditions likely to facilitate the breakthrough and translation of fundamental principles of chronobiology into practical health care include the establishment of universal criteria for data collection and data analysis and interpretation of results thus obtained. Data banks and repositories of chronobiological reference values of physiological variables in clinical health, accounting for gender differences and changes as a function of age should complement existing databases of genetic information. Simple but accurate methods for evaluating the phase of the endogenous circadian rhythm should be developed, standardized, and made widely available. Expert systems using artificial intelligence are being built to help in decision-making in healthcare. Physiological monitoring is now widely used outside the clinic for health surveillance by lay people, and it may soon become feasible to store the data thus collected into individual medical records. Now is the time to incorporate chronobiological principles into these emerging technologies, which will define tomorrow's medical practice. The integration of chronobiological knowledge into these systems should enable the accurate diagnosis of disease and the determination of optimal kind, dose, and timing of pharmacologic or non-pharmacologic intervention to maximize treatment efficacy within the context of personalized medicine.

References

1. D. G. Gubin, *Usp. Fiziol. Nauk*, 2013, **44**, 65.
2. L. S. Mure, H. D. Le and G. Benegiamo, *et al.*, *Science*, 2018, **359**, 6381.
3. S. Panda, The arrival of circadian medicine, *Nat. Rev. Endocrinol.*, 2019, **15**(2), 67.
4. C. Partch, C. Green and J. Takahashi, *Trends Cell Biol.*, 2014, **24**(2), 90.
5. R. Zhang, N. F. Lahens and H. Ballance, *et al.*, *Proc. Natl. Acad. Sci. U. S. A.*, 2014, **111**, 16219.
6. C. R. Cederroth, U. Albrecht and J. Bass, *et al.*, *Cell Metab.*, 2019, **30**(2), 238.
7. C. M. Greco and P. Sassone-Corsi, *J. Exp. Med.*, 2020, **217**(6), e20200702.
8. D. Gubin, D. Weinert and G. Cornelissen, *J. Chronomed.*, 2020, **22**(1), 3–23.
9. S. Vamos and K. Klein, *Glob. Health Promot.*, 2018, **25**(2), 85.
10. G. Cornélissen, R. M. Zaslavskaya, Y. Kumagai, Y. Romanov and F. Halberg, *J. Clin. Pharmacol.*, 1994, **34**, 543–551.
11. L. S. Gaspar, A. R. Álvaro, S. Carmo-Silva, A. F. Mendes, A. Relógio and C. Cavadas, *Br. J. Pharmacol.*, 2019, **176**(16), 2827–2847.

12. G. K. Paschos, J. E. Baggs, J. B. Hogenesch and G. A. FitzGerald, *Annu. Rev. Pharmacol. Toxicol.*, 2010, **50**, 187–214.
13. J. A. Haspel, R. Anafi and M. K. Brown, *et al.*, *JCI Insight*, 2020, **5**(1), e131487.
14. M. Baraldo, *Expert Opin. Drug Metab. Toxicol.*, 2008, **4**(2), 175–192.
15. S. A. Brown, F. Fleury-Olela and E. Nagoshi, *et al.*, *PLoS Biol.*, 2005, **3**(10), e338.
16. A. Reinberg, J. Ghata, F. Halberg, M. Apfelbaum, P. Gervais, P. Boudon, C. Abulker and J. Dupont, *Ann. Endocrinol.*, 1971, **32**, 566–573.
17. K. Kalafatakis, G. M. Russell and C. J. Harmer, *et al.*, *Proc. Natl. Acad. Sci. U. S. A.*, 2018, **115**(17), E4091–E4100.
18. I. Vargas, A. N. Vgontzas and J. L. Abelson, *et al.*, *Sleep Med. Rev.*, 2018, **41**, 234–243.
19. D. G. Gubin, G. Cornelissen and F. Halberg, *et al.*, *Scr. Med. Fac. Med. Univ. Brun. Masaryk.*, 1997, **70**, 207–216.
20. D. Gubin, G. Cornelissen and D. Weinert, *et al.*, *World Heart J.*, 2013, **5**(4), 285–306.
21. W. Arnold, T. Ruf and L. E. Loe, *et al.*, *Sci. Rep.*, 2018, **8**(1), 14466.
22. L. I. Gapon, N. P. Shurkevich, I. M. Tonkonog, T. V. Sereda and D. G. Gubin, *Klin. Med.*, 2004, **82**(1), 31–35.
23. L. I. Gapon, N. P. Shurkevich, A. S. Vetoshkin and D. G. Gubin, *Klin. Med.*, 2006, **84**(2), 39–44.
24. D. G. Hazlerigg and N. J. C. Tyler, *PLoS Biol.*, 2019, **17**(7), e3000360.
25. H. Meng, N. M. Gonzales and D. M. Lonard, *et al.*, *Nat. Commun.*, 2020, **11**(1), 6215.
26. H. Ballance and B. Zhu, *Cell. Mol. Life Sci.*, 2021, **78**(7), 3127–3140.
27. B. Zhu, Q. Zhang and Y. Pan, *et al.*, *Cell Metab.*, 2017, **25**(6), 1305–1319.e9.
28. B. Zhu, C. C. Dacso and B. W. O'Malley, *J. Endocr. Soc.*, 2018, **2**(7), 727–752.
29. G. Andreatta and K. Tessmar-Raible, *J. Mol. Biol.*, 2020, **432**(12), 3525–3546.
30. T. Liu, M. Cavallini, F. Halberg, G. Cornelissen, J. Field and D. E. R. Sutherland, *Experientia*, 1986, **42**, 20–22.
31. V. N. Anisimov, M. A. Zabezhinski and I. G. Popovich, *et al.*, *Am. J. Pathol.*, 2010, **176**(5), 2092–2097.
32. V. N. Anisimov, M. A. Zabezhinski and I. G. Popovich, *et al.*, *Cell Cycle*, 2011, **10**(24), 4230–4236.
33. I. G. Popovich, V. N. Anisimov and M. A. Zabezhinski, *et al.*, *Cancer Biol. Ther.*, 2014, **15**(5), 586–592.
34. M. G. Aspinall and R. G. Hamermesh, *Harv. Bus. Rev.*, 2007, **85**(10), 165.
35. B. Bruguerolle, *Chronobiol. Int.*, 2008, **25**(1), 1–15.
36. M. Hassan, G. Oberg, A. N. Bekassy, J. Aschan, H. Ehrsson, P. Ljungman, G. Lönnerholm, B. Smedmyr, A. Taube and I. Wallin, *et al.*, *Cancer Chemother. Pharmacol.*, 1991, **28**(2), 130–134.

37. Ortiz-Tudela E., Mteyrek A., Ballesta A., Innominato P. F. and Lévi F., *Handb. Exp. Pharmacol.*, 2013, vol. 217, pp. 261–288.
38. S. Giacchetti, P. A. Dugué, P. F. Innominato, G. A. Bjarnason, C. Focan, C. Garufi, S. Tumolo, B. Coudert, S. Iacobelli, R. Smaaland, M. Tampellini, R. Adam, T. Moreau, F. Lévi and ARTBC International Chronotherapy Group, *Ann. Oncol.*, 2012, **23**(12), 3110–3116.
39. M. V. Nascimento Ferreira, L. Goumidi, H. B. Carvalho, A. C. F. De Moraes, A. M. Santaliestra-Pasías, A. Kafatos, D. Molnar, C. P. Lambrinou, S. De Henauw, A. Gutierrez, L. Censi, A. Marcos, K. Widhalm, F. Gottrand, M. Gonzalez-Gross, A. Meirhaeghe, L. A. Moreno and HELENA Study group, *Sleep Med.*, 2018, **46**, 56–60.
40. J. F. Duffy, S. W. Cain and A. M. Chang, *et al.*, *Proc. Natl. Acad. Sci. U. S. A.*, 2011, **108**(Suppl. 3), 15602–15608.
41. D. Gubin and S. N. Kolomeichuk, *J. Chronomed.*, 2019, **21**(2), 14–27.
42. D. Gubin and G. Cornelissen, *J. Chronomed.*, 2019, **21**(2), 8–13.
43. C. Wan, Z. Wang, G. Cornélissen and F. Halberg, *Chronobiologia*, 1992, **19**(3–4), 121–129.
44. D. G. Gubin, D. Weinert and S. V. Solovieva, *et al.*, *Biol. Rhythm Res.*, 2020, **51**(5), 780–793.
45. X. Liang, F. D. Bushman and G. A. FitzGerald, *Proc. Natl. Acad. Sci. U. S. A.*, 2015, **112**(33), 10479–10484.
46. J. N. Cohn, *Herz*, 1991, **16**(Spec No. 1), 267–271.
47. K. N. Maglo, J. Rubinstein, B. Huang and R. F. Ittenbach, *AJOB Empirical Bioethics*, 2014, **5**(4), 37–52.
48. G. Kaur, C. Phillips, K. Wong and B. Saini, *Int. J. Clin. Pharm.*, 2013, **35**, 344–358.
49. G. Sulli, E. N. C. Manoogian, P. R. Taub and S. Panda, *Trends Pharmacol. Sci.*, 2018, **39**(9), 812–827.
50. M. P. Antoch and R. V. Kondratov, *Handb. Exp. Pharmacol.*, 2013, vol. 217, pp. 289-309.
51. R. Von Roemeling, *Ann. N. Y. Acad. Sci.*, 1991, **618**, 292–311.
52. C. Scheiermann, J. Gibbs, L. Ince and A. Loudon, *Nat. Rev. Immunol.*, 2018, **18**(7), 423–437.
53. C. Scheiermann, Y. Kunisaki and P. S. Frenette, *Nat. Rev. Immunol.*, 2013, **13**(3), 190–198.
54. L. C. J. de Bree, V. P. Mourits, V. A. Koeken, S. J. Moorlag, R. Janssen, L. Folkman, D. Barreca, T. Krausgruber, V. Fife-Gernedl, B. Novakovic, R. J. Arts, H. Dijkstra, H. Lemmers, C. Bock, L. A. Joosten, R. van Crevel, C. S. Benn and M. G. Netea, *J. Clin. Invest.*, 2020, **130**(10), 5603–5617.
55. J. E. Long, M. T. Drayson, A. E. Taylor, K. M. Toellner, J. M. Lord and A. C. Phillips, *Vaccine*, 2016, **34**(24), 2679–2685.
56. R. C. Hermida, D. E. Ayala, A. Mojón and J. R. Fernández, *Chronobiol. Int.*, 2013, **30**(1–2), 87–98.
57. R. C. Hermida, J. J. Crespo and M. Domínguez-Sardiña, *et al.*, *Eur. Heart J.*, 2019, ehz754.

58. D. P. Cardinali, G. M. Brown, R. J. Reiter and S. R. Pandi-Perumal, *Sleep Vigilance*, 2020, 1–7.
59. S. Ray and A. B. Reddy, *Nat. Rev. Mol. Cell Biol.*, 2020, **21**(9), 494–495.
60. G. Cornelissen and F. Halberg, Treatment with open eyes: markers-guided chronotheranostics, in *Chronopharmaceutics: Science and Technology for Biological Rhythm-Guided Therapy and Prevention of Diseases*, ed. B. C. Youan, Wiley, Hoboken, NJ, 2009, ch. 11, pp. 257–323.
61. P. Prikryl, G. Cornélissen and J. Neubauer, *et al.*, *Clin. Exp. Hypertens.*, 2005, **27**(2–3), 119–128.
62. Y. Watanabe, F. Halberg, K. Otsuka and G. Cornelissen, *Clin. Exp. Hypertens.*, 2013, **35**, 257–266.
63. G. Cornelissen-Guillaume, F. Halberg, K. Otsuka and R. M. Singh, *World Heart J.*, 2008, **1**(3), 223–232.
64. F. Halberg, D. Powell and K. Otsuka, *et al.*, *Am. J. Physiol. Heart Circ. Physiol.*, 2013, **305**(3), H279–H294.
65. G. Cornélissen, E. Halberg, H. J. Long III, K. Prem, E. Bakken, Y. Touitou, S. Elg, E. Haus and F. Halberg, *Chronobiologia*, 1991, **18**, 153–166.
66. G. Cornelissen and F. Halberg, *Percept. Mot. Skills*, 2000, **90**, 579–586.
67. F. Elrashdy, A. A. Aljaddwi, E. M. Redwan and V. N. Uversky, *J. Biomol. Struct. Dyn.*, 2020, 1–12.
68. F. Halberg, Historical encounters between geophysics and biomedicine leading to the Cornélissen-series and chronoastrobiology, in *Long- and Short-Term Variability in Sun's History and Global Change*, ed. W. Schröder, Science Edition, Bremen, 2000, pp. 271–301.
69. G. Cornélissen, A. Delcourt and G. Toussaint, *et al.*, *Biomed. Pharmacother.*, 2005, **59**(Suppl. 1), S152–S157.
70. S. D. Pierdomenico, A. Bucci, F. Costantini, D. Lapenna, F. Cuccurullo and A. Mezzetti, *J. Am. Coll. Cardiol.*, 1998, **31**(7), 1627–1634.
71. A. J. Viera, F. C. Lin and A. L. Hinderliter, *et al.*, *Hypertension*, 2012, **59**(6), 1157–1163.
72. T. Nabika, H. Ohara, N. Kato and M. Isomura, *Hypertens. Res.*, 2012, **35**(5), 477–484.
73. K. Okamoto, Y. Yamori and A. Nagaoka, Establishment of stroke-prone spontaneously hypertensive rat (SHR), *Circ. Res.*, 1974, **34–35**, I143–I153.
74. R. Braun, W. L. Kath and M. Iwanaszko, *et al.*, *Proc. Natl. Acad. Sci. U. S. A.*, 2018, **115**(39), E9247–E9256.
75. G. A. Bilkey, B. L. Burns and E. P. Coles, *et al.*, *Front. Public Health*, 2019, **7**, 42.
76. C. A. Thaiss, M. Levy and T. Korem, *et al.*, *Cell*, 2016, **167**, 1495–1510.
77. F. Bishehsari, R. M. Voigt and A. Keshavarzian, *Nat. Rev. Endocrinol.*, 2020, **16**, 731–739.
78. R. M. Voigt, C. B. Forsyth, S. J. Green, P. A. Engen and A. Keshavarzian, *Int. Rev. Neurobiol.*, 2016, **131**, 193–205.
79. A. E. Rosselot, C. I. Hong and S. R. Moore, *Curr. Opin. Gastroenterol.*, 2016, **32**, 7–11.

80. W. Feng, J. Liu, H. Ao, S. Yue and C. Peng, *Theranostics*, 2020, **10**(24), 11278–11301.
81. J. K. Paulose, J. M. Wright, A. G. Patel and V. M. Cassone, *PLoS One*, 2016, **11**, e0146643.
82. J. K. Paulose, C. V. Cassone, K. B. Graniczkowska and V. M. Cassone, *iScience*, 2019, **19**, 1202–1213.
83. A. Ballesta, P. F. Innominato, R. Dallmann, D. A. Rand and F. A. Lévi, *Pharmacol. Rev.*, 2017, **69**(2), 161–199.
84. A. Lewy, *J. Clin. Endocrinol. Metab.*, 2010, **95**(7), 3158–3160.
85. H. J. Burgess, V. L. Revell, T. A. Molina and C. I. Eastman, *J. Clin. Endocrinol. Metab.*, 2010, **95**(7), 3325–3331.
86. F. Halberg, E. Jacobson, G. Wadsworth and J. J. Bittner, *Science*, 1958, **128**, 657–658.
87. F. Halberg, E. A. Johnson, B. W. Brown and J. J. Bittner, *Proc. Soc. Exp. Biol.*, 1960, **103**, 142–144.
88. E. Haus and F. Halberg, *J. Appl. Physiol.*, 1959, **14**, 878–880.
89. E. Haus, F. Halberg, M. K. Loken and U. S. Kim, Circadian rhythmometry of mammalian radiosensitivity, in *Space Radiation Biology*, ed. A. Tobias and P. Todd, Academic Press, New York, 1973, pp. 435–474.
90. H. Tamiya, S. Ogawa, Y. Ouchi and M. Akishita, *Sci. Rep.*, 2016, **6**, 32769.
91. P. L. Lowrey and J. S. Takahashi, *Adv. Genet.*, 2011, **74**, 175–230.
92. A. M. Chang, J. F. Duffy and O. M. Buxton, *et al.*, *Sci. Rep.*, 2019, **9**(1), 5350.
93. M. J. Zylka, L. P. Shearman, D. R. Weaver and S. M. Reppert, *Neuron*, 1998, **20**(6), 1103–1110.
94. S. N. Archer, A. U. Viola, V. Kyriakopoulou, M. von Schantz and D. J. Dijk, *Sleep*, 2008, **31**(5), 608–617.
95. S. N. Archer, C. Schmidt, G. Vandewalle and D. J. Dijk, *Sleep Med. Rev.*, 2018, **40**, 109–126.
96. T. D. Viena, C. M. Gobin and A. I. Fins, *et al.*, *J. Circadian Rhythms*, 2016, **14**, 3.
97. A. U. Viola, S. N. Archer and L. M. James, *et al.*, *Curr. Biol.*, 2007, **17**(7), 613–618.
98. A. Hida, S. Kitamura and Y. Katayose, *et al.*, *Sci. Rep.*, 2014, **4**, 6309.
99. M. Turco, A. Biscontin and M. Corrias, *et al.*, *Sci. Rep.*, 2017, **7**(1), 6967.
100. D. J. Dijk and S. N. Archer, *Sleep Med. Rev.*, 2010, **14**(3), 151–160.
101. S. Hasan, N. Santhi and A. S. Lazar, *et al.*, *FASEB J.*, 2012, **26**(6), 2414–2423.
102. S. Hasan, D. R. van der Veen and R. Winsky-Sommerer, *et al.*, *FASEB J.*, 2014, **28**(6), 2441–2454.
103. H. A. Mansour, J. Wood and K. V. Chowdari, *et al.*, *Chronobiol. Int.*, 2017, **34**(5), 624–631.
104. A. Patke, P. J. Murphy and O. E. Onat, *et al.*, *Cell*, 2017, **169**(2), 203–215.e13.
105. S. Miller and T. Hirota, *J. Mol. Biol.*, 2020, **432**(12), 3498–3514.

106. J. P. Etchegaray, K. K. Machida and E. Noton, *et al.*, *Mol. Cell. Biol.*, 2009, **29**(14), 3853–3866.
107. G. Guo, K. Wang and S. S. Hu, *et al.*, *J. Biol. Rhythms*, 2019, **34**(5), 482–496.
108. A. Hirano, G. Shi and C. R. Jones, *et al.*, *eLife*, 2016, **5**, e16695.
109. S. Korge, A. Grudziecki and A. Kramer, *J. Biol. Rhythms*, 2015, **30**(5), 389–395.
110. S. D. Jordan and K. A. Lamia, *Mol. Cell. Endocrinol.*, 2013, **366**(2), 163–169.
111. Q.-J. Meng, L. Logunova and E. S. Maywood, *et al.*, *Neuron*, 2008, **58**, 78–88.
112. J. S. Takahashi, *Nat. Rev. Genet.*, 2017, **18**(3), 164–179.
113. K. L. Toh, C. R. Jones, Y. He and E. J. Eide, *et al.*, *Science*, 2001, **291**(5506), 1040–1043.
114. T. Hirota, T. Okano, K. Kokame, H. Shirotani-Ikejima, T. Miyata and Y. Fukada, *J. Biol. Chem.*, 2002, **277**(46), 44244–44245.
115. K. Nohara, V. Mallampalli and T. Nemkov, *et al.*, *Nat. Commun.*, 2019, **10**(1), 3923.
116. C. A. Thaiss, D. Zeevi and M. Levy, *et al.*, *Cell*, 2014, **159**(3), 514–529.
117. C. Bandín, A. Martinez-Nicolas and J. M. Ordovás, *et al.*, *Int. J. Obes.*, 2013, **37**(8), 1044–1050.
118. F. Benedetti, S. Dallaspezia and M. C. Fulgosi, *et al.*, *Am. J. Med. Genet., Part B*, 2007, **144B**(5), 631–635.
119. M. Garaulet, C. Sánchez-Moreno, C. E. Smith, Y. C. Lee, F. Nicolás and J. M. Ordovás, *PLoS One*, 2011, **6**(2), e17435.
120. D. Katzenberg, T. Young and L. Finn, *et al.*, *Sleep*, 1998, **21**, 569–576.
121. N. L. Barclay, T. C. Eley and J. Mill, *et al.*, *Am. J. Med. Genet., Part B*, 2011, **156B**(6), 681–690.
122. A. M. Chang, A. M. Buch, D. S. Bradstreet, D. J. Klements and J. F. Duffy, *J. Biol. Rhythms*, 2011, **26**(3), 276–279.
123. M. von Schantz and S. N. Archer, Clocks, genes and sleep, *J. R. Soc. Med.*, 2003, **96**(10), 486–489.
124. A. R. Ozburn, K. Purohit and P. K. Parekh, *et al.*, *Front. Psychiatry*, 2016, **7**, 67.
125. A. Bugge, D. Feng and L. J. Everett, *et al.*, *Genes Dev.*, 2012, **26**(7), 657–667.
126. H. Cho, X. Zhao and M. Hatori, *et al.*, *Nature*, 2012, **485**(7396), 123–127.
127. Y. Zhang, B. Fang and M. J. Emmett, *et al.*, *Science*, 2015, **348**(6242), 1488–1492.
128. X. Zhao, T. Hirota and X. Han, *et al.*, *Cell*, 2016, **165**(7), 1644–1657.
129. B. Zhu, L. A. Gates and E. Stashi, *et al.*, *Mol. Cell*, 2015, **60**(5), 769–783.
130. N. A. Agadzhanian and D. G. Gubin, *Usp Fiziol. Nauk*, 2004, **35**(2), 57–72.
131. S. A. Brown, L. Pagani, C. Cajochen and A. Eckert, *Gerontology*, 2011, **57**, 427–434.
132. G. F. Gloston, S. H. Yoo and Z. J. Chen, *Front. Neurol.*, 2017, **8**, 100.

133. G. D. Gubin and A. A. Chesnokov, *Lab. Delo*, 1976, **7**, 430–432.
134. G. D. Gubin and D. Weinert, *Usp. Fiziol. Nauk*, 1991, **22**(1), 77–96.
135. D. G. Gubin, G. D. Gubin, L. I. Gapon and D. Weinert, *Curr. Aging Sci.*, 2016, **9**(1), 5–13.
136. D. G. Gubin, D. Weinert and T. V. Bolotnova, *Curr Aging Sci.*, 2016, **9**(1), 14–25.
137. D. G. Gubin and D. Weinert, *Adv. Gerontol.*, 2015, **5**, 209–218.
138. D. G. Gubin and D. Weinert, *Adv. Gerontol.*, 2016, **6**, 10–20.
139. H. R. Sanghani, A. Jagannath and T. Humberstone, *et al.*, *Mol. Psychiatry*, 2021, **26**(9), 5252–5265.
140. W. Ruan, X. Yuan and H. K. Eltzschig, *Nat. Rev. Drug Discovery*, 2021, **20**(4), 287–307.
141. Y. Wu, D. Tang and N. Liu, *et al.*, *Cell Metab.*, 2017, **25**(1), 73–85.
142. T. Eckle, K. Hartmann and S. Bonney, *et al.*, *Nat. Med.*, 2012, **18**(5), 774–782.
143. B. He and Z. Chen, *Curr. Drug Metab.*, 2016, **17**(5), 503–512.
144. L. A. Solt, Y. Wang and S. Banerjee, *et al.*, *Nature*, 2012, **485**(7396), 62–68.
145. Q. J. Meng, E. S. Maywood and D. A. Bechtold, *et al.*, *Proc. Natl. Acad. Sci. U. S. A.*, 2010, **107**(34), 15240–15245.
146. T. Hirota, J. W. Lee and P. C. St John, *et al.*, *Science*, 2012, **337**(6098), 1094–1097.
147. S. Ray, R. Lach and K. J. Heesom, *et al.*, *Life Sci. Alliance*, 2019, **2**(6), e201900603.
148. T. K. Tamai, Y. Nakane and W. Ota, *et al.*, *EMBO Mol. Med.*, 2018, **10**(5), e8724.
149. N. Xie, L. Zhang, W. Gao, C. Huang, P. E. Huber, X. Zhou, C. Li, G. Shen and B. Zou, *Signal Transduction Targeted Ther.*, 2020, **5**(1), 227.
150. G. Asher, D. Gatfield, M. Stratmann, H. Reinke, C. Dibner, F. Kreppel, R. Mostoslavsky, F. W. Alt and U. Schibler, *Cell*, 2008, **134**(2), 317–328.
151. H. C. Chang and L. Guarente, *Cell*, 2013, **153**(7), 1448–1460.
152. P. T. Foteinou, A. Venkataraman, L. J. Francey, R. C. Anafi, J. B. Hogenesch and F. J. Doyle 3rd, *Proc. Natl. Acad. Sci. U. S. A.*, 2018, **115**(45), 11643–11648.
153. S. Masri, *Curr. Opin. Clin. Nutr. Metab. Care*, 2015, **18**(6), 521–527.
154. D. C. Levine, H. Hong, B. J. Weidemann, K. M. Ramsey, A. H. Affinati, M. S. Schmidt, J. Cedernaes, C. Omura, R. Braun, C. Lee, C. Brenner, C. B. Peek and J. Bass, *Mol. Cell*, 2020, **78**(5), 835–849.e7.
155. S. Kiessling, L. Beaulieu-Laroche and I. D. Blum, *et al.*, *BMC Biol.*, 2017, **15**(1), 13.
156. C. Costantini, G. Renga, F. Sellitto, M. Borghi, C. Stincardini, M. Pariano, T. Zelante, F. Chiarotti, A. Bartoli, P. Mosci, L. Romani, S. Brancorsini and M. M. Bellet, *Front. Cell. Infect. Microbiol.*, 2020, **10**, 30.
157. S. P. Nobs, T. Tuganbaev and E. Elinav, *EMBO Rep.*, 2019, **20**(4), e47129.
158. G. Bordyugov, U. Abraham and A. Granada, *et al.*, *J. R .Soc., Interface*, 2015, **12**(108), 20150282.

159. J. B. Hogenesch and E. D. Herzog, *FEBS Lett.*, 2011, **585**, 1427–1434.
160. C. Schmal, E. D. Herzog and H. Herzel, *J. Biol. Rhythms*, 2018, **33**(1), 84–98.
161. M. Riddle, E. Mezias, D. Foley, J. LeSauter and R. Silver, *Eur. J. Neurosci.*, 2017, **45**(11), 1357–1367.
162. U. Abraham, A. E. Granada and P. O. Westermark, *et al.*, *Mol. Syst. Biol.*, 2010, **6**, 438.
163. M. Doi, A. Ishida and A. Miyake, *et al.*, *Nat. Commun.*, 2011, **2**, 327.
164. D. A. Kalmbach, L. D. Schneider and J. Cheung, *et al.*, *Sleep*, 2017, **40**(2), zsw048.
165. S. E. Jones, V. T. van Hees and D. R. Mazzotti, *et al.*, *Nat. Commun.*, 2019, **10**(1), 1585.
166. J. J. Hughey and A. J. Butte, *J. Biol. Rhythms*, 2016, **31**(6), 588.
167. D. J. Dijk and J. F. Duffy, *J. Biol. Rhythms*, 2020, **35**(5), 421–438.
168. A. Chattoraj, T. Liu, L. S. Zhang, Z. Huang and J. Borjigin, *Rev. Endocr. Metab. Disord.*, 2009, **10**(4), 237–243.
169. N. S. Foulkes, J. Borjigin, S. H. Snyder and P. Sassone-Corsi, *Proc. Natl. Acad. Sci. U. S. A.*, 1996, **93**(24), 14140–14145.
170. D. Zhao, Y. Yu and Y. Shen, *et al.*, *Front. Endocrinol.*, 2019, **10**, 249.
171. P. Kumar Jha, E. Challet and A. Kalsbeek, *Mol. Cell. Endocrinol.*, 2015, **418**(Pt. 1), 74–88.
172. C. Skarke, N. F. Lahens and S. D. Rhoades, *et al.*, *Sci. Rep.*, 2017, **7**(1), 17141.
173. T. W. Hansen, Y. Li, J. Boggia, L. Thijs, T. Richart and J. A. Staessen, *Hypertension*, 2011, **57**(1), 3–10.
174. E. P. Juhanoja, P. J. Puukka, J. K. Johansson, T. J. Niiranen and A. M. Jula, *Blood Press Monit.*, 2016, **21**(2), 63–68.
175. Y. Mochizuki, M. Okutani and Y. Donfeng, *et al.*, *Am. J. Hypertens.*, 1998, **11**, 403–409.
176. D. G. Gubin, D. Weinert, S. V. Rybina, L. A. Danilova, S. V. Solovieva, A. M. Durov, N. Y. Prokopiev and P. A. Ushakov, *Chronobiol. Int.*, 2017, **34**(5), 632–649.
177. D. G. Gubin and G. Cornelissen, *J. Chronomed.*, 2019, **21**(2), 8–13.
178. J. Waterhouse, D. Weinert, D. Minors, S. Folkard, D. Owens, G. Atkinson, I. Macdonald, N. Sytnik, P. Tucker and T. Reilly, *Chronobiol. Int.*, 2000, **17**(4), 539–566.
179. D. Weinert and J. Waterhouse, *Physiol. Behav.*, 2007, **90**(2–3), 246–256.
180. A. J. Lewy, N. L. Cutler and R. L. Sack, *J. Biol. Rhythms*, 1999, **14**(3), 227–236.
181. E. B. Klerman, H. B. Gershengorn, J. F. Duffy and R. E. Kronauer, *J. Biol. Rhythms*, 2002, **17**, 181–193.
182. J. A. Horne and O. A. Ostberg, *Int. J. Chronobiol.*, 1976, **4**(2), 97–110.
183. T. Roenneberg, A. Wirz-Justice and M. Merrow, *J. Biol. Rhythms*, 2003, **18**, 80–90.

184. T. Roenneberg, L. K. Pilz, G. Zerbini and E. C. Winnebeck, *Biology*, 2019, **8**(3), 54.
185. J. F. Duffy, D. W. Rimmer and C. A. Czeisler, *Behav. Neurosci.*, 2001, **115**(4), 895–899.
186. J. S. Emens, K. Yuhas, J. Rough, N. Kochar, D. Peters and A. J. Lewy, *Chronobiol. Int.*, 2009, **26**(3), 474–493.
187. T. Kantermann and C. I. Eastman, *Chronobiol. Int.*, 2018, **35**(2), 280–288.
188. D. G. Gubin, S. N. Kolomeichuk and D. Weinert, *J. Chronomed.*, 2021, **23**(1), 3–15.
189. M. Nováková, M. Sládek and A. Sumová, *Chronobiol. Int*, 2013, **30**(4), 607–617.
190. M. E. Hughes, K. C. Abruzzi and R. Allada, *et al.*, *J. Biol. Rhythms*, 2017, **32**(5), 380–393.
191. S. N. Archer, A. U. Viola, V. Kyriakopoulou, M. von Schantz and D. J. Dijk, *Sleep*, 2008, **31**(5), 608–617.
192. J. M. Lane, I. Vlasac and S. G. Anderson, *et al.*, *Nat. Commun.*, 2016, **7**, 10889.
193. Y. Hu, A. Shmygelska, D. Tran, N. Eriksson, J. Y. Tung and D. A. Hinds, *Nat. Commun.*, 2016, **7**, 10448.
194. F. Agostinelli, N. Ceglia, B. Shahbaba, P. Sassone-Corsi and P. Baldi, *Bioinformatics*, 2016, **32**(12), i8–i17.
195. J. J. Hughey, T. Hastie and A. J. Butte, ZeitZeiger, *Nucleic Acids Res.*, 2016, **44**, e80.
196. J. J. Hughey, *Genome Med.*, 2017, **9**, 19.
197. E. E. Laing, C. S. Möller-Levet and N. Poh, *et al.*, *eLife*, 2017, **6**, e20214.
198. N. Wittenbrink, B. Ananthasubramaniam and M. Münch, *et al.*, *J. Clin. Invest.*, 2018, **128**(9), 3826–3839.
199. R. C. Anafi, L. J. Francey, J. B. Hogenesch and J. Kim, *Proc. Natl. Acad. Sci. U. S. A.*, 2017, **114**(20), 5312–5317.
200. G. Wu, M. D. Ruben and R. E. Schmidt, *et al.*, *Proc. Natl. Acad. Sci. U. S. A.*, 2018, **115**(48), 12313–12318.
201. G. Wu, M. D. Ruben and L. J. Francey, *et al.*, *Genome Med.*, 2020, **12**(1), 73.
202. M. Akashi, H. Soma and T. Yamamoto, *et al.*, *Proc. Natl. Acad. Sci. U. S. A.*, 2010, **107**(35), 15643–15648.
203. G. A. Bjarnason, R. C. Jordan and P. A. Wood, *et al.*, *Am. J. Pathol.*, 2001, **158**(5), 1793–1801.
204. B. A. Katchman, M. Zhu, J. Blain Christen and K. S. Anderson, *Clin. Appl.*, 2018, **12**(6), e1800010.
205. R. D. Munje, S. Muthukumar, B. Jagannath and S. Prasad, *Sci. Rep.*, 2017, **7**(1), 1950.
206. O. Castanon-Cervantes, M. Wu and J. C. Ehlen, *et al.*, *J. Immunol.*, 2010, **185**(10), 5796–5805.
207. A. J. K. Phillips, P. Vidafar and A. C. Burns, *et al.*, *Proc. Natl. Acad. Sci. U. S. A.*, 2019, **116**(24), 12019–12024.
208. R. Dallmann, S. A. Brown and F. Gachon, *Annu. Rev. Pharmacol. Toxicol.*, 2014, **54**, 339–361.

209. F. Kamali, J. R. Fry and G. D. Bell, *Xenobiotica*, 1987, **17**, 635–641.
210. H. Watanabe, S. Nakano, K. Nagai and N. Ogawa, *J. Clin. Pharmacol.*, 1984, **24**, 509–514.
211. E. S. Musiek and G. A. Fitzgerald, *Handb. Exp. Pharmacol.*, 2013, vol. 217, pp. 243-260.
212. X. Pan and M. M. Hussain, *J. Lipid Res.*, 2009, **50**(9), 1800–1813.
213. Y. Wang, Z. Kuang, X. Yu, K. A. Ruhn, M. Kubo and L. V. Hooper, *Science*, 2017, **357**(6354), 912–916.
214. R. Dallmann, A. Okyar and F. Lévi, *Trends Mol. Med.*, 2016, **22**(5), 430–445.
215. H. Miyamoto, E. Nakamaru-Ogiso, K. Hamada and T. K. Hensch, *J. Neurosci.*, 2012, **32**, 14794–14803.
216. W. Pan and A. J. Kastin, *Neuroscientist*, 2017, **23**(2), 124–136.
217. Y. K. Zhang, R. L. Yeager and C. D. Klaassen, *Drug Metab. Dispos.*, 2009, **37**, 106–115.
218. S. Panda, M. P. Antoch, B. H. Miller, A. I. Su, A. B. Schook, M. Straume, P. G. Schultz, S. A. Kay, J. S. Takahashi and J. B. Hogenesch, *Cell*, 2002, **109**, 307–320.
219. M. A. Lazar, Rev-erbs: Integrating Metabolism Around the Clock, in *A Time for Metabolism and Hormones*, ed. P. Sassone-Corsi and Y. Christen, Springer, Cham (CH), 2016.
220. J. J. Gooley and E. C. Chua, *J. Genet. Genomics*, 2014, **41**(5), 231–250.
221. X. Yang, M. Downes, R. T. Yu, A. L. Bookout, W. He, M. Straume, D. J. Mangelsdorf and R. M. Evans, *Cell*, 2006, **126**(4), 801–810.
222. P. Erkekoglu and T. Baydar, *J. Res. Pharm. Pract.*, 2012, **1**, 3–9.
223. B. Lemmer, *Expert Opin. Drug Delivery*, 2005, **2**, 667–681.
224. S. Ohdo, S. Koyanagi and N. Matsunaga, *Adv. Drug Delivery Rev.*, 2010, **62**, 885–897.
225. U. P. Zmrzljak and D. Rozman, *Chem. Res. Toxicol.*, 2012, **25**(4), 811–824.
226. J. G. Johnston and D. M. Pollock, *Free Radical Biol. Med.*, 2018, **119**, 93–107.
227. E. H. Koritzinsky, J. M. Street, R. R. Chari, D. M. Glispie, T. R. Bellomo and A. M. Aponte, *et al.*, *Am. J. Physiol. Renal. Physiol.*, 2019, **317**(5), F1098–F1110.
228. B. Lemmer, *Heart Failure Clin.*, 2017, **13**(4), 739–757.
229. B. Lemmer and M. Middeke, *Chronobiol. Int.*, 2020, **37**(5), 728–730.
230. K. Tamura, I. Kohno, Y. Saito, K. Wakasugi, S. Achiwa, Y. Imanishi, P. Cugini and F. Halberg, Antihypertensive individualized therapeutic strategy, *Difesa Sociale*, 1991, **6**, 109–124.
231. N. Black, A. D'Souza and Y. Wang, *et al.*, *Heart Rhythm*, 2019, **16**(2), 298–307.
232. B. S. P. Fok, G. Cornelissen, F. Halberg, T. T. W. Chu, G. N. Thomas and B. Tomlinson, *J. Cardiol.*, 2004, **97**(Suppl. 2), S14; G. Cornelissen, F. Halberg, E. Bakken, R. M. Singh, K. Otsuka, B. Tomlinson, G. Toussaint, S. Bathina, O. Schwartzkopff, Z. Wang, R. Tarquini, F. Perfetto, G. Pantaleoni, R. Jozsa, P. A. Delmore and E. Nolley, Also in *Biomed. Pharmacother.*, 2004, 58 (Suppl. 1), S69–S86.

233. J. Treadwell and M. McCartney, *Br. J.Gen. Pract.*, 2016, **66**(644), 116–117.
234. J. Brodersen, *Zdr Varst.*, 2017, **56**(3), 147–149.
235. M. S. Kale and D. Korenstein, *BMJ*, 2018, **362**, k2820.
236. P. Ø. Opdal, E. Meland and S. Hjörleifsson, *Scand. J. Primary Health Care*, 2019, **37**(1), 135–140.
237. M. Pausch, A. Schedlbauer, M. Weiss, T. Kuehlein and S. Hueber, *PLoS One*, 2020, **15**(1), e0227457.
238. T. Khoury and Y. Ilan, *Front. Immunol.*, 2019, **10**, 2726.
239. Y. Ilan, *Mol. Ther.-Methods Clin. Dev.*, 2020, **18**, 335–344.
240. S. Ciancia, A. Cafarelli, A. Zahoranova, A. Menciassi and L. Ricotti, *Front. Bioeng. Biotechnol.*, 2020, **8**, 317.
241. Z. Khan, V. Pillay, Y. E. Choonara and L. C. du Toit, *Pharm. Dev. Technol.*, 2009, **14**(6), 602–612.
242. S. Ohdo, S. Koyanagi and N. Matsunaga, *Pharmacol. Ther.*, 2019, **202**, 72–90.
243. N. Reddy Dumpa, S. Bandari and M. A Repka, *Pharmaceutics*, 2020, **12**(1), 52.
244. S. A. Sunil, M. V. Srikanth, N. S. Rao, M. U. Uhumwangho, K. Latha and K. V. Murthy, *Curr. Drug Delivery*, 2011, **8**(6), 622–633.
245. L. Ercolani, A. Ferrari, C. De Mei, C. Parodi, M. Wade and B. Grimaldi, Circadian clock: time for novel anticancer strategies?, *Pharmacol. Res.*, 2015, **100**, 288–295.
246. S. Masri and P. Sassone-Corsi, *Nat. Med.*, 2018, **24**(12), 1795–1803.
247. P. J. DeCoursey, *J. Comp. Physiol.*, 1961, **44**, 331–354.
248. K. Honma, S. Honma and T. Wada, *Experientia*, 1987, **43**(11–12), 1205–1207.
249. S. B. Khalsa, M. E. Jewett, C. Cajochen and C. A. Czeisler, *J. Physiol.*, 2003, **549**(Pt. 3), 945–952.
250. D. S. Minors, J. M. Waterhouse and A. Wirz-Justice, *Neurosci. Lett.*, 1991, **133**, 36–40.
251. R. G. Foster, S. Hughes and S. N. Peirson, *Biology*, 2020, **9**(7), 180.
252. D. P. King and J. S. Takahashi, *Annu. Rev. Neurosci.*, 2000, **23**, 713–742.
253. C. Cajochen, M. Freyburger, T. Basishvili, C. Garbazza, F. Rudzik and C. Renz, *et al.*, *Light Res. Technol.*, 2019, **51**, 1044–1062.
254. O. Stefani, M. Freyburger, S. Veitz, T. Basishvili, M. Meyer, J. Weibel, K. Kobayashi, Y. Shirakawa and C. Cajochen, *J. Pineal Res.*, 2020, e12714.
255. T. Brown, G. Brainard, C. Cajochen, C. Czeisler, J. Hanifin, S. Lockley, R. Lucas, M. Munch, J. O'Hagan, S. Peirson, L. Price, T. Roenneberg, L. Schlangen, D. Skene, M. Spitschan, C. Vetter, P. Zee and K. Wright Jr, Recommendations for Healthy Daytime, Evening, and Night-Time Indoor Light Exposure, Preprints, 2020, 2020120037, DOI: 10.20944/preprints202012.0037.v1.
256. K. W. Houser and T. Esposito, *Front. Neurol.*, 2021, **12**, 630553.
257. A. Lowden, G. Öztürk, A. Reynolds and B. Bjorvatn, *Ind. Health*, 2019, **57**(2), 213–227.

258. S. Soltic and A. Chalmers, *J. Healthcare Eng.*, 2019, **2019**, 5016013.
259. K. Porcheret, L. Wald and L. Fritschi, *et al.*, *Chronobiol. Int.*, 2018, **35**(10), 1365–1374.
260. A. C. Skeldon, A. J. Phillips and D. J. Dijk, *Sci. Rep.*, 2017, **7**, 45158.
261. S. J. Crowley, S. W. Cain, A. C. Burns, C. Acebo and M. A. Carskadon, *J. Clin. Endocrinol. Metab.*, 2015, **100**(11), 4067–4073.
262. V. Daneault, M. Dumont, É Massé, G. Vandewalle and J. Carrier, *J. Physiol. Anthropol.*, 2016, **35**, 9.
263. M. G. Figueiro, *Neurodegener. Dis. Manage.*, 2017, **7**(2), 119–145.
264. R. P. Najjar, C. Chiquet and P. Teikari, *et al.*, *PLoS One*, 2014, **9**(1), e85837.
265. Y. Touitou, *Exp. Gerontol.*, 2001, **36**(7), 1083–1100.
266. R. Hardeland, *Aging Dis.*, 2012, **3**(2), 194–225.
267. I. Misiunaite, C. I. Eastman and S. J. Crowley, *Front. Neurosci.*, 2020, **14**, 99.
268. R. I. Versteeg, D. J. Stenvers and D. Visintainer, *et al.*, *J. Biol. Rhythms*, 2017, **32**(2), 130–142.
269. M. H. Smolensky, L. L. Sackett-Lundeen and F. Portaluppi, *Chronobiol. Int.*, 2015, **32**(8), 1029–1048.
270. A. J. Fields, S. E. Linnville and R. E. Hoyt, *Health Psychol. Open*, 2016, **3**(1), 2055102916648679.
271. L. Coppeta, F. Papa and A. Magrini, *J. Environ. Public Health*, 2018, **2018**, 8468742.
272. R. Nair and A. Maseeh, *Pharmacol. Pharmacother.*, 2012, **3**(2), 118–126.
273. P. Autier, P. Mullie, A. Macacu, M. Dragomir, M. Boniol, K. Coppens, C. Pizot and M. Boniol, *Lancet Diabetes Endocrinol.*, 2017, **5**(12), 986–1004.
274. E. Theodoratou, I. Tzoulaki, L. Zgaga and J. P. Ioannidis, *BMJ*, 2014, **348**, g2035.
275. Y. Zhang, F. Fang and J. Tang, *et al.*, *BMJ*, 2019, **366**, l4673.
276. A. Giustina, R. Bouillon and N. Binkley, *et al.*, *JBMR Plus*, 2020, **4**(12), e10417.
277. D. P. Cardinali and R. Hardeland, *Neuroendocrinology*, 2017, **104**(4), 382–397.
278. D. G. Gubin, G. D. Gubin, J. Waterhouse and D. Weinert, *Chronobiol. Int.*, 2006, **23**(3), 639–658.
279. D. Gubin, V. Neroev, T. Malishevskaya, G. Cornelissen, S. Y. Astakhov, S. Kolomeichuk, N. Yuzhakova, Y. Kabitskaya and D. Weinert, Melatonin mitigates disrupted circadian rhythms, lowers intraocular pressure, and improves retinal ganglion cells function in glaucoma, *J. Pineal Res.*, 2021, **70**(4), e12730.
280. A. J. Lewy, J. S. Emens, R. L. Sack, B. P. Hasler and R. A. Bernert, *Chronobiol. Int.*, 2002, **19**(3), 649–658.
281. A. J. Lewy, *Cold Spring Harbor Symp. Quant. Biol.*, 2007, **72**, 623–636.
282. R. J. Reiter, D. X. Tan and A. Galano, *Physiology*, 2014, **29**(5), 325–333.
283. R. Hardeland, *Diseases*, 2021, **9**(1), 18.

284. S. R. Pandi-Perumal, V. Srinivasan, D. W. Spence and D. P. Cardinali, *CNS Drugs*, 2007, **21**(12), 995–1018.
285. F. Auld, E. L. Maschauer, I. Morrison, D. J. Skene and R. L. Riha, *Sleep Med. Rev.*, 2017, **34**, 10–22.
286. N. Zisapel, *Br. J. Pharmacol.*, 2018, **175**(16), 3190–3199.
287. A. Cagnacci, M. Cannoletta, A. Renzi, F. Baldassari, S. Arangino and A. Volpe, *Am. J. Hypertens.*, 2005, **18**(12 Pt. 1), 1614–1618.
288. F. A. Scheer, G. A. Van Montfrans, E. J. van Someren, G. Mairuhu and R. M. Buijs, *Hypertension*, 2004, **43**(2), 192–197.
289. F. Simko and L. Paulis, *J. Pineal Res.*, 2007, **42**(4), 319–322.
290. E. Grossman, M. Laudon and N. Zisapel, *Vasc. Health Risk Manage.*, 2011, **7**, 577–584.
291. F. Simko, T. Baka, L. Paulis and R. J. Reiter, *J. Pineal Res.*, 2016, **61**(2), 127–137.
292. M. V. Hansen, L. T. Andersen, M. T. Madsen, I. Hageman, L. S. Rasmussen, S. Bokmand, J. Rosenberg and I. Gögenur, *Breast Cancer Res. Treat.*, 2014, **145**(3), 683–695.
293. R. Daneshvar Kakhaki, V. Ostadmohammadi, E. Kouchaki, E. Aghadavod, F. Bahmani, O. R. Tamtaji, R. J. Reiter, M. A. Mansournia and Z. Asemi, *Clin. Neurol. Neurosurg.*, 2020, **195**, 105878.
294. D. P. Cardinali, *Front. Endocrinol.*, 2019, **10**, 480.
295. G. Anderson and R. J. Reiter, *Rev. Med. Virol.*, 2020, **30**(3), e2109.
296. A. O. Sehirli, S. Sayiner and N. Serakinci, *Mol. Biol. Rep.*, 2020, **47**(10), 8229–8233.
297. A. Carrillo-Vico, P. J. Lardone, N. Alvarez-Sánchez, A. Rodríguez-Rodríguez and J. M. Guerrero, *Int. J. Mol. Sci.*, 2013, **14**(4), 8638–8683.
298. V. M. Martín Giménez, F. Inserra and C. D. Tajer, *et al.*, *Life Sci.*, 2020, **254**, 117808.
299. B. S. Alghamdi, *J. Neurosci. Res.*, 2018, **96**(7), 1136–1149.
300. Y. Leng, E. S. Musiek, K. Hu, F. P. Cappuccio and K. Yaffe, *Lancet Neurol.*, 2019, **18**(3), 307–318.
301. V. Neroev, T. Malishevskaya, D. Weinert, S. Astakhov, S. Kolomeichuk, G. Cornelissen, Y. Kabitskaya, E. Boiko, I. Nemtsova and D. Gubin, *Int. J. Mol. Sci.*, 2020, **22**(1), 359.
302. D. Gubin and D. Weinert, *Neural Regener. Res.*, 2022, **17**(8), 1759–1760.
303. M. J. Gerdin, M. I. Masana, M. A. Rivera-Bermúdez, R. L. Hudson, D. J. Earnest, M. U. Gillette and M. L. Dubocovich, *FASEB J.*, 2004, **18**(14), 1646–1656.
304. M. J. Gerdin, M. I. Masana and M. L. Dubocovich, *Biochem. Pharmacol.*, 2004, **67**(11), 2023–2030.
305. I. Prokopenko, C. Langenberg and J. C. Florez, *et al.*, *Nat. Genet.*, 2009, **41**(1), 77–81.
306. M. Huber, A. Treszl, R. Reibis, C. Teichmann, I. Zergibel, J. Bolbrinker, J. Scholze, K. Wegscheider, H. Völler and R. Kreutz, *Eur. J. Intern. Med.*, 2013, **24**(7), 650–655.

307. N. Ferlazzo, G. Andolina, A. Cannata, M. G. Costanzo, V. Rizzo, M. Currò, R. Ientile and D. Caccamo, *Antioxidants*, 2020, **9**(11), 1088.
308. A. Goyal, P. D. Terry, H. M. Superak, C. L. Nell-Dybdahl, R. Chowdhury, L. S. Phillips and M. H. Kutner, *Diabetol. Metab. Syndr.*, 2014, **6**, 124.
309. M. Koziróg, A. R. Poliwczak, P. Duchnowicz, M. Koter-Michalak, J. Sikora and M. Broncel, *J. Pineal Res.*, 2011, **50**(3), 261–266.
310. F. Hong, S. Pan, P. Xu, T. Xue, J. Wang, Y. Guo, L. Jia, X. Qiao, L. Li and Y. Zhai, *Cells*, 2020, **9**(2), 489.
311. J. Yin, Y. Li, H. Han, S. Chen, J. Gao, G. Liu, X. Wu, J. Deng, Q. Yu, X. Huang, R. Fang, T. Li, R. J. Reiter, D. Zhang, C. Zhu, G. Zhu, W. Ren and Y. Yin, *J. Pineal Res.*, 2018, **65**(4), e12524.
312. W. Nelson and F. Halberg, *J. Nutr.*, 1986, **116**(11), 2244–2253.
313. J. P. Chaput, C. Dutil and H. Sampasa-Kanyinga, *Nat. Sci. Sleep*, 2018, **10**, 421–430.
314. M. Schuster, C. Oberlinner and M. Claus, *Chronobiol. Int.*, 2019, **36**(6), 784–795.
315. M. Bailey and R. Silver, *Front. Neuroendocrinol.*, 2014, **35**(1), 111–139.
316. B. Zhang and Y. K. Wing, *Sleep*, 2006, **29**(1), 85–93.
317. H. S. Dashti, S. E. Jones and A. R. Wood, *et al.*, *Nat. Commun.*, 2019, **10**(1), 1100.
318. V. Garfield, *Sleep Med. Rev.*, 2020, **56**, 101413.
319. F. P. Cappuccio, L. D'Elia, P. Strazzullo and M. A. Miller, *Sleep*, 2010, **33**(5), 585–592.
320. O. Itani, M. Jike, N. Watanabe and Y. Kaneita, *Sleep Med.*, 2017, **32**, 246–256.
321. K. M. Full, A. Malhotra and L. C. Gallo, *et al.*, *J. Gerontol., Ser. A*, 2020, **75**(9), 1771–1778.
322. M. A. Grandner and S. P. Drummond, *Sleep Med. Rev.*, 2007, **11**(5), 341–360.
323. S. R. Patel, A. Malhotra, D. J. Gottlieb, D. P. White and F. B. Hu, *Sleep*, 2006, **29**(7), 881–889.
324. M. Korman, V. Tkachev, C. Reis, Y. Komada, S. Kitamura, D. Gubin, V. Kumar and T. Roenneberg, *Sci. Rep.*, 2020, **10**(1), 22225.
325. J. E. Ferrie, M. J. Shipley and F. P. Cappuccio, *et al.*, *Sleep*, 2007, **30**(12), 1659–1666.
326. J. Yin, X. Jin and Z. Shan, *et al.*, *J. Am. Heart Assoc.*, 2017, **6**(9), e005947.
327. K. Okano, J. R. Kaczmarzyk, N. Dave, J. D. E. Gabrieli and J. C. Grossman, *NPJ Sci. Learn.*, 2019, **4**, 16.
328. Z. Weng, W. Ou, J. Huang, M. Singh, M. Wang, Y. Zhu, R. Kumar and S. Lin, *Nat. Sci. Sleep*, 2021, **13**, 103–111.
329. M. F. Borisenkov, E. V. Perminova and A. L. Kosova, *Chronobiol. Int.*, 2010, **27**(6), 1259–1270.
330. A. Adan and V. Natale, *Chronobiol. Int.*, 2002, **19**(4), 709–720.
331. D. Fischer, D. A. Lombardi, H. Marucci-Wellman and T. Roenneberg, *PLoS One*, 2017, **12**(6), e0178782.
332. C. Randler and J. Engelke, *Chronobiol. Int.*, 2019, **36**(7), 888–905.

333. K. V. Allebrandt, M. Teder-Laving, T. Kantermann, A. Peters, H. Campbell and I. Rudan, *et al.*, *Chronobiol. Int.*, 2014, **31**, 731–740.
334. M. J. Leone, M. Sigman and D. A. Golombek, *Curr. Biol.*, 2020, **30**(16), R930–R931.
335. P. M. Wong, B. P. Hasler, T. W. Kamarck, M. F. Muldoon and S. B. Manuck, *J. Clin. Endocrinol. Metab.*, 2015, **100**(12), 4612–4620.
336. E. A. Lucassen, X. Zhao and K. I. Rother, *et al.*, *PLoS One*, 2013, **8**(3), e56519.
337. V. Aho, H. M. Ollila and E. Kronholm, *et al.*, *Sci. Rep.*, 2016, **6**, 24828.
338. A. M. Weljie, P. Meerlo and N. Goel, *et al.*, *Proc. Natl. Acad. Sci. U. S. A.*, 2015, **112**(8), 2569–2574.
339. E. Maury, *Int. J. Mol. Sci.*, 2019, **20**(7), 1597.
340. E. Poggiogalle, H. Jamshed and C. M. Peterson, *Metabolism*, 2018, **84**, 11–27.
341. F. Sciarra, E. Franceschini and F. Campolo, *et al.*, *Int. J. Mol. Sci.*, 2020, **21**(11), 3943.
342. D. H. Loh, J. M. Dragich, T. Kudo, A. M. Schroeder, T. J. Nakamura, J. A. Waschek, G. D. Block and C. S. Colwell, *J. Biol. Rhythms*, 2011, **26**, 200–209.
343. C. Dibner, U. Schibler and U. Albrecht, *Annu. Rev. Physiol.*, 2010, **72**, 517–549.
344. J. A. Mohawk, C. B. Green and J. S. Takahashi, *Annu. Rev. Neurosci.*, 2012, **35**, 445–462.
345. F. Damiola, N. Le Minh, N. Preitner, B. Kornmann, F. Fleury-Olela and U. Schibler, *Genes Dev.*, 2000, **14**, 2950–2961.
346. K. A. Stokkan, S. Yamazaki, H. Tei, Y. Sakaki and M. Menaker, *Science*, 2001, **291**, 490–493.
347. P. Crosby, R. Hamnett, M. Putker, N. P. Hoyle, M. Reed, C. J. Karam, E. S. Maywood, A. Stangherlin, J. E. Chesham and E. A. Hayter, *et al.*, *Cell*, 2019, **177**, 896–909.e820.
348. O. Froy and R. Miskin, *Aging*, 2010, **2**(1), 7–27.
349. C. Escobar, R. Salgado-Delgado, E. Gonzalez-Guerra, A. Tapia Osorio, M. Angeles-Castellanos and R. M. Buijs, *Sleep Disord.*, 2011, **2011**, 964510.
350. U. Albrecht, *Neuron*, 2012, **74**(2), 246–260.
351. B. A. Kent, *Front. Aging Neurosci.*, 2014, **6**, 234.
352. R. E. Mistlberger, *Physiol. Behav.*, 2011, **104**(4), 535–545.
353. J. Song, S. Chu and X. Fang, *et al.*, *Transl. Neurosci.*, 2020, **11**(1), 341–355.
354. S. A. Bae and I. P. Androulakis, *Am. J. Physiol.: Regul., Integr. Comp. Physiol.*, 2019, **317**(1), R98–R107.
355. O. Froy, N. Chapnik and R. Miskin, *Am. J. Physiol.: Endocrinol. Metab.*, 2006, **291**(5), E1017–E1024.
356. S. A. Patel, N. Velingkaar, K. Makwana, A. Chaudhari and R. Kondratov, *Sci. Rep.*, 2016, **6**, 25970.
357. A. Ribas-Latre and K. Eckel-Mahan, *Mol. Metab.*, 2016, **5**(3), 133–152.

358. M. Nováková, M. Sládek and A. Sumová, *J. Biol. Rhythms*, 2010, **25**(5), 350–360.
359. M. Nováková, L. Polidarová, M. Sládek and A. Sumová, *Neuroscience*, 2011, **197**, 65–71.
360. I. Heyde and H. Oster, *Sci. Rep.*, 2019, **9**(1), 20114.
361. J. Husse, A. Leliavski, A. H. Tsang, H. Oster and G. Eichele, *FASEB J.*, 2014, **28**(11), 4950–4960.
362. P. Pevet and E. Challet, *J. Physiol.*, 2011, **105**(4–6), 170–182.
363. M. R. Castillo, K. J. Hochstetler, R. J. Tavernier Jr, D. M. Greene and A. Bult-Ito, *Am. J. Physiol.: Regul., Integr. Comp. Physiol.*, 2004, **287**, R551–R555.
364. E. W. Lamont, L. R. Diaz, J. Barry-Shaw, J. Stewart and S. Amir, *Neuroscience*, 2005, **132**, 245–248.
365. R. Hara, K. Wan, H. Wakamatsu, R. Aida, T. Moriya, M. Akiyama and S. Shibata, *Genes Cells*, 2001, **6**, 269–278.
366. P. L. Haynes, D. Gengler and M. Kelly, *Curr. Psychiatry Rep.*, 2016, **18**(8), 75.
367. D. Minors, G. Atkinson, N. Bent, P. Rabbitt and J. Waterhouse, *Age Ageing*, 1998, **27**(1), 67–72.
368. A. Zisberg, N. Gur-Yaish and T. Shochat, *Sleep*, 2010, **33**(4), 509–514.
369. A. A. de Souza Teixeira, F. S. Lira and J. C. Rosa-Neto, *Life Sci.*, 2020, **261**, 118453.
370. D. Weinert and J. Waterhouse, *Physiol. Behav.*, 2007, **90**(2–3), 246–256.
371. D. Weinert, K. Schöttner, L. Müller and A. Wienke, *Chronobiol. Int.*, 2016, **33**, 1161–1170.
372. A. Martinez-Nicolas, J. A. Madrid and F. J. García, *et al.*, *Sci. Rep.*, 2018, **8**(1), 15027.
373. T. L. Leise, M. E. Harrington and P. C. Molyneux, *et al.*, *Age*, 2013, **35**(6), 2137–2152.
374. A. Power, A. T. Hughes, R. E. Samuels and H. D. Piggins, *J. Biol. Rhythms*, 2010, **25**(4), 235–246.
375. F. van Oosterhout, E. A. Lucassen, T. Houben, H. T. vanderLeest, M. C. Antle and J. H. Meijer, *PLoS One*, 2012, **7**(6), e39693.
376. M. Takahashi, A. Haraguchi and Y. Tahara, *et al.*, *Sci. Rep.*, 2017, **7**, 39771.
377. S. Dupont Rocher, N. Bessot, B. Sesboüé, J. Bulla, D. J. Davenne and A. Gerontol, *Biol. Sci. Med. Sci.*, 2016, **71**(6), 817–822.

Part IV: Methodological Aspects

CHAPTER 22

Chronobiological Study Designs

DIETMAR WEINERT*[a] AND DENIS GUBIN[b]

[a] Institute for Biology/Zoology, Martin Luther University, Halle-Wittenberg, Germany; [b] Department of Biology, Medical University, 625023 Tyumen, Russia
*Email: weinert@zoologie.uni-halle.de

22.1 Introduction

The aim of chronobiological studies is the characterization of time-dependent changes of physiological and behavioural processes. Therefore, it is necessary to compile and analyse time series of adequate measures. These time series are composed of trend(s), rhythmic component(s) and noise, with rhythmic component(s) being the most relevant for chronobiological research. In the present chapter, the focus is placed on 24-hour rhythms and their circadian component. Besides studies on humans, experiments on laboratory animals are included, as some experiments cannot be performed on humans. Focus is also placed on general biological principles.

The design of chronobiological studies depends on their specific focus. One aim of basic research is to characterize the endogenous rhythms of physiological and/or behavioural processes and their interaction with the individual's lifestyle and environmental factors. Overt rhythms consist of endogenous components arising from the body clock or the internal pacemaker and exogenous components, which mask the endogenous rhythm. Therefore, to characterize the endogenous rhythm, particularly its amplitude, phase and period length, masking effects must be eliminated. This can

Chemical Biology No. 23
Chronobiology and Chronomedicine: From Molecular and Cellular Mechanisms to Whole Body Interdigitating Networks
Edited by Germaine Cornelissen and Tsuyoshi Hirota

Published by the Royal Society of Chemistry, www.rsc.org

be done at the source, *i.e.*, while collecting the data, using an appropriate study design, for example, a constant routine protocol. In some cases, it is more convenient to perform studies under everyday life conditions (field studies). Then, the masking effects must be mathematically removed afterwards using so-called purification methods.[1,2]

Besides its importance in basic research, knowledge of biological rhythms is of practical relevance, particularly for medicine. In clinical studies, assessment of circadian rhythms serves several functions. The most important of these functions is to determine the extent of circadian disruption or the misalignment of circadian rhythms with the external environment and the consequences for health and wellbeing (chronodiagnosis). Another practical reason is to precisely time circadian-based treatments such as drug administration or bright light exposure (chronotherapy).

Clinical studies can be performed in the laboratory under controlled environmental conditions to minimize or at least standardize exogenous masking effects. In order to reduce inter-individual variability, homogeneous and well-synchronized samples are used. Sometimes, it is more reasonable to perform clinical studies under every day life conditions, to get an idea of rhythmic changes in normal life. This might be important particularly for personalized chronotherapy. Circadian disruption might also result from different causes, not necessarily originating from the suprachiasmatic nuclei (SCN) or only occurring at an advanced stage. For example, shift-work, transcontinental flights (jet lag), insufficient light during daytime, light at night and acute or chronic diseases may have adverse consequences for the internal and external circadian time order.

22.2 Data Collection: Longitudinal, Transverse, and Hybrid Sampling

To characterize biological rhythms, *i.e.*, their amplitude, phase and/or period length, data must be collected at an appropriate interval over one or several cycles. This is best done at regular, *i.e.*, equally spaced (equidistant) intervals, but can also be achieved based on irregular intervals.

The sampling rate depends on the expected period or on the period that is of immediate interest. Three time points over a single cycle of a sinusoidal rhythm with a known period fully determine its parameters (mean, amplitude, and phase) but are insufficient for rhythm detection and the derivation of uncertainty of parameter estimation. In general, having $2n+1$ equally spaced time points over n cycles is the minimal sampling requirement for characterizing its parameters. However, it is necessary to sample more than three time points per cycle of an assumed period in order to enable statistical testing and the derivation of uncertainty in the estimated parameters (*e.g.*, of a cosine fit). A minimum of six samples per cycle of the highest-order harmonic term qualifying the rhythm's waveform is recommended. The larger the error associated with the measurement and the needed precision in the outcome variable, the larger the sample size should be in order to increase the signal/noise ratio by either using a shorter sampling interval or by increasing the number of replications.

Data can be collected through **longitudinal sampling**, *i.e.*, from an individual over a certain time span. In humans, this is easily achivable, particularly for motor activity, body temperature or cardiovascular variables, as various devices for ambulatory monitoring are available. Animals can be implanted with sensors to monitor their activity, core body temperature, the ECG or blood pressure. In metabolic chambers, additional variables can be measured. Longitudinal studies can also be performed for populations in clinical trials or as part of epidemiological investigations, or for groups of animals when their general activity, food and/or water consumption are recorded.

The length of the time series depends on the longest period to be studied. At least twice the length of this period is necessary in order to show the reproducibility of the pattern, but the frequency resolution increases with the number of cycles covered by the time series. Many circadian studies performed under synchronized conditions sample data over 3 days, but the optimal observation time depends on the signal-to-noise ratio and on the purpose of the experiment. Certain variables like body temperature may be rather stable, but others like blood pressure may be highly variable. If the difference between free days and workdays is of interest or if the weekly variation may add information, then it is advisable to monitor over at least 7 days. It must also be considered that biological variables do not change instantaneously but require some time to undergo a measurable change. In our studies on the effect of daily melatonin dosing on age-dependent changes in circadian body temperature and cardiovascular rhythms, a duration of two weeks for melatonin administration proved sufficient.[3,4] In order to observe the effect of melatonin on ganglion cell function in glaucoma patients, however, an observation period of 90 days was necessary.[5]

In the presence of non-stationarities, *e.g.*, of day-by-day variability in mean, phase and/or amplitude within an individual record, a shorter interval for analysis may be required. In a study comparing body temperature in elderly people to middle-aged individuals,[3] the significantly higher inter-daily phase variability in the elderly population was the major reason for the smaller circadian amplitude when it was estimated over the whole 7-day record.

By choice or necessity, **transverse sampling** can be used. This might be the case in studies on animals when investigating, for example, clock gene expression in the SCN or the concentration of metabolites in a given organ, *i.e.*, if animals can be used only once.[6–8] At each time point, different animals are taken. The number of individuals per timepoint depends on the variability in the sampled variable. The sample should be homogeneous in terms of genetic background, sex, age, *etc.*, and the animals should be synchronized prior to data collection. Adult animals can easily be synchronized using a strong artificial light–dark regimen, while keeping other environmental factors constant. The discrepancy between biological and chronological age in old animals can be problematic. We resolved this issue in our study of age-dependent changes in clock gene expression[8] by selecting old mice of similar biological age on the basis of specific characteristics of their 24-hour activity rhythm (high amplitude but decreased inter-daily stability and a phase advance as compared to younger animals).

Hybrid sampling is a combination of transverse and longitudinal sampling where a given variable is measured repeatedly at intervals from several individuals. If individuals are randomly picked from a homogeneous population, hybrid sampling has the advantage that it allows inferences to be generalized for that population. It is again important to make sure that study participants are well synchronized; phase differences must be considered when summarizing the data for their mean daily pattern. This approach was used to investigate the 24-hour body temperature rhythm of newborns on the second postnatal day. While we could show that all newborns had a pronounced circadian body temperature rhythm, individual acrophases were distributed over the whole 24-hour day, due to the immaturity of photic entrainment mechanisms. Summarizing the data across all babies, the average 24-hour pattern exhibited a very small amplitude, close to zero.[9] Similar results were observed in elderly patients who lived in a veteran house on a self-selected sleep–wake regimen chosen for their own convenience, without any restrictions, except for the scheduled meal times.[3] In a hybrid study of patients suffering from stable or advanced primary open-angle glaucoma, differences in the mean 24-hour patterns of body temperature and intraocular pressure in terms of disease progression and a restorative effect of melatonin administration could be shown.[5,10]

A special category of chronobiological study designs involves **N-of-1 experiments**, clinical trials involving single patients, which are essential for personalized chronotherapy. A widely used procedure is the A^1BA^2 design, where the patient is studied before the treatment (reference) and then again during the treatment and finally when the treatment is terminated. This protocol can be used to determine optimal treatments for diverse disease conditions such as osteoarthritis, chronic neuropathic pain, and attention deficit hyperactivity disorder.[11] For chronobiological/chronotherapeutic studies, another experimental design (A^1-B^1-B^2-B^3-. . .-B^n-A^2) is appropriate, where outcomes from the same treatment administered at different circadian stages are compared.[12] Such studies rely on a marker variable that can be monitored around the clock for several days. For instance, a fixed dose of an anti-hypertensive drug or a combination of drugs is administered at a given circadian stage (*e.g.*, upon awakening); then the timing of its administration is progressively changed to one of several other circadian stages (*e.g.*, delayed by 3 hours until it is taken around bedtime). Monitoring starts before treatment and continues until all treatment times have been implemented. An important feature of such a design is the interval during which treatment is administered at a given circadian stage, and thus the interval between consecutive treatment times, which can be as short as one day or as long as one month or longer. The response to losartan/hydrochlorothiazide was investigated by Watanabe *et al.*[13] who administered the treatment to each of 30 patients at six different times, which progressively varied from the time of awakening to bedtime, each stage lasting for at least one month. At the end of each stage, blood pressure was measured around the clock with an ambulatory blood pressure device for 7 days.

While conventional N-of-1 trials usually compare two treatments, chronobiological treatments focus on optimizing the timing of a specific treatment. However, it is not always feasible to apply the usual randomization and blinding principles. With respect to treatment times, most protocols systematically delay the treatment time by 3 to 4 hours. The timing of the first dose, however, can be randomized. Blinding normally necessitates taking five placebo pills for one active drug pill when six treatment times are tested. This can impose a significant burden on the patient, notably when treatment continues over long periods; in such cases, concern for any effect from the circannual variation may be warranted. For further details concerning experimental N-of-1 designs, applications, and statistical models, refer to Kravitz and Duan 2014.[14]

22.3 Marker Rhythms

Good marker rhythms are physiological variables known to display a prominent circadian rhythm, so that the individual circadian phase can be estimated, which is essential for personalized chronotherapy. These rhythms are also part of the external and internal temporal order, which comprises physiological phase relationships between different rhythmic functions and an appropriate phasing with respect to environmental cycles. Therefore, it might be necessary to record 24-hour rhythms of two or more variables. The appearance of reproducible and stable circadian rhythms of large amplitude with a characteristic phasing is believed to guarantee an optimal functioning of biological systems, including humans, with maximal efficiency, performance and welfare.[15] Accordingly, alterations in these rhythms as compared to clinically healthy people have been used for diagnostic purposes. Circadian disruption might indicate an elevated risk of developing a disease, the presence of an established disease, and even its severity. One must consider, however, that any aberration in the temporal order may have different causes. The source of aberration can be in the central pacemaker, *i.e.*, the SCN itself, in the mechanism of external entrainment, or downstream from the SCN. Examples include deviating endogenous period lengths in patients suffering from advanced or delayed sleep phase syndrome, inadequate photic entrainment in glaucoma, or the malfunctioning of peripheral thermoregulation in diabetes.

Criteria for selecting physiological or behavioural variables as reference differ depending on the aim of the study:

- Their overt rhythm does reflect the endogenous pacemaker.
- The variables are easy to measure and record over long time spans.
- They are of clinical relevance.

Circadian rhythms in mammals are controlled by a central pacemaker, the SCN, which are located in the anterior hypothalamus. They generate self-sustained oscillations and control peripheral rhythms by means of neural and humoral signalling pathways. However, a measured rhythm does not directly reflect the output of this pacemaker. Exogenous factors related to the individual's lifestyle and environment have an effect on the endogenous clock-driven component, which is called masking.[16,17]

The relative size of the exogenous and endogenous components is specific to each variable. Some variables, such as the heart rate and blood pressure have a much larger exogenous component than endogenous component. Accordingly, measurements of these variables normally disclose only little information about the circadian pacemaker. Even though the 24-hour rhythm of heart rate adjusts readily to a shift in the schedule, this fast adjustment only reflects the change in the exogenous rhythmic component. The extent of adjustment of the endogenous rhythm cannot be directly inferred. In contrast, the secretion of melatonin and cortisol into the plasma have a larger endogenous component, so that the circadian rhythms of these hormones reflect the output of the body clock more accurately. Core temperature has endogenous and exogenous components of similar magnitude.[2,18]

22.3.1 Melatonin

The 24-hour melatonin rhythm is considered to be the most robust phase marker, and it is the most commonly used in humans since it best represents the endogenous biological clock.[19–21] Only few exogenous factors are known to affect melatonin concentrations and thus mask its circadian rhythm.[22] Accordingly, the 24-hour melatonin rhythm is a good indicator of the phase of the endogenous oscillator, and to estimate the period length, a minimum of at least two cycles is needed. Besides, it can be of diagnostic value. Any alterations may have consequences for health and welfare, especially considering that the melatonin rhythm is an important internal Zeitgeber.

Melatonin can be assessed by hourly or half-hourly sampling of plasma or saliva. Even though the hormonal concentration is lower in saliva than plasma, with a peak value of merely about 30%, the 24-hour patterns are similar in phase. The urinary concentration of 6-sulphatoxymelatonin (aMT6s), the main metabolite of melatonin, can also be measured, albeit using longer sampling intervals. Since multiple methods exist to measure melatonin, a workgroup was formed to reach a consensus for collecting and analysing melatonin data, thereby helping clinicians and researchers to choose the most appropriate method for their particular needs.[23]

As melatonin synthesis is acutely suppressed by light, and since light exposure can also induce phase responses that differ depending on the circadian stage of exposure,[24,25] collecting plasma or saliva samples must be performed under dim light of <30 lux (some investigators even recommend <10 lux).

Phillips *et al.* investigated the sensitivity to evening light and found huge inter-individual differences.[26] In most sensitive individuals, the effective dose for 50% suppression of melatonin was 6 lux, while in least sensitive individuals, it was 350 lux. The spectral composition of light must also be considered. The suppression of pineal melatonin is mediated by intrinsically photosensitive retinal ganglion cells (ipRGCs) containing melanopsin as photopigment with a maximal sensitivity in the 'blue' region of the visible electromagnetic spectrum.[27]

Posture may also affect the concentration of circulating melatonin; particularly standing or more precisely moving from a supine to a standing position does increase plasma and salivary melatonin concentrations. As this effect can be reversed within 10 minutes, it is recommended to minimize activity and postural changes just prior to and during collection of plasma or saliva samples.[28]

Blood is usually sampled by means of intravenous catheters, allowing frequent sampling over a long span. Catheters should be inserted at least 2 hours before sampling to avoid an increase in adrenergic levels, which may affect the melatonin concentration. As blood can be sampled at frequent intervals throughout the whole day without major disturbance in sleep, plasma melatonin provides highly valuable information. However, this method is invasive, expensive, and requires trained medical personnel.

A more practical and reliable method is saliva sampling,[29] which can be used in field and laboratory studies as well as in clinical routine. Saliva can also be sampled at short intervals. This can be realized at home, although study participants should carefully follow instructions to avoid contamination of samples with food particles or blood. In order to increase the saliva flow rate, it is recommended to gently chew on paraffin film. As melatonin concentrations in saliva are quite stable, samples can be stored at home until they are assayed in the laboratory.

A full 24-hour pattern provides a number of useful markers, such as the times of secretion onset and offset, the duration of secretion, the midpoint and termination of melatonin synthesis (SynOff), and the 24-hour acrophase, among others.[31,33] In practice, however, melatonin is often measured only during several hours before sleep onset under dim light conditions. This way, the so-called "dim light melatonin onset" (DLMO) can be estimated, defining the time at which melatonin secretion starts to rise. The onset times determined in plasma or saliva correlated significantly.[29]

Melatonin onset is defined as the first sample that consistently exceeds a certain threshold. For blood samples, the threshold is 10 pg ml^{-1}. Because of the lower concentrations in saliva, a threshold of 3 pg ml^{-1} is used. Some low-melatonin producers may never exceed a concentration of 10 pg ml^{-1}, *i.e.*, a lower threshold must then be used. It may even be preferable in such cases to use the melatonin synthesis offset (SynOff) because it is not confounded by amplitude and because it is temporally closer to the endogenous mechanism signalling the offset of production.[21] Another objective criterion for determining the onset of secretion is the first value that rises by 2 SD (standard deviations) above the mean of three daytime reference values. Danilenko *et al.*[30] developed a computer-based algorithm and called it the hockey-stick method. It fits the evening melatonin profile using a linear-parabolic function. The point when it transitions from a straight line to a parabola is considered a reliable estimate of DLMO, which does not depend on a threshold value.

An appropriate method for field and laboratory studies is urine sampling and the analysis of aMT6s, the major metabolite of melatonin. Urinary melatonin metabolite excretion has been shown to provide a reliable index of pineal melatonin production.[23] Although the sampling frequency is limited by the ability to void and the compliance with collection, the estimation of a6MTs concentrations in urine is a suitable way to estimate global circadian timing and the amount of melatonin production. Naturally, phase detection is less precise than with more frequent sampling. A phase delay with respect to plasma concentration must also be considered. Benloucif *et al.* obtained the highest aMT6s concentrations between 4:00 and 8:00 h, whereas the midpoint of the plasma melatonin level occurred between 3:00 and 4:00 h.[23] To prevent sleep disruption, overnight aMT6s excretion values can be calculated from the first morning void.[31]

The investigation of the 24-hour melatonin rhythm by measuring its metabolite a6MTs in urine is very convenient for animal studies. This was shown, for example, in studies on Djungarian hamsters, which revealed a clear circadian rhythm, with a delay of 3–4 hours with respect to peak values in pineal melatonin.[32,33] Urine collection over a certain interval may contribute to this phase delay, although on a limited scale. Our studies on Djungarian hamsters revealed clear 24-hour a6MTs profiles reflecting the different circadian phenotypes of investigated animals.[6]

We used a protocol that was first applied in Steinlechner's group in Hannover.[32] Adult hamsters were kept under standardized housing conditions (LD14:10 with "light-on" from 4:00 to 18:00 Central European Time, CET, constant temperature, *ad libitum* food and water) until stable activity patterns were observed. On the day of urine collection, animals were transferred singly to metabolic cages to collect urine every 3 hours over 27 hours. This interval was chosen to give adequate phase resolution, and also to enable sufficient urine volume to be collected for each sample. Food and a mixture of apple juice and water (dilution: 1 : 3) were provided *ad libitum*. The mixture of apple juice and water helped stimulate water consumption and increase urine excretion. Metabolic cages were equipped with a funnel system that allowed separation of urine and faeces. Concentrations of aMT6s were determined using a melatonin sulphate enzyme-linked immunosorbent assay (ELISA) kit. Concentrations were expressed relative to the amount of creatinine in the urine samples, a method necessary to overcome technical difficulties with urine collection.[34]

Melatonin patterns over 24 hours and the DLMO provide the presumed best estimate of the body clock's phase. The DLMO or the 24-hour rhythm of aMT6s is recommended to confirm the diagnosis of various circadian-rhythm sleep–wake disorders, such as delayed or advanced sleep–wake phase disorder and non-24-hour sleep–wake rhythm disorder. Determining the circadian timing of individual patients is also essential to optimize the treatment of these conditions with melatonin or light therapy.[35] The circadian amplitude can also be informative. The circadian amplitude of log-transformed serum melatonin and urinary aMT6s values differed between cancer patients and clinically healthy people[36] and between clinically healthy women at high *vs.* low breast cancer risk,[37] respectively. Remarkably, the MESOR, *i.e.*, the mean value of the approximated cosine function,[38] was not different.

22.3.2 Body Temperature

Core body temperature is also frequently used as a marker of the circadian pacemaker. In studies on animals, body temperature is usually measured by means of implanted transmitters.[39] They allow simultaneous recording of motor activity that is particularly beneficial for unmasking the temperature data. In humans, rectal probes are the preferred method to measure core body temperature. Thermistor pills that are swallowed to pass through the gastrointestinal tract can also be used.[40] Both methods yield similar estimates of acrophases and amplitudes,[41] but the 24-hour mean values are higher in the gut. In contrast, the parallelism between rectal and insulated axillary temperatures is less reliable. The same applies to insulated skin temperature measured at the lateral chest wall.[42] Results showed that gut temperature is a viable alternative to rectal probes, which are not considered to be "user-friendly", particularly for individuals living their usual daily routine. A drawback of gut pills is the short recording time, which depends on gastrointestinal transit.

The characteristics of the circadian rhythm in core body temperature (MESOR, phase, amplitude) are assumed to reflect the corresponding parameters of the endogenous pacemaker. Commonly, the nadir of the rhythm is used as a marker of circadian phase. It occurs during the latter part of the habitual sleep span, *i.e.*, in persons with conventional sleep times around 3–4 am. However, the measured rhythm is a mixture of two components, an endogenous rhythm that reflects the body clock, and a superimposed exogenous ("masking") component arising from a direct effect of lifestyle, mainly the rest activity/sleep cycle.[39,43] Core temperature increases with activity and decreases with sleep. Postural changes, meals, and environmental factors such as ambient temperature, sound, humidity, and bright light also affect body temperature.[44,45] Accordingly, using raw temperature data may yield misleading information about the phase and amplitude of circadian rhythms. Therefore, in order to get a better estimate of the endogenous rhythm, the masking effects are best removed or minimized. This can be achieved by means of constant routine and forced desynchronization protocols. Alternatively, body temperature can be monitored under field conditions in the presence of masking factors, which are removed afterwards mathematically, using the so-called purification methods (see below).

Besides masking factors, modifications of the 24-hour temperature rhythm might be caused by mechanisms downstream from the SCN, as shown, for example, in diabetic patients. Interestingly, changes in MESOR, amplitude and phase were observed already in the pre-diabetic state.[46] In glaucoma patients, the body temperature rhythm was phase-delayed because of a diminished photic synchronization as a consequence of retinal ganglion cell loss. Since the extent of this delay increased with the progression of the disease, it could be used as an indicator of disease severity.[47] These examples show that changes in the 24-hour body temperature rhythm can be of diagnostic value. Moreover, as the temperature rhythm serves as an internal Zeitgeber, aberrations may have the same adverse consequences as described above for melatonin.

In recent years, wrist temperature has often been used as an index of the human circadian system.[48,49] It is a cheap, non-invasive method, which can easily be used to record skin temperature over long time spans. Wireless data loggers (Thermochron iButton, Maxim, San Jose, CA, USA), attached to the inside of the wrist of the non-dominant hand can be used. Other wrist-worn devices are also available that record not only skin temperature but also activity, sleep and environmental temperature (*e.g.*, "Act Trust"; CAT; Condor Instruments, São Paulo, Brazil).

The 24-hour wrist temperature rhythm has a strong endogenous component. It largely reflects vasomotor tone (*i.e.*, vasodilation/vasocontraction of the peripheral blood vessels), which is regulated by the SCN and is related to sleep tendency.[50] Core body temperature is also at least partly regulated by changes in extremities skin blood flow. The decrease in rectal temperature during night time is preceded by an increase in skin blood flow, and its increase in the morning is accompanied by a decrease in skin blood flow, with an average phase difference of 4 hours.[51] According to Sarabia *et al.*,[48] humans' wrist temperature rises sharply during night time, reaching a stable level during sleep and drops in the morning, remaining low during the daytime (*i.e.*, the phase difference with rectal temperature is considerable).

The masking effect of environmental temperature is substantial. To minimize its influence on the measured wrist temperatures, studies are best performed when environmental temperatures are moderate, between 8 °C and 17 °C.[49] Moreover, physical activity, body position, microclimate in bed, food intake and/or taking a nap are factors masking the wrist temperature rhythm. In university students, several peaks due to class attendance appeared in the morning. During holiday weeks, a secondary peak was observed in the afternoon caused probably by an after lunch nap.[48] Nonetheless, when considering that the wrist temperature rhythm is influenced by both endogenous and exogenous factors, it may provide reliable information about the pacemaker's function and prove beneficial for chronodiagnosis and chronotherapy.[49,52,53]

22.3.3 Cortisol

Glucocorticoid secretion is also directly regulated by the circadian pacemaker and can therefore be used in principle as a reference rhythm.[54] Blood concentrations display a circadian rhythm peaking at the onset of activity. In humans, after a sharp rise in the second half of the night, cortisol concentrations reach a maximum in the early morning, a few hours before and just after waking up, declining throughout the day and reaching a nadir within 2 hours after sleep onset. At night, cortisol concentrations are low over several hours. Adrenal hormonal secretion also shows a pulsatile pattern with a primary period of about 60 minutes. This pattern is reflected in the blood as an ultradian rhythm, which has been documented in numerous mammalian species, including humans.[55,56] To detect and characterize the pulsatile behaviour of cortisol, which is of considerable physiological and clinical relevance (see below), dense sampling is needed.

Cortisol concentrations can be measured in blood samples and other body fluids such as saliva, urine and sweat.[57–59] Several factors are known to influence cortisol secretion. Stress does cause bursts of secretory activity.[60] Light raises the morning peak and can cause phase shifts.[61] Sleep onset and deep sleep reduce cortisol secretion, whereas sleep loss, light sleep and awakening stimulate it.[62] The consumption of high-protein meals can also trigger secretory episodes.[63] Accordingly, when investigating the cortisol rhythm, all these factors must be considered, complicating the usage of the cortisol rhythm as a reference for the endogenous pacemaker.

Alterations in the amplitude, phase, and/or waveform of the 24-hour cortisol rhythm can be indicative of illness and survival time, as shown in patients suffering from metastatic colorectal, breast or ovarian cancer.[57,58,64] Cortisol values obtained at appropriate daytimes can also be used to identify disease conditions such as Addison's disease (adrenal hypofunction characterized by low values in the morning) or Cushing's syndrome (adrenal hyperfunction, characterized by very high values at night).

Ultradian cortisol rhythms are important for inter- and intra-cellular signalling through frequency encoding, a common mechanism in biological systems that is particularly important for glucocorticoid-sensitive neural processes.[65] The ultradian pattern changes depending on the pathophysiological state, for example, as a consequence of chronic inflammatory processes.[66,67] Moreover, altered ultradian patterns were linked to psychotic and depressive states.[68,69]

Klerman *et al.* compared the three marker rhythms discussed above. The melatonin rhythm turned out to be the most reliable phase marker for the circadian pacemaker with the lowest variability.[20] DLMO is most widely used as it is easy to determine and interpret. Masking effects are relatively small and can be controlled quite simply. Besides the three mentioned rhythms, further markers do exist that can be measured easily, but may serve only as additional tools.

22.3.4 Physical Activity

Actigraphy is a convenient, non-invasive method to measure the rest/activity cycle in humans or the activity pattern of animals over a long time at short sampling intervals. In humans, activity can be monitored easily under field conditions over many days by means of non-invasive devices worn on the non-dominant wrist.[70] Bellone *et al.* compared the advantages and weaknesses of different portable actigraphy devices in terms of sleep and circadian research.[71] El Allali *et al.* used smartphones as an inexpensive alternative to record motor activity in humans and large mammals and validated activity recordings by means of an actiwatch. The estimated 24-hour rhythms were shown to be similar with both devices.[72]

Actigraphy has been used both in field research and in clinical settings, as it is a simple, inexpensive, and rather accurate method.[73–75] The estimated activity/sleep–wake rhythm is less suitable as a marker of the underlying circadian clock as it is susceptible to masking. However, other rhythm characteristics can be used as indicators of physical fitness, health status, and quality of life of human beings. The inter-daily rhythm stability and the median level of daytime activity were found to be negatively correlated with cognitive decline and depression in elderly women.[74] Hoopes *et al.* used the inter-daily stability and intra-daily variability as biomarkers of cardiometabolic health.[76]

In laboratory animals, motor activity can be monitored using the same transmitters that are implanted to record the core body temperature. Much cheaper, and at the same time most reliable are passive infrared motion detectors. They are mounted above the cage roof in such a way that they detect animal motions in any part of the cage. Both methods record general activity, not just locomotion like infrared barriers, for example.

Running wheels are also quite popular as they produce unusually robust activity rhythms with a clear light–dark distribution of motor activity. The activity onset is tightly coupled to the beginning of the dark span.[77] One has to consider, however, that this is at least partly a consequence of strong photic masking. Thus, activity onset does not necessarily reflect the phase of the endogenous rhythm, particularly following shifts in the light–dark cycle. Running wheels are also used voluntarily and intensively by rodents, thus enhancing the activity level considerably.[78–80] This strengthens the feedback effect onto the central pacemaker of the circadian system. Neurophysiological properties of the SCN and clock gene expression are especially affected.[81–83] As a consequence, the period length of the free-running rhythm[84–87] and the photic phase responses do change.[87–90]

Activity counts can be displayed as double-plotted actograms, which allow a visual inspection of the data. A convenient phase marker is the onset of activity, which characterizes the strength of photic entrainment. In older mice, a phase advance and a decreased phase stability of the activity onset were observed together with an increased inter-daily rhythm variability.[91] These changes have been used to characterize the biological age to help select animals and standardize experimental procedures.[8,92] Interestingly, similar changes were observed in the sleep–wake rhythm of aged people, particularly those suffering from Alzheimer's disease.[74,93]

Besides a visual estimation on screen, activity onsets can be determined mathematically. Ruby *et al.* define activity onset as the first 10-minute interval in a circadian cycle when the number of activity bouts increases above the daily mean activity level and is sustained at or above that level for ≥30 minutes. Activity offset is the time when the number of activity bouts remains below the daily mean for ≥30 minutes.[94] A phase estimation using a one-component cosinor is not appropriate because the activity pattern is far from being sinusoidal. Estimates of phases of maximal and minimal activity can be obtained, however, by fitting a multiple-component model using cosinor. While they may not correspond to the times of activity onset and/or offset, they can be informative, depending on the aim of the study.

22.3.5 Sleep–Wake Pattern

Sleep parameters are less suitable as references because sleep is controlled both by the circadian pacemaker and by homeostatic mechanisms.[95–97] Social factors, the working schedule, individuals′ lifestyle, *etc.* also have an impact on the sleep–wake pattern. However, a few sleep parameters (sleep onset, sleep midpoint, and wake time recorded on sleep logs and verified by actigraphy) are still useful as phase markers, with sleep midpoint being the most reliable one. It correlates best with DLMO,[98,99] especially in "free sleepers" with a self-selected sleep pattern.[100] Assessing the sleep–wake pattern during both work/school and free days, using a sleep diary, or wrist actigraphy together with a sleep diary, is considered a reliable method to assist in the diagnosis of circadian sleep–wake disorders like delayed or advanced sleep phase syndrome and non-24-hour or irregular sleep–wake rhythm (for references, see ref. 70).

22.3.6 Chronotype

The most widely used questionnaire to study the individual timing of activity and sleep was developed by Horne and Ostberg in 1976.[101] Their "Morningness Eveningness Questionnaire" (MEQ) differentiates between morning and evening types and thus reflects some aspects of the circadian rhythm, the circadian chronotype. Later, several questionnaires were designed that were aimed at more specific groups or situations, such as the Munich Chronotype Questionnaire (MCTQ) developed in Roenneberg's lab.[102] While the MEQ contains mostly questions that characterize average preferences, the MCTQ collects information about the actual timing of daily sleep and activity. The Munich questionnaire also assesses working and free days separately, as sleep timing can be rather different on these days.[103] While the sleep–wake cycle closely aligns with the endogenous rhythm on free days, most people have to get up earlier on workdays, *i.e.*, social and biological times are not in phase. This situation was called “social jet lag” by Roenneberg.[104] The consequence of this misalignment is an increasing sleep deficit, especially in evening types, which challenges the body clock and causes circadian disruption.

22.3.7 Gene Expression

In principle, the phase of the circadian pacemaker can be estimated by investigating the expression of clock genes in the SCN, but this approach has only been used in animals thus far. However, the discovery that clock genes are present in almost all cells of the body suggested the existence of autonomous clocks in the periphery. Normally, these peripheral clocks are synchronized with the SCN.[105] With this discovery, the methods for assessing the circadian phase of an organism, including humans, have expanded. One can measure the circadian expression of core clock and other genes in peripheral cells (*e.g.*, in blood mononuclear cells). This method (as those described earlier in this chapter) is time consuming, however, as it requires serial sampling for 24 hours or more, and results are not available in real time.

Recently, several groups are working on developing techniques and analytical methods that would only require a limited number of samples to accurately determine the circadian phase, thus reducing cost and time. These methods use high-dimensional or multivariate material, such as the transcriptome or metabolome that contain a large number of variables, which vary with the circadian phase (for review see Dijk and Duffy[106] and Crnko *et al.*[107]). Machine learning algorithms[108,109] have reduced the number of genes used for these panels from over 100 “time-indicating genes”[110] to about a dozen genes extracted from blood monocytes, for example.[111] Reliable information about the circadian phase can be obtained from the analysis of the activity of a specific set of genes in only one or two samples of blood, skin or other tissues. The method developed by Wittenbrink *et al.*,[111] known as “BodyTime”, reportedly is as accurate as the current gold standard (DLMO). The accuracy of the phase estimation achieved with the described approach, however, has not yet been sufficiently verified. Besides, it should be considered that peripheral oscillators are not necessarily always synchronized with the central pacemaker, a concern in situations when there is circadian disruption like shift work, sleep disorders or various diseases. Accordingly, these new approaches need careful validation to make sure that all parameters are quantified with sufficient precision before using them in basic research or clinical practice.

22.3.8 Cell and Tissue Cultures

Circadian rhythms of clock gene expression can also be investigated using organ explants, cell or tissue cultures. Such studies were first performed on the SCN. In such scenarios, the electrical activity, *i.e.*, the firing rate of neurons could also be measured. Later, cell lines, tissue slices, and explants from peripheral organs were investigated. Fibroblasts have been employed very often because established cell lines are available commercially. It was shown that circadian clocks are present in all cells and that master and subsidiary oscillators use similar molecular mechanisms. However, in fibroblasts and explants of other tissues, the amplitude decreased rapidly, whereas in the SCN, circadian rhythms could be recorded over months. For this reason, it was assumed that only the SCN rhythm is self-sustained or at least stronger.[112] However, experiments with serum shocks revealed robust autonomous and self-sustained circadian oscillators in established cell lines.[113,114] Obviously, circadian oscillators in cultured cells were synchronized by the serum shock, *i.e.*, reset to the same (or a similar) phase.

Serum contains a wide variety of signalling components, such as hormones, growth factors, metabolites, and cytokines. Glucocorticoid hormones are particularly good candidates as synchronizers. They are secreted in circadian cycles, and their receptors are expressed in most peripheral cell types but not in the SCN. According to Balsalobre *et al.*, a 1-hour treatment with the glucocorticoid dexamethasone does induce circadian gene expression in cultured rat-1 fibroblasts.[115] In contrast to a serum shock, however, it does activate only *Per1* and not *Per2* expression. Interestingly, dexamethasone also induces transient phase changes in circadian gene expression in the liver, kidney, and heart *in vivo* but not in SCN neurons. Forskolin, a molecule stimulating cAMP signalling, is also widely used to synchronize cell cultures.[116,117] These substances probably reset the phase by acting through immediate early transcription factors (IETFs), inducing the expression of the period genes *Per1* and/or *Per2*.[118] It must be mentioned, however, that sustained synchronization of clocks in cultured cells over many days could thus far only be accomplished by simulated body temperature rhythms. The heat shock factor HSF1 plays an important role in this process.[119]

The cyclic expression of relevant genes can be quantified *via* the accumulation of their RNAs (quantitative real time PCR) or protein products.[113,117,120] Another method is the luciferase reporter assay. This method is widely used because it is convenient, relatively inexpensive, and gives quantitative measurements of gene expression instantaneously. Cells are transfected with a luciferase reporter gene whose expression is driven by a rhythmically expressed promoter gene (*e.g.*, *Per1*, *Per2*, and *Bmal1*). Bioluminescence cycles are recorded in real time (real-time bioluminescence monitoring).[114,117]

Besides their relevance in basic research, studies of cell cultures are of fundamental practical relevance. The knowledge of specific clock properties of different tissues/organs is important for optimizing tissue engineering and stem cell therapies.[121] The temporal coordination of circadian rhythms and cell division at the single cell level has significant implications. A compromised temporal order is linked to the pathogenesis of various diseases, including cancer.[122,123] A better understanding of mechanisms that synchronize circadian and cell cycles is also highly relevant to optimize the timing of antiproliferative drug treatment.

Ndikung *et al.*[117] used cell cultures as an *in vitro* test system, which may substitute animal testing of potentially hazardous chemicals or drugs. So far, cell-culture-based systems have showed weaker responses to toxic substances as compared to *in vivo* tests. One reason may be that in conventional cell cultures, single cells exhibit their own circadian rhythm with slightly different period lengths. Accordingly, their phases do not match. The authors investigated human colon and breast epithelial cell lines and synchronized them with dexamethasone or forskolin and treated them thereafter with dioxin (TCDD). The obtained results demonstrated that the reconstitution of a synchronous circadian phase could significantly increase the cellular responses. This approach thus improves the physiological relevance of *in vitro* tests and may facilitate their acceptance.

In basic research, Per-luc transgenic mice and rats are also used. In this case, the animals must be killed, and the organs of interest are dissected, after which bioluminescence is measured in tissue slices. However, animals might be treated before. In Gene Block's laboratory, ontogenetic and age-dependent changes in circadian rhythms in different organs of rats and their response to a light–dark shift were investigated with this method.[124–126] Hamaguchi *et al.* showed that the disturbed circadian bioluminescence rhythm in peripheral tissues (liver, kidney, and submandibular salivary gland) of mice can be stabilized through scheduled feeding or exercise.[127]

22.4 Experimental Approaches to Estimate the Endogenous Component of Overt Rhythms

Particularly in the second half of the last century, a number of experiments were performed on humans under temporal isolation in caves and subterranean bunkers. The aim of these studies was to validate the endogenous origin of biological rhythms, particularly the 24-hour rhythm.[128,129] Similar experiments were performed on animals kept under constant environmental conditions. It could be shown that the period length of the endogenous free-running rhythm under these conditions depended on light intensity, was species-specific and changed with age.[128,130] More recently, experiments on humans kept under temporal and social isolation have used ambulatory monitoring to examine not only the circadian rhythm but also the behaviour of other spectral components.[131,132]

These experiments conducted under temporal isolation enable the estimation of the period length of the endogenous rhythm and some of its characteristics but not the phase. The amplitude is also masked, as the study participants maintain their normal lifestyle, except for not being synchronized with the 24-hour day. In order to characterize the endogenous component of overt 24-hour rhythms, driven by the circadian pacemaker, the masking effect of exogenous factors must be minimized by removing its rhythmic effect.

22.4.1 Constant Routine Protocol

In humans, masking effects can be minimized by keeping an individual's environment and lifestyle constant, *i.e.*, all factors that are known to obscure the circadian variable are kept as constant as possible. Such a protocol, known as "constant routine", was first developed by Mills *et al.*, primarily for dealing with temperature and urinary data.[133] This routine has been further developed by others,[134,135] particularly by Czeisler's group.[136] The main changes have been to extend the duration of the routine to 28–40 hours, to eat identical snacks at a defined time interval, for example, every 3 hours, and to engage only in physically undemanding activities, such as listening to music or reading and remaining awake, preferably in the supine position. The constant routine protocol is more than taking a series of measurements under controlled conditions since variables that take longer than a few minutes to stabilize can be measured. This applies, for example, to core body temperature as a heat load takes about 30 minutes to dissipate.

The protocol can be used to separate the endogenous and exogenous components of various 24-hour rhythms affected by activity and sleep. However, some modifications are needed to deal with exogenous factors specific to a particular variable. When the rhythms of core body temperature or blood pressure are investigated, normal domestic lighting levels (about 200 lx) can be applied. But, if the rhythm of melatonin secretion is investigated, the level must be lower (about 15 lx) because more intense light would suppress melatonin secretion.[24] In the case of water or sodium excretion by the kidneys, the intake of these substances needs to be controlled more accurately than can likely be achieved by simply having "a regular intake of identical snacks". When hormones associated with carbohydrate metabolism are investigated, an intravenous infusion of glucose to maintain plasma glucose concentrations constant is required. Eating identical meals regularly does not guarantee a constant uptake of glucose into the bloodstream, since both digestion and absorption from the gut show 24-hour variations.

The constant routine protocol is considered the best way to characterise the endogenous component of a 24 hour rhythm, *i.e.*, the properties of the circadian oscillator. The phase and amplitude thus obtained are considered the "gold standard" because rhythmic effects of the exogenous component have either been removed (sleep disallowed) or made uniform throughout the time of measurement. The endogenous component of body temperature, heart rate and blood pressure closely resembles a sinusoid.[137–139] The cosinor method may thus be applied to provide information about the phase and amplitude of the endogenous component. The MESOR, however, does not reflect the usual mean value due to the specific conditions of the study.

The constant routine protocol also allows estimating the size of the exogenous component. For example, results of body temperature can be compared between a study participant examined under a constant routine protocol and the normal living routine (usual daytime activity and food intake, and nocturnal rest and fasting). The difference between the two

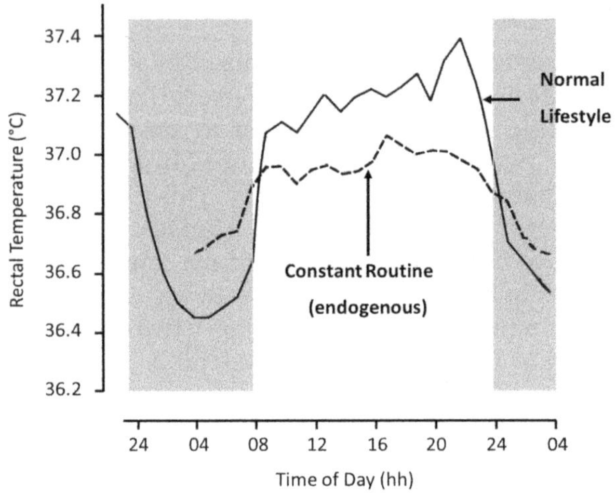

Figure 22.1 Mean body temperature rhythm of a group of eight study participants living their normal lifestyle (solid line) and under constant routine conditions (dashed line). The shaded areas indicate the sleep time under normal life. The difference between the two curves is due to masking effects of the sleep–wake cycle with activity raising temperature and sleep lowering it. Reproduced from ref. 133 with permission from John Wiley & Sons, Copyright 1978 The Physiological Society.

24-hour curves shows the masking effect, rising temperature diurnally and lowering it nocturnally (Figure 22.1).

Under constant routine conditions, different marker rhythms give similar results. In an experiment with a simulated time-zone shift, the rates of adjustment of core temperature and melatonin did not differ during a constant routine protocol.[140] Under usual 24-hour conditions, however, the masked temperature rhythm, with its larger (and fully adjusted) exogenous component, readapted more rapidly than the melatonin rhythm. These results provide evidence that the same central circadian pacemaker generates the endogenous circadian component of both plasma melatonin and body temperature in humans.

Despite the value of the results obtained under constant routine conditions, this protocol suffers several limitations:

(1) It is "laboratory-bound" and unsuitable for field studies. Even in the laboratory, it might cause stress and thus affect the body clock.
(2) Experiments with animals and with individuals who cannot undergo the demanding protocol (babies, or those who are severely ill or mentally fragile) are not possible.
(3) The demands of this protocol are such that it cannot be undertaken on successive days, for example, to assess adjustment to a time zone transition, particularly because sleep is prohibited.
(4) The increasing sleep deprivation may modify circadian patterns. This means that any variable affected by sleep loss or the amount of time awake cannot be assessed satisfactorily.

Item 4 is particularly important when mood and mental performance are considered. It can be tackled using a different protocol, namely, forced desynchronization (see below). In order to avoid the accumulation of sleep pressure, the multiple-nap protocol has also been designed. It is a constant routine with intermittent naps scheduled over a 24-hour or longer day.[141]

The above qualifications do not mean that the constant routine protocols are seriously flawed but rather that they have certain limitations in their execution. Some aspects of the circadian system are also changed. How much the clock itself is affected by the protocols remains to be assessed. Nonetheless, constant routine protocols are superior to the use of raw (masked) data when seeking assessment of the phase and amplitude of the endogenous component of a rhythm.

22.4.2 Forced Desynchronization

In this protocol, the study participants live on non-24-hour "days" that are beyond the range of entrainment of the endogenous circadian oscillator, leading to internal desynchronization.[142] At low light intensities, the circadian pacemaker free-runs with a period close to 24 hours, and episodes of sleep and activity occur at all phases of the endogenous circadian cycle. As a result, masking effects are equally distributed over all rhythm stages. Accordingly, circadian and sleep/wake components of a 24-hour rhythm can be distinguished and studied individually. The method has been used successfully in several studies (for example, ref. 143–146).

The analysis requires the protocol to last for an integer number of "beat periods" between the 24-hour and the imposed non-24-hour period, lasting several days as a result. For example, assuming that the endogenous circadian component has a period of 24 hours, the beat period is 5 solar days with imposed days of 20 hours ($5 \times 24 = 6 \times 20$), and it is 9 solar days with imposed days of 27 hours ($9 \times 24 = 8 \times 27$). Of course, as the period length of the circadian rhythm is slightly longer than 24 hours, an integer number of beat periods can only be approximate.

In our own studies,[143] 27-hour days (with 9 hours of sleep and 18 hours of wakefulness) were undertaken for 17 days, thus covering about two beat cycles (Figure 22.2). Sleep always occurred during the dark span (<10 lux) and was not allowed at other times. During the periods of wakefulness (about 250 lux), study participants were free to choose their activities, which were generally sedentary. The daytime was divided into morning, afternoon, and evening, and meals were taken accordingly. Clocks reading solar time were used to calculate times of retiring and rising as required by this protocol. Core body temperature was measured every 6 minutes using a rectal probe, and movements were recorded every 1 minute using a monitor attached to the non-dominant wrist.

The period length of the body temperature rhythm during the 27-hour days was estimated and a mean rhythm of raw body temperature was educed at this period length (Figure 22.3, top). This way, the effects of sleep–wake

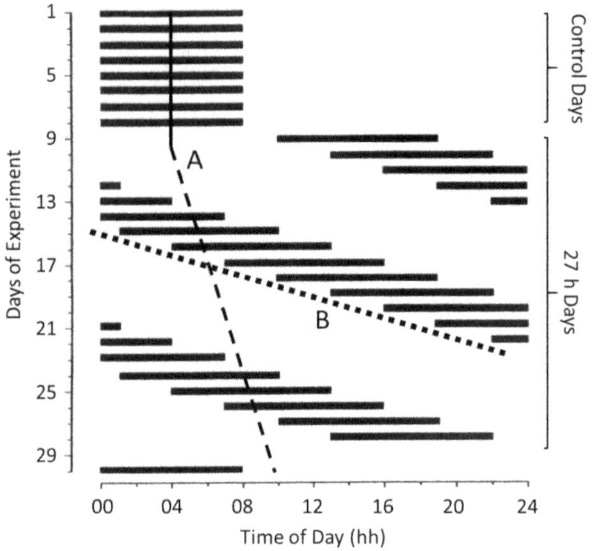

Figure 22.2 The horizontal black bars indicate times allocated for sleep. Control days (1–8): sleep from midnight to 8:00; 27-hour days (9–29): (9 hours asleep and 18 hours awake). Control days were followed by a constant routine protocol lasting 26 hours. Line A connects the times of nadirs of the body temperature rhythm. During control days with period $(\tau) = 24$ hours (solid vertical line), the temperature minimum is entrained to mid-sleep but then free-runs with an average τ of 24.3 hours (oblique dashed line). As a consequence, during the 27-hour days, it no longer coincides with a given phase of the sleep–wake cycle, but drifts from mid-activity to soon after rising, the sleep phase, *etc.* Accordingly, the temperature minimum (and temperature at all other circadian stages) is affected by all stages of the sleep–wake cycle. In other words, the impact of sleep and activity has been averaged out and no longer exerts a rhythmic effect. As a consequence, the mean temperature rhythm, combining all circadian cycles during the 27-hour days, represents the endogenous rhythm (see Figure 22.3, top). Line B (oblique dotted line) represents times just before retiring. These times coincide with different phases of the endogenous temperature rhythm, *i.e.*, the temperatures obtained at these times are affected by all phases of the circadian rhythm. The same applies to all other temperatures obtained at different stages of the 27-hour cycle. Since they have been affected by all phases of the circadian rhythm, the forced desynchronization protocol enables an estimation of the effects of the sleep–wake cycle on temperature by averaging temperatures on a 27-hour scale (see Figure 22.3, bottom). The endogenous effects have been averaged and, therefore, no longer exert a rhythmic effect. Reproduced from ref. 140 with permission from Oxford University Press, Copyright 1991.

were averaged out and therefore no longer exerted a rhythmic effect. In other words, the forced desynchronization protocol enabled an estimation of the endogenous rhythm of temperature. This rhythm is sinusoidal and almost identical to that given by the constant routine (see Figure 22.1).

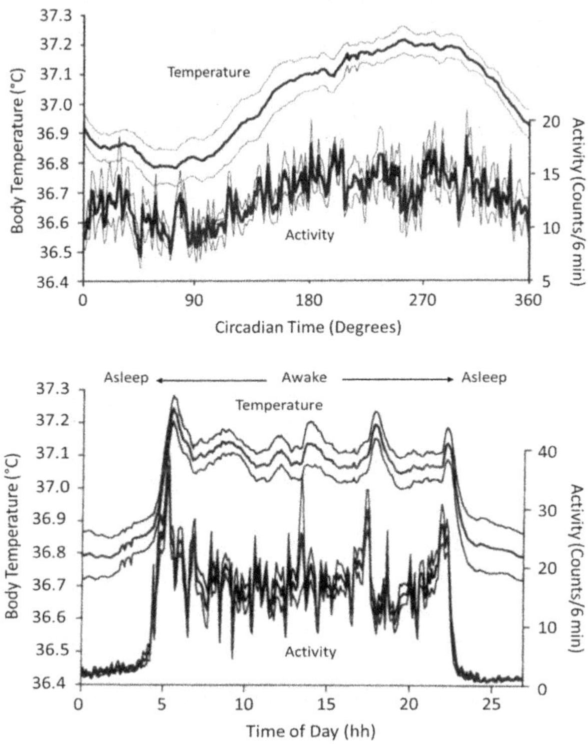

Figure 22.3 Results for core temperature and activity from the forced desynchronization protocol. Top: Circadian (endogenous component); awake and sleep effects have been averaged out. Results are expressed as modulo τ (360° = free-running period, 0° = local midnight on 24-hour control days). Bottom: Effects of the sleep–wake cycle, circadian effects having been averaged out. Results are expressed as modulo 27 hours (0 = midsleep, wake time: 04.5–22.5 decimal hours). For more details, see Figure 22.2. Reproduced from ref. 140 with permission from Oxford University Press, Copyright 1991.

In the next step, mean rhythms of raw temperature and activity during 27-hour days were educed with a period of 27 hours (Figure 22.3, bottom). The activity records after correction for circadian effects reveal a clear dichotomy between "nocturnal" inactivity (asleep) and "daytime" activity (awake). This dichotomy is reflected in the temperature record, differences in heat load produced by activity while awake *vs.* inactivity while asleep being largely responsible for it. This is a clear example of the direct masking effect of the sleep–wake cycle upon the temperature rhythm, an effect confirmed by the observation that regular bouts of activity (rising, preparing meals and retiring) produce transient rises in core temperature.

The forced desynchronization protocol can also be applied to other rhythmic processes, such as mental performance, which is affected by circadian and time-awake factors that can be separated using this protocol.[144]

Similar to constant routines (see above), the length and complexity of the forced desynchronisation protocol prevent its use in field studies and in studies involving children or vulnerable individuals. The circadian system might also be affected by the long-lasting disruption of internal phase relations. Several alternatives have been developed (reviewed in ref. 2 and 43), in which the data are first collected in the presence of masking influences, and then the effects of these influences are removed mathematically. In practice, the effects of the sleep–wake (humans) or rest–activity (animals) cycles are generally removed from body temperature data. The process is termed "purification", the purified data being considered to be a more accurate reflection of the endogenous rhythm than the original (raw) data. These methods are of value in field studies and other circumstances where constant routines and forced desynchronization protocols are not possible or are ethically undesirable. Since such methods do not require the individual's lifestyle to be changed, they can be used to estimate the phase or amplitude of the endogenous component of a circadian rhythm during consecutive days, as we have shown in numerous studies on animals and humans.[39,43,147]

It has been argued that these demasking techniques are unreliable due to nonlinear interactions of endogenous circadian and external influences, and that these methods assume that there is no circadian phase effect on masking.[148,149] The approach we used, however, made no assumptions concerning the shape of the rhythm and the relationship between masking and circadian time. We also used nonlinear functions to describe the interaction between body temperature and motor activity and considered further masking factors in addition to motor activity. The "purified" rhythm was thus shown to be nearly sinusoidal, reflecting the rhythm from the SCN, as expected. We also showed that masking depends distinctly on circadian time, without making any assumption beforehand.[43,147,150] We believe therefore that mathematical demasking techniques are beneficial alternatives in studies where constant routine or forced desynchronization protocols are not applicable.[2]

22.5 Conclusions

Designs of chronobiological studies are very diverse, but none is universal. The specific experimental protocol must be best adapted to the particular study aim, whether it is part of a basic research project or a clinical investigation. Experiments can be performed under laboratory or field conditions, in a highly standardized environment or under everyday life conditions. Investigations of patients following their normal daily routine might be preferred, especially in clinical studies. To characterize the endogenous component of an overt 24-hour rhythm, the impact of exogenous components needs to be assessed by either keeping external factors constant or by recording their effects and eliminating them afterwards.

References

1. D. Minors, J. Waterhouse and W. J. Rietveld, *Biol. Rhythm Res.*, 1996, **27**, 166–174.
2. D. Weinert and J. Waterhouse, in *Biological Timekeeping: Clocks, rhythms and behaviour*, ed. V. Kumar, Springer (India), New Delhi, 2017, pp. 23–45.
3. D. G. Gubin, G. D. Gubin, J. Waterhouse and D. Weinert, *Chronobiol. Int.*, 2006, **23**, 639–658.
4. D. G. Gubin, G. D. Gubin, L. I. Gapon and D. Weinert, *Curr. Aging Sci.*, 2016, **9**, 5–13.
5. D. Gubin, V. Neroev, T. Malishevskaya, G. Cornelissen, S. Y. Astakhov, S. Kolomeichuk, N. Yuzhakova, Y. Kabitskaya and D. Weinert, *J. Pineal Res.*, 2021, **70**, e12730.
6. K. Schöttner, V. Simonneaux, P. Vuillez, S. Steinlechner, P. Pevet and D. Weinert, *Chronobiol. Int.*, 2011, **28**, 873–882.
7. K. Schöttner, P. Vuillez, E. Challet, P. Pévet and D. Weinert, *Chronobiol. Int.*, 2015, **32**, 596–607.
8. H. Weinert, D. Weinert, I. Schurov, E. S. Maywood and M. H. Hastings, *Chronobiol. Int.*, 2001, **18**, 559–565.
9. U. Sitka, D. Weinert, K. Berle, W. Rumler and J. Schuh, *Eur. J. Pediatr.*, 1994, **153**, 117–122.
10. V. Neroev, T. Malishevskaya, D. Weinert, S. Astakhov, S. Kolomeichuk, G. Cornelissen, Y. Kabitskaya, E. Boiko, I. Nemtsova and D. Gubin, *Int. J. Mol. Sci.*, 2021, **22**, 359.
11. P. A. Scuffham, J. Nikles, G. K. Mitchell, M. J. Yelland, N. Vine, C. J. Poulos, P. I. Pillans, G. Bashford, C. del Mar, P. J. Schluter and P. Glasziou, *J. Gen. Intern. Med.*, 2010, **25**, 906–913.
12. F. Halberg, G. Cornélissen and R. B. Singh, *Open Nutraceuticals J.*, 2011, **3**, 100–111.
13. Y. Watanabe, F. Halberg, K. Otsuka and G. Cornelissen, *Clin. Exp. Hypertens.*, 2013, **35**, 257–266.
14. *Design and Implementation of N-of-1 Trials: A User's Guide.*, ed. R. L. Kravitz and N. Duan, Agency for Healthcare Research and Quality, Rockville, MD, 2014.
15. J. Waterhouse and P. J. DeCoursey, in *Chronobiology. Biological Timekeeping*, ed. J. C. Dunlap, J. J. Loros and P. J. DeCoursey, Sinauer, Sunderland, MA, 2004, pp. 325–356.
16. W. J. Rietveld, D. S. Minors and J. M. Waterhouse, *Chronobiol. Int.*, 1993, **10**, 306–312.
17. M. D. Marques and J. M. Waterhouse, *Chronobiol. Int.*, 1994, **11**, 146–155.
18. J. Waterhouse and D. Minors, in *Evolution of the circadian clock*, ed. T. Hiroshige and K. Honma, Hokkaido Univiersity Press, Sapporo, 1994, pp. 275–289.
19. S. Benloucif, M. J. Guico, K. J. Reid, L. F. Wolfe, M. L'Hermite-Baleriaux and P. C. Zee, *J. Biol. Rhythms*, 2005, **20**, 178–188.

20. E. B. Klerman, H. B. Gershengorn, J. F. Duffy and R. E. Kronauer, *J. Biol. Rhythms*, 2002, **17**, 181–193.
21. A. J. Lewy, N. L. Cutler and R. L. Sack, *J. Biol. Rhythms*, 1999, **14**, 227–236.
22. J. Arendt, *J. Biol. Rhythms*, 2005, **20**, 291–303.
23. S. Benloucif, H. J. Burgess, E. B. Klerman, A. J. Lewy, B. Middleton, P. J. Murphy, B. L. Parry and V. L. Revell, *J. Clin. Sleep Med.*, 2008, **4**, 66–69.
24. A. J. Lewy, T. A. Wehr, F. K. Goodwin, D. A. Newsome and S. P. Markey, *Science*, 1980, **210**, 1267–1269.
25. A. J. Lewy, *Prog. Neuropsychopharmacol. Biol. Psychiatry*, 1983, **7**, 551–556.
26. A. J. K. Phillips, P. Vidafar, A. C. Burns, E. M. McGlashan, C. Anderson, S. M. W. Rajaratnam, S. W. Lockley and S. W. Cain, *Proc. Natl. Acad. Sci.*, 2019, **116**, 12019–12024.
27. R. J. Lucas, M. S. Freedman, M. Munoz, J. M. Garcia-Fernandez and R. G. Foster, *Science*, 1999, **284**, 505–507.
28. S. Deacon and J. Arendt, *Neurosci. Lett.*, 1994, **167**, 191–194.
29. A. Voultsios, D. J. Kennaway and D. Dawson, *J. Biol. Rhythms*, 1997, **12**, 457–466.
30. K. V. Danilenko, E. G. Verevkin, V. S. Antyufeev, A. Wirz-Justice and C. Cajochen, *Chronobiol. Int.*, 2014, **31**, 349–355.
31. C. Graham, M. R. Cook, R. Kavet, A. Sastre and D. K. Smith, *J. Pineal Res.*, 1998, **24**, 230–238.
32. A. Stieglitz, F. Spiegelhalter, G. Klante and G. Heldmaier, *J. Pineal Res.*, 1995, **18**, 69–76.
33. M. Niehaus and A. Lerchl, *J. Pineal Res.*, 1998, **25**, 167–171.
34. G. Klante, T. Brinschwitz, K. Secci, F. Wollnik and S. Steinlechner, *J. Pineal Res.*, 1997, **23**, 191–197.
35. H. Keijzer, M. G. Smits, J. F. Duffy and L. M. G. Curfs, *Sleep Med. Rev.*, 2014, **18**, 333–339.
36. B. Tarquini, G. Cornelissen, R. Tarquini, F. Perfetto and F. Halberg, *Neuro Endocrinol. Lett.*, 1999, **20**, 25–28.
37. L. Wetterberg, F. Halberg, B. Tarquini, M. Cagnoni, E. Haus, K. Griffith, T. Kawasaki, L. A. Wallach, M. Ueno, K. Uezo, M. Matsuoka, M. Kuzel, E. Halberg and T. Omae, *Experientia*, 1979, **35**, 416–419.
38. C. Bingham, B. Arbogast, G. C. Guillaume, J. K. Lee and F. Halberg, *Chronobiologia*, 1982, **9**, 397–439.
39. D. Weinert and J. Waterhouse, *Physiol. Behav.*, 1998, **63**, 837–843.
40. J. Waterhouse, B. Drust, D. Weinert, B. Edwards, W. Gregson, G. Atkinson, S. Kao, S. Aizawa and T. Reilly, *Chronobiol. Int.*, 2005, **22**, 207–225.
41. B. Edwards, J. Waterhouse, T. Reilly and G. Atkinson, *Chronobiol. Int.*, 2002, **19**, 579–597.
42. M. Bogh, D. S. Minors and J. M. Waterhouse, *Chronobiol. Int.*, 1994, **11**, 332–339.
43. J. Waterhouse, D. Weinert, D. Minors, S. Folkard, D. Owens, G. Atkinson, I. Macdonald, N. Sytnik, P. Tucker and T. Reilly, *Chronobiol. Int.*, 2000, **17**, 539–566.

44. K. Krauchi, C. Cajochen and A. Wirz-Justice, *J. Appl. Physiol.*, 1997, **83**, 134–139.
45. D. J. Dijk, C. Cajochen and A. A. Borbely, *Neurosci. Lett.*, 1991, **121**, 59–62.
46. D. G. Gubin, A. A. Nelaeva, A. E. Uzhakova, Y. V. Hasanova, G. Cornelissen and D. Weinert, *Chronobiol. Int.*, 2017, **34**, 1136–1148.
47. D. G. Gubin, T. N. Malishevskaya, Y. S. Astakhov, S. Y. Astakhov, G. Cornelissen, V. A. Kuznetsov and D. Weinert, *Chronobiol. Int.*, 2019, **36**, 564–577.
48. J. A. Sarabia, M. A. Rol, P. Mendiola and J. A. Madrid, *Physiol. Behav.*, 2008, **95**, 570–580.
49. M. F. Zeron-Rugerio, A. Diez-Noguera, M. Izquierdo-Pulido and T. Cambras, *Chronobiol. Int.*, 2020, **37**, 1580–1590.
50. K. Krauchi, *Physiol. Behav.*, 2007, **90**, 236–245.
51. J. Smolander, M. Harma, A. Lindqvist, P. Kolari and L. A. Laitinen, *Eur. J. Appl. Physiol. Occup. Physiol.*, 1993, **67**, 192–196.
52. C. G. Scully, A. Karaboue, W. M. Liu, J. Meyer, P. F. Innominato, K. H. Chon, A. M. Gorbach and F. Levi, *Interface Focus*, 2011, **1**, 48–60.
53. M. J. Hasselberg, J. McMahon and K. Parker, *Sleep Med.*, 2013, **14**, 5–11.
54. R. M. Buijs, J. Wortel, J. J. Van Heerikhuize, M. G. Feenstra, G. J. Ter Horst, H. J. Romijn and A. Kalsbeek, *Eur. J. Neurosci.*, 1999, **11**, 1535–1544.
55. M. S. Jasper and W. C. Engeland, *Am. J. Physiol.*, 1991, **261**, R1257–R1268.
56. F. Spiga, J. J. Walker, J. R. Terry and S. L. Lightman, *Compr. Physiol.*, 2014, **4**, 1273–1298.
57. Y. Touitou, A. Bogdan, F. Levi, M. Benavides and A. Auzeby, *Br. J. Cancer*, 1996, **74**, 1248–1252.
58. S. E. Sephton, R. M. Sapolsky, H. C. Kraemer and D. Spiegel, *J. Natl. Cancer Inst.*, 2000, **92**, 994–1000.
59. S. Sheibani, L. Capua, S. Kamaei, S. S. A. Akbari, J. Zhang, H. Guerin and A. M. Ionescu, *Commun. Mater.*, 2021, **2**, 10.
60. F. A. Scheer, K. Hu, H. Evoniuk, E. E. Kelly, A. Malhotra, M. F. Hilton and S. A. Shea, *Proc. Natl. Acad. Sci. U. S. A.*, 2010, **107**, 20541–20546.
61. C. M. Jung, S. B. Khalsa, F. A. Scheer, C. Cajochen, S. W. Lockley, C. A. Czeisler and K. P. Wright Jr, *J. Biol. Rhythms*, 2010, **25**, 208–216.
62. A. Caufriez, R. Moreno-Reyes, R. Leproult, F. Vertongen, E. Van Cauter and G. Copinschi, *Am. J. Physiol.: Endocrinol. Metab.*, 2002, **282**, E1147–E1153.
63. M. F. Slag, M. Ahmad, M. C. Gannon and F. Q. Nuttall, *Metabolism*, 1981, **30**, 1104–1108.
64. M. C. Mormont, A. M. Langouet, B. Claustrat, A. Bogdan, S. Marion, J. Waterhouse, Y. Touitou and F. Levi, *Chronobiol. Int.*, 2002, **19**, 141–155.
65. S. L. Lightman and B. L. Conway-Campbell, *Nat. Rev. Neurosci.*, 2010, **11**, 710–718.
66. M. S. Harbuz, R. J. Windle, D. S. Jessop, D. Renshaw, C. D. Ingram and S. L. Lightman, *Ann. N. Y. Acad. Sci.*, 1999, **876**, 43–52.

67. R. J. Windle, S. A. Wood, Y. M. Kershaw, S. L. Lightman, C. D. Ingram and M. S. Harbuz, *J. Neuroendocrinol.*, 2001, **13**, 905–911.
68. E. A. Young, R. F. Haskett, L. Grunhaus, A. Pande, V. M. Weinberg, S. J. Watson and H. Akil, *Arch. Gen. Psychiatry*, 1994, **51**, 701–707.
69. K. Kalafatakis, G. M. Russell, S. G. Ferguson, M. Grabski, C. J. Harmer, M. R. Munafo, N. Marchant, A. Wilson, J. C. Brooks, J. Thakrar, P. Murphy, N. J. Thai and S. L. Lightman, *Psychoneuroendocrinology*, 2021, **124**, 105096.
70. A. C. Brown, M. H. Smolensky, G. E. D'Alonzo and D. P. Redman, *Chronobiol. Int.*, 1990, **7**, 125–133.
71. G. J. Bellone, S. A. Plano, D. P. Cardinali, D. P. Chada, D. E. Vigo and D. A. Golombek, *Sleep Sci.*, 2016, **9**, 272–279.
72. K. El Allali, H. Farsi, M. Piro, M. Rachid Achaâban, M. Ouassat, E. Challet and P. Pévet, *Chronobiol. Int.*, 2019, 1–11.
73. S. Ancoli-Israel, R. Cole, C. Alessi, M. Chambers, W. Moorcroft and C. P. Pollak, *Sleep*, 2003, **26**, 342–392.
74. S. S. Carvalho-Bos, R. F. Riemersma-van der Lek, J. Waterhouse, T. Reilly and E. J. Van Someren, *Am. J. Geriatr. Psychiatry*, 2007, **15**, 92–100.
75. V. Chevalier, M. C. Mormont, H. Cure and P. Chollet, *Oncol. Rep.*, 2003, **10**, 733–737.
76. E. K. Hoopes, M. A. Witman, M. N. D'Agata, F. R. Berube, B. Brewer, S. K. Malone, M. A. Grandner and F. Patterson, *Chronobiol. Int.*, 2021, **38**, 543–556.
77. D. Weinert, P. Fritzsche and R. Gattermann, *Chronobiol. Int.*, 2001, **18**, 921–932.
78. J. H. Meijer and Y. Robbers, *Proc. R. Soc. B*, 2014, **281**, 20140210.
79. N. Mrosovsky, P. A. Salmon and N. Vrang, *Chronobiol. Int.*, 1998, **15**, 147–158.
80. D. Weinert, K. Schöttner, L. Müller and A. Wienke, *Chronobiol. Int.*, 2016, **33**, 1161–1170.
81. T. Deboer, M. J. Vansteensel, L. Detari and J. H. Meijer, *Nat. Neurosci.*, 2003, **6**, 1086–1090.
82. E. S. Maywood, H. Okamura and M. H. Hastings, *Eur. J. Neurosci.*, 2002, **15**, 216–220.
83. J. Schaap and J. H. Meijer, *Eur. J. Neurosci.*, 2001, **13**, 1955–1962.
84. N. Yamada, K. Shimoda, K. Takahashi and S. Takahashi, *Physiol. Behav.*, 1986, **36**, 357–362.
85. D. M. Edgar, C. E. Martin and W. C. Dement, *J. Biological. Rhythms*, 1991, **6**, 185–199.
86. D. Weinert and T. Weiss, *Biol. Rhythm Res.*, 1997, **28**, 105–120.
87. D. Weinert, K. Schottner, A. C. Meinecke and J. Hauer, *Chronobiol. Int.*, 2018, 1–10.
88. M. C. Antle, R. Sterniczuk, V. M. Smith and K. Hagel, *J. Biological. Rhythms*, 2007, **22**, 524–533.
89. M. R. Ralph and N. Mrosovsky, *J. Biol. Rhythms*, 1992, **7**, 353–359.

90. S. Steinlechner, A. Stieglitz and T. Ruf, *J. Biol. Rhythms*, 2002, **17**, 248–258.
91. H. Weinert, D. Weinert and J. Waterhouse, *Biol. Rhythm Res.*, 2002, **33**, 199–212.
92. J. Sturm, H. Weinert and D. Weinert, *Z. Saugetierkd.-Int. J. Mamm. Biol.*, 2000, **65**, 21–32.
93. E. Werth, E. Savaskan, V. Knoblauch, P. F. Gasio, E. J. van Someren, C. Hock and A. Wirz-Justice, *J. Geriatr. Psychiatry Neurol.*, 2002, **15**, 55–59.
94. N. F. Ruby, N. Joshi and H. C. Heller, *Am. J. Physiol.*, 2002, **283**, R1314–R1320.
95. A. A. Borbely, *Hum. Neurobiol.*, 1982, **1**, 195–204.
96. A. A. Borbely and P. Achermann, *J. Biol. Rhythms*, 1999, **14**, 557–568.
97. S. Daan, D. G. Beersma and A. A. Borbely, *Am. J. Physiol.*, 1984, **246**, R161–R183.
98. S. K. Martin and C. I. Eastman, *Chronobiol. Int.*, 2002, **19**, 695–707.
99. A. M. Reiter, C. Sargent and G. D. Roach, *Chronobiol. Int.*, 2020, 1–13.
100. H. J. Burgess and C. I. Eastman, *J. Sleep Res.*, 2005, **14**, 229–237.
101. J. A. Horne and O. Ostberg, *Int. J. Chronobiol.*, 1976, **4**, 97–110.
102. T. Roenneberg, A. Wirz-Justice and M. Merrow, *J. Biol. Rhythms*, 2003, **18**, 80–90.
103. A. Zavada, M. C. Gordijn, D. G. Beersma, S. Daan and T. Roenneberg, *Chronobiol. Int.*, 2005, **22**, 267–278.
104. M. Wittmann, J. Dinich, M. Merrow and T. Roenneberg, *Chronobiol. Int.*, 2006, **23**, 497–509.
105. C. Dibner, U. Schibler and U. Albrecht, *Annu. Rev. Physiol.*, 2010, **72**, 517–549.
106. D.-J. Dijk and J. F. Duffy, *J. Biol. Rhythms*, 2020, **35**, 421–438.
107. S. Crnko, H. Schutte, P. A. Doevendans, J. P. G. Sluijter and L. W. van Laake, *Physiology*, 2021, **36**, 7–20.
108. F. Agostinelli, N. Ceglia, B. Shahbaba, P. Sassone-Corsi and P. Baldi, *Bioinformatics*, 2016, **32**, i8–i17.
109. J. J. Hughey, T. Hastie and A. J. Butte, *Nucleic Acids Res.*, 2016, **44**, e80.
110. H. R. Ueda, W. Chen, Y. Minami, S. Honma, K. Honma, M. Iino and S. Hashimoto, *Proc. Natl. Acad. Sci. U. S. A.*, 2004, **101**, 11227–11232.
111. N. Wittenbrink, B. Ananthasubramaniam, M. Münch, B. Koller, B. Maier, C. Weschke, F. Bes, J. de Zeeuw, C. Nowozin, A. Wahnschaffe, S. Wisniewski, M. Zaleska, O. Bartok, R. Ashwal-Fluss, H. Lammert, H. Herzel, M. Hummel, S. Kadener, D. Kunz and A. Kramer, *J. Clin. Invest.*, 2018, **128**, 3826–3839.
112. U. Schibler and P. Sassone-Corsi, *Cell*, 2002, **111**, 919–922.
113. A. Balsalobre, F. Damiola and U. Schibler, *Cell*, 1998, **93**, 929–937.
114. E. Nagoshi, C. Saini, C. Bauer, T. Laroche, F. Naef and U. Schibler, *Cell*, 2004, **119**, 693–705.
115. A. Balsalobre, S. A. Brown, L. Marcacci, F. Tronche, C. Kellendonk, H. M. Reichardt, G. Schutz and U. Schibler, *Science*, 2000, **289**, 2344–2347.

116. J. Bieler, R. Cannavo, K. Gustafson, C. Gobet, D. Gatfield and F. Naef, *Mol. Syst. Biol.*, 2014, **10**, 739.
117. J. Ndikung, D. Storm, N. Violet, A. Kramer, G. Schönfelder, N. Ertych and M. Oelgeschläger, *Environ. Int.*, 2020, **134**, 105265.
118. M. Stratmann and U. Schibler, *J. Biol. Rhythms*, 2006, **21**, 494–506.
119. E. D. Buhr, S. H. Yoo and J. S. Takahashi, *Science*, 2010, **330**, 379–385.
120. J. Wuarin and U. Schibler, *Cell*, 1990, **63**, 1257–1266.
121. E. H. Rogers, S. A. Fawcett, V. Pekovic-Vaughan and J. A. Hunt, *Stem Cells Int.*, 2017, **2017**, 2057168.
122. C. Feillet, P. Krusche, F. Tamanini, R. C. Janssens, M. J. Downey, P. Martin, M. Teboul, S. Saito, F. A. Levi, T. Bretschneider, G. T. van der Horst, F. Delaunay and D. A. Rand, *Proc. Natl. Acad. Sci. U. S. A.*, 2014, **111**, 9828–9833.
123. F. Levi, E. Filipski, I. Iurisci, X. M. Li and P. Innominato, *Cold Spring Harbor Symp. Quant. Biol.*, 2007, **72**, 465–475.
124. S. Yamazaki, M. Straume, H. Tei, Y. Sakaki, M. Menaker and G. D. Block, *Proc. Natl. Acad. Sci. U. S. A.*, 2002, **99**, 10801–10806.
125. S. Yamazaki, T. Yoshikawa, E. W. Biscoe, R. Numano, L. M. Gallaspy, S. Soulsby, E. Papadimas, P. Pezuk, S. E. Doyle, H. Tei, Y. Sakaki, G. D. Block and M. Menaker, *J. Biol. Rhythms*, 2009, **24**, 55–63.
126. A. J. Davidson, S. Yamazaki, D. M. Arble, M. Menaker and G. D. Block, *Neurobiol. Aging*, 2008, **29**, 471–477.
127. Y. Hamaguchi, Y. Tahara, M. Hitosugi and S. Shibata, *J. Biol. Rhythms*, 2015, **30**, 533–542.
128. J. Aschoff, *Cold Spring Harbor Symp. Quant. Biol.*, 1960, **25**, 11–28.
129. R. Wever, *The circadian system of man: results of experiments under temporal isolation*, Springer, New York, Heidelberg, Berlin, 1979.
130. D. Weinert, *Chronobiol. Int.*, 2000, **17**, 261–283.
131. G. Cornelissen, F. Halberg, J. Siegelova and A. Galvagno, in *Noninvasive Methods in Cardiology*, ed. F. Halberg, T. Kenner, B. Fiser and J. Siegelova, Faculty of Medicine, Masaryk University, Brno, 2010, pp. 69–74.
132. S. S. d. l. Pena, F. Halberg, A. Galvagno, M. Montalbini, S. Follini, J. Wu, J. Degioanni, F. Kutyna, D. C. Hillman, Y. Kawabata and G. Cornelissen, *Proceedings. Second Annual IEEE Symposium on Computer-based Medical Systems*, 1989.
133. J. N. Mills, D. S. Minors and J. M. Waterhouse, *J. Physiol.*, 1978, **285**, 455–470.
134. D. Dawson, L. Lack and M. Morris, *Chronobiol. Int.*, 1993, **10**, 94–102.
135. J. Foret, A. Daurat, Y. Touitou, A. Aguirre and O. Benoit, *Chronobiol. Int.*, 1996, **13**, 35–45.
136. C. A. Czeisler, J. S. Allan, S. H. Strogatz, J. M. Ronda, R. Sanchez, C. D. Rios, W. O. Freitag, G. S. Richardson and R. E. Kronauer, *Science*, 1986, **233**, 667–671.
137. D. Minors and J. Waterhouse, *Circadian rhythms and the human*, Wright PSG, Bristol, 1981.
138. K. Krauchi and A. Wirz-Justice, *Am. J. Physiol.*, 1994, **267**, R819–R829.

139. S. A. Shea, M. F. Hilton, K. Hu and F. A. Scheer, *Circ. Res.*, 2011, **108**, 980–984.
140. T. L. Shanahan and C. A. Czeisler, *J. Clin. Endocrinol. Metab.*, 1991, **73**, 227–235.
141. C. Cajochen, V. Knoblauch, K. Krauchi, C. Renz and A. Wirz-Justice, *NeuroReport*, 2001, **12**, 2277–2281.
142. R. A. Wever, *Pflügers Arch.*, 1983, **396**, 128–137.
143. J. Waterhouse, D. Weinert, D. Minors, G. Atkinson, T. Reilly, S. Folkard, D. Owens, I. Macdonald, N. Sytnik and P. Tucker, *Chronobiol. Int.*, 1999, **16**, 343–357.
144. D. B. Boivin, C. A. Czeisler, D. J. Dijk, J. F. Duffy, S. Folkard, D. S. Minors, P. Totterdell and J. M. Waterhouse, *Arch. Gen. Psychiatry*, 1997, **54**, 145–152.
145. D. J. Dijk, J. F. Duffy and C. A. Czeisler, *J. Sleep Res.*, 1992, **1**, 112–117.
146. A. E. Hiddinga, D. G. Beersma and R. H. Van den Hoofdakker, *J. Sleep Res.*, 1997, **6**, 156–163.
147. D. Weinert, A. Nevill, R. Weinandy and J. Waterhouse, *Chronobiol. Int.*, 2003, **20**, 249–270.
148. E. B. Klerman, A. Brager, M. A. Carskadon, C. M. Depner, R. Foster, N. Goel, M. Harrington, P. M. Holloway, M. P. Knauert, M. K. LeBourgeois, J. Lipton, M. Merrow, S. Montagnese, M. Ning, D. Ray, F. A. J. L. Scheer, S. A. Shea, D. J. Skene, C. Spies, B. Staels, M.-P. St-Onge, S. Tiedt, P. C. Zee and H. J. Burgess, *Clin. Transl. Med.*, 2022, **12**, e1131.
149. E. B. Klerman, Y. Lee, C. A. Czeisler and R. E. Kronauer, *J. Biol. Rhythms*, 1999, **14**, 260–274.
150. J. Waterhouse, A. Nevill, D. Weinert, S. Folkard, D. Minors, G. Atkinson, T. Reilly, I. Macdonald, D. Owens, N. Sytnik and P. Tucker, *Biol. Rhythm Res.*, 2001, **32**, 511–528.

CHAPTER 23

Identifying and Interpreting Rhythms in Biological Data

ALEXANDER YOO AND RON C. ANAFI*

Department of Medicine, Chronobiology and Sleep Institute, Perelman School of Medicine, University of Pennsylvania, Philadelphia, PA, USA
*Email: ron.anafi@pennmedicine.upenn.edu

23.1 Introduction

The ability to measure and detect rhythms lies at the heart of biological rhythm research. As detailed in the previous chapters, organismal behavior, systems physiology, and molecular biology all undergo regular changes. These periodic cycles can span vastly different periods. The synchronized flashes of fireflies may repeat in seconds.[1] The yeast metabolic cycle[2] and the cell cycle[3] often repeat every few hours. Seasonal changes in physiology, migration, and reproduction are well known across the plant and animal worlds. Driven in part by the experience of the authors, the examples in this chapter focus on *circadian* rhythms, or rhythms that repeat with a period approximating a day. Indeed, all biological rhythms are often categorized in relation to these daily cycles. *Ultradian* rhythms, like the human rapid-eye movement/non rapid eye movement sleep cycle[4] (which repeats every 90–120 minutes), are recurrent cycles that are repeated with a period of less than 24 hours. *Infradian* rhythms, like the human menstrual cycle, have cycles that exceed a day. Nonetheless, the same principles, mathematical techniques, and statistical procedures are often used to identify rhythms regardless of timescale.

Chemical Biology No. 23
Chronobiology and Chronomedicine: From Molecular and Cellular Mechanisms to Whole Body Interdigitating Networks
Edited by Germaine Cornelissen and Tsuyoshi Hirota

Published by the Royal Society of Chemistry, www.rsc.org

Researchers have long studied daily behavioral rhythms in activity and attention, along with rhythms in classical physiological biomarkers like the heart rate, body temperature, or hormone levels. Time-course studies have measured these features over extended durations, identifying biological signals that show regular and repeatable changes with time of day – and sometimes in the absence of environmental stimuli. With the advent of modern molecular techniques, the space of experimental circadian biology, and biological rhythm research in general, has greatly expanded. Researchers can measure gene expression, protein translation, and metabolic function all at the level of individual molecules or at the genome scale. Like more traditional measures of behavior and organ level physiology, each of these data types provides unique opportunities and challenges for researchers interested in biological rhythms.

As molecular biology has evolved, the mathematical and computational challenge to accurately identify and interpret biological rhythms has gained new importance. New technologies are often both technically challenging and expensive. Although impressive in scale, they often suffer from significant technical variability. Moreover, biological experiments typically rely on laboratory animals or human volunteers. Animals must sometimes be killed to obtain these biological data. Finally, the public funds that primarily finance biomedical research are, by necessity, diverted from other societal needs. Thus, both ethical and practical constraints oblige researchers to design experiments and analytic methods that yield the most biological knowledge with the minimum number of new samples collected. While a visual check on selected tracings can be an important first step, different observers often have different assessments of noisy signals. Now, in the face of genome scale data with thousands of experimental measures, even the most dedicated and consistent investigators can no longer "eye-ball" each tracing for evidence of rhythmicity. Rather, computationally tractable methods that can analyze tens of thousands of noisy tracings with mathematical objectivity are required. Like the molecular experimental measures that have spurred their development, the mathematical and statistical approaches to analyze circadian data are evolving quickly.

This chapter will quickly introduce some of the types of errors that complicate some of the core measurements in molecular biology – using the measurement of RNA expression as a representative example. We will then discuss the current techniques to find and describe rhythms in biological data. There is a long and growing list of methods that can be used for this purpose. This chapter will not provide an exhaustive list. Some excellent methods will be excluded due to necessity. Rather, this chapter will emphasize a heuristic understanding of these methods so that readers can better grasp their relative strengths and limitations. In practice, investigators often want to determine if rhythmic biological signals have changed following some perturbation – and we will thus briefly discuss the methods for differential rhythm analysis. The discussion will then focus on methods used to incorporate prior biological knowledge into the interpretation of these data, analyzing rhythmic data at a pathway scale. Finally, we will discuss the growing body of techniques being developed to identify rhythmic signals in the absence of traditional time-course data.

23.2 Errors and Biases in Molecular Measures

The flow of genetic information from DNA to mRNA to functional protein is called the central dogma of molecular biology.[5] Proteins, among other roles, catalyze the metabolic reactions essential for life. Considerable effort has been devoted to measuring the full molecular complement of RNA, proteins, and metabolites – along with their various interactions. These measurements can be obtained at the subcellular, cellular, tissue, organ, and whole organism level. While proteomics and metabolomics are making remarkable strides, it is the technology to measure RNA species that has advanced most quickly. Microarrays are one of the best-established of these genome scale technologies. We will briefly review microarray technology, highlighting the kinds of errors that can influence these measurements and defining key vocabulary as a guide to understanding molecular biology measurements more generally.

Microarrays are slides that contain tens of thousands of labeled fluorescent probes each of which bind to a specific mRNA sequence. The hybridization/binding of the target molecule to its probe increases fluorescence intensity. By scanning the intensity of light emitted at each probe location, the aim of microarray technology is to quantify the expression of thousands of different mRNA transcripts. But all probes will not bind to their targets with equal affinity. Some targets (notably those with more guanine and cytosine residues[6]) are likely to bind more strongly and induce a greater fluorescent output. Some probes may also show non-specific binding and bind several sequences – effectively misidentifying some of our transcripts. These are examples of ***systematic bias*** in our measurements. Of course, small changes in the chemical concentration of the reactants, the temperature of the surroundings, or other changes in processing environment where hybridization occurs can also introduce fluctuations or ***random errors*** in our measurements. Different samples are placed on separate microarrays and groups or batches of arrays are often processed together. These non-biological differences in processing can be shared among groups of samples, leading to systematic influences on thousands of our individual measurements. These errors, shared among batches or groups of samples, are called ***batch effects.*** While informatic processing can aid in downstream processing, a careful researcher will work to make sure that batch processing of samples is split across conditions so that influence of batch can be better distinguished from true experimental differences.[7,8]

There is, unfortunately, no guarantee that the light intensity we observe will linearly increase with transcript binding. Many probes have a baseline level of autofluorescence without any binding of a target transcript. As more and more of a target is present, the probes may begin to saturate and smaller increments in the fluorescent signal might result. Ideally, we would want to take measurements so that each transcript is in the ***dynamic range*** of our probes – that is in the range where our measurements (light intensity) increase linearly with the quantity being measured. Unfortunately, depending on our sample, some transcripts may be near the limit of detection while others could be in a range where our instrument is less sensitive to changes.[7,8]

The fluorescent signal we observe in any probe is affected by the starting quantity of total mRNA. Even if both samples came from cells that were behaving identically, if one sample contained more cells, and more mRNA, we might expect that almost all the probes would grow brighter. We must somehow adjust our measurements to account for these influences. Microarray ***normalization*** is a process that uses information about the intensity of all probes to adjust our measurements and better account for systemic errors, including the total amount of mRNA in the sample. The word "normalization" can, however, have many uses in bioinformatics. Normalization generally refers to adjusting data (perhaps collected at different times, using different tools, or in different conditions) so that they can be analyzed with a common scale. Several well-accepted microarray normalization strategies are used (*e.g.*, MASS, RMA, and GC-RMA). In the case of microarray technology, these normalization strategies result in ***relative measures*** rather than in ***absolute measures*** of transcript abundance. Rather than providing an absolute quantity, like there are 128 molecules of the ARNTL transcript or 1 mM of the ARNTL transcript, they make certain assumptions to provide a relative measure of the amount of ARNTL as compared to all the mRNA in the sample.[7,8]

Our purpose in the above paragraphs was not to denigrate microarray technology. While newer methods have gained popularity, microarrays remain an excellent and cost-effective technology. But regardless of the technology, all biological measurements are subject to random errors and systematic biases. Many also require normalization strategies and provide relative rather than absolute measures of abundance.

Biologists must assume that every measurement reflects a combination of a true biological signal and experimental error and noise. The relative contribution of these elements, the so-called ***signal-to-noise*** ratio, is thus key to any experiment. If the variation that occurs with time, or with an intervention, is relatively small compared to the variation caused by chance, it will be hard to spot. A significant part of the job of the informaticist involves the rigorous discovery of these signals in a sea of noise.

23.3 Methods for Identifying Rhythms in Time Course Data

23.3.1 Describing Rhythms

Regardless of the biological output being assessed or the method used to establish rhythmicity, a few standard terms are used to describe rhythmic

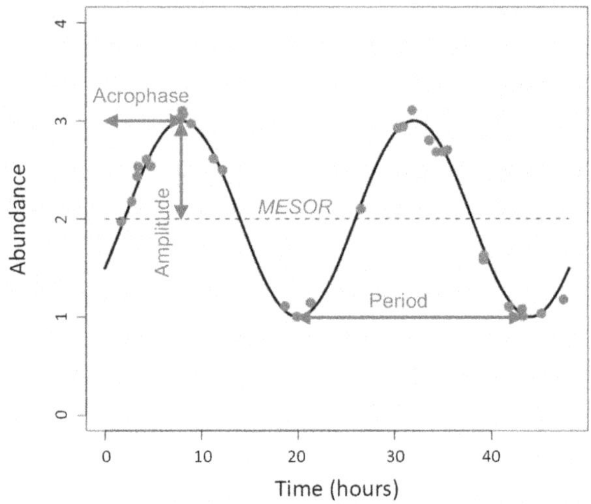

Figure 23.1 Terms used to describe rhythmic biological signals. Idealized data (grey dots) describe the abundance of a rhythmic molecule over time. The data are well fit with a sinusoidal waveform shown in black. The period (*P*) of a waveform is the time it takes for the entire cycle to complete and begin anew. The wave shown has a period of nearly 24 hours and is thus circadian. The amplitude (*A*) is half the difference between the peak and trough values. The acrophase, often denoted by the Greek letter (φ), measures the time at which the waveform reaches its peak relative to a given external standard (*e.g.*, time 0, sunrise, or a new moon). In this example, the acrophase is near 8 h (relative to time 0). The midline estimating statistic of rhythm (MESOR) is the mean or average value across a single evenly sampled idealized period. Note that the MESOR is a rhythm-adjusted mean that might differ from the simple arithmetic mean when the data are not sampled uniformly, as is shown in this example, or do not cover an integer number of cycles. For sinusoidal rhythms, like the one shown, the amplitude of the rhythm is also the difference between the MESOR and the peak or trough value.

variables (Figure 23.1). The *period* of the rhythm is the amount of time it takes to complete one cycle. For circadian rhythms, the period is near (but often not exactly) 24 hours. The same information is alternatively expressed as a *frequency* (*f*), which is simply the inverse of the period (1/period) or the number of cycles that will repeat over a fixed unit of time. The most common unit for describing frequency is Hertz (*Hz*) or the number of cycles per second. The MESOR (midline estimating statistic of rhythm) is the mean or average value across a single evenly sampled idealized period. Note that the MESOR is a rhythm-adjusted mean that differs from the simple arithmetic mean when the data are not sampled uniformly and/or do not cover an integer number of cycles. The *amplitude* (A) of a waveform is half the difference between the peak and trough values, (for sinusoidal waveforms, this is also the difference between the peak or trough of the wave and its MESOR). We can measure the phase of an oscillation relative to a standard reference wave. For periodic signals that are roughly sinusoidal or have a clear, singular peak value, we can define the *acrophase* (φ) of the wave. This is the time in the cycle when the wave reaches its peak.

23.3.2 Finding Rhythms – A Generic Recipe

The various rhythm-finding methods we will discuss each identify some *characteristics* of periodic signals. These characteristics often mirror common intuitions about rhythms, *e.g.*, "rhythmic signals repeat", or "rhythmic signals look like waves". A numerical measure that somehow encapsulates that characteristic, or a *metric*, is then computed. But how high (or low) a metric is required to feel confident that a given biological signal is cycling? If a method gives a score of "11.3" to a given signal, is that good enough?

Without more information, we cannot know the right threshold to use. In answering this question, we use statistical methods that most of us initially find somewhat "backward". We estimate the probability of observing the measured value in a truly *non-rhythmic* signal. More formally, we create a *statistic* by assessing the expected distribution of the metric under the "null hypothesis." If the value of the statistic we observed in our data is expected to occur very rarely under the null hypothesis ("This would be a one in a million event"), we reject the null hypothesis and classify the signal as rhythmic.

Some methods use thousands of random computer simulations to "*bootstrap*" or estimate the expected distribution of metric values if the null hypothesis were true and the various biological signals were arrhythmic. Other methods were developed so that this null distribution can be solved analytically. But what exactly is our null hypothesis? If the signal is not cycling, what *is* it doing? Does it have a constant value? Is it a signal with normally distributed (like a bell curve) noise? Does it monotonically increase or decrease with time? Most commonly, we assume that measurements for a given signal are independent and that they reflect a temporally unchanging Gaussian noise about a fixed value. For some methods the exact nature of the null hypothesis used may not be easy to identify – or may be hard to state. While undeniably useful, these null hypotheses are often imperfect and may not accurately reflect biological reality. The amount of a molecule in my body, or even my temperature, cannot drastically change in an instant. Discrete measurements taken only a short time apart will, by necessity, not be independent from each other. Many signals may change monotonically with age or activity.

23.3.3 Autocorrelation – Looking for Repeating Signals

Autocorrelation[9] is one of the easier rhythmicity tests to understand. Exploring how it works can be instructive in understanding its strengths and weaknesses. Autocorrelation is based on the idea that, regardless of the shape of the underlying waveform, periodic waveforms repeat. If a given biological variable (x) is well described by a function f of time (t), we can write $x \sim f(t)$. But if f is periodic with period (P), the value of x we observe at any time (t) will closely match the value we will observe after another full cycle has completed ($t+P$). So, $f(t) \sim f(t+P)$ for all values of (t). If the level of thyroid hormone in my body has a strong daily rhythm, then the value that we measure at noon today $f(12)$ should be close to the value that will be measured at noon tomorrow $f(12+24)$. Similarly, the value we observe at 3 pm $f(15)$ today should be close to the value we observe at 3 pm tomorrow $f(15+24)$. But how can we create a "metric" to measure this trend?

Correlation is unitless. It measures the strength of the linear association between two variables (Figure 23.2). Simply put, if we plot one variable against another, correlation gives us a number from -1 to 1 that measures how closely these points lie on a straight line. A correlation of $+1$ means that the points lie perfectly along the line. As one variable increases, so does the other. A correlation of -1 also implies a perfect linear trend, but here as one variable increases, the other decreases. For correlations between these values the linear trends become weaker and noisier. A correlation of zero implies that there is no linear trend between the variables.

Autocorrelation uses this well described statistic to quantify our intuition about periodic signals. Instead of comparing two different signals, autocorrelation compares a given biological signal with a time-lagged version of itself (Figure 23.3). We ask if, for every observation we make, does it resemble the observation we make exactly P time units later? We take N evenly spaced measurements of x separated by time interval I, $x_i = x_1, \ldots, x_N$. The auto-correlation function compares the sequence of observations $x_1 \ldots x_{N-L}$ to the series of lagged observations $x_{1+L} \ldots x_N$ for a given time lag $L \times I$. If the time lag $L \times I$ is very close to the period of the signal, then this correlation will be high. In that case, x_1 will be very nearly x_{1+L}, x_2 will be very nearly x_{2+L}, and this pattern will continue through x_{N-L} *and* x_N. Of course, our measurements are likely imperfect and other sources of biological variation might be important. But if we plotted the points (x_1, x_{1+L}), (x_2, X_{2+L}), $(x_3, x_{3+L})\ldots$, our graph looks something like a straight line and the correlation should be relatively high. But how high must it be for us to suspect that there is a rhythm? If we know the distribution of autocorrelation results that would be expected from non-rhythmic signals, we can judge if the autocorrelation result we observed in our data is higher than would be expected from a non-rhythmic signal.

The autocorrelation test has several clear strengths. It makes no assumption about the shape of the underlying periodic waveform. By testing several values for the best lag L, it can also find rhythmic signals without prior knowledge of the underlying period. Of course, this requires that the signal was sampled for a long enough time to observe the repeating pattern and that, given the sampling interval I, there is some lag L such that $L \times I$ is near the period. Moreover, as the linear correlation function is easy to compute, and its statistical properties are very well understood, autocorrelation can be easily computed for thousands of genes or signals, even with low computing power.

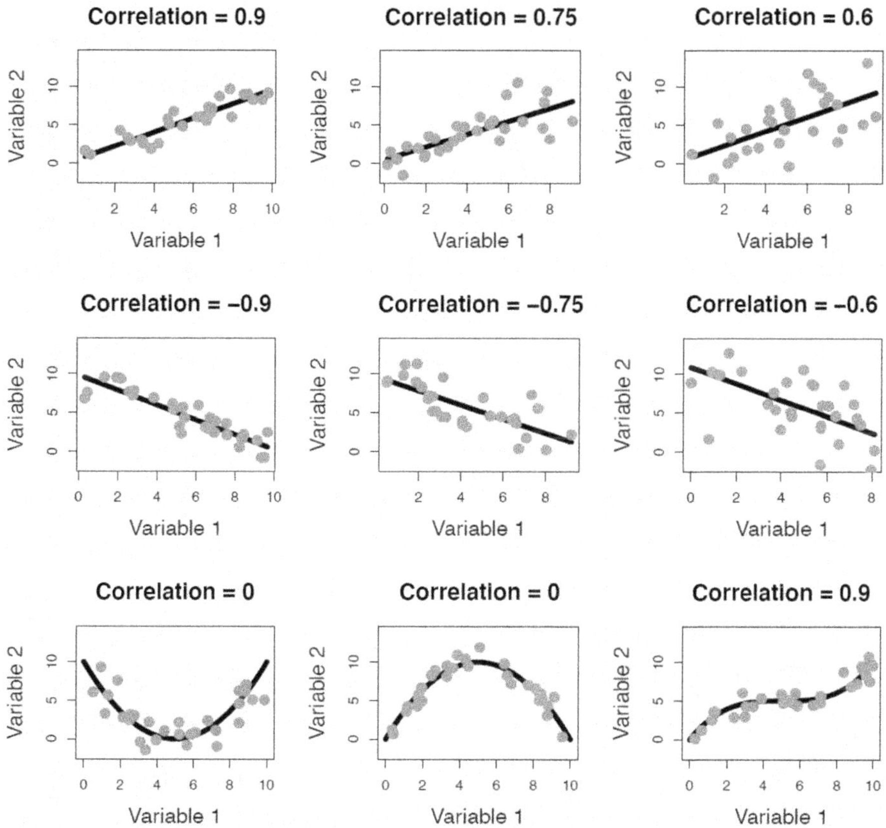

Figure 23.2 Examples of Pearson correlation. Correlation is a unitless number, between −1 and +1, that measures the linear association between two variables. A correlation of +1 implies that the points lie perfectly along a positively sloped line. As one variable increases, so does the other. A correlation of −1 also implies a perfect linear trend, but here as one variable increases, the other decreases. The *top row* of plots displays pairs of variables that are all positively correlated with one another. The correlation between each pair of variables is shown in the plot title. As the correlation decreases, the relationship becomes noisier, and the points spread further from the idealized line. The *middle row* of plots shows examples of variables that are negatively correlated. Correlation measures a *linear* association of variables. As a result, correlation may not reflect the non-linear association between variables. The *bottom row* of plots depicts examples of non-linear associations between variables and the measured linear correlation. Variable 1 can be used to accurately predict variable 2 in the parabolic relationships, but the linear correlation between the variables is ~0.

At the same time, we can also appreciate several key weaknesses of the autocorrelation approach. The method requires evenly spaced samples and has difficulty when dealing with missing samples. While autocorrelation can identify the presence of a repeating signal of a particular period, it does not

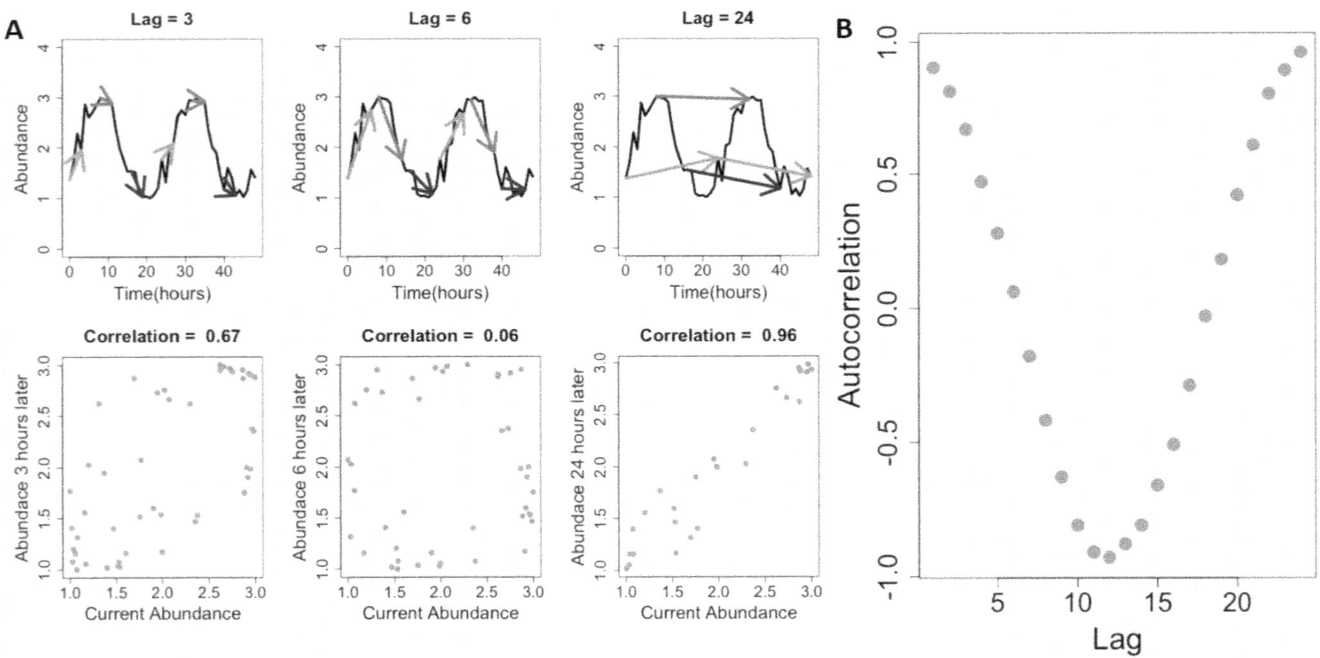

Figure 23.3 Understanding autocorrelation as a test for rhythmicity. Autocorrelation compares the values of a signal at each moment to the value that will be observed after a fixed time lag. If the given signal is rhythmic and the tested lag is near the period of that rhythm, these two signals will be almost identical and will be well correlated. In each of the top figures in A, the same periodic waveform is shown. Arrows connect points lagged by either 3, 6 or 24 hours. For each lag, the bottom row of plots in A compares the values observed at each moment (horizontal axis) to the values that are observed after the fixed lag (vertical axis). With a lag of 3 hours, the plot looks like a narrow ellipse and the two variables are moderately well correlated. With a 6-hour lag, the correlation is near 0. With a lag of 24 hours the correlation is near 1. Figure 23.3B shows a plot of the correlation coefficient as a function of lag. For very short lags, the correlation is high. The signal is continuous and so does not change too rapidly. While the correlation drops as the lag increases, it peaks again at a lag of 24 hours suggesting a circadian rhythm. The null distribution of the expected values of the autocorrelation function in non-rhythmic signals has been well studied. We can use this to identify threshold values of the autocorrelation function that would be very unusual to observe in non-rhythmic data and thus suggest an underlying periodicity.

directly estimate the amplitude, MESOR, or phase of the waveform. Separate methods are often invoked for those estimates after a rhythm is identified. Moreover, the power to detect rhythms drops as the required lag L increases, as fewer and fewer samples are available for the comparison. The linear correlation measure that is at the center of autocorrelation analysis can be very sensitive to noise and extreme data points. Finally, and as we will discuss, newer methods have been shown to have a greater power in identifying rhythms in real world molecular data where noise and sparse sampling are commonplace.

23.3.4 Cosinor Regression – Looking for Sinusoidal Waves

One of the most simple and powerful intuitions underlying the analysis of biological rhythms is that these rhythms are roughly sinusoidal. Dubbed cosinor analysis by Halberg and colleagues in the 1960s, cosinor regression uses curve fitting methods to identify biological features that follow a sinusoidal pattern.[10] Halberg first used non-linear regression to fit observed time course expression data $x(t)$ to a function (Figure 23.4): $x(t) \sim \mathrm{Amp} \times \mathrm{Cos}\left(\frac{2\pi}{P} \times t - \varphi\right) + M + \varepsilon$. In this way, he could estimate the amplitude (Amp), period (P), acrophase (φ) and MESOR (M) of a signal.

The computational ease of the problem and the quality of the available statistical analysis tools are greatly enhanced by taking advantage of some trigonometric identities. Notably a cosine wave of a given period (P) and acrophase (φ) can be expressed as the linear sum of sin and cos terms with the same period, that is, $x(t) \sim \mathrm{Amp} \times \mathrm{Cos}\left(\frac{2\pi}{P} \times t - \varphi\right) + M + \varepsilon$ is mathematically equivalent to $x(t) \sim A \times \mathrm{Cos}\left(\frac{2\pi}{P} \times t\right) + B\,\mathrm{Sin}\left(\frac{2\pi}{P} \times t\right) + M + \varepsilon$, where $\mathrm{Amp} = \sqrt[2]{A^2 + B^2}$ and $\varphi = \mathrm{Atan}\left(\frac{B}{A}\right)$.

While this seems to complicate matters, this formulation of the cosinor regression problem has many advantages. Here the equation is a linear function of all the unknown parameters A, B and M. In the original formulation, the equation was a non-linear function of the parameter φ. The more easily implemented statistical tools of linear regression became available when we use the second formulation. Linear regression also provides well-established statistical tests to evaluate the null hypothesis that the signal is arrhythmic ($A = B = 0$).

In contrast to autocorrelation, cosinor regression does not require that the values are sampled at regular intervals and is not impeded by missing samples. Moreover, by adding terms to our cosinor model that describe group identity or other sources of biological variation, we can account for more complex experimental designs. For example, in some experiments, samples might be repeatedly collected from several subjects (*e.g.*, hourly blood pressure measures over several days in 10 distinct subjects). In these

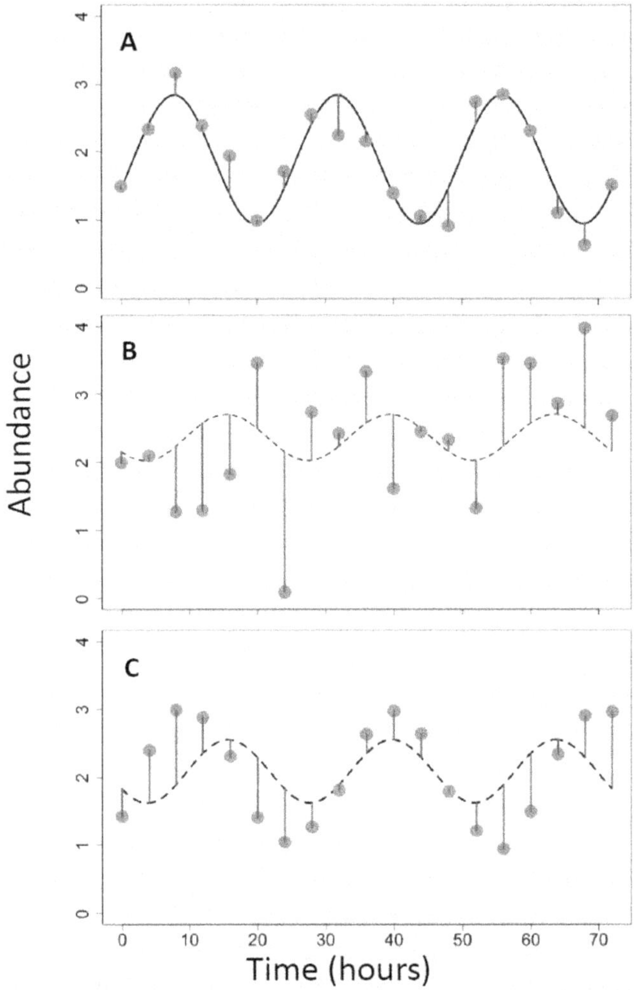

Figure 23.4 Intuitions underlying cosinor regression. Cosinor regression uses curve fitting (linear regression) to find the sinusoidal wave of specified period that best matches the data. (A) Shows observed data (grey dots) well fit with the underlying dark black sinusoid curve. The vertical distances (straight lines) between the curve and each of the data points reflect the discrepancies between the modeled curve and the observed data. These differences, or model errors, are termed the residuals. The best fit curve is chosen so that the sum of the squared errors is minimized. In contrast, the data in (B) have no clear rhythm and even the best fit sinusoidal curve has high residuals/errors. Statistical tests can compare the fit of a model that includes rhythmicity to one that does not. The grey data points shown in (C) are strongly periodic. However, the period of rhythmicity is different from the assumed sinusoidal waveform. As a result, even the best-fit sinusoidal wave has high error. Ideally, cosinor regression requires prior knowledge of the underlying period. Alternatively, waves of many different periods can be tested, but the statistical thresholds must be adjusted accordingly.

cases, a "mixed model" that includes a subject term describing which subject contributed to the sample would be very useful. Although many animal experiments that involve obtaining only one sample from each animal (*e.g.*, collecting different mouse liver samples every few hours) do not require this complexity, the relative ease with which cosinor regression can accommodate hybrid experimental designs is a great advantage. Indeed, established packages for cosinor regression have been developed for many programming languages and have seen widespread use.[10–14] Notably *LimoRhyde*[15] leverages existing methods and packages for transcriptomic data analysis to directly apply a version of cosinor regression to RNA-Seq data. In transcriptome-wide experiments, the expression of thousands of genes is measured. *LimoRhyde* uses the variability of all these different biological signals to help in its estimation of the variability of the signal being studied, constraining the error terms in our model.

Of course, cosinor methods for rhythm detection also have important weaknesses: by assuming a sinusoid shape, cosinor regression has less power to detect non-sinusoidal rhythms. Moreover, it searches for rhythms with a pre-specified period P. This can be difficult if the period of the rhythm is not known prior to the experiment. While one can repeatedly run the analysis using different assumed values for the P, this reduces the statistical power of the test. (See discussion of repeated testing below). Finally, linear regression is quite sensitive to outliers and extreme values. Yet such observations are common in large-scale experiments. More modern methods have demonstrated improved sensitivity and specificity in identifying rhythms in real-world data.

23.3.5 Lomb–Scargle and Fourier Type Methods - Looking for Repeating Signals with More complicated Shapes

Fourier methods build on the strengths of Cosinor models while not requiring a particular waveform shape. One of the primary results of Fourier theory, first developed for the study of heat transfer, is that periodic signals of any shape can be reproduced using the sum of sine and cosine waves of various frequencies.[16] Any signal can be expressed "in frequency space" by providing the required combination of sine and cosine waves rather than the explicit description of the waveform as a function of time (Figure 23.5). The fast Fourier transform (FFT) is a computationally efficient, discrete formulation of Fourier analysis that has allowed the application of these methods to data sampled at regular temporal intervals. Often used in digital video and audio processing, the FFT produces a spectrogram showing the relative contribution of sinusoidal terms required at each frequency to reproduce the observed signal. As you might suspect, signals that are rhythmic at a particular frequency require large contributions from sine and cosines waves at that frequency and spectral "peaks" are observed. Statistical tests can be used to see if the relative contribution of any given frequency (adjusted power) is more than what would be expected given the null hypothesis

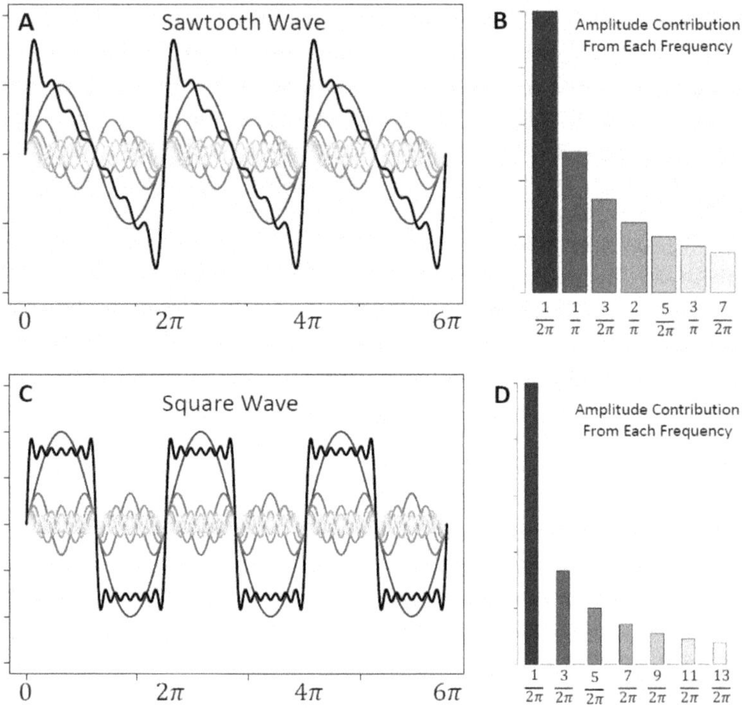

Figure 23.5 Examples of Fourier theory. Fourier theory shows that any "well behaved" periodic waveform can be recreated as the sum of sinusoidal waves. Panel A depicts the first 7 waves (shown in different shades of grey) in the series that converge to form a "sawtooth wave". The summation of these seven waves is shown in black and is beginning to approximate the target sawtooth. The frequencies of the waves plotted in A, and the relative amplitude of each wave is shown in B. The Fourier decomposition of a wave is often shown as a power spectrum (rather than amplitude spectrum) where the power of a wave is related to the square of its amplitude. Panel C depicts the first 7 waves (also shown in different shades of grey) in the Fourier series that converge to a square wave. The frequencies of the waves plotted in C, and the relative amplitude of each wave is shown in D. In C a large "wiggle" is seen near each transition. This is termed "Gibbs phenomenon" and occurs when the target waveform being recreated has discontinuities. As the number of component waves in the Fourier series increases the height of this wiggle will remain relatively unchanged – but they will get progressively narrower.

(arrhythmic data). The expected power spectrogram of pure Gaussian noise, for example, is well established. Like linear regression, Fourier analysis is computationally efficient and well understood. It also overcomes one of the key limitations of cosinor analysis: it does not assume an underlying shape for the biological rhythm. However, these benefits come at a cost. The standard discrete Fourier transform requires regularly sampled data and cannot accommodate missing values. As biological experiments often do not go exactly as planned, missing values are commonplace, and this has been a serious limitation. The Lomb–Scargle[17,18] (LS) periodogram is a generalization of the FFT, developed for astronomy, that can accommodate irregularly sampled data or missing values. LS has experienced increasing adoption in the study of biological rhythms and comparative studies have shown it to be relatively sensitive in detecting rhythms in the setting of real-world noise. Nonetheless, the LS method is less computationally efficient as compared to the methods above and cannot be as easily expanded to accommodate more complex experimental designs. Other methods, which use a more general system to represent signals, called wavelets, have also seen greater use. Wavelet methods[19–21] have been used with particular strength in evaluating more densely sampled signals – like activity patterns, heart rate, temperature, or blood pressure, where subtle differences in the shape of the underlying signal can be reliably estimated and may reflect important physiology.

23.3.6 Non-parametric Methods (JTK and RAIN) – Looking for Blocks of Increasing and Decreasing Abundance

As we have noted, large-scale biological measures are noisy. They often have systematic biases, and the observed biological noise often does not follow a normal distribution (that is, the noise in our measurements or the errors in our regression models often do not follow a traditional "bell curve"). The temporal profiles of biological rhythms are not all perfectly sinusoidal, nor do they clearly follow another specific family of other temporal patterns easily described by varying a few parameters. Non-parametric tests that do not assume a particular form for the experimental noise, specify exact waveform shapes, or show sensitivity to outliers, are well suited for biological rhythm research. As a rule, when compared to parametric methods that make more assumptions, non-parametric statistical methods tend to be more robust. But making fewer assumptions sometimes comes at the cost of decreased sensitivity. In the case of identifying rhythms in genome-scale data, however, some of these non-parametric methods have outperformed parametric approaches in terms of both sensitivity and specificity.

Among the first and most popular of these tests is the JTK_CYCLE[22] (Jonckheere–Terpstra Kendall) algorithm. As a first step, the JTK_CYCLE algorithm replaces the observed values of a biological signal x_i with their rank values (*e.g.*, "this is the smallest value we observed, this is the next smallest"). In this way, one or two extreme measures that might have resulted from an artifact or an outlier subject have only a limited influence on the results. The highest value we observed, for example, is just the highest value. The fact that a single extreme measurement might be 10 times higher than any other measurement is never considered in the analysis.

JTK_CYCLE then combines two established statistical tests to compare the blocks of increasing and decreasing values observed in these ranked data with a library of rhythmic reference waveforms. The Jonckheere–Terpstra test evaluates the medians of several populations (or in this case, blocks of samples) and compares two hypotheses, namely, the null hypothesis that (1) these values are collected from a common population, or (2) that they follow a pre-established order. The Kendall test, on the other hand, analyzes the rank correlations between two variables. By default, the original JTK_CYCLE test uses a reference library of symmetric sine and cosine waves. These reference waves thus all have equally sized blocks of increasing and decreasing values. The newer, BooteJTK[23] (bootstrapped empiric JTK), expands the repertoire of reference waveforms to include a family of asymmetric waves and also improves the statistical correction that is required when testing against several reference waveforms of different shapes, periods, and acrophases.

RAIN[24] is another non-parametric approach that broadens the diversity of waveforms that can be identified. Rather than testing against a specific library of reference waveforms, it uses the non-parametric Mann–Whitney test to see if the rank-ordered data show increasing and decreasing blocks of specified lengths. Thus, like BooteJTK, it is designed to identify a richer repertoire of biological signals.

Notably, as all these methods use rank-ordered data, they cannot directly compute the amplitude or MESOR of the observed pattern. Rather, once the statistical tests establish that a waveform of a particular period and phase is present, other methods (like cosinor regression) are used to estimate these parameters. As in autoregression, methods like RAIN require that the periods being investigated are well matched by the sampling scheme, for example, it might be a little harder to identify 20-hour rhythms with a 6-hour sampling scheme.

23.3.7 Metacycle – Putting it all Together

Knowing that each of the different methods described above has unique strengths and weaknesses, it would be natural to wonder if simply combining them to reach a consensus might be useful. This is the idea underlying Metacycle.[25] Metacycle implements JTK_CYCLE and LombScargle, along with another algorithm, ARSER,[26] that combines features of cosinor regression and autocorrelation. Metacycle first determines if any of the algorithms cannot be applied to the data set to identify rhythms of the desired period – for example, some methods require data from more than one cycle while others cannot accommodate missing data or replicate measures. Each of the applicable tests is then applied to the data and a p-value is computed for each method. Metacycle uses Fischer's method to combine these p-values into a single result. Fischer's method was developed to combine the results of several independent experiments where the same null hypothesis was tested. It provides a recipe for combining the p-values from each of the independent experiments in a single meta-analysis. Notably, this is not how Fischer's test is employed by Metacycle. Metacycle uses Fischer's test to combine p-values generated by applying different, but clearly related, statistical tests to the same data. Moreover, the p-values generated by some of the methods (most notably JTK_CYCLE) do not follow the expected distribution when applied to non-rhythmic data. As a result, the statistical rigor of combining p-values in this way can be questioned and other methods to aggregate testing results can be considered. But empirical testing shows that Metacycle performs well when applied to both real-world and simulated data tests.[27]

23.4 General Comments and Recommendations for Identifying Rhythms

Regardless of the method selected, several factors can influence our ability to accurately detect rhythms. The number of samples that are collected over the course of a single cycle, the number of cycles that are followed, the amplitude of the rhythm, the noise of the individual measurements, and the non-circadian variability of the signal all influence the sensitivity and specificity of our results. Some of the methods discussed above, particularly those based on identifying a particular waveform's shape, can theoretically identify rhythmic waveforms using data from a single cycle. This approach requires strong prior knowledge that the biological signal is indeed rhythmic with an established, pre-defined period. Still most investigators prefer having data spanning at least two complete cycles. In this way, they can confirm that the waveform repeats.[28]

We have tried to emphasize the different intuitions that have guided the development of different rhythm detection methods. Yet, in the end, they all have the same goal: to distinguish between rhythmic and arrhythmic signals. It might then come as no surprise that, for evenly spaced sampling protocols with sufficient sample collection schemes, one can formally prove that several of the above methods give identical or at least mathematically related results. For example, for evenly spaced data the autocorrelation function, Fourier transform, and cosinor regression approaches are all closely related.[29] The choice among these algorithms is thus often related to how easily they can be applied given the constraints of your data or the complexities of the experimental design.

As we noted at the outset, there are several excellent methods that were not reviewed here. Some investigators have also tried to combine aspects of the above methods. For example, rank-based autocorrelation methods have been suggested to enhance the robustness of autocorrelation to experimental noise and extreme values.[30,31] Similarly, cosinor regression models that include more than one frequency can capture signals with more complex waveforms.[10]

A technical but important distinction must also be made between visualization and analysis. While it is a convention in some fields to "double plot" the data – repeating the data when plotting them to emphasize their rhythmic nature – one should never duplicate the data prior to the statistical analysis. This procedure grossly overestimates the statistical confidence in identifying rhythms and leads to many, many false positive results. Simply put, repeating almost any signal over and over will artificially produce a rhythm.[28]

Regardless of the methodology employed, there are clear theoretical limits to our ability to correctly characterize rhythms. Perhaps the most widely known is the Nyquist–Shannon sampling theorem.[32] Accurately identifying features of a signal with a particular frequency requires that samples are collected with at least twice that frequency. For example, to identify and characterize rhythms with a 24-hour period, we must collect samples more than once every 12 hours. The idea behind the Nyquist–Shannon theorem becomes clear if we think of simple systems. Imagine that we are taking snapshots (*e.g.*, pictures) of a very simple oscillating system – a rotating wagon wheel that spins clockwise. Assume that the wheel makes a complete revolution every minute. A single spoke is painted red so that we can easily monitor its position. If we take a snapshot of the wheel every second during its rotation, we will be able to flip through these pictures like a movie and capture the motion. The red spoke will return to its starting position every minute. But now instead imagine that we only take a new picture every 55 seconds. If the red spoke started out at the 12 o'clock position, it will have moved all the way to the 11 o'clock position before the next snapshot is taken. In the next snapshot, it will be at 10 o'clock. If we flip through these snapshots, it will appear as if the wheel is moving counterclockwise $12 \rightarrow 11 \rightarrow 10\ldots$. The wheel would appear to complete a backward rotation after 12 snapshots, or 11 minutes. We would have completely misunderstood the motion of the wheel. This phenomenon – called aliasing – can often be appreciated when watching old western movies where the wagon wheels appear to spin slowly backwards. Of course, the Nyquist–Shannon theory provides only a theoretical limit for the best-case scenario when we are allowed to follow a noiseless signal for as long as we want. The requirements for identifying biological rhythms in real data are often much, much stricter. Experimental studies on *genome scale data* that compared different sampling schemes suggested that balancing concerns of statistical sensitivity and specificity with financial costs and technical difficulty, experimental designs that extract ∼10–12 samples per cycle and cover at least two full cycles are ideal.[33,34] These recommendations were developed with the aim of simply identifying rhythms. A higher sampling density (more frequent sample collection) may be required to precisely characterize rhythm parameters such as the acrophase or amplitude of a waveform.[35] Of course, the "optimal" design for any study depends on financial constraints, the intrinsic variability of the biological signal, and the accuracy of the measurements. These factors are clearly changing as technology improves and depend on the biological signals being studied. If you are focused on a single signal with very high amplitude or when you have strong prior knowledge that a rhythm exists – more restrained experimental designs are often sufficient. Still, these heuristic suggestions provide a good starting point for most researchers.

Newcomers to the field often overlook the need for multiple-test correction when identifying rhythms in large-scale biological data.[36] Statistical tests for rhythm detection are based on observing a measured quantity in our data that would be very unusual in a non-rhythmic signal. A traditional *p*-value cutoff of 0.01 implies that only 1 out of 100 measures would be expected to be as extreme as the value we observed if the null hypothesis was correct. But if tens of thousands of non-rhythmic genes, proteins, metabolites, or other signals are being tested – many of them would likely fit the "1 out of 100" threshold. Multiple-test correction is thus required when a single test is applied to many variables. Corrections are also required when several tests are applied on a single variable – like when a tracing is compared to several reference waveforms.

The simplest approach, termed the Bonferroni correction, creates a stricter threshold by dividing the desired, single *p*-value threshold by the number of tests that are being performed. That is, if one originally would have set a *p*-value threshold of 0.01, but then tests 100 different biological signals, the new Bonferroni corrected threshold required to reject the null hypothesis is $p < 0.01/100 = 0.0001$. In modern molecular experiments tens or hundreds of thousands of measurements are often performed. Bonferroni correction thus sets a very strict threshold – one that is often difficult to meet. In trying to minimize the type I errors (incorrectly rejecting the null hypothesis and classifying non-rhythmic genes as rhythmic), the Bonferroni-corrected threshold will likely have missed many truly rhythmic signals.

Given the strictness of Bonferroni corrections, false discovery rate (FDR) approaches are more commonly applied to genome-scale data.[37] The FDR approach compares how many signals met various significance levels in our real data with how many would have been expected to have met those thresholds if all the signals were non-rhythmic. If we measured 10 000 arrhythmic transcripts in a time course experiment, we would expect to find approximately 1000 signals with a *p*-value < 0.01 and 100 with a *p*-value < 0.001. If we observe 500 transcripts in our experiment that met this $p < 0.001$, we can estimate that $\sim 100/500 = 0.20 = 20\%$ of these might be "false discoveries". Based on our needs we can set the FDR threshold of 5, 10 or 20%, accepting that this fraction of our "hits" will likely be incorrect. Notice that when using the FDR approach, the FDR assigned to a particular biological signal depends not only on that signal but also on all the other signals being tested. Formally, the FDR method also requires that the various variables being tested are all statistically independent. Biological signals operate as part of interconnected networks and the independence of the individual signals is often in doubt. Thus, while FDR approaches are generally accepted as the current standard in analyzing genomic data, statisticians are working to develop methods that better account for biological reality.

Finally, it should be emphasized that statistical significance and biological significance are not one and the same. The amount of a hormone in my blood might regularly and precisely change from 100 to 101 parts per million every day. But depending on the physiology of the hormone, those small but highly statistically significant changes might induce no meaningful physiological changes. On the other hand – there are times when even modest changes can have significant physiological consequences. With experimental noise or technical limits, physiologically critical oscillations might not reach our pre-determined statistical thresholds. Given this concern, researchers sometimes try to impose amplitude thresholds when identifying potentially important biological rhythms. They might, for example, require that a signal has at least a 25% change from peak to trough in addition to meeting a *p*-value threshold. This is a reasonable screening strategy, but it cannot replace biological understanding. In some systems, a 5% change could have very real consequences. But if an enzyme is already saturated, even doubling the amount of its substrate might have no appreciable effect. Simple statistical tests can provide guidance to biological researchers, but they cannot yet replace well-informed biological judgement and the need for experimental follow-up.[38,39]

23.5 Detecting Changes in Rhythms

While the algorithms discussed above can identify periodic signals, researchers often focus on identifying *changes* in rhythmicity. Does the amplitude, acrophase, or period of a particular behavior or biological signal change with drug administration? Do these rhythms differ in the young and old? Do rhythms change with climate? Are certain biological rhythms lost in disease? These topics are among some of the most active areas of biological rhythm research.

Many investigators attempt to answer these questions by simply using the algorithms for rhythm detection: using what has recently been termed the "Venn diagram approach", they separately identify signals that meet a predetermined threshold for rhythmicity in each condition (*e.g.*, $p<0.01$).[40] Then they identify the genes or signals that were identified as rhythmic in one condition (data set) but not the other. But this approach has both high false positive and false negative rates. For example, we could imagine that (in truth) a particular signal is rhythmic and generally remains unchanged by a perturbation. But there may have been fewer samples in one of the two conditions or more experimental noise in one of our trials. In one condition, we reject the null hypothesis of 'no rhythm' with a $p<0.05$ while in the other condition, our metric is right above the required threshold (with a $p=0.052$). The Venn diagram approach would incorrectly conclude that there is a marked change between the two conditions. Similarly, investigators have simply compared the amplitudes and/or periods of rhythmic signals that were estimated in the two different conditions and assumed that any difference reflects a meaningful change in biology. Ultimately, the difficulty with these strategies is that they assess the wrong null hypothesis. Our initial tests for rhythmicity were focused on assessing how confident we were that the different data sets, viewed in isolation, were not "arrhythmic."

But now we are asking something else. If what we now wish to ask is if the rhythms in two conditions are different, the appropriate null hypothesis is that the rhythms in the two conditions are the same. It is this hypothesis that should be directly tested.

Cosinor-type methods were among the first to be used for differential rhythmicity analysis. The statistical methods used more generally for comparing the fit of "nested" regression models are well established and are readily applied to cosinor regression models. Nested model tests can compare how well data are fit by a base linear model and by an expanded linear model that includes everything in the base model – plus additional terms. For example, if one had a base model that predicts an individual's weight as a function of their height, you could compare this with an expanded model that predicts weight as a function of height and waist circumference. Of course, adding flexibility to any model is bound to improve its ability to fit data. An F test (as used in ANOVA) can be used to evaluate the difference in nested model fits and assess if our data support rejecting the null hypothesis and adding these new model terms. In differential rhythms research, the null hypothesis (base model) is to assume that the rhythms are the same in both conditions and are well fitted by a single rhythm waveform. The alternative hypothesis is reflected in an expanded model that assumes different rhythmic waveforms are required to fit data arising from different conditions. This test is often employed after rhythms are first identified in one or both conditions.

Rhythm detection methods that focus on ranked data, like JTK_CYCLE and RAIN, are by their nature, robust to outliers. The robust differential rhythmicity algorithm implemented in DODR[41] uses such a rank-based approach. It uses a cosinor regression-type model on ranked data and replaces the F test used to compare rhythmic models with a dispersion-based measure more appropriate for these data.

An alternative approach is to not define a specific, null hypothesis, but rather use information theory to choose among competing models. The "model selection" approach is also more flexible as it does not require the competing models to be nested. However, it tends to produce more false positive results and does not result in a p-value (as there is no null hypothesis) or confidence measure that is easily interpretable by most biologists.[40] Then again, a good argument can be made that researchers are too reliant on p-values.[38]

Notably all three of the above methods are implemented in the compareRhythms R package and can be applied to any sort of biological data.[40] The LimoRhyde[15] program employs a nested cosinor regression model for differential rhythmicity analysis that specifically uses the Limma statistical package[42,43] designed for application on transcriptomic data and, in contrast to compareRhythms, has the flexibility to incorporate more complicated experimental designs. The recently released Kronos package uses similar methods but aims to support users with only modest programming experience with functions to aid in rhythm visualization.[44]

The above methods based on linear regression can assess if a rhythm differs between two conditions. They do not tell us exactly *what* is different. Has the amplitude of the oscillation changed? The acrophase? In contrast, CircaCompare, uses a non-linear curve fitting procedure. While the statistical properties of non-linear regression are less well established, this unique approach allows for statistical comparison of specific rhythm parameters such as MESOR, amplitude and phase.[44]

23.6 Interpreting Omics Scale Rhythm Data

Regardless of the experiment or hypothesis being tested, one of the biggest challenges in working with genome-scale data is in interpreting the results.[45] Imagine we test the influence of a drug on the liver. We might measure hepatic gene expression in 10 animals that received the drug and in 10 control animals that did not. If the drug influences liver physiology, a statistical comparison might show hundreds or thousands of transcripts exhibiting differential expressions (with higher or lower expression) between the two conditions. But a list of a thousand gene symbols is overwhelming and not immediately interpretable. In the case of a biological rhythm experiment, 100's or 1000's of genes, proteins, or metabolites might vary with the cell cycle phase, circadian time, or lunar phase. Each of these molecules might have different amplitudes and peak expression times. It is often a daunting challenge to understand how these many changes interact to influence physiology. How can researchers interested in biological rhythms interpret these results?

Individually cycling biological macromolecules act as part of a larger framework of pathways and networks to mediate their influence. Molecules that function in the same pathway, share regulatory elements, or are localized to the same cellular locations can often be considered usefully as belonging to sets. For example, metabolites involved in the Krebs cycle might form one set, while genes associated with Parkinson's disease form another. Some molecules may belong to several sets. Prior biological knowledge is often documented in databases of well-curated gene or metabolite sets like Gene Ontology, KEGG,[46] Reactome,[47] and MSigDB.[48]

23.6.1 Overrepresentation Methods

Several tools have been developed to use annotated gene sets to help interpret genome-scale data.[48–53] Most of these were not specifically designed for biological rhythm research. Nonetheless, they have become essential for molecular rhythm research. Some of the most popular tools rely on overrepresentation analysis. Imagine that 500 of the 10 000 (5%) genes expressed in the liver were differentially expressed after administration of the drug. We would expect that by random chance, ~5% of genes involved in various biological pathways would also be differentially expressed. If, in contrast, 45% (90/200) of the genes involved in bile synthesis were differentially expressed, we might suspect that the drug affects bile acid synthesis. Partitioning the members of each gene set into differentially expressed and normally expressed groups, overrepresentation platforms use well-established statistical tests like Fischer's exact test or the chi-square test to identify the overrepresented sets (Figure 23.6A).

In the context of biological rhythm research, this analysis often takes on one of two forms. (1) Rather than separating results into differentially expressed and not-differentially expressed forms – researchers identify rhythmic and non-rhythmic genes (or differentially rhythmic genes). (2) Focusing more on the specifics of the temporal patterns, some researchers bin rhythmic genes into groups that peak in different temporal windows (*i.e.*, those that peak from 2–4 AM, 4–6 AM...), and search for pathways whose activity appears focused at these times.

Although undeniably useful, these approaches also have important limitations and replication studies often show poor overlap.[45,54] Foremost among these disadvantages is the need for strict significance cutoffs. In high throughput studies, experimental noise may push molecules over or below the pre-established statistical threshold. These small p-value shifts can influence many biological signals and markedly change the pathway overrepresentation results. This is compounded when rhythm researchers impose a second cutoff binning data into temporal windows. Depending on how the windows are set up, a given gene set may be concentrated in time but split over adjoining temporal windows. For example, a gene set where many genes peak near 4 AM might have its members split between the 2–4 AM and 4–6 AM windows and thus be missed. Of course, investigators could try many different temporal window lengths and starting times. But the numerous available options can be overwhelming and the need to correct for multiple testing further reduces sensitivity. Equally important, but often overlooked, is the need to choose an appropriate "background list": If we compare the genes that cycle in brain tissue to the whole list of genes that are expressed anywhere in the human body, much of the enrichment we observe may just result from the fact that these rhythmic genes were specifically expressed in the brain. The more appropriate test would be to compare the rhythmic transcripts in the brain to the complement of transcripts expressed in the brain rather than the full human genome. Some,[49] but not all,[50] of the established tools allow users to easily choose or submit background lists that are appropriate for their experimental design.

23.6.2 Aggregate Score Methods

Aggregate score methods were developed to be more robust and overcome some of the difficulties with overrepresentation. Aggregate score methods allow each individual gene to contribute incrementally to the score of the entire set. In this way, aggregate score tools remove the need for strict cutoffs and are more robust against small variations. The most popular method, gene set enrichment analysis[45,48] (GSEA) uses a weighted Kolmogorov–Smirnov (K–S) test

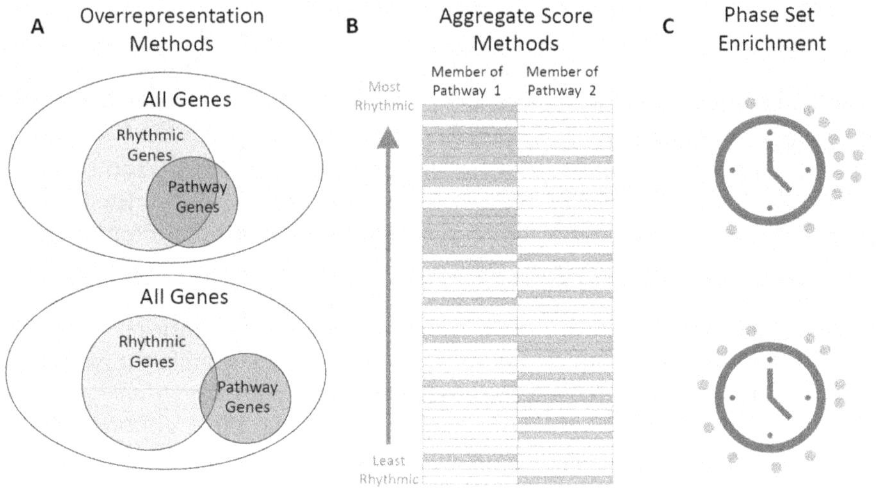

Figure 23.6 Methods to interpret genome scale rhythm experiments. Several techniques have been developed to use prior biological knowledge for interpreting the results of genome scale experiments. Prior knowledge is often summarized as annotated gene, protein, or metabolite sets. Such sets describe molecules that function in the same pathway, share structural features, or are in some other way related. Panel A depicts the overrepresentation approach. A discrete cutoff (often based on *p*-value) is used to classify all tested molecules as either rhythmic or arrhythmic. The set of rhythmic genes is then compared to various annotated gene sets. The significance of the overlap is assessed by a hypergeometric (or related) statistical test. If, for example, 10% of all tested molecules are rhythmic, but 50% of the molecules in a given pathway are rhythmic, cycling molecules are said to be overrepresented in that pathway. The Venn diagram shown in A (*top*) is an example where cycling genes are overrepresented in the given pathway. The Venn diagram shown in A (*bottom*) shows a pathway that is not overrepresented for cycling. Panel B depicts an aggregate score approach implemented by gene set enrichment analysis (GSEA).[48] The application of GSEA for this purpose requires that genes are all ranked by some measure of rhythmicity (*e.g.*, log (*p*-value)). For each gene set, GSEA uses the Kolmogorov–Smirnov test to check if the members of that gene set are concentrated toward one end of the list or if they are distributed at random. Two different pathways are shown in the two columns in B. If a given gene is a member of the pathway, its entry is colored grey. The first column shows a pathway where the grey colored pathway member genes are concentrated at the top of the ranked gene list. The pathway is enriched for high cycling. The second column depicts a gene set where the various members are randomly distributed across the ranked list. Panel C depicts phase set enrichment analysis (PSEA).[56] PSEA is an aggregate score method that focuses on the timing of rhythmic genes or molecules in a pathway rather than the number of cycling molecules in that pathway. PSEA identifies gene sets where the phases of cycling molecules demonstrate some form of temporal coordination (top) or are uniformly distributed across the periodic cycle (bottom).

to identify gene sets whose members are non-uniformly distributed among a ranked list. Metabolite set enrichment takes the same approach.[55] In the context of biological rhythms, all the genes (or metabolites) that were tested can be ranked based on their *p*-value, F-statistic, or other measures used to express our confidence that they are cycling. GSEA then finds gene sets where the members of the set are concentrated toward the front or back of the list, indicating gene sets that are enriched for rhythmic or non-rhythmic molecules (Figure 23.6B). This method also effectively deals with the issues related to choosing the correct background. The enrichment is based on a ranked list – if you did not measure a particular molecule and it is not on the list – it is not included in the calculation.

Although well-suited for identifying biological pathways enriched for cycling components, most aggregate-score tools are poorly suited for studying periodic variables themselves. These tools assume that measurements are monotonic with small values (*e.g.*, 0 h) far away from larger ones (*e.g.*, 24 h). This is not true for phase data. Circadian time (CT) 0 is the same as CT24. The end of one cell cycle is the start of the next. Thus, GSEA and similar methods incorrectly measure the temporal distance between rhythmically expressed molecules. Phase set enrichment analysis (PSEA)[56] is an aggregate score method designed to identify gene sets and biological pathways that show phase coordination. PSEA uses Kuiper's[57] test to find gene sets where the phases of cycling molecules are not uniformly distributed across the periodic cycle but rather show some form of temporal concentration (Figure 23.6C). Kuiper's test can also be used by investigators trying to identify periodic signals in the frequency of other discrete events like daily or weekly rhythms in heart attacks or hospital admissions. Notably PSEA still requires that investigators use a hard cutoff to determine which genes are cycling – even if the phases of those genes are analyzed continuously. PSEA does not solve all the problems with the temporal binning approach.

None of the above methods typically distinguishes between genes that work as activators or repressors in their respective biological pathways. This often reflects the limits of biological annotation. Yet, the expression of pathway repressors in anti-phase to pathway activators may work to augment temporal coordination. While PSEA still often identifies such temporal coordination as a bi-modal (and thus non-uniform) phase distribution, a better approach would be to somehow incorporate the roles of each molecule in the network. Similarly, the above approaches do not distinguish between genes with more or less essential roles in each pathway. Although not discussed here, network enrichment methods have been developed to overcome this limitation.

23.7 Characterizing Biological Rhythms Without Time Course Data

Collecting time course data has been the central experimental method in rhythmic data analysis. Nonetheless, collecting time course data is often difficult, impractical, or even unsafe. This is especially true when the samples we require come from human patients or experimental subjects. If we want to learn the clinical importance of biological rhythms, we will eventually have to characterize them in clinical populations. But can we really ask each patient to stay in a laboratory for repeated sample collection each day? Can we safely and ethically get around-the-clock tumor biopsies? Over the last decade, prompted by the desire to learn about biomedical rhythms at a scale that could support larger biomedical trials, several groups have developed methods to learn about biological rhythms without traditional time course experiments. These generally can be divided into three groups, which we will label (1) supervised time stamping methods, (2) unsupervised ordering methods, and (3) organization assessment methods. To overcome the lack of time course data, all these methods take advantage of the fact that, for many biological rhythms, many different signals share the same oscillatory period. The cell cycle, yeast metabolic cycle, and circadian clock all regulate the expression of hundreds or thousands of transcripts, proteins, and molecules. In each case, almost all those molecules have the same fundamental period but with many different acrophases.

23.7.1 Supervised Time Stamping

Although the idea is very old, *time stamping methods* use what modern researchers would term *supervised* machine learning. Carl Linnaeus, the prolific categorizer of genus and species, is credited for developing one of the earliest time stamping methods. Linnaeus recognized that different flowering plants, like the waterlily and milk thistle, open and close their petals at different times of day.[58] After cataloging the rhythms of many different plants (collecting training data), Linnaeus realized he could carefully select plants (informative features) that had different temporal preferences. If he incorporated those plants in a specially organized garden – he could use his garden to tell time. For example, coat's beards tend to open from 5 am–11 am and cat's ears open from 9 am–1 pm. So, if we take a picture of our garden, and see that the cat's ears were open, but the goat's beards were closed – we can estimate that the picture was taken between 11 am and 1 pm. With the advent of high throughput molecular biology, Hiroki Ueda and colleagues realized that the abundance of different biological molecules could replace the plants in Linnaeus' flower clock.[59,60] Experimenters could carefully collect training data to learn which biomolecular features exhibit cyclic behavior in a given tissue and determine the timing of their peaks and troughs. Then they could compare any new measurement with their library of examples. If, after learning the temporal profile in the murine liver, we see that BMAL is high, CRY1 is low, and DBP is low – we should conclude that the mouse liver was in its biological night. In recent years, several researchers have expanded upon this idea for estimating human circadian phase. Groups have used various biomolecular features, signals arising from different tissues, different "gold standards" for internal circadian phase, and different machine learning algorithms and normalization schemes. Some, for example, use whole blood gene expression data to predict the melatonin phase,[61] a well-established marker for the phase of SCN activity.[62] Others have worked with transcriptomic[63] or metabolic data to predict the time of day ("wall clock time") when a given sample was collected in a "normal" population. Currently, these methods report accuracies in the range of ±3 hours in estimating circadian time. Studies that have used whole blood for this purpose require two samples from each subject, separated by ~12 hours in time.

Like all supervised learning methods, time stamping requires training data that span the tissues and conditions covered in later applications. The flower clock we plant in Panama will likely not even survive, let alone tell time, if planted in Germany. Supervised learning methods are thus ideally suited for tissues where a large collection of temporally annotated, high-quality training data can be obtained – *e.g.*, hair, blood, and skin. Biomarkers that can be recorded non-invasively like the heart rate, body temperature, activity, or other non-invasive biomarkers are also well suited for this approach. In contrast, obtaining molecular signals or gold standard rhythm assessments specific for tissues like the brain and lungs is likely to be more challenging.

There are also many non-circadian differences between individuals (*e.g.*, diet, age, biologic sex, genetic ancestry, and activities). Do some of these variables influence our time-telling molecular signals? The abundance of BMAL in my blood may vary from 5–20 while in yours it may vary from 10–40. A "high level" in my blood may be "low" in yours. Data collected in a lab in Philadelphia may consistently underestimate or overestimate levels as compared to those in a lab in New Orleans. In the same vein, if a particular disease state influences our "time telling molecules", our ability to "tell time" in patients with those diseases would be compromised. Thus, collecting the data and validating these methods on a large and diverse cohort of people will be critical. Nonetheless, by carefully selecting biomolecular features and algorithms, investigators are working to make these supervised methods more robust. Using isolated blood monocytes – as opposed to whole blood – is hypothesized to reduce non-circadian variation. Different people have different fractions of monocytes, neutrophils, and lymphocytes in their blood so that, presumably, analyzing gene expression in purified monocytes will reduce the variability inherent to whole blood measurements. Intriguing work in a smaller number of subjects indicates that only a single sample may be required if isolated monocytes are used in conjunction with more accurate transcriptional measurement technologies.[64] Some research studies suggest that a blood-based metabolomics-approach has similar potential.[65,66] Alternatively, skin samples appear to have higher-amplitude molecular rhythms and are thus a compelling source for single-sample circadian biomarkers.[67] Similarly, different "normalization" or scaling methods have been proposed to reduce some of non-circadian influences.[63] If supervised learning methods and circadian biomarker panels are sufficiently accurate and can be shown to generalize to a large and clinically diverse population, they have the potential to greatly expand the realm of circadian medicine and clinical biological rhythms research.

23.7.2 Unsupervised Ordering

Temporal ordering methods adopt a different approach. They do not require annotated training data. Ordering methods "stich together" many samples – where time is unknown – with the goal of reconstructing a full cycle. Rather than mapping each sample to a specific external time, they order the samples relative to one another (*e.g.*, this sample comes before this one in the cycle – but after that one – in the cycle). These methods use data from an entire population to model the oscillations in any one sample.

The two most prominent unsupervised methods currently in use, OSCOPE[68] and CYCLic Ordering by Periodic Structure (CYCLOPS),[69] both use the idea that, when plotted against each other, out-of-phase signals that share the same period will form an ellipse (Figure 23.6). One could imagine that a particular signal – signal A – is well described by a sine wave with the formula $\sim\sin\left(\frac{2\pi}{P}t\right)$ for some period P. Another signal – signal B – cycles with the same period but is well fit by a cosine wave with the formula $\sim\cos\left(\frac{2\pi}{P}t\right)$. As each new sample is collected, you could plot the observed value of signal A on the horizontal axis, and the value of signal B on the vertical axis. After collecting many samples at many different times, the points would begin to form a circle. The relative position of the samples in that circle would reflect their relative position – or place order – in the underlying biological cycle. While both OSCOPE and CYCLOPS were developed for transcriptomic data, the two methods use this insight differently and were initially developed for studying different biological rhythms.

OSCOPE was developed to order single-cell transcriptomic data by cell-cycle phase. These cells are often grown on the same dish, but each cell may be at a different point in the cell cycle. OSCOPE assumes that enough samples/cells have been collected, so that the samples span all the phases of the cell cycle. OSCOPE then tests every pair of genes, 2 by 2, checking to see, if when plotted against each other, they form an ellipse and thus may share a common biological rhythm (Figure 23.7A and B). It then identifies groups or clusters of genes that all form ellipses with each other, and thus all presumably cycle with the same underlying period. OSCOPE then uses the expression of these genes as if they were spatial coordinates describing the position of each sample. Like a travelling salesman who must visit many cities before returning home, OSCOPE finds a path that connects all the samples in a loop, minimizing the total distance traveled on the road of cycling genes. This oval path is the reconstructed cycle, and once it has been established, the expression of any given gene or other measured quantity can be considered a function of the reconstructed cell cycle phase and tested for rhythmicity.

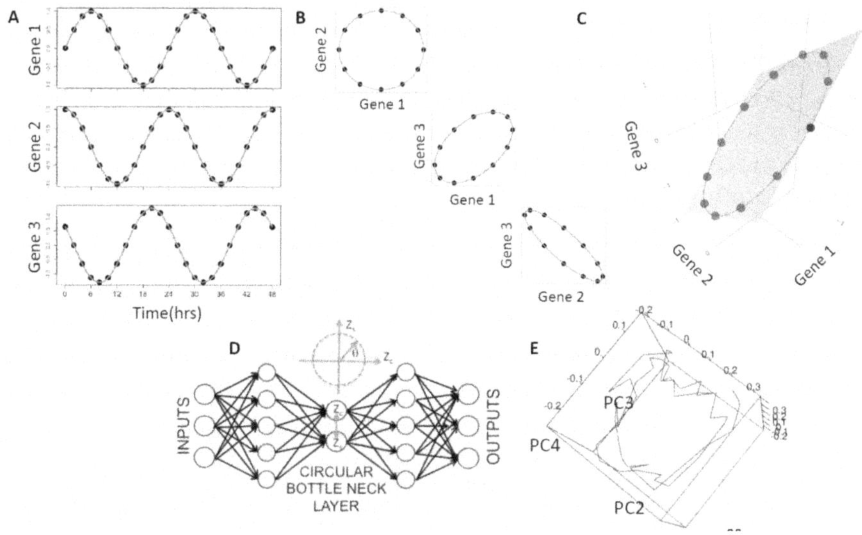

Figure 23.7 Intuitions underlying OSCOPE and CYCLOPS. OSCOPE[68] and CYCLOPS (CYCLic Ordering by Periodic Structure)[69] are two ordering methods that can be used without time course data. Many biological rhythms induce the cycling of hundreds or thousands of molecules all with the same period. Both OSCOPE and CYCLOPS rely on the insight that when plotted against each other, out-of-phase oscillations that share the same period will form an ellipse. Panel A shows the idealized behavior of three such rhythmic molecules – all with a period near 24 hours. In B, the abundance of each of these molecules is plotted as a function of another. All plots form ellipses. OSCOPE identifies clusters of genes that all form ellipses when plotted against each other and thus likely share a common period. OSCOPE uses the expression of these genes in each sample to define that sample's position. OSCOPE then finds the shortest path that connects all the samples and returns to the starting position. This oval path is the reconstructed cycle, and once it has been established, the expression of any given gene or other measured quantity can be evaluated as a function of reconstructed cycle phase. In panel C the expressions of all three genes shown in A are plotted against each other on a 3-dimensional graph, the data trace out a 2-dimensional ellipse embedded in this higher dimensional space. Similarly, a 2-D ellipse will emerge embedded in the higher dimensions if more genes were included. D. In human data there is often non-circadian variability, CYCLOPS uses a neural network to identify the closed 2-d elliptical curve that best captures the variance of noisy high-dimensional data. The CYCLOPS neural network is given gene expression data for each sample. It is trained so that its output best matches this input. The intervening bottle neck layer forces the network to optimally encode these data on a closed elliptical curve. Once the closed curve that best describes the samples is identified, each sample's angular position on the curve reflects its relative position in the molecular cycle. E Principal component analysis was applied to time course transcriptomic data from mouse liver data collected by Hughes and colleagues.[34] A plot of principal components 2, 3, and 4 is shown. When viewed from an appropriate angle, the elliptical structure in these real data becomes evident.

CYCLOPS, on the other hand, was initially developed for analyzing circadian rhythms in transcriptomic data collected from different people. Rather than looking at each pair of genes in isolation to find many different ellipses, CYCLOPS begins with a "seed list" of genes. These are genes that the user selected because they likely cycle as part of the same rhythmic biological process. CYCLOPS does not require that all these seed genes cycle in expression, but it does require that many of them do. CYCLOPS applies singular value decomposition to these seed gene data. This is a mathematical technique to reduce the dimensionality of the data and identify "eigengenes," or characteristic gene expression patterns that describe the seed genes.[3]

As we have already mentioned, plotting the expression of one cycling gene against another out-of-phase cycling gene forms an ellipse. But the same is true if we have 3 cycling genes. If we plot the expression values for all three of these cycling genes on a 3-dimensional graph, a 2-dimensional ellipse will emerge (Figure 23.7C). While we cannot draw a 6- or 7-dimensional graph to plot 6 or 7 cycling genes, in principle the various samples will still form a closed curve lying on a 2-dimensional plane embedded in that space. The coordinates of each point on the plane are a linear combination of all the dimensions. Of course, if there are non-circadian trends in the data, these points will not all lie perfectly on a plane or on an ellipse. CYCLOPS uses a very simple neural network specially constructed to find closed curves in high-dimensional data (Figure 23.7D). It finds the single ellipse that *best* fits all the samples when plotted in this higher-dimensional eigengene space (Figure 23.7E). After finding the ellipse that connects all the samples, CYCLOPS uses the angular position of each sample as a marker of its phase in the circadian cycle. Using these phase assignments as a measure of each sample's internal circadian time, classical rhythm finding methods can be used to identify signals that are rhythmic at the shared base period.

Techniques like OSCOPE and CYCLOPS only provide a *relative* ordering. Additional information is needed to relate our ordering to an external reference (*e.g.*, circadian time 0, or new moon). Even the forward direction of time is unknown. However, additional data and biological knowledge can help anchor these orderings to the external world. Moreover, we can usefully define internal reference points, like the acrophase of a given transcript. CYCLOPS orderings have been aligned to the peak of one or several genes. Even without those anchors, the ordered data can be used to identify oscillatory signals that share a common period.

OSCOPE was designed to analyze data from many nearly identical cells on a dish. Other sources of variation are presumed to have only a small influence on the rhythmic genes. If our two out-of-phase rhythmic signals were also affected by age, temperature, or some other feature that differed markedly among the cells, plotting the signals on a graph might no longer lead to the formation of an ellipse. But the OSCOPE method is particularly well suited for ordering data in controlled settings or when one lacks prior knowledge to create a reasonable seed list.

The concerns about additional sources of variability are more pressing when analyzing clinical data arising from different people – the intended use case for CYCLOPS. Here the choice of seed genes based on prior knowledge, normalization of data, use of eigengenes, and identification of a single best-fit-ellipse are intended to make the technique more robust. Nonetheless, the influence of non-circadian variability on the ordering process remains a very significant issue. Methods that can more effectively order data in the setting of technical and biological covariates are needed – along with more ways to test and assess those methods under realistic constraints.

Both approaches work best when they have samples that span the full biological rhythm being assessed. Without such data, the ellipses they are looking for would not be complete. They also benefit when the amplitudes of the rhythmic signals are large, many genes or signals are involved, and different rhythmic biological outputs peak at a wide variety of distinct phases. These factors may not be known before a study has begun. Thus, a related concern is knowing when to trust the orderings that these methods provide. Both CYCLOPS and OSCOPE have developed metrics to assess the confidence a user should place in the quality of an ordering – but these metrics are far from perfect. Often, the most reliable validation comes in the form of prior biological knowledge or new experiments that can validate aspects of the results.

Finally, there has been a recent explosion of new methods devoted to the ordering problem.[70–73] These methods will continue to improve. However, there is not yet a clear consensus as to how well these methods work in the settings of realistic noise, interindividual variation, sampling bias, and batch processing differences. A more systematic platform for benchmarking new methods will likely be required.

23.7.3 Organization Assessment Methods

Like ordering methods, organization-based methods use data from a population of samples to evaluate oscillator function. However, these methods do not attempt to find oscillatory genes or signals. Thus, they may seem out of place in a chapter devoted to identifying and interpreting biological rhythms. However, these simple, intuitive techniques are finding increasing application in circadian research. The insights that underlie them are also likely easily adapted to other systems. A prime example of this approach is the delta clock correlation distance (ΔCCD), which assesses a simple signature of rhythmic organization that can be computed without time course data.[74]

The core circadian clock is composed of a set of activators (*e.g.*, CLOCK and BMAL/ARNTL) that induce the transcription of a set of repressors (*e.g.*, PER and CRY family genes). This internal architecture imparts a clear temporal organization. The activators tend to peak in expression during one part of the circadian day. As a group, the repressors tend to peak almost 12 hours later. The two groups of genes are generally out of phase with each other. As a result, across many samples, the expression of the activators tends to be positively correlated with each other. In samples where BMAL is high, CLOCK also tends to be high. In contrast, in samples where PER1 is high, PER2 also tends to be high. The two groups of genes tend to be anti-correlated with each other. These strong positive and negative correlations can be easily visualized (Figure 23.8A). Notably we do not need to have temporally annotated data to compute these correlations. If we have collected samples from a wide range of circadian times, we might expect to observe this pattern in correlations. If we take many samples and the expressions of PER1 and BMAL are positively correlated – we might suspect that the circadian clock is not functioning normally in (some of) these samples.

Clock correlation distance (CCD) is a metric designed to assess a distance between two matrixes, quantify this intuition, and more formally measure the functional organization of the core circadian oscillator (Figure 23.8A and B). The underlying hypothesis is that datasets that show a pattern of organization that closely matches a "healthy" reference describe functional clocks. The CCD method compares a matrix of Spearman correlation coefficients between 12 core clock genes to a reference correlation matrix that was identified in healthy tissues. Spearman correlation is much like the linear correlation measure introduced early in this chapter. However, the Spearman correlation is rank-based, and thus like the other non-parametric statistics that were discussed, more robust to outliers.

To compare oscillator organization between two groups, the CCD of one group (*e.g.*, a treatment group) is compared to the CCD of the control group, yielding a ΔCCD. At its core, the ΔCCD is a function of which samples are listed as belonging to group 1 and which samples belong to group 2. Is the clock correlation matrix computed from group 1 samples different from the clock correlation matrix from group 2 samples? Does one group really "looks more like the reference" than the other? This framework makes it easy to assess the significance of this ΔCCD metric by bootstrap. That is, we can now randomly re-assign the samples to two different groups – each time computing the ΔCCD between the two random groups. After repeating this process hundreds or thousands of times, we can assess the ΔCCD we observed in our real groups. Would we expect to see such a big difference between two random groups – or is the ΔCCD we observed "unusual"? We can compute an empirical *p*-value.

It is worth reviewing what the ΔCCD does, and does not, show – and where this analysis can go wrong. Like the ordering and time stamping methods, ΔCCD has had to make many assumptions in drawing conclusions without time course data. As in our other methods that operate without time course data, the most practical concerns relate to non-circadian sources of variability and sampling biases. The simple nature of the ΔCCD test, however, can make it easier to appreciate how these issues can complicate our analysis.

The expected correlation patterns require that we have sampled a wide variety of circadian times – and without too strong temporal bias in our sampling. If almost all our samples come from a limited range of circadian

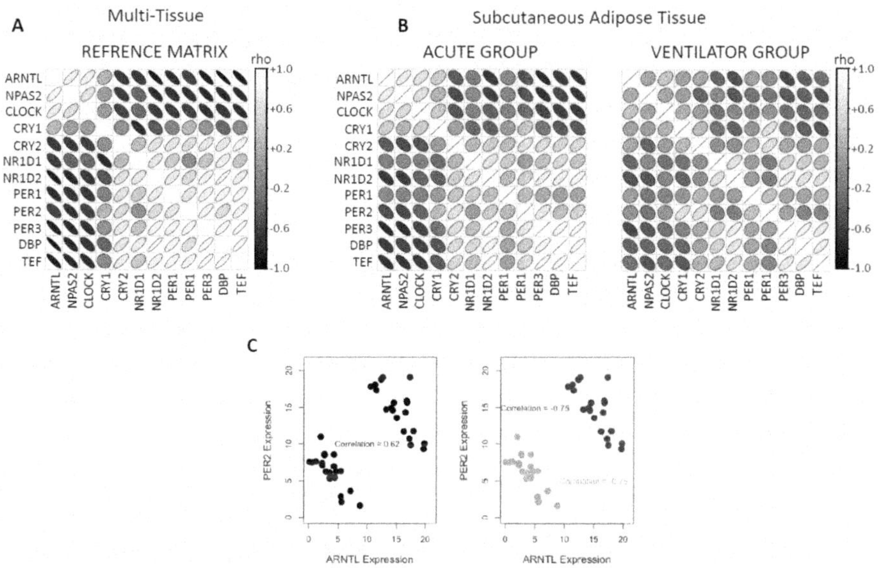

Figure 23.8 Clock correlation distance (CCD) as a measure of circadian organization. The core circadian oscillator is composed of a set of transcriptional activators (*e.g.*, CLOCK, ARNTL, and NPAS2) that induce the transcription of a set of repressors (*e.g.*, PER and CRY family transcripts). The activators tend to be expressed ~12 hours before the repressors. In samples where *ARNTL* expression is high, *CLOCK* expression tends to be high, whereas in samples where *PER2* expression is high, *CRY2* expression also tends to be high. In tissues with a functioning clock the expressions of *ARNTL* and *CLOCK* (and the other activators) tend to be positively correlated across samples while the expressions of *PER* and *CRY* genes also tend to be positively correlated. The two groups of transcripts tend to be negatively correlated with each other. (A) Hughey and colleagues used data from several tissues to create a reference matrix describing the rank correlation pattern between core clock transcripts in samples with functioning circadian clocks.[74] This matrix visually depicts the correlation between each pair of transcripts. Light grey ellipses slanted to the right reflect positive correlations while black ellipses slanted to the left reflect negative correlations. Narrower ellipses represent stronger correlations. The matrixes shown here were plotted with the deltaCCD r package.[74] (B) The genotype expression GTeX dataset uses an autopsy-based sample collection. As an example of the CCD approach, Hollis *et al.* compared the clock correlation matrix observed using samples obtained from individuals who had a sudden or traumatic death with the correlation pattern observed using samples from individuals who died on a ventilator.[77] In many tissues, including adipose tissue, the clock correlation pattern obtained from subjects who died on a ventilator and in intensive care appeared disrupted. (C) The CCD, like other existing techniques that operate without time course data, can be hampered by non-circadian variation or biases in sample collection time. An extreme, idealized example is shown. Combining all data (C *left*), *PER2* and *ARNTL* expression are positively correlated suggesting a departure from reference function and organization. However, these data (C *right*) are composed of data from two different groups (*e.g.*, smokers and non-smokers, males and females, patients with one of two cancer subtypes). In each group the correlation between PER2 and ARNTL is negative and follows the expected pattern.

phases, our rank-correlation measure will not look like our reference. We might have many samples with high ARNTL and low PER with very few samples showing the opposite pattern. This could be likely if samples are collected through clinical practice, where most procedures are performed during a limited range of working hours.

The ΔCCD also ignores the fact that one of the two groups may simply have more non-circadian variability. This variability could reflect random noise, but it could also reflect systematic differences in disease severity or some other non-circadian feature. If other features, like age, diet, or smoking status modulate the expression of the selected clock genes, the expected correlation patterns may appear weaker across the population. Disease states are often, by their very nature, heterogeneous across a population. Indeed, non-circadian confounding can even reverse the sign of the expected correlation patterns in the setting of perfectly normal clock function. Imagine that two types of breast cancers have functioning circadian clock – and in every tumor, ARNTL and PER2 alternate daily with higher and lower expression. However, we could equally imagine that in one type of cancer both ARNTL and PER2 expression are much higher than in the other type of cancer. Even though all the tumors have functioning clocks and, in each tumor, ARNTL and PER2 are negatively correlated, when we put the data from the two cancers together, the correlation between ARNTL and PER2 will appear positive. Relatively low ARNTL and PER2 values are observed together in patients with one cancer type while higher values observed together in patients with the other cancer type (Figure 23.8C). This contrived scenario is an example of Simpson's paradox and one of several situations where organization-based metrics can be misleading.[75,76] Although attempts have been made to address this problem,[77] current methods only account for labeled covariates. If the non-circadian source of variation is not known – it cannot be addressed. It is also theoretically possible, if unlikely, that a dataset could generate a clock correlation matrix that perfectly matches the ideal reference, but that included no samples with a functioning clock. Some samples could remain fixed in time with high ARNTL and low PER2 – while other samples remain fixed in time with low ARNTL and high PER2. Notably, most of these same concerns also apply to the organization and time stamping methods described above.

While organization methods are not perfect, they remain very useful – as long as one remembers their limits. They provide a quick and visually reassuring test that can help experimenters and informaticians determine if more exhaustive studies or more refined analysis techniques are worth pursuing. Informatics researchers are also working to augment correlation-based signatures of clock organization with the requirement that genes in the core circadian clock are highly variable across samples, using this variability as another marker of intact rhythms.[78] More general molecular signatures that predict "healthy circadian function" might be discovered someday. It is equally plausible that informatic signatures will be developed for other, similarly ubiquitous, biological rhythms.

23.8 Conclusions

The tasks of quantitatively evaluating biological data for rhythms, rigorously identifying changes in rhythm parameters, and cogently interpreting rhythmic signals are often daunting challenges – especially for new students. As in all biological sciences, rhythm-focused researchers must work to discern biological signals from the various sources of experimental noise. Designing experiments to minimize errors and account for confounding variables is critical. Once those experiments are completed, however, a growing body of tools and approaches is available to aid the researcher in analyzing these data. While sometimes hard to appreciate, simple intuitions often underly many seemingly complex mathematical approaches. A heuristic understanding of these methods can greatly aid investigators in choosing tests that are appropriate for their data. Best practice guidelines can help guide both the newcomer and experienced researcher and several user-friendly packages make it relatively easy to implement these tests with only modest programing skills. Indeed, new web-based platforms are now offering "one-stop shopping" for many rhythm analysis methods.[79] For more complicated experiments or questions, dedicated bioinformaticians and statisticians who focus on biological rhythms can be an excellent guide.

Like molecular biology and biological rhythms more generally, informatic techniques to identify and interpret rhythms are evolving. While many "classic" approaches remain among the most useful tools, new techniques are also being developed as researchers ask new questions and explore new technologies. Creative solutions are being sought to overcome practical experimental constraints.

Acknowledgements

David Raizen, Germaine Cornelissen, and Nicholas Lahens provided valuable feedback on early drafts of this chapter.

References

1. J. Buck and E. Buck, *Science*, 1968, **159**, 1319–1327.
2. B. P. Tu, A. Kudlicki, M. Rowicka and S. L. McKnight, *Science*, 2005, **310**, 1152–1158.
3. O. Alter, P. O. Brown and D. Botstein, *Proc. Natl. Acad. Sci.*, 2000, **97**, 10101–10106.
4. J. A. Hobson and E. F. Pace-Schott, *Nat. Rev. Neurosci.*, 2002, **3**, 679–693.
5. L. A. Urry, M. L. Cain, S. A. Wasserman, P. V. Minorsky, J. B. Reece and N. A. Campbell, *Campbell Biol.*, 2017, 1123–1134.
6. Z. Wu, R. Irizarry, R. Gentleman, F. M. Murillo and F. Spencer, *J. Am. Stat. Assoc.*, 2004, **99**, 909–917.
7. A. Butte, *Nat. Rev. Drug Discovery*, 2002, **1**, 951–960.

8. D. B. Allison, X. Cui, G. P. Page and M. Sabripour, *Nat. Rev. Genet.*, 2006, **7**, 55–65.
9. P. Dutilleul, *Biol. Rhythm Res.*, 1995, **26**, 173–193.
10. G. Cornelissen, *Theor. Biol. Med. Modell.*, 2014, **11**, 16.
11. M. Moškon, *BMC Bioinf.*, 2020, **21**, 1–12.
12. R. Hou, L. E. Tomalin and M. Suárez-Fariñas, *BMC Bioinf.*, 2021, **22**, 1–7.
13. M. M. Doyle, T. E. Murphy, M. A. Pisani, H. K. Yaggi, S. Jeon, N. S. Redeker and M. P. Knauert, *Comput. Methods Programs Biomed.*, 2021, **209**, 106292.
14. CRAN – Package cosinor, https://cran.r-project.org/web/packages/cosinor/index.html (accessed 10 May 2023).
15. J. M. Singer and J. J. Hughey, *J. Biol. Rhythms*, 2019, **34**, 5–18.
16. C. C. Lin and L. A. Segel, *Mathematics Applied to Deterministic Problems in the Natural Sciences*, Society for Industrial and Applied Mathematics, 1988, p. 609.
17. E. F. Glynn, J. Chen and A. R. Mushegian, *Bioinformatics*, 2006, **22**, 310–316.
18. J. D. Scargle, *Astrophys. J.*, 1982, **263**, 835–853.
19. T. L. Leise, *Methods Enzymol.*, 2015, **551**, 95–119.
20. T. L. Leise and M. E. Harrington, *J. Biol. Rhythms*, 2011, **26**, 454–463.
21. T. S. Price, J. E. Baggs, A. M. Curtis, G. A. Fitzgerald and J. B. Hogenesch, *Bioinformatics*, 2008, **24**, 2794–2795.
22. M. E. Hughes, J. B. Hogenesch and K. Kornacker, *J. Biol. Rhythms*, 2010, **25**, 372–380.
23. A. L. Hutchison, R. Allada and A. R. Dinner, *J. Biol. Rhythms*, 2018, **33**, 339–349.
24. P. F. Thaben and P. O. Westermark, *J. Biol. Rhythms*, 2014, **29**, 391.
25. G. Wu, R. C. Anafi, M. E. Hughes, K. Kornacker and J. B. Hogenesch, *Bioinformatics*, 2016, **32**, 3351–3353.
26. R. Yang and Z. Su, *Bioinformatics*, 2010, **6**, 168–174.
27. W. Mei, Z. Jiang, Y. Chen, L. Chen, A. Sancar and Y. Jiang, *Briefings Bioinf.*, 2021, **22**, 1–13.
28. M. E. Hughes, K. C. Abruzzi, R. Allada, R. Anafi, A. B. Arpat, G. Asher, P. Baldi, C. de Bekker, D. Bell-Pedersen, J. Blau, S. Brown, M. F. Ceriani, Z. Chen, J. C. Chiu, J. Cox, A. M. Crowell, J. P. DeBruyne, D. J. Dijk, L. DiTacchio, F. J. Doyle, G. E. Duffield, J. C. Dunlap, K. Eckel-Mahan, K. A. Esser, G. A. FitzGerald, D. B. Forger, L. J. Francey, Y. H. Fu, F. Gachon, D. Gatfield, P. de Goede, S. S. Golden, C. Green, J. Harer, S. Harmer, J. Haspel, M. H. Hastings, H. Herzel, E. D. Herzog, C. Hoffmann, C. Hong, J. J. Hughey, J. M. Hurley, H. O. de la Iglesia, C. Johnson, S. A. Kay, N. Koike, K. Kornacker, A. Kramer, K. Lamia, T. Leise, S. A. Lewis, J. Li, X. Li, A. C. Liu, J. J. Loros, T. A. Martino, J. S. Menet, M. Merrow, A. J. Millar, T. Mockler, F. Naef, E. Nagoshi, M. N. Nitabach, M. Olmedo, D. A. Nusinow, L. J. Ptáček, D. Rand, A. B. Reddy, M. S. Robles, T. Roenneberg, M. Rosbash, M. D. Ruben, S. S. C. Rund, A. Sancar, P. Sassone-Corsi, A. Sehgal, S. Sherrill-Mix, D. J. Skene,

K. F. Storch, J. S. Takahashi, H. R. Ueda, H. Wang, C. Weitz, P. O. Westermark, H. Wijnen, Y. Xu, G. Wu, S. H. Yoo, M. Young, E. E. Zhang, T. Zielinski and J. B. Hogenesch, *J. Biol. Rhythms*, 2017, **32**, 380–393.

29. N. Wiener, Extrapolation, Interpolation, and Smoothing of Stationary Time Series: With Engineering Applications (M.I.T. Press Series), The MIT Press, 1964.
30. A. Dürre, R. Fried and T. Liboschik, *Wiley Comput. Stat.*, 2015, **7**, 205–222.
31. C. Croux and C. Dehon, *Stat. Methods Appl.*, 2010, **19**, 497–515.
32. C. E. Shannon, *Proc. IRE*, 1949, **37**, 10–21.
33. A. Deckard, R. C. Anafi, J. B. Hogenesch, S. B. Haase and J. Harer, *Bioinformatics*, 2013, **29**, 3174–3180.
34. M. E. Hughes, L. DiTacchio, K. R. Hayes, C. Vollmers, S. Pulivarthy, J. E. Baggs, S. Panda and J. B. Hogenesch, *PLoS Genet.*, 2009, **5**, e1000442.
35. M. E. Hughes, K. C. Abruzzi, R. Allada, R. Anafi, A. B. Arpat, G. Asher, P. Baldi, C. De Bekker, D. Bell-Pedersen and J. Blau, *J. Biol. Rhythms*, 2017, **32**, 380–393.
36. W. S. Noble, *Nat. Biotechnol.*, 2009, **27**, 1135.
37. Y. Benjamini and Y. Hochberg, *J. R. Stat. Soc. B*, 1995, **57**, 289–300.
38. R. L. Wasserstein, A. L. Schirm and N. A. Lazar, *Am. Stat.*, 2019, **73**, 1–19.
39. C. M. Bishop, *Pattern Recognition and Machine Learning*, Springer, 1st edn, 2007.
40. A. Pelikan, H. Herzel, A. Kramer and B. Ananthasubramaniam, *FEBS J.*, 2022, **289**, 6605–6621.
41. P. F. Thaben and P. O. Westermark, *Bioinformatics*, 2016, **32**, 2800–2808.
42. M. E. Ritchie, B. Phipson, D. Wu, Y. Hu, C. W. Law, W. Shi and G. K. Smyth, *Nucleic Acids Res.*, 2015, **43**, e47.
43. B. Phipson, S. Lee, I. J. Majewski, W. S. Alexander and G. K. Smyth, *Ann. Appl. Stat.*, 2016, **10**, 946–963.
44. R. Parsons, R. Parsons, N. Garner, H. Oster and O. Rawashdeh, *Bioinformatics*, 2020, **36**, 1208–1212.
45. N. R. Clark and A. Ma'ayan, *Sci. Signal*, 2011, **4**, tr4.
46. H. Ogata, S. Goto, K. Sato, W. Fujibuchi, H. Bono and M. Kanehisa, *Nucleic Acids Res.*, 1999, **27**, 29–34.
47. D. Croft, A. F. Mundo, R. Haw, M. Milacic, J. Weiser, G. Wu, M. Caudy, P. Garapati, M. Gillespie, M. R. Kamdar, B. Jassal, S. Jupe, L. Matthews, B. May, S. Palatnik, K. Rothfels, V. Shamovsky, H. Song, M. Williams, E. Birney, H. Hermjakob, L. Stein and P. D'Eustachio, *Nucleic Acids Res.*, 2014, **42**, D472–D477.
48. A. Subramanian, P. Tamayo, V. K. Mootha, S. Mukherjee, B. L. Ebert, M. A. Gillette, A. Paulovich, S. L. Pomeroy, T. R. Golub, E. S. Lander and J. P. Mesirov, *Proc. Natl. Acad. Sci. U. S. A.*, 2005, **102**, 15545–15550.
49. D. W. Huang, B. T. Sherman and R. A. Lempicki, *Nat. Protoc.*, 2009, **4**, 44–57.
50. E. Y. Chen, C. M. Tan, Y. Kou, Q. Duan, Z. Wang, G. V. Meirelles, N. R. Clark and A. Ma'ayan, *BMC Bioinf.*, 2013, **14**, 1–14.

51. M. V. Kuleshov, M. R. Jones, A. D. Rouillard, N. F. Fernandez, Q. Duan, Z. Wang, S. Koplev, S. L. Jenkins, K. M. Jagodnik, A. Lachmann, M. G. McDermott, C. D. Monteiro, G. W. Gundersen and A. Ma'ayan, *Nucleic Acids Res.*, 2016, **44**, W90–W97.
52. R. A. Irizarry, C. Wang, Y. Zhou and T. P. Speed, *Stat. Methods Med. Res.*, 2009, **18**, 565–575.
53. J. Reimand, R. Isserlin, V. Voisin, M. Kucera, C. Tannus-Lopes, A. Rostamianfar, L. Wadi, M. Meyer, J. Wong, C. Xu, D. Merico and G. D. Bader, *Nat. Protoc.*, 2019, **14**, 482–517.
54. D. W. Huang, B. T. Sherman and R. A. Lempicki, *Nucleic Acids Res.*, 2009, **37**, 1–13.
55. J. Xia and D. S. Wishart, *Nucleic Acids Res.*, 2010, **38**, W71–W77.
56. R. Zhang, A. A. Podtelezhnikov, J. B. Hogenesch and R. C. Anafi, *J. Biol. Rhythms*, 2016, **31**, 244–257.
57. N. H. Kuiper, *Nederl. Akad. Wetensch. Proc. Ser. A*, 1960, **63**, 38–47.
58. March 2013: Linnaeus' Floral Clock | The Linnean Society, https://www.linnean.org/news/2013/03/04/march-2013-linnaeus-floral-clock (accessed 1 May 2023).
59. Y. Minami, T. Kasukawa, Y. Kakazu, M. Iigo, M. Sugimoto, S. Ikeda, A. Yasui, G. T. J. van der Horst, T. Soga and H. R. Ueda, *Proc. Natl. Acad. Sci. U. S. A.*, 2009, **106**, 9890–9895.
60. H. R. Ueda, W. Chen, Y. Minami, S. Honma, K. Honma, M. Iino and S. Hashimoto, *Proc. Natl. Acad. Sci. U. S. A.*, 2004, **101**, 11227–11232.
61. T. Kasukawa, M. Sugimoto, A. Hida, Y. Minami, M. Mori, S. Honma, K. Honma, K. Mishima, T. Soga and H. R. Ueda, *Proc. Natl. Acad. Sci.*, 2012, **109**, 15036–15041.
62. E. E. Laing, C. S. Möller-Levet, N. Poh, N. Santhi, S. N. Archer and D. J. Dijk, *eLife*, 2017, **6**, e20214.
63. R. Braun, W. L. Kath, M. Iwanaszko, E. Kula-Eversole, S. M. Abbott, K. J. Reid, P. C. Zee and R. Allada, *Proc. Natl. Acad. Sci. U. S. A.*, 2018, **115**, E9247–E9256.
64. N. Wittenbrink, B. Ananthasubramaniam, M. Münch, B. Koller, B. Maier, C. Weschke, F. Bes, J. De Zeeuw, C. Nowozin, A. Wahnschaffe, S. Wisniewski, M. Zaleska, O. Bartok, R. Ashwal-Fluss, H. Lammert, H. Herzel, M. Hummel, S. Kadener, D. Kunz and A. Kramer, *J. Clin. Invest.*, 2018, **128**, 3826.
65. D. Cogswell, P. Bisesi, R. R. Markwald, C. Cruickshank-Quinn, K. Quinn, A. McHill, E. L. Melanson, N. Reisdorph, K. P. Wright and C. M. Depner, *J. Biol. Rhythms*, 2021, **36**, 369–383.
66. T. Woelders, V. L. Revell, B. Middleton, K. Ackermann, M. Kayser, F. I. Raynaud, D. J. Skene and R. A. Hut, *Proc. Natl. Acad. Sci.*, 2023, **120**, e2212685120.
67. G. Wu, M. D. Ruben, R. E. Schmidt, L. J. Francey, D. F. Smith, R. C. Anafi, J. J. Hughey, R. Tasseff, J. D. Sherrill, J. E. Oblong, K. J. Mills and J. B. Hogenesch, *Proc. Natl. Acad. Sci. U. S. A.*, 2018, **115**, 12313–12318.
68. N. Leng, L. F. Chu, C. Barry, Y. Li, J. Choi, X. Li, P. Jiang, R. M. Stewart, J. A. Thomson and C. Kendziorski, *Nat. Methods*, 2015, **12**, 947.

69. R. C. Anafi, L. J. Francey, J. B. Hogenesch and J. Kim, *Proc. Natl. Acad. Sci. U. S. A.*, 2017, **114**, 5312–5317.
70. B. J. Auerbach, G. A. FitzGerald and M. Li, *Nat. Commun.*, 2022, **13**, 6580.
71. Y. Larriba, I. C. Mason, R. Saxena, F. A. J. L. Scheer and C. Rueda, bioRxiv, 2023, preprint, 11, 2022.12.21.519625.
72. L. Talamanca, C. Gobet and F. Naef, *Science*, 2023, **379**, 478–483.
73. S. Liang, F. Wang, J. Han and K. Chen, *Nat. Commun.*, 2020, **11**, 1–8.
74. J. Shilts, G. Chen and J. J. Hughey, *PeerJ*, 2018, **6**, e4327.
75. M. L. Samuels, *J. Am. Stat. Assoc.*, 1993, **88**, 81.
76. Y. Mittal, *J. Am. Stat. Assoc.*, 1991, **86**, 167.
77. H. C. Hollis, J. N. Francis and R. C. Anafi, *Front. Physiol.*, 2022, **13**, 942704.
78. G. Wu, M. D. Ruben, L. J. Francey, Y. Y. Lee, R. C. Anafi and J. B. Hogenesch, *Bioinformatics*, 2022, **38**, 5375–5382.
79. T. G. Brooks, A. Mrčela, N. F. Lahens, G. K. Paschos, T. Grosser, C. Skarke, G. A. FitzGerald and G. R. Grant, *J. Biol. Rhythms*, 2022, **37**, 43.

CHAPTER 24

Quo Vadis

GERMAINE CORNELISSEN*[a] AND TSUYOSHI HIROTA*[b]

[a] Halberg Chronobiology Center, University of Minnesota, Minneapolis, MN, USA; [b] Institute of Transformative Bio-Molecules, Nagoya University, Nagoya, Japan
*Emails: corne001@umn.edu; thirota@itbm.nagoya-u.ac.jp

24.1 Introduction

Keen observers have reported biological cycles with a period of about 24 hours since antiquity.[1,2] While the study of biological rhythms started in earnest only in the 1920s, when the study of rhythms became institutionalized,[3] the discovery of the circadian clock established the study of chronobiology on a solid molecular basis, with the general model for mammalian cellular circadian timekeeping continuing to be refined. As documented and illustrated throughout this book, circadian rhythms play an important role in every aspect of physiology, and their disruption is associated with major disease conditions.

24.2 Interdigitating Networks

Previous chapters have highlighted the complex role played by the circadian system as biological processes usually involve multiple inter-modulating and inter-digitating physiological systems. Since multiple biological domains interact with each other, future studies would greatly benefit from concurrently analyzing their interactions in a systematic manner. The overview of the different physiological systems covered in Part II of this book clearly shows their interconnection as follows:

Chemical Biology No. 23
Chronobiology and Chronomedicine: From Molecular and Cellular Mechanisms to Whole Body Interdigitating Networks
Edited by Germaine Cornelissen and Tsuyoshi Hirota

Published by the Royal Society of Chemistry, www.rsc.org

Already in the suprachiasmatic nuclei (SCN), multiple regional oscillators involving different oscillator cell networks have been reported and documented, as reviewed in Chapter 5. The complete orchestration by the SCN of other brain regions and peripheral organs is, in itself, a complex network that affects physiology as a whole. The evolutionary link between the circadian and hypoxia pathways, both dependent on PAS domains, as light- and oxygen-sensing are interconnected on a molecular level, is reviewed in Chapter 6. The role of inflammation and the immune system in respiratory disorders such as asthma, COPD, pneumonia, and sepsis is illustrated in Chapter 7. Chapter 8, which is related to the renal system, explains how the kidneys serve endocrine functions. Prominent examples include the significance of renin output from the kidney, angiotensin, and aldosterone production in maintaining healthy blood pressure and their relevance to the development of hypertension.

How the microbiome is regulated by the gastrointestinal clock through the presentation of nutrients, hormones, and changes in body temperature is described in Chapter 9, which also shows that some of the bacteria have their own endogenous circadian clocks. The immune system is implicated in some gastrointestinal disorders such as inflammatory bowel disease. As hormones secreted from endocrine glands relay circadian timing signals produced in the SCN to the organs in the whole body, as discussed in Chapter 10, the endocrine system regulates all biological processes in the body, including the development of the brain and nervous system, the growth and function of the reproductive system, and metabolism.

Circadian disruption through environmental or genetic factors often leads to immune system dysfunction. As documented in Chapter 11, failure of the immune system can result in conditions such as allergy, lupus and multiple sclerosis (when the immune system over responds) or HIV and cancer (when the immune system under responds), thus affecting multiple physiological systems and biological processes. The complexity of the haematopoietic system stems from cellular interactions and regulatory mechanisms controlled by the nervous system through circulation, as reviewed in Chapter 12. For instance, among mature cell populations leaving the bone marrow in large numbers are monocytes that further mature into macrophages, considered as the master regulators of local inflammation and fibrosis.

As reviewed in Chapter 13, in addition to providing protection to internal organs and allowing movement, the skeletal system generates blood cells and serves important endocrine functions, such as the regulation of mineral homeostasis. Moreover, skeletal disorders, such as rheumatoid arthritis, again implicate the immune system. Energy homeostasis critically intersects with the skeletal muscle system, since it involves different ways through which muscle cells regulate glucose, lipid, and mitochondrial metabolism on a daily basis, as discussed in Chapter 14. Chapter 15 which focuses on the integumentary system illustrates the important roles played by the cell cycle given the rapid differentiation of skin cells, and by the immune system in view of skin's function to defend the organism against pathogens, chemicals, UV light, and mechanical injury. As reviewed in Chapter 16, reproductive success depends on tightly orchestrated physiological processes. For instance, the timing of ovulation depends both on intact neuronal signaling and coincident estrogen and SCN signals.

24.3 Multi-frequency Rhythms

24.3.1 A 12-hour Clock?

Circadian rhythms are not the only biological cycles. Several other chapters refer to extra-circadian rhythms. The great interest that circadian rhythms generate stems from the discovery of a molecular mechanism underlying their manifestation. Recently, evidence emerged suggesting that oscillations with a period of about 12 hours may be distinct from the circadian rhythm.[4–7] In addition to the 24-hour rhythm, two classes of genes cycling with periods of 12 and 8 hours were originally identified in mouse liver. Distinct 12-hour transcriptional rhythms were also observed in other peripheral tissues such as the heart, kidney, and lungs.[4] The genes cycling with about 12-hour periods in the mouse liver were enriched in the endoplasmic reticulum (ER) and mitochondria-associated metabolism pathways.[5,6] Further studies uncovered prevalent 12-hour gene expression and metabolic rhythms in the mouse liver, coupled with a physiological 12-hour unfolded protein response oscillation. These studies identified a cell autonomous mammalian 12-hour rhythm driven by a dedicated 12-hour pacemaker distinct from the circadian clock that entrains *in vitro* to metabolic and ER stress cues.[6,7] The 12-hour clock is thought to be regulated by the unfolded protein response (UPR) transcription factor XBP1s.[7,8]

Disruption of the 12-hour clock was associated with progression to non-alcoholic fatty liver diseases, while expression of XBP1s in neurons was sufficient to increase the longevity of *C. elegans*.[6] An amplified 12-hour rhythm of heart rate variability helped consolidate the circadian system of astronauts, helping them to adapt to the space environment.[9] Either a too short or too long coupling interval between the 12-hour and 24-hour rhythms in blood pressure resulted in morning or evening hypertension, reduced measures of wellbeing and psychological resilience, and elevated blood pressure, dyslipidemia, arteriosclerosis, and a depressive mood.[10] These examples illustrate the potential roles of the 12-hour clock in regulating human diseases and in guiding chronotherapy in medical practice. As noted, having multiple clock components endows organisms with more flexibility and an enhanced ability to adapt to different environments, thus markedly increasing their survival advantage.[6]

24.3.2 An About-weekly Clock?

Only indirect evidence exists for an endogenous about-weekly rhythm. Apart from socially-synchronized 7-day rhythms characterizing physiological variables[11] as well as pathological conditions, as seen in relation to the incidence of adverse cardiovascular events[12] and suicides,[13] the endogenous nature of about 7-day (circaseptan) rhythms is evidenced from their ability to free-run and their amplification, if not induction, by environmental triggers, which set their phase.

For instance, a circaseptan rhythm in tumor labeling index is apparent in data from H4-IIE-hepatoma-bearing rats after a single i.p. dose of 150 mg kg^{-1} cyclophosphamide.[14] From an epidemiologic viewpoint, Reimann[15] highlighted the dominant incidence of a weekly rhythm in the so-called periodic diseases. Patients with Hodgkin's disease may occasionally exhibit recurrent febrile exacerbations at intervals of 7 or multiples of 7 days.[16] From a clinical viewpoint, Derer[17] described the occurrence of a 6-day rhythm in human blood leukocytes, reticulocytes, excretion of uric acid, and glucocorticoids, among others.

Like circadians, circaseptans have been known since antiquity. Hippocrates (*ca.* 460 BC–*ca.* 370 BC),[18] Galen (*ca.* 129–*ca.* 200),[19] and Avicenna (980–1037)[20] reported the relative incidence of critical days (fever) as a function of days elapsed since disease onset, which exhibits a dampened about 7-day variation.[21] Circaseptan rhythms in DNA synthesis and mitotic activity of the regenerating kidney characterize rats after unilateral nephrectomy, peaking two days after the cytoreductive procedure and at near integer multiples of 7 days thereafter.[22] Similar circaseptan rhythms characterize the regenerating kidney after 20 minutes of contralateral ischemia.[23] Since similar responses may result from the administration of cytoreductive nephrotoxic drugs such as aminoglycoside antibiotics or the anti-cancer drug cisdiamminedichloroplatinum, circaseptans in DNA synthesis and mitotic index may be amenable to new therapeutic approaches.[11]

Rejection episodes after kidney transplantation on different weekdays occurred more frequently about 7, 14, 21 or 28 days post-operation in studies carried out in Milan, Italy,[24] Paris, France,[25] or Minnesota, USA,[24] which involved 212 rejection episodes in 579 patients. Even after a mild stimulus like balneotherapy, about 7-day variations characterize visual reaction time and flicker-fusion frequency recorded longitudinally in 44 patients over four weeks.[26] When referenced to the start of treatment, these variables exhibit a similar phase irrespective of the day of the week when balneotherapy started.[21,26] The first evidence of free-running stems from a 15-year long record, where the circaseptan component of urinary excretion of 17-ketosteroids began to deviate from 7 days after the start of testosterone self-administration, while urine volume remained 7-day synchronized.[27]

Perhaps the most striking demonstration of circaseptans relates to the time course of blood pressure and heart rate monitored during the first few weeks of life, observed most prominently in neonates born prematurely with a low birth weight, as seen in studies carried out in Minnesota, Italy, Spain, the Czech Republic, Germany, Russia, and Japan.[28] On an individual basis, the circaseptan period can deviate significantly from 7 days, indicating free-running. On a population basis, birth synchronizes the phase of the 7-day rhythm, not the day of the week.[28] A small study on twins also found that the circaseptan period, when assessed nonlinearly, was more similar between twins in a pair than among different twin pairs, providing support for a partly endogenous nature of circaseptans.[29]

About-weekly and half-weekly rhythms persist in conditions of social isolation.[30,31] They characterize a number of different variables in different species,[32] including unicells[33,34] and cell cultures.[35] In light transmission studied in *Acetabularia* under constant environmental conditions, removal of the nucleus resulted in the lengthening of the circadian period, the shortening of the circaseptan period, and a decreased circaseptan-to-circadian amplitude ratio.[36] By contrast, when studying dentin formation in rats exposed to constant light, SCN ablation resulted in an increased circaseptan-to-circadian amplitude ratio.[37] These results point to anatomical sites from which circaseptans may have originated.

New opportunities arise to optimize treatment by scheduling its administration according to a circaseptan pattern,[38] as suggested empirically. In a study involving segmental pancreatic transplantation performed on Ma Lewis (RT-1l) recipients with ACI (RT-1a) donor rats, a 7-day modulated daily dose of the immunosuppressant agent cyclosporine prolonged graft function by 30% as compared to rats treated with equal daily doses.[39] Circaseptan scheduling also improved outcomes in studies on the immunomodulation of malignant growth in LOU rats with immunocytoma.[40]

Very little, however, is known about underlying mechanisms. Circaseptans and other multi-day rhythms are known to characterize the recurring growth lines in developing mammalian tooth enamel; these rhythms are strongly associated with all adult tissue, organ, and body masses as well as life history attributes such as gestation length, age at maturity, weaning, and lifespan, particularly among the well-studied primates.[41] While underlying molecular mechanisms remain elusive, metabolomics and genomics pathway analyses from blood plasma in domestic pigs revealed about 5-day rhythms related to growth in which biological functions include cell proliferation, apoptosis, and transcription regulation/protein synthesis, and to degradative pathways.[41] Further investigations aimed at understanding the molecular basis of circaseptans are warranted now since many tools developed to study circadian rhythms could be repurposed to unravel mechanisms underlying the manifestation of about-weekly rhythms.

In contrast to the conventional theory that a biological system is at steady state or cycles with a single frequency, it is now recognized that most biological systems have no homeostatic "set point" but rather oscillate as composite rhythms consisting of superimposed oscillations.[6] While the authors referred to oscillations that often cycle at different harmonics of the circadian rhythm, with the about 12-hour oscillation being the most prevalent, many other rhythms spanning a wide range of frequencies characterize living organisms, as reviewed in Chapter 4.

24.4 Translational Research

24.4.1 Bench Work

New techniques such as genetic modification to reduce the expression of one or more of an organism's genes have permitted the detailed study of physiological processes at the molecular level. By necessity, however, these studies are performed in animal models in the laboratory. Results hence need to be translated for clinical applications, a task that can present challenges beyond accounting for the fact that some animal models are nocturnal while humans are diurnal.

As most research is conducted during regular working hours, one way to facilitate the mapping of circadian rhythms is to first synchronize animals to a fixed lighting regimen of 12 hours of light alternating with 12 hours of darkness (LD12 : 12). By using rooms maintained under opposite lighting regimens, it becomes possible to map the entire circadian profile by collecting samples from both rooms during regular daytime hours. Other lighting regimens such as LD8 : 16 or LD16 : 8 can be used to reflect conditions encountered in winter or summer. Telemetry of whole organisms and bioluminescent probes, in turn, are capable of monitoring certain variables longitudinally, thereby enabling the study of extra-circadian rhythms as well.

24.4.2 Human Studies

The monitoring of marker variables to estimate the internal circadian rhythm of the study participant has long been advocated.[42] It is particularly useful as a guide to treatment timing (chronotherapy). There are several approaches to determine the circadian phase with accuracy. Most of them, however, necessitate the collection of serial samples. Methods have recently been developed and are continuously being improved to estimate the circadian phase from a single or at most very few samples based on the direct assessment of the expression of clock genes and clock-controlled genes, as discussed in several other chapters.

24.4.3 Medications Affecting Circadian Rhythms

As reviewed in several other chapters, several factors can be targeted to modify the period, phase, or amplitude of circadian rhythms. Some of these factors relate to lifestyle changes, such as a regular routine, time-restricted feeding, and exercise. Others, like exposure to bright light in the morning and melatonin administration in the evening have long been known to be useful in the treatment of circadian rhythm sleep disorders, such as delayed sleep phase, advanced sleep phase, free-running, irregular sleep wake, jetlag and shift work.[43,44]

Other factors were approved medications, which were later found to affect circadian rhythms. Among them, lithium continues to be a first-line treatment for bipolar disorder, despite the availability of newer mood stabilizers.[45] The mechanism of action of lithium is not known. Lithium modifies sodium transport in nerve and muscle cells. It alters the metabolism of neurotransmitters, catecholamines and serotonin specifically. Lithium also decreases protein kinase C activity, which alters the genomic expression associated with neurotransmission. Lithium was later found to lengthen the period and increase the amplitude of circadian rhythms.

Metformin, a potent anti-hyperglycemic agent from the biguanide family, is recommended as the first line oral therapy for type 2 diabetes.[46] By decreasing hepatic glucose production, primarily through a mild and transient inhibition of the mitochondrial respiratory-chain complex 1, the AMP-activated protein kinase (AMPK), a cellular metabolic sensor, is activated.[47] Metformin also leads to the activation of liver casein kinase I α (CKIα) and muscle CKIε, known modulators of the feedback loop of the circadian clock.[48] Rapamycin, an mTOR inhibitor and macrolide exhibiting potent antitumor and immunosuppressive activity, is used in clinical settings to prevent rejection in organ transplantation and to treat certain types of cancer.[49] The mammalian target of rapamycin (mTOR) complex 1 (mTORC1) is a master regulator of cell and whole-body energy homeostasis, that serves as an important link between metabolism and the circadian clock, partly by regulating BMAL1 activity.[50] Interestingly, metformin and rapamycin are also considered to be potent geroprotectors with anti-aging effects.[51,52]

24.4.4 Drug Development Targeting the Circadian Clock

The realization that many, if not the majority of, disease conditions are associated with a disrupted circadian system has prompted the development of drugs targeting clock proteins directly, which is now an active area of research. There is indeed an increasing demand to control circadian clock functions in a conditional manner to better understand the circadian system and to discover potential new treatments for clock-related diseases. Small-molecule compounds provide powerful tools to reveal novel functions of target proteins in the circadian clock mechanism, and can be great therapeutic candidates. Chemical screening conducted by measuring cellular circadian rhythms in a high-throughput manner has been identifying compounds that affect circadian rhythms by targeting clock-related proteins.[53] Small-molecule modulators of the circadian rhythm can interact with the clock directly, acting on the clock components of the core loop and secondary stabilization loop (core clock proteins) or with key regulatory mechanisms, which are also capable of significantly altering and controlling the circadian clock (clock related proteins).[54]

A number of kinases, including the serine/threonine kinase, casein kinase 1 (CK1) isoforms CK1δ/ε, regulate the circadian period through post-translational modification of clock proteins.[55] New small-molecule inhibitors of CK1, such as PF-670462, longdaysin, and LH846, have been developed and shown to lengthen the circadian period.[56–59] Modulation of these kinases presents a novel treatment direction for cognitive deficits through circadian modulation. Treatment with PF-670462 improved cognitive performance in mice.[60] CK2, a highly conserved serine/threonine kinase ubiquitously expressed in eukaryotic cells, is another kinase regulating the circadian clock, and a highly specific CK2 inhibitor GO289 was identified from phenotypic screening of circadian clock modulators. GO289 potently inhibits the growth of acute myeloid leukemia cells, establishing a connection between the circadian clock and cancer.[61] Chen *et al.*[62] review strategies for targeting CK2, which is involved in almost all biological processes and regulates various signaling pathways related to cancer progression. Many CK2 inhibitors have been developed. Most of them target the ATP-binding site, such as TBB, CX-4945, and emodin. Thus far, CX4945 is the only small-molecule inhibitor of CK2 that has progressed to clinical trials.[62] While CK2's role is best known in cancer, recent studies have shown its involvement in neurological and psychiatric disorders, infectious diseases, inflammatory and autoimmune diseases, ophthalmic diseases, diabetes, obesity, cardiovascular disease, and cystic fibrosis.[63] Protein kinases are increasingly recognized as valuable targets for clock regulation in cancer.[64]

Nuclear hormone receptors are another important drug target. Many small molecules have been developed to target REV-ERBα (NR1D1) and REV-ERBβ (NR1D2), circadian clock components that function as transcriptional repressors and serve as integrators of cell metabolism with the circadian clock.[65,66] First proposed as a drug target for treating sleep disorders and metabolic syndromes, recent studies found a rather broad role of REV-ERBs in pathological conditions, including local inflammatory diseases, heart failure and cancers. It is also involved in the regulation of circadian drug metabolism that has implications in chronopharmacology.[65,66] Nobiletin is a small molecule shown to activate ROR (retinoid acid receptor-related orphan receptor) nuclear receptors, potentiating circadian oscillation and providing protection against metabolic dysfunctions.[67,68]

Small molecules that modulate clock function may form the basis for drug discovery in clock-related diseases, and an increasing number of compounds that selectively target core clock proteins have been identified. Dysfunction of CRY1 and CRY2 isoforms has been associated with a host of diseases, such as sleep phase disorder and metabolic diseases. The first small molecule identified as a CRY modulator was KL001 that activates both CRY1 and CRY2.[69] KL001 prevents the ubiquitin-dependent degradation of CRY, resulting in the lengthening of the circadian period. Since KL001-mediated CRY stabilization also inhibits glucagon-induced gluconeogenesis in primary hepatocytes,[69] it could help develop novel clock-based treatments for diabetes.[70] KL101 and TH301 are two other small-molecule compounds that lengthen the period of the cellular circadian clock by directly targeting CRYs in mammals.[71] While KL001 targets both CRY1 and CRY2 isoforms, KL101 and TH301 are isoform-selective compounds that require the CRY C-terminal region for their effects[71] and these compounds show no effect on the period in zebrafish, hence illustrating species differences in the effects of isoform-selective CRY-modulating compounds.[72] KL001 derivatives such as KL044[73] and GO200,[74] CRY1-selective compounds such as KL201[75] and TH129,[76] and TH401, which targets both CRY isoforms,[77] were also identified. SHP656, an orally available derivative of KL001, shows efficacy in controlling blood glucose[78] and inhibiting the growth of glioblastoma stem cells (GSC) in animal models.[79] SHP656 and its active isomer SHP1703 were found to be CRY2-selective, suggesting a direct role of CRY2 in glioblastoma anti-tumorigenesis.[80] Intrinsic conformational differences in identical residues of CRY1 and CRY2 contribute to unique interactions with different compound moieties for isoform selectivity.[81,82] In addition to the CRY activators, CRY inhibitors have also been developed and shown to be circadian clock-enhancers.[83] M47 is a small molecule lengthening the circadian period and selectively enhancing the degradation of CRY1; this compound holds promise for treating certain forms of cancer depending on the p53 mutation.[84]

The small molecule CLK8 disrupts the interaction between CLOCK and BMAL1 and interferes with the nuclear translocation of CLOCK. By enhancing the amplitude of the cellular circadian rhythm without affecting the period length, it emerges as a potential candidate for therapeutic development to manage disorders associated with dampened circadian rhythms.[85]

24.5 Concluding Remarks

Looking back at the early days when the terms chronobiology and circadian were first coined and a new discipline was born, it has been a privilege to be a part of the growth, maturation and development of a new science that has such far-reaching applications. The principles underlying every physiological system and biological process reviewed in this book are also applicable in fields other than medicine. Veterinary science,[86] agriculture,[87,88] and pest control[89] are just a few examples. Researchers interested in these areas will also benefit from the information and tools provided in this book.

Technological advances contributed greatly to the breakthroughs and the continued progress made in chronobiology and chronomedicine. Advances made in genetics, cellular and molecular biology help identify critical pathways and their interactions. Without advances in computing and bioinformatics, metabolomics, proteomics and transcriptomics, we would not have information on how organisms make proteins in a global manner and perform other important functions in the cell. Together with advances in chemistry and crystallography, drug development leads to the discovery of many small-molecule compounds targeting the circadian clock, with potential applications in a host of different disease conditions. Newly developed mathematical and statistical approaches benefit the analysis and interpretation of large databases that are now readily obtained, thanks to engineering advances that made data collection possible. Regardless of your area of expertise, much work remains to be done to advance the knowledge of chronobiology quite generally in order to fully realize the potential the field offers for health maintenance and disease prevention as well as treatment when needed.

Among the many questions remaining to be answered, as reviewed throughout this book, one pressing issue challenging the future of the human race relates to space travel. What will time and schedules look like on the Moon or on Mars? In light of the central role played by circadian rhythms regarding health and performance, will our timing system evolve and adapt, and what can we do to be ready when the time comes? With the tools and resources now available, answers to these questions may be sought with confidence. The future of the fields of chronobiology and chronomedicine is bright, with so many prospects for advancing our knowledge and so many opportunities to improve health and combat disease.

References

1. J. Aschoff, Speech after dinner, 1974 Capri Symposium on timing and toxicity, in *Chronobiological Aspects of Endocrinology*, ed. J. Aschoff, F. Ceresa and F. Halberg, Chronobiologia, 1974, vol. 1(Suppl 1), pp. 483–495.
2. K. Otsuka, G. Cornelissen and F. Halberg, in *Chronomics and Continuous Ambulatory Blood Pressure Monitoring, Vascular chronomics: from 7-day/24-h to lifelong monitoring*, Springer, Tokyo, 2016, pp. ix–xx.
3. A. Cambrosio and P. Keating, The disciplinary stake: the case of chronobiology, *Soc. Stud. Sci.*, 1983, **13**, 323–353.
4. M. E. Hughes, L. DiTacchio, K. R. Hayes, C. Vollmers, S. Pulivarthy, J. E. Baggs, S. Panda and J. B. Hogenesch, Harmonics of circadian gene transcription in mammals, *PLoS Genet.*, 2009, **5**, e1000442.
5. M. E. Hughes, H. K. Hong, J. L. Chong, A. A. Indacochea, S. S. Lee, M. Han, J. S. Takahashi and J. B. Hogenesch, Brain-specific rescue of *Clock* reveals system-driven transcriptional rhythms in peripheral tissue, *PLoS Genet.*, 2012, **8**, e1002835.
6. B. Zhu, C. C. Dacso and B. W. O'Malley, Unveiling "Musica Universalis" of the cell: a brief history of biological 12-hour rhythms, *J. Endocr. Soc.*, 2018, **2**, 727–752.
7. B. Zhu, Q. Zhang, Y. Pan, E. M. Mace, B. York, A. C. Antoulas, C. C. Dacso and B. W. O'Malley, A cell-autonomous mammalian 12 hr clock coordinates metabolic and stress rhythms, *Cell Metab.*, 2017, **25**, 1305–1319.e9.
8. Y. Pan, H. Ballance, H. Meng, N. Gonzalez, S-M Kim, L. Abdurehman, B. York, X. Chen, Y. Schnytzer, O. Levy, C. C. Dasco, C. A. McClung, B. W. O'Malley, S. Liu and B. Zhu, 12 h clock regulation of genetic information flow by XBP1s, *PLoS Biol.*, 2020, **18**, e3000580.
9. K. Otsuka, G. Cornelissen, S. Furukawa, K. Shibata, Y. Kubo, K. Mizuno, T. Aiba, H. Ohshima and C. Mukai, Unconscious mind activates central cardiovascular network and promotes adaptation to microgravity possibly anti-aging during 1-year-long spaceflight, *Sci. Rep.*, 2022, **12**, 11862.

10. K. Otsuka, S. Murakami, K. Okajima, K. Shibata, Y. Kubo, D. Gubin, L. A. Beaty and G. Cornelissen, Appropriate circadian-circasemidian coupling protects blood pressure from morning surge and promotes human resilience and wellbeing, *Clin. Invest. Aging*, 2023, **18**, 755–769.
11. F. Levi and F. Halberg, Circaseptan (about 7-day) bioperiodicity—spontaneous and reactive—and the search for pacemakers, *Ric. Clin. Lab.*, 1982, **12**, 323–370.
12. G. Cornelissen, T. K. Breus, C. Bingham, R. Zaslavskaya, M. Varshitsky, B. Mirsky, M. Teibloom, B. Tarquini, E. Bakken and F. Halberg, International Womb-to-Tomb Chronome Initiative Group: Beyond circadian chronorisk: worldwide circaseptan-circasemiseptan patterns of myocardial infarctions, other vascular events, and emergencies, *Chronobiologia*, 1993, **20**, 87–115.
13. G. Cornelissen, B. D. Dimitrov, F. Carandente and F. Halberg, Space and earth weather mirrored in patterns of suicide incidence, *World Heart J.*, 2011, **3**, 31–42.
14. J. V. Moore, R. Rawley, H. A. Hopkins, R. E. Ritenour and W. B. Looney, Cyclophosphamide as an adjuvant to X-rays in treatment of a radioresistant solid tumor of the rat, hepatoma H-4-11-E, *Int. J. Radiat. Oncol., Biol., Phys.*, 1979, **5**, 1471–1474.
15. H. A. Reimann, Periodic disease: observations on old cases and report of new cases and of therapeutic trials, *Arch. Int. Med.*, 1953, **92**, 494–506.
16. H. A. Reimann, Periodic (Pel-Ebstein) fever of lymphomas, *Ann. Clin. Lab. Sci.*, 1977, **7**, 1–6.
17. L. Derer, Rhythm and proliferation with special reference to the six-day rhythm of blood leukocyte count, *Neoplasma*, 1960, **7**, 117–134.
18. Hippocrates, *Die Werke des Hippokrates: die hippokratische Schriftensammlung in neuer deutscher Übersetzung*, ed. R. Kapferer, Hippokrates, Stuttgart, vol. I–V, 1934.
19. C. Galen, in *Clandii Galeni opera anima*, ed. C. G. Kohn, Nachdruck der Ausgabe Leipzig 1825, Hildesheim (zit. n. Bd. 9, S. 769–840), De diebus Decretoriis Lib, 1. 1825/1965, 1965, vol. 1–22.
20. Avicenna/De Venetiis, Über die Seele, Tractatus II, Die Diebus crisis et horus eius 1592, Cap. 1–10.
21. G. Cornelissen and F. Halberg, The biological week and broader time structures (chronomes): In memory of Gunther Hildebrandt, *Percept Mot. Skills*, 2000, **90**, 579–586.
22. K. Hübner, Kompensatorische Hypertrophie, Wachstum und Regeneration der Rattenniere, *Ergeb. Allg. Pathol. Pathol. Anat.*, 1967, **48**, 1–80.
23. F. Halberg, J. Ratte, Kühl JFW, J. Najarian, V. Popovich, R. Shiotsuka, Y. Chiba, L. Cutkomp and E. Haus, Rythmes circaseptidiens—environ 7 jours—synchronisés ou non avec la semaine sociale, *CR Acad. Sci.*, 1974, **278**, 2675–2678.
24. A. de Vecchi, F. Halberg, R. B. Sothern, A. Cantaluppi and C. Ponticelli, Circaseptan rhythmic aspects of rejection in treated patients with kidney transplant, in *Chronopharmacology and Chronotherapeutics*, ed. C. A. Walker,

C. M. Winget and K. F. A. Soliman, Florida A & M University Foundation, Tallahassee, Florida, 1981, pp. 339–353.

25. H. Kreis, M. Lacombe, L. H. Noel, J. M. Descamps, J. Chailley and J. Crosnier, Kidney-graft rejection: has the need for steroids to be re-evaluated?, *Lancet*, 1978, **ii**, 1169.
26. G. Hildebrandt, Circaseptane Reaktionsperiorlik beim Menschen: eine Zeitstruktur von Kratikheit und Heilung, *Therapentikon*, 1990, **4**, 402–413.
27. F. Halberg, M. Engeli, C. Hamburger and D. Hillman, Spectral resolution of low-frequency, small-amplitude rhythms in excreted 17-ketosteroid; probable androgen induced circaseptan desychronization, *Acta endocrinol.*, 1965, **50**(4 Suppl), S5–S54.
28. G. Cornelissen, J. Siegelova and F. Halberg, Blood pressure and heart rate dynamics during pregnancy and early extra-uterine life: Methodology for a chrononeonatology, in *Proceedings, Symposium: The Importance of Chronobiology in Diagnosing and Therapy of Internal Diseases*, ed. F. Halberg, T. Kenner and B. Fiser, Masaryk University, Brno, Czech Republic, 2002, pp. 58–96.
29. G. Cornelissen, M. Engebretson, D. Johnson, K. Otsuka, N. Burioka, J. Posch and F. Halberg, The week, inherited in neonatal human twins, found also in geomagnetic pulsations in isolated Antarctica, *Biomed. Pharmacother.*, 2001, **55**(Suppl 1), 32s–50s.
30. S. Sanchez de la Peña, F. Halberg, A. Galvagno, M. Montalbini, S. Follini, J. Wu, J. Degioanni, F. Kutyna, D. C. Hillman, Y. Kawabata and G. Cornelissen, Circadian and circaseptan (about-7-day) free-running physiologic rhythms of a woman in social isolation, *Proceedings: 2nd Annual IEEE Symposium on Computer-Based Medical Systems*, Computer Society Press, Minneapolis, Washington DC, 1989, pp. 273–278.
31. D. C. Hillman, M. Siffre, G. Milano and F. Halberg, Urinary about-84-hour (circasemiseptan) variations of a woman isolated in a cave and cosmic ray effects, *New Trends Exp. Clin. Psychiatry*, 1994, **10**, 173–178.
32. A. E. Reinberg, L. Dejardin, M. H. Smolensky and Y. Touitou, Seven-day human biological rhythms: An expedition in search of their origin, synchronization, functional advantage, adaptive value and clinical relevance, *Chronobiol. Int.*, 2017, **34**, 162–191.
33. L. N. Edmunds and F. Halberg, Circadian time structure of Euglena: a model system amenable to quantification, in *Neoplasms—Comparative Pathology of Growth in Animals, Plants and Man*, ed. H. Kaiser, Williams and Wilkins, Baltimore, 1981. pp. 105–134.
34. G. Katinas, S. Berger, G. Cornelissen, D. Hillman, J. C. Woolum, M. Engebretson, E. V. Syutkina, A. Masalov, Z. R. Wang, C. M. Wan, E. E. Bakken, O. Schwartzkopff and F. Halberg, Acetabularia acetabulum: Unicellular model of neonatal vascular chronome, *Neuroendocrinol. Lett.*, 2003, **24**(Suppl 1), 201–207.
35. W. Ulmer, G. Cornelissen, M. Revilla, J. Siegelova, J. Dusek and F. Halberg, Circadian and circaseptan dependence of the beta-ATP

peak of four different cancer cell cultures: implications for chrono-radiotherapy, *Scr. Med.*, 2001, **74**, 87–92.
36. J. C. Woolum, G. Cornelissen, D. K. Hayes and F. Halberg, Spectral aspects of Acetabularia's light transmission chronome: effects of enucleation, *Neuroendocrinol. Lett.*, 2003, **24**(Suppl 1), 212–215.
37. H. Shinoda, M. Ohtsuka-Isoya, G. Cornelissen and F. Halberg, Putative circaseptans or other infradians in murine dentin accretion and the suprachiasmatic nuclei, *Neuroendocrinol. Lett.*, 2003, **24**(Suppl 1), 208–211.
38. F. Halberg, G. Cornelissen, F. Halberg, W. Ulmer, S. Sanchez de la Peña, J. Siegelova and O. Schwartzkopff, BIOCOS project. Reasons for a protocol for radiation treatment aimed at exploiting weekly rhythms, ed. F. Halberg, T. Kenner, B. Fiser and J. Siegelova, in *Proceedings, Noninvasive Methods in Cardiology*, Brno, Czech Republic, 2008, pp. 63–73.
39. T. Liu, M. Cavallini, F. Halberg, G. Cornelissen, J. Field and D. E. R. Sutherland, More on the need for circadian, circaseptan and circannual optimization of cyclosporine therapy, *Experientia*, 1986, **42**, 20–22.
40. F. Levi, F. Halberg, G. Chihara and J. Byram, Chronoimmunomodulation: circadian, circaseptan and circannual aspects of immunopotentiation or suppression with lentinan, in *Toward Chronopharmacology*, ed. R. Takahashi, F. Halberg and C. Walker, Pergamon Press, Oxford/New York, 1982, pp. 289–311.
41. T. G. Bromage, Y. Idaghdour, R. S. Lacruz, T. D. Crenshaw, O. Ovsiy, B. Rotter, K. Hoffmeier and F. Schrenk, The swine plasma metabolome chronicles "many days" biological timing and functions linked to growth, *PLoS One*, 2016, **11**, e0145919.
42. F. Halberg, E. Haus, W. Nelson and R. Sothern, Chronopharmacology, chronodietetics and eventually clinical chronotherapy, *Nova Acta Leopold.*, 1977, **225**, 307–336.
43. A. J. Lewy, R. A. Sack and C. L. Singer, Assessment and treatment of chronobiologic disorders using plasma melatonin levels and bright light exposure: the clock-gate model and the phase response curve, *Psychopharmacol. Bull.*, 1984, **20**, 561–565.
44. E. R. Dodson and P. C. Zee, Therapeutics for circadian rhythm sleep disorders, *Sleep Med. Clin.*, 2010, **5**, 701–715.
45. K. Chokhawala, S. Lee and A. Saadabadi, Lithium, in *StatPearls*, StatPearls Publishing, Treasure Island (FL), 2023. Available from: https://www.ncbi.nlm.nih.gov/books/NBK519062/, updated 2023 Feb 26.
46. C. Baker, C. Retzik-Stahr, V. Singh, R. Plomondon, V. Anderson and N. Rasouli, Should metformin remain the first-line therapy for treatment of type 2 diabetes?, *Ther. Adv. Endocrinol. Metab.*, 2021, **12**, 2042018820980225.
47. B. Viollet, B. Guigas, N. Sanz Garcia, J. Leclerc, M. Foretz and F. Andreelli, Cellular and molecular mechanisms of metformin: an overview, *Clin. Sci.*, 2012, **122**, 253–270.

48. M. Barnea, L. Haviv, R. Gutman, N. Chapnik, Z. Madar and O. Froy, Metformin affects the circadian clock and metabolic rhythms in a tissue-specific manner, *Biochim. Biophys. Acta*, 2012, **1822**, 1796–1806.
49. X. Lu, P. Paliogiannis, D. F. Calvisi and X. Chen, Role of the mammalian target of rapamycin pathway in liver cancer: from molecular genetics to targeted therapies, *Hepatology*, 2021, **73**(Suppl 1), 49–61.
50. M. Dadon-Freiberg, N. Chapnik and O. Froy, Rev-Erbα alters circadian rhythms by modulating mTOR signaling, *Mol. Cell. Endocrinol.*, 2021, **521**, 111108.
51. A. Moskalev, E. Chernyagina, V. Tsvetkov, A. Fedintsev, M. Shaposhnikov, V. Krut'ko, A. Zhavoronkov and B. K. Kennedy, Developing criteria for evaluation of geroprotectors as a key stage toward translation to the clinic, *Aging Cell*, 2016, **5**, 407–415.
52. K. Wang, H. Liu, Q. Hu, L. Wang, J. Liu, Z. Zheng, W. Zhang, J. Ren, F. Zhu and G. H. Liu, Epigenetic regulation of aging: implications for interventions of aging and diseases, *Signal Transduction Targeted Ther.*, 2022, **7**, 374.
53. M. Hatori and T. Hirota, Cell-based phenotypic screens to discover circadian clock-modulating compounds, *Methods Mol. Biol.*, 2022, **2482**, 95–104.
54. E. S. Rasmussen, J. S. Takahashi and C. B. Green, Time to target the circadian clock for drug discovery, *Trends Biochem. Sci.*, 2022, **47**, 745–758.
55. R. Narasimamurthy and D. M. Virshup, The phosphorylation switch that regulates ticking of the circadian clock, *Mol. Cell*, 2021, **81**, 1133–1146.
56. L. Badura, T. Swanson, W. Adamowicz, J. Adams, J. Cianfrogna, K. Fisher, J. Holland, R. Kleiman, F. Nelson, L. Reynolds, K. St Germain, E. Schaeffer, B. Tate and J. Sprouse, An inhibitor of casein kinase I epsilon induces phase delays in circadian rhythms under free-running and entrained conditions, *J. Pharmacol. Exp. Ther.*, 2007, **322**, 730–738.
57. Q. J. Meng, E. S. Maywood, D. A. Bechtold, W. Q. Lu, J. Li, J. E. Gibbs, S. M. Dupré, J. E. Chesham, F. Rajamohan, J. Knafels, B. Sneed, L. E. Zawadzke, J. F. Ohren, K. M. Walton, T. T. Wager, M. H. Hastings and A. S. Loudon, Entrainment of disrupted circadian behavior through inhibition of casein kinase 1 (CK1) enzymes, *Proc. Natl. Acad. Sci. U. S. A.*, 2010, **107**, 15240–15245.
58. T. Hirota, J. W. Lee, W. G. Lewis, E. E. Zhang, G. Breton, X. Liu, M. Garcia, E. C. Peters, J. P. Etchegaray, D. Traver, P. G. Schultz and S. A. Kay, High-throughput chemical screen identifies a novel potent modulator of cellular circadian rhythms and reveals CKIα as a clock regulatory kinase, *PLoS Biol.*, 2010, **8**, e1000559.
59. J. W. Lee, T. Hirota, E. C. Peters, M. Garcia, R. Gonzalez, C. Y. Cho, X. Wu, P. G. Schultz and S. A. Kay, A small molecule modulates circadian rhythms through phosphorylation of the period protein, *Angew. Chem. Int. Ed. Engl.*, 2011, **50**, 10608–10611.

60. H. Mahoney, E. Peterson, H. Justin, D. Gonzalez, C. Cardona, K. Stevanovic, J. Faulkner, A. Yunus, A. Portugues, A. Henriksen, C. Burns, C. McNeill, J. Gamsby and D. Gulick, Inhibition of casein kinase 1 δ/ε improves cognitive performance in adult C57BL/6J mice, *Sci. Rep.*, 2021, **11**, 4746.
61. T. Oshima, Y. Niwa, K. Kuwata, A. Srivastava, T. Hyoda, Y. Tsuchiya, M. Kumagai, M. Tsuyuguchi, T. Tamaru, A. Sugiyama, N. Ono, N. Zolboot, Y. Aikawa, S. Oishi, A. Nonami, F. Arai, S. Hagihara, J. Yamaguchi, F. Tama, Y. Kunisaki, K. Yagita, M. Ikeda, T. Kinoshita, S. A. Kay, K. Itami and T. Hirota, Cell-based screen identifies a new potent and highly selective CK2 inhibitor for modulation of circadian rhythms and cancer cell growth, *Sci. Adv.*, 2019, **5**, eaau9060.
62. Y. Chen, Y. Wang, J. Wang, Z. Zhou, S. Cao and J. Zhang, Strategies of targeting CK2 in drug discovery: Challenges, opportunities, and emerging prospects, *J. Med. Chem.*, 2023, **66**, 2257–2281.
63. C. Borgo, C. D'Amore, S. Sarno, M. Salvi and M. Ruzzene, Protein kinase CK2: a potential therapeutic target for diverse human diseases, *Signal Transduction Targeted Ther.*, 2021, **6**, 183.
64. S. Rahman, K. Wittine, M. Sedic and E. P. Markova-Car, Small molecules targeting biological clock; a novel prospective for anti-cancer drugs, *Molecules*, 2020, **25**, 4937.
65. D. J. Kojetin and T. P. Burris, REV-ERB and ROR nuclear receptors as drug targets, *Nat. Rev. Drug Discovery*, 2014, **13**, 197–216.
66. S. Wang, F. Li, Y. Lin and B. Wu, Targeting REV-ERBα for therapeutic purposes: promises and challenges, *Theranostics*, 2020, **10**, 4168–4182.
67. B. He, K. Nohara, N. Park, Y. S. Park, B. Guillory, Z. Zhao, J. M. Garcia, N. Koike, C. C. Lee, J. S. Takahashi, S. H. Yoo and Z. Chen, The small molecule Nobiletint the molecular oscillator to enhance circadian rhythms and protect against metabolic syndrome, *Cell Metab.*, 2016, **23**, 610–621.
68. K. Nohara, V. Mallampalli, T. Nemkov, M. Wirianto, J. Yang, Y. Ye, Y. Sun, L. Han, K. A. Esser, E. Mileykovskaya, A. D'Alessandro, C. B. Green, J. S. Takahashi, W. Dowhan, S. H. Yoo and Z. Chen, Nobiletin fortifies mitochondrial respiration in skeletal muscle to promote healthy aging against metabolic challenge, *Nat. Commun.*, 2019, **10**, 3923.
69. T. Hirota, J. W. Lee, P. C. St John, M. Sawa, K. Iwaisako, T. Noguchi, P. Y. Pongsawakul, T. Sonntag, D. K. Welsh, D. A. Brenner, F. J. Doyle 3rd, P. G. Schultz and S. A. Kay, Identification of small molecule activators of cryptochrome, *Science*, 2012, **337**, 1094–1097.
70. P. S. Humphries, R. Bersot, J. Kincaid, E. Mabery, K. McCluskie, T. Park, T. Renner, E. Riegler, T. Steinfeld, E. D. Turtle, Z. L. Wei and E. Willis, Carbazole-containing sulfonamides and sulfamides: Discovery of cryptochrome modulators as antidiabetic agents, *Bioorg. Med. Chem. Lett.*, 2016, **26**, 757–760.
71. S. Miller, Y. L. Son, Y. Aikawa, E. Makino, Y. Nagai, A. Srivastava, T. Oshima, A. Sugiyama, A. Hara, K. Abe, K. Hirata, S. Oishi, S. Hagihara,

A. Sato, F. Tama, K. Itami, S. A. Kay, M. Hatori and T. Hirota, Isoform-selective regulation of mammalian cryptochromes, *Nat. Chem. Biol.*, 2020, **16**, 676–685.
72. M. Iida, Y. Nakane, T. Yoshimura and T. Hirota, Effects of cryptochrome-modulating compounds on circadian behavioural rhythms in zebrafish, *J. Biochem.*, 2022, **171**, 501–507.
73. J. W. Lee, T. Hirota, A. Kumar, N. J. Kim, S. Irle and S. A. Kay, Development of Small-Molecule Cryptochrome Stabilizer Derivatives as Modulators of the Circadian Clock, *ChemMedChem*, 2015, **10**, 1489–1497.
74. T. Oshima, I. Yamanaka, A. Kumar, J. Yamaguchi, T. Nishiwaki-Ohkawa, K. Muto, R. Kawamura, T. Hirota, K. Yagita, S. Irle, S. A. Kay, T. Yoshimura and K. Itami, C-H activation generates period-shortening molecules that target cryptochrome in the mammalian circadian clock, *Angew. Chem., Int. Ed. Engl.*, 2015, **54**, 7193–7197.
75. S. Miller, Y. Aikawa, A. Sugiyama, Y. Nagai, A. Hara, T. Oshima, K. Amaike, S. A. Kay, K. Itami and T. Hirota, An Isoform-Selective Modulator of Cryptochrome 1 Regulates Circadian Rhythms in Mammals, *Cell. Chem. Biol.*, 2020, **27**, 1192–1198.e5.
76. D. Kolarski, S. Miller, T. Oshima, Y. Nagai, Y. Aoki, P. Kobauri, A. Srivastava, A. Sugiyama, K. Amaike, A. Sato, F. Tama, W. Szymanski, B. L. Feringa, K. Itami and T. Hirota, Photopharmacological Manipulation of Mammalian CRY1 for Regulation of the Circadian Clock, *J. Am. Chem. Soc.*, 2021, **143**, 2078–2087.
77. M. Yagi, S. Miller, Y. Nagai, S. Inuki, A. Sato and T. Hirota, A methylbenzimidazole derivative regulates mammalian circadian rhythms by targeting Cryptochrome proteins, *F1000Research*, 2022, **11**, 1016.
78. P. S. Humphries, R. Bersot, J. Kincaid, E. Mabery, K. McCluskie, T. Park, T. Renner, E. Riegler, T. Steinfeld, E. D. Turtle, Z. L. Wei and E. Willis, Carbazole-containing amides and ureas: Discovery of cryptochrome modulators as antihyperglycemic agents, *Bioorg. Med. Chem. Lett.*, 2018, **28**, 293–297.
79. Z. Dong, G. Zhang, M. Qu, R. C. Gimple, Q. Wu, Z. Qiu, B. C. Prager, X. Wang, L. J. Y. Kim, A. R. Morton, D. Dixit, W. Zhou, H. Huang, B. Li, Z. Zhu, S. Bao, S. C. Mack, L. Chavez, S. A. Kay and J. N. Rich, Targeting Glioblastoma Stem Cells through Disruption of the Circadian Clock, *Cancer Discovery*, 2019, **9**, 1556–1573.
80. S. Miller, M. Kesherwani, P. Chan, Y. Nagai, M. Yagi, J. Cope, F. Tama, S. A. Kay and T. Hirota, CRY2 isoform selectivity of a circadian clock modulator with antiglioblastoma efficacy, *Proc. Natl. Acad. Sci. U. S. A.*, 2022, **119**, e2203936119.
81. S. Miller, A. Srivastava, Y. Nagai, Y. Aikawa, F. Tama and T. Hirota, Structural differences in the FAD-binding pockets and lid loops of mammalian CRY1 and CRY2 for isoform-selective regulation, *Proc. Natl. Acad. Sci. U. S. A.*, 2021, **118**, e2026191118.

82. S. Miller and T. Hirota, Structural and chemical biology approaches reveal isoform-selective mechanisms of ligand interactions in mammalian cryptochromes, *Front. Physiol.*, 2022, **13**, 837280.
83. Y. U. Jeong, H. E. Jin, H. Y. Lim, G. Choi, H. Joo, B. Kang, G. H. Lee, K. H. Liu, H. J. Maeng, S. Chung, G. H. Son and J. W. Jung, Development of non-ethoxypropanoic acid type cryptochrome inhibitors with circadian molecular clock-enhancing activity by bioisosteric replacement, *Pharmaceuticals*, 2021, **14**, 496.
84. S. Gul, Y. K. Akyel, Z. M. Gul, S. Isin, O. Ozcan, T. Korkmaz, S. Selvi, I. Danis, O. S. Ipek, F. Aygenli, A. C. Taskin, B. A. Akarlar, N. Ozlu, N. Ozturk, N. Ozturk, D. O. Unal, M. Guzel, M. Turkay, A. Okyar and I. H. Kavakli, Discovery of a small molecule that selectively destabilizes Cryptochrome 1 and enhances life span in p53 knockout mice, *Nat. Commun.*, 2022, **13**, 6742.
85. Y. U. Doruk, D. Yarparvar, Y. K. Akyel, S. Gul, A. C. Taskin, F. Yilmaz, I. Baris, N. Ozturk, M. Turkay, N. Ozturk, A. Okyar and I. H. Kavakli, A CLOCK-binding small molecule disrupts the interaction between CLOCK and BMAL1 and enhances circadian rhythm amplitude, *J. Biol. Chem.*, 2020, **295**, 3518–3531.
86. H. Li, K. Li, K. Zhang, Y. Li, H. Gu, H. Liu, Z. Yang and D. Cai, The Circadian Physiology: Implications in Livestock Health, *Int. J. Mol. Sci.*, 2021, **22**, 2111.
87. Y. Dakhiya, D. Hussien, E. Fridman, M. Kiflawi and R. Green, Correlations between Circadian Rhythms and Growth in Challenging Environments, *Plant Physiol.*, 2017, **173**, 1724–1734.
88. A. E. Maeda and N. Nakamichi, Plant clock modifications for adapting flowering time to local environments, *Plant Physiol.*, 2022, **190**, 952–967.
89. M. F. Khalid, C.-Y. Lee, S. L. Doggett and G. Veera Singham, Circadian rhythms in insecticide susceptibility, metabolic enzyme activity, and gene expression in Cimex lectularius (Hemiptera: Cimicidae), *PLoS One*, 2019, **14**, e0218343.

Subject Index

www.ingramcontent.com/pod-product-compliance
Ingram Content Group UK Ltd.
Pitfield, Milton Keynes, MK11 3LW, UK
UKHW021850260126
467367UK00003BA/7

9 781839 162008